Gastrointestinal Microbiology

Volume 1
Gastrointestinal Ecosystems and Fermentations

Chapman & Hall Microbiology Series

Physiology/Ecology/Molecular Biology/Biotechnology

C.A. Reddy, Editor-in-Chief
Department of Microbiology
Michigan State University
East Lansing, MI 48824-1101

A.M. Chakrabarty
Department of Microbiology and Immunology
University of Illinois Medical Center
835 S. Wolcott Avenue
Chicago, IL 60612

Arnold L. Demain
Department of Biology, Rm 68-223
Massachusetts Institute of Technology
Cambridge, MA 02139

James M. Tiedje
Center for Microbial Ecology
Department of Crop and Soil Sciences
Michigan State University
East Lansing, MI 48824

Other Publications in the Chapman & Hall Microbiology Series

Methanogenesis; James G. Ferry, ed.
Acetogenesis; Harold L. Drake, ed.

Forthcoming Titles in the Chapman & Hall Microbiology Series

Bacteria in Oligotrophic Environments; Richard Y. Morita
Mathematical Modeling in Microbial Ecology; Arthur L. Koch, Joseph A. Robinson, and George L. Milliken, eds.
Oxygen Regulation of Gene Expression in Bacteria; Rob Gunsalus, ed.

Chapman & Hall Microbiology Series

Gastrointestinal Microbiology

Volume 1
Gastrointestinal Ecosystems and Fermentations

Edited by

Roderick I. Mackie, Ph.D.

Department of Animal Sciences
and Division of Nutritional Sciences
University of Illinois

Bryan A. White, Ph.D.

Department of Animal Sciences
and Division of Nutritional Sciences
University of Illinois

CHAPMAN & HALL

ITP® INTERNATIONAL THOMSON PUBLISHING

New York • Albany • Bonn • Boston • Cincinnati • Detroit • London
Madrid • Melbourne • Mexico City • Pacific Grove • Paris • San Francisco
Singapore • Tokyo • Toronto • Washington

Join Us on the Internet
WWW: **http://www.thomson.com**
EMAIL: **findit@kiosk.thomson.com**

thomson.com is the on-line portal for the products, services and resources available from International Thomson Publishing (ITP). This Internet kiosk gives users immediate access to more than 34 ITP publishers and over 20,000 products. Through *thomson.com* Internet users can search catalogs, examine subject-specific resource centers and subscribe to electronic discussion lists. You can purchase ITP products from your local bookseller, or directly through *thomson.com.*

Visit Chapman & Hall's Internet Resource Center for information on our new publications, links to useful sites on the World Wide Web and an opportunity to join our e-mail mailing list. Point your browser to: **http://www.chaphall.com/chaphall.html** or **http://www.chaphall.com/chaphall/lifesce.html** for Life Science.

Cover design: Curtis Tow Graphics

A service of ITP®

Printed in the United States of America

Chapman & Hall
115 Fifth Avenue
New York, NY 10003

Chapman & Hall
2-6 Boundary Row
London SE1 8HN
England

Thomas Nelson Australia
102 Dodds Street
South Melbourne, 3205
Victoria, Australia

Chapman & Hall GmbH
Postfach 100 263
D-69442 Weinheim
Germany

International Thomson Editores
Campos Eliseos 385, Piso 7
Col. Polanco
11560 Mexico D.F
Mexico

International Thomson Publishing–Japan
Hirakawacho-cho Kyowa Building, 3F
1-2-1 Hirakawacho-cho
Chiyoda-ku, 102 Tokyo
Japan

International Thomson Publishing Asia
221 Henderson Road #05-10
Henderson Building
Singapore 0315

Library of Congress Cataloging-in-Publication Data

Gastrointestinal microbiology.
p. cm.
Contents: v. 1. Gastrointestinal ecosystems and fermentations / edited by Roderick I. Mackie and Bryan A. White. — v. 2. Gastrointestinal microbes and host interactions / edited by Roderick I. Mackie, Bryan A. White, Richard E. Isaacson.
ISBN 0-412-98361-3 (v. 1 : alk. paper) . — ISBN 0-412-98371-0 (v. 2 : alk. paper)
1. Gastrointestinal system—Microbiology. I. Mackie, R. I. (Rod I.) II. White, Bryan A., 1954– . III. Isaacson, Richard E., 1947– . IV. Title: Gastrointestinal ecosystems and fermentations. V. Title: Gastrointestinal microbes and host interactions.
[DNLM: 1. Gastrointestinal System—microbiology. WI 100 G26355 1996]
QR171.G29G37 1996
591.1′3—DC20
DNLM/DLC
for Library of Congress

96-30506
CIP

British Library Cataloguing in Publication Data available

To order this or any other Chapman & Hall book, please contact **International Thomson Publishing, 7625 Empire Drive, Florence, KY 41042.** Phone: (606) 525-6600 or 1-800-842-3636.
Fax: (606) 525-7778. e-mail: order@chaphall.com.

For a complete listing of Chapman & Hall titles, send your request to **Chapman & Hall, Dept. BC, 115 Fifth Avenue, New York, NY 10003.**

Contents

Part I: NUTRITIONAL ECOLOGY

Preface

The gastrointestinal tract is a complex **anaerobic** microbial ecosystem containing a vast assemblage of resident microorganisms performing a multitude of metabolic activities that play a key role in health and disease of humans and animals. Furthermore, the gastrointestinal microbes have a dominant impact on the growth and productivity of both ruminant and non-ruminant animals. This two-volume series on **Gastrointestinal Microbiology** reviews the literature and provides a comprehensive account of the biological significance of the microbiota present in the alimentary tract of a wide range of animals, in terms of their nutritional ecology, biochemical activities, development and composition, interactions and role in host health and disease. Recent developments in the areas of molecular ecology, bacterial genetics, immunological aspects of host microbe interactions at the level of the intestinal mucosa, bacterial translocation and intestinal disease are included.

Although emphasis is placed on domestic ruminants and man, systems which have been extensively researched, this series also provides a full and integrated account of the nutritional ecology and microbial ecology in the gut of many diverse mammals, birds, fish, amphibians, reptiles and insects. This broad perspective allows more realistic interpretation, and better evaluation of, as well as greater insight into, the evolution, ecology, and function of the gastrointestinal ecosystem.

These volumes contain contributions from a multidisciplinary group of internationally recognized authors, all active researchers in their particular fields. These timely volumes emphasize the nutritional, physiological and ecological diversity of the gastrointestinal ecosystem and contribute to a wider understanding of its role not only in normal gastrointestinal function but also in enteric disease.

Contributors

Stephen F. Baron, Biology Department
Bridgewater College
402 East College Street
Bridgewater, VA 22812
USA

Karen A. Bjorndal, Department of Zoology and Center for Sea Turtle Research, University of Florida, Gainesville, Florida 32611, USA

Marvin P. Bryant, Department of Animal Sciences and Department of Microbiology, University of Illinois at Urbana-Champaign, Urbana, Illinois 61801, USA

K.-J. Cheng, Agriculture and Agri-Food Canada, Lethbridge Research Station, Lethbridge, Alberta AB2 9SB, Canada

King-Thom Chung, Department of Microbiology and Molecular Cell Sciences, University of Memphis, Memphis, Tennessee 38152, USA

Kendall D. Clements, School of Biological Sciences, University of Sydney, Sydney, NSW 2006, Australia, Present address: School of Biological Sciences, University of Auckland, Private Bag 92019, Auckland, New Zealand

Michael A. Cotta, United States Department of Agriculture, Agricultural Research Service, National Center for Agricultural Utilization Research, Peoria, Illinois 61604, USA

Burk A. Dehority, Department of Animal Sciences, Ohio Agricultural Research and Development Center, The Ohio State University, Wooster, Ohio 44691-4096, USA

Cecil W. Forsberg, Department of Microbiology, University of Guelph, Guelph, Ontario N1G 2W1, Canada

Glenn R. Gibson, Head, Microbiology Department, Institute of Food Research, Reading Laboratory, Earley Gate, Reading, RG6 6BZ, United Kingdom

Ian D. Hume, School of Biological Sciences, University of Sydney, Sydney, NSW 20006, Australia

Phillip B. Hylemon, Department of Microbiology and Immunology, Medical College of Virginia, Richmond, Virginia 23298-0678, USA

Matthew D. Kane, Laboratory of Molecular Systematics, National Museum of Natural History, Smithsonian Institution, Washington, D.C. 20560, USA

William H. Karasov, Department of Wildlife Ecology, University of Wisconsin, Madison, Wisconsin 53706, USA

George T. Macfarlane, Medical Research Council Dunn Clinical Centre, Hills Road, Cambridge CB2 2DH, United Kingdom

Roderick I. Mackie, Department of Animal Sciences and Division of Nutritional Sciences, University of Illinois at Urbana-Champaign, Urbana, Illinois 61801, USA

Christopher S. McSweeney, CSIRO Division of Tropical Animal Production, Long Pocket Laboratories, Indooroopilly QLD 4068, Australia

Mark Morrison, Department of Animal Sciences, University of Nebraska, Lincoln, Nebraska 68503-0908, USA

James B. Russell, United States Department of Agriculture, Agricultural Research Service, Cornell University, Ithaca, New York 14853, USA

Conrad Vispo, Department of Wildlife Ecology, University of Wisconsin, Madison, Wisconsin 53706, USA

Bryan A. White, Department of Animal Sciences, University of Illinois at Urbana-Champaign, Urbana, Illinois 61801, USA

INTRODUCTION

1

Introduction to Gastrointestinal Microbial Ecology

Marvin P. Bryant

1. Appreciation of Anaerobiosis

Human appreciation of anaerobic life is relatively recent. It was chiefly Pasteur (1858) who alerted the world to the actions of microorganisms in the absence of oxygen. Pasteur described the formation of lactic acid, ethanol, and butyric acid and their specific microbial origins. Fermentation has become the almost universal term to describe these and similar processes occurring in the absence of O_2.

Pasteur's demonstration of life without air soon led to development of many methods for growing such organisms. With all of these methods, many of them quite adequate for excluding O_2 from even the most stringently anaerobic organisms, it is surprising that until about 1940 only spore formers and nonspore formers of clinical importance had been isolated and described (Weiss and Rettger 1937, Prevot 1966). Many of the anaerobes, known from their products to exist in nature, had not been cultured. This was probably due to the great popularity of the Petri dish and the ease of isolating aerobic bacteria on plates. Attempts to inoculate and incubate plates under anaerobic conditions were unsuccessful until the anaerobic glove box technique was perfected (Aranki et al. 1969).

The degree of oxidation or reduction of a chemical system can be defined by the redox potential. In nature, oxygen is the almost universal cause of high redox potential. This raises the question: Does the oxygen or the potential inhibit anaerobes? Hungate (1969) used the Nernst equation to calculate the necessary concentration of O_2 in a culture medium to have a redox potential of -0.33 V (potential required to initiate growth of methanogens). At this potential, this becomes 1.48×10^{-56} molecules O_2/L. This calculation strikingly illustrates that (1) it is difficult to obtain low redox potential for cultivation of strict anaerobes; (2) the permissible concentration of oxygen in solution becomes a statistical function rather than a finite number of molecules; (3) it is impossible to obtain low redox potentials simply by removing oxygen, and the corollary (4) that in order to obtain the required low potential, a reduced system at a lower potential must be added.

This understanding led to development of procedures for media preparation enabling enumeration and isolation of anaerobic bacteria. The roll tube technique, with its numerous modifications and improvements since the original description (Hungate 1950), was considerably superior to other anaerobic methods and contributed much to our knowledge and understanding of anaerobes. Despite the advent of the anaerobic glove box, with its many advantages, modifications of the roll tube technique are still widely used and are standard procedures for anaerobe laboratories. Of importance to the present series on the ecology and physiology of gastrointestinal microbes, these techniques provided the tools that enabled microbial ecologists, particularly those working in the rumen, to advance this field of research significantly.

2. Types of Carbohydrate Fermentation

One of the early problems in the metabolism of anaerobes was to determine the types of fermentation products formed. Many of the enzymes of interest were identified by means of Thunberg (1930) tubes in which cells or cell extracts were analyzed for activity using dyes which changed color with the transfer of H_2. This was followed by the Warburg respirometer in which a gas was liberated or consumed. Acid production could also be measured by including bicarbonate in the medium and calibrating to determine the amount of gas released by a given amount of each acid. Indeed, the pathways of the more important types of anaerobic conversion patterns were firmly established (Wood 1961) through application of these methods. These fermentations have been reviewed by Hungate (1985) and include lactic acid, ethanol, acetate, succinate, propionate, butyrate, and mixed acid fermentations.

The methanogenic fermentation proved more difficult to understand and analyze both biologically and biochemically. Schnellen (1947) was the first to obtain pure cultures of methanogens, isolating *Methanobacterium formicicum* which utilizes formate and *Methanosarcina barkeri* which could use acetate, methanol, CO, and H_2/CO_2. The list of methanogenic substrates has been extended to include methyl-, dimethyl-, and trimethylamines, the methyl groups being converted to methane (Ferry 1993). Although the range of methanogenic substrates is limited, the habitats of methanogens are extremely varied and include sediments, fresh and salt water, thermal hot springs, tree trunks, salt evaporation ponds, geothermal vents, and naturally the rumen and other fermentative gut compartments (Hungate 1985).

The major substrate in anaerobic fermentations is carbohydrate, and since substrate is normally limiting, the pathways yielding the most cell material per unit of substrate will have a competitive advantage and will therefore tend to predominate. This leads to formation of acetic, propionic, and butyric acids as end products, at least in the gastrointestinal tract. However, in sediments and anaerobic digestors the degradation process is extended with the ultimate conversion of these acids to CO_2, CH_4, and H_2S (Hungate 1985). This resulted in the development of the concept of interspecies hydrogen transfer and syntrophy during the anaerobic degradation of organic matter and the discovery and role of obligate proton-reducing fatty acid syntrophs in this process (McInerney and Bryant 1981). The extended transformation is possible because of longer residence time allowing microbial degradation of these and other organic compounds in nature. The unique pathways, enzymes and coenzymes involved in the terminal processes of carbon and electron flow in anaerobic ecosystems, namely methanogenesis and acetogenesis, have recently been documented in exclusive books on these topics (Ferry 1993, Drake 1994).

3. Historical Introduction to Gut Microecology

This monumental two-volume book is concerned with almost all aspects of gastrointestinal microbial ecology from normal to pathologic aspects, from fermentation of the microbiota of protozoa and insects to those of humans and including microbe-microbe and microbe-animal physiologic interactions. Thus, this introductory chapter is mainly limited to my own experience and knowledge in anaerobic microbiology since, inadvertently and with much luck, becoming a part-time undergraduate laboratory glassware washer and very shortly thereafter, a part-time technician, in the laboratory of Robert E. Hungate in June of 1947 at Washington State College (Bryant 1981).

Hungate is recognized by many as the father of modern-day anaerobic microbial ecology. He was the first Ph.D. student of Cornelis Bernardus (C.B. or Kees) Van Niel starting in 1930 after Van Niel came to Hopkins Marine Station of Stanford University in 1929 (Hungate 1979, Van Niel 1967).

Van Niel was a Dutch microbiologist of the Delft School. In the 1670s and '80s Van Leeuwenhoek, the founder of microbiology, made drawings of *Selenomonas* seen tumbling in samples from the human gingival crevice using his crude microscope. He lived in Delft, as did Beijerinck, the great Dutch general microbiologist, whose student A.J. Kluyver, the initiator of comparative biochemistry, was the Ph.D. mentor of Van Niel. Thus, it was almost natural that Van Niel's students—Hungate and H.A. Barker (1978)—were the first two who would emphasize energy metabolism including fermentation products, in their studies on various anaerobic microbial species. Hungate, of more ecologic bent, also emphasized enumeration of species in their natural habitat, fermentation rates, and turnover times of various intermediates in the total microbial ecosystem (Hungate 1960).

Only a few other microbiologists in the era 1920 to 1950 included fermentation products in their descriptions of anaerobic bacterial species; thus, it was very difficult for others to identify strains of these species unless they had rather unique macromorphologies or other unusual features.

Pure cultures of anaerobes were first obtained by the French School of Microbiology (Prévot et al. 1967). In 1861 Louis Pasteur isolated a butyrate-producing spore former, *Clostridium butyricum*. A. Veillon described several pathogenic nonsporing anaerobes, e.g., *Bacillus* (*Bacteroides*) *fragilis*, starting in the 1890s. More modern examples are Andre Prévot, who started his studies about 1930, and Madeleine Sebald, his student, who started working with him at the Pasteur Institute in Paris in the late 1950s. She studied the % G+C of the DNA of many gram-negative anaerobes, their fermentation products, and, more recently, genetics. In 1967, Prévot published his great and massive book with A. Turpin and P. Kaiser, *Les Bactéries Anaérobes*, also containing much of the work of Sebald. This contained 71 genera and over 600 species including many human and animal pathogens and other bacteria from diverse ecosystems.

In Germany, there was much early interest in anaerobes isolated from humans and other animals. Flügge (1886) described *Fusobacterium necrophorum (Bacillus necrophorus)*. Burri and Ankersmit (1906) described *Clostridium (Bacterium) clostridiiforme* and other bacteria from the digestive tract of cattle.

Some great microbiologists in other countries, by no means a complete list, who worked with microbes from anaerobic animal or human ecosystems, includes Noguchi (treponemes) in the early 1900s (Noguchi 1928) and, more recently, Mitsuoka (nonsporing anaerobes) from Japan (Mitsuoka 1980).

In the United Kingdom, Sydney Elsden moved to Cambridge in 1946 (first column chromatography of saturated volatile monocarboxylic fatty acids), where his colleagues A.J.P. Martin and R.L.M. Synge became Nobel laureates for their research on chromatography. On moving to the University of Sheffield, Elsden isolated from the sheep rumen the lactate-fermenting large coccus (LC; *Megasphaera elsdenii*). Elsden and Lewis (1953) and one of his students, Sheila Wilson (1953), briefly reported the isolation and features of four groups of bacteria from the sheep rumen. Elsden sent me Wilson's more complete manuscript which was never published because of his very meticulous standards (see Bryant 1959). I studied bacteria from the bovine rumen and named *Butyrivibrio*, *Prevotella (Bacteroides) ruminicola*, and *Succinivibrio dextrinosolvens*, which were three of the groups isolated by Wilson—the fourth being *Streptococcus bovis*. Later, much study of ruminal nonsporing anaerobes was done by P.N. Hobson and his group at the Rowett Research Institute after S.O. Mann and A.E. Oxford (1954) isolated the anaerobic levorotatory lactic acid–producing *Lactobacillus vitulinus*, typically found in young calves. This long rod species is quite different from the type strain, which is short and found in both calves and adult cattle (Bryant et al. 1958a,b, Sharpe et al. 1973, Sharpe and Dellagio 1977). Collins et al. (1994) showed that *L. vitulinus* should be placed in a new order rather than in *Lactobacillus*. Thus, the species found only in young calves should be renamed ("*Lactobacillus mannii*" would be appropriate).

In the mid-1950s we started a "ruminology" group (term coined by Raymond N. Doetsch) that included the University of Maryland group with Ray and his students in the Microbiology Department, e.g., Ronald J. Gibbons, who has become especially well known for his work in oral microbiology at Forsyth Dental Center in Boston; my group at Beltsville, e.g., Milton Allison and I.M. (Ike) Robinson; and the Virginia Polytechnic Institute group at Blacksburg including W.E.C. Moore and Kendall King (cellulases) and their students. We met about twice a year to present new information and to have intense discussions often with considerable deep philosophic content; e.g., Doetsch said pure cultures were artifacts while I said that the environment normally used for their study was artifactual.

During these informal meetings, Kendall King suggested that I write a comprehensive review of species of rumen bacteria (Bryant 1959) not limited to cellulolytic bacteria (Sijpesteijn 1948, Hungate 1950) while W.E.C. (Ed) Moore suggested that I should broaden my group studies to include anaerobes in man and other mammals. I told Ed that I had plenty to do involving rumen bacteria.

Thus, in 1966 Ed Moore started his group's monumental study of mainly fermentative anaerobes (Moore 1966), which led to their systematic and taxonomic studies of a huge number of anaerobes from many anaerobic habitats. Thus, other researchers could now identify many anaerobes from normal and pathologic systems, especially those from GI tract, (including oral) ecosystems. Of special significance was the standardized study of many previously isolated and maintained strains from many different laboratories. For example, many of the strains from Prévot's culture collection became known much better. Many of our groups of bacteria, especially those from young calves (Bryant et al. 1958a,b) were characterized quite well but could not be identified with known species because of the earlier poor descriptions. After many publications of Moore and L.V. (Peg) Holdeman and their colleagues, many of these and other bacteria were identified.

Eggerth (1935) described several gram-positive, nonsporing rods from human feces—e.g., *Bifidobacterium* spp. and *Butyribacterium rettgeri*, now *E. limosum* (Barker and Haas 1944). In 1952, I published a paper on a rumen spirochete, including the first study of fermentation products of an anaerobic spirochete

(Bryant 1952). It was later named *Treponema bryantii* (Stanton and Canale-Parola 1980).

Bryant and Burkey (1953) published short descriptions and photomicrographs of rumen anaerobes later studied in detail. These included *Ruminococcus flavefaciens*, *R. albus*, *R. bromii*, *Fibrobacter succinogenes*, *Prevotella ruminicola*, *Eubacterium ruminatium* and *Eubacterium* sp., *Selenomonas*, *Lachnospira*, *Butyrivibrio*, *Succinimonas*, *Succinivibrio* and *Treponema*. Bryant et al. (1958a,b) described other anaerobes from young calves, e.g., the long and short species of *Lactobacillus vitulinus*, *Peptostreptococcus productus*, *Megasphaera elsdenii*, *Fusobacterium necrophorum* (shown to degrade lactate), *Eubacterium limosum*, and *Clostridium clostridiiforme*. However, studies on diversity and taxonomy have been greatly enhanced by the application of phylogenetic analysis based largely on small subunit rRNA molecules.

Phylogenetic studies of bacteria using 16S rRNA oligonucleotide analysis started in the laboratory of Carl Woese, University of Illinois, in the 1970s (see Fox et al. 1977b for earlier references). The phylogenetic analysis of anaerobic bacteria, the methanogens, started in 1977 in collaboration between Woese's laboratory and that of R.S. Wolfe (Fox et al. 1977a,b, Balch et al. 1979). This resulted in the discovery of a third kingdom, the Archaebacteria, which included the methanogens. Recently a higher taxonomic level has been proposed which groups organisms into three domains—Archaea, Bacteria, and Eucarya (Woese et al. 1990). With these techniques the phylogeny of many other bacteria, including many from gastrointestinal ecosystems, was published (Fox et al. 1980). 16S rRNA sequencing in phylogenetic studies began later (Woese 1987). Very recently the phylogeny of many bacteria from the gastrointestinal tract and other ecosystems, more or less related to the clostridia, have been studied (Collins et al. 1994). This includes many genera—e.g., *Oxobacter*, *Syntrophomonas*, *Selenomonas*, *Acidaminococcus*, *Megasphera*, *Quinella*, *Sporomusa*, *Ruminococcus*, *Roseburia*, *Peptostreptococcus*, *Lachnospira*, *Fusobacterium*, *Eubacterium*, and anaerobic *Lactobacillus*, and includes 14 or more families, such as Clostridiaceae, Syntrophomonadaceae, Selenomonadaceae, Ruminococcaceae, Fusbacteriaceae, Eubacteriaceae, Heliococcaceae, and more than six unnamed families.

Avgustin et al. (1994 and references therein) determined the genetic diversity and phylogenetic relationships of some rumen gram-negative anaerobes, especially many strains of *Prevotella ruminantium*. Groups represented by Bryant strains 23^T, $GA33^T$, and B_14 are different species, and other species within the group (Bryant et al. 1958a) are sure to be found. Several species are also sure to be found in the *Butyrivibrio* group (Bryant and Small 1956). My idea in naming single species of groups such as *Prevotella* (*Bacteroides*) *ruminicola* and *Butyrivibrio fibrisolvens* was that, if they were not named, other researchers would not further study the groups.

I have covered in this prefatory chapter only a small part of an appropriate introduction to this two-volume book. It is realized that the microbial ecology of the gastrointestinal tract of animals as diverse as humans, termites (as in the studies of Hungate and Breznak), and protozoa are of equal or more importance. However, the best-known microbial ecosystem is still probably that of the rumen.

4. Future Directions in Gut Microecology

Bacteria have traditionally been classified mainly on the basis of phenotypic properties. Despite the vast amount of knowledge generated for the ruminal and other gut ecosystems using traditional techniques, the basic prerequisites for ecological studies, namely, enumeration and identification of all community members, have limitations. The two major problems faced by microbial ecologists include bias introduced by culture-based enumeration and characterization techniques and the lack of phylogenetically based classification scheme (Amann et al. 1990, 1994; Pace et al. 1985). Modern molecular ecology techniques based on sequence comparisons of nucleic acids (DNA or RNA) can be used to provide molecular characterization while at the same time providing a classification scheme that predicts natural evolutionary relationships. These molecular methods provide results that are independent of growth conditions and media used. Also using these techniques, microbes can be classified and identified before they can be grown in pure culture. An example from the rumen is *Quinella ovalis*, a morphologically distinctive but uncultivable organism, which based on 16S rRNA phylogeny was most closely related to the *Selenomonas-Megasphaera-Sporomusa* group in the gram-positive phylum (Krumholz et al. 1993).

Furthermore, in situ hybridization with fluorescently labeled rRNA-targeted nucleic acid probes facilitates in situ identification and phylogenetic placing of uncultured organisms and provides information on three-dimensional relationships in complex microbial populations (Amann 1995). Ultimately genetic capabilities, expression of these capabilities, and taxonomic information are potentially accessible at the individual cell level using nucleic acid (DNA, or mRNA and rRNA) targeted in situ hybridization methods. These nucleic acid–based techniques will enable gut microecologists to answer the most difficult question in microbial ecology: the exact role or function a specific organism plays in its natural environment and its quantitative contribution to the whole (Hungate 1960). Rather than replacing the classical cultural based system, the new molecular-based techniques can be used in combination with the classical approach to improve cultivation, speciation, and evaluation of biodiversity. For example, it is worth noting that the comparative sequence database of 16S and 23S rRNA sequences (more than 4,000 entries covering about 1,800 species) is largely based on pure culture cultivation studies.

It could be argued that the technological impetus for major advances in our knowledge of gastrointestinal microecology during recent decades has been derived from three major sources: the development of anaerobic culture techniques and their application to the study of the rumen ecosystem by Hungate et al.; the use of rodent experimental models to define relationships between gut microbes and the host by Dubos et al.; and the development of gnotobiotic technology by which germ-free or defined-microbiota animals could be derived and maintained. It is likely that the use of molecular ecology techniques based on nucleic acid probes is likely to generate the next major advance in our knowledge and provide for the first time not simply a refinement or increased understanding, but a complete description of the gastrointestinal ecosystem.

References

Amann RI (1995) Flourescently labelled, RNA-targeted oligonucleotide probes in the study of microbial ecology. Mol Ecol 4:543–554.

Amann RI, Krumholz L, Stahl DA (1990) Fluorescent-oligonucleotide probing of whole cells for determinative, phylogenetic, and environmental studies in microbiology. J Bacteriol 172:762–770.

Amann RI, Ludwig W, Schleifer KH (1994) Identification of uncultured bacteria: a challenging task for molecular taxonomists. ASM News 60:360–365.

Aranki A, Syed SA, Kenney EB, Freter R (1969) Isolation of anaerobic bacteria from human gingiva and mouse cecum by means of a simplified gove box procedure. Appl Microbiol 17:568–576.

Avgustin G, Wright F, Flint HJ (1994) Genetic diversity and phylogenetic relationships among strains of *Prevotella (Bacteroides) ruminicola* from the rumen. Int J Syst Bacteriol 44:246–255.

Balch WE, Fox GE, Magrum LJ, Woese CR, Wolfe RS (1979) Methanogens: reevaluation of a unique biological group. Microbiol Rev 43:260–296.

Barker HA (1978) Explorations of bacterial metabolism. Annu Rev Biochem 47:1–33.

Barker HA, Haas V (1944) *Butyribacterium*, a new genus of gram-positive nonsporulating anaerobic bacteria of intestinal origin. J Bacteriol 47:301–305.

Bryant MP (1959) Bacterial species of the rumen. Bacteriol Rev 23:125–153.

Bryant MP (1981) A rumen microbiologist. Paul A. Funk recognition program, University of Illinois as Urbana-Champaign, College of Agriculture Special Publication 62.

Bryant MP, Burkey LA (1953) Cultural methods and some characteristics of some of the more numerous groups of bacteria in the bovine rumen. J Dairy Sci 36:205–217.

Bryant MP, Small N (1956) The anaerobic monotrichous butyric acid-producing curved rod shaped bacteria of the rumen. J Bacteriol 72:16–21.

Bryant MP, Small N, Bouma C, Chu H (1958a) *Bacteroides ruminicola* sp. nov. and *Succinimonas amylolytica* gen. nov., succinic acid–producing bacteria of the bovine rumen. J Bacteriol 76:15–23.

Bryant MP, Small N, Bouma C, Robinson IM (1958b) Studies of the composition of the ruminal flora and fauna of young calves. J Dairy Sci 41:1747–1767.

Burri R, Ankersmit O (1906) *Bacterium clostridiiforme*. In: Ankershmit P, ed. Untersuchunger über die bakterien in Verdauugskanal des Rindes. Zentralbl Bakt Parasitenkd Infektinok Hig Abt I Orig 40:100–118.

Collins MD, Lawson PA, Williams A, et al. (1994) The phyogeny of the genus *Clostridium*: proposal of five new genera and eleven new species combinations. Int J Syst Bacteriol 44:812–826.

Drake HL (1994) Acetogenesis. New York: Chapman and Hall.

Eggerth AH (1935) The gram-positive non-spore-bearing anaerobic bacilli of human feces. J Bacteriol 30:277–290.

Elsden SD, Lewis D (1953) The production of fatty acids by a gram-negative coccus. Biochem J 55:183–189.

Ferry JG (1993) Methanogenesis: ecology, physiology, biochemistry and genetics. New York: Chapman and Hall.

Flügge C (1886) Die Microorganismen. Leipzig: FCW Vogal.

Fox GE, Magrum LJ, Balch WE, Wolfe RS, Woese CR (1977a) Classification of methanogenic bacteria by 16S ribosomal RNA characterization. Proc Natl Acad Sci USA 74: 4537–4541.

Fox GE, Pichman KR, Woese CR (1977b) Comparative catagorizing of 16S ribonucleic acid: molecular approach to procaryotic systematics. Int J Syst Bacteriol 27:44–57.

Fox GE, Stackebrandt E, Hespell RB, et al. (1980) The phylogeny of prokaryotes. Science 209:457–463.

Hungate RE (1950) The anaerobic mesophibic cellulolytic bacteria. Bacteriol Rev 14: 1–49.

Hungate RE (1960) Symposium: selected topics in microbial ecology. I. Microbial ecology of the rumen. Bacteriol Rev 24:353–364.

Hungate RE (1969) A roll tube method for cultivation of strict anerobes. In: Norris JR, Ribbons DW, eds. Methods of Microbiology, Vol. 3B, pp. 177–132. New York: Academic Press.

Hungate RE (1985) Anaerobic biotransformations of organic matter. In: Leadbetter ER, Poindexter JS, eds. Bacteria in Nature, Vol. 1, pp. 39–95. Plenum Publ. Corp, New York: Plenum.

Hungate RE (1979) Evolution of a microbial ecologist. Ann Rev Microbiol 33:1–20.

Krumholz LR, Bryant MP, Brilla WJ, Vicini JL, Clark JH, Stahl DA (1993) Proposal of *Quinella ovalis* gen. nov., sp. nov., based on phylogenetic analysis. Int J Syst Bacteriol 43:293–296.

Mann SO, Oxford AE (1954) Studies on some presumptive lactobacilli isolated from the rumen of young calves. J Gen Microbiol 11:83–90.

McInerney MJ, Bryant MP (1981) Basic principles in the bioconversions in anaerobic digestion and methanogenesis. In: Sofer SS, Zaborsky OR, eds. Biomass Conversion Processes for Energy and Fuels, pp. 277–296. New York: Plenum.

Mitsuoka T (1980) The World of Intestinal Bacteria—The Isolation and Identification of Anaerobic Bacteria; A Color Atlas of Anaerobic Bacteria. Tokyo: Sobunsha (Sobin Press).

Moore WEC (1966) Techniques for routine culture of fastidious anaerobes. Int J Syst Bacteriol 16:173–190.

Noguchi (1928) The spirochetes. In: Jordan, Falk, eds. The Newer Knowledge of Bacteriology and Immunology, pp. 452–497. Chicago: University of Chicago Press.

Pace NR, Stahl DA, Lane DJ, Olsen GJ (1985) Analyzing natural populations by RNA sequences. ASM News 51:4–12.

Pasteur L (1858) Memoire sur la fermentation appelee lactique. Mem Soc Sci Agric Arts Lille 2nd ser 5:13–26.

Prevot AR (1966) Manual for the Classification and Determination of the Anaerobic Bacteria. Fredette V, ed. and transl. Philadelphia: Lea and Febiger.

Prévot AR, Turpin A, Kaiser P (1967) Les Bactéries Anaérobies. Paris: Dunod.

Schnellen CGTP (1947) Onderzoekingen over de methaangistung. Dissertation, Delft.

Sharpe ME, Dellaglio F (1977) Deoxyribonucleic acid homology in anaerobic lactobacilli and in possibly related species. Int J Syst Bacteriol 27:19–21.

Sharpe ME, Latham MJ, Garvie EI, Zingibe J, Kandler O (1973) Two new species of *Lactobacillus* isolated from the bovine rumen, *Lactobacillus ruminis* sp. nov. and *Lactobacillus vitulinus* sp. nov. J Gen Microbiol 77:37–49.

Stanton TB, Canale-Parola E (1980) *Treponema bryantii* sp. nov., a rumen spirochete that interacts with cellulolytic bacteria. Arch Microbiol 127:145–156.

Sijpesteijn AK (1948) Cellulose decomposing bacteria from the rumen of cattle. Ph.D. Thesis, Leiden University, Eduard Ijdo NV, Leiden.

Thunberg T (1930) The hydrogen activating enzymes of the cells. Q Rev Biol 5:318–347.

Van Niel CB (1967) Prefatory chapter: the education of a microbiologist; some reflections. Annu Rev Microbiol 21:1–30.

Weiss JE, Rettger LF (1937) The gram negative *Bacteroides* of the intestine. J Bacteriol 33:423–434.

Wilson SN (1953) Some carbohydrate-fermenting organisms isolated from the rumen of sheep. J Gen Microbiol 9:i–ii.

Woese CR (1987) Bacterial evolution. Microbiol Rev 52:221–271.

Woese CR, Kandler O, Wheelis ML (1990) Towards a natural system of organisms: proposal for the domains Archaea, Bacteria and Eucarya. Proc Natl Acad Sci USA 87: 4576–4599.

Wood WA (1961) Fermentation of carbohydrates and related compounds. In: Gunsalus IC, Stanier RY, eds. The Bacteria, Vol. II. New York: Academic Press.

2

Gut Environment and Evolution of Mutualistic Fermentative Digestion

Roderick I. Mackie

1. Introduction

The gastrointestinal tract is a specialized tube divided into various well-defined anatomical regions extending from the lips to the anus. However, for the purposes of this series concerning the ecology, physiology, metabolism, and genetics of gastrointestinal microbes, discussion is restricted to the stomach (rumen-reticulum, crop, gizzard), small intestine, and large intestine (cecum and colon). Large populations of microorganisms inhabit the gastrointestinal tract of all animals and form a closely integrated ecological unit with the host. This complex, mixed, microbial culture comprising bacteria, ciliate and flagellate protozoa, anaerobic phycomycete fungi, and bacteriophage can be considered the most metabolically adaptable and rapidly renewable organ of the body, which plays a vital role in the normal nutritional, physiological, immunological, and protective functions of the host animal. Development of microbial populations in the alimentary tract of higher animals commences soon after birth. The processes involved in the establishment of microbial populations are complex, involving succession of microorganisms and many microbial and host interactions eventually resulting in dense, stable populations inhabiting characteristic regions of the gut.

2. Intestinal Microbiota

2.1. Conceptual Approaches

The recognition by Dubos et al. (1965) that the microbes in gastrointestinal tracts of animals in a given community can be differentiated into two major groups is consistent with ecological theory (Alexander 1971). Allochthonous as well as autochthonous microbes can usually be found in any gut ecosystem. However, the definition of the autochthonous microbiota (microorganisms that have been present during its evolution) and ''normal'' microbiota (those that are ubiquitous in the given community so that they become established in practically all its members) by Dubos et al. (1965) does not conform strictly to that concept. The designation of autochthonous or normal was based largely on numbers of that organism in adult animals and consequently organisms such as *E. coli* and *S. faecalis* were not included in the autochthonous biota despite the finding that they are prominent during succession in neonatal animals and are always present in the climax biota of adults. Using modern ecological theory, these members are undoubtedly autochthonous members of the gastrointestinal microbiota and not just transient microbes through the ecosystem. In the stable gastrointestinal ecosystem, all available habitats and niches would be occupied by autochthonous microbes.

Savage (1977) considers autochthonous and indigenous as synonyms. Any allochthonous species found in a habitat would not be established (implying colonization and growth) but merely be in passage, being derived from food or water, or from another habitat in the gastrointestinal ecosystem or elsewhere on the host (skin and upper respiratory membranes). Clearly some pathogens are autochthonous to the gastrointestinal ecosystem and can live in harmony with their hosts, becoming pathogenic only when the ecosystem is disturbed in some way. Under such circumstances the microbes may invade a habitat to which they are allochthonous. They may also be pathogenic should they enter alien habitats in a disturbed ecosystem involving an animal species other than their host.

Other pathogens are undoubtedly allochthonous to any gastrointestinal ecosystem, and reside normally in other habitats such as water and soil. Thus, microbes that are potentially pathogenic, as well as those that are nonpathogenic, can be regarded as indigenous to habitats in the gastrointestinal ecosystem (Savage 1977). Also, a particular microbial species may be autochthonous to one habitat in the gastrointestinal tract but allochthonous to another, which it normally passes through after it is shed from its native habitat. The critical distinction to be made is that an autochthonous microbe colonizes the habitat natively, while an allochthonous one cannot colonize it except under abnormal circumstances. Microbes autochthonous to a given habitat can be distinguised from those that are allochthonous using the criteria presented in Table 2.1.

2.2. *Definition of the Environment*

Physical and chemical conditions within the gut of different animals may differ considerably, but are usually relatively constant in a single species on a given diet. This is the case for homeothermic animals, in which, allowing for irregularities in

Table 2.1. Criteria of autochthony for microbes in the gastrointestinal tract

1. Can grow anaerobically
2. Are always found in normal adults
3. Colonize particular areas of the tract
4. Colonize their habitats during succession in infant animals
5. Maintain stable population levels in climax communities in normal adults
6. May associate intimately with the mucosal epithelium in the areas colonized
7. Host acquires immunological tolerance during colonization and succession
8. Not immunogenic in their native host when contained in the gastrointestinal tract

Sources: Dubos et al. 1965, Alexander 1971, Savage 1977.

the intake of food and factors such as temperature, oxygen, acidity, and moisture, vary little with time. In poikilothermic animals, temperature can be a major variable.

The detailed composition of the gut contents of most animals is extremely complex. To date, the microbial environment in the ruminoreticulum has been closely defined and, allowing for variation in the nature and amount of food ingested, serves as a good model for other gut ecosystems, both herbivores and nonherbivores. The hindgut environment is more constant in terms of physical and chemical composition with nutrients for cecocolonic bacteria being provided by undigested dietary polysaccharides and endogenous secretions and tissues such as mucopolysaccharides, mucins, epithelial cells, and enzymes. Many of the properties of rumen contents are listed in Table 2.2, illustrating the complexities that must be considered in media selection and design in order to cultivate, enumerate, and isolate predominant gut bacteria.

2.3. Digestive Physiology of Vertebrate Herbivores

Because of the nature of the plant cell wall and the difficulty in digesting it, herbivores have anatomical and/or physiological adaptations of the digestive tract to compensate for assimilation of this material. Herbivorous reptiles, birds, and mammals usually have enlarged and/or elongated digestive tracts, often including fermentation chambers or sacs in the foregut or hindgut. Cecum-colon (hindgut) fermenters, represent an older differentiation than foregut fermenters, which in turn are older than ruminants (Langer 1991). Bacteria, protozoa, and anaerobic fungi inhabit these enlarged gut compartments as well as other sites in the gastrointestinal tract. The fermentative activity of these microbes results in the production of volatile fatty acids that are absorbed by the host animal, and make a variable and in some cases considerable contribution to its nutritional economy.

There has been considerable debate on the relative merits of foregut and hindgut fermentation in mammals (Janis 1976, Parra 1978, Demment and Van Soest

Table 2.2. Physical, chemical, and microbiological characteristics of mammalian fermentative gut compartments

Characteristic	Property
Physical	
pH	5.5–6.9 (mean 6.4)
Redox potential	− 350 to − 400 mV
Temperature	38–41°C
Osmolality	250–350 mOsmol/kg
Dry matter	10–18%
Chemical	
Gas phase (%)	CO_2, 65; CH_4, 27; N_2, 7; O_2, 0.6; H_2, 0.2
Volatile fatty acids (mM)	Acetate 60–90
	Propionate 15–30
	Butyrate 10–25
	Branched-chain and higher 2–5
Nonvolatile acids (mM)	Lactate < 10
Amino acids and oligopeptides	< 1 mM present 2–3 h postfeeding
Ammonia	2–12 mM
Soluble carbohydrates	< 1 mM present 2–3 h postfeeding
Insoluble polysaccharides	
dietary (cellulose, hemicellulose, pectin)	Always present
endogenous (mucopolysaccharides)	Always present
Lignin	Always present
Minerals	High Na; generally good supply
Trace elements/vitamins	Always present; good supply of B vitamins
Growth factors	Good supply; BCFA, LCFA, purines, pyrimidines, others unknown
Microbiological	
Bacteria	10^{10}–10^{11}/g ($>$ 200 species)
Ciliate protozoa	10^4–10^6/g (25 genera)
Anaerobic fungi	10^3–10^5/g (4 genera)
Bacteriophage	10^7–10^9 particles/mL

1985). The interrelationship between body size, diet, and digestive strategy poses an interesting problem. The relationship between gut capacity and mass of fermentation contents is isometric with increasing body size regardless of whether they are foregut or hindgut fermenting mammals. Since an animal's mass-specific metabolic rate decreases with increasing body size while the ratio of gut capacity to body size remains almost constant, it follows that a large herbivore should have a slower turnover rate of its gut contents than a smaller herbivore. Hence, the mutualistic microbes of a larger herbivore will have longer residence time, allowing greater fermentation or digestion of refractory plant material. Furthermore, the large animal's daily energy requirements can be supplied by fermenting and digesting a smaller fraction of its diet, allowing it to survive on forage of lower quality than a small herbivore. In contrast, a very small herbivore's relatively higher metabolic rate with a rapid turnover makes it difficult for these animals to subsist on a high-fiber diet, and consequently they are typically hindgut fermenters.

Despite the sophisticated anatomical and physiological adaptations for herbivory, these modifications are not essential. Some birds compensate for a lack of structural modification to the intestinal tract by consuming large quantities of grass—e.g., ducks, geese, and the takahe (Morton 1978). Even though emus lack well-developed digestive specializations for microbial fermentation and demonstrate high digesta passage rates, these birds digest plant cell wall constituents moderately well (35% to 45%) with significant levels of VFA production in the intestine with concomitant contribution to the energy economy of the bird. The lack of specialized fermentation chambers is offset by grinding coarse feed particles in the crop and exposure to acid conditions in the stomach prior to fermentation. However, other ratites, such as the ostrich and rhea, which routinely consume high-fiber, low-quality diets, have more elaborate fermentation sites—namely, an elongated colon. Among mammals the giant panda has a slightly modified carnivorelike gut but compensates by consuming large quantities of fodder (Dierenfeld et al. 1982).

2.4. Molecular Approaches

The microbial community inhabiting the gastrointestinal tract is represented by all major groups of microbes and characterized by its high population density and its wide diversity and complexity of interactions. Microbial populations have been described in herbivores, omnivores, and carnivores and in all zoological classes ranging from insects (Chap. 8, Vol. 1); fish (Chap. 6, Vol. 1); reptiles (Chap. 7, Vol. 1); birds (Chap. 5, Vol. 1); rodents (Chap. 4, Vol. 1); lagomorphs (Chap. 4, Vol. 1); marsupials (Chap. 3, Vol. 1); monkeys (Chap. 3, Vol. 1); pigs (Chap. 4, Vol. 1); sheep, goats, cattle, and camels (Chap. 3, Vol. 1); horses (Chap. 4, Vol. 1); antelope, elephants, and even dinosaurs (this chapter). Despite this vast amount of knowledge, the basic prerequisites for ecological studies, namely, enumeration and identification of community members, have tremendous limitations.

These limitations can be overcome using modern molecular ecology techniques based on sequence comparisons of nucleic acids (DNA or RNA) and can be used to provide molecular characterization while at the same time providing a classification scheme that predicts natural evolutionary relationships (Amann et al. 1994, 1995). An example of the power of these modern molecular ecology techniques is provided by the analysis of 16S rRNA sequences (average length 1,500 nucleotides). The highly conserved regions of the rRNA molecule can serve as primer binding sites for in vitro amplification by PCR (Ludwig et al. 1994). The more conserved regions are also useful, serving as targets for universal probes that react with all living organisms or for discriminating between broad phylogenetic groups such as the domains Archaea, Bacteria, and Eucarya (Woese et al. 1990). The more variable sequence regions are more appropriate for genus-, species-, and even strain-specific hybridization probes (Stahl and Amman 1991). Thus the application of molecular ecology techniques based on nucleic acid probes for specific organisms (rRNA) as well as genes (DNA) and their expression (mRNA) will enable scientists to determine the exact role or function a specific organism plays in the gut ecosystem and its quantitative contribution to the whole (Chap. 7, Vol. 2). This is the ultimate goal of the microbial ecologist.

Recently with the advent of molecular biology and the realization of the potential for genetic modification of ruminal bacteria, as well as the discovery of genes of biotechnological importance, a concerted effort of examining the genetics of ruminal bacteria (and fungi) has been established (Chap. 8, Vol. 2). More specifically, it may be possible to utilize recombinant DNA technology to improve cellulose degradation, modify phytotoxins, or produce proteins from synthetic genes that match amino acid requirements of the ruminant under certain production conditions. Initial progress in the area of genetic manipulation to improve or modify rumen microbes or their metabolism has been slow owing to a lack of detailed fundamental knowledge concerning the basic genetics of these strict anaerobes. Researchers have been confronted with problems associated with a lack of native genetic transfer elements appropriate for shuttle vector construction, identifying suitable marker genes, and restriction barriers and genetic diversity within groups of organisms considered to be single species using phenotypic criteria (Chap. 9, Vol. 2). Most progress has been made with the colonic *Bacteroides,* and the related *Prevotella,* where genetic systems have been developed (Chap. 8, Vol. 2). Attention has focused on this genus since *Bacteroides* spp. are largely responsible for fermentation of polysaccharides in the colon, probably play a major role in antibiotic resistance transfer among colonic bacteria, and can be important opportunitistic human pathogens (Chap. 8, Vol. 2).

2.5. *Microbial Interactions*

Microorganisms are most often studied in pure cultures, which provides detailed knowledge regarding their growth and metabolism. Although this has resulted in an enormous body of knowledge, it does not take into account the fundamental importance of the influence that microbes have on each other. Only with a knowledge of both the properties of individual populations and the interactions that occur between them will it be possible to gain some understanding of the structure and function of complex microbial communities. A good example of this is the degradation of organic matter in the gastrointestinal tract and other anaerobic ecosystems where even relatively simple compounds are degraded by the concerted action of several physiologically different anaerobic species. This is reflected in the characteristic pattern of stepwise degradation of organic substrates by a range of organisms with a high degreee of specialization, resulting in different organisms being involved in sequential degradation. The nature of interspecies interactions and the extent of mutual dependence vary enormously, ranging from simple cross-feeding of (essential) nutrients, vitamins, or carbon and energy sources to very specialized total dependence. The best-known of these interactions in anaerobic ecosystems is the removal of electrons formed during fermentation by one species, often in the form of molecular hydrogen, by other species, such as methanogens, sulfate reducers, or hydrogen-oxidizing acetogens. Through this mechanism of interspecies hydrogen transfer, thermodynamically unfavorable reactions can proceed through coupling to exergonic reactions. This contribution (Chap. 10, Vol. 2) highlights the most important aspects of competitive and noncompetitive interactions between microbial species.

The influence of ciliate protozoa on the activity and size of bacterial populations has been studied for many years (Williams and Coleman 1988). However, the observation that fungal zoospores increase in number after defaunation is more recent, providing evidence for a role of protozoa in determining the size of fungal populations. Further research is required to elucidate the exact nature of these predatory interactions, which have been documented by enumeration, substrate disappearance, and electron microscopy. In contrast, good evidence is available for the interaction of both ecto- and endosymbiotic methangens with anaerobic eukaryotic protozoa (Fenchel and Finlay 1995). Of significance is the location of endosymbiotic methanogenic bacteria in close proximity to the hydrogenosome, distinct membrane-bound eukaryotic redox organelles. Thus, it is likely that hydrogen transfer is a significant component of the symbiosis between ciliates and methanogens.

2.6. Host-Microbe Interactions

The association of microbes with tissues of the gastrointestinal tract of animals during evolution has resulted in the development of a "balanced" relationship between microbes and the host. Several biochemical, physiological, and immunological features that are considered intrinsic characteristics of animal species are actually responses by the animal to the physical presence and metabolic activities of the normal microbiota. Host properties influenced by the normal microbiota have been termed "microbiota-associated characteristics" and include altered enzyme activities, bacterial fermentation end products, mucin degradation, elevated mucosal cellularity and enterocyte turnover, more rapid intestinal motility, altered bile acid and cholesterol metabolism, and a multitude of immunological alterations (Chap. 12, Vol. 2).

Although these microbial activities impose a considerable burden on the host in terms of replacement of epithelial cells, detoxification of microbial metabolites, and production of inflammatory or immunological cells, they have a beneficial effect on the host in terms of colonization resistance. Colonization resistance or competitive exclusion includes all factors that hamper colonization of the intestinal tract by exogenous organisms, including pathogens. Some of the regulatory factors are exerted by the animal host; others are mediated by the diet of the host and the gut environment, while other regulatory factors are exerted by the microbes themselves (Chap. 13, Vol. 2). These protective mechanisms are impaired when the normal microbiota is disturbed, such as through the use of antimicrobial agents or before the normal microbiota has a chance to fully develop as in neonatal and immature animals and humans. Further work will be necessary to reveal the relative contributions of the various factors important in controlling the integrity of the intestinal biota.

The gastrointestinal tract provides an extensive surface area in which intimate contact takes place between the host organism and a wide variety of dietary substances and their breakdown products as well as microorganisms, parasites, and exogenous toxins. Accordingly, the intestine must permit the exchange of dietary substances between the gut lumen and the systemic circulation but, at the same time, prevent penetration of pathogenic agents. It is then not surprising that a dense array of mucosal defense mechanisms, both nonspecific and specific, are operative in the intestine. Nonspecific mechanisms include a variety of nonimmunological barriers as well as innate immune components that are operative in the absence of specific, or antigen-driven, immunity. In addition, conventional actively acquired immunological components are well represented in the intestine, both as diffuse cell poplulations and in organized cell aggregates, to ensure the elimination of specific enteric pathogens. Although conveniently categorized as distinct functions, host protection from luminal antigens is ultimately effected by concerted actions of nonimmunologic and immunologic barriers in the intestine (Chap. 14, Vol. 2).

Although the barrier function of the intestinal epithelium serves a key defensive role, mounting evidence indicates that epithelial cells also play a central role in immunological mechanisms operative in the intestine (Berg and Savage 1972, 1975; Foo and Lee 1972). Through the regulated expression of major histocompatibility complex (MHC) genes, epithelial cells initiate local immune responses. Cytokines synthesized and secreted by intestinal epithelial cells serve to integrate the complex population of immune cells resident in the intestine. These integrated functions are required both for maintenance of tolerance to normal luminal antigens and for mounting local inflammatory and immune responses to pathogenic threats encountered in the intestine (Chap. 14, Vol. 2).

Bacterial translocation describes the passage of viable bacteria from the lumen of the gastrointestinal tract to extraintestinal locations in the body, including mesenteric lymph nodes, spleen, livcr, kidney, and the bloodstream. The defense mechanisms preventing bacterial translocation of the indigenous gut microbiata include (1) an ecologically balanced, gastrointestinal microbiota exhibiting bacterial antagonism to prevent bacterial overgrowth; (2) an intact intestinal mucosa providing a physical barrier; and (3) the host immune system (Berg 1992). In the healthy animal, spontaneous bacterial translocation is likely, occurring continuously at a very low rate, but these low numbers of translocating bacteria are killed by the host immune defense, rarely spreading beyond mesenteric lymph nodes. However translocation of enterobacterial pathogens often differs from the indigenous microbiota since they use different mechanisms and approaches to penetrate this barrier. The molecular mechanisms that govern the infection and penetration of the intestinal epithelium by pathogenic bacteria are described in Chap. 15, Vol. 2. Knowledge of these processes facilitates our understanding of the pathological problems and diseases that arise from infection with these organisms, in addition to advancing our understanding of the intimate interactions that occur between bacterial pathogens and their host cells.

Bacterial enteric pathogens are highly adapted organisms capable of producing diarrheal disease in a host. Pathogens have often evolved through lateral transfer of genes associated with increased virulence and/or increased survival. In outbreaks, or individual cases, of enteric infections the causative organisms can often be defined as belonging to distinct pathogenic clones. Thus a common theme emerging from analyses of enteric isolates is that specific bacterial clones are usually responsible for outbreaks of disease and can be distinguished by the possession of distinct combinations of genes which often play a role in virulence (Chap. 16, Vol. 2). These studies rely on the use of descriptive molecular techniques to study population genetics, bacterial virulence, and microbial pathogenesis. It is possible that mutational drift in antigenic composition to completely avoid activation of a host immune response could provide support for the contention that pathogens gcnerally evolve toward avirulence. Regardless, the answer will most likely be provided by detailed consideration of the intimate relationship, signaling, and sensitivity of host-microbe interactions in the gastrointestinal tract.

3. Types of Animal-Microbe Relationships

Animal-microbe relationships are complex and vary tremendously, ranging from competition to cooperation. The animal alimentary tract has evolved as an adaptation enabling the animal to secure food and limit consumption by other animals. This allows the retention and digestion of ingested food, followed by absorption and metabolism of digestion products, while feeding and other activites continue. Since microorganisms grow rapidly under favorable conditions in the gut, they could become serious competitors for the animal's food. This microbial challenge has modified the course of evolution in animals, resulting in selection for varied animal-microbe relationships (Hungate 1976, 1984). The evolutionary strategy in the first case has been to compete with the resident microbes, and in the second to cooperate with them.

3.1. Competition Model

In the competition model, exemplified by carnivorous animals, host and microbe are competitors for the same food. Immunological and other adaptations delay consumption of the host by the resident microbes, and prevent invasion of animal tissues by microorganisms in the gastrointestinal tract. Microbiocidal concentrations of acid are secreted preventing attack of ingested feed, allowing the host's digestive enzymes to act followed by absorption of enzymic digestion products. However, a slower rate of passage, together with rapid growth, results in a large microbial population in the hindgut.

3.2. Cooperation Model

The abundance of carbohydrate in plant cell walls (cellulose and hemicellulose) is the basis for the evolution of the cooperation model. The carbohydrate polymers of plant cell walls are indigestible by most animals (including all mammals), but can be hydrolyzed and fermented by the microbial partner, with the resultant end products of fermentation plus microbial cells being utilized by the host animal. The most widely known and economically important example of a cooperative animal-microbe relationship is found in ruminants in which the capacious, continuously fermenting rumen delays passage of digesta, allowing time for solubilization of fiber components by microorganisms. One cost of this relationship is the breakdown or sacrifice of dietary protein by rumen microorganisms before digestion by the animal's own enzymes. However, poor-quality protein is upgraded through conversion to microbial protein (Tamminga 1979).

3.3. Combination Model

The combined competition-cooperation model of animal-microbe interactions avoids this difficulty since the host enzymatic breakdown products are absorbed before the microbial fermentation takes place. Thus, the host obtains not only the nutrients digested by its own enzymes but also fermentation products from materials its enzymes cannot digest. This type of interaction occurs in horses, elephants, hyraxes, rodents, and lagomorphs (hares and rabbits), but it is probably best exemplified in the termites.

The difference between this and the competition model lies in the extent to which anatomical modifications of the host allow longer retention of the digesta in the hindgut with consequent increased solubilization and fermentation. In both models the microbes act after the host has absorbed the nutrients made available through its own enzymes, and in both models there is marked microbial activity in the hindgut. A disadvantage of the combination model is that, although the host absorbs the fermentation end products, the microbial cells themselves cannot be used as a nutrient source. Some animals have overcome this deficiency by consuming the feces containing the microbes—that is, through coprophagy or cecotrophy. These models of existing animal-microbe relationships are useful when trying to determine the evolution of mutualistic fermentative digestion in the gastrointestinal tract.

4. Evolution of Mutualistic Fermentative Digestion

Much interest has recently been generated in reconstructing the diets of herbivorous dinosaurs and assessing ecological and evolutionary interactions between these reptiles and their food plants. Of considerable importance to this discussion are the physiological correlates of herbivory absent from the fossil record but possibly provided by consideration of herbivory in living vertebrates (Farlow 1987).

4.1. Dentitions and Gastric Mills

Features one would expect to see in herbivorous dinosaurs that provide some evidence of diet include dentition. Herbivorous mammals masticate their fodder to comminute feed into small particles, exposing a large surface area to microbial enzymatic attack. Although some dinosaurs (e.g., ceratopsids and hadrosaurids and to a lesser extent hypsilophodontids and iguanodontids) had rather specialized masticatory dentition, other herbivorous dinosaurs (e.g., prosauropods, sauropods, pachycephalosaurs, stegosaurs, and ankylosaurs) had simple bladelike or peglike teeth adequate for cropping vegetation but not mastication (Farlow 1987).

Herbivorous, granivorous, and insectivorous birds have a highly specialized, muscular gizzard with an inner lining of hard cuticle. In many species the cuticle thickens into hard plates that grind together. In herbivorous birds this action is enhanced by the presence of sand and stones. In ostriches and emus, these consist of pebbles 2 to 3 cm in diameter. Some 200 pebbles, collectively weighing 2.5 kg, were recovered from the gut region of a moa skeleton (extinct large flightless bird) in New Zealand. Although grit is not essential for digestion of plant material, it increases dry matter digestibility and efficiency of digestion. Deliberate stone swallowing and soil swallowing have been reported for modern snakes, lizards, turtles, and crocodilians serving to macerate digesta. In crocodilians, stomach stones most likely serve as ballast (Taylor 1993); in lizards and turtles, pebbles may well aid in comminuting plant tissues and chitinous skeleton of insects. It has been suggested that herbivorous dinosaurs swallowed large stones that collected in a birdlike gizzard, grinding the poorly masticated herbage. Such stomach stones, or gastroliths, have been reported from the gut regions or found nearby of prosauropods, sauropods, and ornithopods (Taylor 1993). Just how widespread this practice was among dinosaurs is unknown, but it could have been fairly common. If this was the case, it would have enhanced their ability to digest coarse herbage. It is worth noting that herbivorous reptiles can attain digestive efficiencies roughly comparable to those found in herbivorous mammals by subjecting feed to digestive processes (including microbial fermentation in some species) for longer periods of time. However, this strategy is not feasible for endothermic herbivores with a rapid metabolic rate and having limited time available to extract energy and nutrients. Thus the main advantage of mastication or gastric grinding is to decrease the length of time needed to attain a digestibility that could be approximated simply by a longer residence time in the animals digestive tract (Farlow 1987).

4.2. Digestive Physiology of Herbivorous Dinosaurs

On the basis of the adaptations commonly found in living vertebrate herbivores, it is likely that herbivorous dinosaurs had relatively long, capacious guts. Furthermore, it is likely that most herbivorous dinosaurs employed a mutualistic gut microbiota. The previously discussed benefits of large size to a mammalian herbivore probably also accrued to large herbivorous dinosaurs. Big dinosaurs probably had low mass-specific metabolic rates, particularly if dinosaurs had lower metabolic rates than birds and mammals, and consequently slow turnover rates of gut contents. Large dinosaurs could have obtained their daily energy requirements from a low-quality fiber diet presumably due to a slow rate of passage and long exposure of digesta to fermentative digestion. As with mammals, herbivorous dinosaurs of different body size probably selected different diets. Thus, smaller herbivorous dinosaurs likely consumed fodder of lower fiber/higher cell–soluble content than their larger relatives in order to sustain their relatively higher metabolic rates resulting from a smaller body size. However, ectothermy may have reduced the dietary differences between large and small herbivorous dinosaurs without eliminating the differences altogether (Farlow 1987).

Discussion of herbivorous dinosaur diets thus far has been based on provision of energy and protein and (most likely) fodder quality in terms of these nutrients. Of equal importance was the allelochemical defenses of plants they ate (see Rosenthal and Janzen 1979 for modern herbivores). Such toxins can markedly reduce the digestibility of plant matter or directly affect the herbivore's metabolism. It has been suggested (Guthrie 1984) that ruminants are adapted to dealing with type 1 defenses (qualitative toxins or allelochemicals) which occur in unpredictable, ephermeral plant species or tissues whereas quantitative (type 2) defenses are typical of more predictable, available, and accessible plant species which are handled better by monogastric hindgut fermenters. Farlow (1987) argues that, given the size of dinosaurs, it is likely that most of them fed mainly on plants that were reasonably predictable, accessible, and available and thus protected by quantitative defenses. Smaller herbivorous dinosaurs, however, may have fed to a greater extent than their larger kin on plants defended by qualitative toxins. Also, if herbivorous dinosaurs had lower metabolic rates than birds and mammals of comparable size, this may have reduced dependence on elaborate detoxification mechanisms.

In summary, it seems likely that most herbivorous dinosaurs, of whatever size, were hindgut fermenters. Finally, the probability that dinosaurs employed microbial fermentation may have implications for their thermal biology. If the relationship between body size and mass of fermentation contents described for mammals applies to herbivorous dinosaurs, then the fermentation mass would have been considerable, allowing the generation of thermoregulatory heat. Thus, they can be considered as having been, to some extent, fermentative endotherms.

4.3. The First Mammalian Herbivores

During the late Cretaceous and early Paleocene periods, plant-eating mammals were frugivores, presumably because fruit can be more easily processed than foliage. Mammals did not invade the herbivorous niche until the middle Paleocene (Collinson and Hooker 1991). Evolution of large size was a prerequisite for the exploitation of leaves because of the need for a longer residence time in the gut for bacterial fermentation to obtain sufficient nutrients from foliage and herbage. In the late Cretaceous, dinosaurs occupied the herbivorous niche although grazers were still absent. The appearance of grazers in the Miocene is coincident with a similar radiation of grassland-forming grasses (Thomasson and Voorhies 1990). Thus herbivore browsers first appear in the middle Paleocene, but they did not become significant until the late Eocene (Collinson and Hooker 1991). Frugivory declined first with the appearance of herbivore browsing followed by an increase in grazers in post-Miocene at the expense of herbivore browsers. The earliest herbivores were ground dwelling (LGMs; large ground mammals) and achieved their dietary specializations largely through evolution from already large, ground-dwelling frugivores or, in the Paleocene, by a size increase from small insectivorous ancestors (Collinson and Hooker 1991). Large size limited them to the ground. Most browsing herbivores in other locomotor niches (SGMs; small ground mammals) changed their diet from frugivory without changing their locomotor adaptation. A period of nearly 30 million years existed in the vertebrate exploitation of leaves after dinosaur extinction and before the first few mammalian herbivores in the middle Paleocene. This was followed by expansion of herbivores in the late Eocene, when climates cooled and more open vegetation became established.

Hume and Warner (1980) published an excellent discussion on the evolution of microbial digestion in mammals. Since the fossil record provides no information on the morphology, physiology, biochemistry, or microbiology of the gut, much of the knowledge must be deduced from what is known about present-day animals coupled with the fossil record of animals and their probable feedstuffs and is therefore highly speculative. Microbial digestion surely arose long before mammals evolved. Microbial habitats exist in all regions of the gut from the mouth to the rectum and have been categorized as luminal, epithelial, and cryptal (Savage 1977). Large luminal populations of microbes develop in regions of the gut with relative stasis where retention time of digesta allows adequate microbial growth. In most deliberations only nutritional contributions to the host animal are considered. These are based on digestion of the plant cell wall by cellulases and hemicellulases provided by the microbial partner, the synthesis of microbial protein from poor-quality dietary proteins and nonprotein nitrogen mainly via ammonia as precursor, and the synthesis of B vitamins and vitamin K. However, little consideration has been given to the protection provided by foregut fermentation resulting in transformation or modification of phytotoxins and mycotoxins in the diet (Mackie 1987; also see Chap. 15, Vol. 1 for review). Other contributions not normally considered in these deliberations are immunologic, physiologic, and protective (see Chaps. 11–16, Vol. 2, for review). These arguments also support the theory that the development of foregut fermentation must have come after an initial development of the hindgut and that all foregut fermenters should have some fermentation in the hindgut (Hume and Warner 1980).

4.4. Special Features of the Ruminant and Molecular Evolution

The ruminant is well adapted to achieve maximal digestion of roughage using the physiological mechanism at the reticuloomasal orifice, which selectively retains large food particles in the reticulorumen. Efficient separation of the fermentative from the acid-secreting region of the stomach may have allowed development of the most obvious special feature of the ruminants, rumination, where foregut digesta is regurgitated, rechewed, and reswallowed in a frequent, regular pattern repeated 500 times per day, occupying a total time of more than 8 hours and involving more than 25,000 chews (Hume and Warner 1980). Rumination occurs in all the Pecora and Tylopoda. In macropods, regurgitation is more irregular and infrequent and involves much less chewing; this has been termed merycism.

Nearly all foregut fermenters have a gastric or ventricular groove (sulcus reticuli) leading directly from the esophagus to the hind stomach (abomasum in ruminants). This ensures that during suckling, milk is channeled directly to the abomasum, bypassing wasteful ruminal fermentation. Foregut fermenting mammals also share interesting and unique features in two enzymes, stomach lysozyme and pancreatic ribonuclease, which accompany this mode of digestion.

At known times on two occasions during mammalian evolution, lysozyme appears to have largely abandoned its usual function, which is to assist animals fight harmful bacteria and acquired a new function as a digestive enzyme in animals with foregut fermentation (Jollés et al. 1989). Its new role is to lyze bacteria in the abomasum for nutritional purposes. On the most recent of these occasions, which began 15 million years ago, leaf-eating monkeys (Colobinae) remolded lysozyme for functioning in stomach fluid (active at low pH and resistant to pepsin breakdown), and the rate of amino acid replacement increased (Dobson et al. 1984, Stewart et al. 1987). By contrast, on the other occasion, as cloven-hoofed animals (ruminants) recruited lysozyme for the same new digestive function, there was no apparent acceleration. Although because this lineage is at least 60 million years old there was an early period of fast evolution, allowing adaptation to functioning in the stomach environment, followed by a period when the rate of amino acid replacement became subnormal (Jollés et al. 1989). Many mammals and birds have a single gene coding for lysozyme. In contrast, ruminants have multiple genes for lysozyme. A traditional explanation for the origin of a gene family is that it provides a means of making more product. This explanation could apply to the ruminant lysozyme case, with the rise in gene number in advanced ruminants being viewed as the result of selection for the high levels of lysozyme in the stomach that may be necessary for efficient digestion of rumen bacteria. Lysozyme accounts for approximately 10% of stomach mucosal protein (Dobson et al. 1984) and 10% of stomach mucosal mRNA (Irwin and Wilson 1989), showing that the stomach genes are highly expressed in cattle and other advanced ruminants, involving gene duplication as well as a change in gene expression. Work in progress on the genomic organization of ruminant lysozyme genes suggests that all of those genes reside on one chromosomal segment (Irwin et al. 1992). The possibility that multigene families can accelerate adaptive evolution, by virtue of their capacity for bringing together functionally coupled substitutions, is emphasized in the review of Irwin et al. (1992).

The content of ribonuclease (RNase) in the pancreas varies greatly among species (Barnard 1969). All ungulates, rodents, and herbivorous marsupials had high amounts with low levels in all other mammals including hindgut fermenters such as equids, elephants, and pigs. Barnard (1969) proposed that RNase developed in ruminants for degradation of bacterial RNA since a large fraction of ingested protein nitrogen is in the form of bacterial protein, which must be digested in the small intestine to be of nutritional benefit. Interestingly, Barnard (1969) proposed a modified version of the nitrogen cycle of the ruminant, placing in context the role of pancreatic nuclease and included the cycling of phosphorus. Because of wide variations in activity and structure, RNase was suggested as a useful source of information on the evolution of a protein and on relationships between enzyme structure and activity. The abundance of RNase sequences from contemporary artiodactyls allows the reconstruction of the RNases that were the evolutionary intermediates in the most recent 40 million years of this evolution. Genes encoding the reconstructed proteins were obtained in the laboratory by site-directed mutagenesis from a synthetic gene for RNase, expressed in *E. coli* and the resulting proteins purified to homogeneity (Stackhouse et al. 1990).

The catalytic activities, substrate specificities, and thermal stabilities of the reconstructed RNases were examined using parsimony analysis to assess the evolution of the reconstructed RNase family. These changes in molecular behavior of reconstructed RNases correspond to a point in the divergent evolution of mammals where digestive physiology of ungulates also underwent substantial changes, ultimately yielding artiodactyls with "true ruminant" foregut digestion (Jermann et al. 1995). Foregut fermentation appears to have substantial adaptive value in many herbivorous environments and may have evolved convergently in marsupial kangaroos, the colobine monkey primates, and more than once within the artiodactyl lineage itself (Jollés et al. 1989).

That a ribonuclease emerged with increased stability, decreased catalytic activity against duplex RNA (increased activity against small RNA substrates and ss RNA), and increased levels of expression at the same time as ruminant digestion emerged, may be a coincidence but it also indicates that the ancestral molecules were not specialized digestive enzymes but played a nondigestive role such as RNase from the brain and seminal plasma (Jermann et al. 1995). This research highlights the possibility that duplication of the RNase gene in the ruminants allowed tissue-specific expression and subsequent specialization of the enzymes, as is seen for ruminant lysozymes, and should encourage more widespread use of evolutionary reconstruction as an experimental tool to direct site-directed mutagenesis.

The transition in the cetaceans (whales, dolphins, and porpoises) from terrestrial life to a fully aquatic existence is one of the most enduring evolutionary mysteries. Previous paleontological and molecular evidence has indicated that cetaceans and artiodactyls constitute a natural clade within the subclass Eutheria (Novacek 1992). Recent phylogenetic analyses of protein (11 nuclear-encoded protein sequences) and mitochondrial DNA (five) sequences indicate that cetaceans are not only intimately associated to the artiodactyls, they are in fact deeply nested within the artiodactyl phylogenetic tree. The results show that Cetacea are more closely related to the Ruminantia than either ruminants or cetaceans are to members of the other two artiodactyl suborders, Suiformes (pigs, peccaries, and hippopotamuses) and Tylopoda (camels and llamas). On the basis of the rate of evolution of mitochondrial DNA sequences and using paleontological reference dates for calibration, the whale lineage branched off a protoruminant lineage 50 million years ago (Graur and Higgins 1994). By implication, the cetacean transition to aquatic life is inferrred to be a relatively recent evolutionary event.

Grauer (1993) reviewed the molecular phylogeny and higher classification of eutherian mammals based on DNA and protein sequences. Phylogenetic trees depicting relationships among 16 eutherian orders are presented (Graur 1993). Evidence from sequence comparisons of mitochondrial DNA suggest that the artiodactyl family Bovidae is monophyletic and that most tribes originated early in the Miocene with all extent lineages present by 16 million to 17 million years ago providing an example of rapid cladogenesis, following the origin of families in the infraorder Pecora (Allard et al. 1992). It has also been shown that Lagomorpha is significantly more closely related to Primates and Scandentia (tree shrews) than it is to Rodentia, invalidating the superordinal taxon Glires (Lagomorpha + Rodentia) (Li et al. 1990, Graur et al. 1996). The question arises, as in bacterial taxonomy and systematics: Will molecular traits replace phenotypic grouping as the main tool of taxonomy and phylogeny in the future? Graur (1993) indicates that such an event would have undesirable consequences, and that a more logical approach would be to utilize both data sets in a phylogenetically meaningful manner to provide insight into the process of evolution at both levels. He suggests using molecular data as the basis for characterization of the dynamics of morphological changes in evolution. Paradoxically, molecular phylogeny may turn out to be the only means by which phenotypic traits retain their scientific value in taxonomy.

Most species of birds have obvious living relatives and are members of well-characterized groups. The hoatzin (*Opisthocomus hoazin*) is one of the few birds that differ in many ways so that its nearest surviving kin are uncertain. Of importance to this discussion is that the hoatzin is the only bird that uses microbial foregut fermentation to convert cellulose into sugars for microbial fermentation. The hoatzin feeds primarily on tender young leaves, twigs, and shoots of trees and marsh plants, which are ingested into a huge muscular crop with a deeply ridged interior lining where active foregut fermentation occurs. The bony sternum and pectoral girdle are modified to accommodate the filled crop, and there is a callosity on the bone skin of the breast where the heavy crop is rested on a branch. The proventriculus and gizzard are small, and the lower esophagus is sacculated, which delays passage of particles into the lower gut, where additional fermentation occurs in the paired ceca. The contents of the crop and esophagus can account for up to 10% of total body weight. The rate of food passage through the digestive system is rapid in many birds, but the hoatzin retains liquids for ca. 18 hours and solids for 24 to 48 hours, similar to retention times in sheep (Grajal et al. 1989, 1991; Dominguez-Bello et al. 1994). These birds also express high levels of a bacteriolytic lysozyme, which is more similar in amino acid sequence to the rock pigeon than that of the domestic fowl (Kornegay et al. 1994). Evolutionary comparison places them among the calcium-binding lysozymes rather than the conventional types. However, biochemical convergence and parallel amino acid replacements have been shown in hoatzin stomach lysozome even though it has a different genetic orgin from the mammalian examples and has undergone more than 300 million years of independent evolution (Kornegay et al. 1994). DNA sequence evidence from the 12S and 16S rRNA mitochondrial genes and from the eye lens protein, α-crystallin, indicate that the hoatzin is most closely related to the typical cuckoos and that divergence occurred at or near the base of the cuculiform phylogenetic tree (Hedges et al. 1995).

Birds have not made as much use of fermentative fiber digestion as have mammals, with only about 3% of existing species regularly consuming herbage. Many avian herbivores consume large quantities of plant material but extract only the readily digestible components; the bulk of cell wall constituents are rapidly expelled without significant microbial fermentation. Even with the family Tetraonidae (grouse and ptarmigan), which utilize enlarged ceca as fermentation chambers, the contribution of fiber degradation to total energy expenditure is low, with less than 20% of basal energy metabolism being derived from this source. It has been suggested that the weight reduction necessary for flight places a constraint on the size and weight of gut compartments. Thus, the birds that should be able to make the most use of fermentative digestion are the large, flightless species such as the emus and ostriches not only because the constraints associated with flying are absent but also because in large animals it is possible to have large fermentation chambers relative to metabolic rate.

Indeed, for the emu (*Dromaius novaehollandiae*) it was found that energy from digestion of neutral detergent fiber contributed up to 63% of standard metabolism and 50% of maintenance requirements (Herd and Dawson 1984). This was achieved despite the fact that rate of passage of feed residues through the tract was rapid and the anatomy of the gastrointestinal tract was simple in structure with small ceca and short colon. Recent research on the ostrich (*Struthio camelus*), the largest extant flightless bird, showed that the hindgut (long paired ceca and elongated colon) contained 58% of the total wet digesta in the entire intestinal tract with high concentrations of VFA (140 to 195 mM).

Furthermore, the long retention times of fibrous feed in the intestinal tract (mean passage rate 40 hours) ensured exposure to microbial digestion for extended periods resulting in high digestibility of fiber and VFA production (Swart et al. 1993a). Theoretical energy contribution of VFA was estimated to be as high as 76% of the metabolizable energy intake in growing chicks (Swart et al. 1993b). These results confirm the importance of gut anatomy and physiology in providing a suitable environment for fermentative digestion and the possible role of evolution in the development of these structures.

5. Synopsis

This two-volume series on gastrointestinal microbiology is structured into six sections. The first section serves as a historical and general introduction to the gut environment, describing the conditions and the concepts involved in gut microecology. It also discusses the evolution of mutualistic gut fermentations and the impact that studies of molecular evolution have on resolving phylogenetic relationships between animals with mutualistic gut fermentation. The following six chapters are concerned with nutritional ecology across a wide range of foregut and hindgut (cecum/colon) fermenters, birds, fish, reptiles, amphibians, and insects. These deliberations focus on the anatomy and physiology of the host and the wide range of nutritional contributions that the resident microbial population makes. The biochemical activities and metabolic transformations that occur in fermentative gut compartments are covered in the following seven chapters. These biochemical activities include polysaccharide degradation; carbohydrate fermentation; nitrogen metabolism; biotransformation of bile acids, cholesterol, and steroids; gastrointestinal toxicology in monogastrics; and ruminal detoxification of phytotoxins and mycotoxins as well as ruminal digestive disorders.

The second volume begins with a section on microorganisms in the digestive tract of humans, ruminants, hindgut fermenters, rodents, and birds, with emphasis on enumeration, taxonomy, and ecology. The aquisition and development of the intestinal microbiota is also reviewed. The nature of interspecies interactions, the extent of mutual dependence, and the fundamental importance of the influence microbes have on each other is covered in Chapter 10 (Vol. 2). Although the normal microbiota is relatively stable in composition, it can be altered by allogenic factors such as diet, stress, antimicrobial agents, and, potentially, probiotics. These factors are reviewed in Chapter 11. Modern approaches that include molecular ecology in the intestinal tract as well as genetic systems and the potential for manipulation of gut anerobes are covered in Chaps. 7–9, Vol. 2.

The final section focuses on the interactions of microbes in the gastrointestinal tract with the host animal. The activities of the normal microbiota have beneficial effects on the host in terms of colonization resistance, but can also impose a considerable burden in terms of replacement of epithelial cells, detoxification of microbial metabolites, and the production of inflammatory and immunological cells. Colonization resistance includes all factors that hamper colonization of the intestinal tract by exogenous organisms including pathogens. Some of the regulatory factors are exerted by the microbes themselves, while others are exerted by the animal host, its diet, and environment (Chap. 13, Vol. 2). In Chapter 14 (Vol. 2), a summary of our current understanding of the immunological aspects of host-microbe interactions in general and specifically at the mucosal surface is provided. A working hypothesis is proposed of host tolerance to indigenous bacteria and their role in stimulating the host's capacity to respond to enteric pathogens is presented (Chap. 14, Vol. 2). Bacterial pathogens can interact with the gastrointestinal mucosal surface and penetrate this barrier, ultimately resulting in disease. The molecular mechanisms by which these pathogens translocate are described in Chap. 15 (Vol. 2). Finally, the relationship of well-defined pathogenic clones of *E. coli* and *V. cholerae* to other pathogenic and commensal strains of the species are described (Chap. 16, Vol. 2). Molecular tools that are highly descriptive and specific are used to describe population genetics and pathogenesis/virulence of their respective variants.

References

Alexander M (1971) Microbial Ecology. New York: John Wiley.

Allard MW, Miyamoto MM, Jarecki L, Kraus F, Tennant MR (1992) DNA systematics and evolution of the artiodactyl family Bovidae. Proc Natl Acad Sci USA 89:3972–2976.

Amann RI, Ludwig W, Schleifer KH (1994) Identification of uncultured bacteria: a challenging task for molecular taxonomists. ASM News 60:360–365.

Amann RI, Ludwig W, Schleifer KH (1995) Phylogenetic identification and in situ detection of individual microbial cells without cultivation. Microbiol Rev 59:143–169.

Barnard EA (1969) Biological function of pancreatic ribonuclease. Nature 221:340–344.

Berg RD (1992) Translocation and the indigenous gut flora. In: Fuller R, ed. Probiotics: The Scientific Basis, pp. 55–58. New York: Chapman and Hall.

Berg RD, Savage DC (1972) Immunological responses and microorganisms indigenous to the gastrointestinal tract. Am J Clin Nutr 25:1364–1371.

Berg RD, Savage DC (1975) Immune response of specific pathogen-free and gnotobiotic mice to antigens of indigenous and non-indigenous microorganisms. Infect Immun 11: 320–329.

Collinson ME, Hooker JJ (1991) Fossil evidence of interactions between plants and plant-eating mammals. Phil Trans R Soc Lond B 333:197–200.

Demment MW, Van Soest PJ (1985) A nutritional explanation for body-size patterns of ruminant and non-ruminant herbivores. Am Nat 125:641–672.

Dierenfeld ES, Hintz, HF, Robertson JB, Van Soest PJ, Oftedal OT (1982) Utilization of bamboo by the giant panda. J Nutr 112:636–641.

Dobson DE, Prager EM, Wilson AC (1984) Stomach lysozymes of ruminants I. J Biol Chem 259:11607–11616.

Dominguez-Bello MG, Michelangeli F, Ruiz MC, Garcia A, Rodriguez E (1994) Ecology of the folivorous hoazin (*Opisthocomus hoazin*) on the Venezuelan plains. Auk 111: 643–681.

Dubos R, Schaedler RW, Costello R, Hoet P (1965) Indigenous, normal and autochthonous flora of the gastrointestinal tract. J Exp Med 122:67–76.

Farlow JO (1987) Speculations about the diet and digestive physiology of herbivorous dinosaurs. Paleobiology 13:60–72.

Fenchel T, Finlay BJ (1995) Ecology and Evolution in Anoxic Worlds. London: Oxford University Press.

Foo MC, Lee A (1972) Immunological response of mice to members of the autochthonous intestinal microflora. Infect Immun 6:525–532.

Grajal A, Strahl SD, Parra R, Dominguez MG, Neher A (1989) Foregut fermentation in the hoatzin, a neotropical leaf-eating bird. Science 245:1236–1238.

Grajal A, Strahl SD (1991) A bird with the guts to eat leaves. Nat Hist 8/91:48–55.

Grauer D (1993) Molecular phylogeny and the higher classification of eutherian mammals. Trends Ecol Evol 8:141–147.

Grauer D, Higgins DG (1994) Molecular evidence for the inclusion of cetaceans within the order Artiodactyla. Mol Biol Evol 11:357–364.

Grauer D, Duret L, Gouy M (1996) Phylogenetic position of the order Lagomorpha (rabbits, hares and allies). Nature 379:333–335.

Guthrie RD (1984) Mosaics, allelochemics, and nutrients: an ecological theory of late Pleistocene megafaunal extinctions. In: Martin DS, Klein RG, eds. Quartenary Extinctions: a Prehistoric Revolution, pp. 259–298. Tucson: University of Arizona Press.

Hedges SB, Simmons MD, Van Dijk MAM, Caspers G-J, De Jong WW, Sibley CG (1995) Phylogenetic relationships of the hoatzin, an enigmatic South American bird. Proc Natl Acad Sci USA 92:11662–11665.

Herd RM, Dawson TJ (1984) Fiber digestion in the emu, *Dramaius novaehollandiae*, a large ratite bird with a simple gut and high rates of passage. Physiol Zool 57:70–84.

Hume ID, Warner ACI (1980) Evolution of microbial digestion in mammals. In: Ruckebusch Y, Thievend P, eds. Digestive Physiology and Metabolism in Ruminants, pp. 665–684. Lancaster: MTD Press.

Hungate RE (1976) Microbial activities related to mammalian digestion and absorption of food. In: Spiller GA, Amen RJ, eds. Fiber in Human Nutrition, pp. 131–149. New York: Plenum Press.

Hungate RE (1984) Microbes of nutritional importance in the alimentary tract. Proc Nutr Soc 43:1–11.

Irwin DM, Wilson AC (1989) Multiple cDNA sequences and the evolution of bovine stomach lysozyme. J. Biol Chem 264:11387–11393.

Irwin DM, Prager EM, Wilson AC (1992) Evolutionary genetics of ruminant lysozymes. Anim Genet 23:193–202.

Janis C (1976) The evolutionary strategy of the Equidae and the origins of rumen and cecal digestion. Evolution 30:757–774.

Jollès J, Jollès P, Bowman BH, Prager EM, Stewart C-B, Wilson AC (1989) Episodic evolution in the stomach lysozymes of ruminants. J Mol Evol 28:528–535.

Jermann TM, Opitz JG, Stackhouse J, Benner SA (1995) Reconstructing the evolutionary history of the artiodactyl ribonuclease superfamily. Nature 374:57–59.

Kornegay JR, Schilling JW, Wilson AC (1994) Molecular adaptation of a leaf-eating bird: stomach lysozyme of the hoatzin. Mol Biol Evol 11:921–928.

Langer P (1991) Evolution of the digestive tract in mammals. Verh Dtsch Zool Ges 84: 169–193.

Li WH, Gouy M, Sharp PM, O'Huigin, Yang YW (1990) Molecular phylogeny of Rodentia, Lagomorpha, Primates, Artiodactyla, and Carnivora, and molecular clocks. Proc Natl Acad Sci USA 87:6703–6707.

Ludwig W, Dorn S, Springer N, Kirchhof G, Schleifer KH (1994) PCR-based preparation of 235 rRNA-targeted, group specific polynucleotide probes. Appl Environ Microbiol 60:3236–3244.

Mackie RI (1987) Microbial digestion of forages in herbivores. In: Hacker JB, Ternouth JH, eds. The Nutrition of Herbivores, pp. 233–265. Sydney: Academic Press.

Morton ES (1978) Avian arboreal folivores: why not. In: Montgomery GG et al., eds. The Ecology of Arboreal Folivores, pp. 123–130. Washington, DC: Smithsonian Institution Press.

Novacek MJ (1992) Mammalian phylogeny: shaking the tree. Nature 356:121–125.

Parra R (1978) Comparison of foregut and hindgut fermentation in herbivores. In: Montgomery GG et al., eds. The Ecology of Arboreal Folivores, pp. 205–229. Washington, DC: Smithsonian Institution Press.

Rosenthal GA, Janzen DH (1979) Herbivores: Their Interaction with Plant Secondary Metabolites. New York: Academic Press.

Savage DC (1977) Interactions between the host and its microbes. In: Clarke RTJ, Bauchop T, eds. Microbial Ecology of the Gut, pp. 277–310. New York: Academic Press.

Stackhouse J, Presnell SR, McGeehan GM, Nambiar KP, Benner SA (1990) The ribonuclease from an extinct bovid ruminant. FEBS Lett 262:104–106.

Stahl DA, Amann RI (1991) Development and application of nucleic acid probes. In: Stackebrandt E, Goodfellow M, eds. Nucleic Acid Techniques in Bacterial Systematics, pp. 205–248. Chichester: John Wiley.

Stewart CB, Schilling JW, Wilson AC (1987) Adaptive evolution in the stomach lysozymes of foregut fermenters. Nature 330:401–404.

Swart D, Mackie RI, Hayes JP (1993a) Influence of livemass, rate of passage and site of digestion on energy metabolism and fiber digestion in the ostrich (*Struthio camelus* var. *domesticus*). S Afr J Anim Sci 23:119–126.

Swart D, Mackie RI, Hayes JP (1993b) Fermentative digestion in the ostrich (*Struthio camelus* var. *domesticus*), a large avian species that utilizes cellulose. S Afr J Anim Sci 23:127–135.

Tamminga S (1979) Protein degradation in the forestomach of ruminants. J Anim Sci 49: 1615–1630.

Taylor MA (1993) Stomach stones for feeding or buoyancy? The occurrence and function of gastroliths in marine tetrapods. Phil Trans R Soc Lond B 341:163–175.

Thomasson JR, Voorhies MR (1990) Grasslands and grazers. In: Briggs DEG, Crowther PR, eds. Paleobiology: A Synthesis, pp. 84–87. Oxford: Blackwell Scientific.

Williams AG, Coleman GS (1988) The rumen protozoa. In: Hobson PN, ed. The Rumen Microbial Ecosystem, pp. 77–128. New York: Elsevier.

Woese CR, Kandler O, Wheelis ML (1990) Towards a natural system of organisms: proposal for the domains Archaea, Bacteria and Eucarya. Proc Natl Acad Sci USA 87: 4576–4579.

I

NUTRITIONAL ECOLOGY

3

Foregut Fermentation

Burk A. Dehority

1. Introduction

Several strategies have evolved over the years that allow mammals to use plants as their sole energy source. The principal carbohydrate in plant materials is cellulose; however, all vertebrate animal species lack the cellulase enzymes needed to utilize this polysaccharide. Thus, they were forced to adopt a microbial population with this capability. Establishment of a viable microbial fermentation somewhere in the gastrointestinal tract depended on the presence of an area or site that was well separated from the acid-secreting portion of the stomach. Movement of digesta through that area would need to be delayed enough to allow adequate time for microbial digestion of the plant material. It is generally believed that microbial digestion of feedstuffs originated in the hindgut (Hume and Warner 1980). As the intake of relatively indigestible fiber increased, there was a selection pressure for an enlargement of the hindgut and a concomitant decrease in rate of passage. However, in those areas of nutritionally low-quality plants, total intake of plant materials increased and an enlargement of the stomach occurred, either by elongation or the formation of diverticula or sacs. Elongation of the body of the stomach was generally accompanied by sacculation of the greater curvature. Diverticula or sacs were derived from the esophagus or in the cardiac region of the stomach with some type of constriction across the gastric canal (Moir 1968).

Three categories of dietary adaptation have been described for mammals, based on the type of food they consume (Chivers and Hladik 1980). In turn, gut structure appears to be directly related to each animal's dietary adaptation. The faunivores (carnivores), animals consuming primarily animal matter, possess a simple undivided or unilocular stomach, a tortuous small intestine, short conical cecum, and simple, smooth-walled colon. Frugivores (omnivores) consume fruits, flowers, seeds, and tubers, and generally supplement their diet with varying amounts of animal matter. This group is comprised mostly of primates with no distinctive structural specialization unique to their gut. The last category, the folivores (herbivores), are animals that consume leaves, grasses, twigs, barks, and gums, all of which can only be digested by symbiotic microorganisms. These animals are characterized by having chambers in their gastrointestinal tract where microbial fermentation can occur. Foregut fermentation occurs in a plurilocular stomach, which is divided into two, three, or four chambers. Fermentation in the hindgut occurs in the cecum or the cecum and colon.

Langer (1986) has expanded the above classification for dietary adaptations into six categories. The first two are the same; (1) omnivores (faunivores) consume considerable amounts of animal material in their food, and (2) frugivores consume fruits, concentrate-type herbage, and some animal material. However, the last category, folivores, has been subdivided into four separate groups. These are: (3) mammals that select concentrate types of herbage (fruits, bulbs, tubers and herbs); (4) animals that consume fresh leaves and only a few seeds; (5) mammals that eat older leaves plus less digestible material such as grass and twigs; and (6) the obligate grazers or grass and bulk roughage eaters. In general, most of the forestomach-fermenting herbivores consume foodstuffs in categories 4, 5, and 6, whereas the hindgut fermenters tend to consume foods classified under categories 2 and 3 (Langer 1986). The exception to this pattern is in the hindgut-fermenting orders Perissodactyla (horses, rhinoceros, etc.) and Proboscida (elephants). Because of their large body size, these hindgut fermenters are able to increase their total intake of high-fiber foodstuffs. By also increasing rate of passage, they can obtain sufficient energy from the smaller proportion of readily available nutrients and subsequent passage of large amounts of undigested residues. This strategy is not applicable in ruminants because the reticuloomasal orifice restricts passage of the larger-size particulate matter (Janis 1976). In other words, fibrous materials are retained in the rumen until particle size can be reduced by rumination and microbial action. A somewhat similar orifice exists in camelids (Vallenas et al. 1971), but not in chevrotains (Langer 1974) or other forestomach fermenters (Janis 1976). Thus, the nonruminant forestomach fermenters, except camelids, would need to select more energy-dense foods which require less fermentation.

2. Foregut Anatomy and Physiology

Possession of a multichambered or plurilocular stomach does not necessarily imply that a pregastric fermentation is taking place. However, volatile fatty acids

(VFA) are produced only by microbial digestion of feedstuffs, and not by mammalian enzymes, so their presence in stomach contents is used as one of the principal criteria to establish the occurrence of a microbial fermentation. In addition, VFA concentrations can provide a rough estimate of the extent of microbial digestion which occurs and reflects on the probable numbers of microorganisms present. Another indicator of forestomach fermentation is the absence of unsaturated fatty acids of food origin in depot fat. These acids are generally hydrogenated by the microorganisms in the foregut (Shorland et al. 1955).

Moir (1965, 1968) lists several herbivorous species in the order Rodentia which have constrictions dividing the pyloric and cardiac portions of the stomach. The cardiac pouch is lined with squamous epithelium and is without glands. These include various species of mice: *Muscardinus* (dormice), *Notomys* (kangaroo mice), *Peromyscus* (deer mice), *Perognathus* (pocket mice); and rats: *Neotoma* (wood rats), *Dipodomys* (kangaroo rats), *Cricetomys* (giant pouched rats); beaver (*Castor*) and voles (*Microtus*). Information about a forestomach fermentation in these animals is limited. However, Camain et al. (1960) have studied in detail the stomach of the Gambian or giant pouched rat (*Cricetomys gambianus*). It consists of two chambers, a larger cardiac chamber (7.25 × 3.0 cm) which connects by a narrow orifice (1 cm) to a smaller pyloric chamber (3.5 × 2.5 cm). The papillated epithelium in the cardiac chamber was covered by a felt or mat of microorganisms, all of which were presumptively identified as *Bacillus* sp., probably *B. cereus*. Since there are no glands or enzymatic activity in the cardiac chamber, it seems likely that some fermentation of foodstuffs would occur; however, it would differ markedly from that in the ruminant or other foregut fermenters which harbor a dense population of many different types of microorganisms.

Marine herbivorous mammals in the order Sirenia (dugongs and manatees) are multilocular, having at least two compartments in their stomachs. However, they contain such highly unsaturated depot fats that any forestomach fermentation is highly questionable. Primarily, the sea cows and manatees are hindgut fermenters (Marsh et al. 1978).

An unusual type of forestomach fermentation is found in animal species that practice coprophagy or reingestion of feces. Although this practice is most prevalent in rodents and lagomorphs (rabbits), it also has been observed in certain primates and marsupials (Bauchop 1977). Essentially, these animals are all hindgut fermenters, and many of the nutritional benefits from a microbial fermentation are lost by excretion in the feces. Reingestion of the feces allows utilization of these fermentation products, especially microbial proteins and vitamins. Griffiths and Davies (1963) have shown that the "soft" fecal pellets reingested by the rabbit are surrounded by a membrane that remains intact in the stomach for as long as 12 hours. Many bacteria are present in these pellets, and fermentation can thus proceed in an otherwise hostile environment. It should be noted, however, that the principal end product of fermentation in the rabbit stomach was found to be lactic acid, not VFA (Alexander and Chowdhury 1958, Griffiths and Davies 1963). The concentration of lactic acid was always quite low (< 5 mmoles/L), casting serious doubt on its contribution as an energy source (Bauchop 1977).

Although the development of pregastric fermentation chambers in the stomach of herbivores is a very successful strategy for the utilization of poor-quality feeds, it appears to be an extremely wasteful process for the utilization of milk (Black and Sharkey 1970). This was demonstrated by Black (1970), where growth rates in young lambs were doubled when milk was infused directly into the abomasum (acid-producing compartment) as compared to infusion into the rumen. Thus, Black and Sharkey (1970) concluded that the reticular groove (sulcus reticuli) was a necessary adaptation in ruminant and ruminantlike herbivores to ensure the efficient utilization of milk by the young animal. Milk is thus channeled past the site of microbial activity and goes directly to the pyloric section of the stomach, being digested by mammalian enzymes in the same way as in animals with a simple stomach. These authors surveyed the literature and found that all animals that utilize a major forestomach fermentation of feedstuffs also possess a reticular groove. Included in this group are species from the orders Marsupiala (kangaroos), Primates (coloboid monkeys), Edentata (sloths), and Artiodactyla (bovids, camelids, deer, hippos). In contrast, most of the herbivorous animals that have only a limited forestomach fermentation do not possess a reticular groove—i.e., lagomorphs, herbivorous rodents, and the hyrax. Possible exceptions to this are the elephant and lemming, both of which possess a rudimentary reticular groove. However, fermentation in the forestomach of these two species is very limited. Thus, it appears that the reticular groove is present in all herbivorous mammals in which ingested milk would otherwise be digested by microorganisms in the forestomach.

Those herbivores with multichambered forestomachs, in which there is some evidence for the occurrence of a microbial fermentation, are listed in Table 3.1. They have been divided into five categories, ranging from animals in which a minimal amount of fermentation occurs, up to the most advanced type, the ruminant.

2.1. Multichambered Forestomach With Minimal Fermentation

The golden hamster (*Mesocricetus auratus*) and rock hyrax (*Procavia habessinica*) both possess two-chambered stomachs, separated into pregastric and gastric sections (Hume and Warner 1980). Stomach contents from the pregastric chamber contain appreciable amounts of volatile fatty acids (VFA), which indicate a microbial fermentation. Mammals in the order Cetacea (whales, porpoises, and dolphins) also possess multichambered stomachs. The number of stomach chambers reported ranges from three to 12; however, most species have a four-

Table 3.1. Classification of the different types of forestomach-fermenting herbivores

Type	Order[a]	Family	Species	Common Name	No. of Stomach Chambers	References[b]
Multichambered forestomach with minimal fermentation	Rodentia	Cricetidae	*Mesocricetus auratus*	Golden hamster	2	1,2
	Hyracoidea	Procaviidae	*Procavia habessinica*	Rock hyrax	2	3
	Cetacea	Balaenidae	*Balaena mysticetus*	Bowhead whale	4	4,5
		Eschrichtidae	*Eschrichtius robustus*	Gray whale	4	4
		Delphinidae	*Neomeris phocaenoides*	Finless porpoise	4	6
			Stenella sp. (3)[c]	Spotted dolphins	4	7,8
Ruminantlike adaptations	Primates	Cercopithecidae	*Presbytis* sp.	Langur monkeys	4	9,11,12
			Colobus sp.	Colobus monkeys	4	11
			Nasalis larvatus	Proboscis monkey	4	11
	Edentata	Bradypodidae	*Bradypus tridactylus*	Three-toed sloth	4	9,10,25
			Choloepus didactylus	Two-toed sloth	4	9,25
	Artiodactyla	Tayassuidae	*Tayassu* sp. (2)[d]	Peccaries	4	13
		Hippopotamidae	*Hippopotamus amphibius*	Hippopotamus	4	10,11,14
	Marsupialia	Macropodidae	*Macropus* sp.[e]	Kangaroos	3	15,19,20
			Thylogale sp.[f]	Wallabies	3	16,19
			Setonix brachyurus	Quokka	3	17,19
			3 general[g]	Rat-kangaroos	3	18,19
	Opisthocomiformes	Opisthocomidae	*Opisthocomus hoazin*	Hoatzin	4	21
Pseudoruminants	Artiodactyla	Camelidae	*Camelus* sp.	Camels	3	9,12,22
			Lama sp.[h]	Llama, alpaca, guanaco	3	9,12,22
			Vicugna vicugna	Vicuña	3	9,12,22
Preruminants	Artiodactyla	Tragulidae	*Tragulus* sp.	Chevrotains	3	12,23
			Hyemoschus aquaticus	Water chevrotain	3	12,23
Ruminants	Artiodactyla	Cervidae	17 genera	Deer, moose, elk	4	12,24
		Giraffidae	*Giraffa camelopardalis*	Giraffe	4	12,24
			Okapia johnstoni	Okapi	4	12,24
		Antilocapridae	*Antilocapra americana*	Pronghorn antelope	4	12,24
		Bovidae	49 genera	Cattle, sheep, goats, buffalo[i]	4	12,24

[a] All orders are in the class Mammalia except for Opisthocomiformes, which is in the class Aves.
[b] (1) Hoover et al. 1969; (2) Ehle and Warner 1978; (3) Clemens 1977; (4) Herwig et al. 1984; (5) Tarpley et al. 1987; (6) Morii and Kanazu 1972; (7) Morii 1972; (8) Morii 1979; (9) Bauchop 1977; (10) Moir 1965; (11) Bauchop 1978; (12) Moir 1968; (13) Langer 1978; (14) Langer 1976; (15) Langer et al. 1980; (16) Langer 1979; (17) Moir et al. 1956; (18) Langer 1980; (19) Hume 1982; (20) Hume and Sakaguchi 1991; (21) Grajal et al. 1989; (22) Vallenas et al. 1971; (23) Hume and Warner 1980; (24) Church 1976; (25) Britton 1941.
[c] *S. attenuata, S. caeruleoalbea,* and *Stenella* sp.
[d] *T. tajacu* (collared peccary), *T. pecari* (white-lipped peccary).
[e] Principal species studied: *M. giganteus* (eastern gray kangaroo, *M. rufus* (red kangaroo), *M. eugenii* (tammar wallaby).
[f] Principal species studied: *T. stigmatica* (red-legged pademelon wallaby), *T. thetis* (red-necked pademelon wallaby).
[g] Species studied: *Bettongia lesueur, Aepyprymnus rufecens, Potorous tridactylus.*
[h] *L. guanacoe* (guanaco), *L. peruana* (llama), *L. pacos* (alpaca).
[i] Additional species: bushbucks, kudus, bongos, elands, bison, duikers, antelope, wildebeests, gazelles, waterbucks, reedbucks, takir, muskoxen.

chambered stomach structure (Tarpley et al. 1987). Relatively high concentrations of VFA have been found in forestomach contents from the bowhead and gray whale, spotted dolphin, and black finless porpoise (Morii 1972, 1979; Morii and Kanazu 1972; Herwig et al. 1984).

GOLDEN HAMSTER

Figure 3.1A is a diagrammatic sketch of the stomach of the golden hamster (*Mesocricetus auratus*). A constriction separates the pregastric and gastric com-

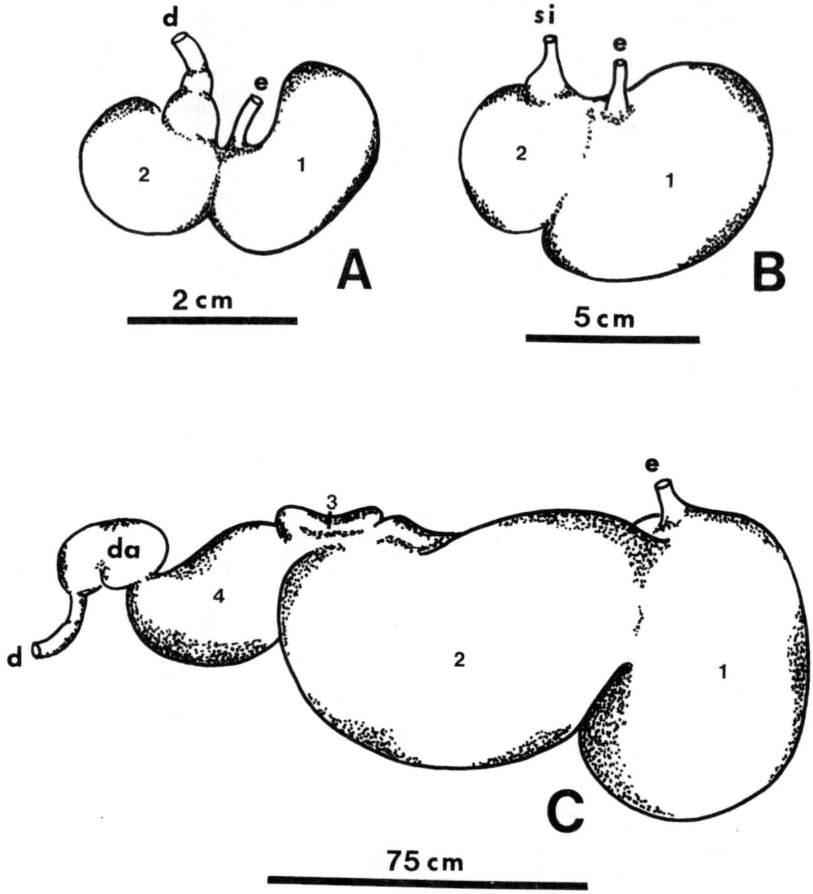

Figure 3.1. Diagrammatic sketches of golden hamster, rock hyrax, and bowhead whale stomachs. (A) Golden hamster (*Mesocricetus auratus*): (1) pregastric pouch; (2) gastric pouch; (e) esophagus; (d) duodenum. Drawn from photographs in Hoover et al. (1969) and Ehle and Warner (1978). (B) Rock hyrax (*Procavia habessinica*): (1) cranial stomach; (2) caudal stomach; (e) esophagus; (si) small intestine. Drawn from Clemens (1977). (C) Bowhead whale (*Balaena mysticetus*): (1) forestomach; (2) fundic chamber; (3) connecting channel; (4) pyloric chamber; (e) esophagus; (da) duodenal ampulla; (d) duodenum. Drawn from Tarpley et al. (1987).

partments, and this area contains a sphincterlike muscular structure which may play a role in the movement of ingesta between the two compartments (Hoover et al. 1969). The esophagus enters the pregastric pouch very close to the sphincter area, and most of the ingested feedstuffs appear to enter directly into the forestomach or pregastric chamber. However, the ingesta moves quite rapidly into the glandular stomach, having a residence time in the forestomach of 1 hour or less (Ehle and Warner 1978). When these authors surgically removed the forestomach, apparent digestibilities of protein, energy, and neutral detergent fiber were not affected.

VFA concentrations in the pregastric pouch are similar to those in the rumen (160 mmol/L) but are markedly lower (45 mmol/L) in the gastric pouch (Hungate 1966, Hoover et al. 1969). This would suggest that some VFA absorption does occur in the pregastric-gastric area. Based on the short residence time of ingesta in the pregastric pouch, Ehle and Warner (1978) concluded that a microbial fermentation in the pregastric chamber would be of limited nutritional benefit. It seems probable that cecal fermentation may be of greater importance in the golden hamster.

ROCK HYRAX

The stomach of the rock hyrax (*Procavia habessinica*) is depicted in Figure 3.1B. Particulate matter was retained for about 4 hours after feeding in the cranial stomach section, and 50% to 80% of the particulate markers were still present in the foregut after 8 hours (Clemens 1977). The rock hyrax digestive tract is unique in that an enlarged chamber, called the midgut sacculation, occurs between the small intestine and the colon. There are three major sites of digesta retention in the rock hyrax, and based on VFA production, a microbial fermentation appears to take place in all three sites—the cranial stomach, the midgut sacculation, and the paired ceca (Clemens 1977). VFA concentrations in the cranial stomach were comparable to those found in the rumen, 125 to 145 mmol/L (Hungate 1966), while lower but still appreciable VFA concentrations were present in the midgut sacculation and ceca, 60 to 80 mmol/L. At this time, the extent of digestion occurring in the cranial section of the stomach is unknown.

BALEEN WHALES

Figure 3.1C presents a diagrammatic sketch of the multichambered stomach of the bowhead whale (*Balaena mysticetus*). The dimensions shown are taken from the stomach of a 9-m bowhead whale (Tarpley et al. 1987). The forestomach is a large, nonglandular sac lined by keratinized stratified squamous epithelium which is connected to the fundic chamber through a rather large orifice (about 10 cm). All three of the latter chambers are glandular. The connecting channel, which connects the fundic and pyloric chambers, is a narrow tubular structure approximately 3.5 × 23 cm with orifices at each end of about 3.0 cm. The pyloric chamber is also a tubular structure, but of considerably larger dimensions (18 × 50 cm). The pyloric sphincter separates the pyloric chamber from the cranial portion of the duodenum, which is a dilated sac called the duodenal ampulla.

VFA concentrations, measured in forestomach samples from seven bowhead whales, averaged 301 mmol/L with a range of 8 to 1,060 mmol/L. Similar samples from four gray whales had an average VFA concentration of 208 mmol/L, ranging from 78 to 558 mmol/L (Herwig et al. 1984). These concentrations of VFA are in a range similar to values reported from rumen contents of both sheep and cattle (Hungate 1966). One major difference in the forestomach fermentation of baleen whales and ruminants is that the land ruminants feed primarily on plant materials while the baleen whales feed mostly on chitinous invertebrates and small fish (Herwig et al. 1984). The absence of papillae in the whale forestomach would also limit the extent of metabolite absorption (Tarpley et al. 1987). Whether a fermentation occurs in the whale forestomach, which is of significance to its nutritional economy, remains to be determined. The available evidence is primarily based on samples collected anywhere from 3 to 12 hours or more after death of the animal, raising the question of postmortem fermentation (Herwig et al. 1984).

2.2. Ruminantlike Adaptations

In contrast to the animals listed in the previous category, which have only a limited fermentation of feedstuffs in their forestomachs, these species rely on microbial activity in the foregut for a major portion of their energy and nutrient requirements. A wide range of animal orders in the class Mammalia are represented in this group, in addition to one representative from the class Aves (Table 3.1). The various animals have forestomachs with either three or four compartments; however, their shape and complexity vary markedly. Classification into this category of ruminantlike adaptations is primarily based on seven criteria (Moir et al. 1956; Moir 1965; Langer 1975, 1979a):

1. Gastric chambers that store the digesta and slow its rate of passage through the stomach.
2. Microorganisms that help to digest the foodstuffs.
3. Microbial fermentation products are absorbed through the walls of the forestomach.
4. Vitamins produced by the microorganisms are available to the host.
5. The microorganisms provide a means for the host to utilize nonprotein nitrogen as a protein source.

6. An effective ventricular sulcus (esophageal groovelike structure) is necessary in the young animals.
7. Microorganisms that can adapt to and metabolize toxic plant materials to harmless substances.

OLD WORLD OR COLOBID MONKEYS

Three genera of monkeys in the family Cercopithecidae, subfamily Colobinae, are unique among primates, based on the size and complexity of their stomach (Hill 1952). All three genera are characterized by having long tails and are vegetarians or herbivores, existing primarily on leaves. This latter characteristic accounts for their common designation as leaf-eating monkeys (Bauchop 1977, 1978). Three species occur in the genus *Colobus*, which are found across central Africa. More than 20 species of *Presbytis*, commonly referred to as langurs, are found in southern Asia from India to China and Indonesia. The third genus, *Nasalis*, contains only a single species, *N. larvatus*, which is found only on the island of Borneo (Walker 1964, Sanderson 1967).

A diagrammatic sketch of the stomach of a langur monkey (*Presbytis* sp.) is shown in Figure 3.2A. Terminology for the four sections is taken from Hill (1952) and Kuhn (1964). An esophageal groove extends from the esophagus along the lesser curvature of the tubular portion down to the pyloric region. Anatomy of the stomach allows distinct separation of the acid-producing pyloric region from the forestomach area (presaccus and saccus). The pH of forestomach contents in langur monkeys was found to range between 5.0 and 6.7 and in colobus monkeys between 5.5 and 7.0 (Kuhn 1964, Bauchop and Martucci 1968, Ohwaki et al. 1974).

A large stomach capacity (11.5% to 31.7% of body weight) ensures a slow passage rate for the ingested plant materials (Kuhn 1964, Bauchop and Martucci 1968, Ohwaki et al. 1974). The three factors, pH, capacity, and reduced rate of passage, provide an ideal environment for growth of anaerobic microorganisms.

VFA concentrations in stomach contents of both colobus and langur monkeys are similar to those found in rumen contents (Bauchop 1977). A high rate of VFA production in vitro, in fresh samples of stomach contents, plus maintenance of a high, relatively constant VFA concentration in stomach contents between 1.5 and 6.5 hours after feeding, would suggest a continuous microbial fermentation (Kuhn 1964, Bauchop and Martucci 1968). Drawert et al. (1962) measured a decrease in the VFA concentration of stomach contents in *Colobus verus* from 230 mmoles/L in the midstomach area to 24 mmoles/L in the pyloric region. This suggests that the VFA are absorbed directly from the forestomach area, similar to absorptions from the rumen.

Bauchop and Martucci (1968) used their data to calculate that a langur monkey could obtain 130% of its daily maintenance energy requirement from VFA pro-

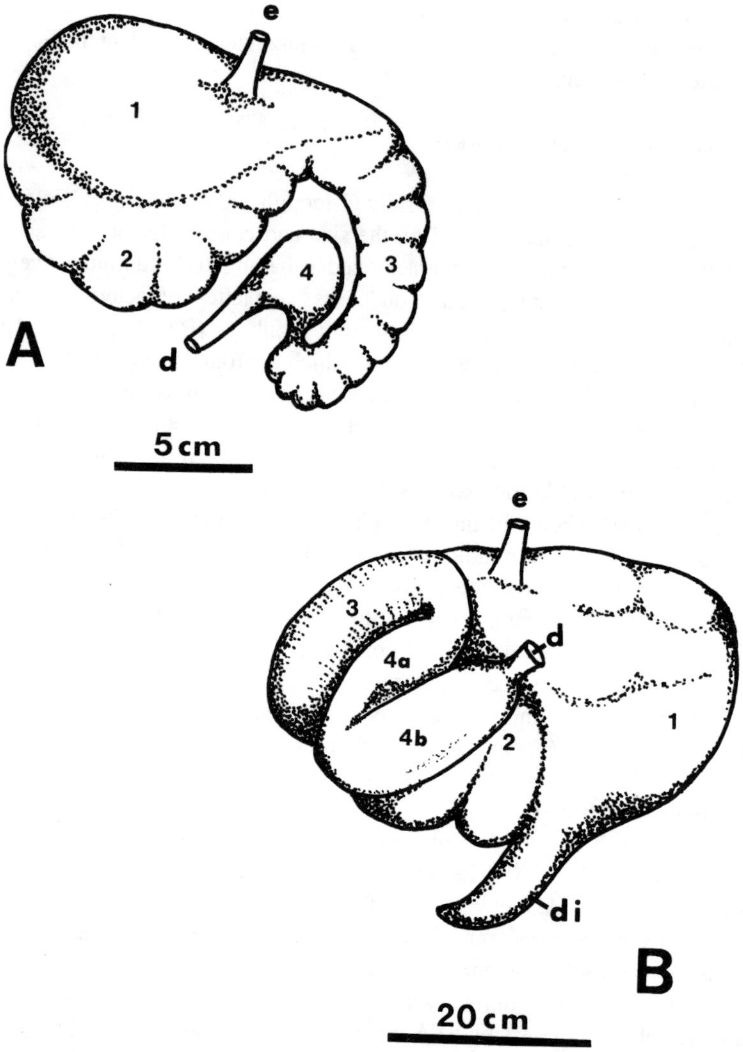

Figure 3.2. Diagrammatic sketches of langur monkey and three-toed sloth stomachs. (A) Langur monkey (*Presbytis* sp.): (1) presaccus; (2) saccus; (3) tubular portion; (4) pyloric region; (e) esophagus; (d) duodenum. Drawn from photographs in Bauchop (1977) and Chivers and Hladik (1980). (B) Three-toed sloth (*Bradypus tridactylus*): (1) right gastric cecum; (2) middle gastric cecum; (3) left gastric cecum; (4a) glandular stomach; (4b) muscular stomach; (e) esophagus; (d) duodenum; (di) diverticulum. Drawn from diagrams in Britton (1941) and Bauchop (1977).

duction. Thus it can be concluded that forestomach microbial fermentation of plant feedstuffs is a major source of energy for coloboid monkeys.

TREE SLOTHS

Two genera of tree sloths, *Bradypus* (three-toed sloths) and *Choloepus* (two-toed sloths), inhabit the tropical forests in central and the upper two-thirds of South America. *Bradypus* species (six or less?) consume only young leaves, tender twigs, and buds of *Cecropia* (large genus of tropical American trees of the mulberry family), while the two species of *Choloepus* eat a variety of leaves, tender twigs, and fruit (Walker 1964). This difference in eating behavior is believed to be the main reason that *Choloepus* is easier to keep in captivity.

Figure 3.2B is a diagrammatic sketch of the stomach of the three-toed sloth *Bradypus tridactylus*, which contains four distinct compartments. The first and largest is termed the right gastric cecum, which is separated from the middle and left gastric ceca by internal pillars. A short esophageal groove extends along the left gastric cecum to the pyloric region. The left gastric cecum leads to the glandular stomach, which connects to the muscular stomach which ends in the pyloric sphincters (Britton 1941, Bauchop 1977). Based on its heavy muscular wall and semidry fibrous and seedy contents, Britton (1941) has suggested that the muscular stomach is a gizzardlike organ that can exert considerable pressure on feedstuffs in its lumen. The stomach and its contents comprise from 20% to 30% of the total body weight (Britton 1941). Rate of passage of foodstuffs appears to be quite slow. A meal of bananas and fruit remained in the stomach for 70 to 90 or more hours after ingestion (Britton 1941). This would imply even longer retention times for less digestible leaves and twigs.

Denis et al. (1967) measured pH of forestomach contents of two *Choloepus* and found them to vary between 5.2 and 5.8 and between 6.4 and 6.7, respectively. When the animals were starved for 1 or 2 days, pH of the contents remained in the range of 7.0 to 7.7, strongly suggesting some mechanisms for maintaining a pH suitable for microbial growth. VFA concentrations were near the bottom of the range that occurs in ruminants (37 to 95 mmoles/L); however, they were shown to be absorbed by the gastric mucosa, as has been found in other ruminantlike animals (Denis et al. 1967, Moir 1965). Percentage composition of VFA in stomach contents of the tree sloth has not been reported to date.

Although information about foregut fermentation in the tree sloth is limited, most observations are indicative of microbial activity, which is important to the overall energy economy of the animal. A large, multichambered forestomach is present in which foodstuffs are retained for extended periods of time. Some mechanism exists for buffering stomach contents, and VFA are produced and absorbed through the gastric mucosa.

PECCARIES

There are three species of peccaries in the family Tayassuidae—the collared peccary (*Tayassu tajacu*), the white-lipped peccary (*Tayassu pecari*), and the recently discovered Chacoan peccary (*Catagonus wagneri*). Distribution of the collared peccary ranges from southern Texas and Arizona southward to Argentina; the white-lipped peccary is found from southern Mexico to southern Brazil and Paraguay; distribution of the Chacoan peccary is limited to Paraguay and northeastern Argentina (Sowls 1984).

The multichambered stomach of the collared peccary is shown diagrammatically in Figure 3.3A. The esophagus opens into the gastric pouch, which is the largest of the four chambers (45% of total stomach volume). A ventricular sulcus (esophageal groovelike structure) runs from the esophageal opening to the glandular stomach. An upper blindsac (11% of total stomach volume) and anterior blindsac (30% of volume) are connected with the gastric pouch at the blindsac junction near the esophageal opening (Langer 1978, 1979a). The pH of contents from the first three compartments of the collared peccary stomach has been reported to range from 5.0 to 6.2 in animals fed experimental diets, and from 5.7 to 6.1 in wild animals (Carl and Brown 1983, Sowls 1984, Lochmiller et al. 1989). VFA concentrations (mmoles/L) have ranged from 81 to 166 in forestomach contents and from 30 to 90 in glandular stomach contents (Langer 1978, Sowls 1984, Lochmiller et al. 1989). Clearly, the concentrations of VFA in the forestomach are similar to those found in the rumen, and they are absorbed through the forestomach wall.

Although several studies indicated that stomach fluids from the peccary were unable to digest cellulose (Sowls 1984, Shively et al. 1985), Carl and Brown (1986) found the mean rate of digesta passage of high-fiber diets to be 52.5 hours in the peccary. Based on an observed 41.4% digestion of cellulose in this diet, they concluded that the peccary may digest fiber as well as other herbivores. Lochmiller et al. (1989) observed an increase in VFA concentrations among the small intestine (14.6 mmoles/L), cecum (61.5 mmoles/L), and large intestine (63.1 mmoles/L). Based on these data and previous observations, they postulated that most of the cellulose is digested in the hindgut. Motile bacteria have been observed in peccary cecal contents (Sowls 1984).

HIPPOPOTAMUS

There are two genera in the family Hippopotamidae, with a single species in each genera. The big or greater hippopotamus (*Hippopotamus amphibius*) is common in most of the great rivers of tropical Africa although it also is found in some small streams and lakes. The second species or pygmy hippopotamus (*Choeropsis liberiensis*) occurs in limited numbers and lives mostly in the

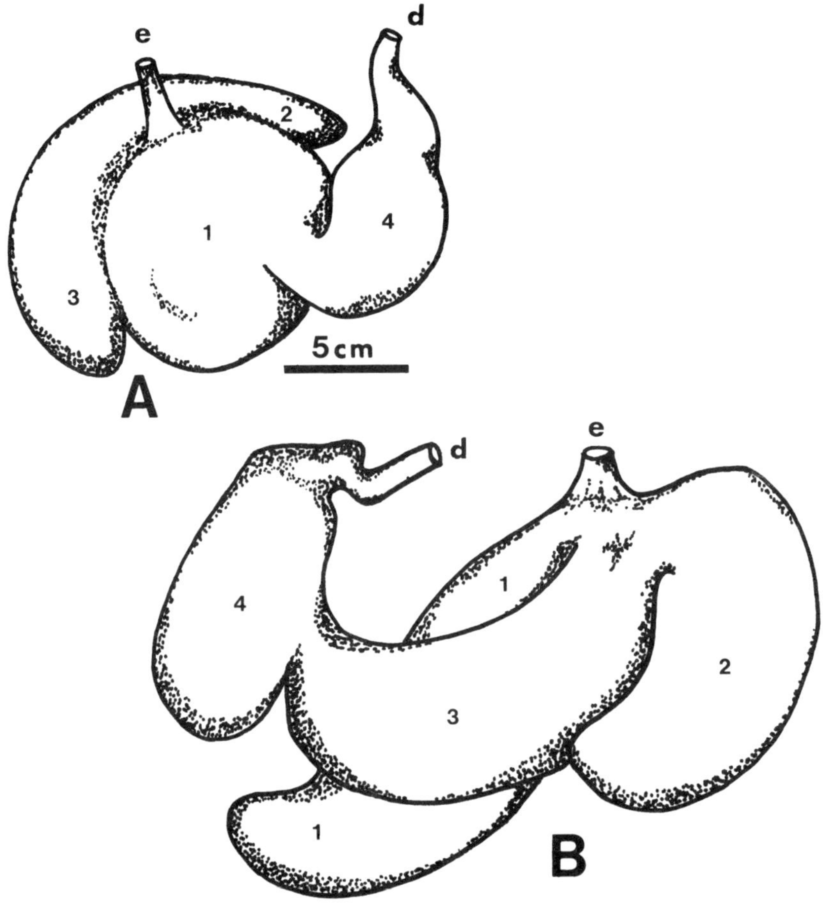

Figure 3.3. Diagrammatic sketches of peccary and hippopotamus stomachs. (A) Collared peccary (*Tayassu tajacu*): (1) gastric pouch; (2) anterior blindsac; (3) upper blindsac; (4) glandular stomach; (e) esophagus; (d) duodenum. Drawn from photographs and diagrams in Langer (1978, 1979a, 1984) and Sowls (1984). Size scale approximated from Langer (1979a). (B) Hippopotamus (*Hippopotamus amphibius*): (1) right diverticulum or visceral blindsac; (2) left diverticulum or parietal blindsac; (3) anterior stomach or connecting compartment; (4) posterior or glandular stomach; (e) esophagus; (d) duodenum. Drawn (not to scale) from photographs and diagrams in Bauchop (1977) and Langer (1976).

streams, wet forests, and swamps of West Africa (Walker 1964, Sanderson 1967). Adult pygmy hippos weigh between 160 and 240 kg, compared to the big hippo which weighs between 3,000 and 4,500 kg (Walker 1964). Almost all of the information available on stomach contents has been collected from the large hippos. In most cases they were shot in the water after grazing overnight. The carcasses rose to the surface about 1 hour after death, due to bloat, at which time samples were collected (Thurston et al. 1968, Van Hoven 1974). It is possible that the various values measured could have been influenced by the postmortem buildup of fermentation products (Thurston et al. 1968).

A diagrammatic sketch of the hippopotamus stomach is shown in Figure 3.3B. The esophagus enters into a small vestibule which opens into two large diverticulae or blindsacs as well as an anterior or connecting stomach compartment, which leads to the glandular stomach. A ventricular sulcus (esophageal groovelike structure) leads from the esophageal opening along the lesser curvature of the connecting stomach compartment (Moir 1968, Langer 1976, Bauchop 1977). In an adult hippo, stomach volume is in the same range as that of adult cattle (Langer 1976). The connecting compartment comprises about 52% of the total stomach volume, with the two blindsacs occupying about 21% each, and the glandular stomach 6%.

The pH of fluid from the forestomach of seven hippos ranged from 5.0 to 5.7; however, the samples were taken at least 2 hours after death. In contrast, pH in the glandular stomach was around 2. VFA concentrations in these same animals were as follows in mmoles/L: diverticulae, 110 to 170; connecting or anterior stomach, 60 to 180; glandular stomach, 25 to 55; small intestine, 6 to 37 (Thurston et al. 1968). Even if postmortem changes slightly decreased pH and increased VFA concentrations, they are certainly in the same range as for ruminants (Hungate 1966). The lower concentration of VFA in the glandular stomach indicates absorption through the forestomach wall. Depot fats in the hippopotamus are highly saturated, which is characteristic of animals with a foregut fermentation (Mattson et al. 1964). Arman and Field (1973) measured similar dry-matter digestibilities in sheep and a hippo, indicating that forestomach fermentation of feedstuffs is an important factor in nutrition of the hippo.

HERBIVOROUS MACROPODID MARSUPIALS

The family Macropodidae, in which all species are forestomach fermenters, is divided into two subfamilies—Macropodinae (kangaroos and wallabies, 37 species), and Potoroinae (rat-kangaroos, nine species). These animals are primarily found in Australia, with lesser numbers in New Guinea and Tasmania (Hume 1982). The habitat of these macropodids ranges from moist forest to desert. Body size (length of body and head), tail length, and weight vary as follows: 235 to 520 mm, 130 to 400 mm, and 0.5 to 3.2 kg for rat-kangaroos; 310 to 1050 mm,

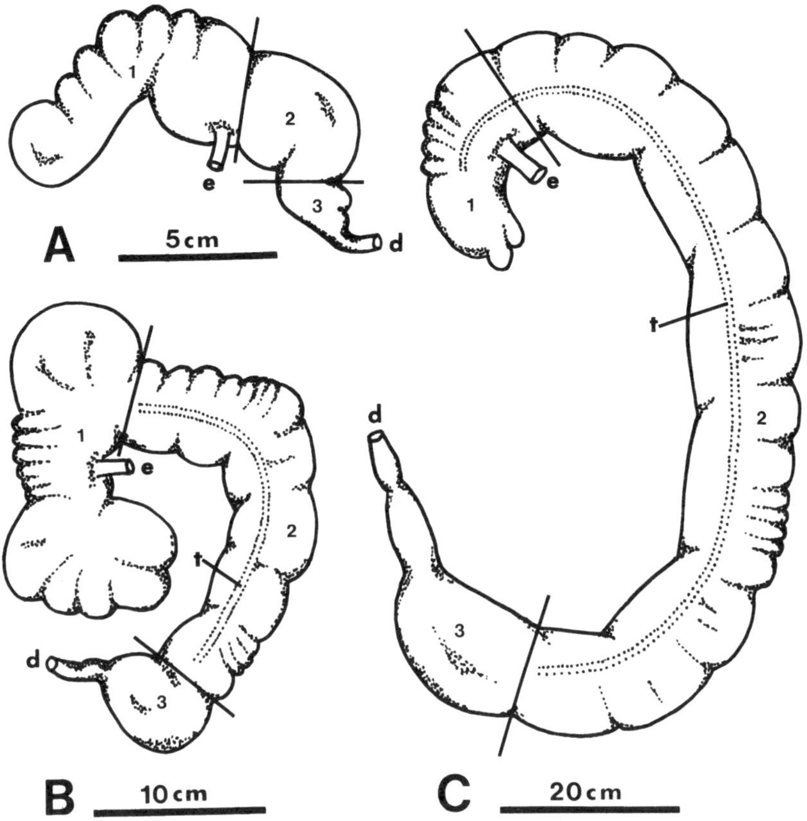

Figure 3.4. Diagrammatic sketches of stomachs of (A) rat-kangaroo (*Bettongia* spp.). (B) Red-necked pademelon wallaby (*Thylogale thetis*). (C) Eastern gray kangaroo (*Macropus giganteus*). (1) Sacciform forestomach; (2) tubiform forestomach; (3) hindstomach; (e) esophagus; (d) duodenum; (t) taenia. Transverse lines show the approximate location of the three stomach sections. Drawn from photographs and diagrams in Hume (1978, 1982) and Dellow (1979).

250 to 750 mm, 1.7 to 24 kg for wallabies; and 800 to 1600 mm, 700 to 1100 mm, and 23 to 70 kg for kangaroos (Walker 1964).

Figure 3.4 presents diagrammatic sketches of representative stomachs in the three macropodid types (kangaroo, wallaby, and rat-kangaroo). The enlarged forestomach is divided into two regions where microbial fermentation of feedstuffs takes place, the sacciform forestomach and the tubiform forestomach (Dellow 1979). Although the opening between these two forestomach regions is wide, the two regions are separated by a permanent ventral fold (Langer et al. 1980), which is adjacent to the cardia (esophageal opening). The third chamber, or hindstomach, is the glandular, acid-secreting region. The overall stomach size varies between the different species; however, relative volumes of the different compartments or regions also vary among the three macropodid types. Relative volumes of the sacciform forestomach in the rat-kangaroo, wallaby, and kangaroo are 75%, 51%, and 23%, respectively; the tubiform forestomach, 14%, 40%, and 70%, respectively; and the hindstomach, 11%, 9%, and 7%, respectively (Dellow 1979, Langer 1980, Hume and Sakaguchi 1991).

In almost all of the macropodids, a gastric sulcus (esophageal groovelike structure) runs from the cardia along the lesser curvature of the tubiform forestomach to the hindstomach (Hume 1982). However, the gastric sulcus is absent from two species of *Thylogale (T. stigmatica* and *T. thetis), Petrogale penicillata,* and *Peradorcas concinna.* In contrast to other forestomach fermenters, the role of the gastric sulcus in the macropodids does not appear to be channeling milk directly to the hindstomach, but rather in assisting caudal movement of liquid digesta. Dellow (1979) found that the initial distribution and subsequent movement of barium sulfate meals was related to the relative sizes of the sacciform and tubiform forestomachs, the presence or absence of a gastric sulcus, and the location of the cardia in relation to the ventral fold between the sacciform and tubiform regions. In the two *Thylogale* species without a gastric sulcus, it was noted that the cardia opens into the sacciform forestomach, whereas in all other macropodids it opens into the tubiform forestomach or on the dividing fold (Dellow 1979, Langer 1979b, Hume 1982).

Moir et al. (1956) measured a wide range of pH values in forestomach contents of the quokka, 4.6 to 8.0. The animals were fed a diet of commercial sheep pellets and chopped oaten hay and sampled at different times after feeding. The highest pH (8.0) occurred in an animal that had been fasted 22 hours. In a later study, Dellow (1979) measured pH in forestomach contents of all three macropodid types. Values ranged from 6.4 to 7.4 in animals eating alfalfa hay ad libitum. Since the chemical composition of parotid and submaxillary saliva was found to be similar among the red kangaroo, sheep, and cattle (Forbes and Tribe 1969, Hume 1982), it would be expected to potentially have a role in buffering the forestomach contents.

VFA concentrations in forestomach contents of the fully fed quokkas ranged from 55 to 147 mmoles/L, and decreased to 7.9 to 13.3 mmoles/L in the hindstomach (Moir et al. 1956). Thus it appears that a fermentation is taking place in the forestomach and that the VFA end products are being absorbed directly from the forestomach. Dellow (1979) made similar measurements in all three macropodid types. VFA concentrations in forestomach contents ranged from 109 to 165 mmoles/L in the sacciform forestomach and from 72 to 131 mmoles/L in the tubiform forestomach. Hindstomach values ranged from 32 to 56 mmoles/L, substantiating the absorption of these acids in the forestomach.

The forestomach appears to play a significant role in digestion of feedstuffs in the macropodids. Between 62% and 65% of OM matter in alfalfa was digested in the stomach. Total acid detergent fiber (cellulose) digestion in the three types ranged from 36.9% to 39.4%; 81.6% to 85% of this occurred in the stomach (Dellow 1979).

HOATZIN

The only animal outside the class Mammalia known to have a significant forestomach fermentation is the hoatzin, in the class Aves (Grajal et al. 1989). The hoatzin (*Opisthocomus hoazin*) is a neotropical leaf-eating bird that is found from the Guianas to Brazil. It is about 750 g in body weight and inhabits swamps and forests. It is estimated that the hoatzin's diet is over 80% green leaves, of which 90% are from only 17 plant species. The hoatzin seems to prefer the new growth.

Figure 3.5 is a diagrammatic sketch of the hoatzin's stomach. The crop and lower esophagus comprise 60.6% and 13.9% of the weight of the total gut, respectively, and are considered to be the areas where a foregut fermentation takes

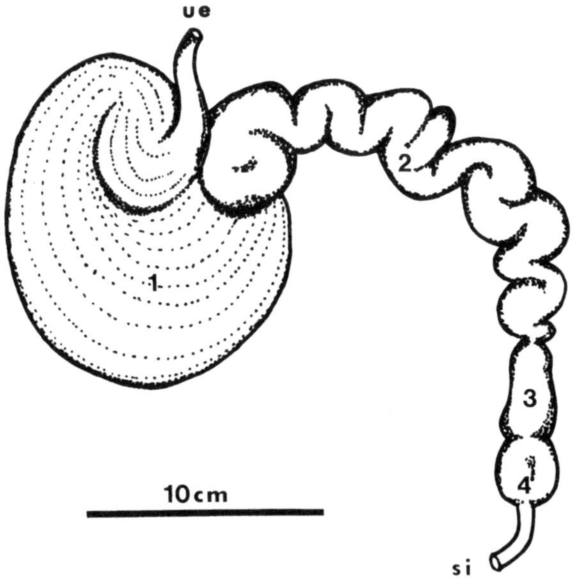

Figure 3.5. Diagrammatic sketch of the hoatzin's foregut: (1) crop; (2) lower esophagus; (3) proventriculus or glandular stomach; (4) gizzard; (ue) upper esophagus; (si) small intestine. Redrawn from Grajal et al. (1989).

place. The proventriculus is the glandular or acid-secreting region, and the gizzard has a grinding and particle size reduction function (Grajal et al. 1989).

Both pH and VFA concentrations in contents from the crop and upper esophagus suggest a microbial fermentation. For the crop and upper esophagus, pH values were 6.3 and 6.7, respectively, and VFA concentrations were 114.5 and 70.3 mmoles/L, respectively. Mean retention times, estimated from pulse doses of plastic markers, indicated that passage rate was slow enough to allow a microbial fermentation. Digestibility trials showed a 34.4% digestibility of cell walls, and the fluid portion from the hoatzin's crop was identical to cattle rumen fluid in digesting plant cell walls in vitro. Thus, all evidence indicates that forestomach fermentation is of significance in the hoatzin.

2.3. Pseudoruminants

Herbivores in the suborder Tylopoda, family Camelidae, are often referred to as pseudoruminants. This designation is based on the fact that like true ruminants (suborder Ruminantia), they ruminate or regurgitate and remasticate ingesta from the forestomach. Included in the family Camelidae are two species of Old World camels (the one-humped camel, *Camelus dromedarius*, and the two-humped camel, *C. bactrianus*) and four species of New World camelids (llama, *Lama peruana*; alpaca, *L. pacos*; guanaco, *L. guanacoe*; and vicuña, *Vicugna vicugna*).

Figure 3.6A shows a diagrammatic three-dimensional sketch of the llama stomach, and 3.6B is a sketch of the llama and guanaco stomach in cross section, extended to show the different compartments. The camelid stomach contains three distinct compartments, the first of which is quite large, containing 83% of the total gastric volume (Vallenas et al. 1971). A ventricular or esophageal groove extends from the cardia to the entrance to the third compartment. Although this first compartment has been compared to the rumen, it differs quite markedly. First, two distinct areas of sacculation are visible, the larger one at the bottom of the caudal sac, and the second at the bottom of the cranial sac. The recessed saccules are lined by glandular mucosa, in contrast to the exposed surfaces which are covered by stratified squamous epithelium. The saccular orifices are partially covered by mucosal diaphragms. During contraction, the saccular mucosa is partially prolapsed or everted through the saccular orifice into the lumen of the compartment. In general, eversion of the saccules is not sustained, and they return to their recessed or inverted form after the contraction. Secondly, the exposed epithelium of the first compartment is nonpapillated. Third, the cardia, or esophageal opening, enters into the upper right part, not into the atrium ventriculi shared by the rumen and reticulum (Figs. 3.6B, 3.8C). And last, the ventricular or esophageal groove in camelids has only a single permanent muscular lip as compared

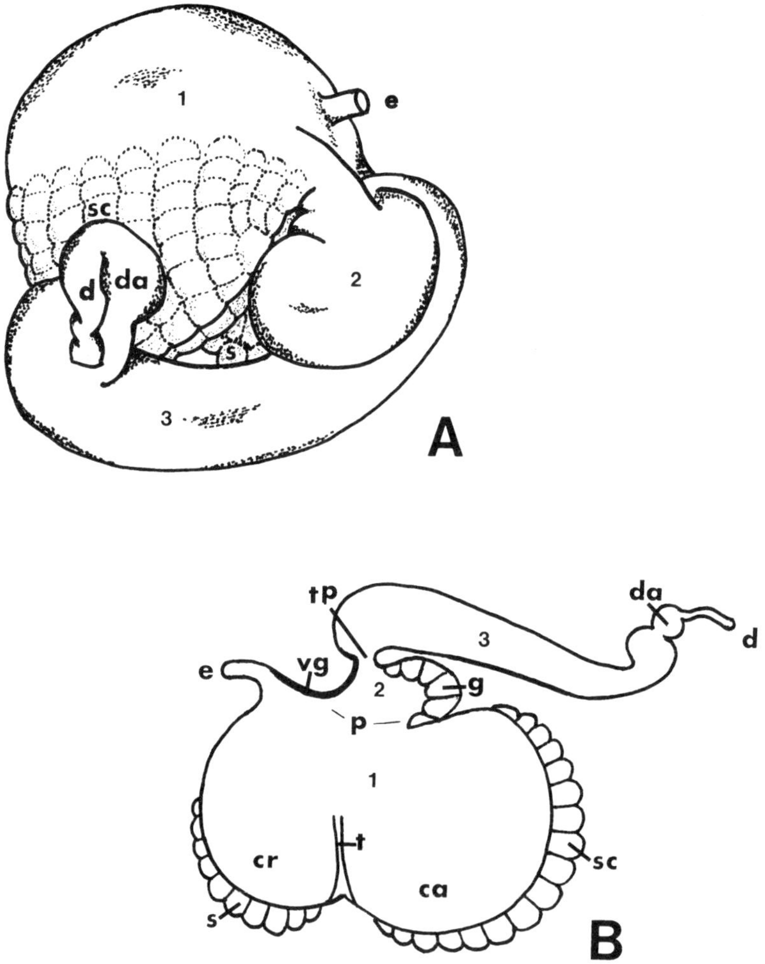

Figure 3.6. (A) Diagrammatic sketch of the llama stomach. (B) Diagrammatic cross section of the llama and guanaco stomach. (1) First compartment; (2) second compartment; (3) third compartment; (da) duodenal ampulla; (d) duodenum; (e) esophagus; (s) sacculated area in cranial sac; (sc) sacculated area in caudal sac; (cr) cranial sac; (ca) caudal sac; (g) glandular cells of second compartment; (t) transverse sulcus; (tp) tubular passage to third compartment; (p) passage to second compartment; (vg) ventricular groove. Redrawn from Vallenas et al. (1971).

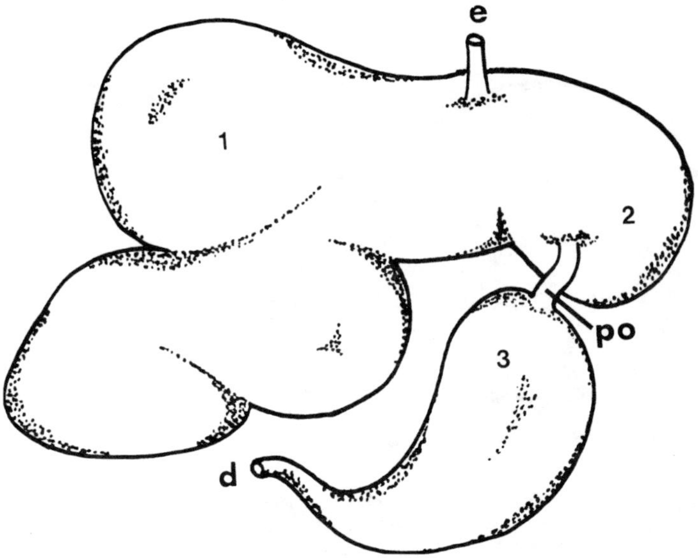

Figure 3.7. Diagrammatic sketch of the chevrotain stomach. (1) Rumen; (2) reticulum; (3) abomasum; (d) duodenum; (e) esophagus; (po) preomasum, tubular passageway connecting the reticulum and abomasum. Redrawn from Moir (1968).

to the ruminant, in which the groove is delineated by two permanent muscular lips.

The second compartment is quite small, containing only 6% of the total gastric volume (Vallenas et al. 1971). Most of this compartment is lined with stratified squamous epithelium, except at the bottom or greater curvature, where deep glandular cells occur. However, unlike the saccules in the first compartment, the glandular mucosa of these cells is papillated, and the basal portion of the cells is not visible on the outer surface (Fig. 3.6B). Despite the many differences, the second compartment has been considered by some to be a homologue to the reticulum in advanced ruminants.

The third compartment is an elongated chamber, connected to the second compartment by a thick-walled narrow tubular passage. It almost appears to be a continuation of the ventricular groove. This compartment contains about 11% of the total gastric volume and is entirely lined with glandular mucosa. The acid-producing cells are located at the terminal end of this compartment. Based on mucosal differences, the proximal four-fifths of this chamber has been suggested as an omasal homologue, and the distal one-fifth as a homologue of the abomasum (Vallenas et al. 1971).

At this time, the exact function of the saccules located in the ventral portion of the first compartment is unknown. Preliminary data indicate that those cells probably secrete both mucous and buffers (Cummings et al. 1972, Eckerlin and Stevens 1973).

Vallenas et al. (1971) were able to obtain the stomach from a 2-week-old camel (*C. bactrianus*) and found fewer but larger individual saccules in the first compartment as compared to the llama and guanaco. Also, the glandular mucosa appeared to be confined to the basal portion of the saccules. Other differences were the possibility of two muscular lips defining the ventricular groove and longitudinal coarse rugae (ridges or folds) in the terminal half of the third compartment. Additional studies, with adult camels, are needed to substantiate these possible differences.

The pH of contents in the caudal sac of the first compartment generally ranged between 6.4 and 7.0 (Vallenas and Stevens 1971), which would allow an active fermentation. VFA concentrations in the first stomach compartment have been reported to range from about 40 to 138 mmoles/L in camelids (Vallenas and Stevens 1971) and from 98 to 185 mmoles/L in camels (Williams 1963), which are similar to those in ruminants (Hungate 1966). Rates of VFA production in camels were found to be equal to those in domestic ruminants (Hungate et al. 1959). It should also be noted that tissue lipids in camels are similar to those found in sheep and cattle—i.e., high in stearic acid and odd-numbered *n*-fatty acids (Bauchop 1977). Hintz et al. (1973) compared digestibility of alfalfa pellets and a complete pelleted hay-grain diet between ponies, sheep, and camelids (one llama and one guanaco). On both diets, the camelids significantly digested more dry matter, neutral detergent fiber, acid detergent fiber, and cellulose than the sheep or ponies.

2.4. Preruminants

Chevrotains, like the pseudoruminants, have only three stomach compartments; however, they differ markedly in that the first two compartments are entirely lined with stratified squamous epithelium and are papillated. The absence of an omasum differentiates them from the true ruminants. There are two genera of Tragulidae; *Hyemoschus*, which contains a single species *H. aquaticus*, is found in Africa and is referred to as the water chevrotain; *Tragulus*, of which there are about six species, are found in India and southeastern Asia and commonly called mouse deer. The type species is *T. javanicus*. The mouse deer are about half as large as the water chevrotains, only weighing 2.5 to 4.5 kg and with an overall length of 0.5 m (Walker 1964).

The chevrotain stomach is shown diagrammatically in Figure 3.7. The ventrocaudal blind sac is extended so that the rumen has an S-shaped appearance, compared to the normal globular, grooved shape seen in the true ruminant (see Fig. 3.8) (Moir 1968). Based on studies of fetal ruminants, Langer (1974) has

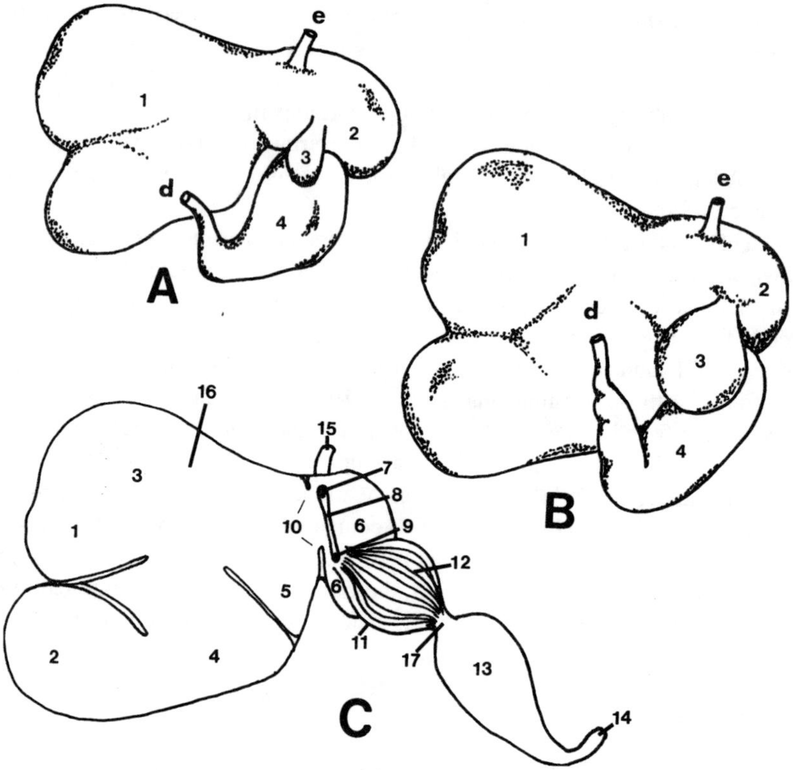

Figure 3.8. Diagrammatic sketches of the ruminant stomach. (A) Stomach of ruminants in the family Cervidae (deer, moose, elk, etc.). (B) Stomach of ruminants in the family Bovidae (cattle, sheep, goats, etc.). Legend for A and B: (1) rumen; (2) reticulum; (3) omasum; (4) abomasum; (d) duodenum; (e) esophagus. (C) Diagrammatic cross section of Bovidae stomach. Legend for C: (1) dorsocaudal blindsac; (2) ventrocaudal blindsac; (3) dorsal sac; (4) ventral sac; (5) cranial sac; (6) reticulum; (7) cardia or esophageal opening; (8) reticular or esophageal groove; (9) reticuloomasal orifice; (10) rumen-reticulum opening; (11) omasum; (12) omasal leaves; (13) abomasum; (14) duodenum; (15) esophagus; (16) rumen; (17) omasal-abomasal orifice. Drawn from photographs and illustrations in Moir (1968), Church (1969), and Hoffman (1988).

concluded that the S-shaped rumen of the chevrotain is a primitive stomach, not simply one reduced in size. He suggests that during the evolutionary process the suborder Ruminantia divided into an infraorder without an omasum (Tragulina) and one with an omasum (Pecora). The reticulum in Tragulidae is large, and the omasum is either absent or present in a rudimentary stage. Two older studies reported an omasum in *Tragulus javanicus* and *Hyemoschus*; while three other investigators (and more recently Langer) could not find this compartment in Tragulina (Langer 1974). If present, it contained no more than two internal leaves and was continuous with the tubular abomasum (Cordier 1894). The ventricular, or esophageal, groove is also primitive in Tragulina (Moir 1968).

Since the reticuloomasal orifice is absent from the tragulids, passage of ingesta into the abomasum or gastric compartment should be considerably faster than in the ruminant (Langer 1974, Janis 1976). The tragulids would also be unable to retain the more fibrous type feedstuffs, which require an extensive fermentation. Although little is known about food selection of the tragulids in the wild, they seem to prefer a low-fiber diet in captivity (Grzimek 1974). Survival of tragulids would be questionable on diets containing appreciable amounts of cellulose (Janis 1976).

2.5. *Ruminants*

The true ruminants are classified into four families, Cervidae, Giraffidae, Antilocapridae, and Bovidae, in the infraorder Pecora, suborder Ruminantia (Romer 1966). The family Cervidae contains 17 genera which include about 53 species. Included in this group are musk deer, muntjacs, fallow deer, axis deer, red deer, Pére David's deer, white-tailed deer, swamp deer, pampas deer, reindeer, moose, and elk. Two species are found in the family Giraffidae, the giraffe and okapi, while only a single species, the pronghorn antelope, occurs in the family Antilocapridae. The family Bovidae contains 49 genera and about 115 species. The most common species in this family are our domesticated cattle, sheep, and goats. A number of the additional species are listed in the footnote to Table 3.1 or can be found in Walker (1964).

Figure 3.8 presents sketches of the stomach in Cervidae and Bovidae plus a diagrammatic cross section of the Bovidae stomach. The ruminant stomach contains four compartments—rumen, reticulum, omasum, and abomasum. The first three compartments are all lined with stratified squamous epithelium and are papillated, while the last compartment, the abomasum, is lined with glandular mucosa. Externally, the stomachs of the four Pecora families differ mainly in overall size, except for the much smaller omasum in Cervidae (Fig. 3.8A). In general, there are differences both within and between families in relative capacity of the various stomach compartments, size and number of omasal leaves, "honeycombed" epithelial cell structure in the reticulum, and papillae in various sections of the rumen (Church 1976). As can be noted in Figure 3.8C, the esophageal groove leads directly from the cardia to the reticuloomasal opening. Closure of this groove in response to suckling provides an efficient means by which the milk from nursing bypasses the rumen and goes directly into the omasum. This allows the milk to be digested by mammalian enzymes, which is a much more efficient process than microbial fermentation in the rumen. Black and Sharkey (1970) have postulated that the esophageal groove was an obligatory adaptation in ruminants and other forestomach fermenters. It allows efficient utilization of milk in the young and permits older animals to make use of the only feeds available in their habitat, poor-quality roughages.

The capacity of the ruminant stomach and its various compartments is difficult to measure. Essentially two procedures have been used, filling the stomach compartments of slaughtered animals with water, addition of markers to the rumen and using several time samples to regress back to zero time. The best estimates from various studies indicate that rumen and abomasum volumes in sheep range from about 3 to 15 L and 2 L, respectively; cattle values range from 35 to 100 L and 3 to 8 L, respectively (Church 1969, 1976; Hoffman 1988a,b). Since there is no distinct separation between the rumen and reticulum, and contents appear to mix freely, most rumen volumes are the combined volumes of these two compartments. Using a fluid marker technique, Lechner-Doll et al. (1990) recently estimated rumen volumes for camels, cattle, sheep, and goats during the dry season in Kenya. Average rumen volume L/100 kg body weight were: camel, 11.3; cattle, 14.9; sheep, 17.4; and goats, 14.5.

Microbial fermentation in the rumen has been extensively documented over the years, beginning with the experiments conducted by Von Tappeiner in 1884 (cited by Hungate 1966). He showed that incubations with fluid from the rumen would not digest cellulose if bacterial growth was inhibited with mild antiseptics which did not prevent hydrolysis. Cellulose disappeared if the antiseptics were omitted from the incubation. Hungate (1966), Clark and Bauchop (1977), and Hobson (1988) provide an excellent overview of our understanding of the rumen microbial fermentation, including the microorganisms involved, extent of digestion, and host utilization of the end products.

3. The Ruminant Animal

Based on a greater efficiency of fiber utilization in ruminants than in other large herbivores, such as horses and elephants, Moir (1965, 1968) postulated that fermentation in the forestomach was superior to that in the cecum and colon. The greater species diversity and higher numbers of ruminants as compared to the large hindgut-fermenting herbivores would indicate an ascending dominance of ruminants since the Oligocene epoch, when the grasslands began to appear. However, Janis (1976) was unwilling to accept this hypothesis without reservation, suggesting that a different digestive strategy is involved. The large hindgut-fermenting herbivores can increase their intake and rate of passage of high-fiber feedstuffs, thereby obtaining adequate nutrients by passing increased amounts of undigested residues through the tract. On the other hand, rate of passage in ruminants is restricted based on the fact that particle size must be reduced enough to allow passage through the reticuloomasal orifice. Although cellulose digestion in the horse is about 70% of that observed in ruminants, digesta passage in the cow and horse are estimated to be 70 to 90 hours and 48 hours, respectively (Phillipson and McAnally 1942, Alexander 1946, Balch and Campling 1965, Heinlein et al. 1966, Vandernoot et al. 1967, Vandernoot and Gilbreath 1970).

Janis (1976) lists several other points that should be considered before concluding that forestomach fermentation is superior. Although the foregut and hindgut microbial fermentations are thought to be very similar (Parra 1978, Moir 1968), the hindgut fermentation is supposedly less efficient because the products of microbial digestion are produced beyond the small intestine, considered to be the usual site of absorption. This problem is alleviated by coprophagy in some hindgut fermentors, but these are generally the smaller animals. Although not well documented, Janis (1976) suggests the possibility that the much larger cecum and colon of the horse could provide adequate surface area for end product absorption. Other differences would be that soluble carbohydrates and proteins are fermented in the rumen, producing volatile fatty acids and ammonia, which may be a less efficient metabolic process for providing glucose and amino acids to the animal. Only one of the volatile fatty acids is glucogenic, propionic acid, and this apparently provides a major portion of the animal's requirements for glucose (Armstrong 1965). The remainder is derived from deamination and metabolism of protein. Ammonia is absorbed, converted to urea, and returned to the rumen via saliva or across the rumen wall. It can then be converted to bacterial protein, which in turn becomes available to the host. This pathway is possibly reflected in the high pancreatic ribonuclease activity observed in ruminants and macropods as compared to hindgut fermenters (Barnard 1969). Apparently this enzyme is essential only in animal species that need to digest large amounts of microbial nucleic acids in the small intestine. Lastly, Janis (1976) suggested that the additional weight of the digestive tract and ingesta may constitute a disadvantage for the forestomach fermentors.

3.1. Unique Characteristics of Digestion in the Ruminant

Among the forestomach-fermenting mammalian herbivores, despite the reservations expressed by Janis (1976), animals in the infraorder Pecora are thought to be the most advanced (Moir 1968, Langer 1974, Hume and Warner 1980). These animals are sometimes referred to as "true" ruminants, having four distinct stomach compartments—the rumen, reticulum, omasum, and abomasum. The ruminant possesses at least one structural (reticuloomasal orifice) and one physiological (rumination) difference, either or both of which set it apart from other forestomach fermenters and place it at the top of the evolutionary scale.

RETICULOOMASAL ORIFICE

The orifice between the reticulum and omasum selectively retains larger food particles, permitting an extended period of microbial fermentation. Only animals in the infraorder Pecora have a reticuloomasal orifice. The tragulids have no apparent constriction separating the reticulum and abomasum (Janis 1976, Hume and Warner 1980). Although a somewhat similar constriction occurs between the second and third stomach compartments in camelids, its potential function in retaining larger particles is unknown (Vallenas et al. 1971). The orifice acts as a sieve, permitting passage of only small particles to the omasum and abomasum. The threshold size of feed particles, above which very few will pass out of the reticulum, appears to be between 1.0 and 2.0 mm in sheep and between 2.0 and 4.0 mm in cattle (Ulyatt et al. 1986). Prigge et al. (1990) estimated threshold particle size in cattle to be between 3 and 5 mm. The majority of particles passing through the reticuloomasal orifice in reindeer were 2 mm or less (Trudell-Moore and White 1983), which, as might be expected, is similar to sizes observed in sheep.

Ulyatt et al. (1986) concluded that particle reduction in the rumen-reticulum is controlled by four processes: (1) chewing associated with eating; (2) chewing associated with rumination; (3) microbial fermentation in the rumen reticulum; and (4) action of rumen contractions. Their studies indicated that chewing associated with eating and rumination was the primary means of reducing particle size. Similar results were obtained in a study by McLeod and Minson (1988). However, even though particle size reduction is required before passage through the reticuloomasal orifice, this does not appear to be the rate-limiting step. The particulate material present in the rumen reticulum at any given time is generally smaller than the determined threshold size. Several factors could be involved—e.g., functional specific gravity of particles as well as the amount of material passed per contraction (Shaver et al. 1986, 1988; Ulyatt et al. 1986).

Functional specific gravity differs from true specific gravity in that air- and gas-filled pockets are not removed in the determination. In the rumen, the specific gravity of particles is obviously influenced by entrapped air and gasses, and when particle size is reduced the entrapped gasses are released (Hooper and Welch 1985). Small particles have a greater functional specific gravity than large particles, and functional specific gravity appears to increase at a faster rate in the small particles. In general, particles small enough to pass through the reticuloomasal orifice, with a functional specific gravity of 1.2 or higher, have the maximum passage rate (Ehle and Stern 1986, Welch 1986, Siciliano-Jones and Murphy 1991). Microbial digestion and ruminal contractions which mix the digesta may be the major factors in increasing functional specific gravity (Hooper and Welch 1985). It is also of interest that the passage rate decreases for particles with functional specific gravities above 2.0 (Ehle and Stern 1986, Welch 1986). This would fit with the 1956 observations of Schels, as cited by Langer (1984), that particles with a high specific gravity become entrapped in the dilated honeycomb epithelial cells of the reticulum. Ulyatt et al. (1986) suggest that microbial fermentation may also contribute to particle size reduction by weakening cell wall structure to the point that chewing during rumination is much more effective. In summary, both particle size and functional specific gravity are critical factors in the passage of particles through the reticuloomasal orifice (Welch 1986).

The other factor to be considered is the actual mechanism by which particles pass through the reticuloomasal orifice. The amount of material that passes with each contraction has been estimated at between 0.25 and 0.5 g DM in sheep (Ulyatt et al. 1984, 1986) and between 1.8 and 3.6 g OM in cattle (Freer et al. 1962). This aspect is discussed at length by Ulyatt et al. (1986); however, how this process is controlled is not well understood at present.

In an overall biological view, rumen particulate matter turnover times in cattle have been observed to range from 1.3 to 3.7 days, with most values being about 2.1 to 2.7 days or 50 to 60 hours. With sheep, turnover times were less, ranging from 0.8 to 2.2 days, with a majority of values being slightly over 1 day (Hungate 1966; Church 1969, 1976; Owens and Goetsch 1988). In this instance, turnover time is defined as the amount of time required for all of the particulate matter present in the rumen at a given time to pass out of the rumen. Level of intake, particle size, specific gravity, and concentration of solids all appear to influence both digestibility and rate of ingesta movement through the reticulorumen. As fiber content of a forage increases and digestibility decreases, passage of particulate matter also decreases (Blaxter et al. 1956, Baile and Forbes 1974). Ruminants appear to voluntarily consume less energy dense feeds to a specific "fill" capacity. If digestibility of the feedstuff is low, it accumulates in the rumen and decreases subsequent food intake. For the ruminant animal to survive, it has been estimated that forage energy digestibility (which approximately equals dry matter digestibility) must be at least 30% to 32% (Blaxter et al. 1961). Thus, with forages of very low energy digestibility, the ruminant animal is actually at a disadvantage because of the sievelike action of the reticuloomasal orifice. Other herbivores, both foregut and hindgut fermenters, can theoretically increase their daily energy intake by increasing both rate of intake and rate of passage.

RUMINATION

All animals in the suborders Ruminantia and Tylopoda ruminate or chew their cud. Bell (1961) has described the process of rumination as a series of four reflex actions—regurgitation, remastication, reinsalivation (mix with saliva), and deglutition (swallowing). He describes the regurgitation phase as a form of controlled vomition that is unique to ruminant digestion. The other phases occur in the digestive strategy of many different species. Rumination generally begins early in a young animal's life, as soon as it begins to consume solid feedstuffs, especially roughages (Gordon 1968). Although not essential for life of the animal, it generally occupies an average of about 8 hours per day on normal diets.

Most evidence seems to support the suggestion that regurgitation is an act of controlled vomition (Moir 1968). Regurgitation of stomach contents has reportedly been observed in other mammals, particularly the macropodids (Moir 1968, Hume 1982). However, regurgitation takes place only intermittently, and the stomach contents appear to be reswallowed without any further chewing. Barker et al. (1963) concluded that this process was not really analogous to rumination in ruminants, and proposed that the term "merycism" be used to describe this activity. This voluntary or controlled vomition, merycism, may occur in many animals including man, and differs from rumination in ruminants. Moir (1968) has suggested that merycism may have actually evolved into the highly coordinated rumination process.

The importance of rumination in decreasing feedstuff particle size and increasing functional specific gravity was discussed in the previous section. Welch and Smith (1968, 1969a,b) have published an excellent series of studies on rumination. In general, they found that (1) the peak of rumination activity takes place at night; (2) fasting causes a rapid decline in rumination, which ceases entirely after about 36 hours; (3) rumination initiates very soon after refeeding; (4) poor-quality roughages with high-fiber content cause an increase in rumination time; and (5) rumination time increased with increased size of single meals and was higher than for an equal weight of hay fed continuously. During an average 24-hour period, they found that sheep fed chopped forage twice a day at 15% excess spent 6 hours eating, 9 hours ruminating, and 9 hours idle. Although both numbers of feedings per day and forage quality varied, somewhat similar rumination times were observed in cattle (Welch and Smith 1970, Welch et al. 1970).

Gordon (1968) estimated that ruminants spend about one-third of their time ruminating, most of which is carried out while lying down. These estimates were well substantiated by the results of Welch and co-workers cited above. Gordon (1968) points out that we do not know for certain why a ruminant carries out this act, and discusses the theories that have been developed to explain rumination.

1. The predator theory. Since ruminants are subject to attack by predators, they eat rapidly without chewing, retreating to a safe place to rest and completely rechew their food. He rejects this theory as unconvincing since a wakeful active flock or herd or even a single animal would be more alert to predators when they are grazing.

2. Addition of oxygen to the ingesta. The original idea was that the addition of oxygen to rumen contents stimulated growth of facultative anaerobic bacteria. However, the redox potential in rumen ingesta is not changed during rumination, and no evidence indicates that facultative anaerobes can perform any function not carried out by the anaerobic bacteria.
3. Rumination increases passage rate of rumen-reticulum contents into the omasum as a result of extrareticular contractions, increased rate of stomach motility, and use of the reticular groove in swallowing remasticated food. However, there appears to be a marked increase in dry matter disappearance during eating, but not during rumination.
4. Decrease in particle size. Although this is a primary function of rumination, it does not seem likely to have been the evolutionary pressure responsible for development of this complex act. More complete primary chewing would probably not reduce the safety of the animals, and it would be a simpler process. Ruckebusch and Marquet (1965) found that animals extensively masticated their food after their rumen was surgically removed.
5. Does rumination increase digestibility? Although this might be expected, particularly with high-fiber roughages, this does not appear to be the case. Pearce and Moir (1964) found that if rumination was prevented by the use of muzzles, digestibility of dry matter, organic matter, nitrogen, and crude fiber all increased. Obviously any decrease in digestibility due to rumination would be affected by increased rate of digestion, as well as increased intake and outflow of food from the rumen.
6. Interaction and summation of activities discussed under (3) and (4) above plus the conservation of energy. Gordon (1968) calculated that a sheep consuming 2,500 kcal/day could save 250 kcal by lying down to ruminate for 8 hours instead of standing and eating. This represents a 10% conservation of energy intake. He further points out that this does not include the energy expended by moving about during grazing.

Regardless of which one or combination of the above reasons is responsible for rumination, it is a biological process that seems to have enabled a large group of mammals to survive and multiply in environments where adequate but only poor-quality forage was available. On the other hand, this may be an energy-wasteful process in commercial animal production, where highly digestible diets are fed.

3.2. Feeding Characteristics of Various Ruminants

On the basis of diet selections in the wild, ruminants have been divided into three general feeding types—grazers (roughage eaters), browsers (concentrate selectors), and intermediate mixed feeders (Hoffman and Stewart 1972). This food preference has resulted in numerous anatomical adaptations in the digestive tract of ruminants. Hoffman (1988a,b) has investigated the morphology of the digestive tract in 58 ruminant species from four continents. They ranged from the grass and roughage eaters (GR) to the concentrate selectors (CS), with a large proportion of intermediate and opportunistic mixed feeders (IM) between the two extremes. Kay et al. (1980) have summarized some of the characteristics that differ between the buffalo (GR) and roe deer (CS). For the buffalo (GR), rumen size, rumen pillars, omasum, food retention time, numbers of cellulolytic bacteria, and diversity of protozoa are relatively larger or more developed. In the roe deer (CS), rumen papillae, reticulum, parotid salivary glands, and fermentation rate were relatively larger or more developed. The adaptable mixed feeders fall in between and are well suited to utilize geographic areas in which vegetation types vary with season. By nature, our domesticated ruminants are either grazers (cattle, sheep, buffalo) or mixed feeders (goat, red deer). However, as mentioned previously, in many instances we now feed these animals energy-dense concentrate diets.

In general, the determining factor in the quantity and quality of food consumed is the requirement of energy. The maintenance requirement for energy of an animal is proportional to the three-quarter power of body weight ($kg^{0.75}$) and its activity or productivity. A small 5-kg productive ruminant would thus require about 3.2 times more metabolizable energy per kg body weight than a 500-kg nonproductive animal (Kay et al. 1980, Van Soest 1982). The small productive ruminant must therefore either consume more feed or a more energy-dense feed. Since the energy-dense diet generally leaves less residue, intake can be increased. The wet weight of rumen-reticulum contents, as a percentage of body weight, decreased from 13.3% for 10 GR species to 11.5% for 14 IM species down to 9.0% for 9 CS species (Kay et al. 1980). Thus, the small ruminants appear to be very successful in selecting concentrate-type diets.

4. Fermentation of Feedstuffs

As discussed earlier, the occurrence of VFA in the stomach contents is indicative of a microbial fermentation. Forages and other feedstuffs are fermented in the forestomach by anaerobic bacteria, protozoa, and fungi. This pregastric fermentation provides at least four major nutritional advantages to the host animal (Moir 1965, Janis 1976, Hume and Warner 1980).

1. Cellulose and other structural polysaccharides, which cannot be degraded by the enzyme systems of the animal, are broken down to a utilizable energy source. This also reduces dry-matter volume which is passed on down the gut.

2. The microorganisms can utilize nonprotein nitrogen for growth. When the microbes pass down the tract, they then serve as a protein source for the animal.
3. Synthesis of vitamins by the microbes alleviates the dietary requirement for all the vitamins except A and D.
4. Detoxification of plant toxins or secondary products by microbes serves a protective function.

4.1. Microbial Populations

Microorganisms that have been observed in stomach contents of the various forestomach-fermenting herbivores are listed in Table 3.2.

GOLDEN HAMSTER

Although information is sparse, two authors have reported the presence of microorganisms in forestomach contents of the golden hamster. Mangold (1929), as cited by both Moir (1965) and Ehle and Warner (1978), states that microorganisms like those in the rumen occur in the forestomach of the golden hamster. Kunstyr (1974) reported the occurrence of several microbial groups in forestomach contents of the golden hamster (coliforms, yeasts, streptococci, lactobacilli and spore formers). The number of microorganisms per gram of contents was much higher in the forestomach than in the glandular stomach.

WHALES

Bacterial concentrations in samples of forestomach contents from the gray and bowhead whales were determined by direct counts (Herwig et al. 1984). Since the extent of dilution of the different samples was unknown, the authors measured chloride concentrations of undiluted stomach contents from a single gray whale and used this to estimate dilution for all their samples.

DOLPHINS

Viable counts of bacteria, pH, and VFA concentrations were measured in fluids collected from the four forestomachs of spotted dolphins (Morii 1972, 1979). Concentrations of viable organisms were quite low (1.8×10^3 to 7.1×10^5), as were VFA concentrations (approx. 0.3 to 1.7 mmoles/L). Samples were obtained from animals caught alive. A number of bacteria were isolated and found to be predominately *Vibrio* species. The author suggested that these bacteria might originate from the intestinal biota of fish that had been ingested by the dolphins,

Table 3.2. Occurrence and numbers of microorganisms in stomach contents of the different forestomach-fermenting herbivores

Animal	Bacteria × 10^8/g	Protozoa × 10^4/mL	Fungi × 10^4/mL	Reference[a]
Golden hamster	+[b]	[c]		1–3
Gray whale	< 1.0–7.7			4
Bowhead whale	5.6–630			4
Spotted dolphin	0.000018–0.0071			5,6
Colobus monkey	26; 50	—[d]		7
Langur monkey	200–1,000	—		8
Two-toed sloth	+	—		9
Collared peccary		30–250; —		10,11
Hippopotamus	+	5.2–15.4		12,13
Quokka	258–296	50–130		1,14
Kangaroo, wallaby	570–7,600	1.5–15	+	15–17
Hoatzin	10–10,000	1.0–10		18,19
Camel	+	7.4–43.7	+	20–22
New World camelids[e]		34.5	+	22–24
Mouse deer (chevrotain)		+		25,26
Ruminants[f]	0.02–314	1–3,370	0.15–150	27–35

[a] (1) Moir 1965; (2) Ehle and Warner 1978; (3) Kunstyr 1974; (4) Herwig et al. 1984; (5) Morii 1972; (6) Morii 1979; (7) Ohwaki et al. 1974; (8) Bauchop and Martucci 1968; (9) Denis et al., 1967; (10) Langer 1978; (11) Carl and Brown 1983; (12) Thurston et al. 1968; (13) van Hoven 1974; (14) Baker 1989; (15) Lintern 1970; (16) Harrop and Barker 1972; (17) Dellow 1979; (18) Grajal et al. 1989; (19) Dominguez-Bello et al. 1991; (20) Hungate et al. 1959; (21) Imai and Rung 1990; (22) Milne et al. 1989; (23) Dehority 1986; (24) Lubinsky 1964; (25) Jameson 1925; (26) Dogiel 1927; (28) Dehority and Orpin 1988; (29) Trinci et al. 1994; (30) Obispo and Dehority 1992; (31) Dehority 1990; (32) Dehority and Varga 1991; (33) Clarke 1977; (34) Williams and Coleman 1992; (35) Ogimoto and Imai 1981.

[b] + = observed, but no quantitative data presented.

[c] Blank space indicates that no information was available on the presence or absence of this microorganism.

[d] — = specifically designated as being absent.

[e] Llama, alpaca, guanaco, vicuña.

[f] Includes almost all the families, genera, and species listed in Table 3.1.

since they resembled the *Vibrio* species commonly inhabiting the gut of various fish (Morii 1972).

COLOBUS MONKEY

Saccular stomach contents from two *Colobus* monkeys contained high concentrations of viable bacteria; however, most isolates were not typical rumen species, and no cellulolytic bacteria were present (Bauchop 1971, Ohwaki et al. 1974). Stomach contents appeared to contain very little leafy material and had a pH of 5.5 or below. Starchy seeds and fruits may have been the main diet, resulting in a low pH with no cellulolytic species or protozoa present (Bauchop 1978).

LANGUR MONKEY

Viable counts of anaerobic bacteria in the forestomach of two langur monkeys (200 to 1,000 $\times$ 10^8/g) were consistently higher than normally observed in rumen contents (100 $\times$ 10^8/g) (Hungate 1966, Bauchop and Martucci 1968, Bauchop 1971). However, cellulolytic bacterial numbers were fairly similar to those in the rumen, ranging from 8 x 10^6 to 1 $\times$ 10^8/g of contents. Two morphological types of cellulolytic bacteria were isolated—one a gram-positive coccus, and the second a *Bacteroides* sp., both of which resemble the major cellulolytic species in the rumen (Hungate 1966).

TWO-TOED SLOTH

A large bacterial population occurs in stomach contents from the two-toed sloth; however, protozoa have not been observed (Denis et al. 1967). Both Jeuniaux (1962) and Denis et al. (1967) were able to demonstrate the presence of cellulolytic bacteria.

COLLARED PECCARY

Information on the microbial population in the forestomach of the peccary is very limited, and what has been reported is somewhat contradictory. No direct observation or proof of the presence of bacteria in the peccary forestomach has been published to date. However, the occurrence of VFA in forestomach contents provides substantial indirect evidence for their presence (Langer 1978, Sowls 1984, Lochmiller et al. 1989). Total tract cellulose digestion is quite high for the peccary (Carl and Brown 1986); however, in a separate study, in vitro cellulose digestion was found to be negligible (Shively et al. 1985). It should be noted that dry matter digestion in the in vitro study was nearly identical to that obtained with steer rumen contents (73% vs. 75%).

Langer (1978) reported that Prins (personal communication) did not find any protozoa in the forestomach contents of three collared peccaries. Carl and Brown (1983) subsequently reported protozoal counts between 0.3 and 2.5 $\times$ 10^6 mL of stomach contents for two collared peccaries that had been kept in captivity. The protozoa species were similar to those of domestic ruminants and other wildlife (white-tailed deer) in the same area. Dehority (unpublished) examined stomach contents from four collared peccary shot in that same geographical area (Texas) and found no protozoa in their stomach contents.

HIPPOPOTAMUS

Thurston et al. (1968) found numerous bacteria and ciliate protozoa in forestomach contents from the hippopotamus. No further reports on the bacteria have been found; however, VFA concentrations are quite similar to those found in rumen contents providing circumstantial evidence for a bacterial fermentation. Mean protozoal concentrations in forestomach contents from 16 hippopotami from South Africa were 15 $\times$ $10^4/cm^3$ for the parietal blindsac, 10.3 $\times$ $10^4/cm^3$ for the visceral blindsac, and 5.2 $\times$ $10^4/cm^3$ for the median or connecting chamber (Van Hoven 1974).

QUOKKA (WALLABY)

Direct counts of bacteria in forestomach contents of the quokka ranged from 232 to 296 $\times$ 10^8 mL, which are quantitatively similar to those in the rumen (Moir 1965). The proportion of cellulolytic bacteria was also of the same order as in ruminants (Moir 1968). It should be kept in mind when comparing different studies that viable counts are generally 10% or less of direct counts (Hungate 1966). Moir et al. (1956) observed about 15 morphological forms of bacteria in quokka forestomach contents, most of which resembled those observed in the rumen of sheep kept under similar conditions. Although the flora was less diverse than in the sheep, in which at least 30 morphological forms were observed, total direct counts were similar.

Other than listing the concentration of total protozoa in the quokka, the only information available is that the fauna was comprised of three unidentified species (Moir 1965). Numbers were highest in the cul-de-sac region of the forestomach (Baker 1989).

KANGAROO AND WALLABIES

From animals shot in the field, direct counts of bacteria in forestomach contents of the eastern gray kangaroo ranged from 570 to 7,600 $\times$ 10^8/g and in the red-necked pademelon from 2,120 to 5,180 $\times$ 10^8. Direct counts in forestomach contents of red-necked pademelon and tammar wallabies kept in captivity and fed chopped alfalfa hay ranged from 660 to 5,110 $\times$ 10^9/g (Dellow 1979). The direct count of bacteria in rumen contents from sheep fed the same diet was 4,430 $\times$ 10^8/g.

Lintern (1970) and Harrop and Barker (1972) observed protozoa in forestomach contents of the tammar wallaby and red kangaroo, respectively. In later studies, Dellow (1979) found ciliate protozoa in the forestomach contents of the eastern gray kangaroo, red-necked pademelon, tammar wallaby, red-necked wallaby, swamp wallaby, and eastern wallaroo. Concentrations ranged from 1.5 to 15 $\times$ 10^4/g. Highest counts were in the sacciform forestomach contents.

Fungal sporangia were observed by Dellow (1979) in all animals except the red-necked pademelon. Bauchop (1983) subsequently reported the occurrence of fungi in the forestomach contents of the gray kangaroo, red-necked wallaby, swamp wallaby and wallaroo.

HOATZIN

Morphologically, the principal anaerobic bacteria isolated from agar roll tubes (Dominguez-Bello et al. 1991) were gram-negative spore-forming rods and gram-positive coccobacilli. Bacterial counts averaged about 11.0 $\times$ 10^8/g of crop contents in adult birds (Grajal et al. 1989), with values up to 1 $\times$ 10^{12} found in young chicks (Dominguez-Bello et al. 1991). Protozoa were also observed in crop contents by Dominguez-Bello et al. (1991), with numbers ranging from 1 $\times$ 10^4 in adults to 1 $\times$ 10^5 in chicks. Only one species of protozoa was observed, which Dominguez-Bello et al. (1991) tentatively suggested to be a species of *Isotricha*. However, Dehority (unpublished) examined these same samples of crop contents and found that although the organism was entirely covered with somatic cilia, it differed from any known rumen protozoal species.

CAMEL

Viable counts of cellulolytic bacteria have been reported from the forestomach of two camels in Africa (Hungate et al. 1959). Concentrations were 8 $\times$ 10^6 and 4.1 $\times$ 10^8/mL of contents in the two animals, respectively. These values were in the same range as those determined on other ruminants in the area. The cellulose-digesting colonies from the camel were similar to those obtained from the rumen cellulolytic species, *Fibrobacter succinogenes*.

Qualitative observations on ciliates in forestomach contents of camels (*Camelus dromedarius*) have been reported by Dogiel (1926, 1928). More recently, Imai and Rung (1990) examined forestomach contents from four bactrian camels (*Camelus bactrianus*) and found ciliate concentrations ranging from 7.4 to 43.7

$\times$ 10^4/mL. In one animal, protozoa concentrations in contents from the first, second, and third stomachs ($\times$ 10^4) were 43.2, 39.6, and 3.3, respectively.

Fungi were qualitatively detected in the feces of the camel (Milne et al. 1989). Since fungi have been found in the rumen, omasum, abomasum, small intestine, cecum, large intestine, and feces of typical ruminants like cattle and sheep, it is suggested that they are carried out of the rumen on particulate matter and pass on down the tract (Grenet et al. 1989, Davies et al. 1993). Concentrations are highest in the reticulorumen and omasum, decrease markedly in the abomasum, remain low in the small intestine, and increase in numbers in the hindgut. Based on these observations, it is assumed that fungi are present in the forestomach of the camel.

NEW WORLD CAMELIDS (LLAMA, ALPACA, GUANACO, AND VICUÑA)

The high VFA concentrations found in stomach contents of the llama and guanaco are certainly indicative of a bacterial fermentation (Vallenas and Stevens 1971). However, no reports on the occurrence of bacteria in forestomach contents of New World camelids have been found.

Occurrence of ciliate protozoa has been qualitatively reported from stomach contents of a single guanaco, two llamas, and three alpacas (Lubinsky 1964, Dehority 1986). The only quantitative data are from a single alpaca, 34.5 $\times$ 10^4 protozoa per mL of forestomach contents (Dehority 1986). This value was quite similar to the protozoa concentration found in rumen contents of a sheep housed in the same area, 36.1 $\times$ 10^4/mL (Dehority, unpublished).

Fungi have been observed in feces from the llama, alpaca, guanaco, and vicuña (Milne et al. 1989). Based on the arguments presented under the section on camels, it is presumed that fungi also occur in the forestomach of New World camelids.

RUMINANTS

All animals in the families Cervidae (deer), Giraffidae (giraffe), Antilocapridae (pronghorn antelope), and Bovidae (cattle, sheep, goats, etc.) are in this classification. Numerous studies have been reported on the bacterial, protozoal, and, more recently, fungal populations occurring in the rumen. References listed in Table 3.2 are for summary and review-type publications, which should identify reports on specific animal species that might be of interest to the reader.

4.2. Comparison Among Animal Species

Although differences may occur in the genera and species of microorganisms that inhabit the forestomach of the various types of foregut fermenters, they all appear capable of digesting the fiber diets of typical herbivores. In addition, particulate matter passage rates also vary between the different forestomach fermenters. Passage rates are slower in cattle than in sheep and faster in nonruminant herbivores (Parra 1978, Hume 1982). The primary cause of slower passage times in the ruminant appears to be the reticuloomasal orifice. Passage rates also tend to be faster for the smaller animals, which, because of their higher energy needs, select more energy dense diets. In contrast, hindgut fermenters have more rapid passage rates since they have no physical restrictions to movement of the digesta (Parra 1978).

It is difficult to compare the efficiency of the forestomach fermentation between the various types of foregut fermenters. Such factors as type of diet, digestibility, intake level, rate of intake, and selectivity can all influence digestibility in the animal. Hintz et al. (1973) compared the digestibility of alfalfa pellets and a complete pelleted ration between camelids (llama and guanaco), ponies, and sheep. Animals were fed equivalent levels of feed based on metabolic body weight. For both diets, dry matter digestibility by camelids was significantly greater than in either sheep and ponies. However, under natural grazing conditions, these animals plus the other types of forestomach-fermenting herbivores probably select markedly different diets.

5. Conclusions

Evolutionary development of multichambered stomachs, with areas adequately separated from the acid-secreting region, has permitted the establishment of a foregut fermentation in numerous animals. Stomach anatomy ranges from a small, two-chambered stomach up to the large, four-chambered ruminant stomach. In general, passage of ingesta through the pregastric chamber(s) is considerably faster in the smaller, simpler multichambered stomachs. These animals tend to feed on concentrate types of plant materials. In contrast, animals whose stomachs have a large pregastric capacity and reduced rate of passage are able to utilize less digestible plant materials such as grasses, twigs, and bulk roughage. The microbial population also becomes more active and complex as the rate of ingesta passage decreases. It is in this latter group of animals that the full potential of the microbial digestion of cellulose for energy is realized. Cellulose is the most abundant polysaccharide on Earth; however, it is unavailable to higher animals because they lack the necessary cellulase enzymes for its utilization.

By far, stomach anatomy and microbial populations have been studied most extensively in the ruminant. Only limited information is available on pseudoruminants and animals with a ruminantlike adaptation, and further studies on this group of animals would seem desirable. This information could provide valuable insight into the factors involved in the evolution of a larger, more complex forestomach and the corresponding changes in the microbial populations to take advantage of this available niche.

References

Alexander F (1946) The rate of passage of food residue through the digestive tract of the horse. J Comp Pathol 56:266.

Alexander F, Chowdhury AK (1958) Digestion in the rabbit's stomach. Br J Nutr 12: 65–73.

Arman P, Field CR (1973) Digestion in the hippopotamus. E Afr Wildl J 11:9–17.

Armstrong DG (1965) Carbohydrate metabolism in the ruminant and energy supply. In: Dougherty RW, ed. Physiology of Digestion in the Ruminant, pp. 272–288. Washington, DC: Butterworths.

Baile CA, Forbes JM (1974) Control of feed intake and regulation of energy balance in ruminants. Physiol Rev 54(1):160–214.

Baker SK (1989) The microbial population of the quokka forestomach. Asian-Aust J Anim Sci 2:458–459.

Balch CC, Campling RW (1965) Rate of passage of digesta through the ruminant digestive tract. In: Dougherty RW, ed. Physiology of Digestion in the Ruminant, pp. 108–123. Washington, DC: Butterworths.

Barker S, Brown GD, Colaby JH (1963) Food regurgitation in the Macropodidae. Aust J Sci 25:430–432.

Barnard EA (1969) Biological function of pancreatic ribonuclease. Nature (Lond) 221: 340–344.

Bauchop T (1971) Stomach microbiology of primates. Annu Rev Microbiol 25:429–436.

Bauchop T (1977) Foregut fermentation. In: Clarke RTJ, Bauchop T, eds. Microbial Ecology of the Gut, p. 223–250. London: Academic Press.

Bauchop T (1978) The significance of microorganisms in the stomach of non-human primates. Wld Rev Nutr Diet 32:198–212.

Bauchop T (1983) The gut anaerobic fungi: colonisers of dietary fibre. In: Wallace G, Bell L, eds. Fibre in Human and Animal Nutrition, pp. 143–148. Wellington: Royal Society of New Zealand.

Bauchop T, Martucci RW (1968) Ruminant-like digestion of the langur monkey. Science 161:698–700.

Bell FR (1961) Some observations on the physiology of rumination. In: Lewis D, ed. Digestive Physiology and Nutrition of the Ruminant, pp. 59–67. London: Butterworths.

Black JL (1970) Nutritional significance of the reticular groove (sulcus reticuli). Aust J Sci 32:332–334.

Black JL, Sharkey MJ (1970) Reticular groove (sulcus reticuli): an obligatory adaptation in ruminant-like herbivores. Mammalia 34:294–302.

Blaxter KL, Graham NMC, Wainman FW (1956) Some observations on the digestibility of food by sheep, and on related problems. Br J Nutr 10:69–91.

Blaxter KL, Wainman FW, Wilson RS (1961) The regulation of food intake by sheep. Anim Prod 3:51–61.

Britton WS (1941) Form and function in the sloth. Q Rev Biol 16:190–207.

Camain R, Quenum A, Kerrest J, Goueffon S (1960) Considerations sur l'estomac de *Cricetomys gambianus*. Bull Mem Ecole Natl Med Pharm Dakar 8:134–142.

Carl GR, Brown RD (1983) Protozoa in the forestomach of the collared peccary (*Tayassu tajacu*). J Mamm 64:709.

Carl G, Brown RD (1986) Comparative digestive efficiency and feed intake of the collared peccary. Southwest Nat 31:79–85.

Chivers DJ, Hladik CM (1980) Morphology of the gastrointestinal tract in primates: comparisons with other mammals in relation to diet. J Morphol 166:337–386.

Church DC (1969) Digestive Physiology and Nutrition of Ruminants, Vol. 1. Digestive Physiology. Corvallis OR: O&B Books.

Church DC (1976) Digestive Physiology and Nutrition of Ruminants, Vol I, 2nd ed. Digestive Physiology. Corvallis OR: O&B Books.

Clarke RTJ (1977) Protozoa in the rumen ecosystem. In: Clarke RTJ, Bauchop T, eds. Microbial Ecology of the Gut, pp. 251–275. London: Academic Press.

Clarke RTJ, Bauchop T (1977) Microbial Ecology of the Gut. London: Academic Press.

Clemens ET (1977) Sites of organic acid production and patterns of digesta movement in the gastrointestinal tract of the rock hyrax. J Nutr 107:1954–1961.

Cordier JA (1894) Recherches sur l'anatomie comparée l'estomac des ruminants. Ann Sci Nat Zool 16:1–128.

Cummings JF, Munnell JF, Vallenas A (1972) The mucigenous glandular mucosa in the complex stomach of two new-world camelids, the llama and guanaco. J Morphol 137: 71–110.

Davies DR, Theodorou MK, Lawrence MIG, Trinci APJ (1993) Distribution of anaerobic fungi in the digestive tract of cattle and their survival in faeces. J Gen Microbiol 139: 1395–1400.

Dehority BA (1986) Protozoa of the digestive tract of herbivorous mammals. Insect Sci Appl 7:279–296.

Dehority BA, Orpin CG (1988) Development of, and natural fluctuations in, rumen microbial populations. In: Hobson PN, ed. The Rumen Microbial Ecosystem, pp. 151–183. London: Elsevier Applied Science.

Dehority BA (1990) Rumen ciliate protozoa in Ohio white-tailed deer (*Odocoileus virginianus*). J Protozool 37:473–475.

Dehority BA, Varga GA (1991) Bacterial and fungal numbers in ruminal and cecal contents of the blue duiker (*Cephalophus monticola*). Appl Environ Microbiol 57:469–472.

Dellow DW (1979) Physiology of digestion in the macropodine marsupials. PhD Thesis, University of New England, Armidale, Australia.

Denis C, Jeuniaux C, Gerebtzoff MA, Goffart M (1967) La digestion stomacalè chez un paresseux: l'unau *Choloepus hoffmanni* Peters. Ann Soc R Zool Belg 97:9–29.

Dogiel VA (1926) Sur quelques infusories nouveaux habitant l'estomac du dromadaire (*Camelus dromedarius*). Ann Parasitol 4:241–271.

Dogiel VA (1928) La faune d'infusoires habitant l'estomac du buffle et du dromadaire. Ann Parasitol 6:323–338.

Dominguez-Bello MG, Michelangeli F, Ruiz MC, Suarez, P (1991) Pregastric fermentation in *Ophistocomus hoazin*, a folivorous bird. 21st Biennial Conference on Rumen Function, Abstracts, Nov. 12–14, 1991, Chicago, p. 13.

Drawert F, Kuhn H-J, Rapp A (1962) Gas chromatographische bestimmung der niederflüchtigen fettsäuren im mager von schlankaffen (Colobinae). Z Physiol Chem 329: 84–89.

Eckerlin RH, Stevens CE (1973) Bicarbonate secretion by the glandular saccules of the llama stomach. Cornell Vet 63:936–945.

Ehle FR, Stern MD (1986) Influence of particle size and density on particulate passage through alimentary tract of Holstein heifers. J Dairy Sci 69:564–568.

Ehle FR, Warner RG (1978) Nutritional implications of the hamster forestomach. J Nutr 108:1047–1053.

Forbes DK, Tribe DE (1969) Salivary glands of kangaroos. Aust J Zool 17:765–775.

Freer M, Campling RC, Balch CC (1962). Factors affecting the voluntary intake of food by cows. 4. The behaviour and reticular motility of cows receiving diets of hay, oat straw and oat straw with urea. Br J Nutr 16:279–295.

Gordon JG (1968) Rumination and its significance. Wld Rev Nutr Diet 9:251–273.

Grajal A, Strahl SD, Parra R, Dominguez MG, Neher A (1989) Foregut fermentation in the hoatzin, a neotropical leaf-eating bird. Science 245:1236–1238.

Grenet E, Fonty G, Jamot J, Bonnemoy F (1989) Influence of diet and monensin on development of anaerobic fungi in the rumen, duodenum, cecum, and feces of cows. Appl Environ Microbiol 55:2360–2364.

Griffiths M, Davies D (1963) The role of the soft pellets in the production of lactic acid in the rabbit stomach. J Nutr 80:171–180.

Grzimek B (1974) Animal Life Encyclopedia, Vol. 13 (Mammals IV). New York: Van Nostrand Reinhold.

Harrop CJF, Barker S (1972) Blood chemistry and gastro-intestinal changes in the developing red kangaroo (*Megaleia rufa* Dermarest). Aust J Expt Biol Med Sci 50:245–249.

Heinlein EFW, Holdren RD, Yoon RM (1966) Comparative responses of horses and sheep to different physical forms of alfalfa hay. J Anim Sci 25:740–743.

Herwig RP, Staley JT, Nerini MK, Braham HW (1984) Baleen whales: preliminary evidence for forestomach microbial fermentation. Appl Environ Microbiol 47:421–423.

Hill WCO (1952) The external and visceral anatomy of the olive colobus monkey (*Procolobus verus*). Proc Zool Soc Lond 122:127–186.

Hintz HF, Schryver HF, Halbert M (1973) A note on the comparison of digestion by New World camels, sheep and ponies. Anim Prod 16:303–305.

Hobson PN (1988) The Rumen Microbial Ecosystem. London: Elsevier Applied Science.

Hoffman RR (1988a) Anatomy of the gastrointestinal tract. In: Church DC, ed. The Ruminant Animal, Digestive Physiology and Nutrition, pp. 14–43. Englewood Cliffs, NJ: Prentice Hall.

Hoffman RR (1988b) Morphophysiological evolutionary adaptations of the ruminant digestive system. In: Dobson A, Dobson MJ, eds. Aspects of Digestive Physiology in Ruminants: Proceedings of a Satellite Symposium of the 30th International Congress of the International Union of Physiological Sciences, pp. 1–20. Ithaca, NY: Comstock Publishing.

Hoffman RR, Stewart DRM (1972) Grazer or browser: a classification based on the stomach-structure and feeding habits of East African ruminants. Mammalia (Paris) 36: 226–240.

Hooper AP, Welch JG (1985) Effects of particle size and forage composition on functional specific gravity. J Dairy Sci 68:1181–1188.

Hoover WH, Mannings CL, Sheerin HE (1969) Observations on digestion in the golden hamster. J Anim Sci 28:349–352.

Hume ID (1978) Evolution of the Macropodidae digestive system. Aust Mammal 2:37–42.

Hume ID (1982) Digestive Physiology and Nutrition of Marsupials. Cambridge: Cambridge University Press.

Hume ID, Sakaguchi E (1991) Pattern of digesta flow and digestion in foregut and hindgut fermenters. In: Tsuda T, Sasaki Y, Kawashima R, eds. Physiological Aspects of Digestion and Metabolism in Ruminants, pp. 427–451. San Diego: Academic Press.

Hume ID, Warner ACI (1980) Evolution of microbial digestion in mammals. In: Ruckebusch Y, Thivend P, eds. Digestive Physiology and Metabolism in Ruminants, pp. 665–684. Lancaster: MTP Press.

Hungate RE (1966) The Rumen and Its Microbes. New York: Academic Press.

Hungate RE, Philips GD, McGregor A, Hungate DP, Buechner HK (1959) Microbial fermentation in certain mammals. Science 130:1192–1194.

Imai S, Rung G (1990) Ciliate protozoa in the forestomach of the bactrian camel in Inner Mongolia, China. Jpn J Vet Sci 52(5):1069–1075.

Jameson AP (1925) A note on the ciliates from the stomach of the mouse deer (*Tragulus meminna* Milne-Edwards) with the description of *Entodinium ovalis* n. sp. Parasitology 17:406–409.

Janis C (1976) The evolutionary strategy of the Equidae and the origins of rumen and cecal digestion. Evolution 30:757–774.

Jeuniaux C (1962) Recherche de polysaccharidases dans l'estomac d'un paresseux (*Choloepus hoffmanni* Pet.). Arch Int Physiol Biochem 70:407–408.

Kay RNB, Engelhardt WV, White RG (1980) The digestive physiology of wild ruminants. In: Ruckebusch Y, Thivend P, eds. Digestive Physiology and Metabolism in Ruminants, pp. 743–761. Lancaster: MTP Press.

Kuhn H-J (1964) Zur kenntnis von bau und funktion des magens der schlankaffern (Colobinae). Folia Primatol 2:193–221.

Kunstyr I (1974) Some quantitative and qualitative aspects of the stomach microflora of the conventional rat and hamster. Zbl Vet Med A21:553–561.

Langer P (1974) Stomach evolution in the Artiodactyla. Mammalia 38:295–314.

Langer P (1975) Macroscopic anatomy of the stomach of the Hippopotamidae Gray, 1821. Zbl Vet Med C4:334–359.

Langer P (1976) Functional anatomy of the stomach of *Hippopotamus amphibius* L. 1758. S Afr J Sci 72:12–16.

Langer P (1978) Anatomy of the stomach of the collared peccary, *Dicotyles tajacu* (L., 1758) (Artiodactyla: Mammalia). Z Säugetierkunde 43:42–59.

Langer P (1979a) Adaptational significance of the forestomach of the collared peccary, *Dicotyles tajacu* (L. 1758) (Mammalia: Artiodactyla). Mammalia 43:235–245.

Langer P (1979b) Functional anatomy and ontogenetic development of the stomach in the macropodine species *Thylogale stigmatica* and *Thylogale thetis* (Mammalia: Marsupiala). Zoomorphologie 93:137–151.

Langer P (1980) Anatomy of the stomach in three species of Potoroinae (Marsupiala: Macropodidae). Aust J Zool 28:19–31.

Langer P (1984) Comparative anatomy of the stomach in mammalian herbivores. J Exp Physiol 69:615–625.

Langer P (1986) Large mammalian herbivores in tropical forests with either hindgut- or forestomach-fermentation. Z Säugetierkunde 51:173–187.

Langer P, Dellow DW, Hume ID (1980) Stomach structure and function in three species of macropodine marsupials. Aust J Zool 28:1–18.

Lechner-Doll M, Rutagwenda T, Schwartz HJ, Schultka W, Engelhardt WV (1990) Seasonal changes of ingesta mean retention time and forestomach fluid volume in indigenous camels, cattle, sheep and goats grazing a thornbush savannah pasture in Kenya. J Agric Sci 115:409–420.

Lintern S (1970) Aspects of nitrogen metabolism in the Kangaroo Island wallaby—*Protemnodor eugenii* (Desmarest). Ph.D. Thesis, University of Adelaide, Adelaide, Australia.

Lochmiller RL, Hellgren EC, Gallagher JF, Varner LW, Grant WE (1989) Volatile fatty acids in the gastrointestinal tract of the collared peccary (*Tayassu tajacu*). J Mamm 70:189–191.

Lubinsky G (1964) Ophryoscolecidae of a guanaco from the Winnipeg Zoo. Can J Zool 42:159.

McLeod MN, Minson DJ (1988) Large particle breakdown by cattle eating ryegrass and alfalfa. J Anim Sci 66:992–999.

Mangold E (1929) Handbuch der Ernahrung and des Stoffwechsel der landwirtshaftlichen Nutztiere. Bd. II. Berlin: Julius Springer.

Marsh H, Spain AV, Heinsohm GE (1978) Physiology of the dugong. Comp Biochem Physiol 61A:159–168.

Mattson FH, Volpenhein RA, Lutton ES (1964) The distribution of fatty acids in the triglycerides of the Artiodactyla (even-toed animals). J Lipid Res 5:363–365.

Milne A, Theodorou MK, Jordan MGC, King-Spooner C, Trinci APJ (1989) Survival of anaerobic fungi in feces, in saliva, and in pure culture. Exp Mycol 13:27–37.

Moir RJ (1965) The comparative physiology of ruminant-like animals. In: Dougherty RW, ed. Physiology of Digestion in the Ruminant, pp. 1–14. Washington, DC: Butterworth.

Moir RJ (1968) Ruminant digestion and evolution. In: Code CF, ed. Handbook of Physiology, Sect. 6, Vol. 5, pp. 2673–2694. Washington, DC: American Physiological Society.

Moir RJ, Somers M, Waring H (1956) Studies on marsupial nutrition. 1. Ruminant-like digestion in a herbivorous marsupial *Setonix brachyurus* (Quoy and Gaimard). Aust J Biol Sci 9:293–304.

Morii H (1972) Bacteria in the stomach of marine little toothed whales. Bull Jpn Soc Sci Fish 38:1117–1183.

Morii H (1979) The viable counts of microorganisms, pH values, amino acid contents, ammonia contents and volatile fatty acid contents in the stomach fluid of marine little toothed whales. Bull Fac Fish Nagasaki Univ 47:55–60.

Morii H, Kanazu R (1972) The free volatile fatty acids in the blood and the stomach fluid from porpoise, *Neomerio phocaenoides*. Bull Jpn Soc Sci Fish 38:1035–1039.

Obispo NE, Dehority BA (1992) A most probable number method for enumeration of rumen fungi with studies on factors affecting their concentration in the rumen. J Microbiol Methods 16:259–270.

Ogimoto K, Imai S (1981) Atlas of Rumen Microbiology. Tokyo: Japan Scientific Societies Press.

Ohwaki K, Hungate RE, Latter L, Hofmann RR, Maloiy G (1974) Stomach fermentation in the East African colobus monkeys in their natural state. Appl Microbiol 27:713–723.

Owens FN, Goetsch AL (1988) Ruminal fermentation. In: Church DC, ed. The ruminant animal, digestive physiology and nutrition, pp. 145–171. Englewood Cliffs, NJ: Prentice Hall.

Parra R (1978) Comparison of foregut and hindgut fermentation in herbivores. In: Montgomery GG, ed. The ecology of arboreal folivores, pp. 205–229. Washington, DC: Smithsonian Institution Press.

Pearce GR, Moir RJ (1964) The influence of rumination and grinding upon the passage and digestion of food. Aust J Agric Res 15:635–644.

Phillipson AT, McAnally RA (1942) Studies on the fate of carbohydrates in the rumen of the sheep. J Exp Biol 19:119–214.

Prigge EC, Stuthers BA, Jacquemet NA (1990) Influence of forage diets on ruminal particle size, passage of digesta, feed intake and digestibility by steers. J Anim Sci 68: 4352–4360.

Romer AS (1966) Vertebrate Paleontology. Chicago: University of Chicago Press.

Ruckebusch Y, Marquet JP (1965) Effets compartementaux de l'ablation du rumen chez les ovins. C R Soc Biol 159:394–396.

Sanderson IT (1967) Living Mammals of the World. Garden City, NY: Doubleday.

Shaver RD, Nytes AJ, Satter LD, Jorgensen NA (1986) Influence of amount of feed intake and forage physical form on digestion and passage of prebloom alfalfa hay in dairy cows. J Dairy Sci 69:1545–1559.

Shaver RD, Nytes AJ, Satter LD, Jorgensen NA (1988) Influence of feed intake, forage physical form, and forage fiber content on particle size of masticated forage, ruminal digesta and feces of dairy cows. J Dairy Sci 71:1566–1572.

Shively CL, Whiting FM, Swingle RS, Brown WH, Sowls LK (1985) Some aspects of the nutritional biology of the collared peccary. J Wildl Manage 49:729–732.

Shorland FB, Weenink RO, Johns AT (1955) Effect of the rumen on dietary fats. Nature (Lond) 175:1129–1130.

Siciliano-Jones J, Murphy MR (1991) Specific gravity of various feedstuffs as affected by particle size and in vitro fermentation. J Dairy Sci 74:896–901.

Sowls LK (1984) The Peccaries. Tucson: University of Arizona Press.

Tarpley RJ, Sis RF, Albert TF, Dalton LM, George JC (1987) Observations on the anatomy of the stomach and duodenum of the bowhead whale, *Balaena mysticetus*. Am J Anat 180:295–322.

Thurston JP, Noirot-Timothee C, Arman P (1968) Fermentative digestion in the stomach of *Hippopotamus amphibius* (Artiodactyla: Suiformes) and associated ciliate protozoa. Nature (Lond) 218:882–883.

Trinci APJ, Davies DR, Gull K, Lawrence MI, Nielsen BB, Rickers A, Theodorou MK (1994) Anaerobic fungi in herbivorous animals. Mycol Res 98:129–152.

Trudell-Moore J, White RG (1983) Physical breakdown of food during eating and rumination in reindeer. Acta Zool Fenn 175:47–49.

Ulyatt MJ, Waghorn GC, John A, Reid CSW, Monro J (1984) Effect of intake and feeding frequency on feeding behaviour and quantitative aspects of digestion in sheep fed chaffed lucerne hay. J Agric Sci 102:645–657.

Ulyatt MJ, Dellow DW, John A, Reid CSW, Waghorn GC (1986) Contribution of chewing during eating and rumination to the clearance of digesta from the ruminoreticulum. In: Milligan LP, Grovum WL, Dobson A, eds. Control of Digestion and Metabolism in Ruminants, pp. 498–515. Englewood Cliffs, NJ: Prentice-Hall.

Vallenas AP, Stevens CE (1971) Volatile fatty acid concentrations and pH of llama and guanaco forestomach digesta. Cornell Vet 61:239–252.

Vallenas A, Cummings JF, Munnell JF (1971) A gross study of the compartmentalized stomach of two new-world camelids, the llama and guanaco. J Morphol 134:399–424.

Vandernoot GW, Gilbreath EC (1970) Comparative digestibility of components of forage by geldings and steers. J Anim Sci 31:351.

Vandernoot GW, Symons L, Lyman R, Fonnesbeck P (1967) Rate of passage of various feed stuffs through the digestive tract of horses. J Anim Sci 26:1309–1311.

Van Hoven W (1974) Ciliate protozoa and aspects of the nutrition of the hippopotamus in the Kruger National Park. S Afr J Sci 70:107–109.

Van Soest PJ (1982) Nutritional Ecology of the Ruminant. Corvallis, OR: O&B Books.

Walker EP (1964) Mammals of the world, Vols. I and II. Baltimore: Johns Hopkins University Press.

Welch JG (1986) Physical parameters of fiber affecting passage from the rumen. J Dairy Sci 69:2750–2754.

Welch JG, Smith AM (1968) Influence of fasting on rumination activity in sheep. J Anim Sci 27:1734–1737.

Welch JG, Smith AM (1969a) Influence of forage quality on rumination time in sheep. J Anim Sci 28:813–818.

Welch JG, Smith AM (1969b) Effect of varying amounts of forage intake on rumination. J Anim Sci 28:827–830.

Welch JG, Smith AM (1970) Forage quality and rumination time in cattle. J Dairy Sci 53:797–800.

Welch JG, Smith AM, Gibson KS (1970) Rumination time in four breeds of dairy cattle. J Dairy Sci 53:89–91.

Williams AG, Coleman GS (1992) The Rumen Protozoa. New York: Springer-Verlag.

Williams VJ (1963) Rumen function in the camel. Nature (Lond) 197:1221.

4

Fermentation in the Hindgut of Mammals

Ian D. Hume

1. Introduction

Terminology in the hindgut is sometimes confusing. The term ''hindgut'' is used in the physiological sense to denote the large intestine (cecum, colon, rectum and anal canal; Lacy 1991). The cecum is a diverticulum located at the junction of the small intestine and colon. The colon varies enormously in length and complexity among mammals, mainly in relation to the natural diet of the animal. In many carnivores the colon is short, straight, and undifferentiated along its length (Hume 1982). In the human, the colon is only 1.0 to 1.5 m in length (Christensen 1989), but can be divided into ascending colon, transverse colon, descending colon, and sigmoid colon; the position or form of each section is described accurately by these terms. In more herbivorous mammals, the colon is much longer, and consequently is often arranged in a spiraling coil (Stevens and Hume 1995). The terms ''ascending,'' ''transverse,'' and ''descending'' colon are not as applicable in these animals; the terms ''proximal colon'' and ''distal colon'' are widely used instead.

The proximal colon is distinguished from the distal colon in the generally smaller diameter and often more segmented appearance of the latter. The cecum, proximal colon, and first part of the distal colon develop from the embryonic midgut; the rest of the distal colon develops from the embryonic hindgut. Thus in the embryological sense, the hindgut is different from its use in the context of this chapter, which deals only with adult animals.

The proximal colon, together with the cecum, functions as the principal site in the hindgut for the retention of food residues and endogenous materials for microbial fermentation. The short-chain fatty acids (SCFA) produced by the fermentation are absorbed throughout the hindgut, but in quantitative terms most is absorbed from the cecum and proximal colon because that is where they are mainly produced. Most SCFA are absorbed as undissociated acid by passive diffusion across the luminal membrane, but the process involves exchange of hydrogen and sodium ions across this membrane (Stevens and Hume 1995). There is also some evidence for a carrier-mediated pathway (Hume et al. 1993), which may achieve prominence at low concentrations of SCFA in the hindgut lumen, as the K_m is low (< 1 mM). Amino acids produced as a result of microbial protein synthesis in the hindgut are apparently not absorbed from that region of the gut by anything but passive, nonmediated diffusion (Hume et al. 1993). Because levels of free amino acids in the luminal contents are kept low by the action of microbial deaminases, uptake of amino acids from the mammalian hindgut is unlikely ever to be of much quantitative importance to the host animal. The same appears to be the case for the B vitamins synthesized during microbial growth. The only way the host animal can access these intracellular microbial nutrients is by ingesting feces (coprophagy).

The distal colon functions as the principal site in the hindgut for net absorption of water and electrolytes, and so retrieves a lot of the water and inorganic ions secreted into the upper digestive tract. In this regard, it is interesting that the distal colon is relatively longer in mammals adapted to arid conditions than in their more mesic counterparts (Woodall and Skinner 1993). Absorption of water can occur against an osmotic gradient, and although the precise mechanism of water transport in the distal colon is unknown, net absorption of water is always coupled with sodium, chloride, and SCFA transport in the same direction, either osmotically or electroosmotically (Reuss 1991).

The rectum functions for the final storage of feces prior to evacuation. Species that defecate at frequent intervals, such as kangaroos, have little in the way of a rectum, but in species that only defecate occasionally, such as the sheep, the rectum and its contents are usually obvious on dissection of the digestive tract (Stevens and Hume 1995).

This chapter concentrates on fermentation in the mammalian hindgut and thus on the cecum and proximal colon, the main sites of microbial fermentation.

2. Anatomy and Physiology of the Cecum and Proximal Colon

There is great diversity among the mammals in hindgut morphology. Most of these variations can be correlated with diet, particularly the ease with which the usual diet of the animal is digested by the animal's own enzymes in the small intestine. Thus in carnivores the hindgut is small, comparatively simple, and only weakly haustrated if at all. A cecum is absent from the carnivorous marsupial family Dasyuridae (Hume 1982), and from several eutherian carnivores such as the mink (Hume and Warner 1980). Hindgut capacity is much greater in herbivores, and reaches its maximum of 12.5% of body mass in large hindgut fermenters such as the horse (Engelhardt et al. 1983).

The relative size and importance of the cecum versus proximal colon vary among hindgut fermenters. In large hindgut fermenters (body mass > 10 kg), the capacity of the proximal colon is greater than that of the cecum, and it is the principal site of microbial fermentation. The cecum in these herbivores appears to function largely as a simple extension of the proximal colon, and there is extensive mixing of contents between the two organs. Hume and Warner (1980) called these hindgut fermenters "colon fermenters." In some colon fermenters such as the marsupial wombats, a cecum is virtually absent. Other colon fermenters include hominoid primates (the apes) including humans; equids (horse, zebra); tapir; rhinos; elephants; dugong; and pig (Hume and Sakaguchi 1991).

The other type of hindgut fermenter is the "cecum fermenter" (Hume and Warner 1980). In these herbivores, nearly all of which weigh less that 10 kg, the principal site of digesta retention and microbial fermentation is the cecum. Fermentation may extend into the proximal colon, the extreme case being the koala, in which the capacities of the proximal colon and cecum are similar (Hume 1982). Cecum fermenters include three other marsupials that feed more or less exclusively on *Eucalyptus* foliage (greater glider, ringtail possums, and brushtail possums); rodents; lagomorphs (rabbit, hare, pika); and some small monkeys (e.g., tamarins, marmosets).

The mammalian hindgut is lined with a columnar epithelium that contains mucus-secreting goblet cells. The hindgut mucosa secretes no enzymes, and the epithelium lacks villi, although there is evidence of microvilli in some species (Lacy 1991). Beneath the mucosa there are two layers of muscle, one circular and one longitudinal. The colon of many mammals, particularly herbivores, is haustrated in appearance. Haustration results from the organization of the longitudinal muscle layer into three thick bands (taeniae). The longitudinal muscle layer is also present between the taeniae but is very thin (Christensen 1989). Contractions of the circular muscle between the taeniae draw in the hindgut wall to form semilunar folds internally, and haustra externally. The semilunar folds form, disappear, and re-form with no constant relationship to the position of fixed points on the hindgut wall. Waves of contraction of the circular muscles move digesta both caudally (peristalsis) and orally (antiperistalsis).

Patterns of digesta transit differ between colon fermenters and cecum fermenters. These are described by reference to indigestible markers given orally and collected and analyzed in the feces (Warner 1981, Stevens and Hume 1995). In colon fermenters, the mean retention time (i.e., the average time for digesta markers to traverse the entire gastrointestinal tract) of particle markers invariably exceeds that of fluid markers (Hume 1989). This separation of digesta phases results from the way in which contractions of the wall of the proximal colon squeeze fluid (and fine particles) through a matrix of large particles. In other words, the haustrations of the colon wall selectively retain the large particles. This maximizes exposure of the most fibrous part of the digesta (in the large particles) to the action of cellulolytic microorganisms in the hindgut, and thus maximizes fiber digestion. Fiber digestibilities in the largest colon fermenters (elephant, rhino) approach those of large ruminants (Hume and Sakaguchi 1991).

In contrast, in cecum fermenters, there is either no significant difference in mean retention times between fluid and particulate markers, or there is selective retention of fluid (and fine particles); that is, the mean retention time of a fluid marker exceeds those of particle markers. As a result, fiber digestibility is often low, but there is great variability among different groups of cecum fermenters. This is because of the complexity and variety of digesta retention mechanisms operating in the hindgut of these small herbivores. The structural and functional bases for the range in digesta marker excretion patterns seen in cecum fermenters are best understood through the application of chemical reactor theory (e.g., Levenspiel 1972) to the digestive system.

3. The Cecum and Colon as Chemical Reactors

Chemical reactor theory has been applied to a wide range of animal digestive systems, both invertebrate (Penry and Jumars 1987) and vertebrate (e.g., Hume 1989, Alexander 1991), in order to clarify the general principles operating in the great diversity of digestive strategies described in the literature. Chemical reactors are classified on the basis of whether input is continuous or discontinuous, and whether reactants are brought together with or without mixing (Penry and Jumars 1987). Three types of chemical reactors are directly applicable to the vertebrate small and large intestines: plug-flow reactors (the small intestine), continuous-flow, stirred-tank reactors (the cecum), and modified plug-flow reactors (the colon; Hume 1989, Hume and Sakaguchi 1991).

3.1. Plug-Flow Reactors

In these reactors there is continuous flow through a usually tubular reaction chamber (Fig. 4.1a). Ideal plug-flow reactors can be described accurately by simple equations. In them, material does not mix along the flow axis, but there is perfect radial mixing. As a result, at steady state, there is a continuous decrease in reactant concentrations and a continuous increase in product concentrations from inlet along the reactor to outlet, but at any point along the flow path composition is uniform in cross section and unchanging with time. Of the three relevant reactor types, plug flow provides the greatest rate of product formation in the

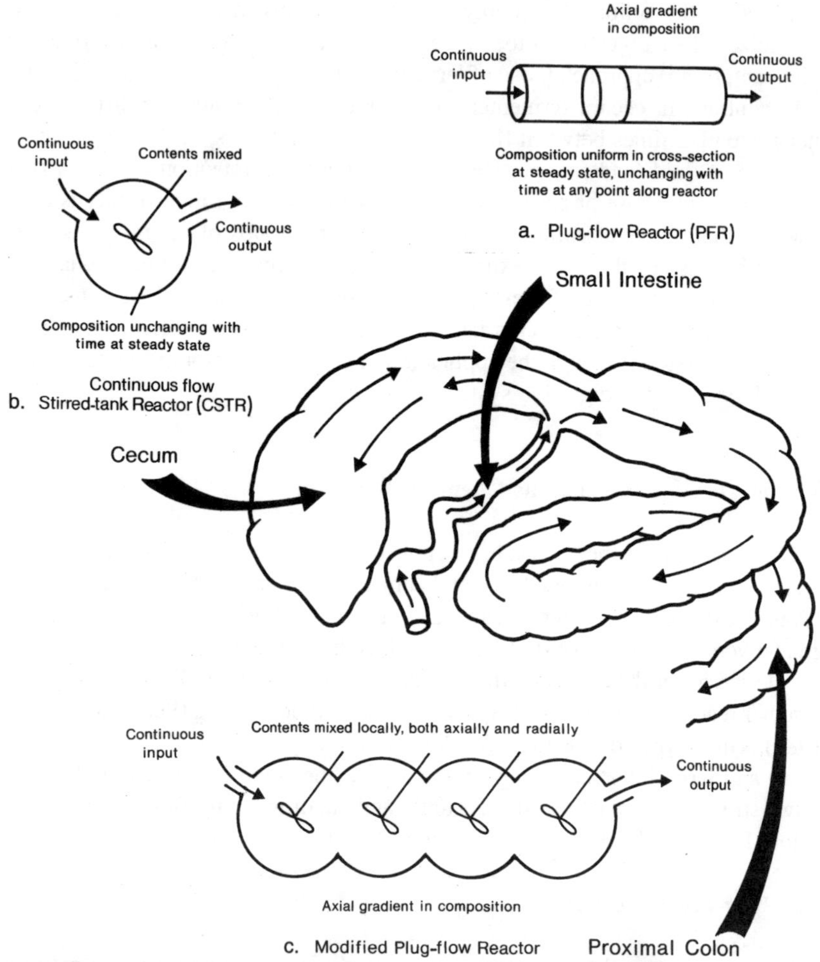

Figure 4.1. Three types of chemical reactors and their analogs in the lower small intestine and upper hindgut of the pig: (a) plug-flow reactor (small intestine); (b) continuous-flow, stirred-tank reactor (cecum); (c) modified plug-flow reactor (proximal colon).

minimum of time and volume under most conditions, although extent of reaction may be low (Penry and Jumars 1987). In the vertebrate gut, plug-flow reactors are best represented by the small intestine, even though flow through this region of the gut is not continuous and even though there is some axial (Macagno and Christensen 1980) as well as radial mixing.

3.2. Continuous-Flow, Stirred-Tank Reactors

These reactors are characterized by continuous flow of materials through a spherical reaction vessel of minimal volume (Fig. 4.1b). The concentration of reactants is diluted immediately upon entry into the vessel by material recirculating in the reactor. This dilution reduces reaction rate, so that rate of digestion is lower than in plug-flow reactors. However, extent of reaction usually exceeds that in plug-flow reactors, and can be very high if material flow is low enough. In an ideal continuous-flow, stirred-tank reactor at steady state, composition is uniform throughout the reactor and unchanging with time. In terms of operating performance, stirred-tank reactors are probably the most flexible of all chemical reactors. In the vertebrate gut, stirred-tank reactors are best represented by diverticulae such as the cecum, even though patterns of digesta flow into and out of this hindgut region are often complex, and rarely continuous (Hume and Sakaguchi 1991).

3.3. Modified Plug-Flow Reactors

When several stirred-tank reactors are connected in series, the flow characteristics of the system are intermediate between those of a stirred-tank reactor and a plug-flow reactor (Fig. 4.1c). Hume (1989) referred to such a hybrid system as a "modified plug-flow reactor." In this system there is substantial axial as well as radial mixing, yet incoming reactants are not immediately mixed with the entire contents of the reactor vessel. As a result, there is a reduction in both the dilution effect seen in a single large stirred tank and the total mixed-flow volume needed to achieve a given extent of conversion of reactants to products (Penry and Jumars 1987). As the number of stirred tanks in series increases, the reactor system becomes more and more like a plug-flow reactor. The proximal colon is best modeled as a modified plug-flow reactor because of the substantial axial mixing caused by antiperistaltic as well as peristaltic contractions (Hume 1989, Hume and Sakaguchi 1991).

4. Digesta Flow and Digestion in Colon Fermenters

The best-studied colon fermenters are the horse (among the Eutheria) and the marsupial wombat. The nutritional niche of these large hindgut fermenters seems to be toward the high-fiber end of the spectrum of plant nutritional quality. Ruminants occupy the central part of the spectrum. Cecum fermenters occupy the low-fiber end of the spectrum (Hume and Sakaguchi 1991). The distribution of these three groups of herbivores across the plant quality spectrum can be understood in terms of chemical reactor theory and body size (and thus the ratio of metabolic requirements to gut capacity, which is high in small animals and low in large animals).

In the horse, Argenzio et al. (1974) found that retention times of particle markers (polyethylene tubing) exceeded those of fluid markers, and that retention times of the particles increased with increasing particle length. This is due to two mechanisms. The first is the resistance to flow of large particles offered by the semilunar folds, as described above. The second is the strong retropulsive waves of contractions of the haustra that originate at a pace maker area located at the large colon pelvic flexure (Sellers et al. 1982). However, the difference in mean retention times of particle and fluid markers in the whole tract of the horse is only 3 to 5 hours (Uden et al. 1982, Orton et al. 1985), much less than would be predicted on the bases of the Argenzio et al. (1974) and Sellers et al. (1982) findings. This is because, opposing the selective retention of particles in the ventral colon, contractions of the muscular wall of the distal colon force fluid back into the right dorsal colon (Björnhag 1987), resulting in selective retention of fluid and fine particles. Thus to model the equine hindgut as a modified plug-flow reactor is an oversimplification of a complex system (Hume and Sakaguchi 1991).

Nevertheless, comparison of plug-flow and stirred-tank reactors offers an explanation for the generally lower-fiber digestibilities by colon fermenters compared with ruminants of similar body size. Plug-flow reactors are characterized by maximal rates of product formation in minimal volume at the expense of maximal extent of conversion. Modified plug-flow reactors share this characteristic (but to a lesser extent). Thus we find that, on similar diets of timothy hay (*Phleum pratense*), colon fermenters such as horses, zebra, elephants, and rhinos exhibit lower digestibilities of organic matter and fiber than do large ruminants (Foose 1982).

In the wombats, Barboza (1993) showed that the mean retention time of a particle marker (57 hours) exceeded that of a fluid marker (33 hours). These values are greater than those of horses (27 hours for particles, 22 hours for fluid; Orton et al. 1985), despite their smaller size (20 to 40 kg). This is related to lower nutrient requirements and, as a consequence, lower voluntary food intakes. The wombat digestive strategy appears to be based on the utilization of forages of low quality and/or limited quantity by a combination of low food intakes, large hindgut volume (Barboza and Hume 1992a), and long mean retention times of digesta. The large difference in retention times of the digesta phases in the wombats is probably due to the lack of any selective retention of fluid in the distal colon, as occurs in the horse.

5. Digesta Flow And Digestion In Cecum Fermenters

The best-studied cecum fermenter is the rabbit (Pickard and Stevens 1972, Björnhag 1981, Sakaguchi et al. 1987), but there is available at least some information on digesta flow and digestion in other cecum fermenters such as other lagomorphs (pika, hare), myomorph rodents (hamster, rat, mouse, vole, lemming), caviomorph rodents (guinea pig, mara, nutria), and some marsupial folivores (koala, greater glider, and ringtail and brushtail possums). Björnhag (1987) has summarized the range of colonic separation mechanisms found in these mammals that result in selective retention of fluid and fine particles in the hindgut. In the rabbit, retropulsive waves of contractions arising from a pacemaker area (the fusus coli) at the junction of the proximal and distal colon lead to retrograde transit of fluid digesta along the wall of the strongly haustrated proximal colon toward the cecum, leaving the bulk of the lumen free for caudal transit of other contents. Fine particles, which include bacteria, tend to be washed along with the flow of fluid. The process is assisted by net secretion of water into the proximal colon, under the control of aldosterone (Clauss 1984), together with net absorption of water from the cecum. This sets up an internal water cycle that operates during hard feces formation but is switched off during soft feces formation. Formation of soft feces is also accompanied by a decrease in motility of the proximal colon, and increased motility of the distal colon.

Other mechanisms that lead to selective retention of fluid and fine particles have been described for other cecum fermenters (Björnhag 1987). In all, the effect of the process is to concentrate digestive effort in the cecum by maintaining higher concentrations of microbes with the most digestible components of the digesta, the solutes and fine particulate matter. It also clears the hindgut of the larger, less digestible particles more quickly that would otherwise be the case. This enables the animal to maintain higher food intakes, although often at the expense of fiber digestibility. Thus the rabbit digests only 10% of the crude fiber of alfalfa, but can survive on much higher-fiber diets than can the guinea pig, which digests 34% of the crude fiber of the same diet (Sakaguchi et al. 1987); the guinea pig has a less effective separation mechanism in the proximal colon, which means that more larger particles are retained in the cecum, and the hindgut is not cleared of large particles as rapidly. Similarly, Foley and Hume (1987) concluded that the main reason why the common brushtail possum (*Trichosurus vulpecula*) cannot subsist on a diet of *Eucalyptus* foliage alone was a lack of a colonic separation mechanism; leaf intake appeared to be constrained by the gut-filling effect of a large mass of indigestible material in the hindgut. In contrast, the common ringtail possum (*Pseudocheirus peregrinus*), which does have a separation mechanism, is easily maintained in captivity on eucalypt leaves as the sole diet (Chilcott and Hume 1985).

Many cecum fermenters ingest feces. When there is little or no difference in the composition of ingested and discarded feces, the practice is referred to as coprophagy. When the ingested feces are of higher nutrient content (i.e., higher in protein and water and lower in fiber), the practice is referred to as cecotrophy (Hörnicke and Björnhag 1980); this term indicates that the soft feces originate from cecal contents. Coprophagy, and more particularly cecotrophy, leads to more effective utilization of nutrients by the animal, especially the essential amino acids and B vitamins synthesized in the cecum by the resident microbes. Chilcott and Hume (1985) calculated that cecotrophy by the common ringtail possum contributed energy equivalent to 58% of digestible energy intake, and nitrogen equivalent to twice the maintenance requirement. In other words, in the absence of cecotrophy the maintenance nitrogen requirement would have been 620 mg/kg$^{0.75}$/day, rather than the value of 290 measured (Hume et al. 1984).

6. Fermentation and Microbiology of the Hindgut

6.1. Environment of the Hindgut

The hindgut environment consists of at least two compartments, the bulk contents of the lumen and the luminal mucous layer. The latter protects the mucosa, lubricates digesta to facilitate their passage, and forms a barrier which modifies the transport of solutes across the epithelium (Filipe and Branfoot 1976). In the cecum of at least three mammals (mice, rats, and guinea pigs), the luminal mucin is well mixed with digesta, and hence there is no definite, continuous mucous layer at the mucosal surface (Sakata and Engelhardt 1981). The lack of a continuous luminal mucous layer probably results from a combination of vigorous mixing of contents and rapid degradation by mucinolytic bacteria. Consequently, food particles are in direct contact with the mucosa.

In contrast, the mucosal surface of the colon is covered almost entirely with a luminal mucous layer. The layer shows marked regional differences in its compactness, thickness, and histochemical composition. In the proximal colon it is spongy, thick (150 μm in rats), and composed mainly of neutral glycoproteins. In the distal colon it is compact, thin (only 16 μm in rats), and composed mainly of acidic glycoproteins (Sakata and Engelhardt 1981). As a result, food particles often penetrate the mucous layer of the proximal colon, but never in the distal colon. The thick, spongy mucous layer in the proximal colon contains a dense population different from the bacterial types of the lumen. In contrast, the thin, compact mucous layer in the distal colon contains considerably lower numbers of bacteria.

The mucin is formed in the Golgi region of the goblet cells which are such a prominant feature of the hindgut epithelium. Within the lumen, the mucin is highly hydrated, containing about 95% water, and thus exists as a gel of glycoprotein mesh held together by hydrogen bonds and a few disulphide bridges (Clamp et al. 1978). The rate of secretion of mucus appears to be under autonomic control, as acetyl choline stimulates mucus release in rabbit and human hindgut mucosa (MacDermott et al 1974).

The mucous layer offers no structural barrier to the movements of small ions, and therefore its ionic composition reflects that of the digesta with which it is in contact. However, macromolecules are excluded from the mucous layer (Edwards 1978). This probably prevents damaging substances having access to the mucosa. Secretory immunoglobulins such as IgA associated with mucus (Kagnoff 1987), which would protect the epithelium against likely pathogens present in the lumen.

Despite the apparently free movement of small ions through the luminal mucous layer, the H ion concentration in the mucous layer is remarkably independent of the H ion concentration in the bulk luminal fluid. The pH in the mucous layer was maintained at 7.08 $\pm$ 0.14 in the proximal colon and at 6.91 $\pm$ 0.14 in the distal colon of guinea pigs when the pH of the luminal solution was varied over the range 5 to 9 (Engelhardt and Rechkemmer 1983). This stable pH microclimate at the luminal epithelial surface explains why SCFA absorption from the colon is largely independent of the pH of the luminal fluid. In normal colonic contents, about 99% of the SCFA is present as anions, but absorption is largely as the undissociated lipid-soluble form (Hume et al. 1993).

Conditions within the lumen of the hindgut are generally conducive to active microbial growth and metabolism. The temperature is constant and close to body temperature, the pH is usually between 6.5 and 7.5 throughout, and the oxygen tension is extremely low. Oxidation reduction potentials (Eh) of -200 mV are often recorded, so that all microorganisms isolated from the hindgut are anaerobes.

Digesta entering the hindgut from the ileum are usually isotonic with plasma. The main cation is sodium (100 to 128 mM), while potassium, although variable between species, is always lower (9 to 57 mM). This variability is the result of high intakes by herbivores. The major cation is usually chloride (25 to 66 mM) in most species, although bicarbonate concentrations of up to 128 mM have been measured in herbivores such as sheep, cattle, and horses (Kay and Pfeffer 1970). Calcium, magnesium, and phosphate concentrations in ileal digesta in sheep and cattle are dependent on dietary intakes, as they are in humans (Ben-Ghedalia et al. 1975). Organic anions in the form of the salts of SCFA and lactate also make a significant contribution to hindgut digesta osmolality in herbivores.

Along the hindgut, the dry matter content increases as a result of net absorption of water. Sodium concentrations decrease along the hindgut as this cation is also retained by net absorption. Potassium concentrations tend to show an inverse relationship to sodium. Chloride concentrations, like those of sodium, decrease along the colon. Compared with the ileum, SCFA concentrations increase in the cecum, remain high in the proximal colon, and decrease through the distal colon (e.g., Barboza and Hume 1992b); this pattern is readily explained as the net result of microbial production, which tends to decrease along the colon, and absorption of SCFA. Differences along the hindgut have beeen found in rates of absorption of the SCFA, but the differences are not consistent between species. For instance, Rechkemmer and Engelhardt (1988) reported that in the guinea pig the proximal colon was more permeable to acetate (the principal SCFA usually produced) than was the distal colon. However, in the prairie vole (*Microtus ochrogaster*), acetate uptake was higher in the distal colon than in either the proximal colon or the cecum (Hume et al. 1993).

6.2. Microbes of the Hindgut

The characteristics of the hindgut microbial ecosystem are not as well understood as those of the rumen. This is partly because of our poor overall understanding of the functions of the mammalian hindgut, and partly because of the preoccupation with the cecum, when in large herbivores the colon is the main site and, in several mammals including humans, virtually the sole site of microbial fermentation in the hindgut. Additional complicating factors are that in the hindgut the microenvironment, and hence perhaps the nature of the microbial fermentation, varies along its length. There may be several different mixing pools of digesta—e.g., in the horse (Argenzio et al. 1974) and wombat (Barboza 1993)—and the most important population of microorganisms may not be the one so easily measured in the lumen, but that attached to the epithelial lining or embedded in the luminal mucous layer associated with the epithelium.

Bacteria and protozoa both occur in the hindgut, but the distribution of protozoa is not universal. The most successful are the ciliates of the equine cecum.

BACTERIA

Bacterial populations as high as 10^{10} to 10^{11}/g contents have been found in the hindgut of most mammals. These are similar to numbers in the forestomach of ruminants, macropod marsupials, and colobid monkeys. In contrast, bacterial numbers in the small intestine and in simple stomachs of mammals are usually less than 10^{8}/g. An average of 55% to 75% of the solids in human feces is bacterial biomass (Wrong et al 1981).

There appears to be no specific hindgut microbiota; neither is the hindgut microbiota uniquely different from that in the rumen. This is reflected in the nature of the end products of the hindgut fermentation. The SCFAs acetate, propionate, and butyrate are formed, along with the gases methane, hydrogen, and carbon dioxide (Wolin 1981). Freter (1983) proposed some general principles relating to bacterial colonization of the hindgut. In the small intestine, the rate of digesta flow, and thus the rate of bacterial washout, exceeds the maximal growth rates of most bacterial species. As a result, this region of the gut can be successfully colonized only by bacteria that adhere to the epithelial lining. In the hindgut, however, digesta flow rates are much slower because of the much greater cross-sectional area of this region of the gut, and because of the dominance of antiperistaltic over peristaltic contractions (Christensen 1989). This means that all but the slowest-growing bacteria can colonize the lumen of the hindgut without having to resort to adhering to the hindgut wall. Thus in the hindgut there are three habitats occupied by bacteria—the lumen, the mucous layer, and the epithelium.

The composition of the hindgut microbiota is remarkably stable. It consists of several hundred anaerobic bacterial species and strains that coexist without one or a few becoming dominant. This stability appears to be a function of inhibition of bacterial multiplication by such compounds as SCFA, hydrogen sulfide, deconjugated bile salts, and bacteriocins. These bacterial inhibitors may prolong the lag phase of invading bacteria sufficiently that they are excreted with the feces before they have a chance to establish themselves (Freter 1983).

However, the observed balance among several hundred bacterial species and strains within the hindgut must be due to a more general mechanism, such as competition for limiting nutrients. The hindgut contains a large number of different substances that serve as bacterial nutrients. The greater the number of limiting nutrients, the greater the diversity of the bacterial population within the ecosystem, since each limiting nutrient will support the one bacterial species or strain that is most efficient in utilizing it.

From these general principles it would be predicted that the bacterial population adhering to the epithelium would be even more stable than that in the lumen. In this case the invader must compete with the resident species or strains not only for limiting nutrients but also for specific adhesion sites. Indeed, the bacteria adhering to the hindgut wall appear to be not only very stable in species composition but very specialized as well, forming dense mats of often virtually pure cultures. This specificity in adhesion sites may not only be physical but also an effect of local immune reactions of the hindgut wall to bacteria.

Most of our knowledge of hindgut bacteria in humans comes from bacterial counts in the feces. Bentley et al. (1972) are one of the few groups to have compared bacterial counts in the terminal ileum, cecum, and transverse colon with stool specimens. This was done with patients undergoing elective cholecystectomy. Bacterial counts were highest in the feces and lowest in the ileum. Despite large numerical differences, however, there did not appear to be marked qualitative differences in the microbiota, at least for the major groups of bacteria.

Table 4.1. Major bacterial groups isolated from the feces of 141 patients

Group	$\log_{10}$ Number of Organisms per g Dry Weight Feces	
	Mean	(Range)
Bacteroides	11.3	(9.2–13.5)
Fusobacterium	8.4	(5.1–11.0
Anaerobic cocci	10.7	(4.0–13.4)
Actinomyces	9.2	(5.7–11.1)
Streptococcus	8.9	(3.9–12.9)
Clostridium	9.8	(3.8–13.1)
Lactobacillus	9.6	(3.6–12.5)
Eubacterium	10.7	(5.0–13.3)
Bifidobacterium	10.2	(4.9–13.4)
Arachnia-Propionibacterium	8.9	(4.3–12.0)
Gram-negative facultative anaerobes	8.7	(4.0–12.4)
Other facultative anaerobes	6.8	(0.7–12.7)

This probably applies to the luminal bacteria only, and not necessarily mucous layer and adherent bacteria.

The major bacterial groups isolated from the feces of 141 patients by Finegold et al. (1983) are shown in Table 4.1. The patients included strict vegetarians, Japanese-Americans on a traditional Japanese diet, and both disease-free individuals and patients with colon cancer on a standard Western diet. There was a wide range of bacterial counts among the 141 specimens, but marked similarity between groups despite the differences in diet and disease state. Of the more than 400 bacterial species so far isolated from the human hindgut, over 95% are obligate anaerobes. The most numerous are *Bacteroides* spp. (gram-negative, nonsporing rods), various gram-positive cocci, *Eubacterium* spp., and *Bifidobacterium* spp., Facultative anaerobes include enterobacteria such as *Escherichia coli* and *Proteus mirabilis*, and some lactobacilli and streptococci.

Information on the hindgut biota of nonhuman primates is limited. The colon of baboons (*Papio* spp.) was found by Uphill et al. (1974) to be heavily populated with lactobacilli, streptococci, staphylococci, and *Clostridium perfringens*. Takeuchi et al. (1974) found the epithelium of the colon of rhesus monkeys (*Macaca mulatta*) to be inhabited by spiral bacteria.

In sheep fed forage diets the main contribution to the biota of the cecum were gram-negative rods of the genera *Bacteroides*, *Butyrivibrio*, and *Fusobacterium* (Mann and Orskov 1973). More minor contributions came from *Streptococcus bovis*, *S. faecalis*, and species of the genera *Peptostreptococcus*, *Micrococcus*, and *Selenomonas*. Cellulolytic activity was high, with counts of cellulolytic bacteria of

10^8/g cecal contents recorded. Anaerobic viable counts can be as high as 9×10^{10}/g contents (Ulyatt et al. 1975).

Gram-negative rods and cocci also predominate in the cecum of the horse, with anaerobic viable counts of 4.9×10^9/g contents (Kern et al. 1974), although cellulolytic counts were low, at 1 to 8×10^7/g contents. About 20% of the isolates were proteolytic.

In contrast to the horse, London (1981) found that the microbiota of the koala cecum was predominantly gram-positive rods (approximately 60%). Presumptive grouping of isolates into genera indicated that *Bacteroides*, *Eubacterium*, *Peptococcus*, *Peptostreptococcus*, and *Propionibacterium* were among the most common genera present in a total viable count of 4×10^{10}/g contents. Osawa et al. (1993) have isolated several bacterial strains from the koala hindgut that are active in breaking down tannin-protein complexes. Partial digestion of tannin-protein complexes in the cecum of another eucalypt foliage feeder, the ringtail possum (*Pseudocheirus peregrinus*), had previously been reported by O'Brien et al. (1986). This may be an important function of the hindgut microbiota of animals that feed almost exclusively on highly tanniniferous *Eucalyptus* foliage.

Considerable attention has been paid to the identification of microbial cells associated with the hindgut wall. Layers of microbial cells on epithelial surfaces tend to be thicker and more complex in the hindgut than in other parts of the digestive tract (Savage 1983). All known forms of bacteria have been found to be associated with the epithelium of the cecum and colon of rats, mice, and other mammals. Most common are fusiform rods and helical bacteria. Some species attach by holdfasts to the epithelial cell membranes, as Takeuchi and Zeller (1972) demonstrated with spirochetes and flagellates in the cecum and colon of the rhesus monkey. In these cases, the structure of the epithelium is markedly altered, with destruction of microvilli and the terminal web of the brush border, as McKenzie (1978) showed with gram-positive bacteria in the cecum of the koala. In other cases the microbes adhere to epithelial surfaces by envelopes composed of complex polysaccharides or other macromolecules on their surfaces, and the epithelium is not structurally altered. This appears to be the case with the tannin-protein complex degrading enterobacteria isolated from the hindgut of the koala by Osawa et al. (1993).

However, most species associate with the epithelium by colonizing the mucous layer overlying it. The mucous layer serves as a structural matrix which contributes to the stability of the microbial communities inhabiting it, and assists with their activities such as motility (Berg and Turner 1979). It serves also as an important source of carbon and energy; various bacterial species isolated from the mammalian digestive tract synthesize enzymes which hydrolyze mucins (Salyers et al. 1977). Estimates of the size of the microbial population associating with the epithelium of the human colon lie between 10^7 and 10^8 cells/g wet tissue (Savage 1983). Hill (1983) found that in the rabbit cecum the total numbers of bacteria recovered on a nonspecific, semidefined medium were 1.7×10^9/g luminal contents versus 2×10^5/g washed, homogenized cecal wall. Twenty-two percent of strains from the latter site were ureolytic, compared with only 5% of strains isolated from luminal contents. In this case it seems clear that the ureolytic bacteria were concentrated at the point of diffusion of urea across the epithelium, as has been shown for the rumen (McCowan et al. 1980, Kennedy et al. 1981) and the human colon (Wolpert et al. 1971).

PROTOZOA

The occurrence of protozoa in the hindgut is not universal. Adam (1951) found ciliate protozoa throughout the equine hindgut, with lowest counts in the cecum and small colon and highest numbers in the ventral and dorsal colon. In addition, there were two distinct populations of ciliates. One, characteristic of the cecum and ventral colon, was dominated by *Blepharcorys uncinata* (particularly in the cecum) and *Cycloposthium bipalmatum* (particularly in the ventral colon). The other, characteristic of the colon caudal to the pelvic flexure to the rectum, was dominated by *Blepharcorys curvigula* and *Bundleia postciliata*. The relative proportions of these dominant species were quite uniform throughout the dorsal and small colon and rectum. Kern et al. (1974) confirmed the presence of protozoa in the cecum of the horse, but they did not examine the colon. Ciliate protozoa have also been reported from the hindgut of the elephant (Kofoid 1935) and pig (Dorst 1973). However, protozoa appear to be absent from the hindgut of foregut fermenters such as ruminants and macropod marsupials. Their metabolic role in those mammals in which they are found in the hindgut is unknown.

FUNGI

Following the discovery of phycomycete fungi in the ovine rumen (Orpin 1975, 1976), examination of the equine cecum by the same author (Orpin 1981) revealed vegetative growth of phycomycete fungi on particles of digesta, as well as uniflagellated motile cells similar to fungal zoospores in the fluid phase. Preliminary tests failed to separate one of three morphologically distinct isolates from the rumen phycomycete *Piromonas communis*. This and one of the other two isolates were shown to digest water-insoluble plant tissues, and were able to use a range of plant carbohydrates for growth, including xylan, cellulose, and starch.

The colon was not investigated by Orpin (1981), but on the basis of the presence of protozoa throughout the equine hindgut it would be surprising if phycomycetes were not present in at least the ventral and dorsal colon. It is also likely that, with further investigation, anaerobic fungi will be shown to be present in other hindgut fermenters. Finegold et al. (1983) found filamentous fungi in the feces of patients on a standard Western diet but not in feces of vegetarians or of Japanese-

Americans on a traditional Japanese diet. Yeasts (including *Candida* spp.) were numerous (5.6×10^{10}/g dry weight) in all six groups of patients examined.

6.3. Rates of Fermentation in the Hindgut

The microbial fermentation in the hindgut appears similar to the forestomach fermentation that has been extensively studied in ruminants. One of the main differences is the production of hydrogen in the colon of some, but not all, humans (Wolin 1981); free hydrogen is never produced in anything but trace amounts in the rumen. Few direct estimates of microbial growth in the hindgut have been published. Instead, authors have assumed tight coupling of microbial growth with fermentation rate. This has allowed measurement of rates of accumulation of fermentation end products as an index of microbial growth rates. The two end products most commonly measured are gases (either total or individual gases such as methane) and SCFA (principally acetic, propionic, and butyric acids).

TOTAL GAS PRODUCTION

Measurements of total gas production are made in vitro. In this procedure the hindgut contents may or may not be buffered, but exogenous substrate should not be added unless the parameter of interest is the potential rate of gas production from a particular substrate rather than actual production rate. Digesta samples are incubated at 1°C or 2°C above body temperature in a sealed container. Gas is allowed to escape into a water-lubricated syringe, and production rate is read directly from the syringe at short (e.g., 1-minute) intervals (El-Shazly and Hungate 1965). The zero-time technique of Carroll and Hungate (1954) can then be used to estimate initial production rates if the rate is not constant over time.

Some estimates of total gas production using these methods are shown in Table 4.2. The most useful values are those from Van Hoven et al. (1981) from the African elephant (*Loxodonta africana*). Fermentation rate was highest in the proximal colon, and declined progressively toward the distal colon. This is consistent with the colon operating as a modified plug-flow reactor; reaction rates are highest at the inlet and progressively decline along the length of the reactor toward the outlet.

METHANE

Van Hoven et al. (1981) found that methane contributed 44% of cecal gas and 52% to 69% of colonic gas in the African elephant. This is much higher than in the rumen, where the gas typically contains about 10% methane (Hoppe et al. 1983). Rates of production of methane in both the rumen and hindgut of sheep have been measured in vivo by Murray et al. (1976, 1978) using an isotope

Table 4.2. Total gas production (μmol/g dry matter/h) in two hindgut fermenters compared with several ruminants

Species	Gas Production			Reference
	Rumen	Cecum	Colon	
Hindgut fermenters				
Cape porcupine (*Hystrix africaeaustralis*)	—	91	75	Van Jaarsveld and Knight-Eloff (1984)
Elephant (*Loxodonta africana*)	—	159	226	Van Hoven et al. (1981)
Foregut fermenters				
(a) Roughage eaters				
Buffalo (*Syncerus caffer*)	203,167	37,79	—	Van Hoven (1980)
Wildebeest (*Connochaetis gnou*)	116,203	37,79	—	Van Hoven and Boomker (1981)
Wildebeest (*C. taurinus*)	181	—	—	Giesecke and Van Gylswyk (1975)
Gemsbok (*Oryx gasella*)	166	—	—	Giesecke and Van Gylswyk (1975)
(b) Concentrate selectors				
Kirk's dikdik (*Madoqua kirki*)	395	123	—	Hoppe et al. (1983)
Suni (*Nesotragus moschatus*)	629	79	—	Hungate et al. (1959)
Gray duiker (*Sylvicapra grimmia*)	370,259	85,103	—	Boomker (1981)

dilution technique, but the method does not appear to have found other applications since that time. Murray et al. (1976) found that methane was produced in the ovine hindgut at about 4.4 mL/kg body mass/h, or 2.6 mL/min; rates in the rumen averaged 18.0 mL/min, or 87% of total production. The two pathways of excretion of methane from the hindgut were via the anus (11%) and through the lungs after absorption into the blood (89%). There was no excretion of methane in the urine.

Methane has also been found to be produced in the hindgut of the rock hyrax (*Procavia habessinica*) by Engelhardt et al. (1978), and in the cape porcupine

(*Hystrix africaeaustralis*) by Van Jaarsveld and Knight-Eloff (1984). The average rate in four rock hyrax was 5.9 mL/kg body mass/h or 0.22 mL/min, similar to the rate in the ovine hindgut. The rate in seven cape porcupines was lower, at 0.08 mL/min.

Some humans are also methane producers. Up to 6% of flatus consists of methane (Cummings and Macfarlane 1991). Bond et al. (1971) found that 34% of a population of over 300 adults produced methane in the colon which was subsequently absorbed into the blood and excreted via the lungs. Maximum production rates were 0.7 mL/min, but they fall to less than 0.03 mL/min on polysaccharide-free diets (Christl et al. 1992).

HYDROGEN

Hydrogen is also produced in the human colonic fermentation, but its rate of production is much more variable than that of methane. Pulmonary excretion of hydrogen is very low in fasting subjects, but increases enormously when nonabsorbable carbohydrates such as lactulose reach the hindgut. Bond et al. (1971) found that ingestion of 10 g lactulose had no effect on pulmonary excretion of methane, which averaged 0.18 mL/min, but hydrogen excretion increased seven fold, to 0.14 mL/min, or 8 mL/g lactulose. Hydrogen production in non-methanogenic humans is about 17 mL/g lactulose (Christl et al. 1992). A number of natural oligosaccharides such as stachyose and raffinose, which are both present in beans, also reach the hindgut and become substrates for microbial production of hydrogen.

In contrast to humans, herbivores appear to produce little hydrogen gas in either the hindgut or the forestomach. In the cape porcupine, only 0.1% of hindgut gas was hydrogen (Van Jaarsveld and Knight-Eloff 1984), compared with 16% in humans (Cummings and Macfarlane 1991).

SHORT-CHAIN FATTY ACIDS

Most reports of SCFA production are based on total SCFA. The contribution made by individual acids to the total SCFA in the hindgut shows the same relationship with fiber content of the diet as is seen in foregut fermenters. In general, as dietary fiber content increases, so does the molar proportion of acetate; concomitantly the molar proportion of propionate decreases (Bergman 1990). The molar proportion of butyrate is less predictable. Proportions of the acids are also influenced by sampling site within the hindgut. In cecum fermenters with a colonic separation mechanism, acetate molar proportions tend to be lower, and propionate higher, in the cecum than the colon. This is the result of selective retention of solutes and fine particles in the cecum, and presumably a faster fermentation. In both cecum and colon fermenters acetate proportions tend to increase along the length of the colon. This is likely to be the result of two factors: the first is greater relative production of acetate than the other acids as the fiber content of digesta progressively increases along the hindgut; the second is the more rapid absorption of acids of greater chain length (Stevens and Hume 1995).

The highest molar proportions of acetate in the hindgut have been reported from two arboreal folivores—the koala (86% to 92%; Cork and Hume 1983), and the howler monkey (94%; Milton and McBee 1983). Most other species have acetate molar proportions in the cecum below 75%, and in the colon below 77%. The reason for the unusually high values in the two folivores is not immediately clear, as the fiber contents of the diets were not unusually high (32% to 37% cell walls for the koala, 28% cell walls for the howler monkeys). However, it may indicate that in both species digestion of the more soluble components of the leaves in the stomach and small intestine is virtually complete, leaving only fiber in the residue to be fermented in the hindgut.

Most estimates of SCFA production rates in the hindgut have been made in vitro rather than in vivo. This is because of incomplete mixing of digesta and markers along the axis of the colon. Estimation of SCFA production in vivo using radioactive isotopic tracers depends on rapid and uniform mixing of infused tracer within a single pool of digesta (Leng and Leonard 1965). This requirement is readily satisfied in sacciform fermentation systems that function as stirred-tank reactors, such as the reticulorumen, but not in tubiform systems that function as modified plug-flow reactors, such as the colon. Representative rates of SCFA production in the hindgut in vitro are given in Table 4.3. Comparisons are made difficult by differences in expression of results. Nevertheless, if results from different workers are recalculated on the assumption that the dry matter content of hindgut digesta is 15%, some comparisons can be made on a common basis.

Among hindgut fermenters there does not seem to be any effect of body size on SCFA production rates. Within the hindgut, fermentation rates are usually highest in the cecum and decline along the length of the colon (Imoto and Namioka 1978, Rose et al. 1986, Barboza and Hume 1992b). This is a reflection of both selective retention of solutes and fine digesta particles in the cecum of some species, and progressive depletion of readily fermentable substrates along the colon that functions as a modified plug-flow reactor.

Most values calculated for the contribution made by total hindgut SCFA production to the estimated maintenance energy requirement of the animal fall between 10% and 30% in hindgut fermenters, but between 1% and 7% for foregut fermenters. In the latter group the contribution from forestomach SCFA production is much more significant, 20% to 67%.

The highest contribution among hindgut fermenters is 36% in the cape hyrax (*Procavia capensis*). This may be related to the unusual morphology of the hyrax hindgut, which includes a large, sacculated cecum separated by a section of proximal colon from a pair of horn-shaped colonic appendages (Rübsamen et al. 1982).

Table 4.3. SCFA production rates in the hindgut of mammals measured in vitro

Species and Body Mass	Diet	Site	SCFA Production mmol/L fluid/h	Contribution to Maintenance Energy Requirement (%)	Reference
(a) Hindgut fermenters					
Greater glider (*Petauroides volans*-Marsupialia), 1.1–1.4 kg	*Eucalyptus* foliage	Cecum	20	8	Foley et al. (1989)
Rock hyrax (*Procavia capensis*), 2–3 kg		Cecum[a]	7–10	?	Rübsamen et al. (1982)
		Colonic appendages[a]	5–6		
Cape hyrax (*Procavia habessinica*), 2–3 kg	Field	Cecum[a]	32–41[b]	36	Eloff and Van Hoven (1985)
		Colonic appendages[a]	28–39[b]		
Brushtail possum (*Trichosurus vulpecula*-Marsupialia), 2–4 kg	*Eucalyptus* foliage	Cecum + proximal colon	19	16	Foley et al. (1989)
Rabbit (*Oryctolagus caniculus*), 3.8 kg	Alfalfa, concentrates	Cecum	41–49[b]	6	Hoover and Heitmann (1972)
Koala (*Phascolarctos cinereus*-Marsupialia), 6.8–7.7 kg	*Eucalyptus* foliage	Cecum	11	9	Cork and Hume (1983)
		Proximal colon	11		
Howler monkey (*Alouatta palliata*), 7.5 kg	Fig leaves, fruit	Cecum	185–263	31	Milton and McBee (1983)
Porcupine (*Erethizon dorsatum*) 8.3 kg	Field	Cecum	28[b]	8	Johnson and McBee (1967)
Cape porcupine (*Hystrix africaeaustralis*), 17–30 kg	Field	Cecum	16	?	Van Jaarsveld and Knight-Eloff (1984)
		Proximal colon	13		
Hairy-nosed wombat (*Lasiorhinus ursinus*-Marsupialia), 24 kg	Chopped barley, straw, corn, casein	Upper proximal colon	12	33	Barboza and Hume (1992b)
		Lower proximal colon	10		
		Distal colon	8		
21 kg	Field	Upper proximal colon	28	?	Barboza and Hume (1992b)
		Lower proximal colon	26		
		Distal colon	16		
Common wombat (*Vombatus ursinus*-Marsupialia), 27 kg	Chopped barley, straw, corn, casein	Upper proximal colon	12	30	Barboza and Hume (1992b)
		Lower proximal colon	7		
		Distal colon	6		
40 kg	Field	Upper proximal colon	16	?	Barboza and Hume (1992b)
		Lower proximal colon	16		
		Distal colon	7		
Pig (*Sus scrofa*), 32–42 kg	Rice bran, alfalfa, concentrates	Cecum	29–33[b]	10–12	Imoto and Namioka (1978)
		Proximal colon	18–32[b]		
		Distal colon	13–14[b]		
Pig (*Sus scrofa*), 39–82 kg	Alfalfa, concentrates	Cecum	82	31	Rose et al. (1986)
		Proximal colon	58		
		Midcolon	47		
Elephant (*Loxodonta africana*)	Field	Cecum	18	?	Van Hoven et al. (1981)
		Midcolon	20		
(b) Foregut fermenters					
Red-necked pademelon (*Thylogale thetis*-Marsupialia), 5.5 kg	Chopped alfalfa hay	Cecum + proximal colon	29	2[c]	Hume (1977)
		Forestomach	39	21[c]	
Red-necked wallaby (*Macropus rufogriseus*-Marsupialia), 11.3 kg	Chopped alfafa hay	Cecum + proximal colon	27	1[c]	Hume (1944)
		Forestomach	52	42[c]	
Grey duiker (*Sylvicapra grimmia*, 15 kg	Field-browse	Cecum + colon	52–113	10–15	Boomker (1983)
		Rumen	40–69	18–38	
Sheep (*Ovis aries*)	Chopped alfafa hay	Cecum + proximal colon	16	7[c]	Hume (1977)
		Rumen	23	29[c]	
Black-tailed deer (*Odocoileus hemionus columbians*), 41 kg	Field	Cecum + proximal colon	102	1	Allo et al. (1973)
		Rumen	138	23	
Wildebeest (*Connochaetes gnou*), 125 kg	Field	Cecum	21–25[b]	6	Van Hoven and Boomker (1981)
		Rumen	20[b]	67	

[a] For terminology of the hindgut of the Hyracoidea, see Rübsamen et al. (1982).

[b] Calculated from results on the basis of a dry matter content of 15%.

[c] On the assumption that ad libitum digestible energy intakes of captive nonreproducing adult animals approximate maintenance.

Table 4.4. SCFA production rates in the hindgut of mammals and the rumen of sheep measured in vivo

Species and Body Mass	Diet	Site	Technique	Net Production mmol/h	Contribution to Maintenance Energy (%)	Reference
(a) Hindgut fermenters						
Rat (*Rattus rattus*), 0.5 kg	Corn, soybean concentrate	Cecum	Postmortem disappearance from cecum	0.42	5	Yang et al. (1970)
Guinea pig (*Cavia porcellus*), 0.6 kg	Laboratory pellets	Cecum + upper proximal colon	Isotope dilution	2.25	17	Sakaguchi et al. (1985)
Rock hyrax (*Procavia habessinica*), 2–3 kg		Cecum + colonic appendages	Isotope dilution	12.5	44	Rübsamen et al. (1982)
Rabbit (*Oryctolagus caniculus*), 2.5–3.0 kg	Laboratory pellets	Cecum	Isotope dilution	6.9	29	Parker (1976)
Pig (*Sus scrofa*), 50–60 kg	Barley, alfalfa	Cecum + proximal colon	Isotope dilution	50.3	9–23	Kennelly et al. (1981)
Human (*Homo sapiens*), 70 kg	British	Cecum + colon	Calculation based on amount of carbohydrate leaving ileum	21–25	6–9	McNeil (1984)
Horse (*Equus caballas*), 162 kg	Grass hay, corn	Cecum + proximal colon	Isotope dilution	362	30	Glinsky et al. (1976)
(b) Foregut fermenters						
Sheep (*Ovis aries*), 47 kg	Dried grass cubes	Cecum + proximal colon	Isotope dilution	18.3	5	Faichney (1969)
Sheep (*Ovis aries*), 47 kg	Alfalfa pellets	Rumen	Isotope dilution	108	34	Faichney (1968)

The effect of this may be an increased total fermentation capacity. Two of the lowest values are from marsupial arboreal folivores—the koala (9%), and the greater glider (*Petauroides volans*) (8%). This is despite very long mean retention times of digesta in both species (Cork and Warner 1983, Foley and Hume 1987), and no doubt reflects the intractability of the highly lignified fiber of *Eucalyptus* foliage.

Despite the problems of incomplete mixing between more than one pool of digesta, several studies based on isotope dilution techniques have resulted in apparently meaningful estimates of SCFA production in the mammalian hindgut in vivo. The results of these studies are summarized in Table 4.4. The contribution to estimated maintenance energy requirements made by SCFA production covers a range similar to that measured in vitro. With two exceptions, the range in hindgut fermenters is 9% to 30%. In the rat, an omnivore, the contribution was low, only 5%. At the other extreme is the rock hyrax (*P. habessinica*) at 44%. This hyrax has a gut morphology similar to that of the cape hyrax, and apparently a similarly large total fermentation capacity.

6.4. Microbial Synthesis

PROTEIN

No estimates of the rates or amounts of microbial protein synthesized in the hindgut appear to have been made for any species. In the rumen, numerous estimates have been published (see Van Soest 1982). Although values as high as 31 g microbial protein per 100 g organic matter (OM) fermented have been reported on purified diets (Ben-Ghedalia et al. 1978), most values on forage diets lie between 16 and 25 g protein per 100 g OM fermented.

It is possible to calculate a microbial protein yield in the hindgut of sheep from the data of Dixon and Nolan (1982, 1983). Ammonia-N incorporated into microbes in the cecum and proximal colon of sheep fed 800 g/day of chopped alfalfa hay averaged 0.6 g/day. On this diet a mean of 62 g OM per day disappeared between the terminal ileum and the end of the proximal colon. From this the microbial protein yield from ammonia-N can be calculated to be only 6.0 g/100 g OM apparently fermented. A further 27.6 g microbial protein was synthesized from other sources of nitrogen. The total, 33.6 g/100 g OM apparently fermented, compares favorably with estimates in the rumen. However, as about 18.8 g microbial protein entered the hindgut from the ileum each day, Dixon and Nolan (1983) concluded that net synthesis of microbial protein in the hindgut was likely to be small.

SOURCES OF AMMONIA FOR MICROBIAL PROTEIN SYNTHESIS

The studies of Dixon and Nolan (1983) suggested that although there were substantial inputs of endogenous nitrogen into the hindgut of sheep, very little entered as urea diffusing across the hindgut wall. Secreted mucin and sloughed epithelial cells were suggested as the main sources of endogenous nitrogen, either entering with ileal digesta or from within the hindgut itself. Whether this is peculiar to high-nitrogen diets (such as the chopped alfalfa hay used by Dixon and Nolan 1983) or is more general in foregut fermenters is not known. In hindgut fermenters fed diets of similar nitrogen content, 38% of endogenous urea was recycled to the whole digestive tract of the rabbit (Regoeczi et al. 1965), 50% to 54% in the horse (Prior et al. 1974), and 63% in the rock hyrax (Hume et al. 1980).

A similar range of values for urea recycling to the whole digestive tract has been reported in foregut fermenters such as ruminants (43% to 56%; Cocimano and Leng 1967, Robbins et al. 1974, Wales et al. 1975), camelids (47% to 74%; Hinderer 1978, Emmanuel et al. 1976), and macropod marsupials (36% to 56%; Dellow and Hume 1982, Chilcott et al. 1985). At least 50% of recycled urea in foregut fermenters is likely to reach the hindgut, both via the small intestine and directly across the hindgut wall (Nolan 1975).

The endogenous urea entering the hindgut would be expected to increase microbial protein synthesis, at least on low-protein diets when the supply of ammonia to the hindgut microbes may be limiting. However, this will only result in increased supplies of essential amino acids to the host animal if the animal is coprophagic; amino acid absorption from the hindgut by active transport has been shown to be negligible (e.g., Hume et al. 1993). In noncoprophagic hindgut fermenters, the main benefit of urea recycling comes from the maintenance of an active microbial fermentation in the hindgut on poor-quality diets, and as a result increased amounts of SCFA, which, in contrast to amino acids, are absorbed directly from the hindgut (Stevens and Hume 1995). This benefit is also shared by coprophagic hindgut fermenters.

VITAMINS

Microbial synthesis of the B vitamins thiamine (B_1), riboflavin (B_2), nicotinic acid, pantothenic acid, biotin, pyridoxine (B_6), folic acid, and cyanocobalamin (B_{12}) in the mammalian hindgut has been amply demonstrated. The effectiveness of this biosynthetic activity in providing for the host animal's requirements depends firstly upon whether the vitamin is found within or outside bacterial cells, and secondly upon whether the vitamin is absorbed across the hindgut wall or must be transferred to the small intestine for absorption.

Most of the thiamine, riboflavin, and nicotinic acid synthesized in the rat cecum is within bacterial cells, whereas the pantothenic acid, biotin, pyridoxine, and folic acid can be isolated from the surrounding medium (Mitchell and Isbel 1942). This will determine the extent to which each particular vitamin is available for absorption from the hindgut. That B vitamins are absorbed from the hindgut, albeit only by passive diffusion, has been demonstrated for at least B_1, B_2, and B_{12} (Sorrell et al. 1971).

The most active site of B vitamin absorption is the small intestine. It is possible that reflux of cecal contents into the ileum (Hörnicke and Björnhag 1980) plays a small but important role in supplying some vitamins synthesized in the hindgut to the host animal tissues. This aspect of the vitamin economy of mammals has received scant attention. Nevertheless, it is likely that the most important means by which hindgut-synthesized vitamins reach the small intestine is by coprophagy or cecotrophy (Hörnicke and Björnhag 1980). For instance, the rabbit is normally independent of a dietary supply of thiamine, riboflavin, pantothenic acid, biotin, and folic acid, but will become deficient in these B vitamins if cecotrophy is prevented and an adequate dietary supply is not provided.

Ascorbic acid (vitamin C) does not appear to be synthesized in the hindgut.

Rather, bacterial degradation of this water-soluble vitamin has been shown to occur (Reid 1948).

Of the fat-soluble vitamins, (A, D, E, and K), only the last mentioned is synthesized by bacteria in the hindgut to any significant extent (Gustafsson et al. 1961). Hollander and Truscott (1974) showed that vitamin K was absorbed from the rat colon by a passive diffusion process they related to its lipid solubility.

7. Summary and Conclusions

The main sites of microbial fermentation in the mammalian hindgut are the cecum and the proximal colon. Hindgut fermenters can be divided into cecum fermenters, in which the main site of digesta retention and fermentation is the cecum, and colon fermenters, in which the proximal colon is the main site. Cecum fermenters are typically small (less than 10 kg body mass), and have relatively high mass-specific metabolic rates and thus high energy and nutrient requirements. They occupy the low-fiber end of the food quality spectrum, but some cecum fermenters can utilize surprisingly high-fiber diets because of a separation mechanism in the proximal colon which results in selective retention of solutes and fine particles (including bacteria) in the cecum (and proximal colon). At the same time, there is relatively rapid elimination of large, hard-to-digest particles. Examples of cecum fermenters include rabbits and other lagomorphs, herbivorous rodents, small primates such as marmots, and several arboreal marsupial species that specialize on *Eucalyptus* foliage. Some (but not all) cecum fermenters ingest a portion of their feces (coprophagy) or special, high-nutrient feces derived from cecal contents (cecotrophy). Coprophagy, and especially cecotrophy, improves nutrient utilization by recycling valuable microbial protein and B vitamins that would otherwise be lost in the feces.

Colon fermenters are typically large ($>$ 10 kg) and include horses and other equids, elephants, rhinos, tapirs, anthropoid primates (including humans), and the marsupial wombats. Colon fermenters occupy the high-fiber end of the food quality spectrum, and are characterized by longer retention times for large particulate digesta than fluid and fine particles. This results in generally higher-fiber digestibilities than in cecum fermenters, and in large colon fermenters fiber digestibility approaches that seen in ruminants of similar body mass.

The dynamics of digesta flow and digestion in hindgut fermenters is best understood by reference to chemical reactor theory. On this basis, the mammalian cecum is best modeled as a continuous-flow, stirred-tank reactor, and the proximal colon as a modified plug-flow reactor (i.e., as a number of stirred tanks arranged in series).

The microbiology of the mammalian hindgut is relatively little studied, but the dense bacterial population (10^{10} to 10^{11}/g contents) seems not to be uniquely different from that found in other fermentative regions of the mammalian gut—namely, the forestomach (of which the reticulorumen has been the best characterized). The hindgut biota is remarkably stable, probably as a result of competition for nutrients in a relatively nutrient-poor environment and the relatively constant composition of food residues entering the hindgut after digestion and absorption in the stomach and small intestine. Over 95% of the bacterial species in the human large bowel are obligate anaerobes, mainly *Bacteroides* spp. Bacteria attached to the hindgut wall have also been characterized, and in humans have been found to be mainly fusiform rods and helical bacteria. Urease activity is largely localized in the bacteria adhering to the wall, the point of diffusion of endogenous urea from the blood into the hindgut lumen. Protozoa and phycomycete fungi are also found in the hindgut of some species, but their occurrence is by no means universal.

Rates of fermentation in the hindgut have usually been measured in vitro because of the difficulty of gaining access to this region of the gut, and also because of the hindgut's mainly tubiform morphology, so that infused isotopes do not mix throughout the contents of the organ. Measurements in vitro have been based on rates of formation of total gas, methane, hydrogen, and SCFA. The latter provide the best estimates of the contribution of microbial fermentation to the energy economy of the host animal. In hindgut fermenters, SCFA contribute between 10% and 30% of the animal's maintenance energy requirement, but estimates as high as 44% have come from the rock hyrax, a small cecum fermenter from East Africa that has an ususually complex gastrointestinal tract. In foregut fermenters there is always a secondary fermentation in the hindgut, but this provides only 1% to 7% of the animal's energy needs, compared with 20% to 67% from the forestomach fermentation.

The SCFAs play other roles, in absorption of other nutrients including water and in the maintenance of the health of the hindgut epithelium. It is this last aspect that is attracting so much attention in humans.

References

Adam KMG (1951) The quantity and distribution of the ciliate protozoa in the large intestine of the horse. Parasitology 41:301–311.

Alexander RMN (1991) Optimization of gut structure and diet for higher vertebrate herbivores. Phil Trans R Soc Lond B333:249–255.

Allo AA, Oh JH, Longhurst WM, Connolly GE (1973) VFA production in the digestive systems of deer and sheep. J Wildl Manage 37:202–211.

Argenzio RA, Lowe JE, Pickard DW, Stevens CE (1974) Digesta passage and water exchange in the equine large intestine. Am J Physiol 226:1035–1042.

Barboza PS (1993) Digestive strategies of the wombats: feed intake, fiber digestion, and digesta passage in two grazing marsupials with hindgut fermentation. Physiol Zool 66: 983–999.

Barboza PS, Hume ID (1992a) Digestive tract morphology and digestion in the wombats (Marsupialia: Vombatidae). J Comp Physiol B162:552–560.

Barboza PS, Hume ID (1992b) Hindgut fermentation in the wombats: two marsupial grazers. J Comp Physiol B162:561–566.

Ben-Ghedalia D, Tagari H, Zamwel S, Bondi A (1975) Solubility and net exchange of calcium, magnesium and phosphorus in digesta flowing along the gut of sheep. Br J Nutr 33:87–94.

Ben-Ghedalia D, McMeniman NP, Armstrong DG (1978) The effect of partially replacing urea nitrogen with protein N on N capture in the rumen of sheep fed a purified diet. Br J Nutr 39:37–44.

Bentley DW, Nichols RL, Condon RE, Gorbach SL (1972) The microflora of the human ileum and intraabdominal colon: results of direct needle aspiration at surgery and evaluation of the technique. J Lab Clin Med 79:421–429.

Berg HC, Turner L (1979) Movement of microorganisms in viscous environments. Nature (Lond) 278:349–351.

Bergman EN (1990) Energy contributions of volatile fatty acids from the gastrointestinal tract in various species. Physiol Rev 70:567–590.

Björnhag G (1981) The retrograde transport of fluid in the proximal colon of rabbits. Swed J Agric Res 11:63–69.

Björnhag G (1987) Comparative aspects of digestion in the hindgut of mammals. The colonic separation mechanism (CSM) (A review). Dtsch Tierarztl Wsschr 94:33–36.

Bond JH, Engel RR, Levitt MD (1971) Factors influencing pulmonary methane excretion in man. J Exp Med 133:572–588.

Boomker EA (1981) A study of the digestive processes of the common duiker, *Sylvicapra grimmia*. MSc thesis, University of Pretoria, Pretoria, South Africa.

Boomker EA (1983) Volatile fatty acid production in the grey duiker, *Sylvicapra grimmia*. S Afr J Anim Sci 13:33–35.

Carroll EJ, Hungate RE (1954) The magnitude of the microbial fermentation in the bovine rumen. Appl Microbiol 2:205–214.

Chilcott MJ, Hume ID (1985) Coprophagy and selective retention of fluid digesta: their role in the nutrition of the common ringtail possum, *Pseudocheirus peregrinus*. Aust J Zool 33:1–15.

Chilcott MJ, Moore SA, Hume ID (1985) Effects of water restriction on nitrogen metabolism and urea recycling in the macropodid marsupials *Macropus eugenii* (tammar wallaby) and *Thylogale thetis* (red-necked pademelon). J Comp Physiol B155:759–767.

Christensen J (1989) Colonic motility. In: Schultz SG, Wood JD, Rauner BB, eds. Handbook of Physiology, Section 6. The Gastrointestinal System, Vol. I, pp. 939–973. Bethesda: American Physiology Society.

Christl SU, Murgatroyd PR, Gibson GR, Cummings JH (1992) Production, metabolism, and excretion of hydrogen in the large intestine. Gastroenterology 102:1269–1277.

Clamp JR, Allen A, Gibbons RA, Roberts GP (1978) Chemical aspects of mucus. Br Med Bull 34:25–41.

Clauss W (1984) Circadian rhythms in Na^+ transport. In: Skadhauge E, Heintze K, eds. Intestinal Absorption and Secretion, pp. 273–283. Lancaster, UK: MTP Press.

Cocimano MR, Leng RA (1967) Metabolism of urea in sheep. Br J Nutr 21:353–371.

Cork SJ, Hume ID (1983) Microbial digestion in the koala (*Phascolarctos cinereus*, Marsupialia), an arboreal folivore. J Comp Physiol B152:131–135.

Cork SJ, Warner ACI (1983) The passage of digesta markers through the gut of a folivorous marsupial, the koala *Phascolarctos cinereus*. J Comp Physiol B152:43–51.

Cummings JH, Macfarlane GT (1991) The control and consequences of bacterial fermentation in the human colon. J Appl Bacteriol 70:443–459.

Dellow DW, Hume ID (1982) Studies on the nutrition of macropodine marsupials. II. Urea and water metabolism in *Thylogale thetis* and *Macropus eugenii*; two wallabies from divergent habitats. Aust J Zool 30:399–406.

Dixon RM, Nolan JV (1982) Studies of the large intestine of sheep. 1. Fermentation and absorption in sections of the large intestine. Br J Nutr 47:289–300.

Dixon RM, Nolan JV (1983) Studies of the large intestine of sheep. 3. Nitrogen kinetics in sheep given chopped lucerne (*Medicago sativa*) hay. Br J Nutr 50:757–768.

Dorst J (1973) Gros intestine—caecum. In: Grasse P-P, ed. Traite de Zoologie, Tome XVI, Fascicule V, Vol. 1, pp. 352–388. Paris: Masson.

Edwards PAW (1978) Is mucus a selective barrier to macromolecules? Br Med Bull 34: 55–56.

Eloff AK, Van Hoven W (1985) Volatile fatty acid production in the hindgut of *Procavia capensis*. Comp Biochem Physiol 80A:291–295.

El-Shazly K, Hungate RE (1965) Method for measuring diaminopimelic acid in total rumen contents and its application to the estimation of bacterial growth. Appl Microbiol 14:27–30.

Emmanuel B, Howard BR, Emady M (1976) Urea degradation in the camel. Can J Anim Sci 56:595–601.

Engelhardt Wv, Rechkemmer G (1983) Absorption of inorganic ions and short-chain fatty acids in the colon of mammals. In: Gilles-Baillien M, Gilles R, eds. Intestinal Transport, pp. 26–45. Berlin: Springer-Verlag.

Engelhardt Wv, Rechkemmer G, Luciano L, Reale E (1983) The hindgut: lessons for man from other species. In: Barbara L, Miglioli M, Phillips SF, eds. New Trends in Pathophysiology and Therapy of the Large Bowel, pp. 3–17. Amsterdam: Elsevier.

Engelhardt Wv, Wolter S, Lawrenz H, Hemsley JA (1978) Production of methane in two non-ruminant herbivores. Comp Biochem Physiol 60:309–311.

Faichney GJ (1968) The production and absorption of volatile fatty acids from the rumen of sheep. Aust J Agric Res 19:791–802.

Faichney GJ (1969) Production of volatile fatty acids in the sheep caecum. Aust J Agric Res 20:491–498.

Filipe MI, Branfoot AC (1976) Mucin histochemistry of the colon. Curr Top Pathol 63: 143–178.

Finegold SM, Sutter VL, Mathisen GE (1983) Normal indigenous intestinal flora. In: Hentges DJ, ed. Human Intestinal Microflora in Health and Disease, pp. 3–32. New York: Academic Press.

Foley WJ, Hume ID (1987) Passage of digesta markers in two species of arboreal folivorous marsupials—the greater glider (*Petauroides volans*) and the brushtail possum (*Trichosurus vulpecula*). Physiol Zool 60:103–113.

Foley WJ, Hume ID, Cork SJ (1989) Fermentation in the hindgut of the greater glider (*Petauroides volans*) and the brushtail possum (*Trichosurus vulpecula*)—two arboreal folivores. Physiol Zool 62:1126–1143.

Foose TJ (1982) Trophic strategies of ruminant versus nonruminant ungulates. PhD thesis, University of Chicago, Chicago, Ill.

Freter R (1983) Mechanisms that control the microflora in the large intestine. In: Hentges DJ, ed. Human Intestinal Microflora in Health and Disease, pp. 33–54. New York: Academic Press.

Giesecke D, Van Gylswyk NO (1975) A study of feeding types and certain rumen functions in six species of South African wild ruminants. J Agric Sci Camb 85:75–83.

Glinsky MJ, Smith RM, Spires HR, Davis CL (1976) Measurement of volatile fatty acid production rates in the cecum of the pony. J Anim Sci 42:1469–1470.

Gustafsson BE, Daft FS, McDaniel EG, Smith JC, Fitzgerald RJ (1961) Effects of vitamin K-active compounds and intestinal microorganisms in vitamin K-deficient germ-free rats. J Nutr 78:461–468.

Hill RRH (1983) Distribution of urease producing bacteria in the rabbit caecum. S Afr J Anim Sci 13:61–62.

Hinderer S (1978) Kinetic des Harnstoff-Stoffwechsels beim Lama bei proteinarmen Diaten. Dissertation, University of Hohenheim, Stuttgart, Germany.

Hollander D, Truscott TC (1974) Colonic absorption of vitamin K-3. Lab Clin Med 83: 648–656.

Hoover WH, Heitmann RN (1972) Effects of dietary fiber levels on weight gain, cecal volume and volatile fatty acid production in rabbits. J Nutr 102:375–380.

Hoppe PP, Van Hoven W, Engelhardt W, Prins RA, Lankhorst A, Gwynne MD (1983) Pregastric and caecal fermentation in dikdik (*Madoqua kirki*) and suni (*Nesotragus moschatus*). Comp Biochem Physiol 75A:517–524.

Hörnicke H, Björnhag G (1980) Coprophagy and related strategies for digesta utilization. In: Ruckebusch Y, Thivend P, eds. Digestive Physiology and Metabolism in Ruminants, pp. 707–730. Lancaster, UK: MTP Press.

Hume ID (1977) Production of volatile fatty acids in two species of wallaby and in sheep. Comp Biochem Physiol 56A:299–304.

Hume ID (1982) Digestive Physiology and Nutrition of Marsupials. Cambridge, UK: Cambridge University Press.

Hume ID (1989) Optimal digestive strategies in mammalian herbivores. Physiol Zool 62: 1145–1163.

Hume ID, Foley WJ, Chilcott MJ (1984) Physiological mechanisms of foliage digestion in the greater glider and ringtail possum (Marsupialia: Pseudocheiridae). In: Smith AP, Hume ID, eds. Possums and Gliders, pp. 247–251. Sydney: Australian Mammal Society.

Hume ID, Karasov WH, Darken BW (1993) Acetate, butyrate and proline uptake in the caecum and colon of prairie voles (*Microtus ochrogaster*). J Exp Biol 176:285–297.

Hume ID, Rübsamen K, Engelhardt Wv (1980) Nitrogen metabolism and urea kinetics in the rock hyrax (*Procavia habessinica*). J Comp Physiol B138:307–314.

Hume ID, Sakaguchi E (1991) Patterns of digesta flow and digestion in foregut and hindgut fermenters. In: Tsuda T, Sasaki Y, Kawashima R, eds. Physiological Aspects of Digestion and Metabolism in Ruminants, pp. 427–451. San Diego: Academic Press.

Hume ID, Warner ACI (1980) Evolution of microbial digestion in mammals. In: Ruckebusch Y, Thivend P, eds. Digestive Physiology and Metabolism in Ruminants, pp. 615–634. Lancaster, UK: MTP Press.

Hungate RE, Phillips GD, McGregor A, Hungate DP, Buecher HK (1959) Microbial fermentation in certain mammals. Science 130:1192–1194.

Imoto S, Namioka S (1978) VFA production in the pig large intestine. J Anim Sci 47: 467–478.

Johnson JL, McBee RH (1967) The porcupine cecal fermentation. J Nutr 91:540–546.

Kagnoff MF (1987) Immunology of the digestive system. In: Johnson LR, ed. Physiology of the Gastrointestinal Tract, 2nd ed., pp. 1699–1728. New York: Raven Press.

Kay RNB, Pfeffer E (1970) Movements of water and electrolytes into and from the intestine of sheep. In: Phillipson AT (ed) Physiology of digestion and metabolism in the ruminant. Oriel Press, Newcastle-upon-Tyne, UK, pp 390–402.

Kennelly JJ, Aherne FX, Sauer WC (1981) Volatile fatty acid production in the hindgut of swine. Can J Anim Sci 61:349–361.

Kern DL, Slyter LL, Leffel EC, Weaver JM, Oltjen RR (1974) Ponies vs. steers: microbial and chemical characteristics of intestinal digesta. J Anim Sci 38:559–564.

Kofoid CA (1935) On two remarkable ciliate protozoa from the caecum of the Indian elephant. Proc Natl Acad Sci 21:501–506.

Lacy ER (1991) Functional morphology of the large intestine. In: Schultz SG, Field M, Frizzell RA, Rauner BB, eds. Handbook of Physiology, Sect. 6: The Gastrointestinal System, Vol. IV, pp. 121–194. Bethesda: American Physiology Society.

Leng RA, Leonard GJ (1965) Measurement of the rates of production of acetic, propionic and butyric acids in the rumen of sheep. Br J Nutr 19:469–484.

Levenspiel O (1972) Chemical Reaction Engineering, 2nd ed. New York: Wiley.

London CJ (1981) The microflora associated with the caecum of the koala (*Phascolarctos cinereus*). MSc Thesis, La Trobe University, Melbourne, Australia.

Macagno EO, Christensen J (1980) Fluid mechanics of the duodenum. Annu Rev Fluid Mech 12:139–158.

MacDermott RP, Donaldson RM Jr, Trier JS (1974) Glycoprotein synthesis and secretion by mucosal biopsies of rabbit colon and human rectum. J Clin Invest 54:545–554.

Mann SO, Orskov ER (1973) The effect of rumen and post-rumen feeding of carbohydrates on the caecal microflora of sheep. J Appl Bacteriol 36:475–484.

McCowan RP, Cheng K-J, Costerton JW (1980) Adherent bacterial populations on the bovine rumen wall: distribution patterns of adherent bacteria. Appl Environ Microbiol 39:233–241.

McKenzie RA (1978) The caecum of the koala, *Phascolarctos cinereus*: Light, scanning and transmission electron microscopic observations on its epithelium and flora. Aust J Zool 26:249–256

McNeil NI (1984) The contribution of the large intestine to energy supplies in man. Am J Clin Nutr 39:338–342.

Milton K, McBee RH (1983) Rates of fermentative digestion in the howler monkey, *Alouatta palliata* (Primates: Ceboidea). Comp Biochem Physiol 74A:29–31.

Mitchell HK, Isbel ER (1942) Intestinal bacterial synthesis as a source of B vitamin for the rat. Univ. of Texas Publ. No. 4237, Part II, pp. 125–134.

Murray RM, Bryant AM, Leng RA (1976) Rates of production of methane in the rumen and large intestine of sheep. Br J Nutr 36:1–14.

Murray RM, Bryant AM, Leng RA (1978) Methane production in the rumen and lower gut of sheep given lucerne chaff: effect of level of intake. Br J Nutr 39:337–345.

Nolan JV (1975) Quantitative models of nitrogen metabolism in sheep. In: McDonald IW, Warner ACI, eds. Digestion and metabolism in the ruminant. Univ. of New England Publ. Unit, Armidale, Australia, pp. 416–431.

O'Brien TP, Lomdahl A, Sanson G (1986) Preliminary microscopic investigations of the digesta derived from foliage of *Eucalyptus ovata* (Labill.) in the digestive tract of the common ringtail possum, *Pseudocheirus peregrinus* (Marsupialia). Aust J Zool 34: 157–176.

Orpin CG (1975) Studies on the rumen flagellate *Neocallimastix frontalis*. J Gen Microbiol 91:249–262.

Orpin CG (1976) Studies on the rumen flagellate *Sphaeromonas communis*. J Gen Microbiol 94:270–280.

Orpin CG (1981) Isolation of cellulolytic phycomycete fungi from the caecum of the horse. J Gen Microbiol 123:287–296.

Orton RK, Hume ID, Leng RA (1985) Effects of exercise and level of dietary protein on digestive function in horses. Equine Vet J 17:386–390.

Osawa R, Bird PS, Harbrow DJ, Ogimoto K, Seynour GJ (1993) Microbiological studies of the intestinal microflora of the koala, *Phascolarctos cinereus*. I. Colonization of the caecal wall by tannin-protein complex degrading enterobacteria. Aus J Zool 41: 599–609.

Parker DS (1976) The measurement of production rates of volatile fatty acids in the caecum of the conscious rabbit. Br J Nutr 36:61–70.

Penry DL, Jumars PA (1987) Modeling animal guts as chemical reactors. Am Nat 129: 69–96.

Pickard DW, Stevens CE (1972) Digesta flow through the rabbit large intestine. Am J Physiol 222:1161–1166.

Prior RL, Hintz HF, Lowe JE, Visek WJ (1974) Urea recycling and metabolism of ponies. J Anim Sci 38:565–571.

Rechkemmer G, Engelhardt W (1988) Concentration- and pH-dependence of short-chain fatty acid absorption in the proximal and distal colon of guinea pig (*Cavia porcellus*). Comp Biochem Physiol 91A:659–663.

Regoeczi E, Irons L, Koj A, McFarlane AS (1965) Isotopic studies of urea metabolism in rabbits. Biochem J 95:521–535.

Reid ME (1948) Gastrointestinal tract of guinea pig and elimination of ascorbic acid given intraperitoneally. Proc Soc Exp Biol Med 68:403–406.

Reuss L (1991) Salt and water transport by gallbladder epithelium. In: Schultz SG, Field M, Frizzell RA, Rauner BB, eds. Handbook of Physiology, Sect. 6: The Gastrointestinal System, Vol. IV, pp. 303–322. Bethesda: American Physiology Society.

Robbins CT, Prior RL, Moen AN, Visek WJ (1974) Nitrogen metabolism of white-tailed deer. J Anim Sci 38:186–191.

Rose CJ, Hume ID, Farrell DJ (1986) Fibre digestion and volatile fatty acid production in domestic and feral pigs. In: Farrell DJ, ed. Recent advances in animal nutrition in Australia. Univ. of New England Press, Armidale, Australia, pp. 347–360.

Rübsamen K, Hume ID, Engelhardt Wv (1982) Physiology of the rock hyrax. Comp Biochem Physiol 72A:271–277.

Sakaguchi E, Becker G, Rechkemmer G, Engelhardt Wv (1985) Volume, solute concentrations and production of short-chain fatty acids in the caecum and upper colon of the guinea pig. Z Tierphysiol Tierernahrg Futtermittelkde 54:276–285.

Sakaguchi E, Itoh H, Uchida S, Horigame T (1987) Comparison of fibre digestion and digesta retention time between rabbits, guinea pigs, rats and hamsters. Br J Nutr 58: 149–158.

Sakata T, Engelhardt W (1981) Luminal mucin in the large intestine of mice, rats and guinea pigs. Cell Tissue Res 219:629–635.

Salyers AA, West SEH, Vercellotti JR, Wilkins TD (1977) Fermentation of mucus and plant polysaccharides by anaerobic bacteria from the human colon. Appl Environ Microbiol 34:529–533.

Savage DC (1983) Associations of indigenous microorganisms with gastrointestinal epithelial surfaces. In: Hentges DJ, ed. Human Intestinal Microflora in Health and Disease, pp. 55–78. New York: Academic Press.

Sellers AF, Lowe JE, Drost CJ, Fendano VT, Georgi, JR, Roberts MC (1982) Retropulsion-propulsion in equine large colon. Am J Vet Res 43:390–396.

Sorrell MF, Frank O, Thomson AD, Aquino H, Baker H (1971) Absorption of vitamins from the large intestine in vivo. Nutr Rep Int 3:143–148.

Stevens CE, Hume ID (1995) Comparative Physiology of the Vertebrate Digestive System. 2nd Ed. Cambridge, UK: Cambridge University Press.

Takeuchi A, Jervis HR, Nakazawa H, Robinson DM (1974) Spiral-shaped organisms on the surface colonic epithelium of the monkey and man. Am J Clin Nutr 27:1287–1296.

Takeuchi A, Zeller JA (1972) Ultrastructural identification of spirochetes and flagellated microbes at the brush border of the large intestinal epithelium of the rhesus monkey. Infect Immun 6:1008–1018.

Uden P, Rounsaville TR, Wiggans GR, Van Soest PJ (1982) The measurement of liquid and solid digesta retention in ruminants, equines, and rabbits given timothy (*Phleum pratense*) hay. Br J Nutr 48:329–339.

Ulyatt MJ, Dellow DW, Reid CSW, Bauchop T (1975) Structure and function of the large intestine of ruminants. In: McDonald IW, Warner ACI, eds. Digestion and Metabolism in the Ruminant. Univ. of New England Printing Unit, Armidale, Australia, pp. 119–133.

Uphill PF, Wilde JKH, Berger J (1974) Repeated examinations, using the laporotomy sampling technique, of the gastro-intestinal microflora of baboons fed a natural or a synthetic diet. J Appl Bacteriol 37:309–317.

Van Hoven W (1980) Rumen fermentation and methane production in the African buffalo (*Syncerus caffer*) in the Kruger National Park. Koedoe 23:45–55.

Van Hoven W, Boomker EA (1981) Feed utilization and digestion in the black wildebeest (*Connochaetes gnou*, Zimmerman, 1780) in the Golden Gate Highlands National Park. S Afr J Wildl Res 11:35–40.

Van Hoven W, Prins RA, Lankhorst A (1981) Fermentative digestion in the African elephant. S Afr J Wildl Res 11:78–86.

Van Jaarsveld AS, Knight-Eloff AK (1984) Digestion in the porcupine *Hystrix africaeaustralis*. S Afr J Zool 19:109–112.

Van Soest PJ (1982) Nutritional Ecology of the Ruminant. Corvallis, Ore.: O&B Books.

Wales RA, Milligan LP, McEwan EH (1975) Urea recycling in caribou, cattle and sheep. Proc. First Int. Reindeer Caribou Symp., Fairbanks, Alaska, pp. 297–307.

Warner ACI (1981) Rate of passage of digesta through the gut of mammals and birds. Nutr Abstr Rev 51:789–820.

Wolin MJ (1981) Fermentation in the rumen and human large intestine. Science 213: 1463–1468.

Wolpert E, Phillips S, Summerskill WHJ (1971) Transport of urea and ammonia production in the human colon. Lancet ii:1387–1390.

Woodall PF, Skinner JD (1993) Dimensions of the intestine, diet and faecal water loss in some African antelope. J Zool Lond 229:457–471.

Wrong OM, Edmonds LJ, Chadwick VS (1981) The Large Intestine: Its Role in Mammalian Nutrition and Homeostasis. Lancaster, UK: MTP Press.

Yang MG, Manoharan K, Mickelsen O (1970) Nutritional contribution of volatile fatty acids from the cecum of rats. J Nutr 100:545–550.

5

The Interaction of Avian Gut Microbes and Their Host: An Elusive Symbiosis

Conrad Vispo and William H. Karasov

1. Introduction

Our goal in this paper is to review information on the role of gastrointestinal (GI) tract microbes in avian nutritional ecology. First, we try to identify the likely forms of interaction between microbes, and host by surveying the avian GI tract as a location/site for microbes, and by summarizing proposed interactions. Second, we consider the impact of possible microbe-host interactions on the bird's nutritional ecology, such as its nutrient requirements and foraging behavior. As avian ecologists, it is this last but difficult question that justifies our interest in the topic. We outline a way of classifying potential microbial effects on avian nutritional ecology and, following this pattern, summarize the scanty information available. We conclude by sketching a study design that forces us to identify the central questions and to propose a means of answering them.

Microbial ecosystems are nearly ubiquitous on this planet, both inside and outside animals. Study of avian gut microbes can provide ornithologists with not only a better understanding of the forces influencing their study taxon's relationship with its environment, but also a reminder that this taxon itself is an environment for other organisms.

Our ultimate interest in this paper is the nutritional ecology of wild birds, yet, for want of information, we often turn to literature on domestic bird species or even mammals, animals whose morphophysiology and gut microbes may be quite distinct from those of wild birds. Hopefully, future studies will be detailed enough to demonstrate the limitations of this overgeneralization.

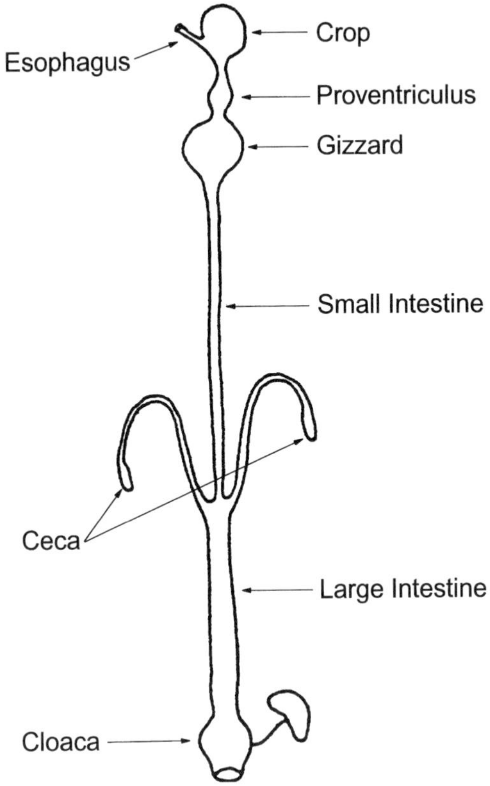

Figure 5.1. Schematic diagram of the avian digestive tract. The sections referred to in the text have been labeled. The organ that drains into the cloaca is the kidney. (Adapted from Stevens 1977.)

2. Overview of the Avian Alimentary System and Its Microbial Habitats

Birds share with all vertebrates the same basic structures for nutrient extraction—a tubular intestine, sometimes with proximal and/or distal fermentation chambers (Fig. 5.1). Special characteristics of the avian digestive tract include a crop in some cases, division of the secretory and triturative functions of the stomach into two distinct compartments (glandular stomach, or proventriculus; and ventriculus, or gizzard), and the development of paired ceca, although some species have only a single cecum and others have none.

Microbes are found throughout the gut, from the mouth to the rectum, attached firmly to the epithelial surfaces, deep in crypts in the mucosa, and free or attached to food particles in the lumen. Of these three types of habitat, major microbial activity in the lumen can only occur in regions of relative stasis, where digesta is retained long enough for there to be significant microbial growth. The only microbes that can remain in regions of high flow are those that attach to epithelial surfaces, as described below for the small intestine.

2.1. Esophagus and Crop

The esophagus of many, perhaps most, species of birds can serve the function of food storage (Ziswiler and Farner 1972). In a wide variety of birds the cervical portion of the esophagus is expanded into a clearly differentiated crop, of which there are a variety of forms. The great diversity of crops is reviewed in McLelland (1979).

The crop serves several functions, including food storage, elaboration of crop "milk" that is fed to young (in pigeons), and perhaps microbial fermentation. Recent study of the pigeon crop has demonstrated the presence of cellulolytic bacteria, and storage and digestion of fiber in the crop was suggested (Shetty et al. 1990). This idea is supported by the relatively high fiber digestibilities reported from pigeons (Hanlan 1949). Scanning electron microscopy (SEM) revealed a large population of bacteria on the wall of the chicken crop and attached to food particles in that organ (Bayer et al. 1975). Some starch digestion occurs in the crop. The amylase activity may arise from a number of possible sources (Ziswiler and Farner 1972). Bacterial starch degradation in the crop has been described (Bolton 1965, Ivorec-Szylit et al. 1965, Champ et al. 1981). Sucrose degradation in the crop is apparently by endogenous enzymes (Pritchard 1972). Because pH may be as low as 4.5 (Ziswiler and Farner 1972), extensive bacterial digestion may not occur in the crops of most birds, although even slight fiber digestion may assist the subsequent work of the gizzard.

In the hoatzin (*Opisthocomus hoazin*), a 750-g folivorous cuculiform from South American Tropical forests, the crop is particularly capacious and muscular (McLelland 1979) and contains a mixed microbial community composed mainly of gram-negative rods at 10^9/mL but also including ciliate protozoa (Grajal et al 1989, Dominguez-Bello et al. 1993a). There is active fermentation throughout the crop and lower esophagus, with substantial digestion of plant cell walls at a pH of 6.3 to 6.7 and short-chain fatty acid (SCFA) concentrations of 115 mM. There is also reduction in particle size of food, probably through trituration by cornified epithelial ridges of the internal ventral surface of the highly muscular crop (Grajal et al 1989). The hoatzin is thus the first bird reported to have a true foregut fermentation similar to that found in several groups of mammalian herbivores.

The very large cervical crop of the kakapo (*Strigops habroptilus*), a rare 1- to 3-kg folivorous parrot from New Zealand, is superficially similar to that of the hoatzin. Its characteristics in relation to microbial fermentation have not been studied, although observations of this bird's diet suggest that it relies largely on extracting cell solubles from ground leaves (Oliver 1955, Williams 1956, Morton 1978). Livezey (1993) raises the possibility that the dodo (*Raphus cucullatus*) had an expanded crop similar to the hoatzin's.

2.2. Stomach

The proventriculus and gizzard tend not to be as easily distinguishable from each other in carnivorous and piscivorous species as in omnivorous, insectivorous, herbivorous, and granivorous species (McLelland 1979). In the former types of feeders both chambers are specialized for storage, and the stomach is saclike. In the latter types of feeders the stomach typically consists of a more distinct, spindle-shaped proventriculus and a larger muscular gizzard. The stomach of granivorous birds has a horny inner lining called koilen that is thick and coarse and aids in grinding; gastroliths (stones, gravel, or sand) perform a similar function in birds (Karasov and Hume 1995). In all cases, this region of the gut would seem to be inhospitable to microbes.

The stomach may retain its contents long enough for microbes to grow in it, but growth would presumably be limited either by a low pH or because the stomach is empty for prolonged periods. The gastric juice secreted in the proventriculus has a pH of 0.2 to 1.2, while pH measured in the gizzard ranges 0.7 to 2.8 in species summarized in Ziswiler and Farner (1972). Even in this acid environment, relatively large numbers of gut wall bacteria are possible (at least in mammals; Kunstyr 1974), perhaps because the attached mucous layer can shelter the participating bacteria somewhat from the ambient pH and because microbial urease can produce NH_3 that counteracts acidity in the immediate microenvironment. Also, uneven mixing of stomach contents allows areas of relatively high pH to persist (Karasov and Hume 1995). Relatively high SCFA concentrations occur in the gizzard and proventriculus of the ostrich (total anaerobes 3×10^6/g ingesta; Swart et al. 1993b), but these are probably not sites of major microbial SCFA production, and the source of the SCFAs is not known (Swart et al. 1993b).

2.3. Intestine

It is often written that the intestines of birds appear to be generally shorter than in mammals, and that they tend to be relatively long in herbivores, granivores, and piscivores and relatively short in carnivores, insectivores, and frugivores (Ziswiler and Farner 1972, McLelland 1979). We have no reason to doubt these generalities, but to our knowledge there has been no systematic analysis of this that carefully controls for body size and phylogenetic affiliations. Within gallinaceous birds the more herbivorous species have longer small intestines than the more granivorous species (Leopold 1953). Intestinal length is modulated within some species, increasing as much as 40% either when switched to high-fiber diet or when acclimated to low temperature (Karasov 1995). A common link in both these manipulations is hyperphagia, and this could be the proximate mechanism for the intestinal enlargement.

The residence time of ingesta is usually short in the small intestine, so in most instances the lumen is unsuitable for colonization by microbes. The epithelium of the small intestine of domestic chickens is colonized by segmented, filamentous bacteria that attach end-on to the apical membranes of enterocytes, beginning in the upper third of the small intestine (Savage 1970). How these components of the normal gut biota influence uptake of nutrients by the small intestine is unknown.

One bird with a small intestinal microbial fermentation that is of some significance is the emu (*Dromaius novaehollandiae*). In this omnivorous bird, the ileum is of sufficient length, diameter, and hence volume (37% of total tract contents) that digesta retention is long enough for fermentation of at least some components of plant cell walls to occur (Herd and Dawson 1984). For birds weighing 38 kg, mean retention times were only 5.5 hours for particles and 4.1 hours for fluid, and the cell wall components that disappeared were hemicelluloses (45%) rather than cellulose (19%) on a cubed diet containing 24% hemicellulose and 10% cellulose. In this regard geese may be similar to the emu (Buchsbaum et al. 1986).

2.4. Ceca

In many birds, the structure of the avian ceca appears different from that in mammals; they are more tubular than saccular, and they contain prominent villi (Planas et al. 1987). Given their higher surface:volume ratio, they seem designed for surface-related processes such as hydrolysis and absorption rather than, or in addition to, a lumen-related process such as fermentation.

There are diverse forms of ceca in birds, and the extent to which they are developed is generally characteristic for each major group of birds (McLelland 1989, Clench and Mathias 1995). Naik and Dominic (1962) surveyed 80 species belonging to 15 orders and 30 families and classified ceca into four main types. The intestinal type is long and histologically resembles the rest of the intestinal tract. The glandular type is also long but contains numerous actively secreting crypts. The lymphoid type, which contains many lymphocytes, and the vestigial type are much reduced in size and probably do not present important microbial environments.

The ceca are the primary site of microbial fermentation in birds such as tinamou, geese, Galliformes (grouse, ptarmigan, domestic chicken, quail), and ratites other than the emu (i.e., ostrich, rhea, cassowary). The predominant organisms are obligately anaerobic bacteria, which occur in the lumen at ca. 10^{10} to 10^{11}/g (wet weight) (Mead 1989). Microbial activity is primarily fermentative, but, as we will discuss below, there has been little evidence of extensive cellulose fermentation (Ziswiler and Farner 1972, Mead 1989). In chickens the cecal bacteria are mainly saccharolytic. Large populations of uric acid-degrading bacteria are commonly observed (McNab 1973, Mead 1989). Adherent bacteria cover the walls of the chicken ceca (Fuller and Turvey 1971).

The functioning of the cecum has been studied primarily in Galliformes. In wild galliformes the ceca are evacuated each morning when the rest of the tract is virtually empty (Farner 1960). The filling of the cecum appears to involve mechanism(s) that selectively retain fluid and small particles (including bacteria). In some birds, fluid (urine) is refluxed by antiperistaltic contractions from the cloaca along the usually short colon and into the ceca. This rinses small particles out of the colonic contents and carries them into the ceca (Bjornhag 1989). Larger particles are left behind to be excreted. Evidence of this process can be seen in the long retention time of liquid ingesta relative to particulate ingesta in the rock ptarmigan (Table 5.1). Fenna and Boag (1974) thought that in galliform birds a meshwork of ridges and villi at the opening into the ceca prevented large particles from entering the ceca at all. Some of these features of cecal anatomy are absent in wild birds kept in captivity on concentrate diets (Hanssen 1979a). The 4.5-cm-long paired ceca of the frugivorous resplendent quetzal are sometimes reported to be packed with fruit skins, possibly suggesting another filling mechanism, at least in this species (Wheelwright 1983).

Table 5.1. Retention time and the components of fiber digestibility for several species of herbivorous birds

Species	Body wt (kg)	Diet	RT(h)[a]		Method[b]	% Digestibilities[c]						Ref
			Liq	Part		NDF	ADF	Holo	Cell	Hemi	Lignin	
Struthioniformes												
Ostrich, *Struthio camelus*	46	ground		40.1	bal	46	39		35	66	8	(1)
Emu, *Dromaius novaehollandae*	38	cubed[d]	4.4	6.2	bal	35	< 0		−3	51	13	(1)
		cubed[e]	3.8	5.3	bal	45	18		19	57	23	(2)
Anseriformes												
Domestic goose, *Anser anser*	5.7	pelleted chow	8	36+								(3)
				4–17								(5)
Brent goose, *Branta bernicla*	1.6	*Spartina patens*			lm		21		31	13		(4)
		S. alterniflora			lm		22		33	39		(4)
Canada goose, *B. canadensis*	4	*Juncus gerardi*			lm		12		18	26		(4)
		S. alterniflora			lm		21		30	23		(4)
Barnacle goose, *B. leucopsis*												
(winter)	1.7	graminoids		1.9	lm		0			20		(17)
(incubation)		graminoids		7.9	lm		26			42		(17)

(*continued*)

Table 5.1. *(continued)*

Species	Body wt (kg)	Diet	RT(h)[a] Liq	Part	Method[b]	% Digestibilities[c] NDF	ADF	Holo	Cell	Hemi	Lignin	Ref
Cape Barren goose, *Cereopsis novaehollandae*	3.7	ground lettuce		1.3	bal		1					(5)
Lesser snow goose, *Chen caerulescens*	2.4	alfalfa pellets			bal	31	16		45	53	8	(18)
Australian wood duck, *Chenonetta jubata*	0.87	intact wild foods			Mn	40	9		11	74	4	(6)
Galliformes												
Ruffed grouse, *Bonasa* umbellus	0.63	ground[f]			bal				20			(7)
		ground[g]			bal				15			(7)
		aspen flower buds			bal	−2	3					(8)
Spruce grouse, *Canachites canadensis*	0.60	pine needles			bal						ca. 0	(9)
Blue grouse, *Dendragapus obscurus*	0.92	conifer needles			bal	7–17[h]						(10)
Red grouse, *Lagopus l. scoticus*												
(wild)	0.65	heather			Mg			38	39		44	(11)
(captive)		heather			Mg			14	11		10	(12)
Willow ptarmigan, *Lagopus* l. lagopus	0.60	willow twigs			bal				10			(13)
		willow twigs, buds			bal				35			(13)
Rock ptarmigan, *Lagopus mutus*	0.45	bulbils			bal			38			4	(14)
		Vaccinium berries			bal			41			0	(14)
		Empetrum berries			bal			−3			13	(14)
		pelleted chow	9.9	1.9								(14)
Western capercaillie, *Tetao urogallus*	2–4	pine needles			bal				12			(13)
Black-billed capercaillie, *T. parvirostris*	2–3.5	larch twigs			bal				40			(13)
Chuckar partridge, *Alectoris* chuckar	0.56	ground[f]			bal				10			(7)
		ground[g]			bal				16			(7)
Bobwhite quail, *Colinus* virginianus	0.18	ground[f]			bal				13			(7)
		ground[g]			bal				22			(7)
Cuculiformes												
Hoatzin, *Opisthocomus hoazin*	0.75	lettuce, chow			bal		77		60			(16)

[a] Retention time. liq = retention time for the liquid portion of the digesta; part = the retention time for the particulate portion of the ingesta.
[b] Method used to calculate digestibilities. bal = mass balance experiments; lm = lignin used as an indigestible marker and digestibilities calculated from its concentration change; Mn = manganese used as an indigestible marker; Mg = magnesium used as an indigestible marker.
[c] NDF = neutral detergent fiber; ADF = acid detergent fiber; Holo = holocellulose, said to equal cellulose plus hemicellulose; Cell = cellulose; Hemi = hemicellulose.
[d] Diet formulated to contain less than 0.1% cell wall material.
[e] Diet formulated to contain 36% cell wall material.
[f] Diet formulated to contain 9.6% cellulose.
[g] Diet formulated to contain 15.4% cellulose.
[h] This range comes from NDF digestibilities calculated separately for needles from four different species of native conifer.

References: (1) Swart et al. 1993a; (2) Herd and Dawson 1984; (3) Clemens et al. 1975; (4) Buchsbaum et al. 1986; (5) Marriott and Forbes 1970; (6) Dawson et al. 1989; (7) Inman 1973; (8) C. G. Guglielmo unpublished data; (9) Pendergast and Boag 1971; (10) Remington 1989; (11) Moss 1977; (12) Moss and Parkinson 1972; (13) Andreev 1988; (14) Moss and Parkinson 1975; (15) Gasaway et al. 1975; (16) Grajal et al. 1989; (17) Prop and Vulink 1992; (18) Sedinger et al. 1995.

2.5. Rectum

The large intestine, or rectum, is a terminal enlargement of the small intestine that extends from the opening of the ceca to the cloaca (Ziswiler and Farner 1972). In the majority of birds it is a relatively short, straight tube. In the ostrich (*Struthio camelus*; a ratite relative of the emu) and the norther screamer (*Chauna chavaria*; an anseriform relative of geese), the rectum is relatively longer (Mitchell 1896). Little seems to be known about the digestive workings of the northern screamer, but recent work (Swart et al. 1987, 1993b) on the domesticated ostrich indicates fermentative activity in its colon may be as extensive as that in the rumen; microbial densities, SCFA production rates and SCFA contribution to metabolizable energy intake were all comparable to values reported for medium to large ruminants. Colon fermentation in the ostrich was, if anything, more active than the fermentation that occurred in its ceca (higher viable counts, greater microbial proteolysis). However, the density of cellulolytic organisms appeared to be much lower than that found in the rumen, and, in contrast to the rumen, protozoa were absent throughout the ostrich GI tract.

Unanswered questions concerning avian anatomy remain; the ceca of most bird species have yet to be described (Clench and Mathias 1995). Furthermore, much of the survey work of avian gut anatomy has been done by zoo pathologists. While zoos provide unparalleled access to an array of species, captivity may drastically alter the natural gut condition. Corroborative information from wild specimens should be collected whenever possible. Nonetheless, thanks to this work from zoological gardens, we know more about the digestive anatomy of many species than we do about their diets in the wild. Unusual anatomies may simply represent the quirks of phylogenetic history, but they may also hint at interesting nutritional ecologies, some involving extensive fermentative digestion. The large crop of the kakapo and the large colon of the northern screamer have already been mentioned. Tinamous (Tinamiformes) are noted for their large and sometimes ornate ceca (Hudson 1920, Beebe 1925, Wetmore 1926). This group offers comparisons in many ways analogous to those made with the much more closely studied tetraonids (e.g., Leopold 1953).

There are also many unanswered questions about fermentative digestion in the avian GI tract. Most of our current understanding arises from study of the galliform and, to some extent, anseriform ceca. There may be variation in cecal function among avian species corresponding to the variation in cecal morphology. Our subsequent discussion of cecal microbial ecology is to some extent handicapped by many unknowns about details of cecal physiology. These include the composition of material entering the ceca and the extent of cecal enzyme secretion and epithelial absorption.

3. Role of Gut Microbes in the Physiology of the Avian Host

Are the intestinal microbes of any benefit at all to the host? Microbial fermentation of some degree probably occurs in the guts of almost all birds, and the resultant SCFAs can be absorbed by the host. However, for many species, one can question both the importance of these SCFAs to the host metabolism and the extent to which microbial digestion is liberating otherwise unaccessible energy. Indeed, many microbes are potential pathogens and nutrient competitors. It has been demonstrated in both gnotobiotic and cecectomized chicken and quail chicks that growth is greater than in their ''conventional'' colleagues (Lewin 1963; Fuller and Coates 1983; Furuse and Yokota 1984, 1985; Muramatsu et al. 1993). The life of the domestic chicken is far removed from that of its wild relatives, and gnotobiotic work on species in the wild has, for obvious reasons, not been done. Hence, while of interest to the poultry industry, the relevance of this work to wild birds is questionable. Nonetheless, as we will reiterate later, only through work combining laboratory study of digestive physiology and field study of nutrition will it be possible to approximate the ecological significance of the gut microbial community.

One line of reasoning used to support the significance of microbial fermentation to hindgut-fermenting birds runs as follows: whatever the function of microbial fermentation, evolution would not have favored fermentation vats as voluminous as the ceca if the fermentation were not of some benefit. This is a weak argument because we cannot say whether the ceca are fermentation vats by design or by circumstance. Indeed, the question of origins may be impossible to answer. Birds have coevolved with their gut biota and whether the present state of that interaction represents avian accommodation to an unavoidable microbial contingent, or a host's welcome to invited guests may simply be unknowable.

The microbiota are diverse, and avian organs may be multifunctional. Below, we do not try to identify *the* function of gut microbes; no single, universally applicable function may exist. We do indicate which functions we believe are most likely and what questions still need to be answered.

4. Microbes and the Functioning of the Gastrointestinal Tract

4.1. Disease Resistance

A range of enteric diseases are potentially fatal to birds. It has been proposed that part of the above-mentioned growth depression in normal chicks compared to gnotobiotic birds is due to sublethal infections by microbes other than the normal gut microbes; putative agents of such infections have been identified (Sieburth et al. 1951, Fuller 1984). Work with chickens has demonstrated that a healthy GI microbiota can reduce the likelihood of invasion by pathogenic bacteria (Barnes 1979, Fuller 1977). The Nurmi principle (in poultry science) proposes that one can improve the survival of chicks by inoculating them early with an ''adult'' gut microbiota (Nurmi and Rantala 1973). Research is now being done to determine the most practical and efficient means of inoculation (Hume et al. 1993b). While reciprocal inoculations between turkeys and chickens appear to provide some health benefits, the benefit tends to be higher when host and inocula source are the same species (Schneitz and Nuotio 1992).

In humans, many studies have investigated the link between the GI biota and GI disorders (Savage 1970, Moore and Holdeman 1975, Ehle et al. 1981). The host-microbe interaction is complex. Factors such as host emotional stress and individual genetics have been shown to alter the microbiota of mammals (Holdeman et al. 1976, Varel et al. 1982), and moulting reportedly affects the GI microbes of birds (Porten and Holt 1993). These alterations may, in turn, affect the susceptibility of the gut to pathogen invasion.

Colonization resistance has two identifiable components: normal, healthy gut microbiota and its associated metabolites which inhibit growth and colonization of pathogens; and the mucosal barrier and its associated immunological components. As regards the first, lactobacilli are one group of bacteria that colonize the GI tract and apparently increase its disease resistance (Sandine et al. 1972, Fuller 1977). They are common to the crop and the entire tract of young birds, and are ubiquitous throughout the small intestine even in adult birds (Barnes 1979, Fuller 1984). These organisms show host-specific attachment to the gut wall. While rarely pathogenic themselves, *Lactobacillus* species produce acid fermentation products which reduce pH to levels unfavorable for pathogenic bacteria. They may also produce a bactericide (Fuller 1977). The proliferation and fermentation of lactobacilli in the avian crop can thus be seen as providing a barrier against pathogenic bacteria (Fuller 1977).

4.2. Adherent Microbes of the Gut Wall

The role of the microbes attached to many portions of the gut deserves further study. Hanssen (1979a) found intracellular amoebae and flagellates closely associated with the cecal epithelia of wild willow ptarmigan (*Lagopus mutus*). Based on his reevaluation of earlier research, he believed such organisms also occurred in the tracts of rock ptarmigan, capercaillie, hazel grouse, and black grouse. Bacteria have been seen associated with microvilli in the ceca of Gambel's quail *Callipepla gambelii* (Strong et al. 1989) and, as noted earlier, in other portions of the gut. Adherent microbes may be present on the surfaces of the more heavily colonized portions of the tract. These can form complex microhabitats (Costerton et al. 1987), and the role of such a lining in the metabolism and absorption of lumen nutrients could be substantial. For instance, the metabolic activity of the microbes of the rumen wall is believed to maintain the urea gradient which allows this important N source to diffuse from the blood stream into the rumen (Owens and Zinn 1988). There may be extensive interaction of host and microbial physiology on these surfaces. Because of its potential effects on luminal biochemical concentrations in the gut, microbial physiology may influence the rate of transport of certain nutrients into or out of the gut blood system and may cause measurements of chemical concentrations in the free lumen to be misleading.

The gut microbes apparently affect not only gut physiology but also epithelial morphology and dynamics (Coates and Fuller 1977, Hanssen 1979b). Among other consequences, elimination of the gut microbiota is accompanied by a change in villi shape and a reduction in the rate of epithelial sloughing (Coates and Fuller 1977). One can speculate that regular sloughing might be a response to reduce the accumulation of attached bacteria. Whether the intestinal microbes benefit the gut lining or not, they are probably an integral part of its dynamics.

We are far from being able to evaluate adequately the significance of GI tract microbes in maintaining gut health and hence enhancing survival of birds in the wild. No doubt both bird and resident microbial community have coevolved at least to the degree that they are mutually tolerant, but it is difficult to believe that the evolutionary raison d'etre for the microbes, if one exists, is their contribution to colonization resistance to pathogenic bacteria; the current microbial community may simply be the lesser evil.

The role of the microbes in gut physiology is more amenable to study, and represents a major research question for those interested in gut processes (Karasov and Hume 1995). The remaining functions proposed for gut microbes generally refer to their activity in one specific organ of the bird—the expanded crop of the hoatzin, or the ceca and colon of others. Aside from the hoatzin, some preprocessing of food may also be carried out by bacteria in the crop of other bird species, but as we have noted, there is little evidence for major microbial fermentation.

5. Contribution of Microbial Fermentation to Host Energetics

Microbial production of SCFAs appears to be of unequivocal nutritional importance to the hoatzin and the ostrich; extensive fiber fermentation occurs in their foreguts and hindguts, respectively (Grajal et al. 1989, Swart et al. 1993b). The contribution of the fermentation products to overall energy requirements has not yet been calculated for the hoatzin, although foregut fermenters, because of the pregastric placement of the fermentation vat, have effectively elected to survive largely from SCFA production. In the ostrich, SCFAs are estimated to provide as much as 75% of the energy intake for growing birds.

The hoatzin is the only known avian foregut fermenter, and the ostrich is unusual in being large and flightless. Such extensive contributions of microbial fermentation to host energetics are not apparent in most other birds. Postmortem measurements of fermentation rates in freshly collected cecal contents from wild Galliformes have indicated that SCFAs may provide from 6% up to 30% of the maintenance energy demands (McBee and West 1969, Thompson and Boag 1975, Gasaway 1976b), but the accuracy of these figures can be questioned. Fermentation rates measured by these methods may underestimate those in the live animals by at least 50% (I.D. Hume, personal communication). On the other hand, the energy demands of wild birds may be substantially higher than those of the captive birds used to estimate maintenance energy requirements. Gasaway (1976b) estimated that field metabolic requirements for the willow ptarmigan were almost three times standard metabolic rates. Finally, the substrate for this fermentation needs to be determined; microbes will readily ferment substrates that the bird would readily absorb for its own use if given the chance. It is not fermentation in general that is of interest but rather SCFAs produced from the microbial metabolism of otherwise undigestible substrates. The most commonly recognized ''undigestible substrate'' is fiber, discussed in the following section.

6. Microbes and Fiber Digestion

Perhaps because of its appeal as adaptive ingenuity and its undeniable importance to ruminant nutrition, the role of gut bacteria in liberating fiber energy that would otherwise be unaccessible to the host has been widely discussed. Yet, at least in the case of most hindgut-fermenting birds, there are, as we will discuss below, several reasons to question this function. Considering the location of the ceca and their filling mechanism, relatively little cellulose may ever reach the ceca. Furthermore, empirical efforts to document extensive cellulose digestion, either through calculated partial digestibilities or attempts to culture cellulolytic bacteria, have rarely revealed substantial activity. The hemicellulose fraction of fiber may be more fully digested; however, microbes may not be central to this process. Suggestions of a direct, cause-and-effect relationship between increased fiber intake and cecal enlargement have been largely discredited. The onus of proof lies on the side of those who would claim that the bacterial breakdown of fiber is of major nutritional importance. Below, we expand and document this interpretation of fiber digestion.

6.1. Fiber

''Fiber'' is a biochemically imprecise term because it generally refers to the variety of polymers that make up the plant cell wall. Some of these components, such as cellulose, can be quite definitively characterized (although even with cellulose, secondary structure may vary). Other components, such as pectin, hemicellulose, and lignin, really represent a range of compounds that have certain characteristics in common which cause them to be grouped together during conventional analysis but that are diverse in other respects (Van Soest 1982, Chesson and Forsberg 1989, Hatfield 1989, Weimer et al. 1990). Furthermore, cellulose, hemicelluloses, and pectins are interwoven with each other in the cell wall, though they are presented as purified substrates in vitro. It is thus not surprising that while microbes are generally recognized as essential for cellulose digestion, their role in the digestion of the other components of fiber is not so clear. One needs to be careful not to attribute differences in ''fiber'' digestibility to differences in the microbial community when, in fact, it may be due to differences in the fiber. Hemicelluloses and pectins are more rapidly and extensively degraded than cellulose, and lignin is recalcitrant to degradation. This caveat is especially applicable to attempts to compare the fiber digestion abilities of wild birds on different diets.

6.2. Birds with Extensive, Microbially Mediated Fiber Digestion

Substantial fiber digestion does occur in the hoatzin crop, and cellulose digestion is extensive (60%; Table 5.1). In vitro, hoatzin crop microbes were found to be as capable of fiber degradation as cow rumen microbes (Grajal et al. 1989). Dominguez-Bello et al. (1993a,b), however, were unable to isolate cellulolytic bacterial colonies, although cellulolytic activity was evidenced in enrichments and under microscopic inspection. The digestion of both cellulose and hemicellulose is important to ostrich nutrition (Swart et al. 1993a). Passage rates can be relatively slow, and the enlarged ceca and proximal colon of this bird provide a suitable fermentation site. Evidence is also beginning to accumulate that appreciable fiber digestion sometimes occurs in geese (Buschbaum et al. 1986, Prop and Vulink 1992, Sedinger et al 1995; see following section).

6.3. Material Entering the Ceca

An accurate knowledge of what materials enter the ceca is crucial to understanding the potential metabolic activities of microbes in most avian hindgut fermenters. Several reports on the composition of cecal droppings exist (e.g., Pendergast and Boag 1971, Moss and Parkinson 1972, Moss 1977, Andreev 1988), but such analyses do not tell what entered the ceca, but rather describe what remained after microbial metabolism.

If we accept several assumptions, we can reason what materials are likely the initial fuels of microbial fermentation in the ceca. This is a weak approach, but, until more direct information is available, it is the best we can do. First, the acid digestion and subsequent nutrient absorption of the upper GI tract may be able to remove much of the readily accessible carbohydrates such as starch and sugars;

proteins and fats from the plant cell cytoplasm are probably also largely absorbed. The acid treatment may also solubilize some fiber such as hemicelluloses (Dawson 1989). Second, most herbivorous birds tend to rapidly process large quantities of food rather than trying to rely on long retention times and extensive digestion of the entire ingesta. This is probably because of allometric considerations we will outline. Rapid transit time and large bulk make it likely that even some theoretically accessible nutrients may escape the upper GI tract. There is indication of this at least in small mammalian herbivores, although mechanisms have also evolved to avoid the loss of energy from cell solubles by fermentation (Cork and Foley 1991). Some forms of starch and protein may also reach the ceca (Nitsan and Alumot 1963, Champ et al. 1981). Third, substantial amounts of endogenous products—enzymes, sloughed cells, bile—are added to the ingesta in the upper tract. Fourth, large particles in the ingesta are excluded from the ceca by a separation mechanism at the cecal-colonic junction (Bjornhag 1989). Lastly, materials are washed into the ceca by the retrograde flow of urine and liquid ingesta from the lower portion of the colon. Based on these assumptions, the probable composition of materials entering the ceca is solutes and small particles from the urine, or the liquid portion of the ingesta with at least traces of escaped digesta, solubilized fiber or endogenous products from the upper GI tract.

In conclusion, cellulose is likely to be a minor component of the material entering the ceca. It is conceivable that endogenous products such as bile, enzymes, or urinary compounds are the major energy and nitrogen sources for the bacteria. Future studies should not only describe the substrate entering the ceca, but also the relative contribution of the substrate components to microbial growth and SCFA production.

6.4. Partial Digestibility of Fiber Fractions and the Importance of Hemicellulose Digestion

In the emu (Herd and Dawson 1984) and the Australian wood duck (Dawson et al. 1989), the importance of cellulose digestion relative to that of hemicellulose may be minor (Table 5.1). The emu GI tract shows little evidence of specialization for fermentation, and the wood-duck, besides being a much smaller bird (870 g) also shows relatively little adaptation for fermentation. Measurable cellulose digestion does occur in the emu but, given the fast transit times of their food, it is not extensive. Yet, high-fiber digestibility values were found for these species (Table 5.1). Much of this fiber digestibility appears to be due to hemicellulose digestion.

Cellulose and hemicellulose digestion accounted for roughly 9% to 31% of extracted dietary energy, with cellulose and hemicellulose making approximately equal contributions in Canada geese studied using lignin as an inert marker

(Buschbaum et al. 1986). In balance trials with alfalfa pellets, Sedinger et al. (1995) concluded that cellulose digestion made an important contribution to overall diet metabolizability. Interesting work with barnacle geese (*Branta leucopsis*) showed that cell wall digestibility increased in concert with retention time as geese went from feeding on agricultural fields during short winter days to feeding on moss during the long days of Arctic summers (Prop and Vulink 1992). Previous studies of geese generally concluded that fiber or cellulose digestion was of minimal importance (Halnan 1949, Marriott and Forbes 1970, Mattocks 1971, Morton 1978), and cellulose concentrations may be comparatively low in the aquatic food plants (Morton 1978). The long retention times recorded by Clemens et al. (1975) would certainly provide sufficient time for substantial fiber fermentation in geese, but those values, measured using short lengths of plastic tubing, need to be corroborated. Geese may modulate their digestion of fiber dependent on the foods and feeding time available.

The distinction between hemicellulose and cellulose digestion is relevant here because features of birds themselves (e.g., comminution, solubilization in acids) apparently result in degradation of at least certain types of hemicellulose (Parra 1978, Robbins 1983, Dawson 1989, Dawson et al. 1989). It may be these avian features, as much as any symbiotes, that enable the emu, the wood duck, and geese to survive on their fibrous diets. Possibly, avian features solubilize hemicelluloses in proximal GI regions, permitting their transport to the ceca where their energy is freed by microbial fermentation (Parra 1978, Dawson 1989).

We have found no calculations of hemicellulose digestion in tetraonids. Existing information (Table 5.1) is variable, with some reports suggesting extensive digestion of fiber including cellulose, and other studies implying that there is almost no digestion of fiber, not even of hemicelluloses. Does this result from methodological inconsistencies among the investigations (e.g., marker vs. mass balance techniques, variation in fiber analysis methods), from differences in fiber digestibilities among the foods studied, or from real differences in the digestive abilities of grouse and their microbes?

6.5. *Occurrence of Cellulolytic Bacteria*

Cellulolytic bacteria have been reported to occur in the ceca of chickens, ducks, geese, turkeys, black grouse, hazel grouse, capercaillie, willow grouse, ruffed grouse, and rock grouse (Suomalainen and Arhimo 1945, Ziswiler and Farner 1972), although McBee (1977) questions the findings of Suomalainen and Arhimo (1975), and other studies have reported little or no cellulose digestion in chickens, sage grouse, Tasmanian native hen, pheasant, and geese (Mattocks 1971, Ridpath 1972, Ziswiler and Farner 1972, Barnes et al. 1973, Remington 1989). However, there is some reason to question the work done to date with wild species. The fiber digestion studies of Gasaway (1976c) and Inman (1973) did provide clear evidence that cellulose degradation occurs in grouse maintained on pelleted diets and fed powdered cellulose, but there is evidence that the ceca of grouse maintained on such unnatural diets are altered and perhaps, unlike the ceca of wild birds, fail to exclude fiber. Furthermore, powdered cellulose has a much smaller particle size than natural fibers and can thus pass into the cecum easily. Feeding grouse unnatural diets with finely ground cellulose might inevitably lead to artificially high-fiber digestibilities.

Isolation of cellulolytic bacteria is not proof that cellulolysis is important to the host; cellulolytic bacteria may occur, and yet the nature of the material entering the ceca and the rate of SCFA production by these organisms may result in little nutritional contribution. Moreover, as pointed out earlier, high rates of SCFA production in the ceca do not necessarily mean that there are high rates of cellulose digestion, since microbial fermentation of uric acid also yields SCFAs (Beck and Chang 1980). Several authors have used the release of labeled CO_2 from fed cellulose as an indicator of host oxidation of the bacterial products of cellulose catabolism (Gasaway 1976c, Duke et al. 1984). Yet, as these authors themselves point out, the CO_2 may have arisen directly from bacterial fermentation, and so, while it does indicate cellulose breakdown, it is not direct evidence of a benefit to the host. Work with turkeys has indicated an increase in cellulolytic bacteria in the ceca of birds fed a diet high in fiber. This was accompanied by an apparent increase in the fiber digestion abilities of these birds (Duke et al. 1984). At the least, such a result suggests that sufficient cellulose is arriving to the ceca to alter the composition of the microbiota, although again the possible effect of captivity and unnatural diets on fiber digestion needs to be recognized.

6.6. Correlations Between Cecal Size and Fiber Intake

Variation in the magnitude of the avian ceca in relation to inter- and intraspecific dietary differences has been cited as evidence of the nutritional importance of fiber, often read as ''cellulose'' digestion. Increase in the size of these organs is often correlated with increased fiber intake (Leopold 1953, Clench and Mathias 1995). Browsing species, for example, have much more extensively developed ceca than granivorous species. More recently this interpretation has been questioned.

Increase in gut dimensions associated with an increase in dietary fiber may be a special case of the more generalized phenomenon we mentioned previously in regard to the intestine; cecal dimensions appear to increase with increased intake. Increasing dietary fiber may have an effect on cecal size only insofar as it reduces diet quality and therefore causes increased intake (Remington 1989). Drobney (1984), for example, found that wood duck ceca enlargement accompanied not only a decrease in diet quality but also an increase in food intake occasioned by reproduction. In an interspecies comparison of cecal development and cellulose digesting ability between ruffed grouse, chukars, and bobwhite quail, it was found that cecal development did not parallel cellulose digestion (Inman 1973). Remington (1989) pointed out that within herbivorous galliformes the ceca may actually be longer among species eating forage with less cell wall and more cell solubles. The strategy of some of these species may not be to derive energy from extensive cellulose digestion, but rather to pass large quantities of fiber through the GI tract and extract the cell solubles.

Experimental work looking at the effects of cecectomy or gnotobiotic treatments on crude fiber digestibility has been equivocal, and there is some indication that reductions in fiber digestibility caused by cecectomy may, in fact, be greater on low-fiber diets (reviewed in Sturkie 1965, McNab 1973, Sibbald 1982, Chaplin 1989).

6.7. *Microbial Digestion of Other Recalcitrant Foods*

Microbial digestion has also been suggested as an explanation for the ability of some birds to survive on waxes. Honey guides consume large quantities of bees wax and derive energy from it (Friedmann and Kern 1956). It was proposed that microbial fermentation was responsible for this digestion, and the ability to digest wax could be conferred on otherwise noncompetent chickens by inoculating with gut bacteria isolated from the honey guides (Friedmann et al. 1957). Some marine birds also consume diets high in waxes. More recently, it has become apparent that, while bacteria exist that can degrade such waxes, and they may occur in the digestive tracts of these birds, the hosts have endogenous enzymes specifically capable of wax digestion (Diamond and Place 1988). The wax-digesting bacteria of their guts, rather than being symbiotes, may be competitors. Likewise, while some bacteria digest chitin, sea birds have endogenous enzymes that are probably more important to the host's degradation of chitin (Jackson et al. 1992). The extent of the degradation, and the post-absorptive nutritional value of the products remain unclear.

6.8. *Issues of Allometry*

Even if one accepts that microbial fermentation may have nutritional benefits for many host species, there are certain a priori constraints that probably have limited the degree to which birds can rely on microbial fermentation for energy. Basically, the relatively small body masses and high metabolic rates of most birds mean that they would starve to death more quickly than a microbial fermentation system could fuel their survival (Morton 1978). Some birds that are large and flightless have been able to escape the mass constraint and can rely more heavily on their fermentation vats (e.g., the ostrich and possibly the rhea). The hoatzin has largely sacrificed flight, and may have relatively low energy requirements, and so is able to survive on the products of microbial fermentation. However, for most of the herbivorous birds major reliance on microbial fermentation may not be a physiologically realistic option.

In conclusion, gut microbes do digest a wider range of nutrients than do vertebrate enzymes. Yet, for most birds, it is unclear that bacteria liberate major quantities of nutrients that would otherwise be inaccessible. SCFAs can provide a relatively large proportion of the energy intake of some species, but for many species, it has not yet been documented that these SCFAs represent energy that would be unavailable to nonmicrobial catabolism. An extensive study following the fates of various labeled substrates, preferably introduced in vivo, would be needed to describe accurately the composition and microbial utilization of the fermentation substrate.

7. Microbes and Nitrogen Digestion

Amino acid digestibility is generally higher in conventional than in cecectomized chickens. However, this result is misleading because the added amino acid that is digested is probably assimilated not by the bird but by its gut microbes (Parsons 1986, Johns et al. 1986). Some forms of protein, e.g., that in raw soybean meal, may not be digested in the upper GI tract and so can reach the ceca in relatively large amounts and be broken down therein by the microbes (Nitsan and Alumot 1963). However, such large-scale escape of protein from the small intestine may not be a common occurrence. To understand the possible contribution of the gut microbes to N balance, we need to follow the fate of the nitrogenous compounds in more detail.

7.1. Contrasting Mammalian and Avian Nitrogen Digestion

Rumen bacteria serve both to increase the biological value of ingested N to the host (i.e., to convert plant amino acid ratios to amino acid ratios more suitable for incorporation into the animal host) and to recycle host nonprotein nitrogen waste (Kay et al. 1980, Van Soest 1982, Wallace and Cotta 1989). Extensive cecotrophy in rodents and lagomorphs permit hindgut-fermenting mammals to realize some of the same benefits (Hornicke and Bjornhag 1980). As with fiber digestion, it has been natural to try to extend this mammalian paradigm to birds. However, the hoatzin is the only known avian foregut fermenter, and there are relatively few records of coprophagy among birds. Whether this dearth of records reflects the actual rarity of coprophagy or an observational bias is not known. Morton (1978) suggested that cecotrophy was rare in birds because, unlike in mammals, cecal droppings would contain urinary material. However, as we noted earlier, the cecal droppings contain the bacterial breakdown products, not the "raw" urinary compounds. Turkeys may at times consume their feces (G. Duke, personal communication), and Gambel's quail are said to be eager to consume their cecal droppings (E.J. Braun, personal communication). In ostrich farming operations, chicks consume fecal pellets of older birds (D. Swart and R.I. Mackie, personal communication). Tracer work with chickens indicated that within flocks significant fecal ingestion occurred (Hornicke and Bjornhag 1980). It is also known that penning chickens in wire cages causes increases in their vitamin requirements, presumably because they are unable to ingest the vitamins found in their cecal droppings (McNab 1973). Tetraonids have been closely observed and apparently are not coprophagic, at least under many conditions (Moss and Hanssen 1980; W. Karasov, personal observation). Whenever coprophagy does occur, birds could benefit from the proteins of their microbes (Mortensen 1984). The mammalian model of microbial nitrogen conversion would not be applicable to those birds that are not coprophagic, but many birds may engage in a type of microbially mediated nitrogen recycling that is not possible in the majority of mammals.

The urinary and digestive tracts of most birds unite in the cloaca and mixing between feces and urine occurs (although some birds, such as the ostrich, do maintain greater separation between feces and urine). Retrograde peristalsis in the colon causes urine to reach the areas of microbial fermentation in the ceca or upper colon. Here, nonprotein nitrogen such as uric acid can be metabolized to ammonia, probably at least in part by the microbes (Featherstone et al. 1962, Blair 1972, Mortensen and Tindall 1981a,b). These products can be absorbed from the gut, and the ammonia can be converted to nonessential amino acids (e.g., glutamine) in the host. Perhaps some amino acid production also occurs in the ceca. While this pathway can be traced, its benefits to the bird are not obvious. The uric acid in the urine originally derives in part from waste ammonia in the bird's bloodstream—ammonia that could have been converted directly to nonessential amino acids without any microbial assistance. Nevertheless, chickens in which urinary reflux has been prevented by colostomy, an operation by which the large intestine is rerouted to an artificial opening that is separate from the cloaca and emptying of the ureters, or in which the microbe community has been subdued by antibiotics may show lower nitrogen retention than control birds when on a low-protein diet (Karasawa and Maeda 1992, Karasawa et al. 1993a; but see also Salter et al. 1974 and Green et al. 1987 for examples of an apparent lack of effect). Bjornhag (1989) suggested that urinary reflux was increased when chickens were fed a low-protein diet. Screech owls which, if they are similar to other owls, have relatively large ceca, reportedly can maintain nitrogen balance on a lower quality diet than kestrels, which have rudimentary ceca (Campbell and Koplin 1986, McLelland 1989).

7.2. *Reflux of Urinary Nitrogen in Birds*

One possible benefit that birds may derive from this recycling of urinary N is based on the role of ammonia in osmoregulation (E.J. Braun, personal communication). The refluxed uric acid travels to the ceca and is converted (bacterially or partially by the host) to glutamine, which is then transported to the kidneys. At the kidneys, ammonia is cleaved from the amino acid and is used to buffer H^+ excretion into the renal tubules. The resultant ammonium is then excreted from the bird. Such an explanation interprets a certain level of ammonia loss as being a requisite part of the avian excretory process, and the reuse of uric acid nitrogen to fill this drain would be advantageous. There is at present no conclusive evidence for this scenario, but it does provide one way of interpreting the outcome of the reported colostomy experiments. Noting the unusually low nitrogen requirements of emus, Dawson and Herd (1983) suggested that nitrogen recycling was occurring. Urinary reflux was observed, but no further details were known.

Poultry workers have exploited the ureolytic abilities of the ceca. Birds cannot, on their own, convert urea to ammonia and thence to amino acids. But, because of the cecal microbes, urea can still be used as a nitrogen source in poultry feeds (Karasawa 1989).

One of the biggest unanswered questions in cecal nitrogen metabolism is the degree to which amino acids are absorbed from the ceca. This is really two questions: are there free amino acids in the cecal lumen? And, are these amino acids absorbed by the bird? Some fiber-associated plant proteins plus endogenous proteins from the lower GI tract arrive to the ceca. Many microbes are probably proteolytic and, because most of the digestion is extracellular, free amino acids are liberated. Host proteases, which could potentially liberate microbial protein, may be present in the avian ceca although their activity is uncertain (Lepkovsky et al. 1964, Barash et al. 1993). However, probably because the free amino acids are rapidly deaminated by other microbes, recorded amino acid concentrations are low. Mortensen (1984) found that the average concentration of seven nonessential amino acids was 0.97 mM while that of the nine essential amino acids was only 0.27 mM. Whether or not the ceca absorb much amino acid depends on the physiological properties of the ceca and the degree to which the cecal tissue can out-compete N-hungry microbes. Although the concentrations may indeed be low, one need not envision passive uptake of amino acids from the ceca. Active amino acid transport from the ceca has been demonstrated in the three bird species studied to date (Obst and Diamond 1989, Moreto et al. 1991). These results are in contrast to the apparent situation in mammals; in both mammal species investigated, there has been no evidence of active amino acid transport from the ceca (Ilundain and Naftelin 1981, Hume et al. 1993a).

Aside from possibly benefiting the bird directly, the microbial breakdown of uric acid may be crucial to the maintenance of the microbial community. Such a role would be valuable to the bird if we accept a positive role of whatever sort for the microbiota. The position of the ceca distal to the small intestine means that relatively little protein may reach the ceca. Uric acid can fertilize the cecal microbiota in a manner similar to the urea in the saliva of ruminants (Mackie 1987; although in ruminants that N may truly be recycled by the host because microbial protein that escapes from the rumen can be digested and absorbed in the small intestine). The ureolytic bacteria could also derive energy from the urea breakdown.

It has also been suggested that the uricolytic activity of the cecal microbes is important to the ceca's function in water reabsorption. If much urinary water is reabsorbed from the ceca (see Skadhauge 1981 for discussion of this role for the ceca), then, without the uricolytic bacteria, uric acid would rapidly accumulate in the ceca, possibly with detrimental consequences (Barnes 1979; but see Anderson and Braun 1984, Campbell and Braun 1986). Ammonia nitrogen recycling may be futile, but the water recycling would not be.

In summary, nonprotein N is entering the ceca and being metabolized by the bacteria; but the benefit of that activity has not yet been clarified and may be indirect.

8. Detoxification of Plant Secondary Metabolites

8.1. Direct Detoxification

Plant foods contain a variety of compounds that can potentially injure the host (e.g., see Cork and Foley 1991). Many of these compounds can be metabolized to less nocuous chemicals by bacteria; unfortunately for the host, innocuous chemicals can also be "activated" by this mechanism (Smith 1992). Chicken feed made from plants in the mustard family contains glucosinolates that become more readily absorbable by the GI tract, and hence more poisonous, when metabolized by the cecal bacteria (Bell 1978; Slominski et al. 1987, 1988; Michaelsen et al. 1994). Some foods, e.g., legume derivatives, may also be toxic because they alter gut chemistry so that, in the competitive gut microbe community, pathogenic bacteria come to dominate (reviewed in Fuller and Coates 1983). Furthermore, reliance on microbes does make the host susceptible to poisoning by bactericides (Cork and Foley 1991). On the other hand, there are documented examples whereby adaptations of the ruminant microbiota allow ruminants to feed on plants that would otherwise not be available to them (Allison and Reddy 1983, Jones and Megarrity 1986, Majak 1991). This function relies largely upon the bacteria intercepting the toxin prior to its absorption by the host, hence it would seem most applicable to foregut fermenters. Plant secondary metabolite detoxification has been suggested as one selective advantage of hoatzin foregut fermentation, because this species consumes a diet that is probably high in secondary metabolites (Dominguez-Bello et al. 1993b). However, the benefits are not always restricted to foregut fermenters. For example, oxalate toxicity occurs because of oxalate absorption in the colon. It is thus likely that the oxalate degradation that has been demonstrated in the colon of some hindgut fermenters does benefit the host (Allison and Cook 1981). Few generalizations can be made, so microbial detoxification mechanisms in hindgut fermenters can neither be assumed nor categorically denied; each case needs to be investigated independently.

8.2. Recovery of Detoxification Costs

Dietary toxins will reach the microbes of avian hindgut fermenters after they enter the host and are excreted into the urine, or if they are unabsorbed by the host and arrive to the ceca with the digesta. In fact, high concentrations of some of these plant compounds have been reported from cecal droppings and may partially explain the hesitance of some birds to ingest cecal droppings. Vertebrates deal with most of the plant secondary metabolites they absorb by making them more water-soluble and then excreting them through the kidneys (Sipes and Gandolfi 1986). The enhancement of water solubility is frequently accomplished by conjugating the toxin to a glucose or amino acid derivative. The conjugated toxin is then excreted in the urine and so may reach the ceca. At this point, bacteria may deconjugate the compound, metabolizing the toxin (as might occur in a foregut fermenter) and returning some of the energy or protein invested in conjugation to the host (i.e., through possible reabsorption of SCFAs or nitrogen compounds from the ceca). All the steps in this sequence are feasible, but work has not yet been done to address this possibility.

Based on measurements of conjugation products in the excreta of captive ruffed grouse feeding on natural foods, Guglielmo et al. (1993) estimated that energy losses via conjugation reactions amounted to 10% to 14% of the daily metabolizable energy intake. These are substantial losses; either the microbiota was not effective in recycling the conjugation products or even more extensive conjugation was occurring but materials were being salvaged. Study of the droppings of wild grouse feeding on some of the same foods indicated that average concentrations of one conjugate, glucuronic acid, were approximately twice those measured in captive grouse (Vispo 1995). This suggests that the apparently high detoxification costs estimated by Glielmo were not artifacts of captivity.

8.3. Avoidance of Toxicants

As the name implies, sage grouse (*Centrocercus urophasianus*) eat large quantities of sage brush (*Artemesia* sp.), a plant that contains high concentrations of monoterpenes. These grouse do not possess grinding gizzards, but do have large ceca. Remington and Braun (unpublished data) suggested that, in a manner analogous to the rapid fiber throughput in other grouse species, sage grouse rapidly pass most of their food through their intestinal tract, funneling soluble nutrients into the ceca but quickly voiding the bulk of the ingesta. The lack of grinding in the gizzard and the swift passage of food might minimize the grouse's exposure to sage secondary compounds. In this case, the microbes of the ceca could enable the bird to cope with a toxic diet, not through any direct metabolic detoxification but by facilitating the host's last-minute physical avoidance of these chemicals.

9. Microbes and Vitamin Synthesis

Many microbes synthesize B-complex vitamins for their own metabolic needs, and some of these may be free in the avian ceca (Coates et al. 1968). However, experimental work indicated that the host did not benefit from these vitamins, perhaps because they could not be absorbed by the cecal wall and/or they were rapidly consumed by the microbes (Coates et al. 1968). Studies of pantothenic acid and biotin indicated that the requirements for these vitamins were *decreased* in gnotobiotic birds (Sunde et al. 1950, Latymer and Coates 1981). In contrast, it was shown that vitamin A demand could be reduced by inoculating chickens with vitamin A-producing bacteria (Pivnyak and Konyakhin 1973). McNab (1973) pointed out that some of the apparent lack of benefit from microbial vitamin synthesis may have occurred because birds were housed in wire-floored cages and hence coprophagy was prevented. Indeed, he noted, it has been realized that housing chickens on hard floors, where coprophagy is possible, reduces their vitamin B_{12} needs (Coates et al. 1968, McNab 1973). In birds that are coprophagic, benefits from microbial vitamin synthesis seem possible.

10. Reciprocal Effects Between Gut Microbes and Avian Foraging Behavior

In this section, we will consider the impact of gut microbe metabolic characteristics on avian (host) feeding ecology, taking the viewpoint of avian, rather than microbial, ecologists. Understanding gut microbial ecology can be relevant to the avian ecologist in at least two general ways. First, the gut microbes can provide static constraints on host foraging. The host/microbe system that is the avian gut has limits on its functioning; understanding the metabolic abilities of the microbes allows one to realize better what bounds birds are being forced to respect during their foraging. Second, the constraints imposed by gut microbes (as with those imposed by host morphophysiology) can be dynamic; understanding the ability of the gut community to adapt to changing diets tells one how rigid the bounds of microbial constraint are and how rapidly they can be changed. In other words, the gut microbiology is relevant not only to the question of what foods a bird can eat today but also how rapidly the bird can adjust if migration or a winter snowfall forces it to consume an alternate diet tomorrow.

Static constraints may be of two sorts. The microbes may determine the range of foods a bird can consume and benefit from because (1) they partially determine the extent to which a host can extract nutrients from a given food; (2) they prevent or facilitate consumption of that food by converting chemicals in the food to more toxic or less toxic substances.

Paralleling the static constraints, dynamic constraints can also be of two types. Gut microbes may determine the rate at which a bird can adapt to new foods because the metabolic properties and taxonomic composition of the microbial community change over time and so alter (1) the realized nutritional value of the food, or (2) the toxicity of the food. Such rate limitation may be important during at least two natural situations—host ontogeny and seasonal (and in some cases, annual) variation in food availability. Host adaptation is determined, as with static constraints, by the interaction of host and microbe characteristics. It would be wrong, for example, to assume that change in the digestive abilities of a herbivorous bird represented solely microbial adaptation given the widely documented morphological and physiological changes known to occur in the gut in response to diet change (Karasov 1995).

Relevant information on birds is scant, and much of it is indirect at best. Below we present the data which we are aware of, including a few references to mammalian work, as ''circumstantial evidence'' for the processes we are discussing.

10.1. Static Constraints

1. To what degree do microbes determine the extent to which a bird can digest a food and so derive nutrients from it? Differences in the digestive abilities of captive and wild individuals do seem to exist (Pendergast and Boag 1971, Moss 1989). Captivity can have a major effect on the gut microbes of wild species. Diet change is one important aspect of captivity, although the effects of associated stress and potentially the exposure to novel microbes must be recognized. Major differences were noted in the microhistology and microbiology of captive and wild willow ptarmigan (Hanssen 1979a). Products of α-ketoglutarate degradation differed between captive and fresh-caught willow ptarmigan although similar rates of uric acid degradation were measured for both (Mortensen and Tindall 1981a,b). Schales et al. (1993) found differences in the aerobic microbiota of captive and wild capercaillie and cautioned that, if these differences were reflected in the rest of the biota, they might threaten the viability of reintroduction efforts.

Interspecies comparisons of the GI tract microbial community are interesting, in part because they may indicate patterns of diet-microbe interaction. Morphological and physiological differences between hosts are no doubt also at least partially responsible for the differences. The aerobic biota of the lower digestive tract of several species of wild birds has been surveyed during investigations of avianborne pathogens (e.g., McClure et al. 1957, Brittingham et al. 1988, Aguirre et al. 1992); inter-species differences have been noted, but their relation to the fermentative microbes is unknown. Barnes (1979) reported observing distinct differences in the cecal microbe composition of ducks as compared with chickens and turkeys. The varied diets of northern tetraonids influences the SCFA profiles in their guts (Gasaway 1976a), possibly reflecting diet-caused differences in the microbiota. A diet effect on cecal SCFA ratios has been demonstrated in mallards (Miller 1976). The occurrence of protozoa provide one marker of microbial community differences. They have been found in hoatzin crops, the ceca of wild quail, Tasmanian native hen, ptarmigan and at least some chickens while they appear to be absent from the guts of ostrich and many domestic species (Lewin 1963, Mattocks 1971, Ridpath 1972, Hanssen 1979a, Bedbury and Duke 1983, Swart et al. 1987, Mead 1989, Dominguez-Bello et al. 1993a). Protozoal protein has a higher biological value than that of bacteria, and the protozoa of the quail ceca, given their coprophagy, and of the hoatzin crop may be directly available to the host.

Differences in gut adaptation for fermentation may be associated with the differences in diet noted between the ostrich and emu. While the former, with its capacity for food retention and extensive fermentation, can consume foods relatively high in cellulose, the emu, with its quick passage rates and unspecialized tract, may rely on taking foods that have lower cellulose contents. The modifications of the sage grouse digestive tract have already been described above; such adaptation may be well suited for the sage grouse's main diet but may limit its ability to consume other foods.

A review of studies that compared in vitro digestibilities using gut microbes from varying mammalian hosts on common substrates suggested that if the substrate was either highly digestible or highly indigestible and host effect was slight, while if foods were of intermediate digestibility there was a larger host effect (Vispo (unpublished data)).

In sum, there does appear to be variation in fermentative gut microbes, at least among species, and, we would assume, also within species whose diets vary geographically, but none of this variation has been formally described so far as we know. There is cause to believe that this variation might result in differences in digestive abilities. However, we still seem very far from being able to explicitly link specific variation in gut microbe composition to corresponding variation in the nutritional constraints experienced by the host.

2. What limits do microbes place on the foods a bird can consume without toxic effect? Little work has been done with birds to isolate microbe-toxin interactions that influence host foraging. Most study of this interaction to date has concerned only the influence of diet on the microbial community. If that microbial community has, in turn, some impact on host digestion, then the microbes' ability to cope with diet toxins is relevant to host nutrition.

Natural dietary toxins have been reported to alter the gut biota of penguins (Sieburth 1959, Soucek and Mushin 1970). Certain penguins were observed to have unusually low gut microbe counts. This reduction was found to be associated with the consumption of crustacea. It was then realized that these crustacea themselves had been eating a certain phytoplankton, and, upon closer inspection, it was found that algae in these plankton contained a bactericidal chemical that was subsequently identified as acrylic acid (Sieburth 1960). Penguin populations that fed on relatively few crustacea appeared to have higher microbial counts, and, between species, those that consumed few crustacea had more gut bacteria. The effect may have extended even one step further up the food chain—predatory birds consuming penguin flesh reportedly had reduced microbes! While this may be a good example of a diet effect on gut microbes, the relevance of those microbes to penguin health and nutrition is unknown.

A bactericidal property of the diet has also been suggested to explain the relative resistance of wild pheasant and ptarmigan carcasses to decomposition (Nordal and Dahle 1971, Raa et al. 1976, Barnes 1979). Hanssen et al. (1984) identified *Vaccinium* as a source of an antimicrobial chemical in the diet of willow ptarmigan and noted that its absence from the diet of captive birds increased their problems with enteric diseases.

Alder is a common food in the habitat of many ruffed grouse, and yet it is rarely eaten. Studies indicated that when cecal content from captive birds that had fed on alder was used as an in vitro substrate, inocula from wild grouse produced significantly lower amounts of gas than when they were raised on a substrate derived from a wild plant that was regularly consumed (Vispo 1995). Alder catkins may contain high levels of secondary compounds (Bryant and Kuropat 1980, Guglielmo and Karasov 1995) and perhaps is avoided in part because of its negative effects on the gut microbes. It is not clear, of course, whether birds avoid alder because it is unacceptable to their bacteria or if the bacteria are susceptible to alder simply because the grouse avoid it for other reasons.

These data suggest the possibility that avian gut microbes may influence avian foraging by affecting diet toxicity, but no studies have directly tested this idea. The hoatzin may be particularly suitable for such study (C. Guglielmo, personal communication). Variation in the metabolic abilities of rumen microbes have been shown to affect the foraging abilities of the host (e.g., Kaufmann et al. 1980), both in terms of what foods it can ingest and in how rapidly it can switch between foods. There is no reason to suppose that the same results would not apply to the hoatzin. This species is found in different areas of the Neotropics where it feeds on an array of different foods, many of which contain secondary compounds (Dominguez-Bello et al. 1993b). One would predict that hoatzins from populations feeding on plants containing different secondary metabolites would have different gut microbe communities with different digestive abilities. Such specialization may be possible because of the relative constancy of the leaf resource as opposed to, for example, the variability of fruit crops. A study of this sort would, given the uniqueness of the hoatzin, have only slight relevance to the nutritional ecology of other birds. Tetraonids may provide good subjects for analogous studies of hindgut fermenters. Within ruffed grouse, for example, some southern populations feed on the leaves of the chemically protected mountain laurel (*Kalmia*), while more northerly populations spend part of the winter consuming browse that may be high in other secondary compounds. A relative, the hazel grouse of Europe (*Bonasa bonasia*) consumes large amounts of alder, albeit a different species from that avoided by ruffed grouse. Potentially, some of this dietary variation reflects gut microbe adjustment.

10.2. Dynamic Constraints

1. How quickly can the avian gut microbe community adapt to changes in diet quality? Increased dietary cellulose or urea has been observed to result in increased cellulolytic or ammoniagenic abilities in the cecal biota of domestic turkeys and chickens respectively (Duke et al. 1984, Karasawa et al. 1993b). These changes may have been slow; improvement in cellulose digestion was still evident after turkeys had been on high-fiber diets for several months, and at least some of this change was associated with increase in cellulolytic gut microbes (Bedbury and Duke 1983, Duke et al. 1984). Microbial succession occurs during the colonization of the young chick, and the process has been described to take a month or more (Mead 1989). The effect of this succession on food digestibility may be relatively slight given the composition of chicken feed, but an inability to fully utilize the diets of adult birds until the adult microbiota is established may partially explain the differences in food choice observed between the young and the adults of some wild species. Milton et al. (1993) suggested that part of the reason young ostriches supplement their diet with insects for several weeks following hatching is because they lack the adult gut microbes necessary to fully utilize the adult, herbivorous diet. Protein demands are also likely higher in growing birds. Inocula from young, wild Tetraonidae reportedly had lower cellulolytic activity than inocula from adults (Suomalainen and Arhimo 1945).

The dynamics of the microbial constraints may also be relevant during seasonal changes in food availability. With wild ruminants, winter/summer differences in the density and composition of the microbial community, occurring in parallel with changes in food quality, have been documented (e.g., Hobson et al. 1975/76a,b, Orpin et al. 1985, Holand and Staalland 1992). However, these changes may not represent adaptation in the sense of changes that improve the microbe community's ability to digest a food. Instead, they may reflect semistarvation of the microbial community during winter. The winter collection of microbes is often not distinctly better at digesting the winter foods than the summer; seasonal differences in the microbial community may result from periods of nutrient scarcity rather than adaptation that enhances digestive ability (Giesecke 1970, Cheng et al. 1993). Results consistent with this hypothesis have been obtained for ruffed grouse; summer microbes appeared to have better growth than winter microbes, regardless of whether they were raised on substrates derived from summer or winter foods (Vispo 1995). Work with rock ptarmigan could detect no seasonal effect on cecal SCFA concentrations, although slight differences in production rates may have occurred (Gasaway 1976a). No seasonal difference was noted in the occurrence of the protozoa embedded in the gut wall of willow ptarmigan (Hanssen 1979a). Until more is known about the effect of changed intake on the composition of material arriving to the ceca, these results will be open to alternative interpretations. The extent and anatomical location of cellulose digestion in geese may vary seasonally (Prop and Vulink 1992) and a study of seasonal variation in the celluloytic activity of ingesta from various portions of the goose digestive tract could be informative.

2. How quickly can the avian gut microbe community adapt to a novel toxin? To our knowledge, there has been no explicit study of this topic, at least in wild species. Bryant and Kuropat (1980) suggested that tetraonid consumption of winter foods in autumn is their way of preadapting their gut microbes to the toxins of the winter diet. They cited reports suggesting that individuals denied an adaptation period were unable to survive when placed on a natural winter diet.

11. Conclusions

By way of conclusion, we will outline the type of study that we believe would provide the most information on the role of the gut microbes in avian nutritional ecology. The study's overall objective would be to determine the nutritional contribution of the gut microbes (in terms of energy or nitrogen) to the host, relative to the host's nutrient demands in the wild. This is not an all-encompassing goal—it would not, for example, directly address the role of gut microbes in host health, and the results from a single species cannot be extrapolated to all other species. However, we believe that what is needed now is an intensive study that can establish a model suitable for subsequent refinement and possible extension. Extensive studies, e.g., ones that compare the microbe communities and digestive abilities of several species, may be suitable for the study of ruminant microbial ecology, but this is only because it is safe to assume that the health of the rumen microbes translates directly into benefits to the host. Such an assumption is backed by a wealth of knowledge on ruminant nutrition—knowledge not available for birds. One extensive study that might be useful, if it is practical, would be a survey of the occurrence of coprophagy in birds. The likelihood of nitrogen and vitamin gains from the cecal microbes depends in large part on the occurrence of this behavior.

We would start our intensive research with a species that is easily studied in captivity and in the wild. For the sake of this example, we will choose the turkey. There is already some knowledge of its nutrition, at least in captivity. It exists in reasonable numbers in the wild, affording adequate sample sizes. It does well in captivity and is large enough to permit operations such as cecal cannulation with relative ease and multiple sampling. The fact that it is hunted potentially provides one with easy access to gut samples from wild birds. Our research questions would be the following.

1. **What is the composition of material arriving at ceca?** This would involve work using captive birds with cecal cannulations (Beattie and Shrimpton 1958) fed known diets simulating winter and summer field diets. Dietary compounds could be radiolabeled so that their fates could be more easily followed (e.g., Hume et al. 1993b). This information should be supplemented by information from wild birds so that the wild condition could also be understood, and the laboratory simulation of it checked. Neutral sugar analysis (e.g., Champ et al. 1989, Buxton 1991), coupled with detergent analysis for the sake of comparability, would be our method of choice for trying to describe the fate of the plant structural polysaccharides.
2. **How is this material processed by the cecal microbes?** Again, work would be with cannulated captive birds. Care would have to be taken to preserve the wild microbiota, perhaps by feeding the captive birds natural foods and/or by beginning the experiments soon after capture. Timed, sequential samplings from the ceca of birds fed labeled substrates could document substrate fates. More precise measurements of specific cecal processes could be made for substrates introduced directly into the ceca via the cannula. Not only the fate of labeled polysaccharides but also the fates of labeled secondary compounds and subsequent conjugates could be followed. Endogenous secretions could be assessed in the ceca of gnotobiotic birds (e.g., Lepkovsky et al. 1964). Model incubation systems have been developed that simulate chicken cecal conditions and may be useful in trying to isolate influential pathways of microbial metabolism (Nuotio and Mead 1993).

3. Which nutrients are absorbed from the ceca? At what rates? Introduction of labeled substrates into the ceca, followed by blood samples from the vessels draining the ceca might be appropriate. In parallel, absorption could be measured as disappearance from the lumen.

4. What is the nutritional importance of this absorption to the host in the wild? What are host field N and energy demands, and how do these compare to estimated absorption rates for these substances from the ceca?

5. How does the natural diet vary (between places, between years)? How does this affect the material arriving to the ceca? How does the gut microbe community with its metabolic processes change, and how, in turn, does the host benefit? Such questions are necessary for detecting potential dynamic constraints imposed by the bacteria. New technology using DNA/RNA probes to follow the occurrence of taxonomic groupings of microbes might provide a more detailed understanding of the relevant microbial ecology (Stahl et al. 1988).

This is not a small, elegant study. It combines both laboratory physiology and field nutritional ecology. We must first understand the physiological contribution of the gut microbes and then measure that contribution against the total demands for those nutrients in wild birds. Without incorporating these two aspects, we will not be able to identify the actual gains birds receive from their gut microbes nor refute those that, as our review shows, have already been proposed.

Acknowledgements

We thank D. Schaefer and C. Knab-Vispo for providing comments and corrections on the manuscript, and C. Guglielmo for sharing some of his ideas and data. Portions of this paper are adapted from a recent review by Karasov and Hume (1995).

References

Aguirre AA, Quan TJ, Cook RS, McLean RG (1992) Cloacal flora isolated from wild black-bellied whistling ducks (*Dendrocygna autumnalis*) in Laguna La Nocha, Mexico. Avian Dis 36:459–462.

Allison MJ, Cook HM (1981) Oxalate degradation by microbes of the large bowel of herbivores: the effect of dietary oxalate. Science 212:675–676.

Allison MJ, Reddy CA (1983) Adaptations of gastrointestinal bacteria in response to changes in dietary oxalate and nitrate. In: Klug MJ, Reddy CA, eds. Current Perspectives in Microbial Ecology, pp. 248–256. Washington: American Society for Microbiology.

Anderson GL, Braun EJ (1984) Cecae of desert quail: importance in modifying the urine. Comp Biochem Physiol 78A:91–94.

Andreev AV (1988) Ecological energetics in palaearctic tetraonidae in relation to chemical composition and digestibility of their winter diets. Can J Zool 66:1382–1388.

Barash I, Nitsan Z, Nir I (1993) Adaptation of light-bodied chicks to meal feeding: gastrointestinal tract and pancreatic enzymes. Br Poult Sci 34:35–42.

Barnes EM (1979) The intestinal microflora of poultry and game birds during life and after storage. J Appl Bacteriol 46:407–419.

Barnes EM, Mead GC, Griffiths NM (1973) The microbiology and sensory evaluation of pheasants hung at 5, 10, and 15 °C. Br Poult Sci 14:229–240.

Bayer RC, Chawan CB, Bird FH (1975) Scanning electron microscopy of the chicken crop—the avian rumen? Poult Sci 54:703–707.

Beattie J, Shrimpton DH (1958) Surgical and chemical techniques for in vivo studies of the metabolism of the intestinal microflora of domestic fowls. Q J Exp Physiol 43: 399–407.

Beck JR, Chang TS (1980) In vitro antibiotic activity on cecal anaerobes with emphasis on uric acid-utilizing bacteria. Poult Sci 59:1197–1202.

Bedbury HP, Duke GE (1983) Cecal microflora of turkeys fed low or high fiber diets: enumeration, identification, and determination of cellulolytic activity. Poult Sci 62: 675–682.

Beebe W (1925) The variegated tinamou *Crypturus variegatus variegatus* (Gmelin). Zoologica 6:195–227.

Bell JM (1984) Nutrients and toxicants in rapseed meal: a review. J Anim Sci 58:996–1010.

Bjornhag G (1989) Transport of water and food particles through the avian ceca and colon. J Exp Zool 3(suppl):32–37.

Blair R (1972) Utilisation of ammonium compounds and certain non-essential amino acids by poultry. World Poult Sci J 28:189–202.

Bolton W (1965) Digestion in the crop of the fowl. Br Poult Sci 6:97–102.

Brittingham MC, Temple SA, Duncan RM (1988) A survey of the prevalence of selected bacteria in wild birds. J Wildl Dis 24:299–307.

Bryant JP, Kuropat PJ (1980) Selection of winter forage by subarctic browsing vertebrates: the role of plant chemistry. Annu Rev Ecol Syst 11:261–285.

Buchsbaum R, Wilson J, Valiela I (1986) Digestibility of plant constituents by Canada geese and Atlantic brant. Ecology 67:386–393.

Buxton DR (1991) Digestibility by rumen microorganisms of neutral sugars in perennial forage stems and leaves. Anim Feed Sci Technol 32:119–122.

Campbell CE, Braun EJ (1986) Cecal degradation of uric acid in Gambel quail. Am J Physiol 251:R59-R62.

Campbell EG, Koplin JR (1986) Food consumption, energy, nutrient and mineral balances in a Eurasian kestrel and a screech owl. Comp Biochem Physiol 83A:249–254.

Champ M, Szylit O, Gallant DJ (1981) The influence of microflora on the breakdown of maize starch granules. Poult Sci 60:179–187.

Champ M, Barry JL, Hoebler C, Delort-Laval J (1989) Digestion and fermentation pattern of various dietary fiber sources in the rat. Anim Feed Sci Technol 23:195–204.

Chaplin SB (1989) Effect of cecetomy on water and nutrient absorption of birds. J Exp Zool 3(suppl):81–86.

Cheng K-J, McAllister TA, Mathiesen SD, Blix AS, Orpin CG, Costerton JW (1993) Seasonal changes in the adherent microflora of the rumen in high-Arctic Svalbard reindeer. Can J Microbiol 39:101–108.

Chesson A, Forsberg CW (1989) Polysaccharide degradation by rumen microorganisms. In: Hobson PN, ed. The Rumen Microbial Ecosystem, pp. 251–284. London: Elsevier Applied Science.

Clemens ET, Stevens CE, Southworth M (1975) Sites of organic acid production and pattern of digesta movement in gastrointestinal tract of geese. J Nutr 105:1341–1350.

Clench MH, Mathias JR (1995) The avian cecum: a review. Wilson Bull 107:93–121.

Coates ME, Ford JE, Harrison GF (1968) Intestinal synthesis of vitamins of the B complex in chicks. Br J Nutr 22:493–500.

Coates ME, Fuller R (1977) The gnotobiotic animal in the study of gut microbiology. In: Clarke RTJ, Bauchop T, eds. Microbial Ecology of the Gut, pp. 185–222. London: Academic Press.

Cork SJ, Foley WJ (1991) Digestive and metabolic strategies of arboreal mammalian folivores in relation to chemical defenses in temperate and tropical forests. In: Palo RT, Robbins CT, eds. Plant Defenses Against Mammalian Herbivory, pp. 133–166. Boca Raton, Fla: CRC Press.

Costerton JW, Cheng K-J, Geesey GG, et al. (1987) Bacterial biofilms in nature and disease. Annu Rev Microbiol 41:435–464.

Dawson TJ (1989) Food utilization in relation to gut structure and function in wild and domestic birds and mammals. Acta Vet Scand 86(suppl):20–27.

Dawson TJ, Herd RM (1983) Digestion in the emu: low energy and nitrogen requirements of this large ratite bird. Comp Biochem Physiol 75A:41–45.

Dawson TJ, Johns AB, Beal AM (1989) Digestion in the Australian wood duck (*Chenonetta jubata*): a small avian herbivore showing selective digestion of the hemicellulose component of fiber. Physiol Zool 62:522–540.

Diamond AW, Place AR (1988) Wax digestion by black-throated honey-guides *Indicator indicator*. Ibis 130:557–560.

Dominguez-Bello MG, Lovera M, Suarez P, Michelangeli F (1993a) Microbial digestive symbionts of the crop of the hoatzin (*Opisthocomus hoazin*): an avian foregut fermenter. Physiol Zool 66:374–383.

Dominguez-Bello MG, Ruiz MC, Michelangeli F (1993b) Evolutionary significance of foregut fermentation in the hoatzin (*Opisthocomus hoazin*; Aves: Opisthocomidae). J Comp Physiol B 163:594–601.

Drobney RD (1984) Effect of diet on visceral morphology of breeding wood ducks. Auk 101:93–98.

Duke GE, Eccleston E, Kirkwood S, Louis CF, Bedbury HJ (1984) Cellulose digestion by domestic turkeys fed low or high fiber diets. J Nutr 114:95–102.

Ehle FR, Robertson JB, Van Soest PJ (1981) Influence of dietary fibers on fermentation in the human large intestine. J Nutr 112:158–166.

Farner DS (1960) Digestion and the digestive system. In: Marshall AJ, ed. Biology and Comparative Physiology of Birds, pp. 411–467. New York: Academic Press.

Featherstone WR, Bird HR, Harper AE (1962) Effectiveness of urea and ammonium nitrogen for the synthesis of dispensable amino acids by the chick. J Nutr 78:198–206.

Fenna L, Boag DA (1974) Adaptive significance of the caeca in Japanese quail and spruce grouse (Galliformes). Can J Zool 52:1577–1584.

Friedmann H, Kern J (1956) The problem of cerophagy or wax-eating in the honey guides. Q Rev Biol 31:19–30.

Friedmann H, Kern J, Rust JH (1957) The domestic chick: a substitute for the honey-guide as a symbiont with cerolytic microorganisms. Am Nat 91:321–325.

Fuller R (1977) The importance of lactobacilli in maintaining normal microbial balance in the crop. Br Poult Sci 18:85–94.

Fuller R (1984) Microbial activity in the alimentary tract of birds. Proc Nutr Soc 43: 55–61.

Fuller R, Turvey A (1971) Bacteria associated with the intestinal wall of the fowl (*Gallus domesticus*). J Appl Bacteriol 34:617–622.

Fuller R, Coates ME (1983) Influence of the intestinal microflora on nutrition. In: Freeman BM, ed. Physiology and Biochemistry of the Domestic Fowl, Vol. 4, pp. 51–61. London: Academic Press.

Furuse M, Yokota H (1984) Protein and energy utilization in germ-free and conventional chicks given diets containing different levels of dietary protein. Br J Nutr 51:255–264.

Furuse M, Yokota H (1985) Effect of the gut microflora on chick growth and utilisation of protein and energy at different concentrations of dietary protein. Br Poult Sci 26: 97–104.

Gasaway WC (1976a) Seasonal variation in diet, volatile fatty acid production and size of the cecum of rock ptarmigan. Comp Biochem Physiol 53A:109–114.

Gasaway WC (1976b) Volatile fatty acids and metabolizable energy derived from caecal fermentation in the willow ptarmigan. Comp Biochem Physiol 53A:115–121.

Gasaway WC (1976c) Cellulose digestion and metabolism by captive rock ptarmigan. Comp Biochem Physiol 54A:179–182.

Gasaway WC, Holleman DF, White RG (1975) Flow of digesta in the intestine and cecum of the rock ptarmigan. Condor 77:467–474.

Giesecke D (1970) Comparative microbiology of the alimentary tract. In: Phillipson AT, ed. Physiology of Digestion and Metabolism in the Ruminant, pp. 306–318. Newcastle-upon-Tyne; Oriel Press.

Grajal A, Strahl SD, Parra R, Dominguez MG, Neher A (1989) Foregut fermentation in the hoatzin, a Neotropical leaf-eating bird. Science 245:1236–1238.

Green S, Bertrand SL, Duron MJC, Maillard R (1987) Digestibility of amino acid in maize, wheat and barley meals, determined with intact and caecectomised cockerels. Br Poult Sci 28:631–641.

Guglielmo CG, Karasov WH (1995) Nutritional quality of winter browse for ruffed grouse. J Wildl Manage. 59:427–436.

Guglielmo CG, Karasov WH, Jakubas WJ (1993) Nutritional costs of a plant secondary metabolite to an avian herbivore. Bull of the Ecol Soc Am 74(2):260.

Hanlan ET (1949) The architecture of the avian gut and tolerance of crude fiber. Br J Nutr 3:245–252.

Hanssen I (1979a) Micromorphological studies on the small intestine and caeca in wild and captive willow grouse (*Lagopus lagopus lagopus*). Acta Vet Scand 20:351–364.

Hanssen I (1979b) A comparison of the microbiological conditions in the small intestine and caeca of wild and captive willow grouse (*Lagopus lagopus lagopus*). Acta Vet Scand 20:365–371.

Hanssen I, Grammeltvedt R, Hellemann A-L (1984) Effects of different diets on viability, and gut morphology and bacteriology in captive willow ptarmigan chicks (*Lagopus l. lagopus*) Acta Vet Scand 25:67–75.

Hatfield RD (1989) Structural polysaccharides in forages and their degradability. Agron J 81:39–46.

Herd RM, Dawson TJ (1984) Fiber digestion in the emu, *Dromaius novaehollandiae*, a large bird with a simple gut and high rates of passage. Physiol Zool 57:70–84.

Hobson PN, Mann SO, Summers R (1975/76a) Rumen micro-organisms in red deer, hill sheep and reindeer in the Scottish highlands. Proc R Soc Edinb 75:171–180.

Hobson PN, Mann SO, Summers R, Staines BW (1975/76b) Rumen function in red deer, hill sheep and reindeer in the Scottish highlands. Proc R Soc Edinb 75:181–198.

Holand O, Staalland H (1992) Nutritional strategies and winter survival of European roe deer in Norway. In: Brown RD, ed. The Biology of Deer, pp. 423–428. New York: Springer Verlag.

Holdeman LV, Good IJ, Moore WEC (1976) Human fecal flora: variation in bacterial composition within individuals and a possible effect of emotional stress. Appl Environ Microbiol 31:359–375.

Hornicke H, Bjornhag G (1980) Coprophagy and related strategies for digesta utilization. In: Ruckebusch Y, Thivend P, eds. Digestive Physiology and Metabolism in Ruminants, pp. 707–730. Lancaster: MTP Press.

Hudson WM (1920) Birds of La Plata. London: JM Dent and Sons Ltd.

Hume ID, Karasov WH, Darken BW (1993a) Acetate, butyrate and proline uptake in the caecum and colon of prairie voles (*Microtus ochrogaster*). J Exp Biol. 176:285–297.

Hume ME, Beier RC, Hinton A Jr, et al. (1993b) In vitro metabolism of radiolabeled carbohydrates by protective cecal anaerobic bacteria. Poult Sci 72:2254–2263.

Ilundain A, Naftelin RJ (1981) Na^+-dependent co-transport of alpha-methyl-D-glucose across the mucosal border of rabbit descending colon. Biochim Biophys Acta 644: 316–322.

Inman DL (1973) Cellulose digestion in ruffed grouse, chukar partridge and bobwhite quail. J Wildl Manage 37:114–121.

Ivorec-Szylit O, Mercier C, Raibuad P, Calet C. 1965. Contribution a l'etude de la degradation des glucides dans le jabot du coq. Influence du taux de glucose du regime sur l'utilization de l'amidon. C R Hebd Seanc Acad Sci Paris 261:3201–3203.

Jackson S, Place AR, Seiderer LJ (1992) Chitin digestion and assimilation by seabirds. Auk 109:758–770.

Johns DC, Low CK, Sedcole JR, James KAC (1986) Determination of amino acid digestibility using caecectomised and intact adult cockerels. Br Poult Sci 27:451–461.

Jones RJ, Megarrity RG (1986) Successful transfer of DHP-degrading bacteria from Hawaiian goats to Australian ruminants to overcome the toxicity of *Leucaena*. Aust Vet J 63:259–262.

Karasawa Y (1989) Effect of colostomy on nitrogen nutrition in the chicken fed a low protein diet plus urea. J Nutr 119:1388–1391.

Karasawa Y, Maeda M (1992) Effect of colostomy on the utilisation of dietary nitrogen in the fowl fed on a low protein diet. Br Poult Sci 33:815–820.

Karasawa Y, Ono T, Koh K (1993a) Relationship of decreased caecal urease activity by dietary penicillin to nitrogen utilisation in chickens fed on a low protein diet plus urea. Br Poult Sci 35:91–96.

Karasawa Y, Umemoto M, Koh K (1993b) Effect of dietary protein and urea on in vitro caecal ammonia production from urea and uric acid in cockerels. Br Poult Sci 34: 711–714.

Karasov WH (1995) Digestive plasticity in avian energetics and feeding ecology. In: Casey C, ed. Avian Energetics and Nutritional Ecology, pp. 61–84. New York: Chapman & Hall.

Karasov WH, Hume ID (1997) Vertebrate gastrointestinal system. In: Dantzler WH, ed. Handbook of Comparative Physiology. Oxford: Oxford University Press. In press.

Kaufmann W, Hagemeister H, Dirksen G (1980) Adaptation to changes in dietary composition, level and frequency of feeding. In: Ruckebush Y, Thivend P, eds. Digestive Physiology and Metabolism in Ruminants, pp. 587–602. Westport, Conn: AVI Publishing Company Inc.

Kay RNB, Engelhardt W, White RG (1980) The digestive physiology of wild ruminants. In: Ruckebusch Y, Thivend P, eds. Digestive Physiology and Metabolism in Ruminants, pp. 743–761. Westport, Conn: AVI Publishing Company.

Kunstyr I (1974) Some quantitative and qualitative aspects of the stomach microflora of the conventional rat and hamster. Zbl Vet Med A21:553–561.

Latymer EA, Coates ME (1981) The influence of microorganisms and of stress on the chick's requirement for pantothenic acid. Br J Nutr 45:441–449.

Leopold AS (1953) Intestinal morphology of gallinaceous birds in relation to food habits. J Wildl Manage 17:197–203.

Lepkovsky S, Wagner M, Furuta F, Ozone K, Koike T (1964) The proteases, amylase and lipase of the intestinal contents of germfree and conventional chickens. Poult Sci 43:722–726.

Lewin V (1963) Reproduction and development of young in a population of California quail. Condor 65:249–278.

Livezey, BC (1993) An ecomorphological review of the dodo (*Raphus cucullatus*) and solitaire (*Pezophaps solitaria*), flightless Columbiformes of the Mascarene Islands. J Zool (Lond) 230:247–292.

Mackie RI (1987) Microbial digestion of forages in herbivores. In: Hacker JB, Ternouth JH, eds. The Nutrition of Herbivores, pp. 233–265. Sydney: Academic Press.

Majak W (1991) Metabolism and absorption of toxic glycosides by ruminants. J Range Manage 45:67–71.

Marriott RW, Forbes DK (1970) The digestion of lucerne chaff by Cape Barren geese, *Cereopsis novaehollandiae* Latham. Aust J Zool 18:257–263.

Mattocks JG (1971) Goose feeding and cellulose digestion. Wildfowl 22:107–113.

McBee RH (1977) Fermentation in the hindgut. In: Clarke RTJ, Bauchop T, eds. Microbial Ecology of the Gut, pp. 185–222. London; Academic Press.

McBee RH, West GC (1969) Cecal fermentation in the willow ptarmigan. Condor 71: 54–58.

McClure HE, Eveland WC, Kase A (1957) The occurrence of certain Enterobacteriaceae in birds. Am J Vet Res 18:207–209.

McLelland J (1979) Digestive system. In: King AS, McLelland J, eds. Form and Function in Birds, pp. 69–181. London: Academic Press.

McLelland J (1989) Anatomy of the avian cecum. J Exp Zool 3(suppl):2–9.

McNab JM (1973) The avian caeca: a review. World Poult Sci J 29:251–263.

Mead GC (1989) Microbes of the avian cecum: types present and substrates utilized. J Exp Zool 3(suppl):48–54.

Michaelsen S, Otte J, Simonsen LO, Sorensen H (1994) Absorption and degradation of individual intact glucosinolates in the digestive tract of rodents. Acta Agric Scand 44: 25–37.

Miller MR (1976) Cecal fermentation in mallards in relation to diet. Condor 78:107–111.

Milton SJ, Dean WRJ, Linton A (1993) Consumption of termites by captive ostrich chicks. S Afr J Anim Sci 23:58–60.

Mitchell PC (1896) On the intestinal tract of birds. Proc Zool Soc 1896:136–159.

Moore WEC, Holdeman LV (1975) Discussion of current bacteriological investigations of the relationships between intestinal flora, diet, and colon cancer. Cancer Res 35: 3418–3420.

Moreto M, Amat C, Puchal A, Buddington RK, Planas JM (1991) Transport of L-proline and alpha-methyl-D-glucoside by chicken proximal cecum during development. Am J Physiol 260:G457-G463.

Mortensen A (1984) Importance of microbial nitrogen metabolism in the ceca of birds. In: Klug MJ, Reddy CA, eds. Current Perspectives in Microbial Ecology, pp. 273–278. Washington DC: American Society for Microbiology.

Mortensen A, Tindall AR (1981a) Caecal decomposition of uric acid in captive and free ranging willow ptarmigan (*Lagopus lagopus lagopus*). Acta Physiol Scand 111: 129–133.

Mortensen A, Tindall AR (1981b) On caecal synthesis and absorption of amino acids and their importance for nitrogen recycling in willow ptarmigan (*Lagopus lagopus lagopus*). Acta Physiol Scand 113:465–469.

Morton ES (1978) Avian arboreal folivores: why not? In: Montgomery GG, ed. The Ecology of Arboreal Folivores, pp. 123–130. Washington DC: Smithsonian Institution.

Moss R (1977) The digestion of heather by red grouse during the spring. Condor 79: 471–477.

Moss R (1989) Gut size and the digestion of fibrous diets by tetraonid birds. J Exp Zool 3(suppl):61–65.

Moss R, Hanssen I (1980) Grouse nutrition. Nutr Abstr Rev Series B 50:555–567.

Moss R, Parkinson JA (1972) The digestion of heather (*Calluna vulgaris*) by red grouse (*Lagopus lagopus scoticus*). Br J Nutr 27:285–298.

Moss R, Parkinson JA (1975) The digestion of bulbils (*Polygonum viviparum* L.) and berries (*Vaccinium myrtillus* L. and *Empetrum* sp.) by captive ptarmigan (*Lagopus mutus*). Br J Nutr 33:197–206.

Muramatsu T, Takasu O, Okumura J (1993) Research note: fructose feeding increases lower gut weights in germ-free and conventional chicks. Poult Sci 72:1597–1600.

Naik DR, Dominic CJ (1962) The intestinal caeca of some Indian birds in relation to food habits. Naturwissenschaften 49:287.

Nitsan Z, Alumot E (1963) Role of the cecum in the utilization of raw soybean in chicks. J Nutr 80:299–304.

Nordal J, Dahle HK (1971) Meat quality of uneviscerated ptarmigans (*Lagopus mutus* Montin) after extended storage. Acta Agric Scand 21:172–177.

Nuotio L, Mead GC (1993) An in vitro model for studies on bacterial interactions in the avian caecum. Lett Appl Microbiol 17:65–67.

Nurmi E, Rantala T (1973) New aspects of *Salmonella* infection in broiler production. Nature 241:210–211.

Obst BS, Diamond JM (1989) Interspecific variation in sugar and amino acid transport in the avian cecum. J Exp Zool 3(suppl):117–126.

Oliver WRB (1955) New Zealand Birds. Wellington, N.Z.: A.H. and A.W. Reed.

Orpin CG, Mathiesen SD, Greenwood Y, Blix AS (1985) Seasonal changes in the ruminal microflora of the high-arctic Svalbard reindeer (*Rangifer tarandus platyrhynchus*). Appl Environ Microbiol 50:144–147.

Owens FN, Zinn R (1988) Protein metabolism of ruminant animals. In: Church DC, ed. The Ruminant Animal. Digestive Physiology and Nutrition, pp. 227–249. Englewood Cliffs, N.J.: Prentice-Hall.

Parra R (1978) Comparison of foregut and hindgut fermentation in herbivores. In: Montgomery GG, ed. The Ecology of Arboreal Folivores, pp. 205–229. Washington: Smithsonian Institution.

Parsons CM (1986) Determination of digestible and available amino acids in meat meal using conventional and caecectomized cockerels or chick growth assays. Br J Nutr 56: 227–240.

Pendergast BA, Boag DA (1971) Nutritional aspects of the diet of spruce grouse in central Alberta. Condor 73:437–443.

Pivnyak IG, Konyakhin AN (1973) Biological value for chickens of live carotene-forming cultures. Nutr Abstr Rev 43:330. Abstract.

Planas JM, Ferrer R, Moretó M (1987) Relation between alpha-methyl-D-glucoside influx and brush border surface area in enterocytes from chicken cecum and jejunum. Pflugers Archiv 408:515–519.

Porten RE Jr, Holt PS (1993) Effect of induced molting on the severity of intestinal lesions caused by *Salmonella enteritidis* infection in chickens. Avian Dis 37:1009–1016.

Pritchard PJ (1972) Digestion of sugars in the crop. Comp Biochem Physiol 43A:195–205.

Prop J, Vulink T (1992) Digestion by barnacle geese in the annual cycle: the interplay between retention time and food quality. Functional Ecol 6:180–189.

Raa J, Moen P, Steen JB (1976) A nutrition dependent antimicrobial principle in tissue of willow ptarmigan (*Lagopus lagopus*). J Sci Food Agric 27:773–776.

Remington TE (1989) Why do grouse have ceca? A test of the fiber digestion theory. J Exp Zool 3(suppl):87–94.

Ridpath MG (1972) The Tasmanian native hen, *Tribonyx mortierii*. III. Ecology. CSIRO Wildl Res 17:91–118.

Robbins CT (1983) Wildlife Feeding and Nutrition. Orlando, Fla: Academic Press.

Salter DN, Coates ME, Hewitt D (1974) The utilization of protein and excretion of uric acid in germ-free and conventional chicks. Br J Nutr 31:307–318.

Sandine WE, Muralidhara KS, Elliker PR, England DC (1972) Lactic acid bacteria in food and health: a review with special reference to enteropathogenic *Escherichia coli* as well as certain enteric diseases and their treatment with antibiotics and lactobacilli. Milk Food Technol 35:691–702.

Savage DC (1970) Associations of indigenous microorganisms with gastrointestinal mucosal epithelia. Am J Clin Nutr 23:1495–1501.

Schales K, Gerlach H, Kosters J (1993) Investigations on the aerobic flora and *Clostridium perfringens* in fecal specimens from free-living and captive capercaillies (*Tetrao urogallus* L., 1758). J Vet Med B 40:469–477.

Schneitz C, Nuotio L (1992) Efficacy of different microbial preparations for controlling *Salmonella* colonisation in chicks and turkey poults by competitive exclusion. Br Poult Sci 33:207–211.

Sedinger JS, White RG, Hupp J (1995) Metabolizability and partitioning of energy and protein in green plants by yearling lesser snow geese. Condor 97:116–122.

Shetty S, Sridhar KR, Shenoy KB, Hegde SN (1990) Observations on bacteria associated with pigeon crop. Folia Microbiol 35:240–244.

Sibbald JR (1982) Measurement of bioavailable energy in poultry feedingstuffs: a review. Can J Anim Sci 62:983–1048.

Sieburth JM (1959) Gastrointestinal microflora of Antarctic birds. J Bacteriol 77:521–531.

Sieburth JM (1960) Acrylic acid, an "antibiotic" principle in *Phaecystis* blooms in Antarctic waters. Science 132:676–677.

Sieburth JM, Gutierrez J, McGinnis J, Stern JR, Schneider BH (1951) Effect of antibiotics on intestinal microflora and on growth of turkeys and pigs. Proc Soc Exp Biol Med 76:15–18.

Sipes IG, Gandolfi AJ (1986) Biotransformation of toxicants. In: Klaassen CD, Amdur MO, Doull J, eds. Casarett and Doull's Toxicology, 3rd ed., pp. 64–98. New York: Macmillan.

Skadhauge E (1981) Osmoregulation in Birds. Berlin: Springer.

Slominski BA, Campbell LD, Slanger NE (1987) Influence of cecectomy and dietary antibiotics on the fate of ingested intact glucosinolates in poultry. Can J Anim Sci 67: 1117–1124.

Slominski BA, Campbell LD, Slanger NE (1988) Extent of hydrolysis in the intestinal tract and potential absorption of intact glucosinolates in laying hens. J Sci Food Agric 42:305–314.

Soucek Z, Mushin R (1970) Gastrointestinal bacteria of certain Antarctic birds and mammals. Appl Microbiol 20:561–566.

Smith GS (1992) Toxification and detoxification of plant compounds by ruminants: an overview. J Range Manage 45:25–30.

Stahl DA, Flesher B, Mansfield HR, Montgomery L (1988) Use of phylogenetically based hybridization probes for studies of ruminal microbial ecology. Appl Environ Microbiol 54:1079–1084.

Stevens CE (1977) Comparative physiology of the digestive system. In: Swenson MJ, ed. Duke's Physiology of Domestic Animals, pp. 216–232. Ithaca, NY: Cornell University Press.

Strong TR, Reimer PR, Braun EJ (1989) Avian cecal microanatomy: a morphometric comparison of two species. J Exp Zool 3(suppl):10–20.

Sturkie PD (1965) Avian Physiology, 2nd ed. Ithaca, NY: Comstock Publishing Associates.

Sunde ML, Cravens WW, Elvehjem CA, Halpin JG (1950) The effect of diet and cecectomy on the intestinal synthesis of biotin in mature fowl. Poult Sci 29:10–14.

Suomalainen H, Arhimo E (1945) On the microbial decomposition of cellulose by wild gallinaceous birds (family Tetraonidae). Ornis Fenn 22:21–23.

Swart D, Mackie RI, Hayes JP (1987) For feathers and leathers. Nuclear Active 36:2–9.

Swart D, Mackie RI, Hayes JP (1993a) Influence of live mass, rate of passage and site of digestion on energy metabolism and fibre digestion in the ostrich (*Struthio camelus* var. *domesticus*). S Afr J Anim Sci 23:119–126.

Swart D, Mackie RI, Hayes JP (1993b) Fermentative digestion in the ostrich (*Struthio camelus* var. *domesticus*), a large avian species that utilizes cellulose. S Afr J Anim Sci 23:127–135.

Thompson DC, Boag DA (1975) Role of the caeca in Japanese quail energetics. Can J Zool 53:166–170.

Van Soest PJ (1982) Nutritional Ecology of the Ruminant. Ithaca, NY: Cornell University Press.

Varel VH, Pond WG, Pekas JC, Yen JT (1982) Influence of high-fiber diet on bacterial populations in the gastrointestinal tracts of obese- and lean-genotype pigs. Appl Environ Microbiol 4:107–112.

Vispo CR (1995) Winter nutrional ecology of ruffled grouse (*Bonasa umbellino* and aspects of digestive fermentation in this species and two other northern herbivores. PhD dissertation University of Wisconsin, Madision.

Wallace RJ, Cotta MA (1989) Metabolism of nitrogen-containing compounds. In: Hobson PN, ed. The Rumen Microbial Ecosystem, pp. 217–249. London: Elsevier Applied Science.

Weimer PJ, Lopez-Guisa JM, French AD (1990) Effect of cellulose fine structure on kinetics of its digestion by mixed ruminal microorganisms in vitro. Appl Environ Microbiol 56:2421–2429.

Wetmore A (1926) Observations on the birds of Argentina, Paraguay, Uruguay, and Chile. Washington DC: Smithsonian Institution.

Wheelwright NT (1983) Fruits and ecology of the resplendent quetzals. Auk 100: 286–301.

Williams GR (1956) The kakapo (*Strigops habroptilus*, Gray): a review and re-appraisal of a near extinct species. Notornis 7:29–56.

Ziswiler V, Farner DS (1972) Digestion and the digestive system. In: Farner DS, King JR, eds. Avian Biology, Vol. II, pp. 343–430. New York: Academic Press.

6

Fermentation And Gastrointestinal Microorganisms In Fishes

Kendall D. Clements

1. Introduction

Microbial fermentation and nutrient synthesis are typically important in organisms with a diet high in fiber (Stevens 1988)—i.e., a diet mainly composed of carbohydrates resistant to endogenous digestive enzymes (Annison 1993). Fermentative digestion thus occurs typically in animals with a diet composed predominantly of plant material (Bergman 1990), and symbioses with microorganisms have been well studied in herbivorous mammals, birds, and reptiles (Stevens 1988). It is therefore surprising that the endosymbiotic communities of the dominant aquatic vertebrate herbivores, the fishes, remain poorly understood. Only recently have diverse microbial communities been reported from the guts of herbivorous fishes (Fishelson et al. 1985, Rimmer and Wiebe 1987, Clements et al. 1989, Clements 1991a), and almost nothing is known of the role of these symbioses in digestion. As a result of this, the material covered in this review will be somewhat different from that covered in other chapters.

The outline of this chapter is as follows. Previous reviews encompassing the subject of fermentative digestion and gastrointestinal microbes in fishes will be examined, and in the process some of the themes that have directed research in this area will be identified. The anatomy and physiology of the gastrointestinal tract of fishes is covered elsewhere, but this work will be briefly discussed to familiarize the reader with the phylogenetic and anatomical diversity of fishes. It will become apparent that although the majority of fish species studied in a physiological context belong to carnivorous, northern-hemisphere taxa, the vast majority of fishes in which fermentation is likely to be important are herbivorous species found in the tropics or the southern hemisphere. The third section of this review will deal with the diversity of endosymbionts found in fishes. Much of this work is purely descriptive and without systematic rigor, yet it serves to indicate directions for future work. The transfer of symbiotic microorganisms between host generations will also be discussed in this section. The fourth section of this chapter will discuss studies of digestive enzyme assays in fishes. Many of these studies seek to assess the importance of exogenous enzymes by the isolated culture of gastrointestinal microorganisms, and rely upon analogy to the characteristic symbioses of terrestrial vertebrates. The utility of this approach will be assessed. The fifth section of this review will cover fermentation in fishes, principally through a description of the levels of short-chain fatty acids (SCFA) in a number of species. The sixth section of the review will discuss these SCFA data in terms of the physiology and diet of the host.

The gastrointestinal biota of fishes has been reviewed from a number of different perspectives in the past, these perspectives naturally reflecting the interests of the reviewer concerned. Physiological reviews such as McBee (1977) and Stevens (1988) and microbiological reviews such as Cahill (1990) and Moriarty (1990) focused on studies of exogenous enzyme activity in fishes. While some of these reviewers suggested that endosymbiotic bacteria may contribute to the nutrition of the host through the supply of enzymes and vitamins, the importance of fermentation to the energy requirements of the host was not considered. McBee (1977) concluded that most fish did not possess the anatomical complexity required for microbial fermentation of cellulose in the gut. Sakata (1990) described the results of a number of studies involving the culture of microbes from fish guts, and considered that anaerobic bacteria contributed to the nutrition of the host fish through the supply of (volatile) fatty acids and vitamins. All of these reviews indicate the fragmentary nature of work in this area, a situation well illustrated by the fact that Cahill's 1990 review of the bacterial biota of fishes does not contain a single reference to the gastrointestinal microbiota of marine herbivorous fishes.

Some recent reviews of fish nutrition have ignored the subject of a gastrointestinal microbiota (e.g., Lovell 1989), while others have hinted at its potential importance (e.g., Helpher 1988). The need for work on the role of gastrointestinal symbionts in fishes has been stressed mainly by biologists interested in marine herbivorous fishes (Horn 1989, 1992; Choat 1991), although until very recently it was assumed that fishes generally lacked a consistent gut microbiota capable of degrading refractory carbohydrates (Montgomery and Gerking 1980). Horn (1989) rated fish/microbe symbioses of particular importance for future research, while Choat (1991) noted our lack of understanding of the biochemical roles of gastrointestinal microbes in fishes. Horn (1992) concluded that microorganisms were likely to become increasingly recognized as being involved in the digestive processes of herbivorous fishes.

The current situation seems to be a lack of communication. While fish ecologists acknowledge the need for an understanding of fermentative processes in fishes, the answers to the major questions lie within the realms of physiology, biochemistry, and microbiology. Those with the skills needed to answer these questions remain unaware of the problem, or lack access to suitable study species. An excellent recent review of energy metabolism in fishes (Christiansen and Klungsøyr 1987) illustrates this point: our knowledge of fish physiology is based largely on studies conducted on carnivorous northern-hemisphere species, and most concern freshwater taxa such as salmonids. A major aim of this review is to bridge the gap between biologists familiar with the physiology and biochemistry of gastrointestinal symbioses in terrestrial herbivores, on the one hand, and biologists familiar with the diversity and ecology of herbivorous fishes, on the other. Attention will therefore focus on those fish taxa most likely to be involved in mutualistic symbioses with gastrointestinal microbes, the marine herbivorous fishes discussed by Horn (1989, 1992) and Choat (1991). Hopefully it will be seen that these aquatic herbivores are as diverse and specialized as their terrestrial counterparts, and are thus worthy of study in their own right.

2. Anatomy and Physiology of the Gastrointestinal Tract

The anatomy and physiology of the feeding apparatus and digestive tract of fishes in general has been dealt with elsewhere in considerable detail (e.g., Kapoor et al. 1975, Fänge and Grove 1979, Helpher 1988, Stevens 1988, Lovell 1989). However, many of the fish taxa important in the context of gastrointestinal symbioses are not well represented in these general accounts. It is these taxa that will be examined in some more detail here.

The anatomy of the gastrointestinal tract is an important consideration in any assessment of microbial fermentation in animals. The low nutritional value of diets high in fiber requires the ingestion and processing of a large volume of food (Stevens 1988). The extra time necessary for microbial digestion of this refractory material requires the retention of food within the gut for extended periods of time. Furthermore, the ability of a herbivore to triturate ingested food influences the rate and efficiency of fermentation (Bjorndal et al. 1990). The ability of fishes to triturate, contain, and retain plant material will therefore have a major bearing on the possible role of gastrointestinal microbes in digestion. These factors will be discussed below.

2.1. Jaws and Teeth

Fishes display tremendous diversity in the form of the skull and jaw elements (Gregory 1933). Most marine herbivorous fishes belong to the order Perciformes, the perchlike fishes. Lists of herbivorous marine species are given in Horn (1989), and genera of tropical marine herbivorous fishes are given in Choat (1991). An examination of these lists of taxa indicates that most species are contained in a small number of families, which are characterized by a number of morphological and structural features (Choat 1991). Herbivorous fishes typically have a small gape, which is suitable for taking rapid, small bites (Choat 1991).

Fish have three kinds of teeth: mandibular, buccal, and pharyngeal (Helpher 1988). Only the mandibular and pharyngeal teeth are important in herbivorous fishes, which do not need to prevent the escape of prey from the mouth. Herbivorous fishes typically have a row of small, closely spaced mandibular teeth which serve to crop plant material or scrape it from the substratum. These teeth are analogous to the incisors of a ruminant, serving to harvest the food without chopping or grinding it. In some groups, notably the Scaridae (parrotfishes) and the Odacidae (butterfish or herring cale), the mandibular teeth have become fused into a parrotlike beak (Clements and Bellwood 1988). This beak serves much the same function as the incisiform teeth of other herbivorous taxa. Other groups of herbivorous fish, such as the freshwater Cyprinidae (carps, goldfishes, etc.), have no teeth in the jaws.

Pharyngeal teeth have been developed to a great extent in some groups of herbivorous fish (Horn 1989, 1992). They are situated on the modified fifth gill arch, which does not carry gills. In cyprinids, pharyngeal teeth can be sharp to cut the food, or rounded to crush the food (Helpher 1988). In some fishes the pharyngeal apparatus has been modified to form a fully functional second set of jaws (Liem and Greenwood 1981). In the herbivorous odacids the pharyngeal apparatus consists of opposable, ridged bones capable of chopping ingested algae into small pieces (Clements and Bellwood 1988). The ultimate development of the pharyngeal apparatus occurs in scarids (Bellwood and Choat 1990). These fishes ingest a considerable amount of inorganic material with their algal diet, and the mixture is triturated to a fine paste between the broad, flat dentigerous surfaces of the pharyngeal jaws.

In summary, the oral jaws contribute little to food preparation in herbivorous fishes. However, the development of the pharyngeal apparatus in some taxa enables ingested plant material to be chopped, shredded, or triturated. Clearly the degree to which ingested food is processed before it enters the stomach depends upon the species involved. For example, the pharyngeal apparatus in aplodactylids (sea carp, marblefish) does little more than convey the food to the stomach, while in scarids it is probably capable of disrupting algal cell walls.

2.2. Stomach

The stomach in fishes can be a straight tube, U-shaped, or Y-shaped with a gastric cecum (Fänge and Grove 1979, Helpher 1988, Stevens 1988). Flow of digesta from the stomach into the intestine is controlled by a muscular valve or sphincter, a fold in the mucous membrane, or both (Stevens 1988). In fishes such as mugilids (mullet) and some acanthurids (surgeonfish), the circular muscle of the stomach is well developed (Figs. 6.1 and 6.2), forming a gizzardlike structure capable of lysing the cells of ingested diatoms and bacteria (Payne 1978, Lobel 1981, Horn 1992). A stomach is absent in many herbivorous fishes, including cyprinids, odacids, and scarids (Figs. 6.3 and 6.4). These stomachless fishes typically have a pharyngeal apparatus for processing food before it enters the intestine (Fänge and Grove 1979, Stevens 1988).

2.3. Intestine

The intestinal tract of fishes varies greatly in both length and arrangement (Fänge and Grove 1979; Stevens 1988). There is no direct correlation between length and feeding habits, although herbivorous species tend to have longer intestines. The intestine can vary from a short, relatively straight tube to a complex of spirals and loops. The length and arrangement of the intestine in fishes is discussed in detail elsewhere (Al-Hussaini 1947, Gohar and Latif 1959, Jones 1968, Mok 1977, Zihler 1982, Hofer 1988, Horn 1989), and examples of 12 species of marine herbivorous fishes are given in the figures.

Many teleost species have from one to 1,000 blind tubes connected with the anterior end of the intestine (Figs. 6.1, 6.2, 6.5 to 6.12). These structures are

Gastrointestinal tracts of marine herbivorous fishes. Anterior end of gut is at top.

Figure 6.1. *Ctenochaetus striatus*, family Acanthuridae. Note muscular stomach.

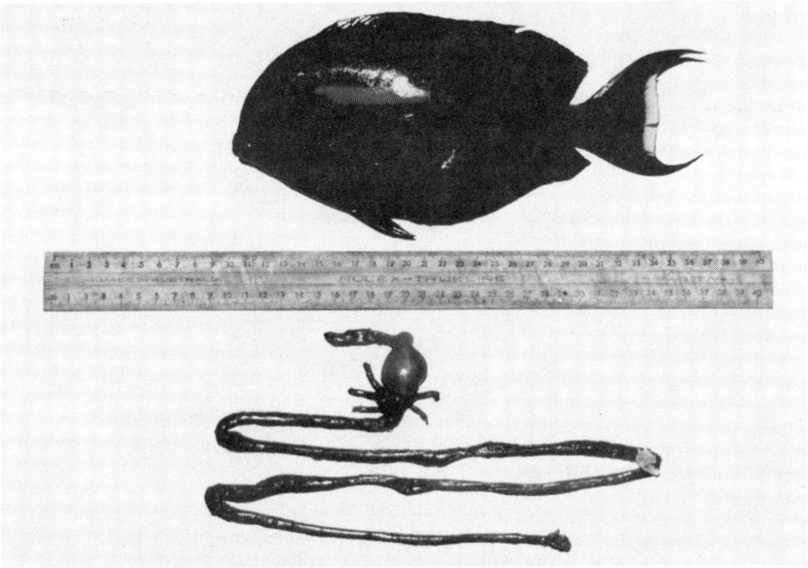

Figure 6.2. *Acanthurus olivaceus*, family Acanthuridae. Note muscular stomach.

Figure 6.3. *Scarus rivulatus*, family Scaridae.

Figure 6.4. *Odax cyanomelas*, family Odacidae. Intervals in scale, 10 mm.

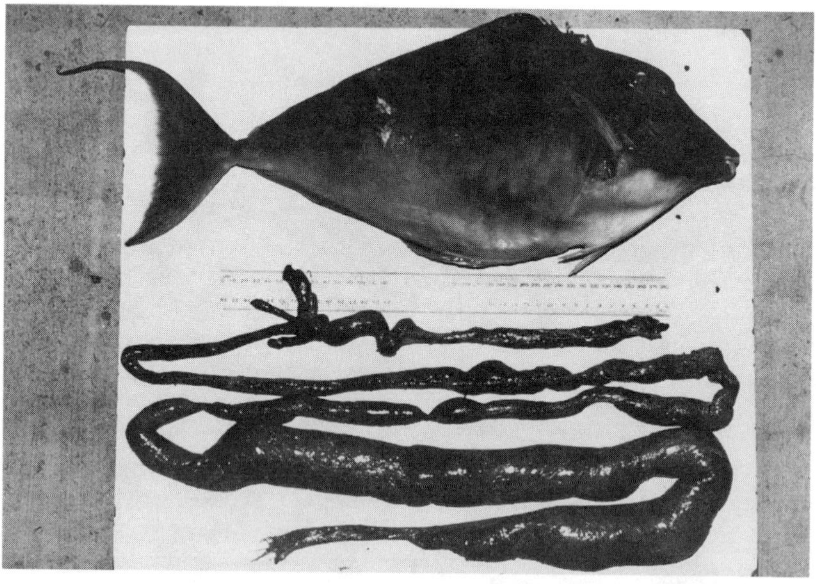

Figure 6.5. *Naso brachycentron*, family Acanthuridae.

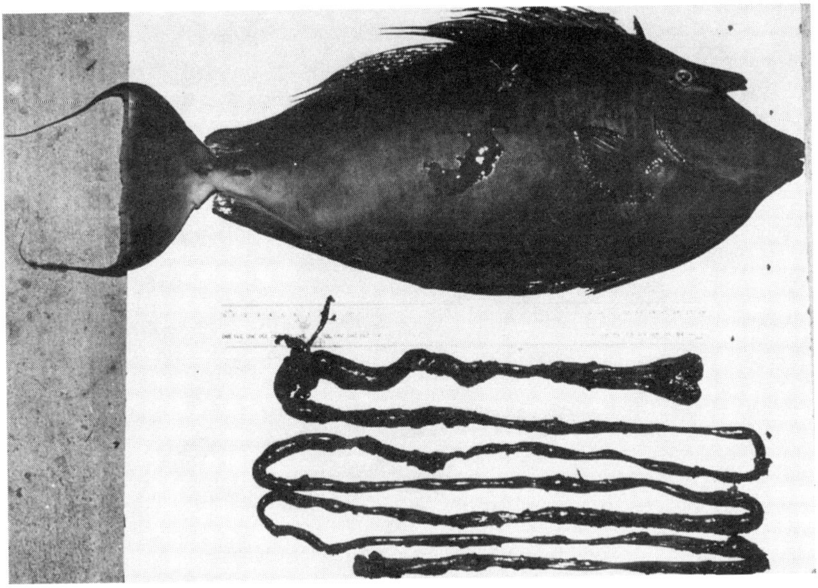

Figure 6.6. *N. unicornis*, family Acanthuridae.

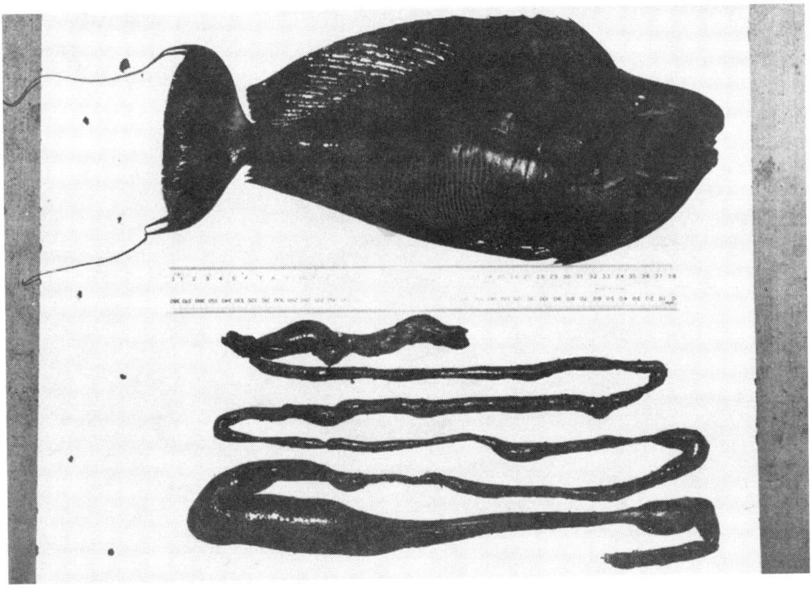

Figure 6.7. *N. vlamingii*, family Acanthuridae.

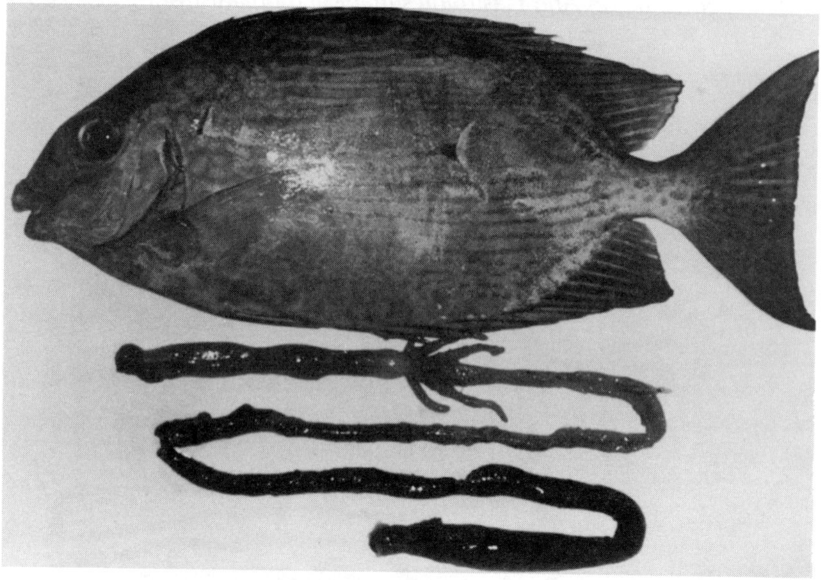

Figure 6.8. *Siganus lineatus*, family Siganidae.

Figure 6.9. *Kyphosus vaigiensis*, family Kyphosidae. Note hindgut chamber, which is separated from the rest of the intestine by a sphincter.

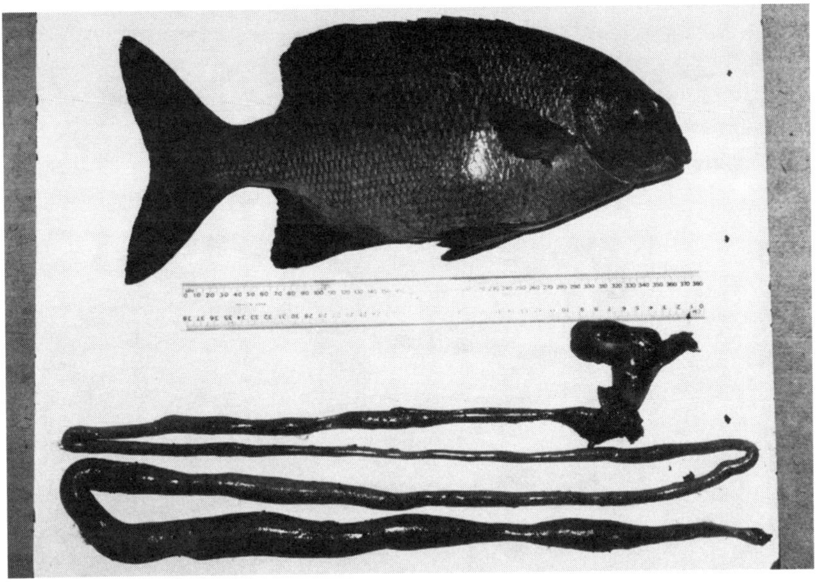

Figure 6.10. *K. cinerascens*, family Kyphosidae.

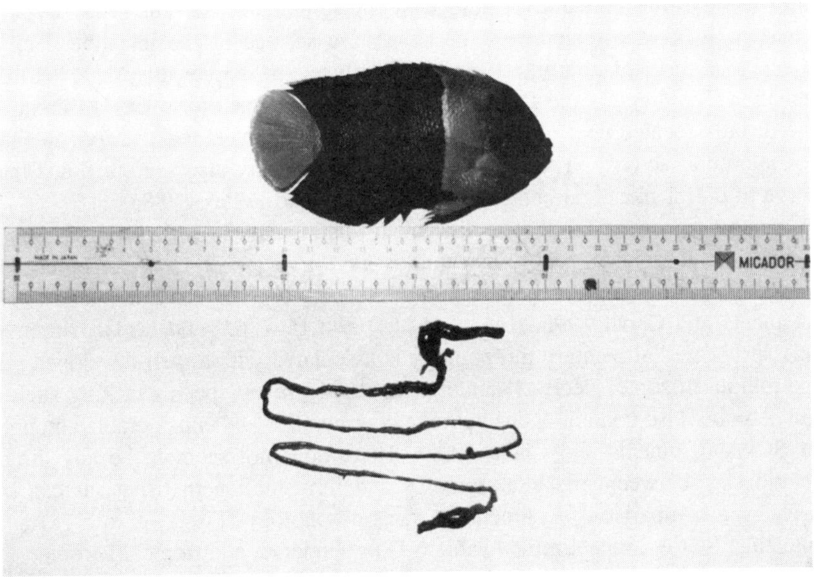

Figure 6.11. *Centropyge bicolar*, family Pomacanthidae. Note hindgut chamber.

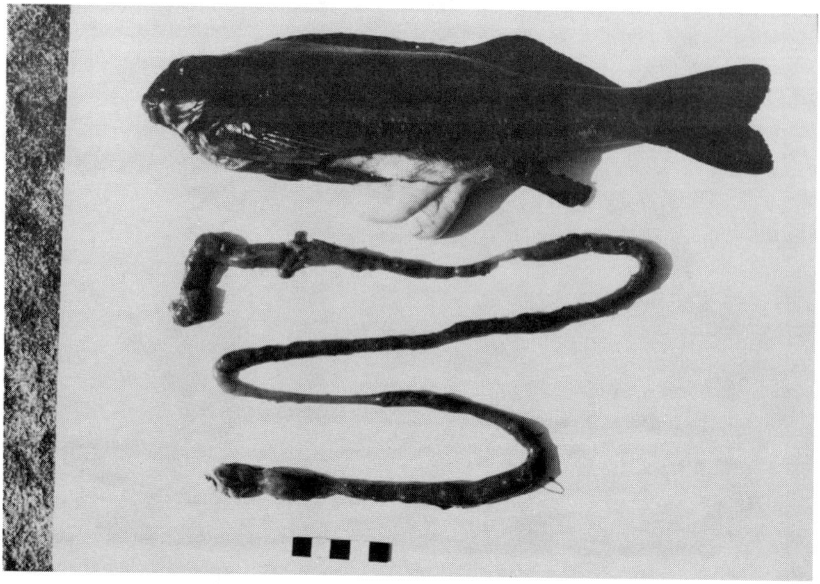

Figure 6.12. *Crinodus lophodon*, family Aplodactylidae. Intervals in scale 10 mm.

known as pyloric cecae and are unique among vertebrates (Stevens 1988). Pyloric cecae in fish serve to increase the absorptive surface of the intestine, lack a specialized resident microbiota, and retain digesta no longer than the proximal intestine (Buddington and Diamond 1987). They are therefore very different to the distally located cecae of reptiles, birds, and mammals, which serve as sites for microbial activity. The pyloric cecae of fishes may also secrete fluids that serve to buffer intestinal contents (Montgomery and Pollak 1988a).

In most fishes the hindgut is difficult to distinguish from the midgut. Some kyphosids, scorpidids, and pomacanthids have a posterior chamber or cecum separated from the rest of the intestine by a sphincter (Rimmer and Wiebe 1987, Clements and Choat 1995, Kandel et al. 1994; Figs. 6.9 and 6.11). Although this specialized alimentary morphology is associated with a well-developed gut microbiota in some species (Kandel et al. 1994), many species lacking such a gut chamber also contain endosymbiotic communities and considerable amounts of SCFA (Clements and Choat 1995). There does not seem to be any direct relationship between intestinal anatomy and fermentation in fishes, as can be seen by a comparison of alimentary tract morphology (Figs. 6.1 to 6.12) and intestinal SCFA concentration (Table 6.1) in temperate and tropical herbivorous fishes.

Table 6.1. Maximum concentration of total SCFA estimated from the intestine of fishes

Family	Species	Salinity	Distn	Diet	N	Gut Section with Max SCFA	Total SCFA ($mmol{\cdot}L^{-1}$)	Source
Acanthuridae	*Acanthurus lineatus*	M	T	FRA	10	PeI	17.7 ± 1.8	1
Acanthuridae	*A. mata*	M	T	P	11	PeI	7.0 ± 1.6	1
Acanthuridae	*A. nigricans*	M	T	FA	12	PoI	29.0 ± 1.7	1
Acanthuridae	*A. nigricauda*	M	T	D, FA	5	PeI	9.7 ± 0.8	1
Acanthuridae	*A. nigrofuscus*	M	T	FA	11	PeI	18.2 ± 1.3	1
Acanthuridae	*A. olivaceus*	M	T	D, FA	10	PeI	11.5 ± 1.4	1
Acanthuridae	*A. triostegus*	M	T	FA	5	PeI	13.8 ± 2.9	1
Acanthuridae	*Ctenochaetus striatus*	M	T	CB, D	10	PoI	9.1 ± 1.6	1
Acanthuridae	*Naso annulatus*	M	T	BI, P	6	PoI	20.0 ± 5.5	1
Acanthuridae	*N. brachycentron*	M	T	A	3	PeI	24.5 ± 1.1	1
Acanthuridae	*N. brevirostris*	M	T	BI, P	16	PoI	17.1 ± 1.9	1
Acanthuridae	*N. hexacanthus*	M	T	P	11	PoI	23.2 ± 3.5	1
Acanthuridae	*N. lituratus*	M	T	BM	12	PoI	39.6 ± 4.8	1
Acanthuridae	*N. tuberosus*	M	T	A, BI	14	PoI	18.8 ± 1.5	1
Acanthuridae	*N. unicornis*	M	T	BM	11	PeI	42.0 ± 4.0	1
Acanthuridae	*N. vlamingii*	M	T	FF, P	14	PoI	24.9 ± 2.8	1
Acanthuridae	*Zebrasoma scopas*	M	T	FRA	8	PoI	31.9 ± 3.6	1
Acanthuridae	*Z. veliferum*	M	T	M	3	PoI	37.0 ± 4.3	1
Aplodactylidae	*Crinodus lophodon*	M	C	FA	12	PeI	27.3 ± 2.9	2
Caesionidae	*Caesio cuning*	M	T	P	6	PI	2.9 ± 0.2	3
Catostomidae	*Carpiodes cyprinus*	F	C	AV, BI, D	2	I	9	7
Centrarchidae	*Micropterus salmoidea*	F	C	C	16	I	18.9	7
Centrachidae	*Pomoxis annularis*	F	C	BI, C	2	I	14	7
Cichlidae	*Oreochromis mossambicus*	E	T	A, D	3	LI	18.2 ± 1.6[a]	8
Clupeidae	*Dorosoma cepedianum*	F	C	BI, D	29	I	2.1	7
Cyprinidae	*Ctenopharyngodon idella*	F	C	AV	2	W	6.8 ± 3.4	5
Cyprinidae	*Cyprinus carpio*	F	C	BI, D	4	MI	18.7 ± 2.1	5
Cyprinidae	*Cyprinus carpio*	F	C	AV, BI, D	29	I	8.1	7
Kyphosidae	*Kyphosus cinerascens*	M	T	A	8	PoI	40.4 ± 6.4	1
Kyphosidae	*K. cornelii*	M	S	MA	2	C	15.5–18.4	6
Kyphosidae	*K. sydneyanus*	M	S	MA	2	C	38.2–38.7	6
Labridae	*Achoerodus viridis*	M	C	BI	5	W	0.8 ± 0.2	3
Labridae	*Notolabrus gymnogenis*	M	C	BI	3	W	0.3 ± 0.3	3
Odacidae	*Odax cyanomelas*	M	C	BM	15	PeI	35.2 ± 2.6	2
Odacidae	*O. pullus*	M	C	BM	9	PoI	37.1 ± 3.3	2
Pomacanthidae	*Centropyge bicolor*	M	T	A, D	4	HC	47.9 ± 4.8	1
Salmonidae	*Salmo gairdneri*	F	C	C	5	PI	14.7	5
Scaridae	*Scarus niger*	M	T	D, FA	4	PeI	9.8 ± 1.7	1
Scaridae	*S. rivulatus*	M	T	D, FA	3	PeI	12.1 ± 2.6	1
Scaridae	*S. schlegeli*	M	T	D, FA	7	PoI	6.2 ± 1.3	1
Scaridae	*S. sordidus*	M	T	D, FA	4	PeI	11.5 ± 1.4	1
Scorpididae	*Medialuna californiensis*	M	S	A	3	HC	24–36	4
Siganidae	*Siganus argenteus*	M	T	A, BI	9	PoI	17.0 ± 1.0	1
Siganidae	*S. corallinus*	M	T	A	4	PoI	10.8 ± 1.9	1
Siganidae	*S. doliatus*	M	T	A	11	PoI	18.4 ± 1.6	1
Siganidae	*S. lineatus*	M	T	A, BI	8	PeI	10.1 ± 1.2	1
Siganidae	*S. puellus*	M	T	A, BI	4	PoI	23.0 + 4.2	1
Siganidae	*S. punctatissimus*	M	T	A	3	PoI	21.9 ± 1.6	1
Siganidae	*S. punctatus*	M	T	A	8	PoI	14.3 ± 3.0	1
Siganidae	*S. vulpinus*	M	T	A	5	PoI	20.7 ± 4.5	1

Values are presented as the mean of the replicates ± SE

Abbreviations: Salinity: E euryhaline, F freshwater, M marine.

Distribution: C temperate, S subtropical, T tropical.

Diet: A algae, AV aquatic vascular plants, BI benthic invertebrates, BM brown macroalgae, C carnivore, CB cyanobacteria, D diatoms and detritus, FA filamentous algae, FF fish feces, FRA filamentous red algae, M macroalgae, P zooplankton.

Gut segment: C cecal pouch, HC hindgut chamber, I total intestine, LI lower third of intestine, MI mid-intestine, PeI penultimate 25% of intestine, PI posterior 50% of intestine, PoI posterior 25% of intestine, W whole gut.

Source: 1 Clements and Choat (1995), 2 Clements et al. (1994), 3 Clements unpubl data, 4 Kandel pers comm, 5 Paris et al. (1977), 6 Rimmer and Wiebe (1987), 7 Smith et al. (1996), 8 Titus and Ahearn (1988).

[a] Value excludes trace propionate presence.

2.4. Gut pH and Redox Potential

The gut pH of herbivorous fishes has been examined in many studies. This follows the suggestion that gastric acidity was one of three mechanisms for plant cell lysis employed by herbivorous fishes (Lobel 1981). Stomach pH in tilapia, a freshwater cichlid, can be as low as 1.3 to 1.5, which is sufficient to lyse algal cells (Moriarty 1973) and detrital bacteria (Bowen 1976). Some pomacentrids (damselfish), pomacanthids (angelfish), stichaeids (pricklebacks), siganids (rabbitfish), and acanthurids are thought to use low pH to lyse algal cells (Lobel 1981, Horn 1992), although Hofer (1988) concluded that a pH below 2 was necessary for this to take place. Values for stomach pH of marine fish range from 2 to 4 in species lacking a trituration mechanism to 6 to 8 in species with a pharyngeal apparatus or a gizzardlike stomach (references in Horn 1989).

The pH of the intestine in acanthurids, pomacentrids, stichaeids, mugilids and scarids ranges from about 6 to 8.5 (references in Horn 1989). *Kyphosus cornelii* and *K. sydneyanus,* which both harbor a diverse gut microbiota, have an intestinal pH of 6 to 7 in the posterior intestine and cecal pouch (Rimmer and Wiebe 1987). Herbivorous odacids have a gut pH of 8 to 9 in the region of the gut densely populated by endosymbionts (Clements 1991a). Endosymbiotic communities in fishes therefore occur over a wide range of pH.

Montgomery and Pollak (1988a) examined the relationship between intestinal pH and presence of the giant bacterial symbiont *Epulopiscium fishelsoni* in the surgeonfish *Acanthurus nigrofuscus.* They found a decline in intestinal pH at the point in the gut where the symbionts occurred in large numbers; this decline did not occur when the symbionts were eliminated by starvation. It is possible that this decrease in pH is caused by the production of SCFA by the bacterial symbionts, although levels of SCFA in this species are low compared to other species of marine herbivorous fish (Clements and Choat 1995).

Redox potential of the gut of fishes has received little attention, presumably because of the lack of work on fermentation in these animals. The entire gut of *Odax cyanomelas* and the intestine of *Crinodus lophodon* were highly anaerobic, a finding also suggested by the large amounts of SCFA in the posterior intestine of these species (Clements et al. 1994). The presence of large amounts of SCFA in many marine herbivorous fish (Rimmer and Wiebe 1987, Clements and Choat 1995) suggests that the intestines of most of these fishes are anaerobic.

2.5. Gut Retention Times

Gut retention time in fishes varies with temperature, frequency of feeding, meal size, biochemical composition of the food, method of feeding, and length and weight of the fish (Fänge and Grove 1979). Total gut emptying time for 69 carnivorous species and three herbivorous species ranges from 10 to 158 hours and 3 to 10 hours, respectively (Fänge and Grove 1979). None of the herbivores in this list are marine species. The gut retention time of the freshwater herbivore *Ctenopharyngodon idella*, the grass carp, is influenced by water temperature (Stevens 1989). Plant material passed through the gut in 18 hours at 10°C, and in 7 to 8 hours at 27°C (references in Stevens 1989). Grass carp feed more selectively at lower temperatures (Stevens 1989), complicating the relationship between temperature and gut retention time. Water temperature also influenced the rate of digesta passage in the detritivorous clupeid *Dorosoma cepedianum*, the gizzard shad (Salvatore et al. 1987). Food passage over time was measured as the distance moved along the intestine by dyed food material. At water temperatures of 10, 15, and 20°C, food passed through the alimentary tract in 8.5, 4.9, and 4.4 hours, respectively (Salvatore et al. 1987).

Gut retention times recorded for marine herbivorous fish are less than 10 hours with three exceptions. The stichaeid *Cebidichthys violaceus*, a cold temperate species with a relatively depauperate gut microbiota (Kandel et al. 1994), can have a gut retention time of greater than 50 hours (Horn 1989). Gut retention time in *C. violaceus* decreases significantly when the fish are fed a low-protein diet (Fris and Horn 1993). The warm temperate species *Kyphosus sydneyanus*, which has a well-developed gut microbiota, has a gut retention time of 21 hours (Rimmer and Wiebe 1987). Gut retention time in the temperate species *Aplodactylus punctatus* is directly proportional to fish length, and ranges from 30 to 52 hours in adults feeding mainly on the kelp *Lessonia trabeculata* (Benavides et al. 1994).

The short retention times recorded for many herbivorous species may be somewhat misleading with respect to maintenance of an endosymbiotic community. Although algae passed through the gut of *Acanthurus triostegus* in 2 hours (Randall 1961), this species consistently contains dense populations of giant bacterial symbionts (Clements et al. 1989). This species must therefore have some mechanism for the selective retention of gut symbionts. The intestinal sacculae characteristic of some species of scarids (Fig. 6.3) increase the holding capacity of the gut, and may also increase retention times of some dietary components (Clements and Bellwood 1988). It is possible that gut retention times of 4 to 6 hours for scarids (Smith and Paulson 1974) reflect only the passage time of particulate material. Many species of herbivorous fishes retain material in the gut overnight (Randall 1961, Fishelson et al. 1985, Montgomery and Pollak 1988a, Montgomery et al. 1989), a factor to be taken into account when considering the impact of retention times on fermentation. A proper comparison of gut retention times between fishes and higher vertebrates must await studies on fish using appropriate particulate and fluid markers.

2.6. Summary and Conclusions

The form of the alimentary tract in fishes is highly diverse, making generalizations difficult. Most species containing high concentrations of SCFA in the gut (Rimmer and Wiebe 1987, Clements et al. 1994, Clements and Choat 1995) lack a mechanism for mechanical trituration of ingested material. Specialized fermentation chambers have been found in a few species, but high SCFA concentrations in the gut of fishes lacking such morphology suggests that material is retained in the intestine long enough for fermentation to take place. The paucity of comparative information on gut retention rates in fishes, particularly herbivores, limits discussion concerning fermentation. The few temperate and subtropical herbivorous taxa studied to date, notably kyphosids and aplodactylids, have lengthy gut retention times.

3. Diversity of Gastrointestinal Symbionts

The microbiology of the intestinal tract of fishes has received considerable attention (Trust and Sparrow 1974, Horsley 1977, Cahill 1990). However, the majority of these studies concern carnivorous, northern-hemisphere fish taxa, which influenced the development of views on the importance of a gut microbiota in fishes. It was considered until recently that the gastrointestinal biota of fishes tended to be similar to that on the skin, gills, and food ingested (Horsley 1977), and that most intestinal bacteria were aerobes or facultative anaerobes (Trust and Sparrow 1974). Conclusions like these led to the assumption that fish lacked a consistent, anaerobic microbiota (Montgomery and Gerking 1980).

More recently, Cahill (1990) concluded that certain genera of facultative and obligately anaerobic bacteria could form large populations in the gut, and that the structure of the alimentary tract influenced the composition of the gut microbiota. She also stated that the role of gut microbiota in fish nutrition had yet to be clearly determined, but that it probably varied with the fish species. It should be noted that neither Horsley (1977) nor Cahill (1990) contains a single reference to marine herbivorous fish, which recently have been found to harbor diverse endosymbiotic communities of anaerobic organisms. Furthermore, the only literature on fermentation in fishes concerns herbivorous species. The following section will therefore focus on herbivorous fishes.

3.1. The Gastrointestinal Symbionts of Herbivorous Fishes

One of the first reports of obligate anaerobes among the resident gut microflora of fishes was that of Trust et al. (1979). Total bacterial numbers cultured in this study ranged from 6×10^4 to 4×10^8g of total alimentary tract plus contents in the herbivorous grass carp, and 1.6×10^8 to 4×10^8 in the omnivorous goldfish. Cultures from the carnivorous trout contained too few bacteria to count. Obligate anaerobes isolated from the alimentary tract of grass carp included species of *Actinomyces*, *Bacteroides*, *Clostridium*, *Eubacterium*, *Fusobacterium*, and *Peptostreptococcus*.

Kamei et al. (1985) isolated two strains of *Bacteroides* from the alimentary tracts of cultured *Sarotherodon niloticus,* a freshwater cichlid which feeds on detritus and microalgae in the wild (Fryer and Iles 1972). The two strains of *Bacteroides* fermented glucose to acetate or acetate and succinate. The authors suggested that these SCFA probably influenced the metabolism of the host fish (Kamei et al. 1985), although this hypothesis was not tested.

A diverse intestinal microbiota including *Spirillum*, trichomonadid flagellates, and an unusual ''protist'' was discovered during the course of a study on the herbivorous surgeonfish *Acanthurus nigrofuscus* in the Red Sea (Fishelson et al. 1985). The protist, which was described subsequently as *Epulopiscium fishelsoni* (Montgomery and Pollak 1988b), ranges in length from 30 to 500 μm and attains densities of 2×10^4 to 1×10^5 cells/mL of gut contents. *Epulopiscium fishelsoni* shows a distinctive pattern of distribution and movement within the intestine, and was found in each of several hundred specimens of *A. nigrofuscus* examined (Fishelson et al. 1985; Montgomery and Pollak 1988b). The report of Fishelson et al. (1985) was the first of a consistent endosymbiosis in the gut of an herbivorous marine fish, and this study was therefore instrumental in generating subsequent research in this area.

Rimmer (1986) found that the development of a gut-resident microbiota was associated with a decrease in the amount of animal material in the diet in juveniles of the subtropical species *Kyphosus cornelii*. Fish greater than 36 mm fork length, which were predominantly herbivorous, contained 3×10^{10} and 7×10^{10} microbial cells per gram wet weight gut contents in the posterior intestine and cecal pouch, respectively. The microbiota of adult *K. cornelii* and *K. sydneyanus* includes rod-shaped bacteria, spirillae, flagellates, and ciliates (Rimmer and Wiebe 1987). Total microbial densities in the cecal pouch of these two species were 28×10^{10} and 50×10^{10} cells/g dry weight, respectively (Rimmer and Wiebe 1987).

The aerobic and facultatively anaerobic bacterial biota of the gastrointestinal tract of the herbivorous surgeonfish *Acanthurus nigrofuscus* was distinct from that of a carnivore, a planktivore, and a detritivore (Sutton and Clements 1988). The stomach contents of *A. nigrofuscus* were devoid of culturable bacteria, and the intestinal tract was dominated by agar-digesting non-*Vibrio* bacteria. Agar-digesting bacteria were also detected from the detritivorous surgeonfish *Ctenochaetus striatus,* but generally comprised less than 8% of the total culturable heterotrophic microflora (Sutton and Clements 1988). The alimentary tracts of *C. striatus,* the carnivore, and the planktivore were dominated by *Vibrio* species, particularly *V. harveyi* and *V. damsela. Vibrio* spp. were detected in the food of the planktivorous species and in the turf algae/substratum over which *A. nigrofuscus* and *C. striatus* fed. Non-*Vibrio* agar-digesting bacteria such as those found in *A. nigrofuscus* were not found in samples of the turf algae/substratum, suggesting that the dominant bacteria cultured from this species were gut-resident.

A range of endosymbionts similar to the giant *Epulopiscium fishelsoni* of the Red Sea occur in the intestinal tract of tropical herbivorous and detritivorous surgeonfish of the Great Barrier Reef (Clements et al. 1989). These microorganisms, subsequently referred to as epulos, were categorized by Clements and coworkers into 10 morphotypes on the basis of cell size and shape, and mode of cell division. The occurrence of epulo morphotypes seemed to be correlated with host feeding ecology (Clements et al. 1989), a relationship that was relatively consistent in samples collected from other tropical and subtropical areas (Clements 1991b). Clements (1991b) suggested that the ubiquitous occurrence of epulos in several species of herbivorous surgeonfish collected from a number of geographical regions raised the possibility that this symbiosis was an obligate relationship. Epulos were not found in any members of the families Kyphosidae, Pomacentridae, Scaridae, Zanclidae, Siganidae, and Blenniidae examined by Clements et al. (1989), but similar organisms were found subsequently in the angelfish *Centropyge bicolor* (Clements 1991b) and herbivorous odacids (Clements 1991a). The largest of the epulo morphotypes was found to be a prokaryote (Clements and Bullivant 1991), closely related to the gram-positive bacterial genus *Clostridium* (Angert et al. 1993). The giant epulos, found only in the gut of herbivorous surgeonfishes, are the largest bacteria known (Angert et al. 1993).

Spirilla, diplomonads, trichomonads, an undescribed species of opalinid, and several unidentified zooflagellate taxa also inhabit the gut of surgeonfishes (Clements 1991b). Surgeonfishes also harbor a diverse intestinal fauna of ciliates, and species of the genera *Balantidium*, *Vestibulongum*, and *Paracichlidotherus* have been described by Grim (1993 and references therein). The presence of microorganisms such as diplomonads, trichomonads, and epulos in the intestine of surgeonfishes indicates that this region of the alimentary tract is anaerobic.

Herbivorous fishes from temperate areas also harbor endosymbiotic communities. Herbivorous odacids, labroid fishes restricted to temperate Australasian waters, contain a diverse microbiota consisting of rod-shaped bacteria, cocci, filamentous bacteria, spirilla, trichomonads, and diplomonads (Clements 1991a). Total microbial numbers in the posterior intestine of the Australian species *Odax cyanomelas* were estimated by epifluorescence microscopy as 5×10^8 cells/g wet weight of gut contents (Clements 1991a). *Odax* species from different geographical areas differ in the composition of endosymbiont communities. The herbivorous aplodactylid *Crinodus lophodon*, which co-occurs with *O. cyanomelas*, also harbors abundant populations of gastrointestinal microorganisms (Clements et al. 1994).

Three microorganisms have been isolated by Mountfort and co-workers from another temperate fish species, the detritivorous mullet *Aldrichetta forsteri.* The facultatively anaerobic marine fungus *Paecilomyces lilacinus* grew well on laminarin and glucose, and fermented hexose to acetate, ethanol, CO_2, and lactate (Mountfort and Rhodes 1991). No growth occurred on cellulose. This work raises the possibility that anaerobic fungi may occur in the gut of other species of marine fish, a possibility not examined by Clements (1991a) in his work on odacids. Two obligately anaerobic, gram-positive bacteria were also isolated from mullet gut contents (Mountfort 1993, Mountfort 1994). The nonsporulating, highly motile rod *Eubacterium* strain P-1 and the sporulating, motile rod *Clostridium grantii* were both present at $>10^5$/mL of mullet gut contents (Mountfort 1993).

Facultative and anaerobic bacteria were isolated from the intestinal tract of the omnivorous pinfish, *Lagodon rhomboides* (family Sparidae), by Luczkovich and Stellwag (1993). The numbers of culturable bacteria varied between different size classes of fish, with the highest numbers occurring in the largest size class. The total number of viable facultative and anaerobic bacteria measured as colony-forming units present per gram of intestinal tract including contents was 2.9×10^7 in the largest size class, which also had the highest dietary proportion of algae and seagrass (Luczkovich and Stellwag 1993). The proportion of the total culturable bacteria capable of metabolising carboxymethylcellulose varied with respect to fish size, although there was no clear trend with fish size.

Finally, the gut-resident microbial flora of four species of herbivorous fishes was discussed by Kandel et al. (1994). The SCFA composition in the hindgut of these fishes ranged from acetate only in the cool temperate stichaeid *Cebidichthys violaceus* to six SCFAs in the warm temperate scorpidid *Medialuna californiensis* and the tropical kyphosids *Kyphosus bigibbus* and *K. vaigiensis.* The presence of six SCFAs, high microbial diversity, and specialized intestinal morphology of the last three species led Kandel and co-workers to suggest that these fishes had well-developed fermentation systems. The authors' assumption that the number of SCFAs present in the gut is related to microbial diversity should be questioned. Acetate dominates the extracellular pool of SCFA in the hindgut fluid of several termite species which contain a very diverse gut microbiota (Odelson and Breznak 1983). The ratios of SCFA present in the gut are influenced by a large number of factors, including the amount and chemical composition of the substrate, the rate at which it is depolymerized, the relative numbers, substrate specificities, preferences and fermentation strategies of individual gut microbes, and the availability of inorganic terminal electron acceptors (Macfarlane and Macfarlane 1993). Furthermore, the hypothesis of Kandel et al. (1994) that warm temperatures and a specialized fermentation chamber are prerequisites for an efficient gut microbiota in fishes should be reconsidered in the light of recent findings on temperate odacids and aplodactylids (Clements et al. 1994).

3.2. Transfer of Endosymbionts in Fishes

Several studies of terrestrial herbivores have emphasized the importance of microbiota acquisition by neonates (e.g., Jones 1984, Troyer 1984). Fishes live in a fluid medium, which presents a different set of problems associated with endosymbiont transfer to those experienced by terrestrial herbivores. The establishment of intestinal microbiota in fishes was reviewed by Cahill (1990), who discussed studies conducted on salmonids and goldfish. She concluded that these fishes derived their intestinal microbiota from bacteria present in the environment, which were able to persist and grow in the alimentary tract. Recent work suggests that there may be more direct mechanisms of symbiont transfer in marine herbivorous fishes.

Fishelson et al. (1985) discussed the pathways of disease and pathways of colonization of the surgeonfish symbiont *Epulopiscium fishelsoni.* These bacteria were not found associated with food or in water collected from aquaria where host surgeonfish *Acanthurus nigrofuscus* were kept (Montgomery and Pollak 1988b), nor were they detected during the course of microbiota surveys in reef sediments over which surgeonfish fed (e.g., Moriarty et al. 1985, Hansen et al. 1987). *Acanthurus nigrofuscus*, like the majority of herbivorous fishes, feeds during daylight hours and rests in reef crevices overnight. Fish collected at night have a bolus of incompletely digested food containing the symbionts in the posterior intestine (Fishelson et al. 1985). At the commencement of feeding in the morning this undigested material is expelled over algal turf-covered surfaces on which the fish feed. The authors suggested that this dispersal of symbiont-laden feces over dietary algae may provide the means by which host fish become colonized. Rimmer (1986) suggested a similar means of symbiont acquisition in juvenile kyphosids, which peck at material suspended in the water column. This material presumably includes fecal fragments from nearby adults containing populations of symbionts.

Clements (1991b) has examined a number of aspects of symbiont biology in surgeonfishes. Most species of surgeonfish settled in areas where adult conspecifics were common, indicating that juveniles had access to adult symbionts at an early stage. Juvenile *Acanthurus nigrofuscus* expelled undigested algae from the posterior intestine at dawn, as did adults of this species (Fishelson et al. 1985), but unlike the adults the juveniles were observed to rapidly ingest this material before it reached the substratum (Clements 1991b). This behavior was interpreted as a mechanism for the retention of symbionts. Coprophagic behavior was not observed in juveniles of the detritivorous species *Ctenochaetus striatus*, which deliberately defecated outside feeding areas. This latter species does not retain a bolus of food in the posterior intestine overnight (Montgomery et al. 1989).

Newly settled surgeonfish did not harbor endosymbionts, but intestinal populations of epulos were rapidly established following settlement (Clements 1991b).

Populations of other endosymbiont taxa, such as zooflagellates and spirilla, took longer to become established in the host gut. An experiment was conducted in which feces from an aquarium containing large juvenile surgeonfishes infected with epulos were added to an aquarium containing newly settled fish lacking epulos. All of the latter juveniles subsequently became infected with epulos, with the exception of one diseased individual. This result strongly suggests that newly settled fish may be colonized by epulos through exposure to fecal material from colonized hosts. The lack of a specific control for this experiment makes it impossible to discount the hypothesis that epulos were introduced into the aquarium during the course of the experiment through the water system or via food. This possibility is however extremely unlikely due to the scarcity of epulos outside the host intestine (Montgomery and Pollak 1988b).

3.3. Summary and Conclusions

Diverse communities of gastrointestinal microorganisms occur in herbivorous members of the families Acanthuridae, Aplodactylidae, Cyprinidae, Kyphosidae, Odacidae, Pomacanthidae, Scorpididae, and Sparidae. Additionally, intestinal symbionts have been reported from stichaeids (Kandel et al. 1994), cichlids (Kamei et al. 1985), mugilids (Mountford and Rhodes 1991; Mountfort et al. 1993, 1994), and pomacentrids, siganids, and zanclids. Representatives of the latter three families contained rod-shaped bacteria and spirilla (Clements 1991b), and these taxa and the speciose blennies require further study. The number of intestinal bacteria in herbivorous taxa such as odacids and kyphosids approach that in the hindgut of terrestrial vertebrate herbivores. The widespread occurrence of abundant populations of gastrointestinal microorganisms in herbivorous taxa suggests that SCFA may be a valuable source of energy in many of these fish species.

The only major taxon of herbivorous fish that does not contain an obvious microbiota is the family Scaridae, the parrotfishes. Over 40 specimens from 13 species of scarids were examined by Clements (1991b), who failed to detect any of the characteristic bacterial rods, spirilla, or zooflagellates found in the majority of tropical herbivorous fish species. Scarids ingest considerable quantities of inorganic sediment, which is triturated in a well-developed pharyngeal mill (Bellwood and Choat 1990). As a result, the alimentary tract contains a slurry of calcium carbonate, creating an environment that may be unfavorable for the development of large populations of microorganisms.

The limited data available suggest that gastrointestinal microorganisms are transmitted in some species by the ingestion of infected fecal material, as occurs in herbivorous reptiles (Troyer 1984), and in other species by the ingestion of infected food or water. Some species of herbivorous fishes appear to have evolved specific mechanisms for the retention and transfer of endosymbionts.

4. Enzyme Activities of the Gastrointestinal Microorganisms

An understanding of the contribution of endosymbionts to digestion requires information on the relative importance of exogenous (produced by gastrointestinal endosymbionts) and endogenous (produced by the host) digestive enzymes. Exogenous cellulases are critical to the nutrition of terrestrial vertebrate herbivores (Stevens 1988) but not to termites and cockroaches (Slaytor 1992). Studies on digestive enzymes have greatly influenced literature opinion on the role of gastrointestinal microorganisms in fishes, and it is therefore appropriate that examples of this research are discussed here. It is beyond the scope of this article to discuss the endogenous enzymes of fishes, which are reviewed elsewhere (Fänge and Grove 1979, Helpher 1988, Stevens 1988, Horn 1989). The following section will therefore focus on studies of the exogenous enzymes of fishes.

4.1. Cellulase Activity in the Alimentary Tract of Fishes

The most important group of exogenous enzymes in symbioses between terrestrial vertebrate herbivores and microorganisms are cellulases, which degrade the cell walls of vascular plants. Vertebrates lack the ability to produce endogenous cellulases, and hence are reliant upon gastrointestinal microorganisms to degrade the cell wall of vascular plants. Many studies have examined cellulase activity in the alimentary tract of fishes, with mixed results.

Stickney and Shumway (1974) investigated cellulase activity in the stomachs of 148 elasmobranch and teleost fishes. Activity was detected in 16 estuarine species and one freshwater species, the catfish *Ictalurus punctatus.* Catfish showed no cellulase activity after streptomycin treatment, while control fish continued to show activity. The authors concluded that cellulase activity derived from gastrointestinal microorganisms rather than the presence of cellulase within the food consumed. This conclusion was challenged by later studies, which found that cellulase activity in fishes was correlated with the ingestion of invertebrate cellulases or cellulolytic bacteria associated with the food (Prejs and Blaszczyk 1977, Lindsay and Harris 1980), and that populations of gastrointestinal microorganisms from fishes exhibit little cellulase activity (Trust et al. 1979, Lésel et al. 1986, Anderson 1991).

The results of these and other studies (reviewed by Horn 1989) led to authors making a series of conclusions about plant digestion in fishes: (1) fishes lack both a resident cellulolytic microbiota and the ability to produce an endogenous cellulase (Lindsay and Harris 1980); (2) a lack of cellulase activity indicates that ingested algae are little utilized by some fishes (Bitterlich 1985); (3) cell walls are degraded by mechanical degradation or low stomach pH (Lobel 1981, Anderson 1987); and (4) fish rely predominantly upon the assimilation of cell storage compounds (Trust et al. 1979). The last of these conclusions was challenged by the finding that cell wall components from the alga *Enteromorpha* were assimilated by the temperate herbivorous fish *Girella tricuspidata*, but the means of cell wall hydrolysis was not determined (Anderson 1987). The most recent chapter in this work was the isolation of cellulolytic microbes from the omnivorous pinfish, *Lagodon rhomboides*, by Luczkovich and Stellwag (1993). The mode of intestinal colonization was not determined, and although the bacteria were absent from sediment samples it was possible that the intestinal colonies were derived from detrital plant material, coprophagy, or the ingestion of invertebrate herbivores containing cellulolytic bacteria (Luczkovich and Stellwag 1993). Nevertheless, the authors conclude that the gastrointestinal microbiota of pinfish may contribute to the breakdown of plant material by pinfish and may be the major source of cellulase.

4.2. Other Exogenous Enzymes in the Alimentary Tract of Fishes

Studies on the digestion of dietary substrates including laminarin, chitin, starch, and lipid have also involved discussions on the role of gastrointestinal microorganisms. Laminarin is a β-1,3-glucan which is a storage polysaccharide in algae, protozoans, and fungi (Piavaux 1977). Laminarinase is therefore a potentially important digestive enzyme in phytophagous, microphagous, and detritivorous fishes. Although endogenous laminarinase activity has been reported from several species of fishes (Piavaux 1977, Sturmbauer et al. 1992), this enzyme can also be produced exogenously (Mountfort and Rhodes 1991). Although the source of laminarinase activity was not determined in the herbivorous rabbitfish *Siganus canaliculatus* (Sabapathy and Teo 1993), the presence of activity in the stomach and pyloric ceca as well as the intestine suggests endogenous production. The failure to detect laminarinase activity in gut fluid extracts does not exclude the possibility of exogenous degradation of this polysaccharide, since the laminarinases of many gut bacteria are cell-associated rather than extracellular (Salyers 1979). Chitinase was not produced by intestinal bacteria isolated from the anchovy *Engraulis capensis* (Seiderer et al. 1987), but is produced by intestinal microflora from the cichlid *Sarotherodon niloticus* (Kamei et al. 1985) and the dover sole *Solea solea* (MacDonald et al. 1986). Pectinolytic and alginolytic activity were exhibited by bacteria isolated from mullet gut contents. *Eubacterium* strain P-1 grew on pectin and *Clostridium grantii* on alginic acid (Mountfort et al. 1993, 1994). Neither isolate grew on laminarin.

Pollak and Montgomery (1994) found that presence of the bacterial symbiont *Epulopiscium fishelsoni* was associated with decreased activity of endogenous amylase, protease, and lipase, possibly through suppression of intestinal pH. However, the results suggested that endosymbionts may contribute to host digestion of lipids by production of exogenous lipase. The effect of host starvation, the treatment used to eliminate epulos from the intestine of experimental fish, on other gastrointestinal microorganisms was not investigated. Therefore the role of epulos cannot be differentiated from that of the rest of the gut microbiota in this study.

4.3. Role of Exogenous Enzymes in Fishes

The amount of research on cellulose digestion in fishes belies the fact that this polysaccharide is an important dietary component in only a small proportion of fish species. Vascular plants, including seagrasses, contain cell walls composed of β-1,4-D-glucose—i.e., α-cellulose. The dominant marine plants are algae, and of the three major groups of marine macroalgae, only chlorophytes contain a significant proportion of α-cellulose in the cell wall (Kloareg and Quatrano 1988). Therefore, while cellulose digestion may be important in some herbivorous freshwater fishes and the few marine fishes that eat seagrass, it is largely irrelevant to the vast majority of marine herbivorous fishes. Since most of the studies on cellulose digestion in herbivorous fishes concern freshwater species, the resulting conclusions have little relevance to plant digestion or the importance of a gut microbiota in marine herbivorous fishes. The chemical composition of the diet must be considered in studies of plant digestion by fishes, as stressed by Montgomery and Gerking (1980).

A more general criticism involves the tendency to speculate on digestive mechanisms in host fish based on the results of isolated bacterial culture experiments. In this respect the work on cellulases in fish is representative of much of the literature on exogenous enzymes in fishes. As pointed out by Luczkovich and Stellwag (1993), much of the controversy concerning the source of cellulase activity in the gut of fishes has been due to the inability to isolate cellulolytic microorganisms. But substrate activity of cultures isolated from fish gut contents can be misleading. For example, a negative result for cellulolysis may be due to an inability to isolate the proportion of the microbiota that degrades cellulose in vivo. Conversely, the detection of cellulase activity in vitro does not demonstrate that cellulose is the preferred substrate in vivo. Many factors could affect the amount of polysaccharide utilized by endosymbionts in the gut, including accessibility of cell wall polysaccharides, substrate competition, low metabolic rates, or the presence of inhibitory substances (Salyers 1979). A demonstration that exogenous cellulolysis occurs in the host requires (1) the detection of exogenous cellulase in vivo, and (2) the detection of the resulting glucose or SCFA in the host tissues or blood. None of the enzyme studies discussed above fulfill these requirements, and therefore they cannot be considered evidence for the degradation by microorganisms of algal polysaccharides in the gut of algal-feeding fish. Future studies should attempt to isolate exogenous and endogenous enzymes from extracts of fish gut contents, and identify the dietary substrates which are fermented by the gut microbiota and assimilated by the host fish.

4.4. Summary and Conclusions

Studies on the enzyme activities of the gastrointestinal symbionts of fishes have done little to determine the role of microorganisms in the nutrition of fishes. Work on freshwater herbivorous fish suggests that the gut microbiota of some species may not degrade refractory polysaccharides, although this has yet to be convincingly demonstrated on populations of mixed gut bacteria. Virtually nothing is known about the metabolic pathways used by the great diversity of microorganisms in the alimentary tracts of marine herbivorous fishes.

5. Gastrointestinal Fermentation

Only very recently has fermentation in the gut of fishes received attention from scientists. Consequently, the number of published studies is limited, particularly in comparison to the amount of literature available on microbial digestion in higher vertebrates. The reasons for this probably include lack of access to specimens, the difficulty of maintaining captive fish compared to reptiles and mammals, and particularly the economic insignificance of most fishes. But research in this area has also been discouraged by the view that fish did not have the anatomical complexity, gastrointestinal microbiota, or pattern of enzyme activity characteristic of the symbioses between terrestrial animals and microorganisms. In part this view was the result of generalizations based on a small number of northern-hemisphere, primarily freshwater, fish taxa. But this view was also the result of the assumption that symbioses between fish and gastrointestinal microorganisms would conform to terrestrial models. The pioneering research of Fishelson et al. (1985) showed not only that fish could harbor a diverse gut microbiota, but also that the endosymbionts could be quite unlike their terrestrial analogs. Subsequently, several studies have investigated the distribution of microorganisms and their metabolic products, SCFA, in the gut of fishes. The following section will focus on SCFA levels in the gut of a variety of fish taxa. The uptake and metabolism of SCFA by the host fish will be discussed in the next section.

5.1. Fermentation in Freshwater Fishes

Paris et al. (1977) examined fermentation in the intestine of three species of freshwater fishes: the common carp *Cyprinus carpio*, the grass carp *Ctenopharyngodon idella*, and the trout *Salmo gairdneri*. The concentration of SCFA was less than 19 mmol L^{-1} in all species (Table 6.1), leading the authors to conclude that fermentation played a minor role in nutrition in these fishes. Interestingly, SCFA concentration in the intestine of the herbivorous grass carp was considerably lower than in the omnivorous common carp. A marked increase in acetate, propionate, and butyrate levels was detected in isolated intestinal preparations incubated at 37°C. This finding suggested that fermentation was potentially important in these species, given the right environmental conditions.

SCFA concentrations of intestinal contents from a euryhaline herbivorous cichlid, *Oreochromis mossambicus*, were analyzed using gas liquid chromatography (GLC) and high-performance liquid chromatography (HPLC) by Titus and Ahearn (1988). The intestine was divided into three sections for sampling, and acetate concentration ranged from 14 to 18 mmol L^{-1} with peak SCFA concentration occurring in the posterior segment (Table 6.1). Trace levels of propionate were detected by GLC throughout the intestine, but no other SCFA were present. These levels of SCFA were described as biologically significant, and the authors concluded that the long intestine in *O. mossambicus* provided an increased retention time, which enhanced the efficiency of fermentation (Titus and Ahearn 1988).

In a detailed study, Smith et al. (1996) analyzed SCFA levels in the intestine of five species of Illinois freshwater fishes. Total SCFA concentrations were similar to background sediment levels in the detritivorous *Dorosoma cepedianum*, while SCFA levels in the omnivores *Cyprinus carpio* and *Carpiodes cyprinus* were about half those recorded by Titus and Ahearn (1988) in *Oreochromis mossambicus* (Table 6.1). The carnivores *Pomoxis annularis* and *Micropterus salmoides* had comparable levels of total SCFA in the intestine to *O. mossambicus* (Table 6.1). Acetate was the predominant SCFA in all five species examined, and the high proportion of this SCFA in *C. carpio* indicated the digestion of fibre (Smith et al. 1996). The proportions of the branched-chain SCFA isobutyrate and isovalerate were highest in the piscivorous species *P. annularis* (29% of total SCFA) and lowest in the omnivore *C. carpio.* An interesting result was the similarity in SCFA concentration between upper and lower intestine samples in all species studied. Marine species investigated to date have shown considerable within-gut variation in SCFA concentration (see below).

Smith et al. (1996) found significant seasonal differences in intestinal SCFA concentration. Mean total SCFA concentrations in spring, summer, and fall samples of *Cyprinus carpio* and *Micropterus salmoides* intestine samples were 6.3 and 6.2, 12.5 and 38.3, and 3.8 and 11.2 mM, respectively. The water temperature in the study area ranged from 10 to 15°C in spring and fall, and from 19 to 24°C in summer. Smith and coauthors concluded that temperature seemed the likeliest cause of these seasonal differences in SCFA concentration, and suggested that low temperatures may impose a constraint on intestinal fermentation as a digestive mechanism.

5.2. Fermentation in Temperate and Subtropical Marine Fishes

Total SCFA concentration in the intestine of two species of subtropical herbivorous kyphosids was measured by Rimmer and Wiebe (1987). SCFA values for the cecal pouch were 15.5 and 18.4 mmol L^{-1} in *Kyphosus cornelii*, and 38.2 and 38.7 mmol L^{-1} in *K. sydneyanus* (Table 6.1). SCFA concentration in the rectum of the latter species was 16 mmol L^{-1}. This study was the first to demonstrate the presence of SCFA in the gut of marine herbivorous fishes, although the conclusion that the *Kyphosus* spp. were able to digest plant cell walls was unsubstantiated.

In a series of papers Mountfort and co-workers report on the fermentation characteristics in vitro of three microorganisms isolated from gut contents of the mullet *Aldrichetta forsteri.* The fermentation of glucose by the facultatively anaerobic marine fungus *Paecilomyces lilacinus* resulted in the formation of ethanol, acetate, and CO_2; lactate was also produced in low quantities (Mountfort and Rhodes 1991). Fermentation of pectin by the obligately anaerobic, gram-positive bacterium *Eubacterium* strain P-1 resulted in the formation of acetate and CO_2, with lesser quantities of ethanol, formate, and methanol (Mountfort et al. 1993). The fermentation of cellobiose and starch by *Eubacterium* strain P-1 produced mainly ethanol and CO_2, and lesser quantities of acetate and formate. Fermentation of alginic acid and hexose mono- and disaccharides by the obligate anaerobe *Clostridium grantii* resulted in the formation of acetate, with lesser quantities of ethanol, formate, and CO_2 (Mountfort et al. 1994). These results raise the question, not investigated by Mountfort and co-workers, of whether the stoichiometry patterns determined in vitro also operate in vivo. Without the selective retention of digesta, it is unlikely that the rapid gut transit times recorded for mugilids (Horn 1989) allow sufficient time for the production of significant quantities of fermentation end products in the gut of *Aldrichetta forsteri.*

Kandel et al. (1994) determined the relative proportions of SCFA in the hindgut of four species of herbivorous fishes by HPLC. Acetate was the only SCFA detected in the temperate stichaeid *Cebidichthys violaceus.* Acetate, propionate, butyrate, isobutyrate, valerate, and isovalerate were each present in the subtropical scorpidid *Medialuna californiensis* and the subtropical/tropical kyphosids *Kyphosus vaigiensis* and *K. bigibbus.* There was a higher concentration of propionate than acetate in each of the last three species, and also more butyrate and isobutyrate than acetate in *K. vaigiensis.* These SCFA proportions are unlike those recorded for any terrestrial vertebrate, where acetate, propionate, and butyrate are produced in a ratio varying from about 75:15:10 to 40:40:20 (Bergman 1990). Three of the eight *M. californiensis* samples examined by Kandel et al. (1994) were analyzed by both HPLC and GLC (J. Kandel, personal communication). The two techniques yielded similar results, which were very different from the other five *M. californiensis* examined, with acetate, propionate, butyrate (plus isovalerate), and valerate (plus isovalerate) in a ratio of 55:21:13:11 (J. Kandel, personal communication). The latter SCFA proportions are much more in line with results from other fishes (e.g., Clements et al. 1995, Clements and Choat 1995) and reptiles (Bjorndal et al. 1991). The total SCFA concentration in the hindgut of the three *M. californiensis* specimens examined by both GLC and HPLC was about 35 mmol L^{-1} (Table 6.1).

Clements et al. (1994) measured SCFA concentration with GLC at five points along the alimentary tract of three temperate Australasian herbivorous fishes—the odacids *Odax cyanomelas* and *O. pullus*, and the aplodactylid *Crinodus lophodon.* SCFAs were distributed predominantly in the posterior half of the intestine of all three species, none of which has a specialized fermentation chamber (Figs. 6.4 and 6.12). Total SCFA concentration was very similar in the two odacids, and slightly lower in the aplodactylid (Table 6.1). These values are about half those reported from the cecum and colon of green turtles feeding on algae (Bjorndal et al. 1991). SCFAs produced in the cecum of green turtles feeding on seagrass provide 15.2% of daily energy requirements (Bjorndal 1979), but a comparison with temperate fish is complicated by the different ambient temperatures experienced by these ectothermic vertebrates.

The ratio of acetate:propionate:butyrate:valerate in the gut section containing the highest SCFA concentration was 83:8:9:1 in *O. cyanomelas*, 64:21:14:1 in *O. pullus*, and 74:17:9:0 in *C. lophodon* (Clements et al. 1994). The SCFA proportions in *O. cyanomelas* and *C. lophodon* are broadly similar to those reported by Bjorndal et al. (1991) for green turtles feeding on algae. The difference in SCFA ratio between the two odacid species is interesting, as both species have similar diets (Choat and Clements 1992). This difference may be due to variation in microbiota composition between the two species (Clements 1991a); another possibility is a difference in the main fermentation substrate. There is a seasonal influence in the selection of specific algal tissues by herbivorous odacids (Clements and Choat 1993); thus, annual periodicity in algal reproductive structures and storage carbohydrates may influence the main substrates available for fermentation in these fishes. Since the degree of oxidation and solubility of the fermentation substrate influence SCFA proportions in other vertebrates (Bjorndal 1979, Macfarlane and Macfarlane 1993), it is likely that SCFA proportions may vary on a seasonal basis in some herbivorous fishes.

Total SCFA concentration is considerably lower in carnivorous labrids than the related herbivorous odacids (Table 6.1). Owing to the very short relative gut length of the labrids, gut volume and, thus, the total amount of SCFA available for host metabolism in these fishes are only a fraction of that of the herbivorous odacids.

5.3. *Fermentation in Tropical Marine Fishes*

Only one study to date has investigated fermentation in the diverse tropical marine taxa Acanthuridae, Scaridae, and Siganidae. Clements and Choat (1995) analyzed with GLC samples of gut contents taken from the stomach and four points along the intestine of 32 species of coral reef fish. The taxa investigated, which were mainly herbivores, included species of the families Acanthuridae, Kyphosidae, Pomacanthidae, Scaridae, and Siganidae. In all species the highest concentration of SCFA occurred in the posterior intestine. Total SCFA concentration in the posterior intestine was lowest in sediment-feeding species, with mean concentrations of 3 to 10 mmol L^{-1} in scarids and detritivorous acanthurids (Table 6.1). All of these species have a mechanism for mechanical trituration of ingested material. The highest concentrations of total SCFA were recorded in *Kyphosus cinerascens*, the pomacanthid *Centropyge bicolor*, and the macroalgal-feeding surgeonfish *Naso lituratus* and *N. unicornis* (Table 6.1). The sediment-feeding species, which contained the lowest concentrations of SCFA, also contained the highest proportions of the branched-chain SCFA isovalerate, a product of leucine catabolism (Massey et al. 1976, Allison 1978). The planktivore *Acanthurus mata* contained the highest proportion of isovalerate, with an acetate:propionate:butyrate:isovalerate ratio of 57:2:3:37. The acetate proportion in this species was the lowest recorded in the study.

Interestingly, substantial amounts of SCFA were detected in some nonherbivorous acanthurids. Both *Naso vlamingii*, which feeds on zooplankton and fish feces, and *N. hexacanthus*, which feeds on salps and other gelatinous zooplankton, contained comparatively high levels of SCFA (Clements and Choat 1995; Table 6.1). The planktivore *Caesio cuning*, which co-occurs with and has a very similar diet to *N. hexacanthus*, contains a far lower concentration of SCFA (Table 6.1). Planktivorous acanthurids are thought to have evolved from herbivorous ancestors (Winterbottom and McLennan 1993), so fermentation in the intestine of species such as *N. hexacanthus* may be a symplesiomorphic feature. Nevertheless, fermentation in the posterior intestine of planktivorous acanthurids may result in more efficient digestion than in non-fermenting planktivores, a possibility also suggested by the fact that *N. vlamingii* eats the feces of planktivorous fish (Robertson 1982). Gut volume in the planktivorous nasine species is not much less than that of congeneric herbivores (Figs. 6.5–6.7).

Variation in SCFA proportions suggests that diet composition has an important influence on fermentation in tropical marine fishes, as it does in terrestrial vertebrates (Stevens 1978, Macfarlane and Macfarlane 1993). Clements and Choat (1995) noted a negative correlation between the amount of highly reduced carbohydrate in the diet and the proportion of acetate in the gut in herbivorous acanthurids. Thus, species such as *Naso unicornis*, which have a macroalgal diet rich in uronic acids, have a very high proportion of acetate to propionate and butyrate. Conversely, species such as *Acanthurus nigricans*, which have a diet of predominantly red algae containing highly reduced galactans and mannans, have a lower proportion of acetate to propionate and butyrate. Carnivorous and microphagous species tend to have a high ratio of isovalerate to other SCFA, which is presumably the result of the amount of leucine in the diet. Protein may therefore be an important substrate for fermentation in some fishes.

5.4. Summary and Conclusions

In general the amount of SCFA in the gut of fishes is lower than that found in vertebrates with foregut or hindgut fermentation (Bjorndal 1979, Stevens 1988, Bjorndal and Bolten 1990). Comparisons with mammals and birds are of limited value owing to the elevated body temperatures in these animals, and many reptiles also function at a temperature well above the maximum experienced by most fishes (Standora et al. 1982, Zimmerman and Tracy 1989). Nevertheless, as ectotherms with hindgut fermentation, herbivorous reptiles provide the best comparison for fishes. Green turtles feeding on seagrass have in excess of 200 mmol L^{-1} SCFA in the colon (Bjorndal et al. 1979); Florida red-bellied turtles, green turtles, feeding on algae, red-footed tortoises, and green iguanas all have have about 60 mmol L^{-1} SCFA in the cecum (Guard 1980, Bjorndal and Bolten 1990, Bjorndal et al. 1991). These values are somewhat higher than those reported from fishes, where SCFA concentration in the gut ranges from less than 1 mmol L^{-1} in carnivorous labrids to over 40 mmol L^{-1} in tropical herbivorous fishes.

The great range of SCFA concentrations in fishes suggests that generalizations on the levels of fermentation in these animals are inappropriate. Species with the highest concentrations of SCFA measured to date are all marine herbivores, although some carnivorous acanthurids appear to ferment components of their diet. Marine carnivorous taxa studied have very low quantities of SCFA, although too few species have been investigated to draw any general conclusions. Summer samples of the freshwater carnivore *Micropterus salmoides* had a relatively high concentration of SCFA in the intestine (Smith et al. 1996), particularly when compared to other species in the same habitat. A comparison of the SCFA concentrations in fish species studied to date indicates that the taxa in which fermentation is most likely to be nutritionally important are the acanthurids, kyphosids, odacids, and pomacanthids.

Many authors have suggested that fermentation in ectothermic vertebrates requires high ambient temperatures (e.g., Zimmerman and Tracy 1989, Lésel 1990, Kandel et al. 1994). The high SCFA concentrations in temperate odacids and aplodactylids (Clements et al. 1994) casts doubt on the generality of this hypothesis. The *Odax cyanomelas* specimens examined by Clements et al. (1994) were collected off Sydney between April and August, when sea temperatures range between 14 and 18°C (Tate et al. 1989). The New Zealand species *O. pullus*, which also contains considerable quantities of SCFA in the intestine (Clements et al. 1994), is subject to water temperatures as low as 10 to 12°C for long periods (Choat and Clements 1992). The presence of elevated SCFA levels in these odacids suggests that low temperature does not prevent the production of substantial quantities of SCFA.

The lack of information on the major substrates fermented by the gastrointestinal microorganisms of fishes makes the importance of microbial digestion in these animals difficult to assess. In terrestrial vertebrate herbivores the cellulolytic activities of gut symbionts allow the host to salvage energy from plant cell walls by the uptake of SCFA. Therefore, the importance of microbial digestion can be estimated by comparing the amounts of energy derived from cell wall and cell contents. In herbivorous fishes the great diversity in the chemical composition of algae complicates the situation, and division of the diet into cell wall and cell content categories is largely uninformative with respect to digestion. Since the chemical composition of many algal species is poorly understood, it is difficult to determine the main dietary substrates of herbivorous fishes without performing carbohydrate analyses on stomach contents. This problem is particularly acute with grazing species, such as some aplodactylids and acanthurids, which eat a great variety of algal taxa. Comparisons of SCFA proportions from fish species for which the diet is known suggest that there is considerable variability in the substrate fermented by the microbiota. There is an urgent need for studies that determine the diet of herbivorous fishes at a biochemical level, and identify the substrates that are utilized at various stages of the digestive process. Information on the rate and pathways of fermentation in fishes is required to understand the relationships among host, diet, and gastrointestinal microorganisms.

6. Uptake and Metabolism of SCFA

As would be expected given the lack of work on fermentation in fishes, very little information is available on SCFA metabolism in these animals. Previous studies assumed that fermentative digestion in the intestine of fishes accrued metabolic benefits to the host (e.g., Rimmer and Wiebe 1987, Horn 1992, Horn and Messer 1992). Although this assumption was not tested, the widespread utilization of SCFA by other vertebrates (Stevens 1978, 1988) suggests that fishes would be unusual if they did not assimilate these end products of microbial metabolism. In mammals 95% of the SCFA produced by microbial fermentation in the large intestine and colon is absorbed, and the rate of SCFA transport is linearly related to lumen concentration (Titus and Ahearn 1992). Transepithelial SCFA transport has been investigated in only one species of fish, while the presence of SCFA in arterial blood provides indirect evidence for the uptake of these metabolites by several other species. Data on the utilization of SCFA by fishes are even more limited, but suggest that intestinal fermentation does contribute to host metabolism.

6.1. Uptake of SCFA

In an excellent series of papers, Titus and Ahearn (1988, 1991, 1992) investigated the mechanism of acetate transport in the intestine of the cichlid *Oreoch-*

romis mossambicus. SCFA uptake occurs by electroneutral carrier-mediated exchange, in which acetate is exchanged specifically for bicarbonate originating in the blood or produced intracellularly. This process results in a net movement of acetate from the intestinal lumen, across the intracellular space, and into the blood. Titus and Ahearn calculated that, at a luminal concentration of 15 mmol L^{-1}, 60% of acetate uptake across the brush border membrane occurs by passive diffusion, while 21% of acetate transport across the basolateral membrane occurs by passive diffusion (Titus and Ahearn 1991). Carrier-mediated transport is more important at low luminal SCFA concentrations, and acts as an accessory to diffusion at high luminal SCFA concentrations (Titus and Ahearn 1992). This suggests that the SCFA transport mechanism in *O. mossambicus* allows the the fish to maximize the uptake of SCFA at low luminal concentrations. Combined with the relatively high levels of SCFA detected in many species of herbivorous fish (Rimmer and Wiebe 1987, Clements et al. 1994, Clements and Choat 1995), this result indicates that SCFA are an important source of nutrients for many species of fishes. Another direct demonstration of SCFA uptake by fishes is provided by Smith et al. (1996), who orally dosed *Cyprinus carpio* with 2-^{14}C acetate. The label was subsequently detected in liver, muscle, and blood.

Indirect evidence for the uptake of fermentation products is provided by the presence of SCFA in the blood of fishes (Clements et al. 1994, Clements and Choat 1995). The ability of fish to produce acetate endogenously (see below) complicates the assumption that SCFA in the blood is derived from fermentation in the gut. Titus and Ahearn (1991) assumed that blood acetate concentration in *Oreochromis mossambicus* would be about 1 mmol L^{-1}, similar to levels in other vertebrates. Literature values estimated subsequently by GLC from a range of temperate and tropical fishes support this assumption. The concentration of acetate in arterial blood of the temperate herbivores *Odax cyanomelas* and *Crinodus lophodon* is 1.74 $\pm$ 0.17 and 1.79 $\pm$ 0.20 mmol L^{-1} respectively (Clements et al. 1994). The concentration of acetate in arterial blood was 1.13 $\pm$ 0.20 mmol L^{-1} in the carnivorous labrid *Notolabrus gymnogenis*, and 1.22 $\pm$ 0.28 mmol L^{-1} in the carnivorous odacid *O. acroptilus*. Acetate was undetected in arterial blood of the labrid *Achoerodus viridis* (Clements, unpublished). The levels of acetate in the blood of *N. gymnogenis* and *O acroptilus*, which have very low levels of SCFA in the gut, are similar to although lower than those in the herbivorous species. This result indicates that the concentration of SCFA in the blood of temperate species is not proportional to the concentration in the gut, an expected result. The concentration of SCFA in the blood would depend on its rate of uptake from the gut, the rate of SCFA metabolism by enterocytes, the rate of SCFA transport across the gut epithelium, and the rate of blood flow. The unexpectedly high concentration of acetate in the blood of carnivorous species may reflect a higher rate of endogenous acetate production in these taxa, or a lower rate of acetate utilization in carnivorous species compared to herbivorous species.

The concentration of acetate in the arterial blood of 32 species or tropical fishes ranged from 0.45 ± 0.11 mmol L^{-1} in the parrotfish *Scarus niger* to 3.80 ± 1.89 mmol L^{-1} in the surgeonfish *Naso brachycentron* (Clements and Choat 1995). As with the temperate species, plasma concentration of acetate was poorly predicted by intestinal concentration. The planktivorous caesionid *Caesio cuning* had an arterial acetate concentration of 1.08 ± 0.03 mmol L^{-1} (Clements, unpublished), similar to values from herbivorous species with much higher levels of SCFA in the gut.

In addition to acetate, the branched-chain SCFA isovalerate was detected in the plasma of the detritivorous surgeonfish *Ctenochaetus striatus,* and all *Acanthurus* and *Siganus* spp. examined (Clements and Choat 1995). Comparable levels of isovalerate in humans lead to severe acidosis (Tanaka et al. 1966), so the presence of this SCFA in the blood of fishes is surprising. There was a strong relationship between the levels of isovalerate in the blood and gut of the surgeonfish species examined, suggesting that this SCFA was derived from fermentation in the gut. Other groups of species—e.g., for example scarids and some of the nasine surgeonfish—contained isovalerate in the gut but not in the plasma. The fact that only a proportion of the species that contain isovalerate in the gut also contain this SCFA in the plasma suggests interspecific variation in its transport or metabolism, urging caution in the extension of results to other species.

6.2. Metabolic Utilisation of SCFA

If acetate produced in the gut by gastrointestinal microorganisms is utilized by the host fish for energy purposes, then one would expect to find the enzyme necessary for the immediate utilization of this SCFA, acetyl CoA synthetase, in fish tissues. Acetyl CoA synthetase activates acetate to acetyl CoA, which can then be used as a substrate in the citric acid cycle or for lipid synthesis (Ballard 1972). The distribution of acetyl CoA synthetase activity in five tissues from five species of temperate fishes is presented in Table 6.2. In the herbivorous species

Table 6.2. Activity of acetyl CoA synthetase in temperate fishes

			Tissue				
Species	Family	Diet	Liver	Heart	Muscle	Gut	Kidney
Crinodus lophodon	Aplodactylidae	herbivore	1.20 ± 0.43 (8)	1.33 ± 0.84 (8)	0.16 ± 0.07 (8)	1.77 ± 0.24 (8)	6.48 ± 3.18 (10)
Odax cyanomelas	Odacidae	herbivore	2.74 ± 0.44 (22)	1.79 ± 0.19 (22)	0.47 ± 0.07 (18)	1.82 ± 0.23 (18)	3.55 ± 0.51 (12)
O. acroptilus	Odacidae	carnivore	0.50 ± 0.29 (3)	1.26 ± 0.51 (2)	0.16 ± 0.07 (3)	2.25 ± 1.07 (3)	—
Achoerodus viridis	Labridae	carnivore	1.46 ± 0.17 (5)	3.33 ± 1.13 (5)	0.25 ± 0.03 (4)	4.95 ± 0.91 (5)	31.13 ± 9.54 (5)
Notolabrus gymnogenis	Labridae	carnivore	1.06	1.73	0.17	0.81	3.58

Results are expressed as $nmol{\cdot}sec^{-1}{\cdot}g^{-1}$ tissue. Results are means ± SE of the mean. Values in parentheses indicate number of independent experiments. Assay method is described in Clements et al. (1994). Data for *O. cyanomelas* and *C. lophodon* from Clements et al. (1994).

Odax cyanomelas and *Crinodus lophodon*, activity is lowest in the muscle and highest in the kidney, with intermediate levels of activity in the liver, heart, and gut. The results in Table 6.2 show that acetyl CoA synthetase activity is present in both herbivorous and nonherbivorous species. This finding compares with those from mammals, in which nonherbivores clear acetate from the blood faster than herbivores. This suggests that in fish as in mammals acetyl CoA synthetase is a constitutive enzyme—i.e., that its synthesis in fish is not strongly regulated by the amount of acetate that has to be metabolized. Clearly, many tissues in these fishes have the ability to use acetate, and the high activity in the kidney may allow the fish to salvage energy from acetate in the blood. The distribution of activity in these fishes was similar to that reported from ruminant and hindgut fermenting mammals (Clements et al. 1994). Further evidence that fish are capable of activating acetate is provided by studies on european eels, in which C^{14} acetate added to aquaria and injected directly into fish tissues was incorporated in vivo into tissue lipids (Hansen and Abraham 1983, Hansen 1987).

Clements et al. (1994) also examined the ability of *Odax cyanomelas* and *Crinodus lophodon* to produce acetate endogenously, by the hydrolysis of acetyl CoA to acetate with the enzyme acetyl CoA hydrolase. The highest activity of acetyl CoA hydrolase in both species was found in the liver, with lower levels of activity present in the heart, muscle, and gut. Only trace levels of acetyl CoA hydrolase activity were detected in the kidney of *O. cyanomelas*, while in *C. lophodon* levels were lower than but comparable to levels in other tissues. Acetyl CoA, unlike acetate, cannot diffuse across the plasma membrane, and therefore the presence of acetyl CoA hydrolase in fish tissues allows the redistribution of an oxidizable substrate throughout the body.

6.3. Summary and Conclusions

The limited data available at present suggest that fish have the ability to transport acetate from the intestinal lumen into the blood and then use it for energy purposes or in lipid synthesis. Titus and Ahearn (1992) proposed that carrier-mediated transport in fishes became less important relative to diffusion as luminal SCFA concentration rose, suggesting that these metabolites are utilised even in fishes with very low levels of SCFA in the gut. The carrier-mediated acetate transport system described by Titus and Ahearn (1988, 1991) for *Oreochromis mossambicus* is similar to that found in rats (Mascolo et al. 1991, Reynolds et al. 1993), suggesting that the mechanisms of SCFA transport may be similar in all vertebrates.

There appears to be no clear relationship between the level of acetate in the blood and that in the gut of a variety of fish species. The presence of acetate in the blood of carnivorous species with low levels of fermentation in the gut indicates that SCFA may be an important blood fuel in many fishes irrespective of diet. The widespread activity of acetyl CoA hydrolase in fish tissues makes it impossible at present to interpret the contribution of intestinal fermentation to the blood SCFA pool, although the presence of isovalerate in the blood of several species is strong evidence that some of the SCFA in the blood is produced by gastrointestinal microorganisms. The absence of propionate and butyrate from the arterial blood of species with high levels of these SCFA in the gut is not surprising. In mammals, most propionate is removed from portal blood by the liver (Bergman 1990); much of the butyrate is metabolized by enterocytes, for which it provides a major source of energy. Therefore only trace amounts of these SCFA are present in the arterial circulation. In fishes it is likely that both of these SCFA are present in portal blood, and therefore contribute to the metabolic requirements of both the liver and gut.

7. Conclusions

The main site of fermentation in fishes seems to be the posterior intestine. Although hindgut chambers separated from the remainder of the gut by a sphincter are present in some species, high concentrations of SCFA are also found in species lacking such specialized morphology. Information on digesta flow and retention is extremely limited for herbivorous taxa, and progress in this area has been largely confined to theoretical considerations (e.g., Horn and Messer 1992). There is a great need for detailed studies on the anatomy and physiology of the gastrointestinal tract of herbivorous fishes.

A number of recent studies have found diverse endosymbiotic communities within the gut of temperate, subtropical, and tropical marine herbivorous fishes. This descriptive work has laid the foundation for more detailed microbiological research (e.g., Mountfort et al. 1993, 1994). Some of the organisms discovered to date have proved to be of general biological significance, such as the giant epulo bacteria inhabiting the intestines of some surgeonfish species. These spectacular examples aside, very little is known of the anaerobic microbiota of marine herbivorous fishes. Particular attention is drawn to the lack of information on the role and identity of intestinal zooflagellates and fungi.

A great proportion of the work on the role of endosymbionts in fishes has focused on enzyme production. Much of this work involves in vitro experiments on isolated bacterial strains; as a consequence, the results are inconclusive with respect to in vivo conditions. Future studies must take into account the substrate requirements of the microbiota as a whole, in conditions that approximate as closely as possible the environment of the gut. Since carbohydrase activities are found both extracellularly and within the bacterial fraction of gut contents (Cummings and Macfarlane 1991), attempts to delineate exogenous and endogenous enzymes will require considerable sample preparation.

The substantial levels of SCFA found in some of the few taxa investigated suggest that fermentation is a potential source of energy for many species of fishes. Most fishes examined have a relatively high concentration of acetate in the plasma compared to mammals, and this SCFA presumably serves as a readily oxidizable source of energy. The contribution of fermentation by gastrointestinal microorganisms to the plasma SCFA pool is uncertain and must be a priority for future studies. Experiments using ^{14}C-labeled substrates with perfused gut preparations or whole fish appear to be the most suitable methods, since SCFA production rates estimated by incubation of gut contents in vitro are potentially unreliable (Mathers and Annison 1993). Good information on gut retention time, for both fluid and particulate digesta, and gut volume relative to body mass, is also required to establish the importance of fermentation in fishes. It should be noted that there are many potential roles for gastrointestinal microorganisms in fishes other than the supply of nutrients in the form of SCFA. Production of exogenous digestive enzymes, retention of water, and conservation of nitrogen are just a few of the possible roles played by the microbial endosymbionts of fishes.

Although data are very limited, information on marine herbivorous fishes suggests that intestinal fermentation may contribute to host metabolism in these animals. Given the amount of information on fermentation available for other groups of animals, why have fishes been so sadly neglected? As suggested earlier, part of the problem is lack of access to specimens: most marine herbivorous fishes occur in either tropical or southern-hemisphere waters. Although many of these fishes are abundant and important to the ecology of coral reefs, they are generally only important as food for subsistence fisheries in developing countries. The economic imperative that has driven research on animals of great domestic importance such as the ruminants has been lacking for fishes. But it seems also that research on fermentation in fishes has been delayed by the perpetuation in the literature of a misconception: that cellulose digestion and therefore microbial fermentation are extremely rare in marine herbivores. This misconception was repeated as recently as 1992 in a review of plant-herbivore interactions (Hay and Steinberg 1992), and results from a failure to understand the biochemical composition of the diet of the majority of marine herbivores. With the exception of animals feeding upon seagrasses, such as some green turtles, most marine herbivores encounter little cellulose. The cell walls of marine algae comprise a great variety of polysaccharides such as alginic acid, mannans, xylans, and galactans. It will be these compounds and the cell storage polysaccharides of algae, such as laminarin and floridean starch, that will serve as the substrates for fermentation in the majority of marine herbivores.

Research on fermentation in fishes needs to address the fundamental differences between the diets of terrestrial and aquatic herbivores, for it seems that part of the problem in the past has been a failure to identify the real dietary substrates

Table 6.3. Parameters of diet, digestion, feeding behavior, and endosymbionts in some herbivorous surgeonfishes

	Naso unicornis *N. lituratus*	*Acanthurus lineatus* *A. nigrofuscus* *A. triostegus*
Diet	brown macroalgae	turfing red and green algae
Growth rate of dietary algae	slow?	rapid
Major dietary polysaccharides	β-linked	α-linked
Dietary protein	moderate	high
Dietary fiber	high	low
Chemical defenses in diet	moderate to high	low
Readily accessible calories in diet	low	moderate to high
Feeding periodicity	episodic	continuous
Gut passage time	slow?	rapid
Defecation rate	low	high
Rate of digestion	slow?	rapid
Hindgut swelling	present	absent
Size of bacterial symbionts	small to medium ($< 100\ \mu m$)	very large ($> 300\ \mu m$)
Locomotory ability of symbionts	low to average	high
Primary digestive carbohydrases	laminarinase, β-glucosidase, alginase?	amylase, galactase
Source of digestive carbohydrases	exogenous?	endogenous?
Concentration of SCFA in gut	$\approx 40\ mmol \cdot L^{-1}$	$\approx 17\ mmol \cdot L^{-1}$
Contribution of SCFA to host metabolism	high?	low?

of fishes. Recent work on marine herbivorous fishes suggests that there may be great diversity in the mechanics of fermentation processes in different species. A speculative example of this diversity is provided in Table 6.3, where two groups of tropical herbivorous fishes are contrasted on the basis of parameters of diet, digestion, feeding behavior, and endosymbionts. This scheme is modeled on a comparison between two primate species (Milton 1993), and is intended to illustrate the variation inherent within even a single family of fishes. Many of these fishes have remained largely unchanged since the Eocene (Choat 1991, Winterbottom and McLennan 1993), so symbiotic relationships between host and microorganism have the potential to be ancient. It is hoped that this review will encourage physiologists, microbiologists, and biochemists to investigate these symbioses, which may help to shed light on the evolution of fermentative processes in terrestrial vertebrates.

Acknowledgments

This review was substantially improved by the advice and criticism of Howard Choat, Ian Hume, Michael Slaytor, Ed Stevens, and Pam Veivers. I am grateful to Mike Horn, Judy Kandel, Doug Mountfort, and Timothy Smith for providing unpublished manuscripts, and Graham Bailey for access to GLC facilities. I thank the Australian Research Council for support in the form of a postdoctoral fellowship and a small grant through the University of Sydney.

References

Al-Hussaini AH (1947) The feeding habits and the morphology of the alimentary tract of some teleosts living in the neighbourhood of the Marine Biological Station, Ghardaqa, Red Sea. Publ Mar Biol Stn Al Ghardaqa 5:1–61.

Allison MJ (1978) Production of branched-chain volatile fatty acids by certain anaerobic bacteria. Appl Environ Microbiol 35:872–877.

Anderson TA (1987) Utilization of algal cell fractions by the marine herbivore the luderick, *Girella tricuspidata* (Quoy and Gaimard). J Fish Biol 31:221–228.

Anderson TA (1991) Mechanisms of digestion in the marine herbivore, the luderick, *Girella tricuspidata* (Quoy and Gaimard). J Fish Biol 39:535–547.

Angert ER, Clements KD, Pace NR (1993) The largest bacterium. Nature 362:239–241.

Annison G (1993) The chemistry of dietary fibre. In: Samman S, Annison G, eds. Dietary Fiber and Beyond—Australian Perspectives, Vol. 1, pp. 1–18. Sydney: Nutrition Society of Australia, Occasional Publications.

Ballard FJ (1972) Supply and utilisation of acetate in mammals. Am J Clin Nutr 25: 773–779.

Bellwood DR, Choat JH (1990) A functional analysis of grazing in parrotfishes (family Scaridae): the ecological implications. Environ Biol Fish 28:189–214.

Benavides AG, Cancino JM, Ojeda FP (1994) Ontogenetic changes in gut dimensions and macroalgal digestibility in the marine herbivorous fish, *Aplodactylus punctatus*. Funct Ecol 8:46–51.

Bergman EN (1990) Energy contributions of volatile fatty acids from the gastrointestinal tract in various species. Physiol Rev 70:567–590.

Bitterlich G (1985) Digestive enzyme pattern of two stomachless filter feeders, silver carp, *Hypophthalmichthys molitrix* Val., and bighead carp, *Aristichthys nobilis* Rich. J Fish Biol 27:103–112.

Bjorndal K (1979) Cellulose digestion and volatile fatty acid production in the green turtle, *Chelonia mydas*. Comp Biochem Physiol [A] 63:127–133.

Bjorndal KA, Bolten AB (1990) Digestive processing in a herbivorous freshwater turtle: consequences of small-intestine fermentation. Physiol Zool 63:1232–1247.

Bjorndal KA, Bolten AB, Moore JE (1990) Digestive fermentation in herbivores: effect of food particle size. Physiol Zool 63:710–721.

Bjorndal KA, Suganuma H, Bolten AB (1991) Digestive fermentation in green turtles, *Chelonia mydas,* feeding on algae. Bull Mar Sci 48:166–171.

Bowen SH (1976) Mechanism for digestion of detrital bacteria by the cichlid fish *Sarotherodon mossambicus* (Peters). Nature 260:137–138.

Buddington RK, Diamond JM (1987) Pyloric ceca of fish: a ''new'' absorptive organ. Am J Physiol 252:G65-G76.

Cahill MM (1990) Bacterial flora of fishes: a review. Microb Ecol 19:21–41.

Choat JH (1991) The biology of herbivorous fishes on coral reefs. In: Sale PF, ed. The Ecology of Fishes on Coral Reefs. pp. 120–155. San Diego: Academic Press.

Choat JH, Clements KD (1992) Diet in odacid and aplodactylid fishes from Australia and New Zealand. Aust J Mar Freshwat Res 43:1451–1459.

Christiansen DC, Klungsøyr L (1987) Metabolic utilization of nutrients and the effects of insulin in fish. Comp Biochem Physiol 88B:701–711.

Clements KD (1991a) Endosymbiotic communities of two herbivorous labroid fishes, *Odax cyanomelas* and *O. pullus.* Mar Biol 109:223–229.

Clements KD (1991b) Gut microorganisms of surgeonfishes (family Acanthuridae). Unpublished PhD thesis, James Cook University of North Queensland, Townsville, Australia.

Clements KD, Bellwood DR (1988) A comparison of the feeding mechanisms of two herbivorous labroid fishes, the temperate *Odax pullus* and the tropical *Scarus rubroviolaceus.* Aust J Mar Freshwat Res 39:87–107.

Clements KD, Bullivant S (1991) An unusual symbiont from the gut of surgeonfishes may be the largest known prokaryote. J Bacteriol 173:5359–5362.

Clements KD, Choat JH (1993) Influence of season, ontogeny and tide on the diet of the temperate marine herbivorous fish *Odax pullus* (Odacidae). Mar Biol 117:213–220.

Clements KD, Choat JH (1995) Fermentation in tropical marine herbivorous fishes. Physiol Zool 68:355–378.

Clements KD, Sutton DC, Choat JH (1989) Occurrence and characteristics of unusual protistan symbionts from surgeonfishes (Acanthuridae) of the Great Barrier Reef, Australia. Mar Biol 102:403–412.

Clements KD, Gleeson VP, Slaytor M (1994) Short-chain fatty acid metabolism in temperate marine herbivorous fish. J Comp Physiol B. 164:372–377.

Cummings JH, Macfarlane GT (1991) The control and consequences of bacterial fermentation in the human colon. J Appl Bacteriol 70:443–459.

Fänge R, Grove D (1979) Digestion. In: Hoar WS, Randall DJ, Bretts JR, eds. Fish Physiology, Vol. VIII, pp. 161–260. Energetics and Growth. New York: Academic Press.

Fishelson L, Montgomery WL, Myrberg AA (1985) A unique symbiosis in the gut of tropical herbivorous surgeonfish (Acanthuridae: Teleostei) from the Red Sea. Science 229:49–51.

Fris MB, Horn MH (1993) Effects of diets of different protein content on food consumption, gut retention, protein conversion, and growth of *Cebidichthys violaceus* (Girard), an herbivorous fish of temperate zone marine waters. J Exp Mar Biol Ecol 166:185–202

Fryer G, Iles TD (1972) The Cichlid Fishes of the Great Lakes of Africa. Neptune City, NJ: TFH Publications.

Gohar HAF, Latif AFA (1959) Morphological studies on the gut of some scarid and labrid fishes. Publ Mar Biol Stn Al-Ghardaqa, Red Sea. Publ Mar Biol Stn Al Ghardaqa 10: 145–189.

Gregory WK (1933) Fish skulls. A study of the evolution of natural mechanisms. Trans Am Phil Soc 23:75–481.

Grim JN (1993) Description of somatic kineties and vestibular organization of *Balantidium jocularum* sp. n., and possible taxonomic implications for the class Litostomatea and the genus *Balantidium.* Acta Protozool 32:37–45.

Guard CL (1980) The reptilian digestive system: general characteristics. In: Schmidt-Nielsen K, Bolis L, Taylor CR, Bentley PJ, Stevens CE, eds. Comparative Physiology: Primitive Mammals, pp. 43–51. Cambridge: Cambridge University Press.

Hansen HJM (1987) Comparative studies on lipid metabolism in various salt transporting organs of the European eel (*Anguilla anguilla*). Mono-unsaturated phosphatidylethanolamine as a key substance. Comp Biochem Physiol 88B:323–332.

Hansen HJM, Abraham S (1983) Influence of temperature, environmental salinity and fasting on the patterns of fatty acids synthesized by gills and liver of the european eel (*Anguilla anguilla*). Comp Biochem Physiol 75B:581–587.

Hansen JA, Alongi DM, Moriarty DJW, Pollard PC (1987) The dynamics of benthic microbial communities at Davies Reef, central Great Barrier Reef. Coral Reefs 6:63–70.

Hay ME, Steinberg PD (1992) The chemical ecology of plant-herbivore interactions in marine versus terrestrial communities. In: Rosenthal GA, Berenbaum MR, eds. Herbivores: Their Interactions with Secondary Plant Metabolites, Vol. 2, 2nd ed, pp. 371–413. New York: Academic Press.

Helpher B (1988) Nutrition of Pond Fishes. Cambridge: Cambridge University Press.

Hofer R (1988) Morphological adaptations of the digestive tract of tropical cyprinids and cichlids to diet. J Fish Biol 33:399–408.

Horn MH (1989) Biology of marine herbivorous fishes. Oceanogr Mar Biol Annu Rev 27:167–272.

Horn MH (1992) Herbivorous fishes: feeding and digestive mechanisms. In: John DM, Hawkins SJ, Price JH, eds. Plant-Animal Interactions in the Marine Benthos. Systematics Assosciation Special Volume No. 46, pp. 339–362. Oxford: Clarendon Press.

Horn MH, Messer KS (1992) Fish guts as chemical reactors: a model of the alimentary canals of marine herbivorous fishes. Mar Biol 113:527–535.

Horsley RW (1977) A review of the bacterial flora of teleosts and elasmobranchs, including methods for its analysis. J Fish Biol 10:529–553.

Jones CG (1984) Microorganisms as mediators of plant resource exploitation by insect herbivores. In: Price PW, Slobodchikoff CN, Gaud WS, eds. A New Ecology: Novel Approaches to Interactive Systems. pp. 53–99. New York: John Wiley.

Jones RS (1968) Ecological relationships in Hawaiian and Johnston Island Acanthuridae (surgeonfishes). Micronesica 4:309–361.

Kamei Y, Sakata T, Kakimoto D (1985) Microflora in the alimentary tract of *Tilapia:* characterization and distribution of anaerobic bacteria. J Gen Appl Microbiol 31: 115–124.

Kandel JS, Horn MH, Van Antwerp W (1994) Volatile fatty acids in the hindguts of herbivorous fishes from temperate and tropical marine waters. J Fish Biol. 45:527–529.

Kapoor BG, Smit H, Verighina IA (1975) The alimentary canal and digestion in teleosts. Adv Mar Biol 13:109–230.

Kloareg B, Quatrano RS (1988) Structure of the cell walls of marine algae and ecophysiological functions of the matrix polysaccharides. Oceanogr Mar Biol Annu Rev 26: 259–315.

Lésel R (1990) Thermal effect on bacterial flora in the gut of rainbow trout and African catfish. In: Lésel R, ed. Microbiology in Poecilotherms, pp. 33–38. Amsterdam: Elsevier Science.

Lésel R, Fromageot C, Lésel M (1986) Cellulose digestibility in grass carp, *Ctenopharyngodon idella* and in goldfish, *Carassius auratus.* Aquaculture 54:11–17.

Liem KF, Greenwood PH (1981) A functional approach to the phylogeny of the pharyngognath teleosts. Am Zool 21:83–101.

Lindsay GJH, Harris JE (1980) Carboxymethylcellulase activity in the digestive tracts of fish. J Fish Biol 16:219–233.

Lobel PS (1981) Trophic biology of herbivorous reef fishes: alimentary pH and digestive capabilities. J Fish Biol 19:365–397.

Lovell T (1989) Nutrition and Feeding of Fish. New York: Van Nostrand Reinhold.

Luczkovich JJ, Stellwag EJ (1993) Isolation of cellulolytic microbes from the intestinal tract of the pinfish, *Lagodon rhomboides:* size-related changes in diet and microbial abundance. Mar Biol 116:381–388.

MacDonald NL, Stark JR, Austin B (1986) Bacterial microflora in the gastro-intestinal tract of Dover sole (*Solea solea* L.), with emphasis on the possible role of bacteria in the nutrition of the host. FEMS Microbiol Lett 35:107–111.

Macfarlane GT, Macfarlane S (1993) Factors affecting fermentation reactions in the large bowel. Proc Nutr Soc 52:367–373.

Mascolo N, Rajendran VM, Binder HJ (1991) Mechanism of short-chain fatty acid uptake by apical membrane vesicles of rat distal colon. Gastroenterology 101:331–338.

Massey LK, Sokatch JR, Conrad RS (1976) Branched-chain amino acid catabolism in bacteria. Bacteriol Rev 40:42–54.

Mathers JC, Annison EF (1993) Stoichiometry of polysaccharide fermentation in the large intestine. In: Samman S, Annison G, eds. Dietary Fibre and Beyond—Australian Perspectives, Vol. 1, pp. 123–135. Sydney: Nutrition Society of Australia, Occasional Publications.

McBee RH (1977) Fermentation in the hindgut. In: Clarke RTJ, Bauchop T, eds. Microbial Ecology of the Gut, pp. 185–222. London: Academic Press.

Milton K (1993) Diet and primate evolution. Sci Am 269:70–77.

Mok H-K (1977) Gut patterns of the Acanthuridae and Zanclidae. Jpn J Ichthyol 23: 215–219.

Montgomery WL, Gerking SD (1980) Marine macroalgae as foods for fishes: an evaluation of potential food quality. Env Biol Fish 5:143–153.

Montgomery WL, Pollak PE (1988a) Gut anatomy and pH in a Red Sea surgeonfish, *Acanthurus nigrofuscus*. Mar Ecol Prog Ser 44:7–13.

Montgomery WL, Pollak PE (1988b) *Epulopiscium fishelsoni* N.G., n. sp., a protist of uncertain taxonomic affinities from the gut of an herbivorous reef fish. J Protozool 35: 565–569.

Montgomery WL, Myrberg AA Jr, Fishelson L (1989) Feeding ecology of surgeonfishes (Acanthuridae) in the northern Red Sea, with particular reference to *Acanthurus nigrofuscus* (Forsskal). J Exp Mar Biol Ecol 132:179–207.

Moriarty DJW (1973) The physiology of digestion of blue-green algae in the cichlid fish, *Tilapia nilotica*. J Zool (Lond) 171:25–39.

Moriarty DJW (1990) Interactions of microorganisms and aquatic animals, particularly the nutritional role of the gut flora. In: Lésel R, ed. Microbiology in Poecilotherms, pp. 217–222. Amsterdam: Elsevier Science.

Moriarty DJW, Pollard PC, Alongi DM, Wilkinson CR, Gray JS (1985) Bacterial productivity and trophic relationships with consumers on a coral reef (MECOR I). Proc 5th Int Coral Reef Congress, Tahiti, Vol. 3, pp. 457–462.

Mountfort DO (1993) Ecophysiological significance of anaerobes in the gastro-intestinal tracts of marine fish. In: Guerrero R, Pedrós-Alió C, eds. Trends in Microbial Ecology, pp. 213–216. Barcelona: Spanish Society for Microbiology.

Mountfort DO, Grant WD, Morgan H, Rainey FA, Stackebrandt E (1993) Isolation and characterization of an obligately anaerobic, pectinolytic, member of the genus *Eubacterium* from mullet gut. Arch Microbiol 159:289–295.

Mountfort DO, Rainey FA, Burghardt J, Stackebrandt E (1994) *Clostridium grantii* sp. nov., a new obligately anaerobic, alginolytic bacterium isolated from mullet gut. Arch Microbiol 162:173–179.

Mountfort DO, Rhodes LL (1991) Anaerobic growth and fermentation characteristics of *Paecilomyces lilacinus* isolated from mullet gut. Appl Environ Microbiol 57: 1963–1968.

Odelson DA, Breznak JA (1983) Volatile fatty acid production by the hindgut microbiota of xylophagous termites. Appl Environ Microbiol 45:1602–1613.

Paris H, Murat JC, Castilla C (1977) Etude des acides gras volatils dans l'intestin de troit especes de poissons Teleosteens. C R Seances Soc Biol Fil 171:1297–1301.

Payne AI (1978) Gut pH and digestive strategies in estuarine grey mullet (Mugilidae) and tilapia (Cichlidae). J Fish Biol 13:627–629.

Piavaux A (1977) Distribution and localization of the digestive laminarinases in animals. Biochem Syst Ecol 5:231–239.

Pollak PE, Montgomery WL (1994) Giant bacterium (*Epulopiscium fishelsoni*) influences digestive enzyme activity of an herbivorous surgeonfish (*Acanthurus nigrofuscus*). Comp Biochem Physiol, 108A:657–662.

Prejs A, Blaszczyk M (1977) Relationships between food and cellulase activity in freshwater fishes. J Fish Biol 11:447–452.

Randall JE (1961) A contribution to the biology of the convict surgeonfish of the Hawaiian Islands, *Acanthurus triostegus sandvicensis.* Pac Sci 15:215–272.

Reynolds DA, Rajendran VM, Binder HJ (1993) Bicarbonate-stimulated [^{14}C] butyrate uptake in basolateral membrane vesicles of rat distal colon. Gastroenterology 105: 725–732.

Rimmer DW (1986) Changes in diet and the development of microbial digestion in juvenile buffalo bream, *Kyphosus cornelii.* Mar Biol 92:443–448.

Rimmer DW, Wiebe WJ (1987) Fermentative microbial digestion in herbivorous fishes. J Fish Biol 31:229–236.

Robertson DR (1982) Fish feces as fish food on a Pacific coral reef. Mar Ecol Prog Ser 7:253–265.

Sabapathy U, Teo LH (1993) A quantitative study of some digestive enzymes in the rabbitfish, *Siganus canaliculatus* and the sea bass, *Lates calcarifer.* J Fish Biol 42: 595–602.

Sakata T (1990) Microflora in the digestive tract of fish and shell-fish. In: Lésel R, ed. Microbiology in Poecilotherms, pp. 171–176. Amsterdam: Elsevier Science.

Salvatore SR, Mundahl ND, Wissing TE (1987) Effect of water temperature on food evacuation rate and feeding activity of age-0 gizzard shad. Trans Am Fish Soc 116: 67–70.

Salyers AA (1979) Energy sources of major intestinal fermentative anaerobes. Am J Clin Nutr 32:158–163.

Seiderer LJ, Davis CL, Robb FT, Newell RC (1987) Digestive enzymes of the anchovy *Engraulis capensis* in relation to diet. Mar Ecol Prog Ser 35:15–23.

Slaytor M (1992) Cellulose digestion in termites and cockroaches: what role do symbionts play? Comp Biochem Physiol 103B:775–784.

Smith RL, Paulson AC (1974) Food transit times and gut pH in two Pacific parrotfishes. Copeia 1974:796–799.

Smith TB, Wahl DH, Mackie RI (1996) Volatile fatty acids and anaerobic fermentation in temperate piscivorous and omnivorous freshwater fish. J Fish Biol 48:829–841.

Standora EA, Spotila JR, Foley RE (1982) Regional endothermy in the sea turtle, *Chelonia mydas*. J Therm Biol 7:159–165.

Stevens CE (1978) Physiological implications of microbial digestion in the large intestine of mammals: relation to dietary factors. Am J Clin Nutr 31:S161-S168.

Stevens CE (1988) Comparative Physiology of the Vertebrate Digestive System. Cambridge: University Press.

Stevens CE (1989) Evolution of vertebrate herbivores. ACTA Vet Scand, suppl 86:9–19.

Stickney RR, Shumway SE (1974) Occurrence of cellulase activity in the stomachs of fishes. J Fish Biol 6:779–790.

Sturmbauer C, Mark W, Dallinger R (1992) Ecophysiology of aufwuchs-eating cichlids in Lake Tanganyika: niche separation by trophic specialization. Env Biol Fish 35:283–290.

Sutton DC, Clements KD (1988) Aerobic, heterotrophic gastrointestinal microflora of tropical marine fishes. Proc 6th Int Coral Reef Symp 3:185–190. Choat JH et al., eds. Sixth International Coral Reef Symposium Executive Committee, Townsville, Australia.

Tanaka K, Budd MA, Efron ML, Isselbacher KJ (1966) Isovaleric acidemia: a new genetic defect of leucine metabolism. Proc Natl Acad Sci USA 56:236–242.

Tate PM, Jones ISF, Hamon BV (1989) Time and space scales of surface temperatures in the Tasman Sea, from satellite data. Deep-Sea Res 36:419–430.

Titus E, Ahearn GA (1988) Short-chain fatty acid transport in the intestine of a herbivorous teleost. J Exp Biol 135:77–94.

Titus E, Ahearn GA (1991) Transintestinal acetate transport in a herbivorous teleost: anion exchange at the basolateral membrane. J Exp Biol 156:41–61.

Titus E, Ahearn GA (1992) Vertebrate gastrointestinal fermentation: transport mechanisms for volatile fatty acids. Am J Physiol 262:R547-R553.

Troyer K (1984) Microbes, herbivory and the evolution of social behaviour. J Theor Biol 106:157–169.

Trust TJ, Sparrow RAH (1974) The bacterial flora in the alimentary tract of freshwater salmonid fishes. Can J Microbiol 20:1219–1228.

Trust TJ, Bull LM, Currie BR, Buckley JT (1979) Obligate anaerobic bacteria in the gastrointestinal microflora of the grass carp (*Ctenopharyngodon idella*), goldfish (*Carassius auratus*), and rainbow trout (*Salmo gairdneri*). J Fish Res Board Can 36: 1174–1179.

Winterbottom R, McLennan DA (1993) Cladogram versatility—evolution and biogeography of acanthuroid fishes. Evolution. 47:1557–1571.

Zihler F (1982) Gross morphology and configuration of digestive tracts of Cichlidae (Teleostei, Perciformes): phylogenetic and functional significance. Neth J Zool 32: 544–571.

Zimmerman LC, Tracy CR (1989) Interactions between the environment and ectothermy and herbivory in reptiles. Physiol Zool 62:374–409.

7

Fermentation in Reptiles and Amphibians

Karen A. Bjorndal

1. Introduction

As ectotherms, amphibians and reptiles do not have to support the metabolic expense of endothermy and are characterized by low rates of energy flow and high efficiencies of biomass conversion, relative to birds and mammals (Pough 1983). The low energy requirements of amphibians and reptiles have important ramifications for digestive processing and, in herbivorous species, for the level of energy that must be generated by fermentations in the gastrointestinal tract.

The digestive physiology of herbivorous amphibians and the ecology and physiology of their symbiotic gut microbes are the poorest-studied of any group of vertebrate herbivores. Some amphibians—such as adult *Hyla truncata*, a frog that feeds on fruit (Da Silva et al. 1989, Fialho 1990); *Rana hexadactyla*, a frog that feeds on aquatic vegetation (Das and Coe 1994); some salamander larvae; the toad *Bufo marinus* (Alexander 1964); and *Siren lacertina* (Ultsch 1973)—ingest sufficient vegetation to be considered omnivorous or herbivorous. However, the major group of herbivorous amphibians is the anuran tadpoles. Many, but not all, anuran tadpoles are herbivorous. They feed on epiphytic, epibenthic and suspended algae, algae in their fecal pellets, and detritus (Duellman and Trueb 1986, Seale 1987).

Herbivorous reptiles are limited to two groups: the chelonia (turtles and tortoises) and the sauria (lizards). Substantially more research has been conducted on digestion in herbivorous reptiles than in herbivorous amphibians. However, as will become apparent in this chapter, there are still large gaps in our knowledge.

2. Gastrointestinal Tract Morphology of Herbivorous Reptiles

Herbivorous lizards have teeth that are adapted for shearing vegetation. Within the Iguanidae, herbivorous species have pleurodont teeth that are multicusped, strongly compressed laterally, sharp, and bladelike (Hotton 1955, Montanucci 1968, Throckmorton 1976). In a comparison of the herbivore *Ctenosaura pectinata* and the insectivore *Anolis carolinensis*, Ray (1965) found that in *C. pectinata* maxillary and dentary tooth rows have continuous dental palisades in which the broadly expanded, strongly cusped crowns closely succeed one another and tend to overlap. These continuous rows provide excellent shearing blades, unlike the row of teeth in *A. carolinensis* that are characterized by gaps of variable size. The genus *Liolaemus* includes a large number of relatively small lizards (5 to 20 g). The more herbivorous species within this group are characterized by dentition that is more blade-like and cuspate (Troyer 1988; 1991). The agamid herbivore *Uromastyx aegyptius*, in contrast, has acrodont teeth, and the cheek teeth are massive and have no cusps (Throckmorton 1976).

Turtles lack teeth and rely instead on a keratinized beak or rhamphotheca to grasp and shear vegetation. The green turtle, *Chelonia mydas*, is the only herbivorous sea turtle and the only sea turtle with a serrated rhamphotheca, presumably to aid in clipping vegetation (Balazs 1980). Other relationships of rhamphotheca structure to diet apparently have not been studied.

The jaw structure in extant reptiles precludes mastication of food (Throckmorton 1980; Norman and Weishampel 1985). This lack of particle size reduction results in ingestion of large food particles, such as entire leaves (Iverson 1982, Foley et al. 1992), which may have significant effects on both the extent and rate of digestion and fermentation (Bjorndal et al. 1990).

Within lizards and turtles, there is a general trend for herbivorous species to have longer digestive tracts, and more specifically longer and more capacious large intestines, than do omnivorous or carnivorous species (Lonnberg 1902, Guard 1980, Thompson 1980, Iverson 1982, Bjorndal 1985a, Troyer 1988, 1991, Dearing 1993). Digestive tracts of herbivorous species are illustrated in Lonnberg (1902), El-Toubi and Bishae (1959), Guard (1980), Iverson (1980; 1982), Stevens (1988), and Troyer (1991).

Herbivorous lizard species in the Iguanidae, Agamidae, and Scincidae have colic structures that act to slow the passage of digesta through the colon and to increase the surface area of that region (Iverson 1980, 1982). These species are characterized by relatively large adult body size and no ontogenetic change in their herbivorous diet. All species in the Iguaninae except the marine iguana, *Amblyrhynchus cristatus*, have one to 11 transverse valves in the proximal colon. The valves are either circular or semilunar, and, if the former are present, they are proximal to the semilunar valves. There is no significant change in the number of valves with age or body size within a species. However, among species, there is a significant, positive relationship between the number of colic valves and adult body size (Iverson 1982). The proximal colon in the marine iguana is not greatly enlarged but has many irregular transverse folds (Iverson 1980). This morphological difference may reflect differences in diet; marine iguanas feed on marine algae whereas other iguanines feed on terrestrial vegetation. Lizards in the genus *Uromastyx* (Agamidae) and *Corucia* (Scincidae) have colic structures similar to those in iguanines (Iverson 1980).

Other herbivorous lizards lack the structured colons described above. Most of these species are less specialized folivores. Their diets may contain a higher proportion of low-fiber plants parts (e.g., fruits) and animal matter, and they may undergo an ontogenetic shift in diet from insectivory to herbivory. Also, many of these species have relatively smaller adult body sizes. Morphological adaptation in these species is limited to an enlarged colon relative to that of less herbivorous relatives. Examples are in the genera *Agama*, *Angolosaurus*, *Anolis*, *Basiliscus*, *Egernia*, *Gerrhosaurus*, *Macroscincus*, *Phymaturus*, *Physignathus*, *Tiliqua*, and *Trachydosaurus* (Iverson 1982); *Liolaemus* (Troyer 1988, 1991); and *Cnemidophorus* (Dearing 1993).

No species of turtle is known to have partitioned colons (Troyer 1991). In tortoises (terrestrial turtles), all of which are herbivorous, the large intestine is approximately equal in length to the small intestine and there is an eccentric dilation of the proximal colon (termed a cecum by Guard, 1980) in those species that have been examined (Guard 1980, Bjorndal 1989, Troyer 1991).

The only herbivorous marine turtle is the green turtle, *Chelonia mydas*, in which the large intestine is approximately 2.5 times the length of the small intestine (Thompson 1980, Bjorndal 1985a). The proximal colon is expanded into what has been termed a ''functional'' cecum (Bjorndal 1979). During the 3- to 4-month reproductive season in which female green turtles fast, the digestive tract is empty except for the ''cecum,'' which contains approximately 1 L of a dark green fluid that may serve as a refugium for gut microbes until feeding resumes (Bjorndal 1979).

In herbivorous freshwater turtles that have been examined, the length of the small intestine is about four to six times that of the large intestine (Bjorndal and Bolten 1990 and unpublished data). The proximal colon has an eccentric dilation in *Pseudemys nelsoni* and *Trachemys scripta* (Bjorndal and Bolten 1990 and unpublished data). As discussed below, the short hindgut does not necessarily reflect a lesser reliance on gut fermentation in freshwater turtles.

3. Fermentation in Herbivorous Reptiles

Although the role of gut fermentations in the nutrition of herbivorous reptiles was slow to be recognized (Bjorndal 1979, Troyer 1982), fermentation of complex

Table 7.1. Reptile species known or presumed to harbor microbial cellulolytic fermentations based on studies of fermentation end products (Products), cell wall digestibility (CWD), and/or identification of microbes (Microbes).

Species	Evidence			References
	Products	CWD	Microbes	
Turtles				
Chelonia mydas	X			Bjorndal 1979
		X		Bjorndal 1980
Geochelone carbonaria	X			Guard 1980
		X		Bjorndal 1989
Geochelone denticulata		X		Bjorndal 1989
Geochelone gigantea		X		Hamilton and Coe 1982
Gopherus polyphemus	X	X		Bjorndal 1987
Xerobates agassizii		X		Meienberger et al. 1993
Pseudemys nelsoni	X	X		Bjorndal and Bolten 1990
Trachemys scripta		X		Bjorndal and Bolten 1993
Lizards				
Conolophus subcristatus		X		Christian et al. 1984
Dipsosaurus dorsalis		X		Zimmerman and Tracy 1989
Iguana iguana	X			Guard 1980
	X	X		Troyer 1984a
	X		X	McBee and McBee 1982
Sauromalus hispidus		X		Voorhees 1981
Sauromalus obesus		X		Zimmerman and Tracy 1989
Sauromalus varius		X		Voorhees 1981
Uromastyx aegyptius	X	X		Foley et al. 1992
Egernia cunninghami		X		Andrews 1984, cited in Foley et al. 1992

carbohydrates has now been demonstrated in a number of reptiles, including lizards, freshwater and marine turtles, and tortoises (Table 7.1). The presence of fermentations in several species has been confirmed by analysis of end products—primarily short-chain fatty acids (SCFAs). In only a few species have microbes been identified. In many species, a fermentation is assumed to be present because of the significant degradation of complex carbohydrates in the digesta as it passes through the digestive tract. In addition to the species listed in Table 7.1, Dearing (1993) suggested a fermentation may be present in the hindgut of the lizard *Cnemidophorus murinus* based on the odor of fermentation and large numbers of nematodes in the region.

3.1. Fermentation Regions

No reptile is known to harbor a foregut fermentation. In marine turtles, tortoises, and lizards, the location of the fermentation is in the large intestine. In *Pseudemys*

Table 7.2. Concentrations of fermentation end products along the digestive tracts of reptiles. Values are mean (SD)

		Small Intestine				Colon				
Species	Stomach	Anterior	Mid	Posterior	Cecum	Anterior	Mid	Posterior	Feces	References
Chelonia mydas										Bjorndal 1979
Seagrass diet										
(N = 2)										
SCFA[a]	8	—	58	—	156	191	207	62	—	
Lactic acid[a]	0.7	—	0.9	—	2.8	2.8	2.0	0.6	—	
Algae diet										Bjorndal et al. 1991
(N = 4)										
SCFA[b]	4	—	13	—	67	66	63	17	—	
	(4)		(20)		(22)	(18)	(30)	(5)		
SCFA[c]	164	—	203	—	1174	1220	870	134	—	
	(154)		(106)		(339)	(319)	(521)	(118)		
Geochelone carbonaria										Guard 1980
SCFA[b]	1	—	7	—	63	—	23	—	9.1	
(N = 2)										
Gopherus polyphemus										Bjorndal 1987
SCFA[b]	—	—	—	—	—	—	—	—	25	
(N = 5)									(8)	
Pseudemys nelsoni	N = 12	N = 1	N = 11	N = 3	N = 12	—	N = 10	N = 9	—	Bjorndal and Bolten 1990
SCFA[b]	2	11	55	41	60	—	61	36	—	
	(0.4)	—	(14)	(7)	(12)		(15)	(9)		
SCFA[c]	46	289	1131	1059	1282	—	1091	520	—	
	(11)	—	(347)	(117)	(306)		(258)	(218)		
Iguana iguana										
SCFA[b]	1	—	10	—	51	—	16	—	13	Guard 1980
(N = 2)										
SCFA[b]	200	—	200	—	807	—	463	—	200	Troyer 1984a
(N = 3)										
Uromastyx aegyptius										Foley et al. 1992
(N = 4)										
SCFA (mM)	13	10	8	6	120	76	—	60	—	
(SE)	(4)	(2)	(1)	(1)	(3)	(6)		(9)		
Ethanol (mM)	1.5	1.7	1.9	1.9	1.7	2.5	—	1.2	—	
(SE)	(0.3)	(0.7)	(0.5)	(0.7)	(0.4)	(0.8)		(0.3)		

[a] μmol/mL strained fluid.
[b] μmol/mL total sample.
[c] μmol/g dry matter.

nelsoni, the only freshwater turtle species examined, the fermentation is in the small and large intestines (Bjorndal and Bolten 1990). The region of fermentation has been confirmed by analyses of SCFA concentrations along digestive tracts (Table 7.2) in the marine turtle *Chelonia mydas*, in the tortoise *Geochelone carbonaria* (Guard 1980), in the freshwater turtle *Pseudemys nelsoni* (Bjorndal and Bolten 1990), and in the lizards *Iguana iguana* (Guard 1980, Troyer 1984a) and *Uromastyx aegyptius* (Foley et al. 1992). *Chelonia mydas* feeding on seagrasses (which are angiosperms) and on marine algae had similar patterns of SCFA concentrations along their digestive tracts (Bjorndal et al. 1991).

In herbivorous mammals, wet mass of contents from the fermentation region of the gut is isometric with body mass for both foregut fermenters and hindgut fermenters (Parra 1978, Justice and Smith 1992). Exponents in regression equations for mammals with foregut fermentations and mammals with hindgut fermentations based on data from Parra (1978) and Justice and Smith (1992) were not significantly different ($\alpha = 0.05$, t-test for comparison of 2 slopes, Zar 1984) nor significantly different from an exponent of 1 (t-test for significance of regression; Zar 1984). The regression line for mammalian hindgut fermenters is shown in Fig. 7.1.

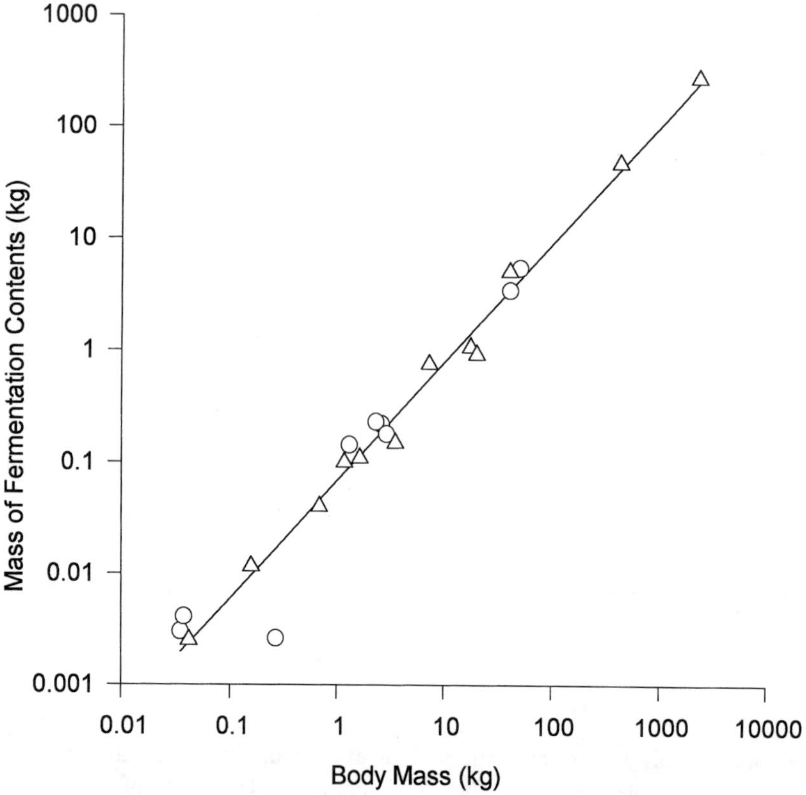

Figure 7.1. Relation of wet mass of fermentation contents to body mass in mammals with hindgut fermentations (△) and in reptiles (○). Line is regression line for mammalian hindgut fermenters (equation $Y = 1.17X^{1.06}$; range in body mass 0.04 to 2337 kg; n = 12) and for reptiles (equation $Y = 1.15X^{1.06}$; range in body mass 0.04 to 50.5 kg; n = 9). See text for statistical comparison between mammals and reptiles and for sources of data.

Capacity of the fermentation region has the same relation to body mass in reptiles as in mammals (Fig. 7.1). Capacities of the fermentation regions of the reptiles *Iguana iguana* (Troyer 1984a), *Chelonia mydas* and *Geochelone carbonaria* (Bjorndal and Bolten, unpublished data), *Pseudemys nelsoni* (Bjorndal and Bolten 1990), *Uromastyx aegyptius* (Foley et al. 1992), and *Cnemidophorus murinus* (Dearing 1993) are plotted against body mass in Figure 7.1. Exponent (1.06) of the regression equation is not significantly different from those of the mammalian regression equations (t-test for comparison of 2 slopes; Zar 1984) or from an exponent of 1 (t-test for significance of regression; Zar 1984). The point that falls below the line in Figure 7.1 represents the value for one individual *Geochelone carbonaria*. This value is the only one available for tortoises; more data are needed to determine whether tortoises consistently fall below predicted values.

The regressions presented here are based on interspecific comparisons. Meienberger et al. (1993) stressed the need for intraspecific comparisons. The only analysis of scaling of gut capacity to body mass over a wide range of body sizes within a reptile species suggested that the relationship was also isometric (Troyer 1984a). More studies are needed to establish the scaling of fermentation capacity to body mass both among and within species of herbivorous reptiles. Also, data are needed to assess whether percent dry matter of the contents of fermentation chambers is inversely related to body size in reptiles, as it apparently is in mammals (Justice and Smith 1992).

The small size of the hindgut in herbivorous freshwater turtles suggested that these species may not rely on fermentation of their plant diet. However, the herbivorous *Pseudemys nelsoni* does rely on fermentation, and the fermentation region has expanded into the small intestine (Table 7.2) (Bjorndal and Bolten 1990). This record is the only one of which I am aware of a significant fermentation in the small intestine of a vertebrate. There was no possibility of contamination of the small intestine contents with colon contents. In the 11 turtles sampled, the mid-small intestine samples, which had the highest SCFA concentrations, were taken 0.5 m from the ileocolic valve, and the entire intestinal tract was full, preventing backflow of digesta.

In comparisons of hindgut fermentation and foregut fermentation, one major advantage of hindgut fermentation is that the host animal is able to digest and absorb the high quality dietary protein and soluble carbohydrates before the more refractory diet components enter the hindgut for fermentation (Janis 1976). Because fermentation is shifted into the small intestine in *P. nelsoni*, endogenous enzymes in the small intestine are in competition for the high quality substrate with the microbiota, which will attack soluble carbohydrates and protein more rapidly than refractory cell walls. Thus, by having a fermentation in the small intestine, it appears that *P. nelsoni* has lost the major advantage of hindgut fermentation without gaining any of the many advantages of foregut fermentation.

In *P. nelsoni*, fermentation may have expanded into the small intestine to ensure a sufficiently large fermentation vat. Based on the equation for mammals with hindgut fermentations (Fig. 7.1), a 2.6-kg herbivore should have 186 g wet mass of fermentation contents. For *P. nelsoni* with an average mass of 2.6 kg, fermentation contents equaled 221 g wet mass. If small intestine contents were not included, fermentation contents would be only 69 g. Thus, without including the small intestine region, the volume of the fermentation region in *P. nelsoni* would be well below that of other herbivores.

To enlarge the volume of the fermentation vat, the size of the hindgut could be increased or the fermentation could move into the small intestine. Because space inside a turtle shell is limited, hindgut expansion would require a corresponding decrease in the small intestine. If small *P. nelsoni* are omnivorous (there is no information on diet for this size class), a long small intestine may be necessary to ensure high digestive efficiency at a life stage in which rapid growth is critical for survival and future reproductive success. There may also be phylogenetic constraints. The relative roles of phylogeny and diet in determining gut morphology in turtles needs further study.

3.2. Gastrointestinal Motility and Passage Rates

In addition to the size of the fermentation region and morphological structures designed to slow the flow of digesta in the area of fermentation, gut motility controls the movement of digesta through fermentation regions. Gut motility is poorly studied in reptiles. Antiperistalsis—peristaltic movements originating near the coprodeum and moving along the hindgut toward the ileocolic valve—has been described in the turtle *Chinemys reevesii* (= *Geoclemys reevesii*) (Hukuhara et al. 1975). Antiperistaltic waves occurred at intervals of 18 to 25 seconds and moved about 1 mm/sec; it is not clear whether turtles were anesthetized when these measurements were made.

Antiperistalsis may play two critical roles in fermentation in reptiles. First, antiperistalsis may act to prolong the stay of digesta in the fermentation region. This delay may be particularly important in tortoises in which the fermentation vat may be small and in those reptiles that lack morphological structures to retain digesta. By relying on antiperistalsis to delay passage of digesta rather than on morphological structures, many herbivorous reptiles benefit by having more flexible digestive processing. Greater flexibility allows more efficient processing of the wide range of diet items (fruits, seeds, foliage) ingested by many reptile herbivores (Bjorndal 1989, Bjorndal and Bolten 1993) and accommodate the ontogenetic shift from insectivory to herbivory that some herbivorous reptiles undergo (Iverson 1982).

Second, antiperistalsis may move urinary nitrogen from the cloaca to the proximal hindgut and provide an important source of nitrogen to the hindgut fermentation (Stevens 1989). This movement has been well documented in birds (references in Stevens 1988). In an adult *Iguana iguana*, $BaSO_4$ moved from the coprodeum to the proximal colon in about 1 hour (Guard 1980).

Movement of digesta through the gut of reptiles is generally slow relative to that of birds and mammals of similar size. Data on transit times for free-feeding reptiles (i.e., not force-fed) on foliage diets range from 2.8 to 21.2 days (Table 7.3). Transit times of particles in most mammals with hindgut fermentations range from 1 hour to 48 hours, although sloths have transit times of 96 to 144 hours (Warner 1981).

Limited data on differential movement of liquid and particulate fractions indicate that the liquid fraction moves significantly faster than does the particulate fraction in *Geochelone carbonaria* and *Iguana iguana* (Guard 1980). Values for transit time in Table 7.3 were obtained by the rather crude method of recording the time required for a marker (e.g., plastic flagging or beads) introduced in the food to appear in the feces. If Meienberger et al. (1993) are correct that movement of digesta in desert tortoises is similar to that in a plug-flow reactor (Penry and Jumars 1987), then this method of measuring transit time may be relatively accurate for the particle fraction of digesta. However, the method may be less reliable in reptiles with more complex gut structure (e.g., iguanine lizards).

Transit times can vary greatly with diet. In the tortoises *Geochelone carbonaria* and *G. denticulata*, transit times averaged 3 days on a diet of guava fruit and 9 days on *Lantana* foliage (Bjorndal 1989). In *Iguana iguana*, mean transit times varied from 3.8 to 8.5 days with a general pattern of fruits passing more rapidly than flowers which passed more rapidly than foliage (Marken Lichtenbelt 1992). This variation illustrates the flexibility in digestive processing discussed above.

3.3. Nematodes in Reptile Digestive Tracts

A recurring question that has not yet been addressed is the role of nematodes that are found in huge numbers in the fermentation regions of the digestive tracts of many herbivorous reptiles. Schad et al. (1964) were apparently the first to suggest that the large numbers of nematodes in the colon of *Testudo graeca* may affect cellulose digestion in that region by consuming cellulolytic bacteria. Dubuis et al. (1971) hypothesized that nematodes were important in digestive processing in *Uromastyx acanthinurus*, and Nagy (1977) suggested that nematodes may be the source of cellulase in the hindgut of the herbivorous lizard *Sauromalus obesus*. Iverson (1982) proposed three possible roles for the nematodes: 1. mixing and mechanical breakdown of digesta, thus increasing surface area; 2. production of waste products useful to the host; and/or 3. regulation of microbial populations in the colon.

Nematode populations can reach large numbers. Iverson (1982) estimated that a healthy adult *Cyclura carinata* harbored more than 15,000 nematodes (families

Table 7.3. Transit time in days (TT), mass-specific intake in g DM/kg body mass per day (I), and percent cell wall digestibility (CWD) in herbivorous reptiles free-feeding on foliage diets. N is sample size, %CW is cell walls (as percent dry matter) in diet, values are means. For each reptile species, each line represents a different diet.

Species	N	TT	I	CWD[a]	%CW[b]	Reference
Turtles						
Chelonia mydas	12	—	3.2	86A	44	Bjorndal 1980
Pseudemys nelsoni	5	2.8	2.4	86B	37	Bjorndal and Bolten 1993
	6	3.4	2.6	16B	39	Bjorndal and Bolten 1993
Geochelone carbonaria	4	9.5	2.1	37	49	Bjorndal 1989
Geochelone denticulata	5	8.7	2.3	41	49	Bjorndal 1989
Geochelone gigantea	—	10.3	3.5[c]	38–45	61–66A	Hamilton and Coe 1982
Gopherus polyphemus	7	13	—	73	40	Bjorndal 1987
Xerobates agassizii	19	17.8	—	37	26	Meienberger et al. 1993
	20	21.2	—	59	72	Meienberger et al. 1993
Lizards						
Conolophus pallidus	9	4.5	—	—	≈ 21B	Christian et al. 1984
Iguana iguana	4	7.0	2.8	64	42	Marken Lichtenbelt 1992
	4	8.5	1.0	52	36	Marken Lichtenbelt 1992
	4	6.1	5.2	76	24	Marken Lichtenbelt 1992
	10	5.4	4.4	81	34	Marken Lichtenbelt 1992
Uromastyx aegyptius	4	—	10[d]	69C	44	Foley et al. 1992

[a] Unless noted method = total collection; A = acid detergent lignin as indigestible marker; B = in vitro indigestible cell walls as indigestible marker; C = manganese as indigestible marker.

[b] Unless noted CW = neutral detergent fiber; A = holocellulose; B = cellulose by Crampton & Maynard procedure.

[c] Calculated for 30-kg tortoise feeding on tortoise turf in the wet season.

[d] Estimated from water turnover, N = 3.

Atractidae and Oxyuridae) in its colon. The mid and posterior small intestine in the freshwater turtle *Pseudemys nelsoni* had very dense populations of the nematode *Spironoura procera* (superfamily Oxyuroidea) (Bjorndal and Bolten 1990). Nematodes appear to be most common in the regions of initial fermentation. In *P. nelsoni*, which has a small-intestine fermentation, boli in the small intestine were composed of approximately 30% to 100% nematodes. Numbers of nematodes decreased greatly in the colon, and those present appeared less active (Bjorndal and Bolten 1990). In iguanines, nematodes are most abundant in the proximal hindgut, the region of initial fermentation (Iverson 1982).

If a mutualism exists between the reptilian host and its nematode populations, the most likely benefit to the reptile host is reduction of digesta particle size (Bjorndal and Bolten 1990), which would increase fermentation rate (Bjorndal et al. 1990). A similar mutualism has been suggested for gastric nematodes in the horned lizard, *Phrynosoma cornutum* (Waldschmidt et al. 1987). The horned lizard is an ant specialist; nematodes may disarticulate the ants and facilitate digestion in the lizard's stomach.

3.4. Identification and Enumeration of Microbes

Few studies have identified gut microbes in herbivorous reptiles. *Lampropedia* sp. was identified in the colon contents of the tortoise *Testudo graeca*, and in the feces of the tortoise *Gopherus polyphemus* and the lizard *Iguana iguana* (Schad et al. 1964). Schad et al. (1964) also noted the presence of large numbers of nematodes, bacteria, and occasionally ciliates in the colon of *T. graeca*. Examination of posterior rectum contents from three tortoises (*Testudo* sp.) revealed *Escherichia coli*, Streptococci, and *Bacteroides*. Log_{10} viable counts per gram for the three organisms were 3.0, 4.4, and 4.8, respectively (Smith 1965).

Troyer (1982) identified *Lampropedia merismopedioides* and a large ciliate protozoan, similar to holotrichs in the rumen, in the feces of young *Iguana iguana*. Using direct counts of formalin-preserved colon contents, Troyer (1984b) recorded bacterial concentrations from 6.9 to 16.1 $\times$ 10^{10} cells/g dry mass in young *Iguana iguana*.

In the most thorough study to date, McBee and McBee (1982) cultured contents from the hindgut of eleven *Iguana iguana* and made direct clump counts on formalin-preserved contents. Species in the genus *Leuconostoc* were dominant in eight of the iguanas; species in the genus *Clostridium* were dominant in the other three iguanas. Five species in the genus *Lachnospira* were present; no species in the genera *Bacteroides*, *Fusobacterium*, or *Ruminococcus* were found. Direct microscopic counts yielded 30 $\times$ 10^9 clumps of bacterial cells per gram content. Colony counts yielded values from 3.3 to 23.5 $\times$ 10^9/g content.

3.5. *Acquisition of Microbes*

Because *Leuconostoc* and *Clostridium* are common in the environment, McBee and McBee (1982) suggested that each *Iguana iguana* establishes its own microbiota as hatchlings by ingesting these bacteria with either soil or feces from adult iguanas. Troyer (1982; 1984b) demonstrated that hatchlings first consume soil within the nest chamber, which establishes populations of soil microbes in the hindgut. For the first few weeks of life, iguanas actively seek and preferentially feed on fresh feces from older iguanas, and a more complex microbe community is established. Stomach pH is neutral for the first week of life, which facilitates establishment of microbial populations. After one week, the stomach becomes acidic (pH = 1 to 2). Hatchling iguanas isolated from soil or feces of other iguanas initially grew more slowly than did hatchlings that were given access to either soil or feces of conspecifics (Troyer 1982, 1984b). Hatchlings of other herbivorous lizard species (*Amblyrhynchus cristatus* and *Cyclura carinata*) ingest adult feces (personal communications cited in Troyer 1984c).

The situation is more complex for the herbivorous green turtle, *Chelonia mydas*. Posthatchling green turtles apparently spend their first few years of life feeding as carnivores or omnivores on the ocean surface. Only when they reach a carapace length of about 25 cm do they leave the pelagic habitat and settle on benthic feeding grounds; at this time they shift to a herbivorous diet and must acquire an appropriate microbial community (Bjorndal 1985a). How they acquire their gut microbiota is not known. When several green turtles that had been raised to a size of 25 cm carapace length on a diet of fish and invertebrates were released into an enclosed area of seagrass (*Thalassia testudinum*) pasture, they immediately began to feed on the seagrass, which is the major diet species of green turtles in the Caribbean. At first, the plant blades passed through the digestive tract with little sign of digestion. Only after two months, did the feces resemble the well digested feces of older green turtles (Bjorndal, unpublished data). The area in which the young turtles were held was adjacent to feeding pastures of older green turtles. The young turtles may have acquired microbes from feces of the other turtles carried on water currents or from the thick detritus layer present in *T. testudinum* pastures.

Many herbivorous reptiles undergo long periods of time in which they do not feed. Reptiles do not feed during periods when sources of water or food are insufficient or during periods of cool temperatures. *Dipsosaurus dorsalis* hibernate from October to March (Mautz and Nagy 1987), and *Sauromalus obesus* do not feed from November to March (Nagy and Shoemaker 1975). *Xerobates agassizii* hibernate from December through February (Nagy and Medica 1986), and *Gopherus polyphemus* retreat to their burrows during cool weather and may remain there several months without feeding. Female *Chelonia mydas* do not feed during periods of reproductive activity, which can last several months (Bjorndal

1982). It is not known whether microbial populations are maintained in the guts of reptiles during these extended fasts, or whether reptiles must reinoculate their digestive systems when feeding resumes.

More studies are needed on the acquisition of microbes in reptile species. Troyer (1984c) offers an intriguing discussion on the possible role of microbial transfer between and within generations of herbivores in the evolution of social behavior, particularly in species that lack parental care, such as herbivorous reptiles.

3.6. Fermentation End Products

Fermentation end products have been quantified in four species of turtles and in two lizard species (Table 7.2). The end products generated from fermentations in reptiles are the same as those that have been identified in the guts of birds and mammals: SCFA (primarily acetic, propionic, and butyric acids), ethanol, lactic acid, methane, carbon dioxide, and hydrogen gas (Stevens 1988). In reptiles, fermentation end products were first characterized in *Chelonia mydas* feeding on the seagrass *Thalassia testudinum* (Bjorndal 1979). As has been found in all reptiles studied to date, SCFAs were the primary end products. Lactic acid was present at low levels, ethanol was not detected, and hydrogen and carbon dioxide were the major gases evolved during in vitro fermentation. In *Pseudemys nelsoni* on a diet of the aquatic plant *Hydrilla verticillata*, carbon dioxide was the primary gas produced in vitro, with small amounts of methane and hydrogen gas also produced (Bjorndal and Bolten 1990).

Although fermentation end products have been quantified in only two species of lizards, there have been three studies on *Iguana iguana* (Table 7.2). Acetic, propionic, and butyric acids were the primary end products with lesser amounts of lactic acid and some of the higher volatile fatty acids in wild-caught *I. iguana*; ethanol was not found in measurable amounts (McBee and McBee 1982). In addition to high concentrations of SCFA, ethanol was present in low concentrations in wild-caught *Uromastyx aegyptius* (Foley et al. 1992).

Concentrations of SCFA along the gut clearly indicate that the fermentation region is the hindgut in all reptiles except in *P. nelsoni* which has an active fermentation in the small intestine and large intestine (Table 7.2). Tests of significance for the differences in SCFA concentrations were reported in three studies. In *Chelonia mydas* feeding on algae, SCFA concentrations in the cecum, anterior colon and mid colon are significantly higher than those in the esophagus, stomach, small intestine, and posterior colon (Bjorndal et al. 1991). In *P. nelsoni*, SCFA concentrations in the stomach are significantly lower than those in the posterior colon, which are significantly lower than those in the small intestine, cecum and mid colon (Bjorndal and Bolten 1990). In *U. aegyptius*, the concentration of SCFA is significantly lower in the stomach and small intestine than in the hindgut,

Table 7.4. Relative concentrations of short chain fatty acids in the fermentation contents in reptiles

acetic > propionic > butyric > valeric	
	Gopherus polyphemus on legume leaf diet (Bjorndal 1987)
	Pseudemys nelsoni on aquatic plant diet (Bjorndal and Bolten 1990)
	Chelonia mydas on algae diet (Bjorndal et al. 1991)
acetic > propionic > butyric	
	Iguana iguana (Troyer 1984a)
acetic > propionic > valeric > butyric	
	Gopherus polyphemus on mixed grazing diet (Bjorndal 1987)
acetic > butyric > propionic > valeric	
	Uromastyx aegyptius on natural plant diet (Foley et al. 1992)
acetic > butyric > propionic	
	Chelonia mydas on seagrass diet (Bjorndal 1979)

and concentrations in the posterior colon and rectum are significantly lower than in the cranial parts of the cecum (Foley et al. 1992). The low values in the posterior colon in all species reflect the depletion of fermentable substrate in the digesta in that region.

Comparisons of SCFA concentrations are limited because of the different techniques and units employed (Table 7.2). However, the values are quite similar considering the variation in diet, species, and habitats. These values fall within the range of concentrations (160 to 1156 μmol/g dry matter) reported for foregut and hindgut fermentations in herbivorous mammals (Parra 1978).

Relative concentrations of SCFA in reptiles are given in Table 7.4. The sequence most frequently reported for reptiles—acetic > propionic > butyric > valeric—is the most common in gut fermentations in vertebrates (Van Soest 1982). The effect that diet can have on relative concentrations of SCFA is demonstrated by the shifts in relative concentrations found in *Chelonia mydas* on two diets and in *Gopherus polyphemus* on two diets (Table 7.4). In all species for which relative concentrations were reported for sequential samples along the gut, the relative concentrations remained the same throughout the fermentation region (Bjorndal 1979, Troyer 1984a, Bjorndal and Bolten 1990, Bjorndal et al. 1991, Foley et al. 1992).

3.7. Rates of Fermentation

In addition to the rates of SCFA production in Table 7.5, McBee and McBee (1982) reported a rate of 440.5 μmol/h for the total hindgut contents from two adult male *Iguana iguana* at 27° to 36°C. Rates of fermentation are difficult to

Table 7.5. Rates of production of short chain fatty acids (SCFAs) and pH along the digestive tracts of reptiles. Values are mean (SD)

Species	Stomach	Small Intestine Anterior	Small Intestine Mid	Small Intestine Posterior	Cecum	Colon Anterior	Colon Mid	Colon Posterior	References
Chelonia mydas									
Seagrass diet (N = 2)									Bjorndal 1979
SCFA[a] at 30°C	—	—	—	—	12	—	—	—	
Lactic acid[a]	—	—	—	—	0.5	—	—	—	
pH	4	—	7	—	6	6	6	7	
Algae diet (N = 4)									Bjorndal et al. 1991
pH	2.6 (1.3)	—	6.6 (0.5)	—	6.4 (0.5)	6.8 (0.8)	6.4 (0.5)	6.2 (0.4)	
Pseudemys nelsoni									Bjorndal and Bolten 1990
Sample size for SCFA	—	N = 1	N = 7	N = 2	N = 11	—	N = 9	N = 7	
SCFA[b] at 30°C	—	10	16 (8)	11 (7)	12 (9)	—	11 (9)	12 (8)	
SCFA[c] at 30°C	—	244	388 (137)	372 (207)	238 (135)	—	241 (202)	163 (97)	
Sample size for pH	N = 12	N = 7	N = 12	N = 7	N = 12	—	N = 12	N = 12	
pH	1.8 (0.8)	7.1 (0.3)	6.8 (0.3)	7.3 (0.2)	6.8 (0.2)	—	6.8 (0.3)	7.2 (0.3)	
Iguana iguana									Troyer 1984a
pH (N = 18)	1.5	—	7.5	—	7.5	—	7.5	—	
Uromastyx aegyptius (N = 4)									Foley et al. 1992
SCFA[a] at 40°C	—	—	—	—	31.1	31.1	—	—	
SCFA[c] at 40°C	—	—	—	—	148	148	—	—	
pH (SE)	1.8 (0.1)	6.9 (0.1)	7.0 (0.1)	7.1 (0.1)	6.9 (0.1)	7.0 (0.1)	—	7.1 (0.1)	

[a] μmol/mL strained fluid per hour.
[b] μmol/mL total sample per hour.
[c] μmol/g dry matter per hour.

compare because of differences in temperature at which the determinations were made and in units in which the values are expressed (Table 7.5). Rates of fermentation reported for reptiles fall within the range of values (32 to 629 μmol/g dry matter per hour) summarized by Parra (1978) for mammals with foregut or hindgut fermentations. Although fermentation rate tends to decrease along successive sections of the fermentation region as fermentable substrate is consumed, the pattern is not significant in *Pseudemys nelsoni* (Bjorndal and Bolten 1990), the only species in which the pattern was analyzed.

Relative rates of evolution of individual SCFA are acetic > butyric > propionic for *Chelonia mydas* (Bjorndal 1979) and *Uromastyx aegyptius* (Foley et al. 1992), and acetic > propionic > butyric > valeric in *P. nelsoni* (Bjorndal and Bolten 1990). The large proportion of acetic generated in these reptiles is similar to that in mammals on foliage diets (Stevens 1988). However, in two *Iguana iguana*, the relative production rates were butyric > acetic > propionic (McBee and McBee 1982).

Based on several lines of evidence, it is believed that reptiles absorb and use SCFA generated by gut fermentations as an energy source. First, pH levels are near neutral in fermentation regions (Table 7.5). Second, SCFA concentrations are low near the cloaca and in the feces of herbivorous reptiles (Table 7.2). Third, about 10 μmol SCFA/mL blood was measured in efferent venous blood and less than 4 μmol SCFA per ml of blood was found in the arterial blood supply of *Iguana iguana* (McBee and McBee 1982). Fourth, $^{14}CO_2$ was detected in respired air from *Uromastyx aegyptius* following oral doses of [^{14}C]cellulose (Foley et al. 1992).

Chelonia mydas does not absorb SCFA relative to their production rates, but in the order: butyric > propionic > acetic. This order is the most common in herbivorous mammals (Stevens 1988) and ptarmigan (*Lagopus* spp.) (Gasaway 1976a,b). In contrast, SCFA were absorbed relative to the rates at which they were produced in *Uromastyx aegyptius*: acetic > butyric > propionic (Foley et al. 1992). The authors suggested that formation of butyrate from acetate may be responsible for this deviation from the normal pattern.

3.8. Extent of Fermentation

Cell wall digestibility (CWD) in herbivorous reptiles free-feeding on foliage diets are presented in Table 7.3. The range of values reflect differences in diets as can be seen in the variation in CWD for different diets within a reptile species. Marken Lichtenbelt (1992) reported a range of CWD from 52% to 81% for *I. iguana* on four foliage diets. A wide range of 16% and 86% CWD was measured in *P. nelsoni* on two diets (Bjorndal and Bolten 1993).

It is not clear what proportion of the variation in CWD results from differences in the fermentation capacity of the herbivores. The only herbivorous species to be compared on the same diet are the tortoises *Geochelone carbonaria* and *G. denticulata*. CWD on a diet of *Lantana* foliage was not significantly different for the two tortoise species (Bjorndal 1989). In a comparison of herbivorous (*Pseudemys nelsoni*) and omnivorous (*Trachemys scripta*) freshwater turtles on two plant diets, the herbivore had a significantly greater energy gain from cell wall fermentation than did the omnivore on one diet and there was no significant difference in energy gain from cell wall fermentation on the other diet (Bjorndal and Bolten 1993). Studies on different species feeding on the same diet are needed to compare the fermentation capacities in herbivorous reptiles.

Microbial populations in herbivorous reptiles can attain very extensive fermentation of plant cell walls (Table 7.3), equivalent to that in mammals on diets of similar cell wall content (Bjorndal 1979, Troyer 1984d). However, these high levels of cell wall fermentation are achieved at a much slower rate in reptiles than in mammals. Although the capacity of the fermentation chambers apparently are equivalent for reptiles and mammals (Fig. 7.1), the long transit times and low intakes (Table 7.3) in reptiles indicate that a smaller quantity of digesta is fermented during a given period of time in reptiles.

3.9. Contribution of Fermentation to Energy Balance

Fermentations provide a substantial portion of the energy budget in herbivorous reptiles. Studies on three species of reptiles have estimated the contribution of SCFA to the animal's energy budget. In all studies, the animals were free-feeding on a natural diet. McBee and McBee (1982) calculated that SCFA provided 30% and 38% of the energy requirement of two adult male *Iguana iguana*. The energy requirement of the iguanas was estimated from Nagy's (1982) equation for field metabolic rates of iguanid lizards.

Two other studies calculated the energy generated by fermentation as a percent of digestible energy intake (DEI) on the same diet. In *P. nelsoni*, energy from SCFA approached 100% of DEI on a diet of an aquatic plant *Hydrilla verticillata* (Bjorndal and Bolten 1990). SCFA provided the energy equivalent of 47% of DEI in *U. aegyptius* on a natural mixed plant diet (Foley et al. 1992). All authors cautioned that the estimates were calculated based on a constant rate of fermentation at a constant temperature. There have been no studies on variation in rates of fermentation throughout the day or on the effect of temperature on fermentation rate, but one would predict that fermentation rates would decrease with lower body temperatures.

The portion of the energy balance not met by fermentation products must come from digestion of cell contents. Because reptiles do not chew their food, it is not clear how access to cell contents is gained. SCFA must supply a higher percentage of energy in *P. nelsoni* than in other reptiles because the fermentation in the small intestine decreases the digestion by endogenous enzymes and absorption of other energy sources such as dietary starches and sugars. The values for reptiles either exceed or fall within the upper range of estimates for small mammals with hindgut fermentations (5% to 39%) and are within the range for mammals with foregut fermentations (25% to 100%) (Parra 1978).

Most of the studies of cell wall fermentation in reptiles are based on single-species diets. To elucidate the role that fermentation plays in the energy balance of reptiles, studies are needed that explore the interaction of diet items—or associative effects—in herbivorous reptiles. In the omnivorous turtle *Trachemys scripta*, there was a positive associative effect between duckweed (*Spirodela polyrhiza*) and insect larvae (*Tenebrio*) (Bjorndal 1991). Duckweed was fermented to a significantly greater extent and turtles gained significantly more energy from the fermentation of duckweed when insect larvae were included in the diet.

Microbial populations in reptiles may provide benefits other than energy to their hosts, such as production of vitamins, improved nitrogen and mineral uptake, and conservation of water (McBee 1989). However, no studies have been conducted in these areas. Another benefit from microbial activities has been demonstrated for the green turtle, *Chelonia mydas*. The seagrass *Thalassia testudinum* comprises 80% of the diet of the green turtle in the Caribbean (Mortimer 1982). Microbial activity in the hindgut of the green turtle results in production of feces with small particle size and a lower carbon to nitrogen ratio than the fresh blades of *T. testudinum*. Thus, the microbes significantly shorten decomposition and enrichment cycles in seagrass beds and increase productivity of the green turtle's major food plant (Thayer et al. 1982).

3.10. Effect of Body Size

The constraint that small body size may impose on the ability of reptiles to utilize a plant diet has been a recurring theme in the study of herbivorous reptiles because of early observations that herbivorous reptiles appeared to be limited to animals of relatively large size. The question of body size and herbivory in reptiles was first raised by Szarski (1962) who suggested that many herbivorous lizards were large in order to avoid predation; he hypothesized that there are few herbivorous reptile species because they are vulnerable to predation. Sokol (1967) then suggested that herbivorous lizards were generally of large body size because smaller lizards lacked the strength to "reduce vegetation."

In 1973, Pough published a paper in which he suggested that below a body mass of 50 to 100 g, reptiles could not meet the relatively high mass-specific energy requirements on a plant diet. When Pough developed this hypothesis, the study of herbivory in reptiles was in its infancy. At that time it was assumed that reptiles were very inefficient herbivores that had no adaptations for utilization of plant cell walls and that all reptilian herbivores went through an early stage of insectivory before switching to an herbivorous diet. Termed "an enduringly provocative paper" by Troyer (1991), Pough's work stimulated interest in the field of herbivorous reptiles. Further studies revealed that there are a number of reptile species that are herbivorous at a small size, less than 50 to 100 g (reviews in Greene 1982, Iverson 1982, Troyer 1991). These small reptiles either are juveniles of large species, such as 3.5-g hatchling *Dipsosaurus dorsalis* (Mautz and Nagy 1987) and 12-g *Iguana iguana* (Troyer 1984d), or are species with small adult size, such as 20- to 50-g *Cnemidophorus murinus* (Dearing 1993) and 20-g *Liolaemus* spp. (Troyer 1991).

Although it is now clear that very small reptiles can exist on a plant diet, the question that Pough raised is still valid. How do small reptilian herbivores meet their high mass-specific energy needs? Apparently, small reptiles do not meet their higher mass-specific energy needs by increasing their relative gut capacity (Fig. 7.1) (Troyer 1984d). Studies indicate that they meet their higher energy demands in several ways.

First, small reptiles may feed more selectively, ingesting plant matter that can be more rapidly fermented. Small *Iguana iguana* select younger foliage with higher protein and less lignin (Troyer 1984d). Young *Dipsosaurus dorsalis* feed primarily on flowers (Mautz and Nagy 1987). Hatchling *P. nelsoni* ingest only leaves of *Hydrilla verticillata*, whereas adults ingest the entire fronds (leaves and stems), which are higher in lignin and lower in protein (Bjorndal and Bolten 1992). Small reptiles can efficiently harvest the more digestible plant parts because they require less absolute volume of food than do larger conspecifics and their small mouth size facilitates selective feeding.

Second, the small mouth size of the smaller reptiles also results in smaller bite size, and thus smaller particle size of digesta. Because reptiles cannot chew their food, smaller bite size can make significant differences in penetration of structural barriers to digestion and in increasing surface to volume ratio of digesta particles. Hatchling *P. nelsoni* physically damaged the thick cuticle of duckweed (*Spirodela polyrhiza*) fronds as they ingested the fronds one at a time. This structural damage to the plant allowed hatchlings to process the diet at significantly higher rates and to the same extent as did adults that swallowed many duckweed fronds simultaneously without penetrating the cuticles (Bjorndal and Bolten 1992). Higher surface to volume ratios in digesta particles result in higher fermentation rates in herbivores (Bjorndal et al. 1990).

Third, small reptiles may select higher body temperatures that may increase the rate of digestive processing. Captive juvenile *Amblyrhynchus cristatus* and *Conolophus pallidus* maintained higher (about 1.4°C) temperatures than did captive adults (Wilhoft 1958), but this study was based on a small number of animals, several of which were in poor health. Young *A. cristatus* in the wild also maintain higher temperatures than do adults and have shorter transit times than adults (Wikelski et al. 1993). However, there was no significant difference in body temperatures of adult and hatchling *Dipsosaurus dorsalis* in the field (Mautz and Nagy 1987). Troyer (1984d) found that small *Iguana iguana* had significantly shorter transit times than did large iguanas, but she could not identify the cause of the more rapid processing.

3.11. Effect of Temperature

Temperature has a major effect on physiological processes in ectotherms. It is therefore expected that temperature would influence digestive processing in reptiles, and a number of studies have addressed this question. These studies have yielded conflicting results. In some studies, digestibility of dry matter or energy increased with temperature (Harlow et al. 1976, Troyer 1987, Zimmerman and Tracy 1989 [for *Dipsosaurus dorsalis*]) whereas in other studies, digestibility did not vary with temperature (Ruppert 1980, Christian 1986, Karasov et al. 1986, Zimmerman and Tracy 1989 [for *Sauromalus obesus*], Baer 1992, Marken Lichtenbelt 1992, Wikelski et al. 1993). The different conclusions may be a result of the methods employed. Many of these studies relied on force-feeding reptiles a diet with small particle size. The problems of relying on force-fed animals have been discussed (Bjorndal et al. 1990, Baer 1992, Meienberger et al. 1993). For example, Harlow et al. (1976) is the study most commonly cited for illustrating a positive relationship between temperature and digestibility in an herbivorous reptile. In that study, *Dipsosaurus dorsalis* were force-fed the same amount of rabbit pellets at four temperatures (28°, 33°, 37°, and 41°C). However, Zimmerman and Tracy (1989) showed that if *D. dorsalis* were force-fed rabbit pellets at 33°, 37°, and 41°C, based on the predicted intake needed to balance energy requirements at each temperature (lower intake at lower temperatures), temperature had a significant, but much smaller, effect on digestibility.

From more recent studies based on free-feeding reptiles, it appears that intake and passage rate are positively related to temperature, whereas digestibility is not (Christian 1986, Baer 1992, Marken Lichtenbelt 1992). However, since more food is digested to the same extent in less time, the rate of digestion must also increase with temperature to maintain the same extent of digestion.

Further research is needed on the effect of temperature on digestive processing. No study has attempted to separate the temperature effects on digestion between endogenous enzymatic digestion and microbial activity. Studies of the effect of temperature on rates of fermentation and production of SCFA in vitro with inoculum from reptiles are needed. Also, no study has yet included the heat generated by fermentation in the temperature relations of herbivorous reptiles.

4. Gastrointestinal Tract Morphology of Anuran Tadpoles

Many species of anuran tadpoles rely on filter feeding, and the elaborate feeding structures associated with filter feeding have been described in detail (references cited in Seale 1987, Wassersug and Heyer 1988). Filter feeding requires specialization of the branchial skeleton and gills. Tadpoles can capture particles from 0.126 to 200 μm in diameter (Wassersug 1972, Seale 1980). Many species have keratinized mouthparts that are used to create suspensions by scraping attached algae or to allow macrophagous feeding. Filter feeding is replaced by macrophagous feeding during development in many anuran tadpoles (Burggren and Just 1992). Filter-feeding tadpoles are generally considered to be non-selective feeders within a range of particle sizes (Farlowe 1928, Jenssen 1967, Seale and Beckvar

1980, Sekar 1992) and, unlike most herbivores, do not avoid cyanobacteria (Seale and Beckvar 1980). However, Wagner (1986) described behavioral selection of conifer pollen by *Hyla regilla* tadpoles.

The digestive tract of anuran tadpoles is divided into a foregut, midgut and hindgut. The foregut includes the esophagus; stomach, which includes a glandular structure known as the manicotto glandulare; and a short, ciliated region that leads to the midgut (Griffiths 1961). In herbivorous tadpoles, the stomach is poorly differentiated, has a neutral pH, and serves primarily as a storage site for food with limited digestive activity (Fox 1984). In carnivorous tadpoles, the stomach may be more differentiated with low pH and high pepsin activity (Ruibal and Thomas 1988, Carroll et al. 1991).

The midgut is very long, narrow in diameter, and tightly coiled in herbivorous tadpoles (Noble 1931, Toloza and Diamond 1990a). Most enzymatic digestion and absorption occurs in this region (Reeder 1964). The hindgut is a relatively short, straight tube leading to the cloaca.

Gut morphology can change with diet within a species. Babak (1903, 1906; cited in Altig and Kelly 1974) found that when the same species of tadpole was fed plant or animal diets, the tadpoles on the plant diet developed longer intestines. Tadpoles of *Gastrotheca gracilis* maintained on a plant diet had greater development of the posterior intestine and fewer opalinid protozoans compared with tadpoles fed a carnivorous diet (Michel de Cerasuolo and Teran 1991). Gut mass relative to body mass was higher in *Rana catesbeiana* tadpoles fed a lettuce diet than in tadpoles fed ground beef (Toloza and Diamond 1990b). Some species of spadefoot toads (*Scaphiopus* spp.) have distinct omnivorous and carnivorous tadpole morphs. The carnivorous morphs have shorter intestines than do the omnivorous morphs (Seale 1987).

Altig and Kelly (1974) reported a similar pattern of diet and gut morphology among species. In their study of gut morphology in 13 species of tadpoles, those species known to be herbivorous had longer relative gut lengths than did carnivores.

The morphology and physiology of the digestive tract of herbivorous tadpoles undergo radical changes at metamorphosis. Mouthparts are remodeled, and intestines of herbivorous tadpoles shorten dramatically in preparation for the carnivorous diets of adult forms (Dodd and Dodd 1976, Fox 1984, Burggren and Just 1992). Average mass of stomach and intestine tissue of *Rana pipiens* was reduced by 56.5% (Kuntz 1924). However, at least in *Rana catesbeiana*, the amount of body mass invested in the digestive tract does not change significantly at metamorphosis; the intestines are shorter but thicker-walled in adult forms (Toloza and Diamond 1990a). Ciliated epithelial cells disappear from the lining of the gut and major histological changes occur (Bonneville 1963, Fox 1981, Dauca and Hourdry 1985). The occurrence and activities of digestive enzymes change

(Altig et al. 1975, Reeder 1964, Burggren and Just 1992), as do nutrient transport systems (Toloza and Diamond 1990a,b).

5. Microbes in the Gastrointestinal Tract of Anuran Tadpoles

Most research on gut microbiota in anuran tadpoles has involved a very interesting unicellular organism (referred to here as the unicell) that has been identified as the alga *Prototheca* sp. (Richards 1958, 1962), the yeast *Candida humicola* (Steinwascher 1979), and, most recently, *Prototheca richardsi* (Beebee 1991, Beebee and Wong 1992), a unicellular, unpigmented, heterotrophic alga. For many years, studies reported that under conditions of high population density and low food resources, large tadpoles appeared to have a competitive advantage over small tadpoles, as evidenced by more rapid growth and eventual metamorphosis in the large tadpoles and slow growth and failure to metamorphose in the small tadpoles. These effects were observed in both mixed species and single species groups of tadpoles. Richards (1958, 1962) suggested that *Rana pipiens* tadpoles deposited a growth-inhibiting chemical onto the unicell, and that this chemical then slowed the growth of tadpoles that later ingested the unicell. Studies by West (1960) and Rose and Rose (1961) demonstrated that the unicell did not have to be present for the effect of the chemical to occur. Thus, ingestion of the unicell was not essential, which refuted Richards' hypothetical mechanism. Steinwascher (1978) demonstrated that the effect only occurred at low food levels in *Rana utricularia* tadpoles and hypothesized that the secretion of the growth inhibitor was stimulated by competition for food between the unicell and the tadpole. But in further studies, Steinwascher (1979) found that in large *R. clamitans* tadpoles that eventually metamorphosed and reproduced, the unicell was a mutualistic symbiont that enhanced growth, whereas it retarded growth and decreased probability of metamorphosis in smaller tadpoles. Also, at low concentrations of unicells, individual tadpoles grew more rapidly than did control tadpoles, but at higher unicell concentrations, individual tadpoles grew more slowly than did controls (Steinwascher 1979). Steinwascher suggested that growth was retarded through exploitative competition between the unicells and the tadpole for the digesta in the intestines of the tadpole.

In a recent study, Beebee and Wong (1992) provided evidence for a mechanism of interference competition between size classes of tadpoles mediated by *Prototheca richardsi*. Small tadpoles (*Bufo calamita*) were attracted to *Prototheca*-rich feces and away from higher quality food. Production of protothecans in feces of large tadpoles (*Rana temporaria*) was inversely related to food resources. Thus, under low food regimes, small tadpoles were largely coprophagous and left the high quality food for the large tadpoles. Growth inhibition resulting from *Prototheca* apparently is not species specific (Beebee 1991).

All of the above unicell studies were conducted with tadpoles housed in aquaria. The role of unicells in tadpole nutrition under natural conditions has not been determined. Petranka (1989) found growth inhibition to be a rare event of low magnitude in a field study with *Rana utricularia* tadpoles, but Beebee (1991) suggested that Petranka's methods were inappropriate for detection of unicell-mediated effects.

Few studies have investigated the potential role of gut microbes in the nutrition of anuran tadpoles. Reeder (1964, p. 140) stated that there was a possibility that tadpoles might benefit from bacterial fermentation of complex carbohydrates. However, many authors have dismissed microbial fermentation as a significant source of nutrition because of the short passage time of digesta (Savage 1952, 1961; Reeder 1964, p. 135). Passage times that have been reported for tadpoles are relatively short. Passage time was from 4 to 8 hours in *Rana temporaria* and from 4 to 6 hours in *Bufo bufo* at 16° to 18°C (Savage 1952); from 2 to 6 hours in *Bufo calamita* and > 6 hours in *Rana temporaria* (Beebee and Wong 1992); 6 hours in *Xenopus laevis* (Ueck 1967, cited in Altig and McDearman 1975); and 3 to 6 hours in *Rhacophorus cruciger* (Costa and Balasubramanian 1965). Mean clearance times for five species of tadpoles ranged from 0.5 to 1.7 hours on a diet of commercial pelleted rabbit food at 22°C (Altig and McDearman 1975), but Steinwascher (1978) has suggested that the method used may have underestimated passage rates in that study.

Examination of intestinal contents of *Bufo regularis* tadpoles demonstrated that they ingested microorganisms as they ingested the nutrient-rich surface of the substrate (Nathan and James 1972). Nathan and James identified several genera of protozoa in the intestines of the tadpoles and suggested they may play a role similar to that of protozoa in termites. However, their study demonstrated that protozoans were not necessary to the growth and development of the tadpoles. Of the several species of protozoa that were found in the digestive tracts of the tadpoles while they were feeding on mud, only the euglenoid Mastigophora maintained its populations in the gut when mud was removed from the diet.

Identification of bacteria in the digestive tract of anuran tadpoles has been limited to pathogenic forms (Hird et al. 1981, 1983). I found no study that analyzed digesta for fermentation end products, or that evaluated the degradation of cell walls in the digestive tract. Such studies are needed. There are few digestibility measurements available for anuran tadpoles on plant diets, but those suggest a relatively high digestibility of plant matter, which in turn suggest degradation of cell wall material. Mean digestibility of organic matter in four species of anuran tadpoles on a diet of commercial rabbit pellets ranged from 54% to 86% at 22°C (Altig and McDearman 1975). However, because ash was used as an indigestible diet marker in this study, the results may not be reliable (Bjorndal 1985b). Seale (1987) cited a study by Marian (1982) that reported an absorption efficiency of about 80% for tadpoles fed plant material. Without information on the composition of these plant diets, it is difficult to assess the probability of cell wall degradation from the digestibility data.

Passage times of a few hours do not necessarily preclude significant cell wall fermentation, as evidenced by studies on emu, *Dromaius novaehollandiae* (Herd and Dawson 1984), and the Australian wood duck, *Chenonetta jubata* (Dawson et al. 1989). However, this rapid cell wall degradation apparently relies on acid hydrolysis of hemicellulose in the gizzard of the birds, followed by rapid fermentation of the solubilized hemicellulose in the cecum (Dawson 1989). Because the foregut in herbivorous tadpoles is not acidic, such a mechanism would not be possible. Also, hemicellulose is not a major cell wall constituent of algae. Algal complex carbohydrates may or may not be susceptible to similar digestive processing.

Anuran tadpoles may be able to benefit nutritionally from the actions of bacteria that continually pass through their digestive tract, rather than harboring permanent populations of microbes. Characteristics of many anuran tadpoles make a reliance on allochthonus—or transient—microbes not unlikely: 1. the extent to which many species feed on detritus and practice coprophagy would provide a reliable source of such microbes; 2. the lack of acidic pH throughout the digestive tract (Reeder 1964) and the low or absent pepsin activity would promote passage of live bacteria through the intestines; 3. filter-feeding and scraping feeding mechanisms result in ingestion of a constant supply of small particles that could be rapidly attacked by microbes; and 4. the lack of any apparent morphological structure that could harbor gut microbes.

Clearly, research is needed on digestive processing in anuran tadpoles. Studies on almost any aspect in herbivorous forms would yield important information in this little-known group.

6. Conclusions

Any attempt to synthesize the field of digestive processing in herbivorous reptiles is hampered by the methods employed in many of the studies. Because herbivorous reptiles are often disinclined to feed on natural diets in captivity, researchers have resorted to force-feeding experimental animals and/or feeding them unnatural diets. Often, these diets are low in fiber and/or have small particle size. McBee (1989) cautions that these feeding practices tend to change hindgut fermenters into omnivores and alter hindgut processes. He states that feeding finely ground food to iguanas changes "their thin-walled, turgid, hyaline hindgut filled with a slightly acidic content into a thick-walled opaque and flaccid organ half-filled with a stinking mess." Clearly, studies on fermentation in such animals would yield questionable results.

Much research is needed to improve our understanding of even the most basic elements of fermentation in amphibians and reptiles. For amphibians, we must establish whether fermentations exist and whether anuran tadpoles receive any benefit from gut microbes. For reptiles, research needs have been identified in the appropriate sections throughout this chapter. Fermentations in herbivorous lizards with simple hindguts—such as *Cnemidophorus* or *Liolaemus* species—have not been studied and may provide insightful comparisons with fermentation processes in lizards with elaborated hindguts or in turtles with simple hindguts.

Acknowledgments

I am grateful to John E. Moore, a dedicated animal scientist who was sufficiently open-minded to encourage and support a student interested in turtle nutrition. I benefited from his guidance during my postdoctoral studies at the Animal Nutrition Laboratory, University of Florida. Alan Bolten has continually shared his insights and improved the experimental designs of many of the studies on which this chapter is based. John Iverson has repeatedly and generously shared his extensive knowledge of the literature. I thank Indraneil Das for sharing his unpublished manuscripts. Alan Bolten, Martha Crump, and Frank Hensley provided constructive criticism on the manuscript. I thank Harold Avery, Perry Barboza, Martha Crump, Ken Nagy, Olav Oftedal, Warren Porter, Rodolfo Ruibal, and Linda Zimmerman for providing literature. I thank Peter Eliazar for his assistance with preparation of this publication.

References

Alexander, TR (1964) Observations on the feeding behavior of *Bufo marinus* (Linne). Herpetologica 20:255–259.

Altig R, Kelly JP (1974) Indices of feeding in anuran tadpoles as indicated by gut characteristics. Herpetologica 30:200–203.

Altig R, Kelly JP, Wells M, Phillips J (1975) Digestive enzymes of seven species of anuran tadpoles. Herpetologica 31:104–108.

Altig R, McDearman W (1975) Percent assimilation and clearance times of five anuran tadpoles. Herpetologica 31:67–69.

Andrews SP (1984) Aspects of fibre digestion in Cunningham's skink, *Egernia cunninghami*. Unpublished thesis, University of New England, Armidale, Australia.

Babak E (1903) Ueber den Einfluss der Nahrung auf die Lange des Darmkanals. Biol Zentr 22:477–483, 519–528.

Babak E (1906) Ueber die morphogenetische Reaktion des Darmkanals des Froschlarvae auf Muskelprotein verschiedener Tierklassen. Beitr Chem Physiol Pathol 7:323–330.

Baer DJ (1992) Effects of diet composition and ambient temperature on digestive function and bioenergetics of the green iguana (*Iguana iguana*). Ph.D. dissertation, Michigan State University, East Lansing.

Balazs GH (1980) Synopsis of biological data on the green turtle in the Hawaiian Islands. NOAA Tech Memo NOAA-TM-NMFS-SWFC-7.

Beebee TJC (1991) Purification of an agent causing growth inhibition in anuran larvae and its identification as a unicellular unpigmented alga. Can J Zool 69:2146–2153.

Beebee TJC, Wong ALC (1992) Prototheca-mediated interference competition between anuran larvae operates by resource diversion. Physiol Zool 65:815–831.

Bjorndal KA (1979) Cellulose digestion and volatile fatty acid production in the green turtle, *Chelonia mydas*. Comp Biochem Physiol 63A:127–133.

Bjorndal KA (1980) Nutrition and grazing behavior of the green turtle *Chelonia mydas*. Mar Biol 56:147–154.

Bjorndal KA (1982) The consequences of herbivory for the life history pattern of the Caribbean green turtle, *Chelonia mydas*. In: Bjorndal KA, ed. Biology and Conservation of Sea Turtles, pp. 111–116. Washington: Smithsonian Institution Press.

Bjorndal KA (1985a) Nutritional ecology of sea turtles. Copeia 1985:736–751.

Bjorndal KA (1985b) Use of ash as an indigestible dietary marker. Bull Mar Sci 36: 224–230.

Bjorndal KA (1987) Digestive efficiency in a temperate herbivorous reptile, *Gopherus polyphemus*. Copeia 1987:714–720.

Bjorndal KA (1989) Flexibility of digestive responses in two generalist herbivores, the tortoises *Geochelone carbonaria* and *Geochelone denticulata*. Oecologia 78:317–321.

Bjorndal KA (1991) Diet mixing: nonadditive interactions of diet items in an omnivorous freshwater turtle. Ecology 72:1234–1241.

Bjorndal KA, Bolten AB (1990) Digestive processing in a herbivorous freshwater turtle: consequences of small-intestine fermentation. Physiol Zool 63:1232–1247.

Bjorndal KA, Bolten AB, Moore JE (1990) Digestive fermentation in herbivores: effect of food particle size. Physiol Zool 63:710–721.

Bjorndal KA, Bolten AB (1992) Body size and digestive efficiency in a herbivorous freshwater turtle: advantages of small bite size. Physiol Zool 65:1028–1039.

Bjorndal KA, Bolten AB (1993) Digestive efficiencies in herbivorous and omnivorous freshwater turtles on plant diets: do herbivores have a nutritional advantage? Physiol Zool 66:384–395.

Bjorndal KA, Suganuma H, Bolten AB (1991) Digestive fermentation in green turtles, *Chelonia mydas*, feeding on algae. Bull Mar Sci 48:166–171.

Bonneville MA (1963) Fine structural changes in the intestinal epithelium of the bullfrog during metamorphosis. J Cell Biol 18:579–597.

Burggren WW, Just JJ (1992) Developmental changes in physiological systems. In: Feder ME, Burggren WW, eds. Environmental Physiology of the Amphibians, pp. 467–530. Chicago: University of Chicago Press.

Carroll EJ Jr, Seneviratne AM, Ruibal R (1991) Gastric pepsin in an anuran larva. Dev Growth Differ 33:499–507.

Christian KA (1986) Physiological consequences of nighttime temperature for a tropical, herbivorous lizard (*Cyclura nubila*). Can J Zool 64:836–840.

Christian KA, Tracy CR, Porter WP (1984) Diet, digestion, and food preferences of Galapagos land iguanas. Herpetologica 40:205–212.

Costa HH, Balasubramanian S (1965) The food of the tadpoles of *Rhacophorus cruciger* (Blyth). Ceylon J Sci 5:105–109.

Das I, Coe M (1994) Dental morphology and diet in anuran amphibians from south India. J Zool 233:417–427.

Da Silva HR, De Britto-Pereira MC, Caramaschi U (1989) Frugivory and seed dispersal by *Hyla truncata*, a neotropical treefrog. Copeia 1989:781–783.

Dauca M, Hourdry J (1985) Transformations in the intestinal epithelium during anuran metamorphosis. In: Balls M, Bownes M, eds. Metamorphosis: The Eighth Symposium of the British Society for Developmental Biology, pp. 36–58. Oxford: Clarendon Press.

Dawson TJ (1989) Food utilization in relation to gut structure and function in wild and domestic birds and mammals. Acta Vet Scand Suppl 86:20–27.

Dawson TJ, Johns A, Beal AM (1989) Digestion in the Australian wood duck (*Chenonetta jubata*): a small avian herbivore showing selective digestion of the hemicellulose component of fiber. Physiol Zool 62:522–540.

Dearing MD (1993) An alimentary specialization for herbivory in the tropical whiptail lizard, *Cnemidophorus murinus*. J Herpetol 27:111–114.

Dodd MHI, Dodd JM (1976) The biology of metamorphosis. In: Lofts B, ed. Physiology of the Amphibia, pp. 467–599. New York: Academic Press.

Dubuis AL, Faurel L, Grenot C, Vernet R (1971) Sur le regime alimentaire du lezard saharien *Uromastyx acanthinurus* Bell. C R Acad Sci (Paris) Ser D 273:500–503.

Duellman WE, Trueb L (1986) Biology of Amphibians. New York: McGraw-Hill.

El-Toubi MR, Bishai HM (1959) On the anatomy and histology of the alimentary tract of the lizard *Uromastyx aegyptia* (Forskal). Bull Fac Sci Cairo Univ 34:13–50.

Farlowe V (1928) Algae of ponds as determined by an examination of the intestinal contents of tadpoles. Biol Bull 55:443–448.

Fialho RF (1990) Seed dispersal by a lizard and a treefrog—effect of dispersal site on seed survivorship. Biotropica 22:423–424.

Foley WJ, Bouskila A, Shkolnik A, Choshniak I (1992) Microbial digestion in the herbivorous lizard *Uromastyx aegyptius* (Agamidae). J Zool Lond 226:387–398.

Fox H (1981) Cytological and morphological changes during amphibian metamorphosis. In: Gilbert LI, Frieden E, eds. Metamorphosis: A Problem in Developmental Biology 2nd ed, pp. 327–362. New York: Plenum Press.

Fox H (1984) Amphibian Morphogenesis. Clifton, NJ: Bioscience, Humana Press.

Gasaway WC (1976a) Seasonal variation in diet, volatile fatty acid production and size of the cecum of rock ptarmigan. Comp Biochem Physiol 53A:109–114.

Gasaway WC (1976b) Volatile fatty acids and metabolizable energy derived from cecal fermentation in the willow ptarmigan. Comp Biochem Physiol 53A:115–121.

Greene HW (1982) Dietary and phenotypic diversity in lizards: why are some organisms specialized? In: Mossakowski D, Roth G, eds. Environmental Adaptation and Evolution, pp. 107–128. Stuttgart: Gustav Fischer.

Griffiths I (1961) The form and function of the fore-gut in anuran larvae (amphibia, salientia) with particular reference to the manicotto glandulare. Proc Zool Soc Lond 137:249–283.

Guard CL (1980) The reptilian digestive system: general characteristics. In: Schmidt-Nielsen K, Bolis L, Taylor CR, Bentley PJ, Stevens CE, eds. Comparative Physiology: Primitive Mammals. pp. 43–51. Cambridge: Cambridge University Press.

Hamilton J, Coe M (1982) Feeding, digestion and assimilation of a population of giant tortoises (*Geochelone gigantea* (Schweigger)) on Aldabra atoll. J Arid Environ 5: 127–144.

Harlow HJ, Hillman SS, Hoffman M (1976) The effect of temperature on digestive efficiency in the herbivorous lizard, *Dipsosaurus dorsalis*. J Comp Physiol 111:1–6.

Herd RM, Dawson TJ (1984) Fiber digestion in the emu, *Dromaius novaehollandiae*, a large bird with simple gut and a high rate of passage. Physiol Zool 57:70–84.

Hird DW, Diesch SL, McKinnell RG, et al. (1981) *Aeromonas hydrophila* in wild-caught frogs and tadpoles (*Rana pipiens*) in Minnesota. Lab Anim Sci 31:166–169.

Hird DW, Diesch SL, McKinnell RG, et al. (1983) Enterobacteriaceae and *Aeromonas hydrophila* in Minnesota frogs and tadpoles (*Rana pipiens*). Appl Environ Microbiol 46:1423–1425.

Hotton N III (1955) A survey of adaptive relationships of dentition to diet in the North American Iguanidae. Am Midl Nat 53:88–114.

Hukuhara T, Naitoh T, Kameyama H (1975) Observations on the gastrointestinal movements of the tortoise (*Geoclemys reevesii*) by means of the abdominal-window-technique. Jpn J Smooth Muscle Res 11:39–46.

Iverson JB (1980) Colic modifications in iguanine lizards. J Morphol 163:79–93.

Iverson JB (1982) Adaptations to herbivory in Iguanine lizards. In: Burghardt GM, Rand AA, eds. Iguanas of the World: Their Behavior, Ecology and Conservation. pp. 60–76. Park Ridge, NJ: Noyes Publications.

Janis C (1976) The evolutionary strategy of the equidae and the origins of rumen and cecal digestion. Evolution 30:757–774.

Jenssen TA (1967) Food habits of the green frog, *Rana clamitans*, before and during metamorphosis. Copeia 1967:214–218.

Justice KE, Smith FA (1992) A model of dietary fiber utilization by small mammalian herbivores, with empirical results for *Neotoma*. Am Nat 139:398–416.

Karasov WH, Petrossian E, Rosenberg L, Diamond JA (1986) How do food passage rate and assimilation differ between herbivorous lizards and nonruminant mammals? J Comp Physiol B156:599–609.

Kuntz A (1924) Anatomical and physiological changes in the digestive system during metamorphosis in *Rana pipiens* and *Amblystoma tigrinum*. J Morphol 38:581–598.

Lonnberg E (1902) On some points of relation between the morphological structure of the intestine and the diet of reptiles. Bihang Till K Svenska Vet-Akad Handlingar Band 28 Afd IV (No. 8):3–53.

Marian MP (1982) Ecophysiological studies in frog culture (*Rana tigrina* Daud). Ph.D. thesis, Madurai Kamaraj University, Madurai, India.

Marken Lichtenbelt WD (1992) Digestion in an ectothermic herbivore, the green iguana (*Iguana iguana*): effect of food composition and body temperature. Physiol Zool 65: 649–673.

Mautz WJ, Nagy KA (1987) Ontogenetic changes in diet, field metabolic rate, and water flux in the herbivorous lizard *Dipsosaurus dorsalis*. Physiol Zool 60:640–658.

McBee RH (1989) Hindgut fermentation in nonavian species. J Exp Zool Suppl 3:55–60.

McBee RH, McBee VH (1982) The hindgut fermentation in the green iguana, *Iguana iguana*. In: Burghardt GM, Rand AA, eds. Iguanas of the World: Their Behavior, Ecology and Conservation. pp. 77–83. Park Ridge, NJ: Noyes Publications.

Meienberger C, Wallis IR, Nagy KA (1993) Food intake rate and body mass influence transit time and digestibility in the desert tortoise (*Xerobates agassizii*). Physiol Zool 66:847–862.

Michel de Cerasuolo A, Teran HR (1991) Aspectos histoquimicos del tracto digestivo larval de *Gastrotheca gracilis* Laurent, en relacion con la alimentacion. Acta Zool Lilloana 40:69–81.

Montanucci RR (1968) Comparative dentition in four iguanid lizards. Herpetologica 24: 305–315.

Mortimer JA (1982) Feeding ecology of sea turtles. In: Bjorndal KA, ed. Biology and Conservation of Sea Turtles. pp. 103–109. Washington: Smithsonian Institution Press.

Nagy KA (1977) Cellulose digestion and nutrient assimilation in *Sauromalus obesus*, a plant eating lizard. Copeia 1977:355–362.

Nagy KA (1982) Energy requirements of free-living iguanid lizards. In: Burghardt GM, Rand AA, eds. Iguanas of the World: Their Behavior, Ecology and Conservation, pp. 49–59. Park Ridge, NJ: Noyes Publications.

Nagy KA, Medica PA (1986) Physiological ecology of desert tortoises in southern Nevada. Herpetologica 42:73–92.

Nagy KA, Shoemaker VH (1975) Energy and nitrogen budgets of the free-living desert lizard *Sauromalus obesus*. Physiol Zool 48:252–262.

Nathan JM, James VG (1972) The role of protozoa in the nutrition of tadpoles. Copeia 1972:669–679.

Noble GK (1931) The Biology of the Amphibia. New York: McGraw-Hill.

Norman DB, Weishampel DB (1985) Ornithopod feeding mechanisms: their bearing on the evolution of herbivory. Am Nat 126:151–164.

Parra R (1978) Comparison of foregut and hindgut fermentation in herbivores. In: Montgomery GG, ed. The Ecology of Arboreal Folivores, pp. 205–229. Washington: Smithsonian Institution Press.

Penry DL, Jumars PA (1987) Modeling animal guts as chemical reactors. Am Nat 129: 69–96.

Petranka JW (1989) Chemical interference competition in tadpoles: does it occur outside laboratory aquaria? Copeia 1989:921–930.

Pough FH (1973) Lizard energetics and diet. Ecology 54:837–844.

Pough FH (1983) Amphibians and reptiles as low-energy systems. In: Aspey WP, Lustick SI, eds. Behavioral Energetics, pp. 141–188. Columbus: Ohio State University Press.

Ray CE (1965) Variation in the number of marginal tooth positions in three species of iguanid lizards. Breviora 236:1–15.

Reeder WG (1964) The digestive system. In: Moore JA, ed. Physiology of the Amphibia, pp. 99–149. New York: Academic Press.

Richards CM (1958) The inhibition of growth in crowded *Rana pipiens* tadpoles. Physiol Zool 31:138–151.

Richards CM (1962) The control of tadpole growth by alga-like cells. Physiol Zool 35: 285–296.

Rose SM, Rose FC (1961) Growth-controlling exudates of tadpoles. Symp Soc Exp Biol 15:207–218.

Ruibal R, Thomas E (1988) The obligate carnivorous larvae of the frog, *Lepidobatrachus laevis* (Leptodactylidae). Copeia 1988:591–604.

Ruppert RM (1980) Comparative assimilation efficiencies of two lizards. Comp Biochem Physiol 67A:491–496.

Savage RM (1952) Ecological, physiological and anatomical observations on some species of anuran tadpoles. Proc Zool Soc Lond 122:467–514.

Savage RM (1961) The Ecology and Life History of the Common Frog (*Rana temporaria temporaria*). London: Sir Isaac Pitman & Sons.

Schad GA, Knowles R, Meerovitch E (1964) The occurrence of *Lampropedia* in the intestines of some reptiles and nematodes. Can J Microbiol 10:801–804.

Seale DB (1980) Influence of amphibian larvae on primary production, nutrient flux, and competition in a pond ecosystem. Ecology 61:1531–1550.

Seale DB (1987) Amphibia. In: Pandian TJ, Vernberg FJ, eds. Animal Energetics, Vol. 2, pp. 467–552. Bivalvia Through Reptilia, San Diego: Academic Press.

Seale DB, Beckvar N (1980) The comparative ability of anuran larvae (genera: *Hyla*, *Bufo* and *Rana*) to ingest suspended blue-green algae. Copeia 1980:495–503.

Sekar AG (1992) A study of the food habits of six anuran tadpoles. J Bombay Nat Hist Soc 89:9–16.

Smith HW (1965) Observations on the flora of the alimentary tract of animals and factors affecting its composition. J Pathol Bacteriol 89:95–122.

Sokol OM (1967) Herbivory in lizards. Evolution 21:192–194.

Steinwascher K (1978) Interference and exploitation competition among tadpoles of *Rana utricularia*. Ecology 59:1039–1046.

Steinwascher K (1979) Host-parasite interaction as a potential population-regulating mechanism. Ecology 60:884–890.

Stevens CE (1988) Comparative Physiology of the Vertebrate Digestive System. Cambridge: Cambridge University Press.

Stevens CE (1989) Evolution of vertebrate herbivores. Acta Vet Scand Suppl 86:9–19.

Szarski H (1962) Some remarks on herbivorous lizards. Evolution 16:529.

Thayer GW, Engel DW, Bjorndal KA (1982) Evidence for short-circuiting of the detritus cycle of seagrass beds by the green turtle, *Chelonia mydas* L. J Exp Mar Biol Ecol 62: 173–183.

Thompson SM (1980) A comparative study of the anatomy and histology of the oral cavity and alimentary canal of two sea turtles: the herbivorous green turtle *Chelonia mydas* and the carnivorous loggerhead turtle *Caretta caretta* (includes discussion of diet and digestive physiology). M.S. thesis, James Cook University, North Queensland, Townsville, Australia.

Throckmorton GS (1976) Oral food processing in two herbivorous lizards, *Iguana iguana* (Iguanidae) and *Uromastix aegyptius* (Agamidae). J Morphol 148:363–390.

Throckmorton GS (1980) The chewing cycle in the herbivorous lizard *Uromastix aegyptius* (Agamidae). Arch Oral Biol 25:225–233.

Toloza EM, Diamond JM (1990a) Ontogenetic development of nutrient transporters in bullfrog intestine. Am J Physiol 258:G760-G769.

Toloza EM, Diamond JM (1990b) Ontogenetic development of transporter regulation in bullfrog intestine. Am J Physiol 258:G770-G773.

Troyer K (1982) Transfer of fermentative microbes between generations in a herbivorous lizard. Science 216:540–542.

Troyer K (1984a) Structure and function of the digestive tract of a herbivorous lizard *Iguana iguana*. Physiol Zool 57:1–8.

Troyer K (1984b) Behavioral acquisition of the hindgut fermentation system by hatchling *Iguana iguana*. Behav Ecol Sociobiol 14:189–193.

Troyer K (1984c) Microbes, herbivory and the evolution of social behavior. J Theor Biol 106:157–169.

Troyer K (1984d) Diet selection and digestion in *Iguana iguana*: the importance of age and nutrient requirements. Oecologia 61:201–207.

Troyer K (1987) Small differences in daytime body temperature affect digestion of natural food in a herbivorous lizard (*Iguana iguana*). Comp Biochem Physiol 87A:633–636.

Troyer K (1988) Morphological specializations for herbivory in small lizards: the genus *Liolaemus*. Am Zool 28:197A.

Troyer K (1991) Role of microbial cellulose degradation in reptile nutrition. In: Haigler CH, Weimer PJ, eds. Biosynthesis and Biodegradation of Cellulose, pp. 311–325. New York: Marcel Dekker.

Ueck M (1967) Der Manicotto glandulare (Drusenmagen) der Anuranlarvae in Bau, Funktion and Beziehung zure Gesamtlange des Darmes. Sonderdruck Zeit Wissen Zool 176: 173–270.

Ultsch, GR (1973) Observations on the life history of *Siren lacertina*. Herpetologica 29: 304–305.

Van Soest PJ (1982) Nutritional Ecology of the Ruminant. Corvallis, OR: O & B Books.

Voorhees ME (1981) Digestive efficiency of *Sauromalus varius*. M.S. thesis, Colorado State University, Fort Collins.

Wagner WE (1986) Tadpoles and pollen: observations on the feeding behavior of *Hyla regilla* larvae. Copeia 1986:802–804.

Waldschmidt SR, Jones SM, Porter WP (1987) Reptilia. In: Pandian TJ, Vernberg FJ, eds. Animal Energetics, Vol. 2, pp. 553–619. Bivalvia Through Reptilia. San Diego: Academic Press.

Warner ACI (1981) Rate of passage through the gut of mammals and birds. Nutr Abstr Rev 51B:789–820.

Wassersug R (1972) The mechanism of ultraplanktonic entrapment in anuran larvae. J Morphol 137:279–288.

Wassersug RJ, Heyer R (1988) A survey of internal oral features of leptodactyloid larvae (Amphibia: Anura). Smithson Contrib Zool 457:1–99.

West LB (1960) The nature of growth inhibitory material from crowded *Rana pipiens* tadpoles. Physiol Zool 33:232–239.

Wikelski M, Gall B, Trillmich F (1993) Ontogenetic changes in food intake and digestion rate of the herbivorous marine iguana (*Amblyrhynchus cristatus*). Oecologia 94: 373–379.

Wilhoft DC (1958) Observations on preferred body temperature and feeding habits of some selected tropical iguanas. Herpetologica 14:161–164.

Zar JH (1984) Biostatistical Analysis, 2nd ed. Englewood Cliffs, NJ: Prentice Hall.

Zimmerman LC, Tracy CR (1989) Interactions between the environment and ectothermy and herbivory in reptiles. Physiol Zool 62:374–409.

8

Microbial Fermentation in Insect Guts

Matthew D. Kane

1. Introduction

If biological success is measured either by numbers of individuals or diversity of species, insects are undoubtedly the most successful group of animals in the history of life on earth. There are more species of insects than all other animal species combined (Wilson 1992). This observation is especially remarkable in light of the fact that insects do not colonize marine habitats, a limitation that restricts them to approximately 30% of the planet's surface area.

The extreme phylogenetic radiation exhibited by insects has been accompanied by extensive diversification in dietary habit and nutritional physiology. Many intriguing behavioral and physiological adaptations have been employed by insects to assist them in this dietary and nutritional diversification (Slansky and Rodriguez 1987, Hunt and Nalepa 1994), including the establishment of nutritionally based symbiotic relationships with microorganisms (Buchner 1965, McBee 1977, Anderson et al. 1984, Barbosa et al. 1991). Insects have supplemented the nutritive potential of their food by establishing relationships with protozoa (Honigburg 1970, Yamin 1979, Fenchel 1987, Radek et al. 1992, Kirby and Margulis 1994), fungi (Sands 1969, Martin 1987, Darlington 1994), and/or bacteria (Breznak 1982, 1990, 1994; Cruden and Markovitz 1987; Breznak and Brune 1994). While microbial communities in the guts of termites represent the most dramatic example of this phenomenon (Prins and Kreulen 1991, Breznak and Brune 1994), digestive nutritional relationships with microbes have been discovered among members of a variety of other insect orders (Table 8.1). However,

Table 8.1. Insects for which digestive nutritional relationships with microbes have been examined.

Insect Group (order)	General References
Termites (Isoptera)	Breznak 1982, 1984; Martin 1987; Breznak and Brune 1994
Cockroaches (Blattaria)	Cruden and Markovitz 1987, Kane and Breznak 1991b, Gijzen and Barugahare 1992
Beetles (Coleoptera)	Bayon 1980, 1981a,b; Bayon and Mathelin 1980; Martin 1987
Crickets (Orthoptera)	Ulrich et al. 1981, Kaufman et al. 1989, Kaufman and Klug 1990, 1991
Grasshoppers (Orthoptera)	Mead et al. 1988
Locusts (Orthoptera)	Charnley et al. 1985
Craneflies (Diptera)	Klug and Kotarski 1980, Lawson et al. 1984, Lawson and Klug 1989
Fruitflies (Diptera)	Drew and Lloyd 1991, Daser and Brandl 1992
Ants (Hymenoptera)	Martin 1987, Caetano 1989
Bees (Hymenoptera)	Gilliam et al. 1990, Gilliam and Taber 1991, Cano et al. 1994
Wasps (Hymenoptera)	Kukor and Martin 1983, Martin 1987

even among comparatively nonspeciose groups such as termites (2,200 species), the diversity of insects makes comparative studies a daunting prospect. With this in mind, many other insect-microbe nutritional interactions undoubtedly remain to be discovered, and even those that have been identified need to be examined more closely to have a better understanding of the underlying mechanisms that govern the ecology and evolution of these fascinating interactions.

This chapter will primarily be concerned with microorganisms that reside in the guts of insects, and how the metabolic activities of these microbial populations and communities contribute to insect digestion, nutrition, and dietary diversification. Considerable evidence that "mycetocyte" bacteria (intracellular bacterial endosymbionts present in fat body and other tissues of a variety of insect species) are important to host nutrition underscores the view that important nutritional interactions between insects and microbes are not restricted to the gut environment (Douglas 1989). Nevertheless, microbial fermentation in the guts of insects, especially those that consume a refractile, nutrient-poor diet consisting largely of lignocellulosic materials (McBee 1977, Martin 1991, Prins and Kreulen 1991), is an important component of the global carbon cycle and represents the most eccentric modification of digestive processes among members of the Earth's most numerous and diverse class of animals (Chapman 1982).

2. Morphology and Physiology of the Insect Gut

2.1. Morphology and Structure

While insects represent the most diminutive animals considered in this volume, sizes and weights of the group actually vary over five orders of magnitude, from tiny parasitic wasps such as *Trichogramma minutum* (length, 0.37 mm; live weight, 4 μg) to the monstrous Goliath Beetle, *Goliathus goliathus* (length, 12.5 cm; live weight, 42 g; Metcalf et al. 1962). Not surprisingly, the length of the alimentary canal is usually a reflection of, and at least equivalent to, the length of the individual, although in some insects, it can be stretched out to several times the length of the body, and the gut and its contents can account for 40% or more of the live weight.

The consensus gut of insects consists of three main sections, the fore-, mid-, and hindgut, which, when removed and unraveled, are configured as a straight passage from mouth to anus (Figure 8.1A). The outer lobed edges of the midgut form ceca just below where the foregut meets the midgut. The alimentary canal is further distinguished by the presence of Malpighian tubules, which extend out from the junction between the mid- and hindguts. Relative lengths of different sections can vary and may be correlated with the feeding habits of the insect. For example, in most insects the hindgut ileum is a comparatively narrow tube leading to the rectum and anus, but in insects which depend on resident gut microbes to assist them with the digestion of plant matter, the most common sites of attachment, colonization, and proliferation are regions of the hindgut, and so this structure has become significantly more extended and/or bulbous in such taxa (Figure 8.1B). This anatomical adaptation is most apparent and well studied in termites, in which an enteric valve separates the midgut from a capacious hindgut microbial fermentation chamber known appropriately as the "paunch." The ileum and paunch are elongated and further partitioned in the higher termite subfamilies Termitinae, Apicotermitinae, and Nasutitermitinae (Bignell 1994). Some insects also possess spines or bristles protruding inward from the hindgut epithelium, and upon which bacterial cells may colonize, or filaments may become entrained (Bignell 1984).

Cells of the fore- and hindgut tissue deposit a chitinous cuticular intima not unlike that covering the body wall. Insects experience developmental cycles in which they molt (shedding of the outer epidermis accompanied by the synthesis of fresh cuticle), and the fore- and hindgut cuticle undergoes a similar molting process, which can result in the temporary removal of symbiotic microbiota from the hindgut. By contrast, midgut cells do not produce cuticle, but are lined with a chitinous, fibrillate and delicate peritrophic membrane (Bignell 1984, Chapman 1985). Synthesis of the peritrophic membrane appears to occur continuously, with portions of membrane sloughing off with the passage of food. While it is relatively

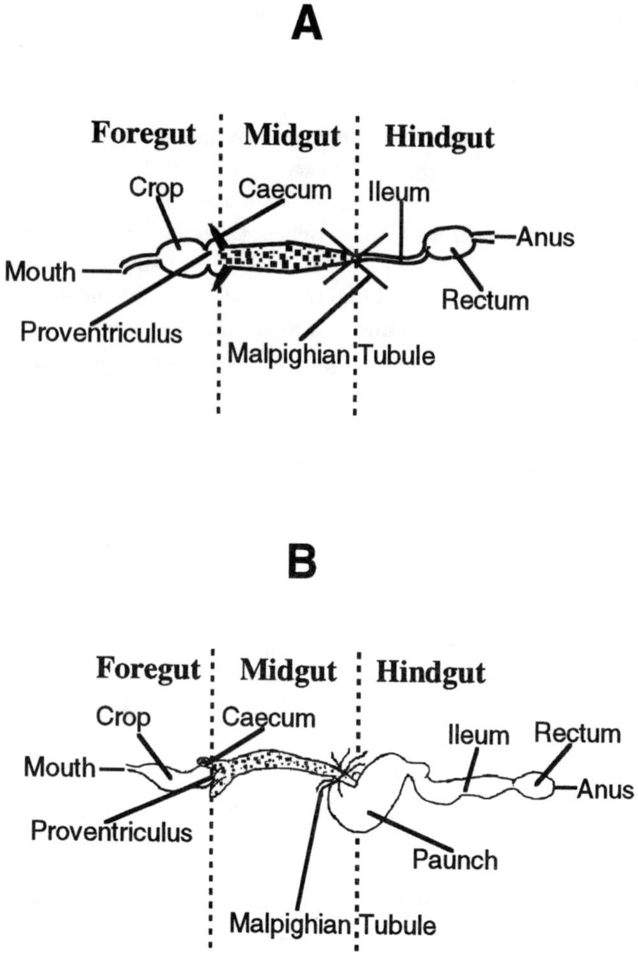

Figure 8.1. Diagram of the anatomy of the digestive tract typical of most insects (A) and that of the wood-feeding termite, *Reticulitermes flavipes* (B).

uncommon for microorganisms to attach to midgut tissue, it is unclear if this is due to the rapid turnover of the peritrophic membrane, or if the membrane itself is unsuitable for microbial colonization.

2.2. Gut Movements and Activities

As food is ingested, extrinsic muscles of the pharnyx and esophagus are responsible for contractions of the foregut, which serve to move liquids and food particles from the crop (sometimes a storage site) through the proventriculus into the midgut. Minimally, the proventriculus serves as a valve between fore- and midgut, but in some insects (e.g., cockroaches, crickets, and some wood-feeding beetles) the region has differentiated into grinding plates, presumably to help break up food particles. Tissue of the midgut serves the dual function of digestive enzyme secretion and nutrient absorption. Continuous synthesis and posterior movement of peritrophic membrane helps transport food through the midgut to the hindgut. In most insects, food is passed backward to the rectum and anus via a narrow hindgut ileum. However, those insects for which microbial fermentation is an important feature of their digestive processes use their expanded hindguts as digestive pouches from which significant nutrient absorption also occurs (McBee 1977). The rectum functions principally as a site for water and ion reclamation prior to excretion at the anus.

Food transit times in the alimentary canal of insects vary from less than 1 hour to more than 48 hours, depending on the insect and the method of measurement used (Bignell 1984). In cockroaches and termites, passage of food is relatively rapid through the fore- and midgut ($<$ 3 hours), whereas food and food residue tends to be retained for longer periods in the hindgut and rectum. Some examples of food retention times in the hindguts of insects exhibiting microbial fermentations include: the scarab beetle, *Oryctes nasicornis*, 6 hours (Bayon 1980a); the American cockroach, *Periplaneta americana*, 5 hours (Snipes and Tauber 1937); and the termite, *Reticulitermes flavipes*, 26 hours (Breznak 1982).

2.3. pH and Redox Potential

Physicochemical conditions inside the insect gut have been difficult to study due to the tiny size of the environment and the incursive nature of sampling procedures. Comparisons are further hampered by differences in methodology (e.g., use of pH or redox dyes vs. microelectrodes). Accordingly, pH measurements seem to have varied greatly among different taxa and in different regions of the gut, although a general tendency toward increasing pH values from fore- to hindgut has been documented by Bignell (1984). When present, the fermentative regions of insect hindguts usually exhibit pH values between 6 and 7.5, but in some cases values of 10 or higher have been recorded (Bignell 1984, Brune et al. 1995a).

By definition, fermentative processes are anaerobic and require at least a localized low (negative) redox potential. The fore- and midguts of insects generally lack resident populations of microorganisms and are usually aerobic, although E_h measurements in the microbe-free midguts of lepidopteran larvae varied from a low of −188 (*Manduca sexta*), to as high as +238 mV (*Lymantria dispar*; Appel and Martin 1990). By contrast, anaerobic regions of low redox potential commonly occur in hindguts of those insects with sites of significant microbial fermentation activity, even though the volume of this habitat is often on the order of 1 to 2 μL, and it is surrounded by aerobically respiring insect tissue. For example, typical measurements of E_h values have ranged from −50 to −270 mV (termites; Breznak and Brune 1994), +157 to −173 mV (cockroaches; Bignell 1984), and −80 to −100 mV (scarab beetles; Bayon 1980b). Often, the observation of anaerobic microbial activities such as nitrogen fixation, fermentation, and CO_2 reduction to either acetate and/or methane, and the enrichment and isolation of strictly anaerobic microbes including (depending on the insect species) cellulolytic, flagellate protozoa, CO_2-reducing acetogenic bacteria and methanogic archaea (Hungate 1939, 1943, 1946; Yamin 1980; Brauman et al. 1992; Cruden and Markovitz 1987; Breznak and Brune 1994; Breznak 1994) has been taken to be de facto evidence of the anaerobic nature of certain insect gut habitats.

Nevertheless, a variety of actinomycetes and other aerobic and facultative bacteria have also been isolated from the hindguts many of these same insects (Pasti and Belli 1985, Bignell et al. 1991, Brune et al. 1995b), suggesting that labeling the hindgut of a termite as "highly anaerobic" is an oversimplification. While such observations appear to be contradictory, recent studies have helped to increase our understanding of the situation. Experiments using redox dyes demonstrated differences in redox potential between the epithelium and the hindgut interior (Kuhnigk et al. 1994), and microelectrodes have now been used to provide an oxygen profile of the gut tract, showing that the hindguts of termites are capable of forming steep oxygen gradients (Brune et al. 1995a). The emerging picture is that oxygen diffuses across the hindgut lumen, but is also then quickly metabolized, leaving the center 25 to 40% of the cavity as an highly reduced anaerobic habitat (experiments with *Nasutitermes lujae* and *Reticulitermes flavipes*; Brune et al. 1995a). Oxygen consumption by aerobic and facultative microbes is responsible for creating anoxic regions of the gut in which fermentative activities occur, but the relative contributions of aerobic and anaerobic microbial metabolism to termites remains to be clarified.

3. Fermentation in Guts of Termites and Cockroaches

3.1. Nutritional Biology of Termites

Two aspects of the ecology of termites stand out above all others: their dramatically successful exploitation of lignocellulose as a food resource, and the extent to which this success is tied to symbiotic interactions with microorganisms. Nutritionally based insect-microbe symbiotic interactions involving termites (Isoptera) with protozoa, fungi, and bacteria have been the subject of intensive study, as summarized previously (Breznak 1982, 1984b, 1990, 1994; Martin 1987; Darlington 1994; Breznak and Brune 1994). There are over 2,200 living species of termites, one or more of which can be found in various habitats over two-thirds of the earth's land surface (Wood and Johnson 1986). They cause considerable damage to agriculture and manmade structures, but as decomposers of large amounts of biomass, termites are also important to the recycling of nutrients, particularly in tropical regions (Wood and Sands 1978, Edwards and Mill 1986).

Termites are eusocial insects (Wilson 1971), and in the colonies of all species, morphologically differentiated castes perform different functions. Food resources are typically collected by termites belonging to a ''worker'' caste, which then feed themselves as well as other members of the termite society (Wood 1978, 1986; Waller and La Fage 1987). One can easily imagine that relationships between termites and their gut microbiota probably originated in an ancestral form via the consumption of microbes that were initially ingested as the animals foraged. In extant species, by contrast, newly hatched, or freshly molted termites acquire an appropriate microbial inoculum by ingesting a fecal droplet from another colony member, a behavior known as ''anal trophallaxis.'' Cleveland et al. (1934) first proposed the intriguing hypothesis that the beginnings of termite social behavior must have been closely correlated with the origins of their associations with gut microbes inasmuch as termites require social interaction for cross-inoculation (reviewed by Nalepa 1994).

The food of termites consists of living, dead, decomposing, or highly decomposed plant matter, i.e., substances that can generally be described as lignocellulosic in composition (Wood 1978). Although a few termites are polyphagous, and eat a variety of living or dead plant materials, termite biologists recognize at least three main categories of specialization with regard to feeding behavior (Wood 1978, Wood and Johnson 1986, Waller and La Fage 1987):

1. Termites that eat living, dead, or decomposing vegetation (trees, grass, and/or leaves). Wood-feeding species are the best-studied termites of this group because of their economic impact (Edwards and Mill 1986).
2. Termites that cultivate aerobic, cellulolytic basidiomycete fungi of the genus *Termitomyces*, which, in addition to plant materials, are consumed by the termite. This habit is limited to, but pervasive among, members of the subfamily Macrotermitinae.
3. Termites that eat soil and presumably derive their nutrition from digestion of soil organic matter (humus). Although their nutrition is understood the least, they constitute approximately 45% of all termite species (Wood and Johnson 1986) and are especially numerous and active in tropical habitats.

Higher-level systematic relationships within the Isoptera still require resolution. Most classifications of termites recognize at least five families of "lower," or more primitivelike termites (Masto-, Kalo-, Hodo-, Rhino-, and Serritermitidae) and one family of "higher," or more derived, termites (Termitidae; Krishna 1970). The lower termite families together constitute about 25% of all living species; the remainder belong to the Termitidae. There is some correlation between phylogenetic classification and diet in that the lower termites are almost exclusively wood-feeders, whereas the higher termites include plant- (wood-, grass-, and/or leaf-) feeding, fungus-cultivating, and soil-feeding representatives (Wood 1978).

Despite intensive research on certain taxa, only a small fraction of the more than 2,200 termite species have been studied with respect to nutritional ecology and symbiotic interaction with microorganisms, so it is difficult to generalize about such relationships. However, one property apparently shared by all termites, regardless of feeding behavior or phylogenetic classification, is the presence of an anaerobic microbial community in the hindgut region of their alimentary tract, a community which can be densely situated, and extraordinarily diverse (Bignell et al. 1980, Breznak 1982). Lower termites harbor both anaerobic prokaryotes and protozoa, including cellulolytic flagellates, in their hindguts. By contrast, the hindguts of higher termites lack permanent populations of protozoa, and contain only bacteria and archaea (Cleveland 1924, Breznak 1982, Wood and Johnson 1986). This feature begs a critical question: What was the relationship between the loss of gut protozoa in the Termitidae, and the extensive species and dietary diversification exhibited by members of this family? Intuition suggests that initial radiation of the Isoptera was correlated with their unique ability to exploit wood as a source of carbon and energy, due, in part, to the acquisition of gut microbes. The most species-rich group of extant termites, however, are those whose dietary habits have diversified to include soil, grass or other resources. Yet, these are the same termites that have not retained an association with cellulolytic flagellates. Is it possible that during one period of their evolution, the presence of protozoan gut symbionts was essential to the early stages of termite diversification, but at a later stage, the loss of this same trait facilitated even greater phylogenetic radiation, as evidenced by the Termitidae?

One of the most conspicuous groups of bacteria observed in wet mounts of gut contents from both higher and lower termites is spirochetes (Breznak and Pankratz 1977, To et al. 1978, Breznak 1984a, Czolij et al. 1985). Their presence is very common in the guts of freshly collected specimens, and they are often abundant therein, so it seems unlikely that spirochetes are pathogenic to termites. Unfortunately, no termite gut spirochetes have ever been obtained in pure culture, so their function(s) with respect to gut processes, and any benefit which they might confer upon their termite hosts are still a mystery (Breznak 1984a, Bermudes et al. 1988). The role of other termite gut microbes in host nutrition is discussed in detail below.

3.2. Role of Microorganisms in the Nitrogen Metabolism of Termites

Wood-feeding termites digest a resource that contains up to 100 times less N (on a dry-weight basis) than that exhibited by termite tissues (Breznak 1984b). Consequently, termites have evolved ways of utilizing the activities of their gut microbiota to acquire and/or conserve combined N. One source of termite N acquisition is through N_2 fixation activities of gut-associated bacteria (Fig. 8.2) (Breznak 1973, French et al. 1976, Potrikus and Breznak 1977, Rohrmann and Rossman 1980, Prestwich and Bentley 1982). Results show that N_2 fixation rates

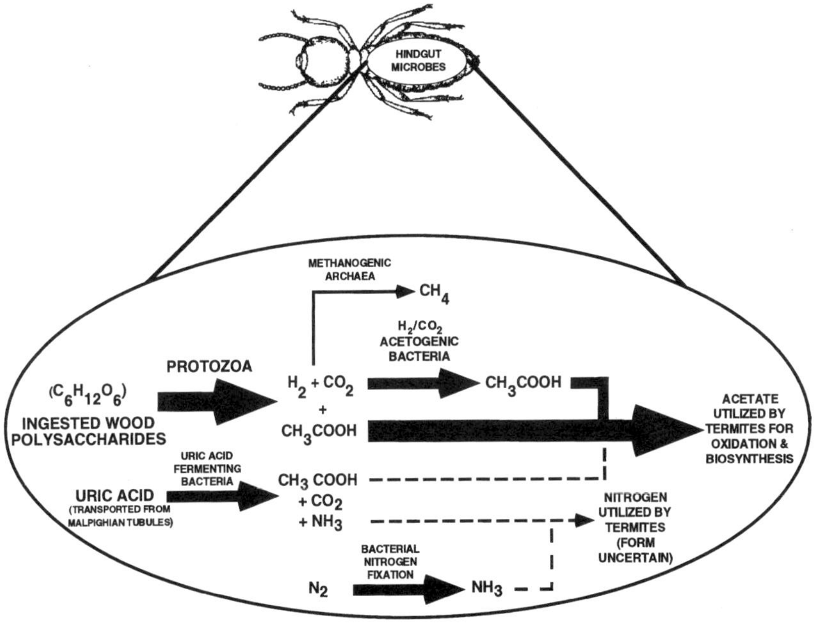

Figure 8.2. Model for metabolic reactions mediated by microorganisms in the hindguts of lower, wood-feeding termites. The illustration is a composite; not all reactions have been demonstrated to be significant in the hindguts of a particular species, and different reactions (e.g., nitrogen fixation vs. uric acid fermentation) may be more or less important to the nutrition of the host, depending on the taxon to be considered. Solid lines indicate reaction steps for which there is strong evidence; reaction steps indicated by dashed lines are likely to occur, but their significance is still uncertain.

for some wood-feeding species are significant enough to support up to 50% of the termite's N requirements (Rohrmann and Rossman 1980, Breznak1984b). By contrast, most of the dietary nitrogen of fungus-cultivating termites is obtained directly from digestion of ingested fungal tissue (Collins 1983). It is possible that termites might also acquire N by digestion of part of their gut microbiota, or through assimilation of combined N (e.g., amino acids) excreted by gut bacteria (Mauldin et al. 1978).

Mechanisms for nitrogen conservation involving gut microbes can also be important to termite N metabolism. In studies with the lower, wood-feeding termite *Reticulitermes flavipes*, it was found that, although this species has the ability to synthesize uric acid (a common nitrogenous excretory product of insects), the compound is not voided in the termite's feces (Potrikus and Breznak 1980). Instead, uric acid is transported via Malpighian tubules from its site of synthesis (fat body tissue) to the gut, and then is immediately and completely fermented to acetate, CO_2, and NH_3 by anaerobic bacteria (Fig. 8.2) (Potrikus and Breznak 1981). Uric acid nitrogen liberated by hindgut bacteria is ultimately assimilated back into termite tissues, although the principle mechanism for this last recycling step remains to be established.

Soil-feeding termites can be extremely abundant in remote, tropical regions. Attempts to relocate their nests to other sites that would be more easily amenable to experimental manipulation succeed only when substantial amounts of nearby soil are also relocated, and even then colonies usually only survive for a month or two at most. This suggests that factors such as soil nitrogen content are probably very important to their viability (Collins 1983). Unfortunately, so little is understood about the nutrition of soil-feeding termites, that any statement concerning the role of their gut microbes in nitrogen metabolism would be purely speculative.

3.3. *Role of Microorganisms in the Digestion of Carbohydrates by Termites*

Termites are faced with the formidable task of digesting relatively refractory lignocellulosic substances as their major carbon and energy source. For plant-feeding and fungus-cultivating termites, cellulose is the main component which must be digested, and their symbiotic interactions with microorganisms, to various degrees, help them to do this.

Even though some lower wood-feeding termites appear to be able to synthesize cellulase components (reviewed in Slaytor 1992, Breznak and Brune 1994), considerable evidence indicates that these termites depend largely on anaerobic, cellulolytic protozoa to hydrolyze ingested cellulose (Cleveland 1924; Hungate 1939, 1943; Breznak 1984b). An important chapter in the elucidation of this relationship was provided by the isolation and culture of *Trichomitopsis termopsidis* and *Trichonympha sphaerica*, dominant anaerobic protozoa from the hindgut of the lower termite, *Zootermopsis angusticollis*, and subsequent demonstration that these two species converted crystalline cellulose to acetate, H_2 and CO_2 (Yamin 1980, 1981; Odelson and Breznak 1985b). Cellulase and other polymer-hydrolyzing enzyme activities (e.g., amylase, xylanase, and protease) have also been detected in crude extracts of *T. termopsidis* (Yamin and Trager 1979, Odelson and Breznak 1985a).

Our current understanding of the overall utilization of cellulose by lower, wood-feeding termites is based largely on Yamin's work and studies with the gut microbiota of another lower termite, *Reticulitermes flavipes*. The dissimilation of much of wood polysaccharides in the hindgut of *R. flavipes* is essentially a homoacetic fermentation of cellulose, as depicted in Figure 8.2. This diagram is based on a model that has received substantial experimental support (Odelson and Breznak 1983, Breznak and Switzer 1986). Initially, wood polysaccharides (principally cellulose) are endocytosed by protozoa, which then hydrolyze the cellulose and ferment the liberated glucosyl units to acetate, H_2 and CO_2. H_2 and CO_2 produced by protozoa are subsequently consumed, mainly by H_2-oxidizing, CO_2-reducing, acetogenic bacteria to form additional acetate (although some H_2/CO_2 is converted to methane by methanogenic archaea). Acetate is then used by the termite as an important source of carbon and energy (O'Brien and Breznak 1984). In fact, the entire respiratory requirement of *R. flavipes* can be met by the oxidation of microbially produced acetate, which is present in *R. flavipes* hindguts at a concentration of about 80 mM (compared to a concentration of 45 to 49 mM in bovine rumen fluid; Odelson and Breznak 1983). According to this scheme, roughly two-thirds of hindgut acetate is derived from activities of protozoa, and one-third is derived from bacterial H_2/CO_2 acetogenesis. The novel bacterial component of this model will be discussed in detail in Section 3.5. Uric acid fermentation represents an additional source of acetate, although its contribution to hindgut acetate pools is not clear.

Although this simplified model may be conceptually appealing, it obscures the fact that the hindgut microbial community of lower termites is extremely diverse (Breznak and Pankratz 1977, To et al. 1980, Breznak 1982) and remains largely uncharacterized due to the limitations of culture methods (McBee 1977, Bull and Hardman 1991). The specific contribution of some of the other abundant organisms in the gut to host nutrition, or to the functional stability and activity of the total hindgut microbial community, may also be of critical importance to the overall health and development of the insect (e.g., Guo et al. 1990).

For species of Termitidae that have retained the ancestral wood-feeding habit, cellulose digestion must be different than that proposed for lower termites because of the absence of hindgut protozoa. After considerable controversy, Slaytor and co-workers have provided a solid body of evidence which supports the view that the origin of enzymes that comprise the cellulase repertoire in wood-feeding Termitidae is primarily the termite itself (Slaytor 1992) For example, *Nasuti-*

termes species that have been examined secrete cellulase enzymes in the midgut (OBrien et al. 1979, McEwen et al. 1980, Hogan et al. 1988). Treatment of *Nasutitermes walkeri* and *N. exitiosus* with antibacterial drugs had little or no immediate deleterious effect on cellulase activity (Hogan et al. 1988). Thus, in such *Nasutitermes* species, cellulose hydrolysis does not depend on the presence of cellulolytic microorganisms at all. Nevertheless, cellulose hydrolysis is only the first step in the catabolism of this polymer, and the above scenario as suggested by Slaytor (1992) still provides little foundation to the notion that higher termites do not require their gut microbes to thrive on cellulose. On the contrary, presence of a morphologically diverse bacterial community in the hindgut of wood-feeding representatives of the Termitidae (e.g., Czolij et al. 1985), and demonstration that their hindgut contents reduce CO_2 to acetate (Breznak and Switzer 1986, Brauman et al. 1992, Breznak 1994, Williams et al. 1994) suggests that extensive bacterial metabolism occurs there. Fermentation byproducts such as acetate and other compounds resulting from bacterial metabolism (e.g., combined forms of N, vitamins, and other biosynthetic precursors) are undoubtedly also important to the nutrition of wood-feeding termitids. Lack of comparative data simply highlights the many details concerning nutritional interactions between termites and their gut bacteria which remain to be clarified.

A good case in point concerns the role of aerobic and facultative bacteria which have also been isolated from the guts of wood-feeding termites. These include actinomycetes and other bacteria capable of aerobic degradation of cellulose (Pasti and Belli 1985, Bignell et al. 1991), and other substances such as monoaromatic compounds that are typical subunits within polymeric lignin (Pasti et al. 1990, Kuhnigk 1994, Brune et al. 1995b). For example in studies with *Reticulitermes flavipes* (lower) and *Nasutitermes lujae* (higher) termites, aerobic bacteria capable of mineralizing benzoic or cinnamic acid were present at levels of 10^5 cells per gut (Brune et al. 1995b). Whether the aerobic metabolism of cellulose and/or aromatic compounds by bacteria are significant with respect to the mass balance of carbon flow in termites is still unknown, and it will be important for future studies to quantitatively assess this potential contribution to termite carbohydrate metabolism.

Diverse dietary habits observed among different members of the single, higher termite family, Termitidae, are derived when compared with those of the lower termite families. In the Termitidae subfamily Macrotermitinae, a benefit obtained from the cultivation of *Termitomyces* fungal gardens is the acquisition, by ingestion of fungal tissue, of some components of the cellulase enzyme complex not made by the termite itself (Martin 1987, Rouland et al. 1988a,b; but see also Veivers et al. 1991). As with the higher, wood-feeding termites, fungus-cultivating termites harbor a community of bacteria in their hindguts. However, almost no information exists concerning the role of bacteria in the nutrition of Macrotermitinae. It is likely that hindgut bacterial fermentation is part of the digestive strategy of members of the Macrotermitinae, but much work along these lines remains to be done.

While the ability of termites to feed on soil has evolved only within the Termitidae, it occurs in all three non-fungus-cultivating termitid subfamilies and is pervasive among members of the subfamily Apicotermitinae and most of the Termitinae as well. This remarkable dietary adaptation has apparently arisen independently several times (Noirot 1992) and is coincidental with characteristic changes in gut morphology (Bignell 1994). Soil-feeding termites also have an anaerobic prokaryotic community in their elongated hindguts and harbor bacterial populations in the so-called mixed segment (Bignell et al. 1980, 1983). The food of such termites has not been well defined in chemical terms, but probably consists of residues of lignins and tannins that comprise the organic component of soil referred to as humus. Evidence suggests that anaerobic, methanogenic, bacterial consortia capable of degrading aromatic compounds from ingested soil could be important to carbohydrate metabolism in the guts of soil-feeding termites (Brauman et al. 1990, 1992), or such consortia may be interact syntrophically with aerobic and facultative gut bacteria in decomposing soil organics. However, fundamental aspects of the nutrition and digestive processes of these fascinating insects have not been investigated in detail, and attempts at doing so have been greatly hampered by the lack of success in transferring colonies of soil feeders to the laboratory. Inasmuch as most (75%) termite taxa are higher termites and a majority of those species (67%; Noirot 1992) feed on soil, the lack of knowledge about this group represents the most neglected area of research on termite digestion, nutrition, and symbiosis. The importance of overcoming limitations to their study cannot be stressed enough.

3.4. Cockroach-Microbe Interactions, and Their Nutritional and Evolutionary Implications

Cockroaches and mantids are believed to share common ancestry with termites in a monophyletic grouping known as the Dictyoptera. Some have considered cockroaches to be the termites closest relatives (e.g., Krishna 1970), although a recent cladistic analysis based on morphological, developmental, and behavioral characters concluded that cockroaches and mantids are closer to one another than either are to termites (Thorne and Carpenter 1992). Most cockroaches are generally considered to be omnivorous, but members of one genus, *Cryptocercus*, are strictly wood feeders (Cleveland et al. 1934) and nutritional similarities between these large, slow moving roaches and lower, wood-feeding termites have clouded issues concerning dictyopteran phylogeny. Like lower termites, the roach *Cryptocercus punctulatus* harbors a diverse population of anaerobic prokaryotes and cellulolytic protozoa in its hindgut (Cleveland et al. 1934; Honigburg 1970; Nalepa 1984, 1990), and the metabolic interactions that occur between *C. punctulatus* and its gut microbes are analogous to those of lower termites. However, the traditional view that the microbiota of lower termites and those of *C. punctulatus* were inherited from a common ancestor has been challenged, most recently by B.L. Thorne. Extant *C. punctulatus* roaches and *Zootermopsis angusticollis* (lower) termites can be found in the same decaying logs, and it has been known for sometime that protozoa can be transferred between defaunated individuals of the two species (Cleveland et al. 1934). Moreover, when individual nymphs of *C. punctulatus* or *Z. angusticollis* were placed into one another's nests, they were killed and eaten by members of the other group (Thorne 1990). This evidence, taken together with results of the aforementioned systematic study, particularly the lack of shared, derived characters between *Cryptocercus* roaches and lower termite taxa (Thorne and Carpenter 1992), has provided reasonable support for the hypothesis that transfer or exchange of gut microbes may have occurred between ancestral members of these two groups (Thorne 1990 1991, but see also Nalepa 1991).

Another investigator who has addressed the evolution of microbe-mediated cellulose digestion in cockroaches and termites is M. M. Martin (1991), who suggested that the initial acquisition of cellulolytic microbes by wood-feeding ancestors of cockroaches and/or termites occurred prior to, and subsequently facilitated their specialization on a diet high in cellulose. This hypothesis outlined two separate, critical steps: acquisition of cellulolytic microbes by a detritus-feeder or scavenger already dependent upon other hindgut microbes for nutritional support, and then expansion of food resources to include those high in cellulose, upon which the insect ultimately becomes specialized. An alternative to the scenario suggested by Martin is that major components of a microbial community capable of efficiently fermenting substrates high in cellulose first assembled on woody particles, which are widely distributed in nature, rather than inside the gut of a detritus-feeding or scavenging insect, even one that had already become nutritionally dependent upon gut bacteria. This speculation leads to a one-step hypothesis that termite and/or roach ancestors sampled wood particles covered with "biofilm" communities of cellulose-digesting microbiota, and that their specialization on a diet high in cellulose coincided with acquisition of microbial consortia associated with their food.

In Australia, there is another wood-feeding cockroach, *Panesthia cribrata*, and although bacteria and protozoa are present in its hindgut, cellulolytic protozoa are conspicuously absent. Elimination of all of the protozoa and most of the bacteria from hindguts of *P. cribrata*, by feeding the animals a diet of filter paper treated with an antibiotic solution, had no effect on their respiration rate or quotient or on the general health of the animals over a 12-week period (Scrivener et al. 1989). While this demonstrated that *P. cribrata* is able to digest cellulose independent of its gut microbiota, results of a 12-week experiment do not preclude microbial contributions to the growth and development of the insect, as was suggested by studies with another cockroach species (below).

Knowledge of the nutritional relationship between microbes and cockroaches other than wood-feeding specialists such as *C. punctulatus* or *P. cribata* is mostly based on studies with two omnivorous species, *Periplaneta americana* and *Eublaberus posticus*. These two cockroaches have dense, anaerobic bacterial populations in their hindguts (Cruden and Markovitz 1987). In addition, natural populations of *P. americana* in Tanzania contain significant populations of *Nyctotherus ovalis*, an anaerobic cilliated protozoan with endosymbiotic, methanogenic archaea (Gijzen et al. 1991), but such protists appear to be absent from roach colonies maintained for long periods under laboratory conditions in the United States (Kane and Breznak 1991b).

Although bacteria have been isolated from the hindguts of *P. americana* and *E. posticus* (Cruden and Markovitz 1987), their contribution to cockroach nutrition is unclear, in part because of the omnivorous and complex nature of the insects' diet. Laboratory specimens are usually reared on a diet of dog chow and water, and under such conditions, bacterial fermentation products (acetate, lactate) have been detected in different regions of the gut and can be transported across the hindgut epithelium (Bracke et al. 1978, Hogan et al. 1985). Concentrations of acetate in the fore-, mid-, and hindgut fluids of dog chow-fed *P. americana* were 17, 14, and 7 mM, respectively, while those for lactate were 6, 8, and 5 mM (Kane and Breznak 1991). However, when specimens of *P. americana* were fed a diet high in fiber (e.g., milled cereal leaves, milled corncob, or bran cereal), production of acetate and lactate by bacteria in the foregut diminished (Kane and Breznak 1991b). (Dietary changes also affected methane emission by hindgut microbiota, as discussed further, below.) In another study, *P. americana* cockroaches were administered metronidazole, a drug that essentially eliminated all anaerobic bacteria from the alimentary tract. The health of adult cockroaches was unaffected, but the growth of immature *P. americana* was retarded. Moreover, the guts of metronidazole-fed immature specimens were degenerate (Bracke et al. 1978). This evidence should serve as a reminder that contributions of gut microbes to the nutrition and vitality of wood-feeding cockroaches (i.e., *P. cribrata*) may be significant even if the insect can digest cellulose independent of its gut microbes for periods of up to 12 weeks.

It is also worth mentioning that a system analogous to that of wood-feeding termites, involving the conservation of uric acid nitrogen through the activities of bacterial symbionts, probably exists for most cockroaches as well, although it involves intracellular "mycetocyte" bacteria present in fat body tissue rather than in the hindgut (Mullins and Cochran 1972, 1975a,b; Downer 1982; Cochran 1985). However, studies of this system have been obstructed by the unavailability of authentic mycetocyte bacteria in pure culture. Until direct evidence for bacterial uricolysis is forthcoming, the nature of the symbiosis between cockroaches and their mycetocyte bacteria will remain uncertain.

3.5. Competition for H_2 in the Guts of Termites and Cockroaches

In previous sections, it was evident that the contribution of gut microorganisms to the carbohydrate metabolism of termites and cockroaches is best understood for lower, wood-feeding termites such as *R. flavipes*. A novel aspect of the model describing symbiotic degradation of cellulose by *R. flavipes* was the role accorded to H_2/CO_2 acetogenic bacteria as consumers of H_2 produced by protozoa (Fig. 8.2). To appreciate why bacterial H_2/CO_2 acetogenesis is such a novel aspect of this model, it is worthwhile to briefly discuss the role of H_2-consuming bacterial reactions in the terminal steps of anaerobic microbial decomposition processes.

Anaerobic decomposition of plant polymers occurs in a wide variety of aquatic and terrestrial habitats (Hungate 1985), and in the digestive tracts of many animals (Wolin 1974; other chapters in this volume) including insects (Prins and Kreulen 1991). During intermediate steps of this process, fermentative and fatty acid- and alcohol-oxidizing microorganisms may produce large amounts of H_2. However, H_2-consuming bacteria effectively remove H_2, keeping H_2 partial pressures in natural environments quite low (Wolin 1982). Removal of H_2 is a crucial component of anaerobic microbial decomposition processes, because the intermediate, H_2-producing reactions of anaerobic microbial food webs will proceed less efficiently, or not at all, if H_2 is allowed to accumulate (Wolin 1982, Dolfing 1988).

In environments that are low in other inorganic electron acceptors such as NO_3^{-1} and SO_4^{-2} (as is the case with most animal gastrointestinal tracts), CO_2 reduction is the de facto terminal, H_2-consuming (electron sink) process. Two types of H_2-consuming, CO_2-reducing, energy-yielding, microbial processes are known to occur in anaerobic habitats: H_2/CO_2 methanogenesis, and H_2/CO_2 acetogenesis. When other inorganic electron accepting compounds are limited or absent, CO_2 reduction to methane, rather than acetate, is usually the main, terminal, H_2-consuming reaction (Table 8.2, reaction No. 1). There are >50 species described that can grow via this reaction, all of which belong to a physiologically and phylogenetically coherent group of strictly anaerobic archaea whose growth is obligately methanogenic (Boone and Mah 1988, Ferry 1993). Most methanogenic archaea can grow on $H_2 + CO_2$, and some can also produce methane from acetate, formate, methanol, ethanol, CO, or methylamines (Whitman 1985, Boone and Mah 1988, Ferry 1993).

A taxonomically broad sampling of arthropods was recently assayed for methane emission, including representatives of many different insect groups. Of 110 taxa surveyed, methane was detected only consistently from various species of termites, cockroaches, scarab beetles, and millipedes, and was always restricted to the hindgut region of their alimentary canal (Hackstein and Stumm 1994).

Table 8.2. Comparative energetics for the reduction of CO_2 to 1 mol methane or acetate.

Reaction No.	Reaction	$\Delta G^{0'}$ (kJ/reaction)	$\Delta G^{0'}$ (kJ/H_2 consumed)
1	$4\ H_2 + CO_2 \rightarrow CH_4 + 2\ H_2O$	−135.6	−33.9
2	$4\ H_2 + 2\ CO_2 \rightarrow CH_3COOH + 2\ H_2O$	−104.6	−26.2
3	$CH_3OH + H_2 + CO_2 \rightarrow CH_3COOH + H_2O$	−81.1	−81.1
4	$0.5\ C_7H_4O_3(OCH_3)_2 + CO_2 + H_2 \rightarrow 0.5\ C_7H_4O_3(OH_2 + CH_3COOH$	−92.8	−92.8

Sources: Thauer et al. 1977, Breznak and Blum 1991, Liu and Suflita 1993.

Methanogens were observed in wet mounts (by F_{420} fluorescence microscopy; Doddema and Vogels 1978) as endosymbionts of hindgut protozoa, as attached populations associated with hindgut cuticular structures (bristles, spines; Bignell 1984), and/or as free-living populations within microbial communities in the hindgut cavity. Previous photomicrographic and physiological studies have provided some details of the associations between methanogenic archaea and insect gut protozoa, including the flagellate *Trichomitopsis termopsidis* in the hindgut of *Z. angusticolis* termites (Lee et al. 1987, Messer and Lee 1989) and the ciliate *N. ovalis* in the hindgut of the cockroach *Periplaneta americana* (Gijzen et al. 1991). These delightful examples of microbial interactions are fascinating to observe, but specific benefits to the insect from such associations have yet to be clearly demonstrated.

In their recent survey, Hackstein and Stumm (1994) emphasized the role of methanogenesis as a qualitative evolutionary trait of the animal rather than one associated with its dietary habits, because arthropods with similar diets emitted methane based on their taxonomic affiliation, and not what kind of diet they consumed. Perhaps the hindguts of non-methane-emitting arthropod groups with diets similar to those of their methane-emitting counterparts are not suitable environments for the growth of methanogens in general, or possibly some arthropods may produce antibiotic substances effective against methanogens. Whatever the case, in assessing the nutritional and evolutionary significance of the phenomenon of methane emission from insects, it is also important to consider quantitative aspects of methane emission, especially in relationship to the animal's carbohydrate metabolism. If rates of methane emission are insignificant compared to host metabolic processes, an insect-methanogen association is of dubious evolutionary significance, and may simply be an artifact of microbial activity associated with ingested biomass. It is also worth noting that molecular systematic studies of endosymbiotic associations of methanogens with protozoa indicate that such partnerships have formed repeatedly and independently and thus reflect a general ecological rather than evolutionary phenomenon (Embly and Finlay 1994).

The above discussion is worth considering because other studies indicate that rates of methane emission by hindgut archaea can be significantly related to an insect's diet. For example, in *P. americana* cockroaches in American, laboratory-reared colonies, rates of methane production by hindgut microbes were found to increase when individuals were fed diets high in fiber such as milled cereal leaves, milled corncob, or bran cereal rather than the typical diet of dog chow (Kane and Breznak 1991b). Similarly, in Tanzanian colonies, which, in addition to hindgut prokaryotes, harbor significant populations of the cilliate *N. ovalis* and its associated methanogenic endosymbionts, rates of methane emission varied with diet (Gijzen et al. 1991). In this instance, changes in methane production rates were specifically attributed to the effect of diet on populations of *N. ovalis*, and in a follow-up paper it was shown that removal of protozoa had a negative effect on body weight, generation time, and methane production rates, whereas direct inhibition of methanogenesis, by adding bromoethanesulfonic acid to drinking water, resulted in a 98% decrease in methane emission but had no effect on protozoa numbers or insect body weight (Gijzen and Barugarahare 1992). Other evidence of the relationship between insect diet and methanogenesis comes from a survey of 24 diverse termite taxa representing different feeding guilds, in which mean rates of methane emission from soil-feeding termites were observed to be seven- to 10-fold higher than those of wood-feeding termites (Brauman et al. 1992). Although this is an example of the effect of host diet on microbial methane production, it also illustrates the degree to which insect diet may affect competitive interactions between methanogens and other bacteria. This point is elaborated further, in the passages that follow.

An alternative to CO_2 reduction to methane is CO_2 reduction to acetate by H_2/CO_2 acetogenic bacteria (Table 8.2, reaction No. 2). The biochemistry of H_2/CO_2 acetogenesis has been thoroughly investigated (Wood et al. 1986a,b, Fuchs 1986, Dolfing 1988, Drake 1994). Although most biochemical studies have been done by using a few selected species (*Clostridium thermoaceticum* and *Acetobacterium woodii*), many new H_2/CO_2 acetogenic bacteria have been described in the past 10 years (Drake 1994). Unlike H_2/CO_2 methanogenesis, H_2/CO_2 acetogenesis is not restricted to a phylogenetically coherent group, but is a property of several distantly related taxa of anaerobic bacteria that include both gram-positive and gram-negative genera, several endospore-forming representatives, and mesophilic and thermophilic species. As a physiological group, H_2/CO_2 acetogenic bacteria can metabolize more than 60 different compounds including sugars, organic and amino acids, and alcohols (Ljungdahl 1986, and references contained therein). Most species convert carbohydrates to acetate as the principal fermentation product and therefore have been called "homoacetogens." However, a few others (e.g., *Eubacterium limosum*) form both acetate and butyrate (Sharak Genthner et al. 1981). It is also curious that when H_2/CO_2 acetogens are growing heterotrophically, some can be H_2 producers rather than H_2 consumers (Dolfing 1988). For example, one H_2/CO_2 acetogen can perform reaction No. 2 (Table 8.2) in the reverse direction when grown in coculture with an H_2-consuming methanogen (Lee and Zinder 1988).

Despite recent advances in our understanding of the microbiology and biochemistry of H_2/CO_2 acetogenic bacteria, almost no information exists concerning their ecology or their role as H_2 consumers in nature. They have been isolated from a wide variety of anaerobic habitats but it is not yet clear whether H_2/CO_2 acetogens that have been isolated using $H_2 + CO_2$ as the carbon and energy source, are, in fact, H_2 consumers in situ. Theoretical considerations suggest that H_2/CO_2 methanogenesis should always outcompete H_2/CO_2 acetogenesis for H_2, because CO_2 reduction to methane (Table 8.2, reaction No. 1) is thermodynamically more favorable than CO_2 reduction to acetate (Table 8.2, reaction No. 2) (Thauer et al. 1977, Cord-Ruwisch et al. 1988). This is especially true under conditions involving ''typical'' natural concentrations of substrates and products found in anoxic habitats (Conrad et al. 1986, Dolfing 1988). Moreover, methanogenic archaea have a 10- to 100-fold lower threshold for H_2 than do their acetogenic counterparts (Breznak and Kane 1990) despite the fact that the H_2-V_{max}/K_m values of these two groups of hydrogenotrophic bacteria are in the same range (Breznak and Switzer 1986). However, there are reports of a few habitats where rates of H_2/CO_2 acetogenesis compare favorably to, or even exceed rates of H_2/CO_2 methanogenesis. These include certain anaerobic freshwater lakes and sediments (Phelps and Zeikus 1984, Jones and Simon 1985, Conrad et al. 1989), and the gastrointestinal tracts of humans (Lajoie et al. 1988), rodents (Prins and Langhorst 1977), baleen whales (Wilmarth et al. 1985), and, most spectacularly, wood-feeding termites (Breznak 1994 and references therein).

The Odelson-Breznak model of symbiotic cellulose utilization by lower, wood-feeding *R. flavipes* termites (Fig. 8.2) suggested that H_2/CO_2 acetogenesis outcompetes H_2/CO_2 methanogenesis for H_2 in hindguts of this species. The reasoning that led to the initial hypothesis, and subsequent experimental confirmation of its accuracy, have been recounted in a recent review (Breznak 1994). Given theoretical considerations, and results of in vitro studies favoring the opposite, the discovery that CO_2 reduction to acetate almost totally outcompeted CO_2 reduction to methane in *R. flavipes* was surprising, and it inspired an in-depth examination into the prevalence of the phenomenon and the agents responsible. To summarize, results similar to those obtained for *R. flavipes* were observed for five of six other wood-feeding species representing three families of lower termites, and for the wood-feeding cockroach *C. punctulatus*, but also for eight wood-feeding and one grass-feeding species representing the Termitidae (Breznak and Switzer 1986, Brauman et al. 1992). Protozoa may have been the principle source of H_2

in the guts of lower termites and *C. punctulatus*, but their absence in the guts of termitids indicates that they were not the only factor contributing to the dominance of acetogenesis over methanogenesis, and that reducing equivalents generated by gut bacteria in wood-feeding termites also can provide a suitable substrate for CO_2 reduction to acetate.

By contrast, rates of methane emission were significantly greater than rates of H_2/CO_2 acetogenesis for five of six soil-feeding termites, and for three species of fungus-cultivating termites (Brauman et al. 1990, 1992). Although little is known about the role of bacterial fermentation in the nutrition of soil-feeding, and fungus-cultivating termites, this is clear evidence that fermentation does occur in their hindguts, and that it is different from that of their wood-feeding relatives. It should be emphasized that acetogenesis and methane production were far from mutually exclusive, yet results pointed to the inverse relationship between the two processes: when rates of methanogenesis where high, acetogenesis rates were low, and vice versa. Another important conclusion that can be infered from the above is that dominance of bacterial acetogenesis over methanogenesis in the termite hindgut is, along with wood-feeding, an ancestral trait, whereas gut fermentation favoring methanogenesis is, like soil-feeding and fungus-cultivating, a derived trait.

In order to better understand the mechanisms by which CO_2-reducing acetogenic bacteria outcompete methanogenic archaea for H_2 in guts of wood-feeding termites, pure cultures of acetogens were pursued. Unfortunately, and for reasons that are not entirely understood, repeated attempts to isolate H_2/CO_2 acetogenic bacteria from gut contents of *R. flavipes* met with a discouraging lack of success. However, enrichments of gut contents from other termites were more fruitful and led to the isolation and characterization of three novel acetogenic taxa. *Sporomusa termitida*, was described from the higher, wood-feeding termite *Nasutitermes nigriceps* (Breznak et al. 1988), and *Acetonema longum* from the lower, wood-feeding, *Pterotermes occidentis* (Kane and Breznak 1991a). In addition, *Clostridium mayombeii* was obtained from hindgut preparations of *Cubitermes speciousus,* an abundant soil-feeding termitid from the Congo's Mayombei rainforest, despite that fact that H_2/CO_2 acetogenesis rates were insignificant for this species (Kane et al. 1991).

S. termitida and *A. longum*, the two acetogenic isolates from wood-feeding termites, were both gram-negative, yet formed heat-resistant endospores. These two properties occur together in very few bacteria that have been described to date, and comparative sequence analysis of the 16S ribosomal RNAs of *S. termitida* and *A. longum* helped to identify them as members of a small and unusual group of bacteria with gram-negative cell walls that, nevertheless, seem to be more closely related to gram-positive bacteria (Kane and Breznak 1991a). Aside from sharing this unusual ancestry, other traits exhibited by these two isolates, such as the spectrum of organic substrates which they could use for growth, were as different from one another as they were from *C. mayombeii* (a gram-positive endospore former), or from acetogens isolated from other habitats. So, while isolation of acetogenic bacteria confirmed that they were, in fact, present in termite guts, it presented no obvious explanation for their competitiveness versus methanogens. The characterization of H_2/CO_2 acetogenic bacteria from gut contents of other termites may provide a more complete database from which to help reveal a common property related to their ability to compete for H_2.

As indicated above, acetogens are generally more physiologically versatile than methanogens, and it has been suggested that this versatility, along with an ability to utilize H_2 and organic substrates simultaneously (mixotrophy), may give acetogens a competitive advantage in the guts of wood-feeding termites (Breznak and Blum 1991). Mixotrophy was demonstrated for *S. termitida*; the organic substrates used in combination with H_2 were methanol, mannitol, lactate, or glycine. Of these, mixotrophic, acetogenic growth on methanol and H_2 (Table 8.2, reaction No. 3) appeared to be the most advantageous, since it was three times more energetically favorable than methanogenesis on a per H_2-utilized basis (Breznak and Blum 1991). Recent studies with an H_2/CO_2 acetogen isolated from subsurface sediments are also relevant in this regard. Subsurface isolate SS1 grows mixotrophically by using $H_2 + CO_2$ and the O-methyl groups of syringate (a lignin monoaromatic model compound) simultaneously (Table 8.2, reaction No. 4) (Liu and Suflita 1993). When considered on a per-mol-H_2-consumed basis, the Gibbs free energy yield ($\Delta G^{\circ\prime}$) calculated for this mixotrophic reaction (No. 4) is also more energetically favorable than that of H_2/CO_2 methanogenesis (reaction No. 1). Future studies might involve obtaining methanogenic isolates from termite guts and studying their competition against *S. termitida* and *A. longum* for H_2. Naturally, these should also examine competition for H_2 in the presence of methanol or methoxylated aromatics like syringate, both of which, as constituents of lignocellulose, might be representative of what they encounter in the hindguts of wood-feeding termites.

Other explanations for the observed competitiveness of acetogens for H_2 in the guts of wood-feeding termites also deserve attention. For example, it is possible that specific compounds produced by wood-feeding termites or released from ingested woody particles either are of benefit to the growth of acetogens or inhibit the growth of methanogens. Conversely, general chemical and physical parameters (pH, osmolarity, microbial aggregation, etc.) inside the guts of wood-feeding termites may favor one group over the other.

4. Fermentation in Guts of Other Insects

Metabolism by microbes in the guts of termites and cockroaches has played a greater role in the digestion, nutrition, and dietary diversification of these insects than it has for other groups examined thus far, and this is coincidental with the high degree to which members of these two orders utilize various forms of lignocellulose as their principle source of carbon and energy. Many other insect species feed on various plant structures but usually void the more refractile components (i.e., lignocellulose) in their feces. Nevertheless, significant microbial populations and/or communities have been observed in the guts of a variety of other insects, especially those that feed on lignocellulosic resources. The remainder of this chapter shall be concerned with summarizing these studies.

Certain members of the scarab beetles (Coleoptera: Scarabaeidae) feed on wood and/or sawdust. The role of microbial metabolism in the digestion of one of these, *Oryctes nasicornis*, has been examined. Wood consumed by this species has apparently first undergone a certain amount of destruction by soil fungi, which are also ingested (Bayon 1981a). *O. nasicornis* larvae have a hindgut "proctodeal dilation" which appears to be anoxic, and in which they harbor a diverse community of bacteria (Bayon 1981a,b). Some microbial populations are also associated with the midgut. Although enzyme assays could detect no cellulase activity in any portion of the gut (Bayon 1980b), experiments using ^{14}C-cellulose showed that significant cellulose degradation occurred in the hindgut and, to a lesser extent, in the midgut. Decomposition of cellulose was found to be associated with bacterial fermentation resulting in significant volatile fatty acid (VFA) production (Bayon and Mathelin 1980). The VFA pool was dominated by acetate (90%), which was highest in the midgut. Methane production localized within the proctodeal dilation has also been detected from *O. nasicornis*, but only at about the same rate (0.05 $\mu M\ g^{-1}\ h^{-1}$; Bayon and Etiévant 1980) as that of wood-feeding termites for which H_2/CO_2 acetogenesis is the dominant H_2-consuming process. However, it is not known whether H_2/CO_2 acetogenesis occurs in the guts of *O. nasicornis*; the extent to which this species depends on microbial metabolism for augmentation of its nutritional requirements is also not known.

Some members of the insect order Orthoptera cause significant damage to crops. The cricket *Achaeta domesticus*, the grasshopper *Melanoplus sanguinipes*, and the desert locust *Schistocerca gregaria* each have substantial populations of gut bacteria. Bacterial cells were observed throughout the gut tract, but greatest numbers were found in the hindgut region, and it has been possible to create germ-free cultures of *A. domesticus* and *S. gregaria* in which the insects appear to develop and reproduce in a manner that is generally similar to that of their conventional counterparts (Charnley et al. 1985, Mead et al. 1988, Kaufman et al. 1989). However, when *A. domesticus* were fed a relatively refractile diet of alfalfa, they were able to digest plant polysaccharides contained in their food more efficiently than were germ-free crickets (Kaufman and Klug 1991). The central products of bacterial metabolism were acetate and propionate in molar ratios of approximately 4:1, respectively. Very low levels of methane production were also detected from live *A. domesticus* specimens (M.G. Kaufman, personal communication). Since crickets are somewhat omnivorous, it was suggested that their bacterial biota may help them adjust to the sorts of fluctuations in diet content that they encounter in the wild (Kaufman and Klug 1991).

Zoraptera are a group of tiny, poorly known insects that have been found in dead trees in tropical and subtropical environments and that have protozoa and bacteria associated with wood particles found in their guts, but investigations are only just beginning into the nature of the association (Choe 1992). Bacteria have also been found in the guts of cephalotine ants (Hymenoptera: Formicidae: tribe Cephalotini), bees (Hymenoptera: Apidae), tephritid fruit flies (Diptera: Tephritidae), and other representative terrestrial insects (Table 8.1), but the role they play in host digestion and nutrition is probably not very significant (Buchner 1965, Caetano 1989, Gilliam et al. 1990, Gilliam and Taber 1991, Drew and Lloyd 1991, Daser and Brandl 1992). By contrast, fermentation by gut bacteria appears to be more important for certain aquatic insects involved in the decomposition of detritus (coarse, particulate organic matter, mostly of plant origin). Insect detritivores which attack allochthonous plant materials in woodland streams are sometimes known as "shredders" and play an important role in the flow of carbon and energy in these habitats (Cummins and Klug 1979). Stream detritivores associated with the decomposition of allochthonous plant matter include the larvae of caddisflies (Trichoptera), stoneflies (Plecoptera), and craneflies (Diptera: Tipulidae) (Cummins and Klug 1979, Dudley and Anderson 1982). Cellulose hydrolysis was detected in anterior portions of the gut by larvae of the stonefly, *Pteronarcys proteus* and the caddisfly *Pycnosyche luculenta*, and in the hindgut of larvae of the cranefly *Tipula abdominalis*, the latter of which also exhibited a higher assimilation efficiency of ^{14}C-cellulose than either *P. proteus* or *P. luculenta* (Sinsabaugh et al. 1985). Tipulid larvae harbor a diverse assemblage of bacteria, particularly in the hindgut region (Klug and Kotarski 1980), and are commonly found in woody debris (Dudley and Anderson 1982). Moreover, wood and leaf particles are usually found inside their guts (Pereira et al. 1982). Accordingly, *T. abdominalis* has been studied as a model system to examine the role that gut bacteria may play in the digestion and nutrition of stream detritivores.

In field studies, growth of *T. abdominalis* larvae was greatest when consuming hickory leaves that had already undergone decomposition in a stream for periods of between 3 and 5 weeks, and in complementary laboratory feeding studies, insect growth and assimilation efficiency were greater for insects fed leaves colonized by a mixture of fungi and bacteria than for insects fed leaves colonized by fungi alone (Lawson et al. 1984). However, increased growth and assimilation efficiency did not appear to be the result of digestion of significant microbial biomass, but rather were associated with some sort of preconditioning of leaf litter by colonizing microbes that enabled the insect to thrive on the colonized and conditioned material (Lawson et al. 1984, Findlay et al. 1986). Bacterial fermentation occurred in the hindguts of *T. abdominalis*, resulting primarily in the production of acetate, which could be used by larvae for oxidation and biosynthesis. The utilization of acetate was confirmed by verifying transport of ^{14}C-acetate across the gut wall into the hemolymph, and detecting its subsequent incorporation into intermediates of the tricarboxylic acid cycle. It was estimated that oxidation of acetate produced by hindgut bacteria of *T. abdominalis* could supply between 5% and 16% of the insect's respiratory demand (Lawson and Klug 1989). The importance of specific bacterial populations in the hindgut fermentation of *T. abdominalis* remains to be investigated.

5. Concluding Remarks

For insects examined so far, substantial microbial fermentation is generally restricted to the hindguts of species such as termites, cockroaches, and others that thrive on a diet of some form of lignocellulose. Moreover, in these insects, there is strong evidence of a correlation between diet and gut microbial activities, such as bacterial CO_2 reduction to acetate or methane. Undoubtedly, many other submerged logs in rainforest soils or woodland streams are awaiting some inquisitive biologist to roll them over and reveal new insect-microbe digestive associations that have yet to be discovered.

Surprisingly, it now appears that the ability of insects themselves to hydrolyze cellulose with enzymes secreted from salivary glands or midgut tissue is an ancient trait whose evolution may have coincided with, or even predated, the establishement of interactions with fermenting consortia of microorganisms in their digestive tracts (Slaytor 1992, Treves and Martin 1994), but this finding does not diminish the importance of gut microbial fermentation to host survival and competitiveness. To understand the role of microbial fermentation in insect guts, it is important to distinguish between the origin of cellulolytic activity on the one hand, and the ability of an insect to thrive by digestion and assimilation of naturally occurring lignocellulosic substrates, on the other. Evidence that lignocellulose-consuming insects obtain benefit from the fermentative activities of their gut microbes is well established from experiments that have shown that some species, such as termites, cannot survive on their natural diet without their gut symbionts (Cleveland 1924), while the health, vitality, and digestive activities of others, such as cockroaches and crickets, are diminished when deprived of their full complement of gut microbes (Cruden and Markovitz 1987, Kaufman and Klug 1991). Furthermore, with respect to termites, intensive biochemical and microbiological studies have been successful in elucidating the roles of specific gut microbial populations in critical aspects of their carbon and nitrogen metabolism (Breznak 1984a, Breznak and Brune 1994).

It is fascinating to speculate whether coevolutionary interactions between insects and members of their gut microbial communities exist and, if so, whether they may be related to dietary diversification of the host. The role of gut microbes in the evolutionary diversification of higher termites has been dismissed as insignificant (Bignell 1994), but the dearth of nutritional and phylogenetic studies on this most speciose termite family (especially on soil feeders), combined with difficulties in identifying and enumerating bacterial populations within complex microbial communities (Bull and Hardman 1991), means that it is premature to assess the role and quantitative significance of bacteria in their nutrition and dietary diversification. Molecular methods and advances in phylogenetic inference should prove helpful in examining the ecology and evolution of microbial communities in insect hindguts and correlating it with that of their hosts (Kane and Pierce 1994). For example, while the nature of associations between bacteria and bees are still unclear, the recovery of 16S rDNA fragments representing *Bacillus* species from extinct bees preserved in Dominican amber confims that insect-microbe associations can be very ancient (Cano et al. 1994).

The focus of this chapter has been on the role of gut microorganisms in insect nutrition, digestion, and nutritional diversification, but it is worth mentioning in closing that gut microbes may also play a role in detoxifying plant-produced defensive toxins (Dowd 1991) and that insect guts may also serve as a source of novel microbes with biotechnologically useful metabolic activities (Bull et al. 1992, Dowd 1992, Taguchi et al. 1992).

Acknowledgments

I thank Alain Brauman, John Breznak, Andreas Brune, Jae Choe, Michael G. Kaufman, and Michael M. Martin for sharing unpublished data and/or discussing aspects of their research, and Margaret Collins and Rod Mackie for reading an earlier version of the manuscript.

References

Anderson JM, Rayner ADM, Walton DWH (1984) Invertebrate-Microbial Interactions. Cambridge: Cambridge University Press.

Appel HM, Martin MM (1990) Gut redox conditions in herbivorous lepidopteran larvae. J Chem Ecol 16:3277–3290.

Barbosa P, Krischik VA, Jones CG (1991) Microbial Mediation of Plant-Herbivore Interactions. New York: John Wiley and Sons.

Bayon C (1980a) Transit des aliments et fermentations continues dans le tube digestif d'une larve xylphage d' Insecte: Oryctes nasicornis (Coleoptera: Scarabaeidae). C R Acad Sci Paris 290:1145–1148.

Bayon C (1980b) Volatile fatty acids and methane production in relation to anaerobic carbohydrate fermentation in *Oryctes nasicornis* larvae (Coleoptera: Scarabaeidae). J Insect Physiol 26:819–828.

Bayon C (1981a) Modifications ultrastructurales des parols végétales dans le tube digestif d'une larve xylophage *Oryctes nasicornis* (Coleoptera: Scarabaeidae): role des bactéries. Can J Zool 59:220–229.

Bayon C (1981b) Ultrastructure del'epithelium intestinal et flore parietale chez larve xylophage d'*Oryctes nasicornis* (Coleoptera: Scarabaeidae). Int J Insect Morphol Embryol 10:359–371.

Bayon C, Etiévant P (1980) Methanic fermentation in the digestive tract of a xylophagous insect: *Oryctes nasicornis* L. larva (Coleoptera: Scarabaeidae). Experientia 36:154–155.

Bayon C, Mathelin JA (1980) Carbohydrate fermentation and byproduct absorption studied with labeled cellulose in *Oryctes nasicornis* (Coleoptera, Scarabeidae). J Insect Physiol 26:833–840.

Bermudes D, Chase D, Margulis L (1988) Morphology as a basis for taxonomy of large spirochetes symbiotic in wood-eating roaches and termites: *Pillotina* gen. nov., nom rev.; *Pillotina calotermitidis* sp. nov., nom rev.; *Diplocalyx* gen. nov., nom rev.; *Diplocalyx calotermitidis* sp. nov., nom rev.; *Hollandina* gen. nov., nom rev.; *Hollandina pterotermitidis* sp. nov., nom rev.; and *Clevelandina reticulitermitidis* gen. nov., sp. nov. Int J Syst Bacteriol 38:291–302.

Bignell DE (1984) The arthropod gut as an environment for microorganisms. In: Anderson JM, Rayner ADM, Walton DWH, eds. Invertebrate-Microbial Interactions, pp. 205–228. Cambridge: Cambridge University Press.

Bignell DE (1994) Soil-feeding and gut morphology in higher termites. In: Hunt JH, Nalepa CA, eds. Nourishment and Evolution in Insect Societies. pp. 131–158. Boulder: Westview Press .

Bignell DE, Anderson JM, Crosse R (1991) Isolation of facultatively aerobic actinomycetes from the gut, parent soil and mound materials of the termites *Procubitermes aburiensis* and *Cubitermes severus*. FEMS Microbiol Ecol 85:151–160.

Bignell DE, Oskarsson H, Anderson JM (1980) Distribution and abundance of bacteria in the gut of a soil-feeding termite, *Procubitermes aburiensis* (Termitidae, Termitinae). J Gen Microbiol 117:393–403.

Bignell DE Oskarsson H, Anderson JM, Ineson P, Wood TG (1983) Structure, microbial associations and function of the so-called "mixed segment" of the gut in two soil-feeding termites, *Procubitermes aburiensis* and *Cubitermes severus* (Termitidae, Termitinae). J Zool Lond 201:445–480.

Boone DR, Mah RH (1988) Group I. Methanogenic archaeobacteria. In: Staley JT, Bryant MP, Pfennig N, Holt JG eds. Bergey's Manual of Systematic Bacteriology, Vol. 3. Baltimore: Williams and Wilkins.

Bracke JW, Cruden DL, Markovetz AJ (1978) Effect of metronidazole on the intestinal microflora of the American cockroach, *Periplaneta americana* L. Antimicrob Agents Chemother 13:115–120.

Bracke JW Markovetz AJ (1980) Transport of bacterial endproducts from the colon of *Periplaneta americana*. J Insect Physiol 26:85–89.

Brauman A, Kane MD, Labat M, Breznak JA (1990) Hydrogen metabolism by termite gut microbes. In: Belaich JP, Bruschi M, Garcia JL, eds. Microbiology and Biochemistry of Strict Anaerobes Involved in Interspecies Hydrogen Transfer, pp. 369–371. New York: Plenum Publishing.

Brauman A, Kane MD, Labat M, Breznak JA (1992) Genesis of acetate and methane by gut bacteria of nutritionally diverse termites. Science 257:1384–1387.

Breznak JA (1982) Intestinal microbiota of termites and other xylophagous insects. Annu Rev Microbiol 36:323–343.

Breznak JA (1984a) Biochemical aspects of symbiosis between termites and their intestinal microbiota. In: Anderson JM, Rayner ADM, Walton DWH, eds. Invertebrate-Microbial Interactions, pp. 173–203. Cambridge: Cambridge University Press.

Breznak JA (1984b) Hindgut spirochetes of termites and *Cryptocercus punctulatus*. In: Krieg NR, ed. Bergey's Manual of Systematic Bacteriology, Vol. 1, pp. 68–70. Baltimore: Williams and Wilkins.

Breznak JA (1990) Metabolic activities of the microbial flora of termites. In: Lesel R, ed. Microbiology of Poecilotherms, pp. 63–68. Amsterdam: Elsevier.

Breznak JA (1994) Acetogenesis from carbon dioxide in termite guts. In: Drake HL, ed. Acetogenesis. pp. 303–330. New York: Chapman and Hall.

Breznak JA, Blum JS (1991) Mixotrophy in the termite gut acetogen, *Sporomusa termitida*. Arch Microbiol 156:105–110.

Breznak JA, Brill WJ, Mertins, JW, Coppel HC (1973) Nitrogen fixation in termites. Nature 244:577–580.

Breznak JA, Brune A (1994) Role of microorganisms in the digestion of lignocellulose by termites. Annu Rev Entomol 39:453–487.

Breznak JA, Kane MD (1990) Microbial H_2/CO_2 acetogenesis in animal guts: nature and nutritional significance. FEMS Microbiol Rev 87:309–314.

Breznak JA Pankratz HS (1977) In situ morphology of the gut microbiota of wood-eating termites [*Reticulitermes flavipes* (Kollar) and *Coptotermes formosanus* (Shiraki)]. Appl Environ Microbiol 33:406–426.

Breznak JA, Switzer JS (1986) Acetate synthesis from H_2 plus CO_2 by termite gut microbes. Appl Environ Microbiol 52:623–630.

Breznak JA, Switzer JM, Seitz H-J (1988) *Sporomusa termitida* sp. nov., an H_2/CO_2-utilizing acetogen isolated from termites. Arch Microbiol 150:282–288.

Brune A, Emerson D, Breznak JA (1995a) The termite microflora as an oxygen sink: microelectrode determination of oxygen and pH gradients in guts of lower and higher termites. Appl Environ Microbiol 61:2681–2687.

Brune A, Miambi E, Breznak JA (1995b) Roles of oxygen and the intestinal microflora in the metabolism of lignin-derived phenylpropanoids and other monoaromatic compounds by termites. Appl Environ Microbiol 61:2688–2695.

Buchner P (1965) Endosymbiosis of Animals With Plant Microorganisms. New York: Interscience.

Bull AT, Goodfellow M, Slater JH (1992) Biodiversity as a source of innovation in biotechnology. Annu Rev Microbiol 46:219–252.

Bull AT, Hardman DJ (1991) Microbial diversity. Curr Opin Biotechnol 2:421–428.

Caetano FH (1989) Endosymbiosis of ants with intestinal and salivary gland bacteria. In: Schwemmler W, Gassner G, eds. Insect Endocytobiosis: Morphology, Physiology, Genetics, Evolution, pp. 57–76. Boca Raton, Fla: CRC Press.

Cano RJ, Borucki MK, Higby-Schweitzer M, Poinar HN, Poinar GO, Pollard KJ (1994) Bacillus DNA in fossil bees: an ancient symbiosis? Appl Environ Microbiol 60: 2164–2167.

Chapman RF (1982) The Insects: Structure and Function. Cambridge: Harvard University Press.

Chapman RF (1985) Structure of the digestive system. In: Kerkut GA, Gilbert LI, eds. Comprehenisive Insect Physiology, Biochemistry and Pharmacology, Vol. 4, pp. 165–212. Oxford: Pergamon Press.

Charnley AK, Hun J, Dillon RJ (1985) The germ-free culture of desert locusts *Schistoccerca gregaria*. J Insect Physiol 31:477–485.

Choe JC (1992) Zoraptera of Panama with a review of the morphology, systematics and biology of the order. In: Quintero D, Aiello A, eds. Insects of Panama and Mesoamerica: Selected Studies pp. 249–246. Oxford: Oxford University Press.

Cleveland LR (1924) Symbiosis between termites and their intestinal protozoa. Proc Natl Acad Sci USA 19:424–428.

Cleveland LR, Hall SR, Sanders EP, Collier J (1934) The wood-feeding roach *Cryptocercus*, its protozoa, and the symbiosis between protozoa and roach. Mem Am Acad Arts Sci 17:184–342.

Cochran DG (1985) Nitrogen excretion in cockroaches. Annu Rev Entomol 30:29–49.

Collins NM (1983) The utilization of nitrogen resources by termites (Isoptera). In: Lee JA, McNeill S, Rorison IH, eds. Nitrogen as an Ecological Factor, pp. 381–412. Oxford: Blackwell.

Conrad RB, Schink B, Phelps TJ (1986) Thermodynamics of H_2-consuming and H_2-producing metabolic reactions in diverse methanogenic environments under in situ conditions. FEMS Microbiol Ecol 38:353–360.

Conrad R, Bak F, Seitz H-J, Thebrath B, Mayer HP, Schutz H (1989) Hydrogen turnover by psychrotrophic homoacetogenic and mesophilic methanogenic bacteria in anoxic paddy soil and ake sediment. FEMS Microbiol Ecol 62:285–294.

Cord-Ruwisch R, Seitz H-J, Conrad R (1988) The capacity of hydrogenotrophic anaerobic bacteria to compete for traces of hydrogen depends on the redox potential of the terminal electron acceptor. Arch Microbiol 149:350–357.

Cruden DL, Markovetz AJ (1987) Microbial ecology of the cockroach gut. Annu Rev Microbiol 41:617–643.

Cummins KW, Klug MJ (1979) Feeding ecology of stream invertebrates. Annu Rev Ecol Syst 10:147–172.

Czolij RM, Slaytor M, O'Brien RW (1985) Bacterial flora of the mixed segment and the hindgut of the higher termite *Nasutitermes exitiosus* Hill (Termitidae, Nautitermitinae). Appl Environ Microbiol 49:1226–1236.

Darlington JPEC (1994) Nutrition and evolution in fungus-growing termites. In: Hunt JA, Nalepa CA, eds. Nourishment and Evolution in Insect Societies, pp. 105–130. Boulder: Westview Press.

Daser U, Brandl R (1992) Microbial gut floras of eight species of tephritids. Biol J Linn Soc 45:155–165.

Doddema HJ, Vogels GD (1978) Improved identification of methanogenic bacteria by fluorescence microscopy. Appl Environ Microbiol 36:752–754.

Dolfing J (1988) Acetogenic bacteria. In: Zehnder AJB, ed. Biology of Anaerobic Microorgansisms, pp. 417–468. New York: John Wiley and Sons.

Douglas A (1989) Mycetocyte symbiosis in insects. Biol Rev 64:409–434.

Dowd PF (1991) Symbiont-mediated detoxification in insect herbivores. In: Barbosa P, Krischik VA, Jones CG, eds. Microbial Mediation of Plant-Herbivore Interactions, pp. 411–439. New York: John Wiley and Sons.

Dowd PF (1992) Insect fungal symbionts: a promising source of detoxifying enzymes. J Indust Microbiol 9:149–161.

Downer RGH (1982) Fat body and metabolism. In: Bell WJ, Adiyodi KG, eds. The American Cockroach, pp. 151–174. New York: Chapman and Hall.

Drake HL (1994) Acetogenesis. New York: Chapman and Hall.

Drew RAI, Lloyd AC (1991) Bacteria in the life cycle of tephritid fruit flies. In: Barbosa P, Krischik VA, Jones CG, eds. Microbial Mediation of Plant-Herbivore Interactions, pp. 441–465. New York: John Wiley and Sons.

Dudley T, Anderson NH (1982) A survey of invertebrates asssociated with wood debris in aquatic habitats. Melanderia 39:1–22.

Edwards R, Mill AE (1986) Termites in Buildings. W. Sussex: Rentokil Ltd.

Embly TM, Finlay BJ (1994) The use of small subunit rRNA sequences to unravel the relationships between anaerobic ciliates and their methanogen endosymbionts. Microbiology 140:225–235.

Fenchel T (1987) Ecology of protozoa: the biology of free-living phagotrophic protists. Madison, Wisc.: Science Tech, Inc.

Ferry JG (ed) (1993) Methanogenesis. New York: Chapman and Hall.

Findlay S, Meyer JL, Smith PJ (1986) Incorporation off microbial biomass by *Peltoperla* sp. (Plecoptera) and *Tipula* sp. (Diptera). J North Am Benthol Soc 4:306–310.

French JRJ, Turner GL, Bradbury JF (1976) Nitrogen fixation by bacteria from the hindgut of termites. J Gen Microbiol 95:202–206.

Fuchs G (1986) CO_2 fixation in acetogenic bacteria: variations on a theme. FEMS Microbiol Rev 39:181–213.

Gijzen HJ, Broers CAM, Barugahare M, Stumm CK (1991) Methanogenic bacteria as endosymbionts of the ciliate *Nyctotherus ovalis* in the cockroach hindgut. Appl Environ Microbiol 57:1630–1634.

Gijzen HJ, Barugahare M (1992) Contribution of anaerobic protozoa and methanogens to hindgut metabolic activities of the american cockroach, *Periplaneta americana*. Appl Environ Microbiol 58:2565–2570.

Gilliam M, Buchman SL, Lorenz BJ, Schmalzel RJ (1990) Bacteria belonging to the genus *Bacillus* associated with three species of solitary bees. Apidologie 21:99–105.

Gilliam M, Taber S (1991) Diseases, pests, and normal microflora of honeybees, *Apis mellifera*, from feral colonies. J Invert Pathol 58:286–289.

Guo L, Quilici DR, Chase J, Blomquist GJ (1991) Gut tract microorganisms supply the precursors for methyl-branched hydrocarbon biosynthesis in the termite *Zootermopsis nevadensis*. Insect Biochem 21:327–333.

Hogan ME, Slaytor M, O'Brien RW (1985) Transport of volatile fatty acids across the hindgut of the cockroach, *Panesthia cribrata* Saussure and the termite, *Mastotermes darwiniensis* Froggatt. J Insect Physiol 31:587–591.

Hogan M, Veivers PC, Slaytor M,Czolij RT (1988) The site of cellulose breakdown in higher termites (*Nasutitermes walderi* and *Nasutitermes exitiosus*). J Insect Physiol 34: 891–899.

Honigburg, BM (1970) Protozoa associated with termites and their role in digestion. In: Krishna K, Weesner FM, eds. Biology of Termites, Vol. 1, pp. 1–36. New York: Acedemic Press.

Hungate RE (1939) Experiments on the nutrition of Zootermopsis. III. The anaerobic carbohydrate dissimilation by the intestinal protozoa. Ecology 20:230–245.

Hungate RE (1943) Quantitative analyses on the cellulose fermentation by termite protozoa. Ann Entomol Soc Am 36:730–739.

Hungate RE (1946) The symbiotic utilization of cellulose. J Elisha Mitchell Soc 62:9–24.

Hungate RE (1985) Anaerobic biotransformations of organic matter. In: Leadbetter ER, Poindexter JS, eds. Bacteria in Nature, Vol. 1, pp. 39–95. New York: Plenum.

Hunt JH Nalepa CA (1994) Nourishment and Evolution in Insect Societies. Boulder: Westview Press.

Jones JG, Simon BM (1985) Interactions of acetogens and methanogens in anaerobic freshwater sediments. Appl Environ Microbiol 49:944–948.

Kane MD, Brauman, A Breznak JA (1991) *Clostridium mayombeii*, sp. nov. an acetogenic bacterium from the guts of the African soil-feeding termite, *Cubitermes speciousus*. Arch Microbiol 156:99–104.

Kane MD, Breznak JA (1991a) *Acetonema longum*, gen. nov. sp. nov., an H_2/CO_2 acetogenic bacterium from the termite, *Pterotermes occidentis*. Arch Microbiol 156:91–98.

Kane MD, Breznak JA (1991b) Effect of host diet on the production of organic acids and methane by cockroach gut bacteria. Appl Environ Microbiol 57:2628–2634.

Kane MD, Pierce NE (1994) Diversity within diversity: molecular approaches to studying microbial interactions with insects. In: Schierwater B, Streit B, Wagner G, DeSalle R, eds. Molecular Methods in Ecology and Evolution, pp. 509–524. Berlin: Berkhauser Verlag. In press.

Kaufman MG, Klug MJ (1990) Microbial community metabolism in the digestive tract of crickets (Orthoptera: Gryllidae): Implications for omnivorous insects. In: Lesel R, ed. Microbiology of Poecilotherms, pp. 69–72. Amsterdam: Elsevier.

Kaufman MG, Klug MJ (1991) The contribution of hindgut bacteria to dietary carbohydrate utilization by crickets (Orthoptera: Gryllidae). Comp Biochem Physiol 98:117–123.

Kaufman MG, Klug MJ, Merrit RW (1989) Growth and food utilization parameters of germ-free house crickets, *Acheta domesticus*. J Insect Physiol 35:957–967.

Kirby H, Margulis L (1994) Harold Kirby's symbionts of termites: karymastigont reproduction and calonymphid taxonomy. Symbiosis 16:7–16.

Klug MJ, Kotarski S (1980) Bacteria associated with the gut tract of larval stages of the aquatic cranefly *Tipula abdominalis* (Diptera; Tipulidae). Appl Environ Microbiol 40: 408–416.

Krishna K (1970) Taxonomy, phylogeny, and distribution of termites. In: Krishna K, Weesner FM, eds. Biology of Termites, Vol. 2, pp. 127–152. New York: Academic Press.

Kuhnigk T, Borst E-M, Ritter A, et al. (1994) Degradation of lignin monomers by the hindgut flora of xylophagous termites. Sys Appl Microbiol 17:76–85.

Kukor JJ, Martin MM (1983) Aquisition of digestive enzymes by siricid woodwasps from their fungal symbiont. Science 220:1161–1163.

Lajoie, SF, Bank S, Miller TL, Wolin MJ (1988) Acetate production from hydrogen and [^{13}C]carbon dioxide by the microflora of human feces. Appl Environ Microbiol 54: 2723–2727.

Lawson DL, Klug MJ (1989) Microbial fermentation in the hindguts of stream detritivores. J North Am Benthol Soc 8:85–91.

Lawson DL, Klug MJ, Merrit RW (1984) The influence of the physical, chemical, and microbiological characteristics of decomposing leaves on the growth of the detritivore *Tipula abdominalis* (Diptera: Tipulidae). Can J Zool 62:2339–2343.

Lee MJ, Schreurs PJ, Messer AC, Zinder SH (1987) Association of methanogenic bacteria with flagellated protozoa from a termite hindgut. Curr Microbiol 15:337–341.

Lee MJ, Zinder SH (1988) Isolation and characterization of a thermophilic bacterium which oxidizes acetate in syntrophic association with a methanogen and which grows autotropphically on H_2-CO_2. Appl Environ Microbiol 54:124–129.

Liu S, Suflita JM (1993) H_2-CO_2-dependent anaerobic O-demethylation activity in subsurface sediments and by an isolated bacterium. Appl Environ Microbiol 59:1325–1331.

Ljungdahl LG (1986) The autotrophic pathway of acetate synthesis in acetogenic bacteria. Annu Rev Microbiol 40:415–450.

Martin MM (1987) Invertebrate-Microbial Interactions: Ingested Fungal Enzymes in Arthropod Biology. Ithaca: Cornell University Press.

Martin MM (1991) The evolution of cellulose digestion in insects. Phil Trans R Soc Lond B 333:281–288.

Mauldin JK, Rich NM, Cook DW (1978) Amino acid synthesis from ^{14}C-acetate by normally and abnormally faunated termites, *Coptotermes formosanus*. Insect Biochem 8:105–109.

McBee RH (1997) Fermentation in the hindgut. In: Clarke RTJ, Bauchop T, eds. Microbial Ecology of the Gut, pp. 185–222. London: Academic Press.

McEwen SE, Slaytor M, Obrien RW (1980) Cellobiase activity in three species of Australian termites. Insect Biochem 10:563–567.

Mead LJ, Khachatourians GG, Jones GA (1988) Microbial ecology of the gut in laboratory stocks of the migratory grasshopper, *Melaoplus sanguinipes* (Fab.) (Orthoptera: Acridae). Appl Environ Microbiol 54:1174–1181.

Messer AC, Lee MJ (1989) Effect of chemical treatments on methane emission by the hindgut microbiota in the termite *Zootermopsis angusticolis*. Microb Ecol 18:275–284.

Metcalf CL, Flint WP, Metcalf RL (1962) Destructive and Useful Insects, 4th ed. New York: McGraw-Hill.

Mullins DE, Cochran DG (1972) Nitrogen excretion in cockroaches: uric acid is not a major product. Science 177:699–701.

Mullins DE, Cochran DG (1975a) Nitrogen metabolism in the American cockroach. I. An examination of positive nitrogen balance with respect to uric acid stores. Comp Biochem Physiol 50A:489–500.

Mullins DE, Cochran DG (1975b) Nitrogen metabolism in the American cockroach. II. An examination of negative nitrogen balance with respect to mobilization of uric acid stores. Comp Biochem Physiol 50A:501–510.

Nalepa CA (1984) Colony oviposition, protozoan transfer and some life history characteristics of the woodroach *Cryptocercus punctulatus* Scudder. Behav Ecol Sociobiol 14: 273–279.

Nalepa CA (1990) Early development of nymphs and establishment of the hindgut symbiosis in *Cryptocercus punctulatus* Scudder (Dictyoptera: Cryptocercidae). Ann Entomol Soc Am 81:637–641.

Nalepa CA (1991) Ancestral transfer of symbionts between cockroaches and termites: an unlikely scenario. Proc R Soc Lond B 246:185–189.

Nalepa CA (1994) Nourishment and the origin of termite eusociality. In: Hunt JH, Nalepa CA, eds. Nourishment and Evolution in Insect Societies, pp. 57–104. Boulder: Westview Press.

Noirot C (1992) From wood- to humus-feeding: an important trend in termite evolution. In: Billen J, ed. Biology and Evolution of Social Insects, pp. 107–119. Leuven, Belgium: Leuven University Press.

O'Brien RW, Breznak JA (1984) Enzymes of acetate and glucose metabolism in termites. Insect Biochem 14:639–643.

O'Brien GW, Veivers PC, McEwen SE, Slaytor M, O'Brien RG (1979) The origin and distribution of cellulase in the termites, *Nasutitermes exitiosus* and *Coptotermes lacteus*. Insect Biochem 9:619–625.

Odelson DA, Breznak JA (1983) Volatile fatty acid production by the hindgut microbiota of xylophagous termites. Appl Environ Microbiol 45:1602–1613.

Odelson DA, Breznak JA (1985a) Cellulase and other polymer-hydrolyzing activites of *Trichomitopsis termopsidis*, a symbiotic protozoan from termites. Appl Environ Microbiol 49:622–626.

Odelson DA, Breznak JA (1985b) Nutrition and growth characteristics of *Trichomitopsis termopsidis*, a cellulolytic protozoan from termites. Appl Environ Microbiol 49: 614–621.

Pasti MB, Belli ML (1985) Celluloytic activity of actinomycetes isolated from termites (Termitidae) gut. FEMS Microbiol Lett 26:107–112.

Pasti MB, Pometto AL III, Nuti MP, Crawford DL (1990) Lignin-solubilizing ability of actinomycetes isolated from termite (Termitidae) gut. Appl Environ Microbiol 56: 2213–2218.

Pereira CRD, Anderson NH, Dudley T (1982) Gut content analysis of aquatic insects from wood substrates. Melanderia 39:23–33.

Phelps TJ, Zeikus JG (1984) Influence of pH on terminal carbon metabolism in anoxic sediments from a mildly acidic lake. Appl Environ Microbiol 48:1088–1095.

Potrikus CJ, Breznak JA (1977) Nitrogen-fixing *Enterobacter agglomerans* isolated from the guts of wood-eating termites. Appl Environ Microbiol 33:392–399.

Potrikus CJ, Breznak JA (1980) Uric acid in wood-eating termites. Insect Biochem 10: 19–27.

Potrikus CJ, Breznak JA (1981) Gut bacteria recycle uric acid nitrogen: a strategy for nutrient conservation. Proc Natl Acad Sci USA 78:4601–4605.

Prestwich GD, Bentley BL (1982) Ethylene production by the fungus comb of Macrotermitines (Isoptera, Termitidae): a caveat for the use of the acetylene reduction assay for nitrogenase activity. Sociobiology 7:145–152.

Prins RA, Kreulen DA (1991) Comparative aspects of plant cell wall digestion in insects. Anim Feed Sci Technol 32:101–118.

Prins RA, Lankhorst A (1977) Synthesis of acetate from CO_2 in the cecum of some rodents. FEMS Microbiol Lett 1:255–258.

Radek R, Hausmann K, Breunig A (1992) Ectobiotic and endycytobiotic bacteria associated with the termite flagellate *Joenia annectens*. Acta Prorozool 31:93–107.

Rouland C, Civas A, Renoux J, Petek F (1988a) Purification and properties of cellulases from the termite *Macrotermes mulleri* (Termitidae, Macrotermitinae) and its symbiotic fungus *Termitomyces* sp. Comp Biochem Physiol 91B:449–458.

Rouland C, Civas A, Renoux J, Petek F (1988b) Synergistic activities of the enzymes involved in cellulose degradation, purified from *Macrotermes mulleri* and from its symbiotic fungus *Termitomyces* sp. Comp Biochem Physiol 91B:459–465.

Rohrmann GF Rossman AY (1980) Nutrient strategies of *Macrotermes ukuzii* (Isoptera: Termitidae). Pedobiologia 20:61–73.

Sands WA (1969) The association of termites and fungi. In: Krishna K, Weesner FM, eds. Biology of Termites, Vol. 1, pp. 495–524. New York: Academic Press.

Scrivener AM, Slaytor M, Rose HA (1989) Symbiont-independent digestion of cellulose and starch in *Panesthia cribrata* Saussure, an Australian wood-eating roach. J Insect Biochem 35:935–941.

Sharak Genthner BR, Davis CL, Bryant MP (1981) Features of rumen and sewage sludge strains of *Eubacterium limosum*, a methanol- and H_2-CO_2-utilizing species. Appl Environ Microbiol 42:12–19.

Sinsabaugh RL, Linkins AE, Benfield EF (1985) Cellulose digestion and assimilation by three leaf-shredding aquatic insects. Ecology 66:1464–1471.

Slansky F, Rodriguez JG (1987) Nutritional Ecology of Insects, Mites, Spiders and Related Invertebrates. New York: John Wiley and Sons.

Slaytor M (1992) Cellulose digestion in termites and cockroaches: what role do symbionts play? Comp Biochem Physiol 103B:775–784.

Snipes BT, Tauber OE (1937) Time required for food passage through the alimentary tract of the cockroach, *Periplaneta americana* Linn. Ann Entomol Soc Am 30:277–284.

Taguchi F, Chang JD, Takiguchi S, Morimoto M (1992) Efficient hydrogen production from starch by a bacterium isolated from termites. J Ferment Bioeng 73:244–245.

Thauer RK, Jungermann KK, Decker K (1977) Energy conservation in chemolithotrophic bacteria. Bacteriol Rev 41:100–180.

Thorne BL (1990) A case for ancestral transfer of symbionts between cockroaches and termites. Proc R Soc Lond B 241:37–41.

Thorne BL (1991) Ancestral transfer of symbionts between cockroaches and termites: an alternative hypothesis. Proc R Soc Lond B 246:191–195.

Thorne BL, Carpenter JM (1992) Phylogeny of the Dictyoptera. Syst Entomol 17:253–268.

To L, Margulis L, Cheung ATW (1978) Pillotinas and hollandinas: distribution and behavior of large spirochaetes symbiotic in termites. Microbios 22:103–133.

To LP, Margulis L, Chase D, Nutting WL (1980) The symbiotic microbial community of the sonoran desert termite: *Pterotermes occidentis*. BioSystems 13:109–137.

Treves DS, Martin MM (1994) Cellulose digestion in primitive hexapods: effect of ingested antibiotics on gut microbial populations and gut cellulase levels in the firebrat, *Thermobia domestica* (Zygentoma, Lepismatidae). J Chem Ecol 20:2003–2020.

Ulrich RG, Buthala DA, Klug MJ (1981) Microbiota associated with the gastrointestinal tract of the common house cricket, *Acheta domestica*. Appl Environ Microbiol 41: 246–254.

Veivers PC, Muhlemann R, Slaytor M, Leuthold RH, Bignell DE (1991) Digestion, diet and polytheism in two fungus-growing termites: *Macrotermes subhyalinus* Rambur and *M. michaelseni* Sjostedt. J Insect Physiol 37:675–682.

Waller DA, La Fage JP (1987) Nutritional ecology of termites. In: Slansky F, Rodriguez J, eds. Nutritional Ecology of Insects, Mites, Spiders and Related Invertebrates, pp. 487–532. New York: John Wiley and Sons.

Whitman WB (1985) Methanogenic bacteria. In: Woese CR, Wolfe RS, eds. The Bacteria: A Treatise on Structure and Function, Vol. VIII: Archaebacteria pp. 3–84. New York: Academic Press.

Williams CM, Veivers PC, Slaytor M, Cleland SV (1994) Atmospheric carbon dioxide and acetogenesis in the termite *Nasutitermes walkeri* (Hill). Comp Biochem Physiol 107A:113–118.

Wilmarth KR, Boone DR, Mah RH (1985) Hydrogen utilizing bacteria in the colon of cetaceans. Abstr Ann Mtg Am Soc Microbiol, p. 164.

Wilson EO (1971) The Insect Societies. Cambridge: Harvard University Press.

Wilson EO (1992) The Diversity of Life. Cambridge: Harvard University Press.

Wolin MJ (1974) Metabolic interactions among intestinal microbes. Am J Clin Nutr 27: 1320–1328.

Wolin MJ (1982) Hydrogen transfer in microbial communities. In: Bull AT, Slater JH, eds. Microbial Interactions and Communities, pp. 323–356. London: Academic Press.

Wood HG, Ragsdale SW, Pezacka E. (1986a) The acetyl-CoA pathway a newly discovered pathway of autotrophic growth. Trends Biochem Sci 11:1–5.

Wood, HG, Ragsdale SW, Pezacka E (1986b) The acetyl-CoA pathway of autotrophic growth. FEMS Microbiol Rev 39:345–362.

Wood TG (1978) Food and feeding habits of termites. In: Brian MV, ed. Production Ecology of Ants and Termites, pp. 55–80. Cambridge: Cambridge University Press.

Wood TG, Johnson RA (1986) The biology, physiology, and ecology of termites. In: Vinson SB, ed. Economic Impact and Control of Social Insects, pp. 1–68. New York: Praeger.

Wood TG, Sands WA (1978) The role of termites in ecosystems. In: Brian MV, ed. Production Ecology of Ants and Termites, pp. 245–293. Cambridge: Cambridge University Press.

Yamin MA (1979) Flagellates of the orders Trichomonadida Kirby, Oxymonadida Grassä, and Hypermastigida Grassi and Foa reported from lower termites (Isoptera families Mastotermitidae, Kalotermitidae, Hodotermitidae, Termopsidae, Rhinotermitidae and Serritermitidae) and from the wood-feeding roach *Cryptocercus* (Dictyoptera: Cryptocercidae). Sociobiology 4:1–120.

Yamin MA (1980) Cellulose metabolism by the termite flagellate *Trichomitopsis termopsidis*. Appl Environ Microbiol 39:859–863.

Yamin MA (1981) Cellulose metabolism by the flagellate *Trichonympha* from a termite is independent of endosymbiotic bacteria. Science 211:58–59.

Yamin MA, Trager W (1979) Cellulolytic activity of an axenically-cultivated termite flagellate, *Trichomitopsis termopsidis*. J Gen Microbiol 113:417–420.

II

BIOCHEMICAL ACTIVITIES AND METABOLIC TRANSFORMATIONS

9

Carbohydrate Fermentation, Energy Transduction and Gas Metabolism in the Human Large Intestine

George T. Macfarlane and Glenn R. Gibson

1. Introduction

The principal sources of carbon and energy for bacteria growing in the human large intestine are resistant starches, plant cell wall polysaccharides, and host mucopolysaccharides, together with various proteins, peptides, and other lower-molecular-weight carbohydrates that escape digestion and absorption in the small bowel (Cummings and Macfarlane 1991, Macfarlane and Cummings 1991). These complex polymers are degraded by a wide range of bacterial polysaccharidases, glycosidases, proteases, and peptidases to smaller oligomers and their component sugars and amino acids. Intestinal bacteria are then able to ferment these substances to short chain fatty acids (SCFAs), hydroxy and dicarboxylic organic acids, H_2, CO_2, and other neutral, acidic, and basic end products. Carbohydrate metabolism is quantitatively more important than amino acid fermentation in the human large intestine, particularly in the proximal colon, where substrate availability is greatest. The hydrolysis and metabolism of carbohydrates in the large intestine are influenced by a variety of physical, chemical, biological, and environmental factors, some of which are shown in Table 9.1.

The major SCFAs produced during breakdown of carbohydrates by intestinal bacteria are acetate, propionate, and butyrate. Other fermentation products include ethanol, lactate, and succinate (Fig. 9.1), but these metabolites are further oxidized by cross-feeding species in the large gut, and, with the occasional exception of the proximal bowel, they do not accumulate to any significant extent (Macfarlane et al. 1992). Other SCFA present in lower amounts include formate, valerate, caproate, and the branched chain fatty acids (BCFAs) isobutyrate, 2-methyl-butyr-

Table 9.1. Some factors affecting fermentation of carbohydrates in the human large intestine

Chemical composition of the substrate
Amount of substrate available for fermentation
Physical form of the substrate including particle size, solubility, and association with undigestible complexes such as lignins, tannins and silica
Colonic transit time
Composition of the gut microbiota with respect to species diversity and relative numbers of different types of bacteria
Ecological factors including competitive and cooperative interactions among bacteria
Rates of depolymerization of substrates
Substrate specificities and catabolite regulatory mechanisms of individual gut species
Fermentation strategies of individual substrate utilizing bacteria
Availability of inorganic electron acceptors
pH of gut contents
Antibiotic therapy

ate, and isovalerate, which are formed by fermentation of branched chain amino acids. SCFA concentrations vary in different anatomical regions of the large bowel, but are highest in the proximal colon (Fig. 9.2). This is mainly due to higher levels of substrate availability in this region of the gut. Colonic pH is also more acidic in the proximal bowel, as a result of higher SCFA production rates. Cummings et al. (1987) showed that molar ratios of acetate : propionate : butyrate are similar in different regions of the large intestine (proximal colon 57 : 22 : 21; distal colon 57 : 21 : 22); however, these data probably reflect the relative absorptive abilities of the proximal and distal bowels, and do not mean that the fermentative activities of intestinal bacteria are the same throughout this organ (Macfarlane et al. 1994).

SCFAs are rapidly absorbed from the large bowel and metabolized at various sites in the body. Quantitation of fermentation in the large intestine is difficult and must be made using indirect methods. However, some appreciation of the scale of bacterial activities in this organ can be obtained from autopsy studies, which show that relatively high concentrations of SCFA (288 μM) are present in portal blood, with A/V differences of 192 μM, giving an approximate colonic production rate of 277 mmol/d (Cummings et al. 1987, 1989). Estimates indicate that 95% to 99% of SCFAs formed by intestinal bacteria are absorbed (Engelhardt et al. 1991), suggesting that these fermentation products could, under certain circumstances, provide a potentially useful source of energy for the host. Theoretical calculations show that up to 540 kcal/d might be obtained from colonic fermentation, possibly contributing as much as 10% of the hosts daily energy requirements (McNeil 1984, Wisker and Feldheim 1994).

Either through SCFA formation, or gas production, the fermentative activities

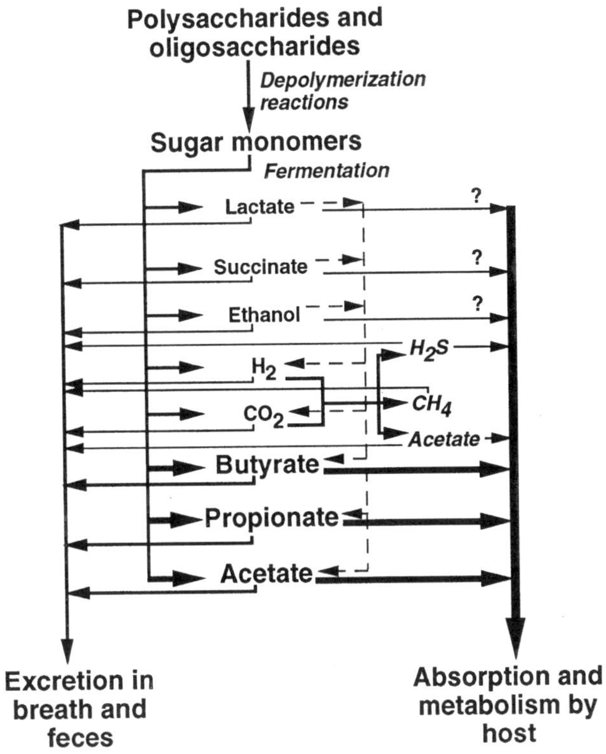

Figure 9.1. Overview of carbohydrate fementation in the human large intestine showing the fates of the major products of bacterial metabolism.

of intestinal bacteria impact on many facets of host metabolism. It is therefore important to understand how the gut microbiota functions and interacts with the host, especially in relation to substrate availability. Carbohydrate fermentation can be considered to be the major force driving the activities of the microbiota, and this chapter discusses the acquisition and metabolism of these substances by intestinal bacteria. Because little detailed information is available concerning the metabolism of carbohydrates in human gut species, particularly with respect to transport mechanisms, studies made on their rumen counterparts are discussed where appropriate.

2. Carbohydrate Utilization by Intestinal Bacteria

2.1. General Transport Mechanisms

Prokaryotes assimilate growth substrates by either passive or active mechanisms, depending on whether or not energy is expended during nutrient transloca-

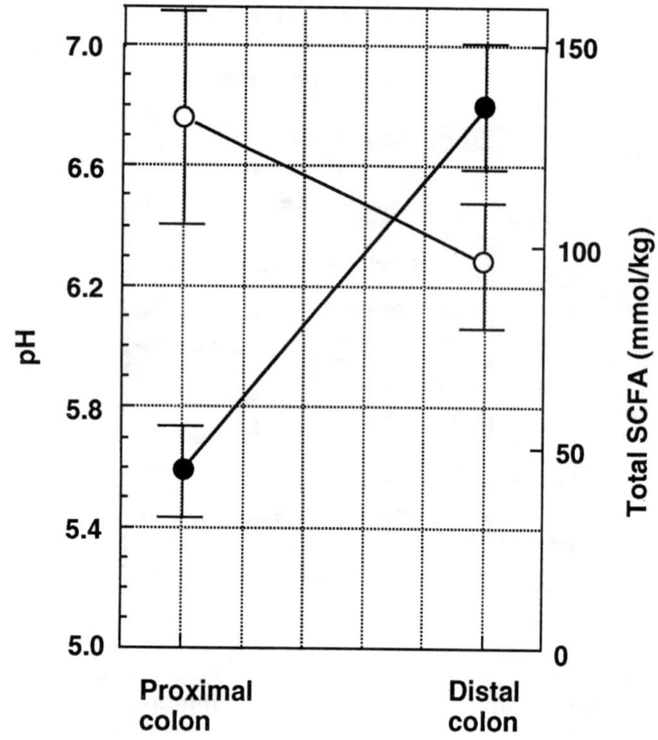

Figure 9.2. Total SCFA concentrations (○) and pH (●) in different regions of the large intestine. Error bars show SEM (Cummings et al. 1987).

tion across the bacterial membrane (Booth 1988). Passive diffusion does not require an energy generating step, because biological membranes are selectively permeable to the solutes. In gram-negative bacteria, where there are two lipid bilayers of the outer and cytoplasmic membranes, it is thought that solute size is important in regulating the uptake of nutrients. Schlegel (1986) observed that a molecular mass of over 600 Da is an effective cutoff point in limiting this form of substrate uptake in gram-negative organisms. During passive diffusion simple permeation of lipid soluble cations can occur, or transmembrane protein complexes may facilitate diffusion by selectively recognizing molecules that may be accumulated intracellularly, with a pore often being formed to effect uptake. Examples include the outer membrane porins Omp F, Omp C, and Lam B (Booth 1988).

Some obligate anaerobes use ion pumps located in the cytoplasmic membrane to couple electron transport-linked phosphorylation to ion translocation. Bacterial cell membranes are characteristically maintained in an energized state by a transmembrane H^+ electrochemical potential, which can be used for maintaining ion gradients and substrate transport (Morris 1986, Jones 1988). This proton motive force (Δp) consists of a chemical component, which is the pH gradient (ΔpH), and an electrical component (membrane potential, $\Delta\psi$ as shown below

$$\Delta p = \Delta\psi - 2.3RT/F(\Delta pH)$$

where R, T, and F represent the gas constant, the absolute temperature, and the Faraday constant, respectively. In fermentative species, the transmembrane potential difference is generated by membrane-associated H^+ ATPases which are driven by the hydrolysis of ATP formed in fermentation reactions. The F_o elements provide channels for driving H^+ outward through the cell membrane (Saier et al. 1993). This results in the interior of the bacteria becoming alkaline and negatively charged compared to the external environment.

Active transport systems are energy dependent and are usually ATP driven (Saier et al. 1993). Two of the most common and widely studied processes are the following.

PERIPLASMIC BINDING PROTEIN DEPENDENT SYSTEMS (PBPDS)

These uptake mechanisms possess solute binding proteins which are located in the periplasm that are responsible for transporting carbohydrates in an unmodified state. PBPDS have an energy requirement which is probably provided by ATP hydrolysis catalyzed by one of the periplasmic proteins (Silhavy et al. 1978, Higgins et al. 1985, Ferro-Luzzi Ames 1990). The mechanism is sensitive to osmotic shock which causes loss of periplasmic binding sites (Neu and Heppel 1965). In *E. coli*, PBDS are known to be responsible for transport of arabinose, maltose, ribose, xylose, and methyl β-galactose. In the case of maltose, it is thought that five proteins arc involved in substrate uptake by this mechanism (Dills et al. 1980).

ION-LINKED TRANSPORT SYSTEMS

Booth (1988) observed that ion-linked transport systems in prokaryotes have several unique features which have facilitated molecular studies on their mechanisms of action: reversibility—in the absence of an energy source they may be driven by artificially generated ion and charge gradients; their location in membrane vesicles; each product of the system is encoded by a single gene.

As defined by Mitchell (1963), three classes of ion-linked transport system

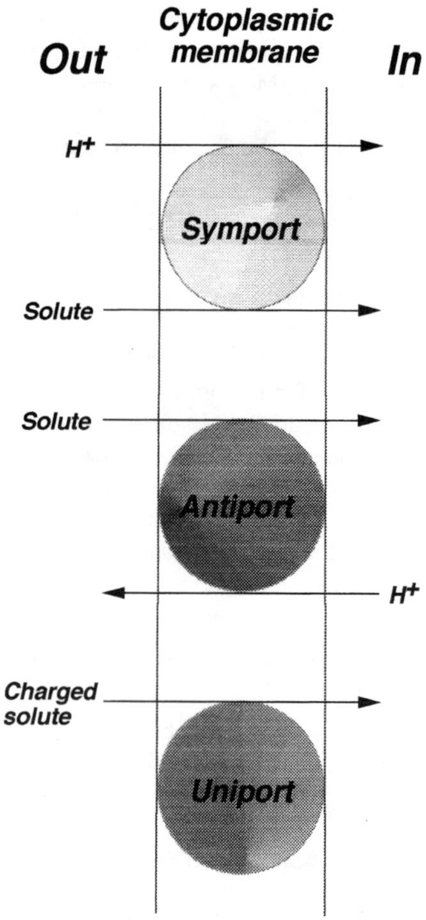

Figure 9.3. Simplified scheme showing the direction of movement of solutes and coupling ions in ion-linked bacterial secondary transport systems.

exist (Fig. 9.3). These secondary active transport processes include various forms of symports, antiports, and uniports. Symports have been fairly well investigated and consist of a solute and coupling ion traversing the membrane in the same direction. A typical model is the sugar:H^+ symport where solute accumulation is energized by a transmembrane proton gradient (Mitchell 1963, West 1970, West and Mitchell 1973). Protons are pumped out of the bacterial cell by electron flow between carriers or by proton translocating ATPases, and a proton motive force (pmf) is generated. Proton symports have been shown to be used for the uptake of carbohydrates by *Clostridium pasteurianum* (Mitchell et al. 1987) and the rumen bacterium *Selenomonas ruminantium* (Martin and Russell 1988, Russell et al. 1990).

Antiport systems consist of the solute and coupling ion moving in opposite directions; for example, Na^+/H^+ antiports function with Na^+ as the solute leading to its expulsion from the cell at the expense of a H^+. The formation of a transmembrane Na^+ gradient may then be used to drive the uptake of other solutes by Na^+ symports. Such a system is used for melibiose uptake in *E. coli* (Wilson and Wilson 1987).

Uniports are transport systems in which there is no coupling ion and where the driving force for uptake is provided by the charge on the substrate molecule itself, which establishes a membrane potential. A characteristic of energy coupling in ion-linked systems is that it usually results from a transmembrane pmf, as described earlier (Jones 1988).

PHOSPHOENOLPYRUVATE (PEP)-DEPENDENT GROUP TRANSLOCATION

This mechanism (Fig. 9.4) catalyzes the uptake of sugars, low-molecular-weight oligosaccharides, and sugar alcohols at the expense of PEP, as follows (Kundig 1974):

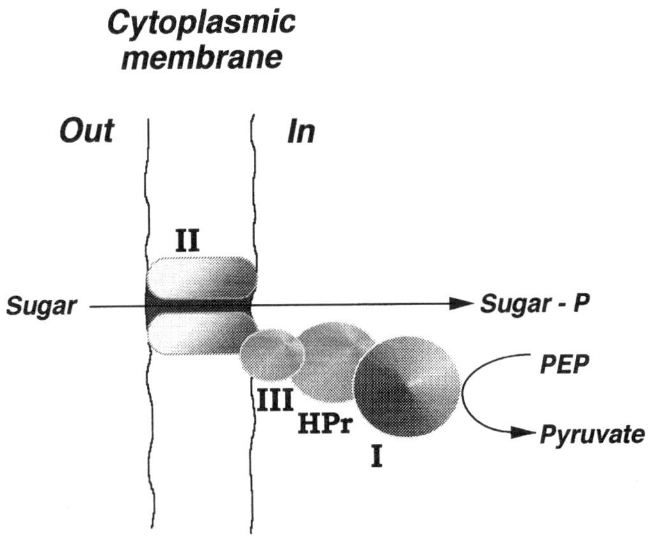

Figure 9.4. Outline of group translocation processes showing the spatial location of various components of the phosphoenolpyruvate : phosphotransferase system (PEP : PTS). I, II, and III refer to enzymes I, II, and III, respectively. HPr = heat-stable protein. Modified from Dills et al. 1980.

$$\text{sugar} + \text{PEP} \rightarrow \text{sugar phosphate} + \text{pyruvate}$$

Simultaneous translocation and phosphorylation of the sugar occurs as it traverses the cytoplasmic membrane (Postma and Roseman 1976, Postma and Lengeler 1985). A group of soluble and membrane bound proteins are involved which are collectively known as the phosphoenolpyruvate:phosphotransferase (PEP:PTS) system (Kundig 1974). Of these, at least two general soluble proteins (enzyme I and HPr) and a membrane-bound, substrate-specific protein (enzyme II) are essential. Enzyme I catalyzes phosphorylation of the heat stable protein (HPr) at the expense of PEP:

$$\text{PEP} + \text{enzyme I} \rightleftharpoons \text{phosphoenzyme I} + \text{pyruvate}$$

$$\text{phosphoenzyme I} + \text{HPr} \rightleftharpoons \text{enzyme I} + \text{phospho-HPr}$$

The phosphorylated solute is then transported through the cytoplasmic membrane in a reaction facilitated by enzyme II:

$$\text{phospho-HPr} + \text{enzyme II} \rightleftharpoons \text{HPr} + \text{phosphoenzyme II}$$

$$\text{phosphoenzyme II} + \text{sugar}_{out} \rightleftharpoons \text{enzyme II} + \text{sugar-phosphate}_{in}$$

The phosphoryl group of phospho-HPr is concomitantly transferred to the sugar forming a sugar phosphate. In some systems, enzyme III acts as a phospho carrier between HPr and enzyme II:

$$\text{phospho-HPR} + \text{enzyme III} \rightleftharpoons \text{HPr} + \text{phosphoenzyme III}$$

$$\text{phosphopnzyme III} + \text{enzyme II} \rightleftharpoons \text{enzyme III} + \text{phosphoenzyme II}$$

Enzyme III is located on the inner surface of the cytoplasmic membrane. However, because this protein may also be involved in catabolite repression, it is able to diffuse freely through the cell, enabling it to interact with soluble enzymes such as adenyl cyclase (Postma and Lengeler 1985).

The PEP:PTS system primarily operates in bacteria that use the glycolytic pathway of metabolism. It is energetically efficient in these organisms because two moles of PEP are produced from one hexose. One PEP is used to translocate another hexose molecule, while the other can be utilized for energy transduction (Saier and Chin 1990). Intestinal bacteria in which this process has been extensively studied include *Escherichia coli* (Booth 1988), lactobacilli (Chassey and Thompson 1983, Veyrat et al. 1994), *Streptococcus bovis, Fibrobacter (Bacteroides) succinogenes, Prevotella (Bacteroides) ruminicola, Megasphaera elsdenii* (Martin and Russell 1986, Russell et al. 1990), and *Clostridium thermocellum*

(Patni and Alexander 1971), where sugars and sugar alcohols such as glucose, mannose, mannitol, fructose, and galactose may be translocated.

2.2. *Carbohydrate Uptake*

Many different types of carbohydrate are present in the large intestine at any given moment, with the concentrations of individual substrates continuously changing as they are broken down, replenished, or replaced. Consequently, nutritionally versatile bacteria, which can grow on a variety of carbon sources and are able to adapt rapidly to changing nutritional conditions, are likely to have a considerable competitive advantage.

Regulation of carbohydrate transport in many prokaryotes is affected by catabolite regulatory mechanisms, including catabolite repression, catabolite inhibition, and inducer exclusion. In catabolite repression, the site of control is at the level of transcription; for example, synthesis of proteins involved in transport of substrate into the cell (Magasanik 1970). Catabolite inhibition and inducer exclusion are more responsive and rapid-acting mechanisms, in which the activities of existing enzymes involved in either transport or the initial stages of metabolism are regulated (McGinnis and Paigen 1969, 1973; Saier and Roseman 1976).

One characteristic of catabolite regulation is the manifestation of substrate preferences in bacteria grown in the presence of a mixture of carbon sources. This is often dependent on the relative availabilities of the substrates, and can be seen when bacteria are cultured on either polysaccharides or sugars. For example, experiments using carbon-limited continuous cultures of *Bacteroides ovatus* grown on either guar gum (mannose backbone, galactose side chains) or xylan (xylose backbone, arabinose side chains) showed that the backbone and side chain sugars were simultaneously utilized (Macfarlane et al. 1990). Similarly, when the same organism was grown on a mixture of starch and arabinogalactan, both polymers were co-utilized during carbon-limited growth; however, in carbon-excess chemostats, starch was preferentially used. This was not because synthesis of hydrolytic enzymes involved in depolymerizing the substrates was repressed, as arabinose, and to a lesser extent galactose accumulated in the cultures (Macfarlane and Gibson 1991). This indicated that regulation was occurring at the level of substrate transport.

The results of a series of investigations on sugar utilization in both human intestinal and rumen bacteria are shown in Table 9.2. It can be seen that substrate preferences are highly variable; even among the bifidobacteria, where the eight species studied exhibited very different patterns of substrate utilization. The mechanisms of sugar uptake in bifidobacteria are largely unknown, although experiments with *B. breve* showed that phosphorylation was involved in glucose transport (Degnan and Macfarlane 1993), but it was unclear from these studies

Table 9.2. Examples of substrate preferences exhibited in saccharolytic intestinal bacteria

Bacteria	Examples	References
Bifidobacterium[a]		Degnan and Macfarlane (1991)
pseudolongum	Sequential uptake of galactose, followed by glucose and xylose	
adolescentis	Glucose and galactose co-utilized, mannose and arabinose uptake repressed	
catenulatum	Only utilized glucose and galactose. Galactose taken up after glucose depletion	
angulatum	Glucose and galactose co-utilized, xylose uptake repressed	
infantis	Galactose repressed mannose and galactose utilization	
bifidum	Sequential uptake of glucose > galactose > mannose	
breve	Glucose and arabinose co-utilized, mannose and galactose uptake repressed	
longum	Glucose and xylose co-utilized, galactose uptake repressed	
Bacteroides[a]		Degnan (1992)
ovatus	Glucose, galactose, mannose, xylose co-utilized; arabinose uptake repressed	
thetaiotaomicron	Glucose, galactose, xylose, arabinose co-utilized. Mannose repressed glucose, galactose and arabinose uptake, but not xylose	
Ruminococcus albus[b]	Cellobiose preferentially used over glucose, due to repression of glucose transport systems	Thurston et al. (1993)
	Cellobiose preferred to either xylose or arabinose, through repression of transport systems and various catabolic enzymes. Glucose and xylose co-utilized	Thurston et al. (1994)
Selenomonas ruminantium[b]	Both glucose and maltose used in preference to either arabinose or xylose. Effect on xylose utilization resulted from catabolite inhibition by glucose PTS	Strobel (1993)
	Glucose, sucrose, xylose co-utilized; maltose, cellobiose, lactate uptake repressed by glucose and sucrose	Russell and Baldwin (1978)[c]
Megasphaera elsdenii[b]	Lactate used in preference to glucose	Hino et al. (1994)
	Glucose, maltose, lactate simultaneously utilized; sucrose uptake repressed by glucose	Russell and Baldwin (1978)[c]
Prevotella ruminicola[b]	Maltose used preferentially. Utilization of cellobiose, sucrose, and xylose occurred after maltose depletion	Russell and Baldwin (1978)[c]

[a] Grown on mixtures of glucose, galactose, mannose, xylose, arabinose.

[b] Rumen strains.

[c] Grown on various mixtures of glucose, cellobiose, maltose, sucrose, xylose, lactate.

whether this was mediated by a PEP:PTS. In contrast, arabinose uptake appeared to occur by facilitated diffusion.

As with bifidobacteria, little is known concerning sugar transport in human colonic bacteroides. However, Hylemon et al. (1977) showed that *B. thetaiotaomicron* did not assimilate glucose by a PEP:PTS, but their data suggested that sugar uptake might occur by facilitated diffusion in a carrier-mediated reaction. In these studies, mannose was found to inhibit glucose transport. Degnan (1992) obtained evidence that glucose, galactose, xylose, and arabinose transport systems were inducible in *B. thetaiotaomicron*, and that mannose was the preferred carbon source. This work also indicated the absence of a PEP:PTS for either mannose, xylose, or galactose.

Saccharolytic clostridia occur in comparatively high numbers in the human colon. A wide range of carbohydrate transport processes have been identified in these organisms, including facilitated diffusion, ATP or pmf-dependent uptake systems, and PEP:PTS, while there is evidence that control of inducible transport systems is effected through catabolite regulation and catabolite inhibition mechanisms (reviewed in Booth and Mitchell 1987).

3. Fermentation Reactions in Gut Anaerobes

3.1. Fermentation Strategies

Although a wide range of carbohydrates are potentially available for fermentation in the colon by several hundred different species of anaerobic bacteria, these substances are catabolized in a relatively small number of metabolic pathways. A highly simplified overview of the routes of hexose and pentose utilization in the large bowel is shown in Fig. 9.5.

The majority of intestinal bacteria use the glycolytic or Embden Meyerhof Parnas (EMP) pathway to ferment carbohydrates. Glucose is initially phosphorylated by hexokinase, or, if the organism has a PTS, it is phosphorylated during transport into the cell. The carbohydrate is converted to pyruvate in a series of reactions, which can be summarized as:

$$\text{glucose} + 2\,\text{NAD}^+ + 2\,\text{ADP} + 2\,\text{Pi} \rightarrow 2\,\text{pyruvate} + 2\,\text{NADH} + 2\,\text{ATP}$$

As will subsequently be seen, pyruvate formed in glycolysis may then participate in a variety of metabolic reactions.

The pentose phosphate pathway, which is also known as the hexose monophosphate shunt, occurs in many intestinal bacteria and can be used in dissimilatory metabolism of pentoses (see Fig. 9.5) or to produce NADPH and pentoses for biosynthetic purposes. In hexose metabolism, glucose-6-phosphate is oxidatively

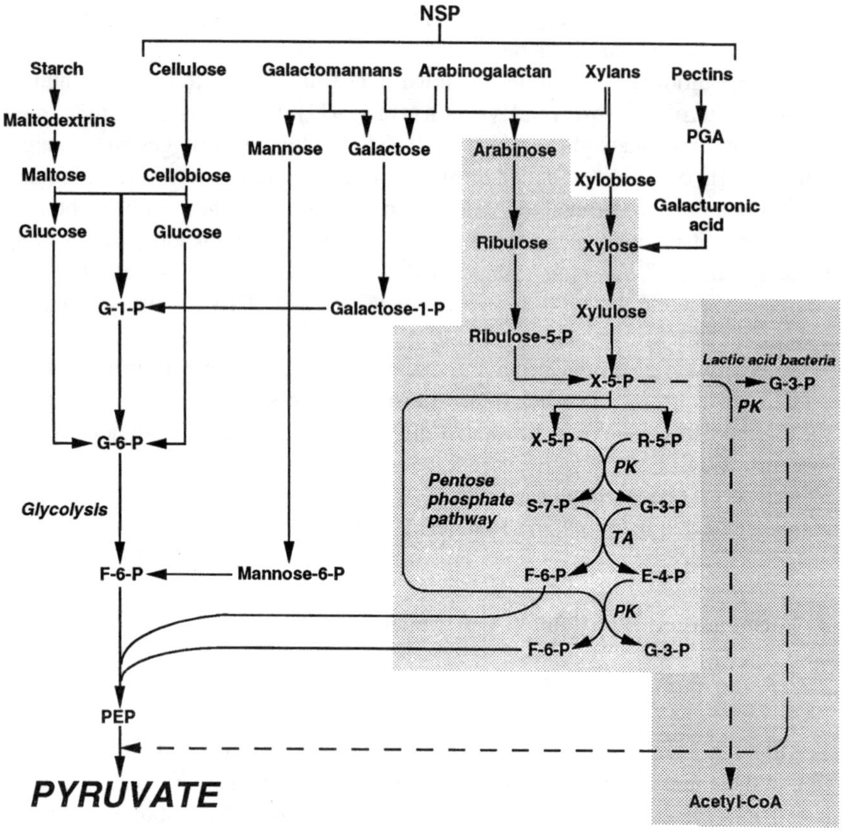

Figure 9.5. Simplified scheme of hexose and pentose utilization by intestinal bacteria. The shaded area delineates pentose metabolism. PK, phosphoketolase; TA, transaldolase; PGA, polygalacturonic acid; X-5-P, xylulose-5-P; R-5-P, ribose-5-phosphate; S-7-P, sedoheptulose-7-phosphate; G-3-P, glyceraldehyde-3-phosphate; F-6-P, fructose-6-P; E-4-P, erythrose-4-phosphate; G-1-P, glucose-1-phosphate; G-6-P, glucose-6-phosphate, PEP, phosphoenolpyruvate.

decarboxylated to ribulose-5-phosphate via 6-phosphogluconate, which subsequently undergoes a series of nonoxidative interconversions mediated by transketolases and transaldolase, in which triose and hexose phosphates are formed. In dissimilatory metabolism, these intermediates can then feed into various levels of the EMP pathway, leading ultimately to pyruvate synthesis.

In the Entner-Doudoroff (ED) pathway, glucose-6-phosphate is dehydrogenated to 6-phosphogluconate, from which glyceraldehyde-3-phosphate and pyruvate are produced. The ED pathway has a comparatively restricted taxonomic distribution in prokaryotes, and it is used to metabolize hexoses in some facultative anaerobes. For example, in *Escherichia coli*, it is used as an auxiliary pathway in the utilization of aldonates. However, the ED pathway has a low ATP yield and it is not important in strictly anaerobic colonic anaerobes.

Anaerobic chemotrophs growing in the large intestine can be differentiated into two broad groups—species that are able to carry out anaerobic respiration, and those that are restricted to fermentative metabolism and ATP formation through substrate level phosphorylation (SLP) reactions. Fermentations carried out by intestinal bacteria do not require respiratory chains that use molecular O_2 or other inorganic ions as terminal electron acceptors, since the electron acceptors are metabolic products derived from the original substrate. Therefore, with respect to the formation and consumption of reducing power, fermentation reactions are self-balancing, with the magnitude of the redox difference between substrates and products determining the amount of energy produced. Fermentations in which high levels of acetate are formed, can, in general, produce higher levels of ATP. Compared to oxidative metabolism, fermentations are energetically inefficient and have characteristically low ATP yields. Large amounts of substrate are required for growth in fermentative bacteria, which in turn leads to large quantities of metabolic end products being formed.

Energy transduction in the majority of human intestinal microorganisms, including the clostridia and lactic acid bacteria, occurs through SLP in a series of reactions involving dehydrogenases and lyases, although in some species (e.g., members of the *B. fragilis* group and propionibacteria) electron transport-linked phosphorylation is associated with the reduction of fumarate to succinate. In propionibacteria growing on lactate, three lactates are converted to two propionates and one acetate, thus forming three ATPs—two from electron transport-linked phosphorylation, and one from SLP.

Thauer et al. (1977) listed the $\Delta G^{0'}$ values of hydrolysis of the most common high-energy compounds involved in energy transduction in gut anaerobes as: Acetyl phosphate (51.6 kJ/mol); butyryl phosphate (44.8 kJ/mol); diphosphoglycerate (51.9 kJ/mol); phosphoenolpyruvate (51.6 kJ/mol). Release of energy from these compounds is catalyzed by a variety of kinases, including acetate kinase, butyrate kinase, phosphoglycerate kinase, and pyruvate kinase.

Fermentations are regulated by the need to maintain redox balance, principally through the reduction and oxidation of ferredoxins, flavins, and pyridine nucleotides. To a large degree, this affects the flow of carbon through the bacteria, the energy yield obtained from the substrate, and the fermentation products formed. Synthesis of reduced products including H_2, lactate, succinate, butyrate, and ethanol is used to effect redox balance during fermentation, whereas production of more oxidized substances, such as acetate is associated with ATP generation. Conversely, more reduced fermentation products result in comparatively low ATP yields. Many gut anaerobes take advantage of the metabolic flexibility offered by branched fermentation pathways compared to linear reaction sequences, since this allows them to modulate the thermodynamic efficiency of substrate catabolism through regulating ATP formation and redox balance. The central roles of pyruvate and acetyl-CoA at branch points in fermentation are shown in Figure 9.6, together with the various metabolic strategies that have evolved in different groups of intestinal bacteria.

Arguably, one of the most important factors affecting fermentation is substrate availability. For example, an increase in glucose concentration in cultures of *Clostridium bifermentans*, *C. sporogenes*, and *Peptostreptococcus anaerobius* resulted in comparative increases in ethanol and acetate synthesis, with, depending on species, various reductions in butyrate, butanol and BCFA (Turton et al. 1983). Table 9.3 shows the significance of substrate availability and physiologic and environmental factors such as dilution rate, pH, CO_2, and the effect of vitamin B_{12} on fermentation product formation in continuous cultures of a selected group of intestinal bacteria. Some of these studies will be addressed in more detail later.

3.2. Electron Sinks

Succinate, ethanol, lactate, and molecular H_2 variously function as electron sinks in fermentation reactions which serve to oxidize and recycle reduced coenzymes and ferredoxins formed in the initial stages of substrate catabolism (Fig. 9.6), thereby enabling fermentation pathways to continue in operation.

Succinate is a major end product of metabolism in relatively few human gut species, but many intestinal bacteria form ethanol as a predominant end product of metabolism in mixed fermentations. This product is a highly reduced metabolite which consumes 2 mol NADH/mol ethanol formed in bacteria. Unlike the ethanol fermentation in yeasts, which oxidizes only one NADH, and where the alcohol is produced from pyruvate by pyruvate decarboxylase and alcohol dehydrogenase (ADH), the vast majority of bacteria form ethanol from acetyl-CoA through the activities of acetaldehyde dehydrogenase and ADH. In ethanol-producing species that ferment carbohydrates by the glycolytic pathway, acetate must also be formed to maintain redox balance, because this pathway can only supply half of the NADH that would be required if ethanol was the only fermentation product.

Most lactate-producing bacteria in the large intestine form the L-isomer (Fig.

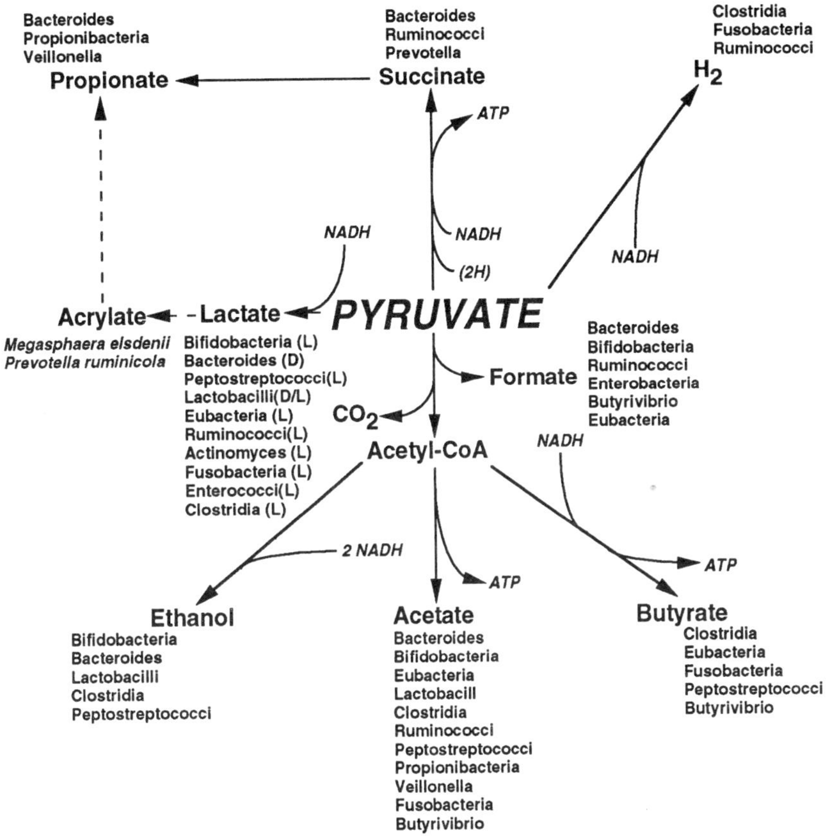

Figure 9.6. Overview of pyruvate metabolism in the large intestine in relation to ATP formation, regeneration of oxidized pyridine nucleotides, and the formation of various SCFA and electron sinks. The major types of bacteria associated with these fermentation products are also shown. In the lactate producing organisms, the isomeric forms of the acid are shown in parenthesis.

9.6). Studies have shown that lactate (and ethanol) are mainly present in the proximal colon (Cummings et al.1987, Macfarlane et al. 1992). For reasons discussed earlier, this primarily reflects substrate availability and shows that as far as carbohydrate fermentation is concerned, the majority of energy-spilling reactions in colonic bacteria occur in this region of the bowel. In some species, lactate formation is pH-dependent (Macfarlane et al. 1994), while both in vitro and in vivo studies indicate that its production in the colon is particularly associated with starch fermentations (Macfarlane and Englyst 1986, Etterlin et al. 1992).

Table 9.3. Effect of carbon availability, dilution rate, and other physiological and nutritional factors on fermentation product formation in continuous cultures of intestinal bacteria. (Molar ratios of principal fermentation products)

Bacteria	Growth Conditions	D (h^{-1})[a]	A	P	B	L	S	F	E	References
Bifidobacterium	Carbon (glucose)-limited	0.10	59	—	—	—	—	32	8	Degnan and
breve		0.45	60	—	—	—	—	29	10	Macfarlane
	Carbon (glucose) excess	0.09	70	—	—	25	—	—	3	(1994)
		0.60	68	—	—	25	—	—	6	
Ruminococcus	Carbon (cellobiose)-limited	0.06	47	—	—	27	21	6	—	Pettipher and
flavefaciens[b]		0.15	35	—	—	11	25	29	—	Latham (1979)
	Carbon (cellobiose) excess	0.06	33	—	—	10	24	33	—	
		0.15	33	—	—	16	19	32	—	
Ruminococcus	Carbon (cellulose)-limited									Shi and Weimer
flavefaciens[b]	pH 7.05	0.019	52	—	—	—	35	13	—	(1992)
	pH 7.08	0.059	36	—	—	—	34	30	—	
	Carbon (cellulose)-limited									
	pH 6.02	0.019	49	—	—	—	29	23	—	
	pH 6.16	0.059	38	—	—	—	32	30	—	
Ruminococcus	Carbon (glucose)-limited	0.05	50	—	—	4	—	33	14	Thurston et al.
albus[b]		0.39	39	—	—	—	—	40	20	(1993)
	Carbon (cellobiose)-limited	0.04	67	—	—	2	—	15	16	
		0.35	42	—	—	—	—	37	20	
Bacteroides	Carbon (arabinogalactan)-	0.025	50	32	—	—	18	—	—	Salyers et al.
thetaiotaomicron	limited	0.20	40	8	—	—	53	—	—	(1981)
Bacteroides fragilis	Carbon (glucose)-limited	0.04	44	50	—	1	3	2	—	Caspari and
	N_2 atmosphere	0.15	18	12	—	52	8	10	—	Macy
										(1983)
	Carbon (glucose)-limited	0.04	55	31	—	2	11	1	—	
	CO_2 atmosphere	0.16	46	13	—	4	33	4	—	
Bacteroides ovatus	Carbon (starch/	0.06	58	33	—	—	9	—	—	Macfarlane and
	arabinogalactan)-limited	0.19	53	8	—	—	39	—	—	Gibson (1991)
	Carbon (starch/arabinogalactan)	0.06	49	15	—	5	31	—	—	
	excess	0.19	49	11	—	4	36	—	—	
Prevotella	Carbon (xylose)-limited	0.025	23	—	—	69	8	—	—	Williams and
ruminicola[b]		0.15	21	—	—	75	4	—	—	Withers (1985)
Prevotella	Carbon (glucose)-limited with	0.05	26	73	—	—	1	—	—	Strobel (1992)
ruminicola[b]	vitamin B_{12}	0.35	35	29	—	—	36	—	—	
	Carbon (glucose)-limited no	0.05	30	3	—	—	67	—	—	
	vitamin B_{12}	0.36	24	5	—	—	71	—	—	
Propionibacterium	Carbon (lactate)-limited	0.02	41	59	—	—	—	—	—	Allison and
acnes		0.10	30	70	—	—	—	—	—	Macfarlane
Clostridium	Carbon (glucose)-limited	0.04	74	—	17	3	6	—	—	(1989b)
perfringens		0.16	80	—	8	12	—	—	—	Allison and
	Carbon (glucose) excess	0.04	49	—	12	35	5	—	—	Macfarlane
		0.16	18	—	3	80	—	—	—	(1989a)
Clostridium	Carbon (sucrose)-limited	0.05	32	—	68	—	—	—	—	Keith et al. (1982)
butyricum	Carbon (sucrose) excess	0.05	20	—	80	—	—	—	—	

A = acetate, P = propionate, B = butyrate, L = lactate, S = succinate, F = formate, E = ethanol.

[a] Approximate values.

[b] Rumen strains.

Measurements of lactate in gut contents also show a positive correlation with increased levels of residual starch in the cecum, confirming this relationship (Macfarlane and Gibson 1994). Lactate metabolism in individual groups of bacteria is discussed in more detail in subsequent sections of this chapter.

3.3. *Other Forms of Energy Transduction*

Some fermentative anaerobes can generate energy by establishing transmembrane electrochemical gradients using Na^+-translocating ATPases. In these systems, free energy is conserved as an electrochemical Na^+ potential, which can be converted into a $\Delta\psi$ if the cell membrane contains a Na^+/H^+ antiporter. Na^+ pumps are biotin-dependent decarboxylases, and the reactions they catalyze are highly exergonic. A good example occurs in *Veillonella alcalescens* where a Na^+-translocating decarboxylase (methylmalonyl-CoA decarboxylase) produces an electrochemical Na^+ gradient using an electroneutral Na^+/H^+ antiport, which enables a Δp to be generated. It is estimated that this allows the bacterium to produce an extra 0.3 ATP per substrate molecule decarboxylated (Hamilton 1988).

In some lactic acid bacteria, the efflux of lactate together with H^+ generates a Δp thereby contributing to energy production (Michels et al. 1979). In their energy-recycling model, ATP produced by substrate catabolism occurs via SLP, but is also dependent on the stoichiometry of H^+/lactate excretion (Driessen and Konings 1990).

3.4. *Fermentation in* **Bacteroides fragilis**-*type Bacteria*

Species belonging to this group are numerically predominant in the large intestine and play an important role in polysaccharide breakdown and carbohydrate fermentation (Finegold et al. 1983, Salyers 1984). The initial stages of glucose catabolism in these organisms occurs via the EMP pathway. PEP and pyruvate produced in these reactions may then undergo a variety of metabolic fates, as shown in Figure 9.7. Acetate, succinate, propionate, and to a lesser degree lactate are major products of fermentation, but as can be seen from Table 9.3, the relative amount of each metabolite formed is strongly influenced by nutritional and environmental factors.

ATP is formed by SLP in the conversion of glucose to pyruvate, but energy transduction also occurs by a pmf generated during the reduction of fumarate to succinate:

$$\text{fumarate}^{2-} + H_2 \rightarrow \text{succinate}^{2-} \quad (\Delta G^{0'} = -86.0 \text{ kJ/mol})$$

Although succinate formation by the fumarate reductase system is energetically advantageous, in that it enables the bacteria to form extra ATP, the product is

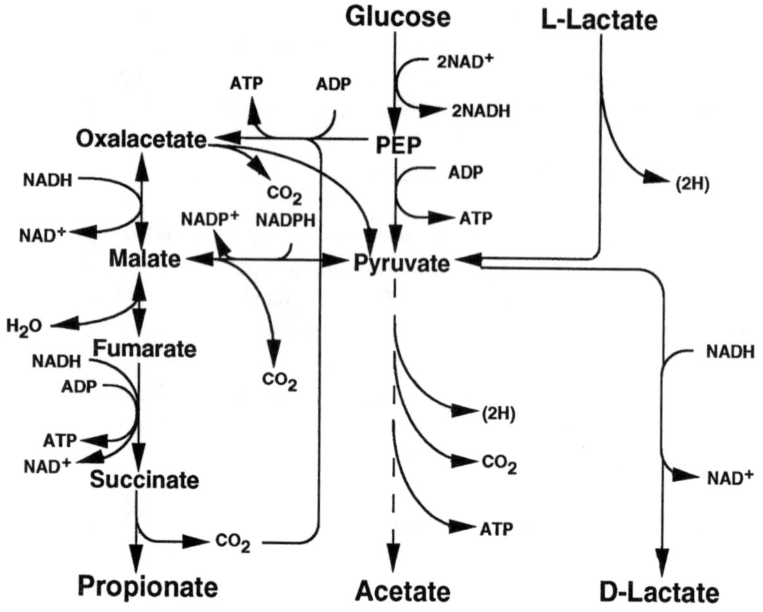

Figure 9.7. Pathways of glucose and L-lactate fermentation in *Bacteroides fragilis*, showing succinate decarboxylation to propionate and the significance of subsequent CO_2 cycling in converting C3 acids into C4 acids that can enter the succinate pathway. Adapted from Macy et al. (1978).

also highly reduced and serves as a sink for the disposal of reducing equivalents formed during glycolysis.

The metabolic significance of the succinate pathway in *B. fragilis* was demonstrated by Macy et al. (1975), who demonstrated that the bacterium requires hemin for synthesis of (NADH) fumarate oxidoreductase and a particulate *b*-type cytochrome. Growth rates were shown to be low in the absence of hemin and the main fermentation products were fumarate and lactate. In contrast, cell yields were nearly three-fold higher and growth rates four times faster when hemin was available.

Carbohydrate metabolism in *B. fragilis*-type bacteria is CO_2-dependent in that CO_2 is required to convert the 3C-compound PEP into the C4-compound oxaloacetate by PEP carboxykinase, in the initial reaction of the succinate pathway (see Fig. 9.7). Most fermentative anaerobes produce CO_2 by oxidative decarboxylation of pyruvate, but in *B. fragilis* this may not provide sufficient substrate for the succinate pathway to operate. Succinate is therefore decarboxylated to produce propionate and CO_2, which can then be recycled. Other bacteria that form propionate via the succinate pathway include the propionibacteria (Cummins and Johnson 1992) and veillonella (Ng and Hamilton 1973). The acrylate pathway is used by a few species such as *Prevotella ruminicola* (Wallnofer and Baldwin 1967) and *Megasphaera elsdenii* (Hino and Kuroda 1993) to form propionate. Although this route becomes quantitatively significant in ruminants under certain feeding regimes (Hungate 1966), it is unlikely to be important in propionate production in the human colon.

The effect of CO_2 availability on fermentation product formation in *B. fragilis* was demonstrated by Caspari and Macy (1983) using glucose-limited chemostats (see Table 9.3). They found that cultures grown under N_2 mainly produced acetate and propionate at low dilution rates, and that lactate became more important as dilution rates were increased. In cultures grown under CO_2, acetate and propionate predominated at low dilution rates, changing to acetate and succinate at high dilution rates. The regulatory effect of CO_2 on the succinate pathway occurs through PEP carboxykinase, which is induced by low pCO_2 but repressed by high pCO_2.

As Table 9.3 shows, a similar relationship between dilution rate and propionate and succinate formation occurred in carbon-limited cultures of *B. thetaiotaomicron* and *B. ovatus*. However, in carbon-excess chemostats, *B. ovatus* formed high levels of acetate and succinate independent of dilution rate, suggesting that when sufficient carbohydrate was available for the bacteria, oxidative decarboxylation of pyruvate provided enough CO_2 for the succinate pathway to operate, and that decarboxylation of succinate to propionate was inhibited.

3.5. Fermentation in Saccharolytic Clostridia

Saccharolytic clostridia almost exclusively use the EMP pathway for fermentation (Woods 1961). One hexose molecule is metabolized to two pyruvates with the production of NADH and ATP, as outlined in Figure 9.8. Acetate, lactate, and butyrate are the major organic acids produced from pyruvate. The fermentation pathways in saccharolytic clostridia are usually branched, providing a high degree of metabolic flexibility during substrate catabolism. The majority of pyruvate formed during carbohydrate fermentation is converted by pyruvate:ferredoxin oxidoreductase to acetyl-CoA and CO_2 with concomitant formation of reduced ferredoxin. Acetyl-CoA is the key intermediate in fermentation reactions mediated by saccharolytic clostridia and is an important precursor of acetate, ethanol and butyrate (Rogers 1986).

In lactate-producing clostridia such as *C. perfringens*, high substrate concentrations increase the amounts of lactate produced but reduce acetate and H_2 formation (Table 9.3). In these organisms, regulation of fermentation is effected at the level of pyruvate. Lactate dehydrogenase (LDH) activity is dependent on high intracellular levels of fructose-1,6-bisphosphate (F-1,6-P_2), which occur only

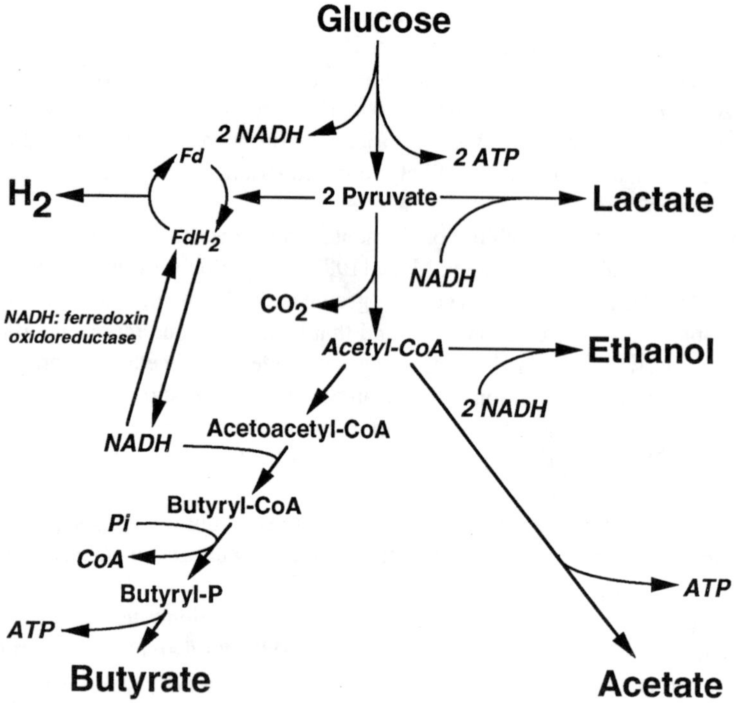

Figure 9.8. Simplified pathways of glucose fermentation in saccharolytic clostridia, showing the interrelationships of glycolysis, energy generation and the various roles of butyrate, ethanol, H_2 and lactate as electron sinks. FdH_2, reduced ferredoxin. Fd, oxidized ferredoxin; Pi, inorganic phosphate; CoA, coenzyme A.

under certain nutritional conditions—for example, Fe limitation, or growth in the presence of high concentrations of fermentable substrate (Andreesen et al. 1989). In contrast to carbon-limited growth, where energy gain is maximized by producing ATP from acetyl phosphate, a reduction in the efficiency of energy transduction occurs when carbon is in excess, and lactate becomes the principal electron sink. As occurs in many fermentative bacteria grown under carbon excess, the reduction in energy generating efficiency is compensated by the increase in carbon flux through the cells.

Few species carry out pure butyrate fermentations of the type outlined below, in which H_2 evolution enables H^+ to be used as electron acceptors:

$$\text{glucose} + 3\ \text{ADP} + 3\ \text{Pi} \rightarrow \text{butyrate} + 2\ CO_2 + 2H_2 + 3\ \text{ATP}$$

Instead, ethanol and acetate are often formed by butyrate producing clostridia (Keith et al. 1982). In these bacteria, fermentation is controlled to some degree by

CoA/acetyl-CoA and NAD^+/NADH ratios in the cell. Acetyl-CoA is an allosteric activator of NADH:ferredoxin oxidoreductase, whereas free CoA inhibits the enzyme (Thauer et al. 1977). Regulation of the efficiency of free energy conversion is exerted through the activity of NADH:ferredoxin oxidoreductase, because this enzyme influences the dissimilation of electrons to either H_2 or butyrate (Fig. 9.8). The conversion of acetyl-CoA to butyrate depends on NADH concentration, and under conditions where substrate is present in excess, the bacteria produce higher levels of butyrate, as shown in the chemostat studies in Table 9.3. However, under carbon limitation, the thermodynamic efficiency of fermentation is increased if acetate is also produced. This is because the conversion of one acetyl-CoA to acetate yields one ATP, whereas two acetyl-CoAs are needed to form butyrate. Thus, twice as much ATP can be generated if acetate is produced. When acetate and butyrate are made, reducing equivalents must be disposed of via hydrogenase, thereby increasing H_2 formation.

Ethanol may also be produced in mixed fermentations carried out by clostridia, in which case acetyl-CoA is used to form acetaldehyde, which is then reduced to ethanol. These two reactions each require NADH, and the bacteria must transfer electrons from reduced ferredoxin via ferredoxin:NAD oxidoreductase to NADH, which then allows alcohol formation.

3.6. Lactic Acid Bacteria Fermentations

Like ethanol, lactate is used as an electron sink by many gut species (see Fig. 9.6) and can be either a major or minor product of fermentation. The taxonomically diverse assemblage of organisms collectively known as the lactic acid bacteria form lactate as a major end product of metabolism and, from the viewpoint of the large intestine, can be taken to include enterococci, lactobacilli, and bifidobacteria.

Lactobacilli and enterococci can be differentiated into species that carry out homolactic-type fermentations and those that are heterofermentative. In homolactic metabolism glucose is converted into two lactates through the glycolytic pathway, with concomitant generation of 2 mol ATP/mol glucose, which are formed in SLP reactions. The conversion of two pyruvates into two lactates enables redox balance to be maintained.

Lactate is not necessarily always the only fermentation product formed by homofermentative lactic acid bacteria. Chemostat studies show that under some circumstances carbohydrate metabolism is affected by dilution rate, and hence, substrate availability. DeVries et al. (1970) demonstrated that *Lactobacillus casei* grown in glucose-limited continuous culture completely fermented the carbohydrate to acetate, formate, and ethanol at low dilution rates, whereas at high dilution rates only trace amounts of these metabolites were produced, with lactate as the principal fermentation product. They suggested that the intracellular concentration of F-1,6-P_2 regulated pyruvate metabolism, because LDH in this bacterium has an obligate requirement for F-1,6-P_2 for its activity. Thus, at high dilution rates the augmentation in carbon flow through the cells increased F-1,6-P_2 availability and hence LDH activity. Conversely, at low dilution rates, the phosphoroclastic enzyme was more active than LDH allowing the bacterium to produce more ATP under these energy-limiting growth conditions.

Heterofermentative lactic acid bacteria produce lactate, CO_2, and ethanol by the hexose monophosphate shunt, in which the metabolism of one glucose generates one ATP in SLP reactions, which is half of the energy yield obtained in the homolactic fermentation. An outline of the heterolactic fermentation pathway is shown in Figure 9.9. The key enzyme in these reactions is a phosphoketolase that

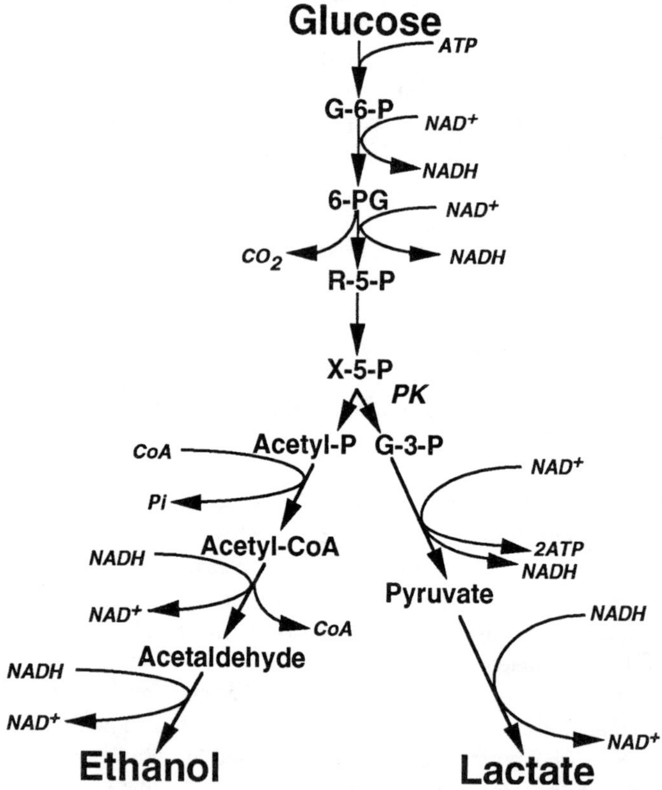

Figure 9.9. Heterolactic fermentation pathway as found in some lactic acid bacteria. G-6-P, glucose-6-phosphate; 6-PG, 6-phosphogluconate; R-5-P, ribulose-5-phosphate; X-5-P, xylulose-5-phosphate; PK, phosphoketolase; G-3-P, glyceraldehyde-3-P; Pi, inorganic phosphate; CoA, coenzyme A.

hydrolyzes xylulose-5-phosphate to acetyl-CoA and glyceraldehyde-3-phosphate. Both lactate and ethanol serve as electron sinks, and ATP is formed in the conversion of glyceraldehyde-3-phosphate to pyruvate. Heterofermentative lactic acid bacteria are able to ferment pentoses in the phosphoketolase pathway, as outlined in Figure 9.5, which is an inducible system; however, many homofermentative species are unable to ferment pentoses or are facultatively heterofermentative with respect to pentose metabolism (Kandler 1983).

Some lactic acid bacteria are aerotolerant and can oxidize NADH with O_2, through the activities of superoxide dismutase and peroxidases. This allows pyruvate to be converted to acetyl-CoA instead of lactate, thereby enabling more acetate and ATP to be produced (Hamilton 1988).

3.7. *Fermentation in Bifidobacteria*

In some individuals, bifidobacteria can constitute up to 25% of the total culturable gut microbiota (Mitsuoka 1982, Hidaka et al. 1986). Consequently, like *Bacteroides*, these organisms are likely to play, by virtue of their numbers, a major role in carbohydrate fermentation in the large bowel. The principal products of fermentation were originally believed to be acetate and lactate in a 3:2 ratio (Rasic 1983) according to the reaction:

$$2 \text{ glucose} + 5 \text{ ADP} + 5 \text{ Pi} \rightarrow 3 \text{ acetate} + 2 \text{ lactate} + 5 \text{ ATP}$$

Hexoses are fermented by the fructose-6-phosphate shunt (DeVries et al. 1967), in which they undergo a series of cleavage and isomerization reactions that produce pentose phosphates, which are subsequently metabolized to acetyl phosphate and glyceraldehyde-3-phosphate (G-3-P). ATP is generated from acetyl phosphate and G-3-P is converted to pyruvate in reactions similar to those found in the EMP pathway (Modler et al. 1990). Under certain growth conditions, however, ethanol and formate are also produced (DeVries and Stouthamer 1968). This is because pyruvate may be metabolized by two routes: It can be reduced by NADH to produce lactate, or alternatively, pyruvate can undergo phosphoroclastic cleavage to formate and acetyl phosphate (Fig. 9.10). Half of the resulting acetyl phosphate must be reduced to ethanol to facilitate oxidation of NADH produced earlier in the metabolic pathway; the remainder is available to make extra ATP through the formation of acetate (DeVries and Stouthamer 1968). In addition, ATP may also be obtained by end-product efflux-type mechanisms which were described earlier (Bezkorovainey 1989).

Chemostat studies with *Bifidobacterium breve* have shown that carbon availability has a major effect on fermentation product formation (Table 9.3). Degnan and Macfarlane (1994) demonstrated that acetate, and formate were the principal metabolic products in carbon-limited cultures, whereas under nitrogen limitation,

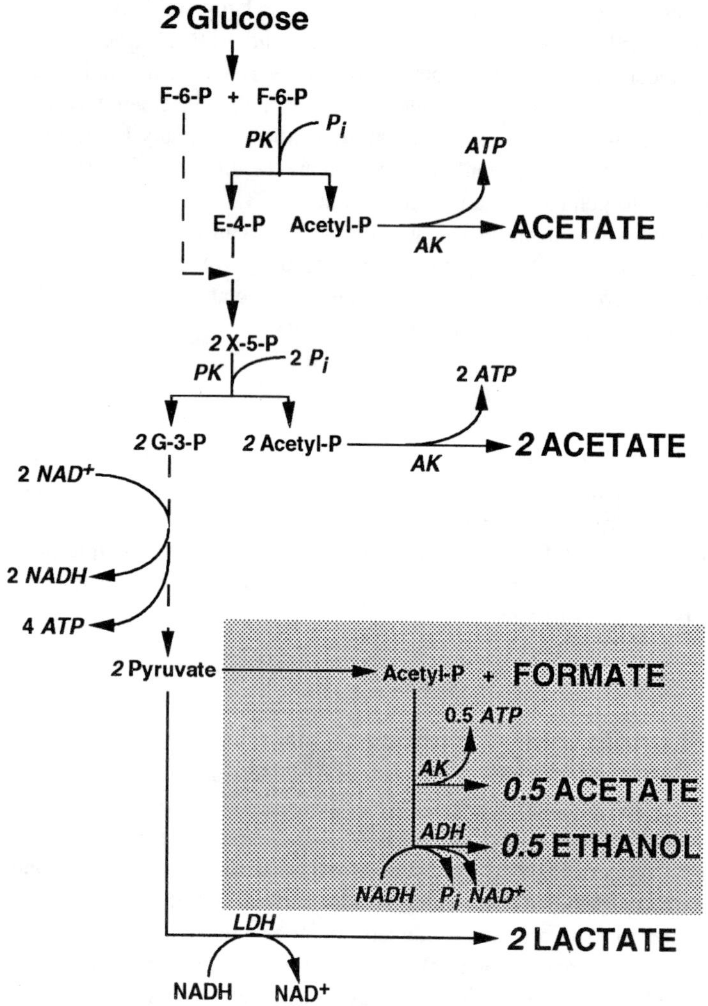

Figure 9.10. Pathways of glucose fermentation in *Bifidobacterium breve* showing the different metabolic fates of pyruvate (Degnan and Macfarlane 1994). F-6-P, fructose-6-phosphate; Pi, inorganic phosphate; E-4-P, erythrose-4-phosphate; X-5-P, xylulose-5-phosphate; G-3-P, glyceraldehyde-3-phosphate; PK, phosphoketolase; ADH, alcohol dehydrogenase; LDH, lactate dehydrogenase; AK, acetate kinase.

acetate, and ethanol predominated. These data show that the metabolic fate of pyruvate was dependent on carbon availability. Under carbon limitation, pyruvate was preferentially cleaved to acetate and formate instead of being reduced to lactate, enabling the bacterium to generate extra ATP. In contrast, the increase in substrate availability in nitrogen-limited chemostats provided sufficient ATP for growth; under these circumstances, reduction of pyruvate to lactate served as an uncoupling mechanism, which facilitated carbon flow through the cells without concomitant generation of energy by SLP.

Three hypotheses have been proposed to explain the regulation of pyruvate dissimilation in lactic acid bacteria (DeVries and Stouthamer 1968): In some species, it has been shown that LDH has an obligate requirement for F-1,6-P_2 for activity; therefore, by regulating the levels of this molecule in the cell, possibly through the activity of phosphofructokinase, it would be possible to directly affect lactate synthesis. Alternatively, the relative amounts of LDH and the phosphoroclastic enzyme may vary under different growth conditions, while their relative affinities for pyruvate will influence its dissimilation. Enzyme measurements in *B. breve* suggested that the latter explanation was most likely (Degnan and Macfarlane 1994).

The chemical composition of the fermentable substrate also influences the types of fermentation product formed by bifidobacteria. For example, DeVries and Stouthamer (1968) showed that *B. bifidum* fermented both glucose and xylose to acetate and formate, whereas the principal end products of lactate and mannitol fermentations were acetate and lactate and acetate, ethanol, and formate, respectively.

4. Hydrogen Metabolism In The Colon

4.1. Hydrogen Production

Together with ethanol, lactate, and succinate, H_2 is an important fermentation intermediate in the large intestine. These metabolites are used as sinks for electron disposal in anaerobic bacteria in catabolic reactions involving both sugars and amino acids. Mechanisms by which H_2 can be produced by carbohydrate fermenting bacteria include (Macfarlane and Gibson 1994):

1. Cleavage of pyruvate to formate which is then metabolized to H_2 and CO_2 by formate hydrogen lyase—e.g., enterobacteria.
2. Generation from pyruvate through the activity of pyruvate:ferredoxin oxi-

doreductase and hydrogenase—e.g., some clostridia. This reaction is exergonic ($\Delta G^{0\prime} = -19.2$ kJ/mol) and occurs as follows:

$$\text{pyruvate} + \text{Fd}_{\text{ox}} \rightleftharpoons \text{Fd}_{\text{red}} + \text{acetyl-CoA} + \text{CO}_2$$

Pyruvate probably has a central role in H_2 metabolism in the colon, because it is energetically most efficient to produce H_2 from this metabolite (Thauer 1977).

3. Formation from oxidation of the pyridine and flavin nucleotides $NADH_2$ and $FADH_2$ to NAD^+ and FAD^+—e.g., ruminococci and some clostridia. The production of reduced ferredoxin from NADH is an endergonic reaction ($\Delta G^{0\prime} = +18.8$ kJ/mol) catalyzed by NADH: ferredoxin oxidoreductase as follows:

$$\text{NADH} + \text{Fd}_{\text{ox}} \rightleftharpoons \text{Fd}_{\text{red}} + \text{NAD}^+ + \text{H}^+$$

Hydrogen production by mechanisms 1 and 2 depends less on pH_2 than on its formation from NADH, which is only thermodynamically possible at low pH_2 values. Thus, high pH_2 strongly inhibits the formation of reduced ferredoxin from pyridine nucleotides (Gottschalk and Peineman 1992). In some species, such as *Ruminococcus albus* and *Bacteroides xylanolyticus*, maintenance of low pH_2 is dependent on interspecies H_2 transfer reactions with H_2-consuming syntrophs. The physiological significance of these interactions is that they enable the H_2-producing species to generate more oxidized fermentation products and extra ATP through disposing of excess reducing power via NADH:ferredoxin oxidoreductase, instead of using ethanol as an electron sink (Thauer et al. 1977, Biesterveld et al. 1994).

In the human large bowel, physical removal of H_2 gas occurs through its excretion in breath or flatus and by the activities of H_2 consuming syntrophs. The relative spatial proximity of bacteria which produce and/or utilize H_2 is probably an important influence on overall metabolism of the gas. In this respect it has been shown, using sewage sludge, that over 90% of H_2 transfer reactions occur between bacterial populations existing in close proximity or in microniches (Conrad et al. 1985). Moreover, mass transfer resistance of H_2 between gas-liquid interfaces is also likely to affect its rate of utilization, since the gas is poorly soluble. Strocchi and Levitt (1992) showed that the extent of H_2 removal, and to some degree, its net production, was influenced by the extent of stirring of gut contents which was directly related to the partial pressure of the gas. Stirring allows movement of H_2 from colonic contents to the adjacent gas space at a more rapid rate than by diffusion alone, thereby elevating pH_2 in the gaseous phase. At high pH_2 values, human intestinal bacteria were able to oxidize H_2 at a rapid rate.

Cummings (1995) has proposed the following stoichiometry of carbohydrate fermentation in the large intestine:

$$59\ C_6H_{12}O_6 + 38\ H_2O \rightarrow 60\ \text{acetate} + 96\ CO_2 + 256\ H^+ + 22\ \text{propionate} + 18\ \text{butyrate}$$

Effectively, the above equation means that considerable potential for gas production exists in the large bowel. Theoretically, well in excess of 1 L of H_2 can be produced daily from an average dietary input of 40 to 50 g carbohydrate into the colon (Levitt et al. 1995). From the host viewpoint, this is occasionally of medical significance owing to symptoms of abdominal discomfort resulting from excessive flatulence. In reality, however, the total flatus volume of an average Western adult rarely exceeds 1 L/day, with H_2 contributing only about 16% of the total (Kirk 1949; Levitt 1969, 1971). Human calorimetry studies have confirmed that in healthy volunteers, the ingestion of nonabsorbable carbohydrates results in the excretion of only 2.5% to 14% of predicted H_2 production (Christl et al. 1992). It is believed that H_2 consuming microoganisms, such as methanogens, are primarily responsible for the discrepancy between theoretical and actual levels of H_2 excretion. However, CH_4 is only a fermentation end product in some individuals (see later), whereas H_2 utilization occurs even in persons who do not harbor a methanogenic biota, suggesting that other groups of bacteria are involved in H_2 metabolism in the large gut. In fact, many anaerobic bacteria are able to use molecular H_2 as an electron donor in anaerobic metabolism. Table 9.4 summarizes some of the mechanisms whereby this can occur.

4.2. Hydrogen Utilization

METHANOGENESIS

A taxonomically diverse range of methanogenic organisms exist in many different anaerobic ecosystems such as sewage sludgc; soils; marine, estuarine, and freshwater sediments; and the rumen and large intestines of various animals (Balch et al. 1979). These include the horse, sheep, cow, rat, pig, goose, chicken, and turkey. In the turkey and chicken, *Methanogenium* is thought to be a major component of the methanogenic biota, whereas in other animals, species belonging to the genera *Methanobrevibacter, Methanobacterium, Methanomicrobium,* and *Methanosarcina* are numerically predominant (Archer and Harris 1986). Considerable inter- and intraindividual variability in CH_4 excretion is seen in humans, where it is estimated that some individuals can produce between 40 mL and 4 L CH_4/day (Bond et al. 1971). However, only two species of methanogen have so far been identified as being part of the normal gut biota (Table 9.4). They are *Methanobrevibacter smithii*, which reduces CO_2 with H_2, and *Methanosphaera*

Table 9.4. Principal mechanisms involved in hydrogen disposal in human colonic microorganisms

Mechanism	Equation	Free Energy Change (kJ/mol)	Main Organisms Involved	Substrate Affinity (μmol/L)
Methanogenesis	$4H_2 + CO_2 \rightarrow CH_4 + 2H_2O$	−131	*Methanobrevibacter smithii*	6
	4 formate $\rightarrow CH_4 + 3CO_2 + 2H_2O$	−130.1	*Methanobrevibacter smithii*	NA
	$H_2 + CH_3OH \rightarrow CH_4 + H_2O$	−112.5	*Methanosphaera stadtmaniae*	NA
Dissimilatory sulfate reduction	$4H_2 + SO_4^{2-} + H^+ \rightarrow HS^- + 4H_2O$	−152.2	*Desulfovibrio* spp.	1
Acetogenesis	$4H_2 + 2CO_2 \rightarrow CH_3COOH + 2H_2O$	−95	Clostridia, eubacteria, peptostreptococci	NA
Dissimilatory nitrate reduction	$H_2 + NO_3^- \rightarrow NO_2^- + H_2O$	−163.2	NA (*Escherichia coli,* veillonella, some saccharolytic clostridia)	NA

NA = Data not available.

stadtmaniae, which combines methanol with H_2 to form CH_4 (Whitman et al. 1992). The latter organism is found only in relatively low numbers and appears to have a more restricted distribution than *M. smithii* in CH_4 excreting individuals. Both species grow relatively slowly and have an obligate growth requirement for H_2. Cell population densities of *M. smithii* in the colon range from 10^8 to 10^{10}/g contents (Miller and Wolin 1982), but significant numbers of other methanogenic organisms, such as species able to utilize acetate or other methanogenic precursor compounds, have not been detected.

In ruminant animals, methanogenesis is ubiquitous, but in humans only about 35% to 40% of persons in Western populations are methanogenic, although carriage rates are higher than this in Africa (Bond et al. 1971, McKay et al. 1985, Segal et al. 1988, Gibson et al. 1993). Infants below 2 years of age are usually CH_4-negative, although an age-related increase occurs as the children get older (Peled et al. 1985). It is possible that this observation is connected to competition for reducing equivalents with H_2-consuming bacteria, since acetogenesis has been detected in newborn lambs but disappears on the development of active methane production (Dore et al. 1992). However, the relative substrate affinities of these microorganisms for H_2, together with energetic considerations (Thauer et al.

1977), ought to favor methanogenesis in a competitive environment in which H_2 is limiting. It is possible that acetogens grow better in the underdeveloped colon because many species are also saccharolytic or amino acid fermenting.

In ruminants, the majority of H_2 produced during fermentation is used in CH_4-producing reactions, but in humans, H_2 as well as CH_4 may be excreted. This may be explained in part by varying retention times and differences in the relative rates of mixing of contents in the colon and rumen. Moreover, the organisms themselves may manifest adaptive nutritional differences.

The reduction of CO_2 to CH_4 (Fig. 9.11) is carried out by an electron transport chain consisting of various dehydrogenases, reductases and electron carriers (Thauer et al. 1977). Factor$_{420}$ is a fluorescent, low-molecular-weight electron carrier that is specific to methanogens and is often used to detect their presence in mixed bacterial suspensions. A hydrogenase is used to catalyze the reduction of F_{420} with H_2, while CO_2 reduction is mediated by a variety of cofactors, of which coenzyme M is the most studied (McBride and Wolfe 1971, Taylor and Wolfe 1974).

Initially, CO_2 is bound to methanofuran (MF) and reduced to formylmethanofuran (formyl-MF):

$$2H^+ + CO_2 + MF \rightarrow \text{formyl-MF} + H_2O$$

The formyl moeity is then transferred to tetrahydromethanopterin (H_4MPT):

$$\text{formyl-MF} + H_4MPT \rightarrow MF + \text{formyl-}H_4\text{-MPT}$$

H_2O is cleaved from the formyl-H_4MPT to give methenyl-H_4-MPT:

$$\text{formyl-}H_4MPT \rightarrow \text{methenyl-}H_4\text{-MPT} + H_2O$$

This is subsequently reduced to methylene-H_4-MPT in conjunction with F_{420}:

$$\text{methenyl-}H_4\text{-PMT} + F_{420} + H_2 \rightarrow F_{420} + \text{methylene-}H_4\text{-MPT}$$

This compound is then reduced to methyl-H_4-MPT:

$$\text{methylene-}H_4\text{-PMT} + 2H^+ \rightarrow \text{methyl-}H_4\text{-MPT}$$

The methyl group is transferred to coenzyme M in a transferase reaction:

$$\text{methyl-}H_4\text{-MPT} + \text{HS-CoM} \rightarrow \text{methyl-CoM} + H_4MPT$$

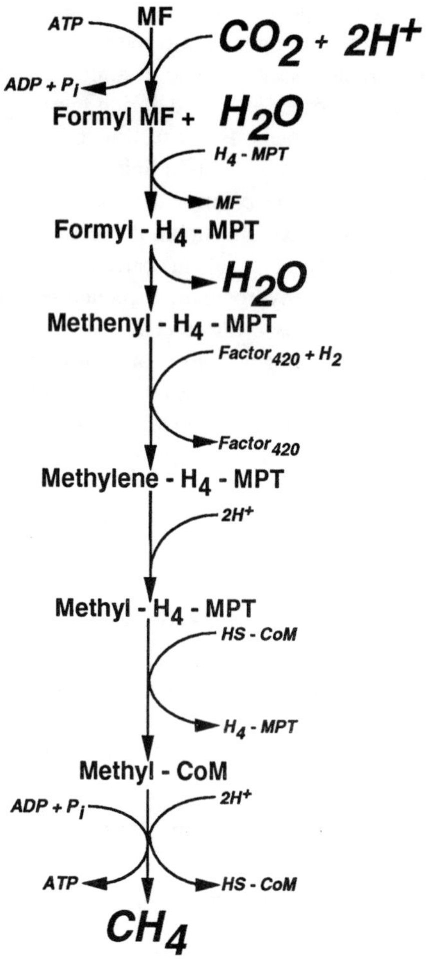

Figure 9.11. Pathway of methane formation from hydrogen and carbon dioxide. Information from Rouviere and Wolfe (1988), Blaut et al. (1990), and Whitman et al. (1992). MF, methanofuran; formyl MF, formylmethanofuran; H_4-MPT, tetrahydromethanopterin; CoM, coenzyme M.

Finally, methyl-coenzyme M is reduced to CH_4 in a methylreductase reaction (Blaut et al. 1990):

$$\text{methyl-CoM} + 2H^+ \rightarrow \text{HS-CoM} + CH_4$$

In the rumen, the composition of fermentation products is strongly dependent on methanogenesis, but in humans no significant differences in short chain fatty acid production patterns have been found between methanogenic and non-methanogenic persons (Macfarlane and Cummings 1991).

DISSIMILATORY SULFATE REDUCTION

One alternative to methanogenesis as a process whereby H_2 can be disposed of in the colon is through the activities of sulfate-reducing bacteria (SRB). These organisms are more frequently detected in feces of healthy persons living in the United Kingdom than are methanogens, occurring in approximately 65% of the population (Gibson et al. 1993). It has been shown that colonic SRB are able to outcompete methanogens for the mutual growth substrate H_2, an observation that may explain the absence of significant methanogenesis in some individuals (Gibson et al. 1988a,c). In fecal slurries, methanogenesis was only an important H_2 sink when SRB were either absent or inhibited with molybdate (Gibson et al. 1988a). In a study of 87 healthy volunteers, SRB occurred in high numbers, up to 10^{11}/g wet weight, in feces of nonmethanogenic persons (Gibson et al. 1993). This compares with counts of between 10^2 and 10^7/mL rumen fluid (Coleman 1960, Huisingh et al. 1974). As shown in Table 9.5, three broad-based population

Table 9.5. Methanogenesis and dissimilatory sulfate reduction in healthy individuals in the United Kingdom[a]

	Population Groupings					
	Group 1 (N = 21)		Group 2 (N = 9)		Group 3 (N = 57)	
Measurement	Range	Mean	Range	Mean	Range	Mean
Breath CH_4[b]	3.2–90.2	24.7 ± 10.19	3.2–40.0	23.5 ± 6.24	0	0
CH_4 production rates[c]	8.0–886	120 ± 97	5.0–634	101 ± 49.6	0	0
SRB counts[d]	ND	ND	3.1–5.5	4.1 ± 1.2	7.9–11.4	9.3 ± 3.3
Sulfate reduction rates[c]	T	T	T	T	3.4–374	82.0 ± 34.1

[a] 87 persons were studied. Data are mean values ± SEM.

[b] ppm.

[c] nmol (g wet wt. feces)$^{-1}$ h^{-1}.

[d] $\log_{10}$ (g wet wt. feces)$^{-1}$.

ND = not detected; T = trace amounts detected; SRB = sulfate-reducing bacteria.

groupings could be recognized in human volunteers: Group 1 consisted of 21 persons who were strong CH_4 producers in which SRB were completely absent from their stools. In Group 2 (n = 9) methanogenesis occurred and low numbers of SRB (ca. 10^5/g wet weight feces) were detected, although their metabolic activities were negligible. The final group consisted of 57 volunteers who had high counts of fecal SRB and complete absence of methanogenesis. The numerically predominant SRB were desulfovibrios which accounted for 67% to 91% of total SRB counts. Species belonging to the genera *Desulfobacter* (9% to 16%), *Desulfobulbus* (5% to 8%), and *Desulfotomaculum* (2%) were present in considerably lower numbers.

Because H_2-utilizing SRBs, such as desulfovibrios, have a growth requirement for sulfate, variations in the availability of this anion may select for methanogenesis (limiting SO_4^{2-}) or sulfate reduction (excess SO_4^{2-}). Sulfate is present in the diet and studies with ileostomists indicate that dietary sulfate reaching the colon may range from 2 to 9 mmol/day (Florin et al. 1991). Moreover, depolymerization and fermentation of host-produced glycoproteins that have a high sulfate content, such as mucins and chondroitin sulfate, appear to strongly stimulate sulfide production in mixed cultures of colonic bacteria containing SRB, due to release of the anion in free form (Gibson et al. 1988a,b). A carbohydrate sulfatase that has activity toward sulfated linkages in mucin oligosaccharides has been found in human intestinal isolates of *Bacteroides fragilis* and *B. thetaiotaomicron* (Tsai et al. 1992). In addition, preliminary results in our laboratory have indicated that batch culture incubations of these species in the presence of mucin and chondroitin sulfate results in the release of large amounts of free sulfate.

In an environment in which H_2 is limiting, and where SRB, methanogens, and acetogenic bacteria are all present, then given sufficient sulfate availability, SRB should theoretically outcompete the others for H_2. The explanation for this lies in the relative affinities of SRB and methanogens for H_2, and possibly to a lesser degree, energetic considerations (Table 9.4).

Although H_2 is not the only electron donor that SRB are able to use for dissimilatory sulfate reduction in the gut, stoichiometric considerations indicate that H_2 is likely to be important, because 4 moles of this gas can be converted to H_2S by only 1 mole of sulfate requires higher concentrations of sulfate per mole of electron donor oxidized (Sorensen et al. 1981, Parkes et al. 1989), as shown below:

$$4H_2 + SO_4^{2-} + H^+ \rightarrow HS^- + 4H_2O$$

In contrast, the oxidation of other common electron donors by SRB in the gut such as acetate:

$$\text{Acetate} + SO_4^{2-} \rightarrow HS^- + 2HCO_3^- \text{ (\textit{Desulfobacter} spp.)}$$

ethanol:

$$2 \text{ ethanol} + SO_4^{2-} \rightarrow 2 \text{ acetate} + HS^- + H^+ + 2H_2O \text{ (\textit{Desulfovibrio} spp.)}$$

lactate:

$$2 \text{ lactate} + SO_4^{2-} + H^+ \rightarrow 2 \text{ acetate} + 2CO_2 + 2H_2O + HS^- \text{ (\textit{Desulfovibrio} spp.)}$$

propionate:

$$4 \text{ propionate} + 3SO_4^{2-} \rightarrow 3HS^- + H^+ + 4HCO_3^- + 4 \text{ acetate (\textit{Desulfobulbus} spp.)}$$

and butyrate:

$$2 \text{ butyrate} + 3SO_4^{2-} \rightarrow 3HS^- + H^+ + 4HCO_3^- + 2 \text{ acetate (\textit{Desulfotomaculum} spp.)}$$

The major end product of sulfate reduction is H_2S, when H_2 is the electron donor (Fig. 9.12). Sulfate can be replaced by sulfite, thiosulfate, tetrathionate, or sulfur as an electron acceptor (Postgate 1984). The initial step involves transport of sulfate into the bacterial cell (Cypionka 1989), which is then reduced by combination with ATP to produce adenosine phosphosulfate (APS) and pyrophosphate. APS is then converted to sulfite (Stille and Truper 1984), which is subsequently reduced via a range of intermediates to form the sulfide ion.

Although there is debate concerning the identity of many of the intermediates involved in dissimilatory sulfate reduction, the most widely accepted concept is the cyclic mechanism described by Postgate (1984), in which sulfite is dehydrated to metabisulfite, followed by conversions to dithionite and trithionate. The trithionate is split into thiosulfate (and sulfite), which is converted into sulfite and sulfide. These reactions are mediated by a variety of hydrogenases and reductases. At least three classes of hydrogenase are thought to exist in SRB (Fauque et al. 1988, Widdel and Hansen 1992). They contain either (1) iron and sulfur; (2) nickel, iron, and sulfur, or (3) nickel, selenium, iron, and sulfur. In desulfovibrios, hydrogenases are often periplasmic, although some species may have cytoplasmic enzymes (Bell et al. 1974, Fauque et al. 1991). Cytochrome C_3 is an important electron carrier in SRB, but ferredoxins, flavodoxins, menaquinones, rubredoxins and other cytochromes also have a role (Peck and LeGall 1982, Peck 1993).

The predominant H_2 utilizing SRB in the human colon belong to the genus

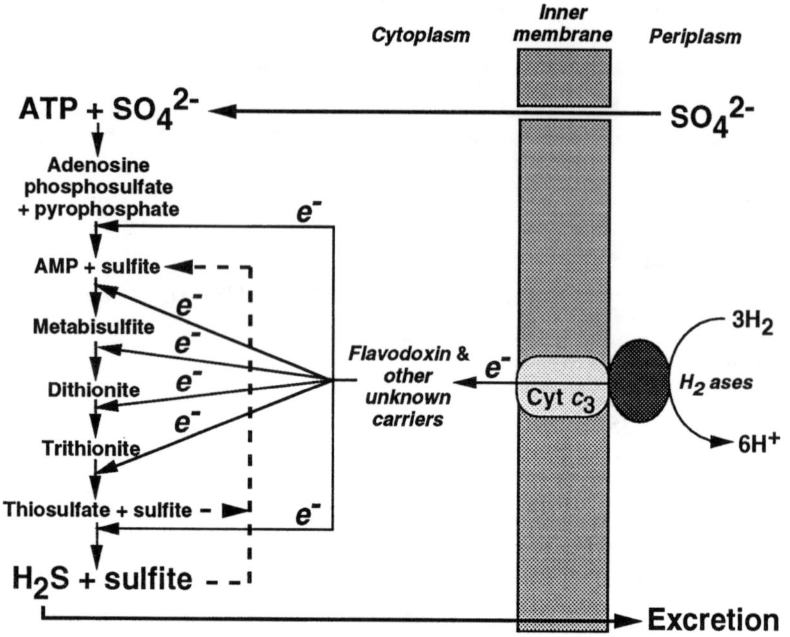

Figure 9.12. Generalized pathway of dissimilatory sulfate reduction by *Desulfovibrio* spp. utilizing H_2 as the electron donor. Adapted from Postgate (1984), Widdel and Hansen (1992), and Peck (1993). The hydrogenases are usually located in the periplasm, and they can, depending on species, be one of three types: 12Fe, NiFe, or NiFeSe. These hydrogenases carry out similar biochemical reactions, but variations in electron acceptor specificities allow the enzymes to feed electrons from H_2 into different electron transfer sequences. Cytochrome C_3 is required as a specific cofactor for transferring electrons to soluble carriers. These electron carriers are responsible for transport of electrons to enzymes involved in sulfate reduction. Some have not yet been fully identified, but include a flavodoxin (transports an electron to bisulfite reductase in *Desulfovibrio vulgaris*). Details of the transmembrane electron transfer have not been completely resolved, but it is thought that a proton motive force is set up across the bacterial membrane. The H_2 oxidation pathway generates 3 ATP from the reduction of sulfite to sulfide. However, as 2ATP are consumed during sulfate activation, the net gain of this reaction is 1 ATP.

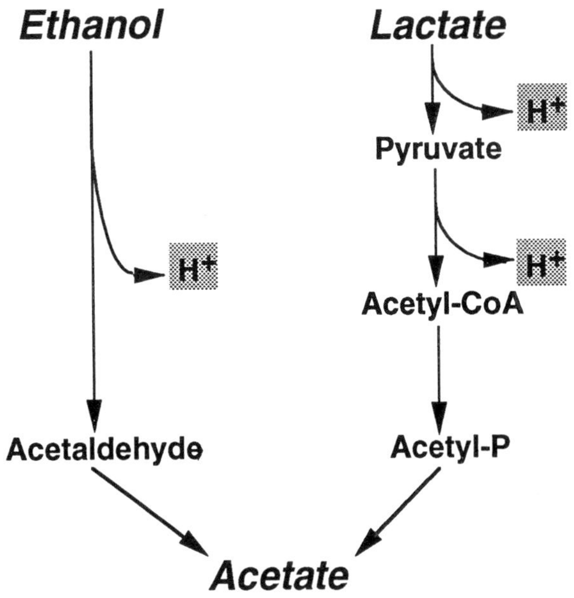

Figure 9.13. Fermentative growth of *Desulfovibrio* spp. on ethanol and lactate demonstrating how H^+ might be generated for subsequent interspecies H_2 transfer reactions.

Desulfovibrio. The hydrogenases in marine strains of these bacteria have received some interest because of their reversible nature. This enables the organisms, under certain conditions such as low electron acceptor availability, to carry out fermentative rather than oxidative metabolism. Substrates such as lactate, pyruvate, and ethanol serve as electron donors to yield H_2 using the mechanism shown in Figure 9.13 (Peck 1993). Subsequently, the H_2 produced may be metabolized by syntrophic bacteria such as methanogens. It is unclear whether this particular type of interspecies H_2 transfer occurs to a significant extent in the human gut, although recent data have shown that in approximately one-third of methanogenic persons tested, SRB exist in the presence of active methanogenesis and are not reducing sulfate (e.g., group 2 subjects described earlier). It is possible that these bacteria were existing by fermentative metabolism.

A further question that may be addressed when discussing human intestinal SRB concerns the fate of H_2S that they produce. This product is extremely toxic toward mammalian cells and is currently receiving interest as a possible cofactor in the initiation and/or maintenance of the inflammatory bowel disease, ulcerative colitis (Gibson et al. 1991, Roediger et al. 1993). Potential mechanisms for disposal of H_2S include:

1. Utilization as an S source in amino acid biosynthesis by other bacterial components of the gut microbiota.

2. Conversion into less toxic forms in the intestinal mucosa, such as volatile mercaptides or methanediol (Weiseger et al. 1980), e.g.:

$$H_2S + \text{S-adenosylmethionine} \rightarrow CH_3SH \text{ (methanethiol)} + \text{S-adenosylhomocysteine}$$

3. HS^- oxidation by colonocytes.

4. Conversion to thiocyanate in the colonic mucosa (Roediger et al. 1993).

At present, it is unclear whether either elevated SRB activities or a defect in the mechanism of sulfide detoxification may be involved in the pathogenesis of ulcerative colitis.

ACETOGENESIS

Acetate production from H_2 and CO_2 has been demonstrated to occur in the cecum of rat (Prins and Lankhorst 1977), termite gut (Breznak and Switzer 1986), and pig bowel (Greening and Leedle 1989) as well as feces from cattle, horses, and rabbits (Demeyer and DeGraeve 1991, Stevani et al. 1991). The studies of Lajoie et al. (1988) indicated that in some individuals, the activities of acetogenic bacteria may be of quantitative significance in terms of H_2 disposal, although the process is thermodynamically unfavorable at low pH_2—e.g., $<10^{-4}$ atm (Diekert 1992). Both dissimilatory sulfate reduction and methanogenesis should usually dominate H_2-dependent acetate production in anaerobic ecosystems. Indeed, Cord-Ruwisch et al. (1988) showed that desulfovibrios were particularly adept at scavenging H_2 at very low pH_2 values ($<10^{-4}$ atm).

Other studies with humans suggest that acetogenesis is relatively insignificant, in terms of H_2 metabolism, when either dissimilatory sulfate reduction or methanogenesis are occurring (Gibson et al. 1990). Thus, in the large intestine, acetogenesis should only become important when either sulfate reduction or methanogenesis are not active, or in specialized microniches in large-bowel contents that may contain active acetogens. Because both SRB and methanogens are relatively more acid-sensitive than acetogens (Fig. 9.14), reduction of CO_2 by H_2 to form acetate could potentially occur in the lower pH environment of the proximal colon. Macfarlane et al. (1992) showed, using gut contents from human sudden death victims, that both methanogenesis and sulfate reduction were quantitatively more important in the distal bowel than their activities in the cecum or ascending colon.

Virtually nothing is known about the types of acetogenic bacteria that are

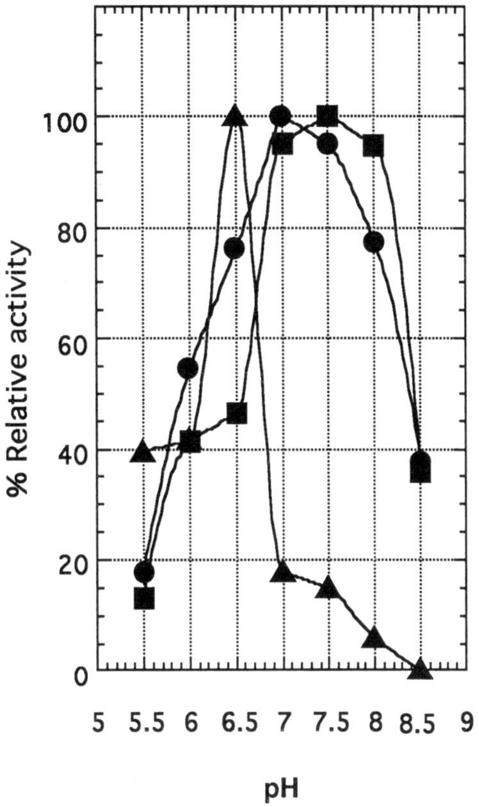

Figure 9.14. Effect of pH on methanogenesis, dissimilatory sulfate reduction, and acetogenesis by mixed populations of human intestinal bacteria (Gibson et al. 1990). ●, CH_4 production; ■, H_2S formation; ▲, acetogenesis.

present in the human large gut, or of their activities. Species belonging to the genera *Clostridium, Peptostreptococcus*, and *Eubacterium* are acetogenic; however, many of these bacteria are also able to ferment carbohydrates or amino acids, either of which is likely to be an energetically preferential strategy for growth in the colon.

Some clostridia can ferment hexoses to 3 moles of acetate, one of which is formed exclusively from CO_2 and requires ATP (Andreesen et al. 1973). The initial oxidation of glucose is carried out by a dehydrogenase, followed by CO_2-dependent acetogenesis involving a series of electron carriers and reductases (Thauer et al. 1977).

In the mechanism given by Diekert (1992), and shown in Figure 9.15, during homoacetogenesis, CO_2 is first reduced to formate, a reaction catalyzed by formate dehydrogenase:

$$CO_2 + 2H^+ \rightarrow HCOO^- + H^+$$

Formate is activated by binding to tetrahydrofolate (THF) at the expense of ATP:

$$\text{formate} + \text{THF} + \text{ATP} \rightarrow \text{ADP} + \text{Pi} + \text{formyl-THF}$$

A cyclohydrogenase then catalyzes the reaction:

$$\text{formyl-THF} \rightarrow OH^- + \text{methenyl-THF}^+$$

Similar to the methanogenesis reaction, methenyl-THF is further reduced:

$$\text{methenyl-THF}^+ + \text{NAD(P)H} \rightarrow \text{NAD(P)}^+ + \text{methylene-THF}$$

Methylene-THF reductase then catalyzes the reaction:

$$\text{methylene-THF} + 2H^+ \rightarrow \text{methyl-THF}$$

The methyl group is transferred to an enzyme bound corrinoid, E-[Co]:

$$\text{methyl-THF} + \text{E-[Co]} \rightarrow \text{THF} + CH_3\text{-[Co]-E}$$

The carboxyl group of acetate may then be formed from CO_2, by the activity of carbon monoxide dehydrogenase:

$$CO_2 + 2H^+ \rightarrow CO + H_2O$$

Synthesis of acetyl-CoA then occurs from the carbonyl group, corrinoid bound methyl group and CoA:

$$CO + CH_3\text{-X} + \text{CoA} \rightarrow \text{X} + \text{acetyl-CoA}$$

Finally, acetyl CoA acts as a precursor for acetate synthesis via acetyl phosphate, which is catalyzed by a phosphotransferase and acetate kinase:

$$\text{acetyl-CoA} + \text{Pi} \rightarrow \text{CoASH} + \text{acetyl-P} + \text{ADP} \rightarrow \text{acetate} + \text{ATP}$$

It has been estimated that over 25% of the total acetate formed during fermentation

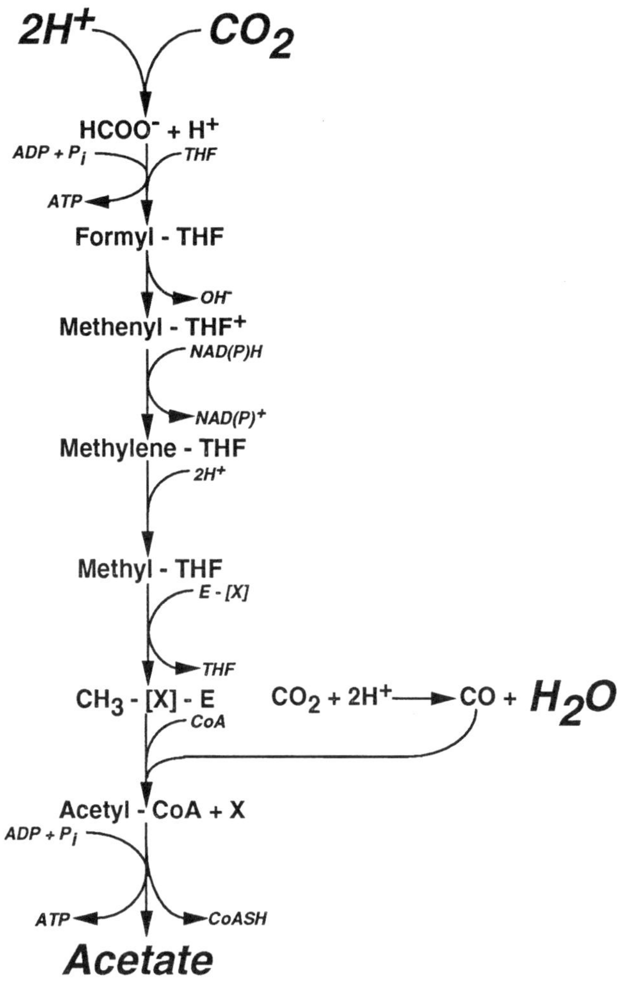

Figure 9.15. Pathway of acetate formation from hydrogen and carbon dioxide. Information from Fuchs (1986), Wood et al. (1986), and Diekert (1992). THF, tetrahydrofolate; E-[X], enzyme-linked corrinoid.

is due the activities of acetogenic bacteria (Duncan and Henderson 1990), although this does not explain the quantitative and qualitative lack of variability in the profile of fermentation end product formation seen in CH_4 excretors and non-producers. Future studies should further clarify the role of acetogenesis in the large intestine. In particular, the identity, physiology and ecology of these nutritionally versatile organisms will be of interest.

DISSIMILATORY NITRATE REDUCTION

From purely energetic considerations, it might be expected that H_2 disposal using NO_3^- as an electron acceptor would be preferred to either methanogenesis, dissimilatory sulfate reduction or acetogenesis, since the $\Delta G^{0\prime}$ of the reaction:

$$NO_3^- + H_2 \rightarrow NO_2^- + H_2O$$

is -163.2 kJ/mol (Thauer et al. 1977). However, the availability of NO_3^- as a terminal electron acceptor in the human large intestine is unclear. Florin et al. (1990) using ileostomists concluded that little NO_3^- entered the colon from dietary sources, but this work did not take into account endogenous NO_3^- synthesis and secretion by activated macrophages in the colonic mucosa or its transport to the the large bowel in blood (Beeken et al. 1987, Iyengar et al. 1987). Despite the uncertainty over the availability of this anion in the gut, studies have shown that NO_3^- or its metabolites almost completely inhibited both H_2 production and methanogenesis by mixed populations of human gut microbes (Allison and Macfarlane 1988). These studies also showed that H_2 was the preferred electron donor for dissimilatory NO_3^- reduction and that the process enabled the bacteria to produce more oxidized fermentation products during starch fermentation.

Nitrate reduction is widespread in many facultative and obligately anaerobic bacteria. The reaction in enterobacteria and anaerobes such as *Propionibacterium acnes* (Allison and Macfarlane 1989b) is catalyzed by membrane associated electron transport systems involving nitrate reductase (NaR) (Thauer et al. 1977), but in saccharolytic clostridia such as *C. perfringens* and *C. butyricum* (Keith et al. 1982) the enzyme is soluble. In *C. butyricum* grown in nitrogen-limited chemostats, it was shown that NaR served as an electron sink and that electrons were preferentially used to reduce NO_3^- to NO_2^- instead of reducing H^+ to molecular H_2. Nitrate reduction also enabled the bacterium to increase ATP production by SLP, thereby producing more acetate and less butyrate and ethanol (Keith et al. 1982).

OTHER POTENTIAL HYDROGEN CONSUMING PROCESSES

In addition to the H_2-utilizing reactions described above, a wide range of other microbiological sinks are potentially available for disposing of this gas in the large bowel. Examples include pyrimidine reduction in some clostridia (Hilton et al. 1975), its use as an electron donor in amino acid fermentations (Barker 1981), and its use in the reduction of fumarate in organisms such as *Wollinella succinogenes* (Bronder et al. 1982). Hydrogen is also used in the reduction of 2-enoates in amino acid catabolism in some proteolytic clostridia, which are formed as fermentation intermediates in dissimilatory reactions involving leucine, lysine, and phenylalanine (Bader and Simon 1983). However, virtually nothing is known about these processes, and it is therefore not possible to assess their physiologic and ecologic significance.

References

Allison C, Macfarlane GT (1988) Effect of nitrate on methane production by slurries of human faecal bacteria. J Gen Microbiol 134:1397–1405.

Allison C, Macfarlane GT (1989a) Influence of pH, nutrient availability, and growth rate on amine production by *Bacteroides fragilis* and *Clostridium perfringens*. Appl Environ Microbiol 55:2894–2898.

Allison C, Macfarlane GT (1989b) Dissimilatory nitrate reduction by *Propionibacterium acnes*. Appl Environ Microbiol 55:2899–2903.

Andreesen JR, Scaupp A, Neurater C, Brown A, Ljungdahl LG (1973) Fermentation of glucose fructose, and xylose by *Clostridium thermoaceticum*: effects of metals on growth yield, enzymes, and the synthesis of acetate from CO_2. J Bacteriol 114:743–751.

Andreesen JR, Bahl H, Gottschalk G (1989) Introduction to the physiology and biochemistry of the genus *Clostridium*. In: Minton NP, Clarke DJ, eds. Clostridia, Biotechnology Handbooks 3, pp. 27–62. New York: Plenum Press.

Archer DB, Harris JE (1986) Methanogenic bacteria and methane production in various habitats. In: Barnes EM, Mead GC, eds. Anaerobic Bacteria in Habitats Other Than Man, pp. 185–223. Oxford: Blackwell Scientific.

Bader J, Simon H (1983) ATP formation is coupled to the hydrogenation of 2-enoates in *Clostridium sporogenes*. FEMS Microbiol Lett 20:171–175.

Balch WE, Fox GE, Magrum LJ, Woese CL, Wolfe RS (1979) Methanogens: reevaluation of a unique biological group. Microbiol Rev 43:260–296.

Barker HA (1981) Amino acid degradation by anaerobic bacteria. Annu Rev Biochem 50:23–40.

Beeken W, Northwood I, Beliveau C, Gump D (1987) Phagocytes in cell suspensions of human colonic mucosa. Gut 28:976–980.

Bell GR, LeGall J, Peck HD (1974) Evidence for the periplasmic location of hydrogenase in *Desulfovibrio gigas*. J Bacteriol 120:994–997.

Bezkorovainy A (1989) Nutrition and metabolism of bifidobacteria. In: Bezkorovainy A, Miller-Catchpole R, eds. Biochemistry and Physiology of Bifidobacteria. Boca Raton, Fla.: CRC Press.

Biesterveld S, Zehnder AJB, Stams AJM (1994) Regulation of product formation in *Bacteroides xylanolyticus* X5–1 by interspecies electron transfer. Appl Environ Microbiol 60:1347–1352.

Blaut M, Muller V, Gottschalk G (1990) Energetics of methanogens. In: Krulwich TA, ed. Bacterial Energetics, pp. 505–537. London: Academic Press.

Bond JH, Engel RR, Levitt MD (1971) Factors influencing pulmonary methane excretion in man. An indirect method of studying the in situ metabolism of the methane-producing colonic bacteria. J Exp Med 133:572–588.

Booth IR (1988) Bacterial transport: energetics and mechanisms. In: Anthony C, ed. Bacterial Energy Transduction, pp. 377–429. London: Academic Press.

Booth IR, Mitchell WJ (1987) Sugar transport and metabolism in clostridia. In: Reizer J, Peterkofsky A, eds. Sugar Transport and Metabolism in Gram-Positive Bacteria, pp. 165–185. Chichester: EllisHorwood.

Breznak JA, Switzer JM (1986) Acetate synthesis from H_2 and CO_2 by termite gut microbes. Appl Environ Microbiol 52:623–630.

Bronder M, Mell H, Stupperich E, Kroger A (1982) Biosynthetic pathways of *Vibrio succinogenes* with fumarate as terminal electron acceptor and sole carbon source. Arch Microbiol 131:216–223.

Caspari D, Macy JM (1983) The role of carbon dioxide in glucose metabolism of *Bacteroides fragilis*. Arch Microbiol 135:16–24.

Chassey BM, Thompson J (1983) Regulation of lactase-phosphoenol-pyruvate dependent phosphotransferase system and β-D-phosphogalactoside and galactohydrolase activities in *Lactobacillus casei.* J Bacteriol 154:1195–1203.

Christl SU, Murgatroyd PR, Gibson GR, Cummings JH (1992) Quantitative measurement of hydrogen and methane from fermentation using a whole body calorimeter. Gastroenterology 102:1269–1277.

Coleman GS (1960) A sulfate reducing bacterium from the sheep rumen. J Gen Microbiol 22:423–436.

Conrad R, Phelps TJ, Zeikus JG (1985) Gas metabolism evidence in support of the juxtaposition of hydrogen-producing and methanogenic bacteria in sewage sludge and lakesediments. Appl Environ Microbiol 50:595–601.

Cord-Ruwisch R, Sietz HJ, Conrad R (1988) The capacity of hydrogenotrophic anaerobic bacteria to compete for traces of hydrogen depends on the redox potential of the terminal electron acceptor. Arch Microbiol 149:295–299.

Cummings JH (1995) Short chain fatty acids. In: Gibson GR, Macfarlane GT, eds. Human Colonic Bacteria: Role in Nutrition, Physiology and Health, pp. 101–130. Boca Raton: CRC Press.

Cummings JH, Gibson GR, Macfarlane GT (1989) Quantitative estimates of fermentation in the hindgut of man. In: Skadhauge E, Norgaard P, eds. Proceedings of the International Symposium on Comparative Aspects of the Physiology of Digestion in Ruminant and Hindgut Fermenters. Acta Vet Scand Suppl 86:76–82.

Cummings JH, Macfarlane GT (1991) The control and consequences of bacterial fermentation in the human colon. J Appl Bacteriol 70:443–459.

Cummings JH, Pomare EW, Branch WJ, Naylor CPE, Macfarlane GT (1987) Short chain fatty acids in human large intestine, portal, hepatic and venous blood. Gut 28: 1221–1227.

Cummins CS, Johnson JL (1992) The genus *Propionibacterium*. In: Balows A, Truper HG, Dworkin M, Harder W, Schliefer KH, eds. The Prokaryotes, 2nd Ed. A Handbook on the Biology of Bacteria: Ecophysiology, Isolation, Identification, Applications, pp. 834–849. New York: Springer-Verlag.

Cypionka H (1989) Characterization of sulfate transport in *Desulfovibrio desulfuricans*. Arch Microbiol 152:237–243.

Degnan BA (1992) Transport and Metabolism of Carbohydrates by Anaerobic Gut Bacteria. Ph.D. thesis, University of Cambridge.

Degnan BA, Macfarlane GT (1991) Comparison of carbohydrate substrate preferences in eight species of bifidobacteria. FEMS Microbiol Letts 84:151–156.

Degnan BA, Macfarlane GT (1993) Transport and metabolism of glucose and arabinose in *Bifidobacterium breve*. Arch Microbiol 160:144–151.

Degnan BA, Macfarlane GT (1994) Effect of dilution rate and carbon availability on *Bifidobacterium breve* fermentation. Appl Microbiol Biotechnol 40:800–805.

Demeyer DI, DeGraeve K (1991) Differences of stoichiometry between rumen and hindgut fermentation. Adv Animal Physiol Nutr 22:50–61.

DeVries W, Stouthamer, AH (1968) Fermentation of glucose, lactose, galactose, mannitoland xylose by bifidobacteria. J Bacteriol 96:472–478.

DeVries W, Gerbrandy SJ, Stouthamer, AH (1967) Carbohydrate metabolism in *Bifidobacterium bifidum*. Biochim Biophys Acta 136:415–425.

DeVries W, Kapteijn WMC, Van der Beek EG, Stouthamer AH (1970) Molar growth yields and fermentation balances of *Lactobacillus casei* L3 in batch cultures and in continuous cultures. J Gen Microbiol 63:333–345.

Diekert G (1992) The acetogenic bacteria. In: Balows A, Truper HG, Dworkin M, Harder W, Schleifer KH, eds. The Prokaryotes, 2nd Edn. A Handbook on the Biology of Bacteria: Ecophysiology, Isolation, Identification, Applications, pp. 517–533. New York: Springer-Verlag.

Dills SS, Apperson A, Schmidt MR, Saier MH (1980) Carbohydrate transport in bacteria. Microbiol Rev 44:385–418.

Dore J, Rieu-Lesme F, Fonty G, Gouet P (1992) Preliminary study of non-methanogenic hydrogenotrophic microflora in the rumen of new-born lambs. Ann Zootech 41:82.

Driessen AJM, Konings WN (1990) Energetic problems of bacterial fermentations: Extrusion of metabolic products. In: Krulwich TA, ed. Bacterial Energetics, pp. 449–478. London: Academic Press.

Duncan AJ, Henderson C (1990) A study of the fermentation of dietary fibre by human colonic bacteria grown in vitro in semi-continuous culture. Microbial Ecol Health Dis 3:87–93.

Engelhardt W, Busche R, Gros G, Rechkemmer G (1991) Absorption of short-chain fatty acids: Mechanisms and regional differences in the large intestine. In: Cummings JH, Rombeau, JL, Sakata T, eds. Short-chain Fatty Acids: Metabolism and Clinical Importance, pp. 60–62. Columbus: Ross Laboratories Press.

Etterlin C, McKeowen A, Bingham SA, Elia M, Macfarlane GT, Cummings JH (1992). D-lactate and acetate as markers of fermentation in man. Gastroenterology 102:A551.

Fauque G, Peck HD, Moura JJG. et al. (1988) The three classes of hydrogenases from sulfate-reducing bacteria of the genus *Desulfovibrio*. FEMS Microbiol Rev 54:299–344.

Fauque G, LeGall J, Barton LL (1991) Sulfate-reducing and sulfur-reducing bacteria. In: Shively JM, Barton LL, eds. Variations in Autotrophic Life, pp. 271–337. New York: Academic Press.

Ferro-Luzzi Ames, G (1990) Energetics of periplasmic transport mechanisms. In: Krulwisch TA, ed. Bacterial Energetics, pp. 225–246. San Diego: Academic Press.

Finegold SM, Sutter VL, Mathisen GE (1983) Normal indigenous intestinal flora. In: Hentges DJ, ed. Human Intestinal Microflora in Health and Disease, pp. 3–31. London: Academic Press.

Florin THJ, Neale G, Gibson GR, Christl SU, Cummings JH (1991) Metabolism of dietary sulphate: absorption and excretion in humans. Gut 32:766–773.

Florin THJ, Neale G, Cummings JH (1990) The effect of dietary nitrate on nitrate and nitrite excretion in man. Br J Nutr 64:387–397.

Fuchs G (1986) CO_2 fixation in acetogenic bacteria: variations on a theme. FEMS Microbiol Rev 39:181–213.

Gibson GR, Cummings JH, Macfarlane GT (1988a) Competition for hydrogen between sulphate-reducing bacteria and methanogenic bacteria from the human large intestine. J Appl Bacteriol 65:241–247.

Gibson GR, Cummings JH, Macfarlane GT (1988b) Use of a three-stage continuous culture system to study the effect of mucin on dissimilatory sulfate reduction and methanogenesis by mixed populations of human gut bacteria. Appl Environ Microbiol 54: 2750–2755.

Gibson GR, Macfarlane GT, Cummings JH (1988c) Occurrence of sulphate-reducing bacteria in human faeces and the relationship of dissimilatory sulphate reduction to methanogenesis in the large gut. J Appl Bacteriol 65:103–111.

Gibson GR, Cummings JH, Macfarlane GT, et al. (1990) Alternative pathways for hydrogen disposal during fermentation in the human colon. Gut 31:679–683.

Gibson GR, Cummings JH, Macfarlane GT (1991) Growth and activities of sulphate-reducing bacteria in gut contents of healthy subjects and patients with ulcerative colitis. FEMS Microbiol Ecol 86:103–112.

Gibson GR, Macfarlane S, Macfarlane GT (1993) Metabolic interactions involving sulphate-reducing bacteria and methanogenic bacteria in the human large intestine. FEMS Microbiol Ecol 12:117–125.

Gottschalk G, Peinemann S (1992) The anaerobic way of life. In: Balows A, Truper HG, Dworkin M, Harder W, Schleifer KH, eds. The Prokaryotes, 2nd Ed. A Handbook on the Biology of Bacteria: Ecophysiology, Isolation, Identification, Applications, pp. 301–311. New York: Springer-Verlag.

Greening RC, Leedle JAZ (1989) Enrichment and isolation of *Acetitomaculum ruminis,* gen. nov., sp. nov. acetogenic bacteria from the bovine rumen. Arch Microbiol 151: 399–405.

Hamilton WA (1988) Energy transduction in anaerobic bacteria. In: Anthony C, ed. Bacterial Energy Transduction, pp. 83–149. London: Academic Press.

Hidaka H, Eida T, Takizawa T, Tokunaga T, Tashiro Y (1986) Effects of fructooligosaccharides on intestinal flora and human health. Bifid Microflora 5:37–50.

Higgins CF, Hiles ID, Whalley K, Jamieson DK (1985) Nucleotide binding by membrane components of bacterial periplasmic binding protein-dependent transport systems. EMBO J 4:1033–1040.

Hilton MG, Mead GC, Elsden SR (1975) The metabolism of pyrimidines by proteolytic clostridia. Arch Microbiol 102:145–149.

Hino T, Kuroda S (1993) Presence of lactate dehydrogenase and lactate racemase in *Megasphaera elsdenii* grown on glucose or lactate. Appl Environ Microbiol 59: 255–259.

Hino T, Shimada K, Maruyama T (1994) Substrate preference in a strain of *Megasphaera elsdenii*, a ruminal bacterium, and its implications in propionate production and growth competition. Appl Environ Microbiol 60:1827–1831.

Huisingh J, McNeill JJ, Matrone G (1974) Sulfate reduction by a *Desulfovibrio* species isolated from sheep rumen. Appl Microbiol 28:489–497.

Hungate RE (1966) The Rumen and Its Microbes, pp. 8–90. New York: Academic Press.

Hylemon PB, Young JL, Roadcap RF, Phibbs PV (1977) Uptake and incorporation of glucose and mannose by whole cells of *Bacteroides thetaiotaomicron.* Appl Environ Microbiol 34:488–494.

Iyengar R, Stueke DJ, Mareletta MA (1987) Macrophage synthesis of nitrite, nitrate and N-nitrosamines: precursors and role of the respiratory burst. Proc Natl Acad Sci USA 84:6369–6373.

Jones CW (1988) Membrane associated energy conservation in bacteria: a general introduction. In: Anthony C, ed. Bacterial Energy Transduction, pp. 1–82. London: Academic Press.

Kandler O (1983) Carbohydrate metabolism in lactic acid bacteria. Anton van Leeuwen 49:209–224.

Keith SM, Macfarlane GT, Herbert RA (1982) Dissimilatory nitrate reduction by a strain of *Clostridium butyricum* isolated from estuarine sediments. Arch Microbiol 132:62–66.

Kirk E (1949) The quantity and composition of human colonic flatus. Gastroenterology 12:782–794.

Kundig W (1974) Molecular interactions in the bacterial phosphoenolpyruvate-phosphotransferase system (PTS). J Supramol Struc 2:695–714.

Lajoie SF, Bank S, Miller TL, Wolin MJ (1988) Acetate production from hydrogen and [^{13}C]carbon dioxide by the microflora of human feces. Appl Environ Microbiol 54: 2723–2727.

Levitt MD (1969) Production and excretion of hydrogen gas in man. N Engl J Med 281: 122–127.

Levitt MD (1971) Volume and composition of human intestinal gas determined by means of an intestinal washout technique. N Engl J Med 284:1394–1398.

Levitt MD, Gibson GR, Christl SU (1995) Gas metabolism in the large intestine. In:

Gibson GR, Macfarlane GT, eds. Human Colonic Bacteria: Role in Nutrition, Physiology and Health, pp. 131–154. Boca Raton: CRC Press.

Macfarlane GT, Cummings JH (1991) The colonic flora, fermentation, and large bowel digestive function. In: Philips SF, Pemberton JH, Shorter RG, eds. The large intestine: physiology, pathophysiology and disease. New York: Raven Press, pp. 51–92.

Macfarlane GT, Englyst HE (1986) Starch utilization by the human large intestinal microflora. J Appl Bacteriol 60:195–201.

Macfarlane GT, Gibson GR (1991) Co-utilization of polymerized carbon sources by *Bacteroides ovatus* grown in a two-stage continuous culture system. Appl Environ Microbiol 57:1–6.

Macfarlane GT, Gibson GR (1994) Metabolic activities of the normal colonic flora. In: Gibson SAW, ed. Human health: the contribution of microorganisms. London: Springer Verlag, pp. 17–52.

Macfarlane GT, Gibson GR, Cummings JH (1992) Comparison of fermentation reactions in different regions of the human colon. J Appl Bacteriol 72:57–64.

Macfarlane GT, Gibson GR, Macfarlane S (1994) Short chain fatty acid and lactate production by human intestinal bacteria grown in batch and continuous culture. In: Binder HJ, Cummings JH, Soergel KH, eds. Short Chain Fatty Acids. Lancaster: Kluwer Academic Publishers, pp. 44–60.

Macfarlane GT, Hay S, Macfarlane S, Gibson GR (1990) Effect of different carbohydrates on growth, polysaccharidase and glycosidase produduction by *Bacteroides ovatus*, in batch and continuous culture. J Appl Bacteriol 68:179–187.

Macy JM, Ljungdahl LG, Gottschalk G (1978) Pathway of succinate and propionate formation in *Bacteroides fragilis*. J Bacteriol 134:84–91.

Macy JM, Probst I, Gottschalk G (1975) Evidence for cytochrome involvement in fumarate reduction and adenosine 5′-triphosphate synthesis by *Bacteroides fragilis* grown in the presence of hemin. J Bacteriol 123:436–442.

Magasanik B (1970) Glucose effects: inducer exclusion and repression. In: The Lactose Operon, pp. 189–319. Cold Spring Harbor, NY: Cold Spring Harbor Laboratories.

Martin SA, Russell JB (1986) Phosphoenolpyruvate-dependent phosphorylation of hexoses by ruminal bacteria: evidence for the phosphotransferase transport system. Appl Environ Microbiol 52:1348–1352.

Martin SA, Russell JB (1988) Mechanisms of sugar transport in the rumen bacterium *Selenomonas ruminantium.* J Gen Microbiol 134:819–828.

McBride BC, Wolfe RS (1971) Biochemistry of methane formation. In: Gould RF, ed. Advances in Chemistry Series 105, pp. 11–22. Washington: American Chemical Society.

McGinnis JF, Paigen K (1969) Catabolite inhibition: a general phenomenon in the control of carbohydrate utilization. J Bacteriol 100:902–913.

McGinnis JF, Paigen K (1973) Site of catabolite inhibition of carbohydrate metabolism. J Bacteriol 114:885–887.

McKay FL, Eastwood MA, Brydon WG (1985) Methane excretion in man: a study of breath, flatus and feces. Gut 26:69–74.

McNeil NI (1984) The contribution of the large intestine to energy supplies in man. Am J Clin Nutr 39:338–342.

Michels PAM, Michels JPJ, Boonstra J, Konings WN (1979) Generation of an electrochemical gradient in bacteria by excretion of metabolic end products. FEMS Microbiol Lett 5:357–364.

Miller TL, Wolin MJ (1982) Enumeration of *Methanobrevibacter smithii* in human feces. Arch Microbiol 131:14–18.

Mitchell P (1963) Molecule, group, and electron translocation through natural membranes. Biochem Soc Symp 22:142–168.

Mitchell WJ, Sadegh Roolin M, Mosely MJ, Booth IR (1987) Regulation of carbohydrate utilization in *Clostridium pasteurianum.* J Gen Microbiol 133:31–36.

Mitsuoka T (1982) Recent trends in research on intestinal flora. Bifid Microflora 1:3–24.

Modler HW, McKellar RC, Yaguchi M (1990) Bifidobacteria and bifidogenic factors. Can Inst Food Sci Technol J 23:29–41.

Morris JG (1986) Anaerobiosis and energy-yielding metabolism. In: Barnes EM, Mead GC, eds. Anaerobic Bacteria in Habitats Other Than Man, pp. 1–21. Oxford: Blackwell Scientific Publications.

Neu HC, Heppel LA (1965) The release of enzymes from *Escherichia coli* during theformation of spheroplasts. J Biol Chem 240:3685–3692.

Ng SKC, Hamilton IR (1973) Carbon dioxide fixation by *Veillonella parvula* M4 and its relation to propionic acid formation. Can J Microbiol 19:715–723.

Parkes RJ, Gibson GR, Mueller-Harvey I, Buckingham WJ, Herbert RA (1989) Determination of the substrates for sulphate-reducing bacteria within marine and estuarine sediments with different rates of sulphate reduction. J Gen Microbiol 135:175–187.

Patni NJ, Alexander JK (1971) Catabolism of fructose and mannitol in *Clostridium thermocellum:* presence of phophoenolpyruvate : fructose phosphotransferase, fructose-1-phosphate kinase, phosphoenolpyruvate : mannitol phosphotransferase, and mannitol-1-phosphate dehydrogenase in cell extracts. J Bacteriol 105:226–231.

Peck HD (1993) Bioenergetic strategies of the sulfate-reducing bacteria. In: Odom JM, Singleton R, eds. The Sulfate-Reducing Bacteria: Contemporary Perspectives, pp. 41–76. New York: Springer Verlag.

Peck HD, LeGall J (1982) Biochemistry of dissimilatory sulphate reduction. Phil Trans R Soc Lond B298:443–466.

Peled Y, Gilat T, Libermann T, Bujanover Y (1985) The development of methane production in childhood and adolescence. J Pedriatr Gastroenterol Nutr 4:575–579.

Pettifer GL, Latham MJ (1979) Production of enzymes degrading plant cell walls and fermentation of cellobiose by *Ruminococcus flavefaciens* in batch and continuous culture. J Gen Microbiol 110:29–38.

Postgate JR (1984) The Sulphate-Reducing Bacteria, 2nd Ed. Cambridge: Cambridge University Press.

Postma P, Roseman S (1976) The bacterial phosphoenolpyruvate:sugar phosphotransferase system. Biochim Biophys Acta 457:213–257.

Postma P, Lengeler LW (1985) Phosphoenolpyruvate: carbohydrate phosphotransferase system of bacteria. Microbiol Rev 49:232–269.

Prins RA, Lankhorst A (1977) Synthesis of acetate from CO_2 in the cecum of some rodents. FEMS Microbiol Lett 1:255–258.

Rasic JL (1983) The role of dairy foods containing bifido- and acidophilus bacteria in nutrition and health? North Eur Dairy J 48:80.

Roediger WEW, Duncan A, Kapaniris O, Millard S (1993) Reducing sulfur compounds of the colon impair colonocyte nutrition: implications for ulcerative colitis. Gastroenterology 104:802–809.

Rogers P (1986) Genetics and biochemistry of *Clostridium* relevant to development of fermentation processes. Adv Appl Microbiol 3:1–60.

Rouviere PE, Wolfe RS (1988) Novel biochemistry of methanogenesis. J Biol Chem 263: 7913–7916.

Russell JB, Strobel HJ, Martin SA (1990). Strategies of nutrient transport by ruminal bacteria. J Dairy Sci 73:2996–3012.

Russell JB, Baldwin RL (1978) Substrate preferences in rumen bacteria: evidence of catabolite regulatory mechanisms. Appl Environ Microbiol 36:319–329.

Saier MH, Chin M (1990) Energetics of the bacterial phosphotransferase system in sugar transport and the regulation of carbon metabolism. In: Krulwich TA, ed. Bacterial Energetics, pp. 273–299. San Diego: Academic Press.

Saier MH, Fagan MJ, Hoischen C, Reizer J (1993) Transport mechanisms. In: Sonenshein A, Hoch JA, Losick R, eds. *Bacillus subtilus* and Other Gram Positive Bacteria, pp. 133–156. Washington: American Society for Microbiology.

Saier MH, Roseman S (1976) Sugar transport. Inducer exclusion and regulation of the melibiose, maltose, glycerol, and lactose transport systems by the phosphoenolpyruvate: sugar phosphotransferase system. J Biol Chem 251:6606–6615.

Salyers AA (1984) *Bacteroides* of the human lower intestinal tract. Annu Rev Microbiol 38:293–313.

Salyers AA, Arthur R, Kuritza A (1981) Digestion of larch arabinogalactan by a strain of human colonic *Bacteroides* growing in continuous culture. J Agric Food Chem 29: 475–480.

Schlegel HG (1986) General Microbiology, 6th Ed. Cambridge: Cambridge University Press.

Segal I, Walker ARP, Lord S, Cummings JH (1988) Breath methane and large bowel cancer risk in contrasting African populations. Gut 29:608–613.

Shi Y, Weimer PJ (1992) Response surface analysis of the effects of pH and dilution rate on *Ruminococcus flavefaciens* FD-1 in cellulose-fed continuous culture. Appl Environ Microbiol 58: 2583–2591.

Silhavy TJ, Ferenci T, Boos W (1978) Sugar transport systems in *Escherichia coli.* In: Rosen BP, ed. Bacterial Transport, pp. 127–169. New York: Marcel Dekker.

Sorensen J, Christensen D, Jorgensen BB (1981) Volatile fatty acids and hydrogen as substrates for sulfate-reducing bacteria in anaerobic marine sediments. Appl Environ Microbiol 42:5–11.

Stevani J, Durand M, DeGraeve K, Demeyer D, Grivet JP (1991) Degradative abilities and metabolism of rumen and hindgut microbial ecosystems. In: Sakata T, Snipes RL, eds. Hindgut '91, pp. 123–135. Tokyo: Senshu University Press.

Stille W, Truper HG (1984) Adenylylsulfate reductase in some new sulfate-reducing bacteria. Arch Microbiol 41:1230–1237.

Strobel HJ (1992) Vitamin B_{12}-dependent propionate production by the ruminal bacterium *Prevotella ruminicola* 23. Appl Environ Microbiol 58:2331–2333.

Strobel HJ (1993) Evidence for catabolite inhibition in regulation of pentose utilization and transport in the ruminal bacterium *Selenomonas ruminantium.* Appl Environ Microbiol 59:40–46.

Strocchi A, Levitt MD (1992) Factors affecting hydrogen production and consumption by human fecal flora: The critical role of hydrogen tension and methanogenesis. J Clin Invest 89:1304–1311.

Taylor CD, Wolfe RS (1974) Structure and methylation of Coenzyme M, ($HSCH_2CH_2SO_3$). J Biol Chem 240:4879–4885.

Thauer R, Jungermann K, Decker K (1977) Energy conservation in anaerobic bacteria. Bacteriol Rev 41:100–180.

Thurston B, Dawson KA, Strobel HJ (1993) Cellobiose versus glucose utilization by the ruminal bacterium *Ruminococcus albus.* Appl Environ Microbiol 59:261–2637.

Thurston B, Dawson KA, Strobel HJ (1994) Pentose utilization by the ruminal bacterium *Ruminococcus albus.* Appl Environ Microbiol 60:1087–1092.

Tsai HH, Sunderland D, Gibson GR, Hart CA, Rhodes JM (1992) A novel mucin sulphatase from human faeces: its identification, purification and characterization. Clin Sci 82: 447–454.

Turton LJ, Drucker DB, Ganguli LA (1983) Effect of glucose concentration in the growth medium upon neutral and acidic fermentation end-products of *Clostridium bifermentans, Clostridium sporogenes* and *Peptostreptococcus anaerobius.* Med Microbiol 16:61–67.

Veryrat A, Monedero V, Perez-Martinez G (1994) Glucose transport by the phosphoenol-pyruvate :mannose phosphotransferase system in *Lactobacillus casei* ATCC 393 and its role in carbon catabolite repression. J Gen Microbiol 140:1141–1149.

Wallnofer P, Baldwin RL (1967) Pathway of propionate formation in *Bacteroides ruminicola*. J Bacteriol 93:504–505.

Weiseger RA, Pinkus LM, Jakoby WB (1980) Thiol-S-methyltransferase: suggested role in detoxification of intestinal hydrogen sulphide. Biochem Pharmacol 29:2885–2887.

West IC (1970) Lactose transport coupled to proton movements in *Escherichia coli.* Biochem Biophys Res Commun 41:655–661.

West IC, Mitchell P (1973) Stoichiometry of lactose-proton symport across the plasma-membrane of *Escherichia coli.* Biochem J 132:587–592.

Whitman WB, Bowen TL, Boone DR (1992) The methanogenic bacteria. In: Balows A, Truper HG, Dworkin M, Harder W, Schleifer KH, eds. The Prokaryotes, 2nd Ed. A Handbook on the Biology of Bacteria: Ecophysiology, Isolation, Identification, Applications, pp. 719–767. New York: Springer-Verlag.

Widdel F, Hansen TA (1992) The dissimilatory sulfate- and sulfur-reducing bacteria. In: Balows A, Truper HG, Dworkin M, Harder W, Schleifer KH, eds. The Prokaryotes, 2nd Ed. A Handbook on the Biology of Bacteria: Ecophysiology, Isolation, Identification, Applications, pp. 582–624. New York: Springer-Verlag.

Williams AG, Withers SE (1985) Formation of polysaccharide depolymerase and glycoside hydrolase enzymes by *Bacteroides ruminicola* subsp. *ruminicola* grown in batch and continuous culture. Curr Microbiol 12:79–84.

Wilson DM, Wilson TH (1987) Cation specificity for sugar substrates of melibiose carriers in *Escherichia coli.* Biochim Biophys Acta 904:191–200.

Wisker E, Feldheim W (1994) Energy value of fermentation. In: Binder HJ, Cummings JH, Soergel KH, eds. Short Chain Fatty Acids, pp. 20–28. Lancaster: Kluwer Academic Publishers.

Wood HG, Ragsdale SW, Pezacka E (1986) The acetyl-CoA pathway: a newly discovered pathway of autotrophic growth. Trends Biochem Sci 11:14–18.

Woods WA (1961) Fermentation of carbohydrates and related compounds. In: Gunsalus IC, Stanier RY, eds. The Bacteria: A Treatise on Structure and Function, pp. 59–100. New York: Academic Press.

10

Polysaccharide Degradation in the Rumen and Large Intestine

Cecil W. Forsberg, K.-J. Cheng, and Bryan A. White

1. Introduction

The rate and extent of degradation of plant structural tissues during digestion depends on a combination of animal, plant, and microbial factors. The major animal effect is the nature of the digestive tract itself. Ruminants with pregastric ruminal fermentation generally have a lower rate of passage than is seen in monogastric animals, and rumination (regurgitation and remastication) of large particles permits them to macerate fibrous materials more completely than do monogastric animals. Ruminants are thus more efficient at digesting fibrous plant material than are monogastric animals, in which fermentation occurs only in the large intestine. In contrast to ruminants, monogastric animals often tend to consume more readily digestible feeds. For example, cereals are a major component of the diets commonly fed to domestic animals. Thus, there is often a quantitative compositional difference in the substrates available to the digestive microbiota present in ruminants and monogastric animals. Cellulolytic organisms which are of pivotal importance in the rumen are generally viewed to have a lesser role than other fermentative organisms in the postgastric intestinal fermentation of domestic animals; however, there are exceptions—for example, the horse. Undoubtedly the presence or absence of particular gut species is directly related to the diet consumed and the retention time of the system.

Fermentation in the lower gut is often considered unimportant for the animal's well-being; however, short-chain fatty acids arising from this fermentation are utilized as energy and carbon sources by the epithelial cells lining the gut as well as being absorbed into the bloodstream and metabolized further in the liver (Macfarlane and Macfarlane 1993).

On entry to the ruminant digestive system, plant materials are masticated, which opens the tissues to microbial degradation. The hydrolytic process is accomplished in the rumen by a limited number of bacterial, fungal, and protozoal species. Similarly, in the large intestine only a few species are responsible for hydrolysis of polysaccharides and oligosaccharides that bypass the small intestine. This chapter focuses on the unique hydrolytic capabilities of these organisms from a physiological and genetic perspective. The challenging issue that faces us is the extent to which physiological or genetic characteristics of these microbes can be exploited to enhance or perhaps modify fiber degradation toward improvements in animal production. Potential areas for application include acceleration of digestive processes which are limiting steps in milk production in high-producing dairy cattle and in weight gain of beef animals, or establishment and maintenance of a selective fermentation, such as a *Bifidobacterium* predominance in the large intestine, to promote the health of the animal.

There are a number of excellent recent reviews dealing with aspects of plant cell wall biodegradation in the rumen (Cheng et al. 1991b, Forsberg et al. 1993, Forsberg and Cheng 1992, Malburg et al. 1992, Thomson 1993, White et al. 1993b, Wubah et al. 1993), and others which deal specifically with postgastric intestinal fermentations (Cummings and Macfarlane 1991, Macfarlane and Macfarlane 1993). This chapter provides an update with recent molecular information integrated with the established knowledge of the ruminal and intestinal fermentations.

2. The Plant Cell Wall and Its Polymers

A discussion of the mechanisms of fiber degradation should be prefaced by a review of the structure and composition of the plant cell wall. The structure of grasses and legumes has been reviewed by Wilson (1993). The typical organization of the stem of a grass from the outside in is: waxy cuticular layer, epidermis, sclerenchyma, mesophyll, vascular lignified zone and pith parenchyma in the center (Fig. 10.1A). The cuticle is resistant to microbial degradation, but mastication disrupts the barrier making internal tissues broadly accessible to microbial colonization. The sclerenchymal tissue is colonized primarily by ruminal fungi, while mesophyll and starch-granule enriched parenchymal tissue are readily colonized by the major fibrolytic bacteria, fungi, and protozoa. Accessibility is an important factor in determining digestibility, and tissues with thicker cell walls, including cells of vascular tissues and sclerenchyma bundles, are poorly digested. Another perspective is cell wall surface area versus wall volume, where the same relationship can be observed; in this respect, mesophyll and parenchymal cell walls are thinner than walls of vascular tissue and sclerenchyma cells and are more readily digested.

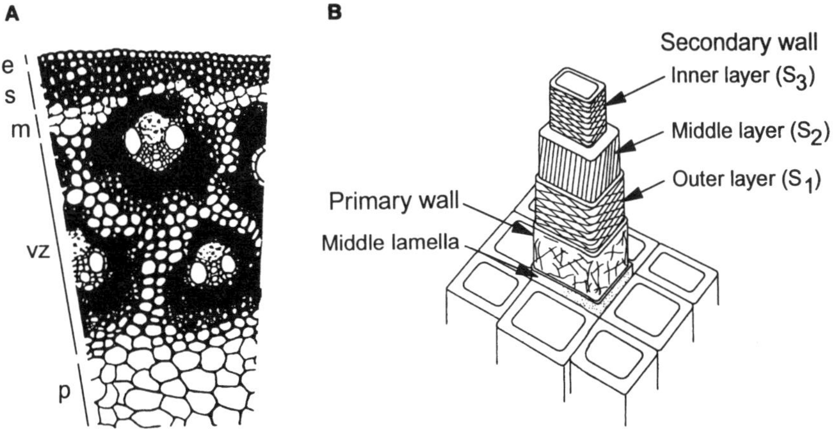

Figure 10.1. (A) Cross section of a mature stem of the tropical grass sorghum. Solid black areas indicate high lignification. e, epidermis; s, sclerenchyma; m, mesophyll; vz, lignified vascular zone; p, pith parenchyma. From Wilson (1993), with permission. (B)Schematic structure of a coniferous gymnosperm tracheid illustrating the orientation of cellulose microfibrils in the primary and secondary (S_1, S_2, S_3) wall layers. The orientation of cellulose microfibrils in the walls of sclerenchyma fibers from dicotyledon woods is similar. Redrawn from Dinwoodie (1975) after Wardrop and Bland (1959).

Plant cells are composed of an outer primary cell wall comprising a base structure of cellulose microfibrils which is laid down while cells are dividing and expanding (Fig. 10.1B). In some cell types—e.g., the chlorenchyma and parenchyma in the midrib area of grasses—the primary cell wall is the only wall formed. The primary cell walls of contiguous cells are separated by a middle lamella. In cells containing only the primary cell wall, neither it nor the separating middle lamella become lignified (Wilson 1993). The secondary cell wall is laid down inside the primary cell wall after cell expansion is complete. It is comprised of up to three layers of cellulose microfibrils organized in parallel, each layer having a different orientation (Fig. 10.1) (Dinwoodie 1975). Rye grass sclerenchyma fibers, for example, have secondary walls with two layers (Harris 1990).

The plant cell wall is a complex of cellulosic microfibrils embedded in a matrix of interwoven hemicellulosic polymers and insoluble proteins. The major cell wall components of mesophyll from a typical monocot, rye grass, and parenchyma from a typical dicot, kale, are shown in Table 10.1. Gross composition is generally quite similar for monocotyledonous and dicotyledonous cell walls from different tissues and different species (Chesson 1993). However, some characteristic differences have been described. The primary walls of monocot (Gramineae) grass species have a substantially lower proportion of pectic substances as compared

Table 10.1. Calculated polymer composition of primary cell walls extracted from monocotyledonous (perennial rye grass) and dicotyledonous (Kale stem parenchyma cells) plants

Component	Ryegrass Mesophyll	Kale Parenchyma
Cellulose	44.3	35.4
Mixed-link glucan	3.0	—
Xyloglucan	11.6	7
Arabinoxylan	14	2
Arabinogalactans	4.6	10.2
Rhamnogalacturonans	4.6	21.4
Lignin	1.9	3.8
Acetyl groups	9.0	—

Sources: Data for rye grass taken from Chesson et al. (1985); that for Kale from Wilson et al. (1988, 1989).

to dicot species. In dicots, pectic materials are located in the middle lamella (Moore and Staehelin 1988) and have a major role in cementing the cells together (Ishii 1984). In the grasses (monocots), a combination of pectic materials and glucuronoarabinoxylan cement cell walls together in the leaves (Ishii 1984). Monocots also have xyloglucan as a minor component. A distinct mixed-linked glucan is present, and there is a high content of extensively acetylated arabinoxylan as compared to dicotyledonous plants. There are also more phenolic residues in monocots than in dicots (Monties 1989, Chesson 1993).

Cellulose is the major component of the plant cell wall. It is composed of β-1,4-linked glucose residues in chains aligned and hydrogen bonded to form microfibrils. The microfibrils are hydrogen bonded to arabinoxylans, xyloglucans, and glucomannans, which are celluloselike in their conformations (Carpita and Gibeaut 1993, Talbott and Ray 1992) and presumably interact with more than one cellulose microfibril to form noncovalent crosslinks. At completion of the expansion phase of cell growth, covalent crosslinks form in nonlignified primary and secondary cell walls. These include glycosidic crosslinks between oligosaccharides, direct ester links between uronic acid residues of one polysaccharide and hydroxyls on a neighbouring polysaccharide, and ester linkages of hydroxycinnamic acids to arabinoxylans (Iiyama et al. 1994, Lam et al. 1990). Once the secondary cell wall begins to form, polymerization of lignin precursors occurs in primary walls (Iiyama et al. 1994). Lignin is a highly branched, random polymer generated by the free-radical condensation of aromatic alcohols (Gold and Alic 1993). Lignin deposition is initiated in cell corners, continues in the middle lamella, and proceeds through the primary cell wall into the secondary cell wall. As it progresses, it encapsulates polysaccharides and proteins. Because of the close association of polysaccharides and lignin, crosslinks occur including (1) ester linkages between uronic acids of glucuronoxylan or rhamnogalacturonans and hydroxyl groups on the lignin surface to give α or γ esters on monolignol side chains (Lam et al. 1990); (2) direct ether linkages, possibly involving glucose or mannose residues, between polysaccharides and lignin (Lam et al. 1990); and (3) glycosidic links from the carbohydrate to terminal phenolic acids on lignin (Bacic et al. 1988, Lam et al. 1990).

Proteins are present in plant cell walls as enzymes and as structural components. The structural proteins include the extensins (glycine-rich proteins), proline-rich proteins, and hydroxyproline-rich proteins. The hydroxyproline-rich proteins are crosslinked by isodityrosine bridges (Fry 1988). Many of the other structural proteins contain tyrosine-rich repeated sequences that could also involve isodityrosine crosslinks (Showalter 1993). Additionally, tyrosine and cysteine residues in wall proteins may form crosslinks with hydroxycinnamic acids esterified to walls (Bacic et al. 1988). There is evidence that both hydroxyproline-rich proteins and glycine-rich proteins are associated with lignin and may act as a foci for lignin polymerization (Iiyama et al. 1994).

Obviously the cell wall is a formidable structure which restricts rapid biodegradation of plant material. The low porosity mesh of noncovalently and covalently crosslinked polymers in the wall determines that a successful assault on this barrier can be made only by those organisms possessing a battery of hydrolytic enzymes sufficient for simultaneous cleavage of the polymers involved.

3. Polysaccharides and Monosaccharides Metabolized by Ruminal Microorganisms

The structural and soluble carbohydrates fermented by some major ruminal bacteria and fungi are presented in Tables 10.2, 10.3 and 10.4. This information provides a guide to the hydrolytic enzymes and substrate uptake systems expected to be found in the ruminal microorganisms. The predominant cellulolytic bacteria are *Fibrobacter succinogenes*, *Ruminococcus albus*, *R. flavefaciens*, and *Butyrivibrio fibrisolvens*, although only a few strains of *B. fibrisolvens* are able to extensively degrade cellulose in plant cell walls (Dehority 1993). The major hemicellulose utilizers include *Ruminococcus* sp., *Prevotella ruminicola*, and *B. fibrisolvens*. Pectin is degraded primarily by *Lachnospira multiparus*, *B. fibrisolvens*, *Succinovibrio dextrinosolvens*, and *Treponema bryantii* (Dehority 1993). Major starch degraders include *Ruminobacter amylophilus*, *Streptococcus bovis*, and *B. fibrisolvens* (Cotta 1988, Kotarski et al. 1992).

The gut fungi generally have a broader fermentative capability than do the bacteria (Table 10.4). The capacity to degrade cellulose and xylan is nearly ubiquitous among them, and all will grow on a broad range of simple disaccharide and monosaccharide carbohydrates (Wubah et al. 1993). Because of their broad array

Table 10.2. Structural and soluble carbohydrates fermented by gram-negative ruminal bacteria

Substrate	*Fibrobacter succinogenes*	*Prevotella ruminicola*	*Ruminobacter amylophilus*	*Succinivibrio dextrinosolvens*	*Succinimonas amylolytica*	*Treponema bryantii*	*Selenomonas ruminantium*
Cellulose	+	—	—	—	—	—	—
Xylan	—	d		—	—		—
Pectin	d	d		+		+	—
Starch	d	d	+	—	+	—	d
Cellobiose	+	+	—	d	—	+	+
Maltose	d	d	+	+	+	—	+
Sucrose	—	d	—	d	—	+	d
Lactose	d	+	—	+	+	+	+
Glucose	+	+	—	+	+	+	+
Fructose	—	+	—	d	—	—	+
Galactose	—	+	—	+		+	+
Mannose	—	+	—	d		+	+
Xylose	—	d	—	+	—	+	+
Arabinose	—	d	—	d	—	+	+
Mannitol	—	—	—	d	—	—	+

Source: Extracted from Stewart and Bryant (1988).

Abbreviations: +, positive reaction; —, negative reaction; d, reaction varies among strains; blank, not tested.

All sugars are the D isomer except for arabinose, which is the L isomer.

Table 10.3. Structural and soluble carbohydrates fermented by gram-positive ruminal bacteria

Substrate	*Ruminococcus albus*	*Ruminococcus flavefaciens*	*Lachnospira multiparus*	*Butyrivibrio fibrisolvens*	*Streptococcus bovis*	*Eubacterium ruminantium*	*Megasphera elsdenii*
Cellulose	+	+	—	d	—	—	—
Xylan	+	+	—	+		d	
Pectin			+	+	d	d	—
Starch	—	—	d	d	+	—	—
Cellobiose	+	+	+	d	+	+	+
Maltose	—	—	d	d	+	d	+
Sucrose	d	d	+	d	+	d	d
Lactose	d	d	d	d	+	d	—
Glucose	d	d	+	+	+	+	+
Fructose	d		+	+	+	+	+
Galactose	—	—	d	+	+	—	—
Mannose	d	d	d	+	+	—	—
D-Xylose	d	d	d	d	d	d	—
L-Arabinose	d	d	—	+	d	d	—
Mannitol	d	—	—	—	d	—	+

Source: Extracted from Stewart and Bryant (1988).

Abbreviations: +, positive reaction; —, negative reaction; d, reaction varies among strains; blank, not tested.

All sugars are the D isomer except for arabinose, which is the L isomer.

Table 10.4. Structural and soluble carbohydrates fermented by gut fungi

	Isolate[a]						
Substrate	Nh	Nt	Np	Ns	Cs	Ps	Nj
Cellulose	+ +[b]	+ +	+ +	+ +	—	+ +	+ +
Xylan	+ +	nd	+ +	+ +	+ +	+ +	+ +
Pectin	—	—	—	—	—	—	+
Starch	nd	nd	+ +	+ +	—	—	+ +
Pullulan	+ +	nd	nd	+ +	—	—	nd
Inulin	+ +	nd	nd	+ +	—	—	nd
Cellobiose	+ +	+ +	+ +	+ +	+ +	+ +	+ +
Maltose	+ +	+ +	+ +	+ +	—	—	+ +
Sucrose	+ +	+ +	+ +	+ +	—	—	nd
Raffinose	+ +	+	nd	+ +	—	—	+ +
Lactose	+	+ +	nd	+ +	+ +	+ +	+ +
D-Glucose	+ +	+ +	+ +	+ +	+ +	+ +	+ +
L-Fructose	+ +	+ +	—	+ +	+ +	+ +	+ +
D-Galactose	—	—	+ +	—	—	—	nd
D-Mannose	—	—	—	—	—	—	—
D-Xylose	+ +	+ +	+ +	+ +	+ +	+	nd
L-Arabinose	—	—	nd	—	—	—	—
L-Rhamnose	nd	—	nd	—	—	—	nd
D-Ribose	—	—	nd	—	—	—	nd

Sources: Breton et al. 1989, Gordon and Phillips, 1989, Mountfort and Asher, 1985, Phillips and Gordon, 1988, Theodorou et al. 1988.

[a] Nh, *Neocallimastix hurleyensis;* Nf, *Neocallimastix frontalis;* Np, *Neocallimastix patriciarum;* Ns, *Neocallimastix* sp.; Cs, *Caecomyces* sp.; Ps, *Piromyces* sp.; Nj, *Neocallimastix joyonii.*

[b] + +, good growth; +, poor growth; —, no growth; nd, not determined.

of cell wall hydrolytic enzymes, gut fungi colonize both leaf tissue and highly lignified stem cell wall tissue. However, they are restricted in the ability to use pectin (Wubah et al. 1993), and some strains do not digest starch (Table 10.4) (McAllister et al. 1993).

Several species of ciliate protozoa of the genera *Diplodinium* and *Eudiplodinium* are able to use cellulosic plant fragments (Orpin 1988). Since protozoa ingest cellulose by phagocytosis, the question arises whether the hydrolytic enzymes are produced by the protozoa or by cellulolytic bacteria coingested with the substrate.

The mechanism by which gut microorganisms colonize and degrade plant cell walls has been the subject of extensive research. Figure 10.2 illustrates sites of entry that permit access to plant cell walls. Following penetration of the cutinous outer layer and the epidermis, colonization of the vegetative tissue continues by hydrolysis of pectin in the middle lamella by pectinolytic species or species

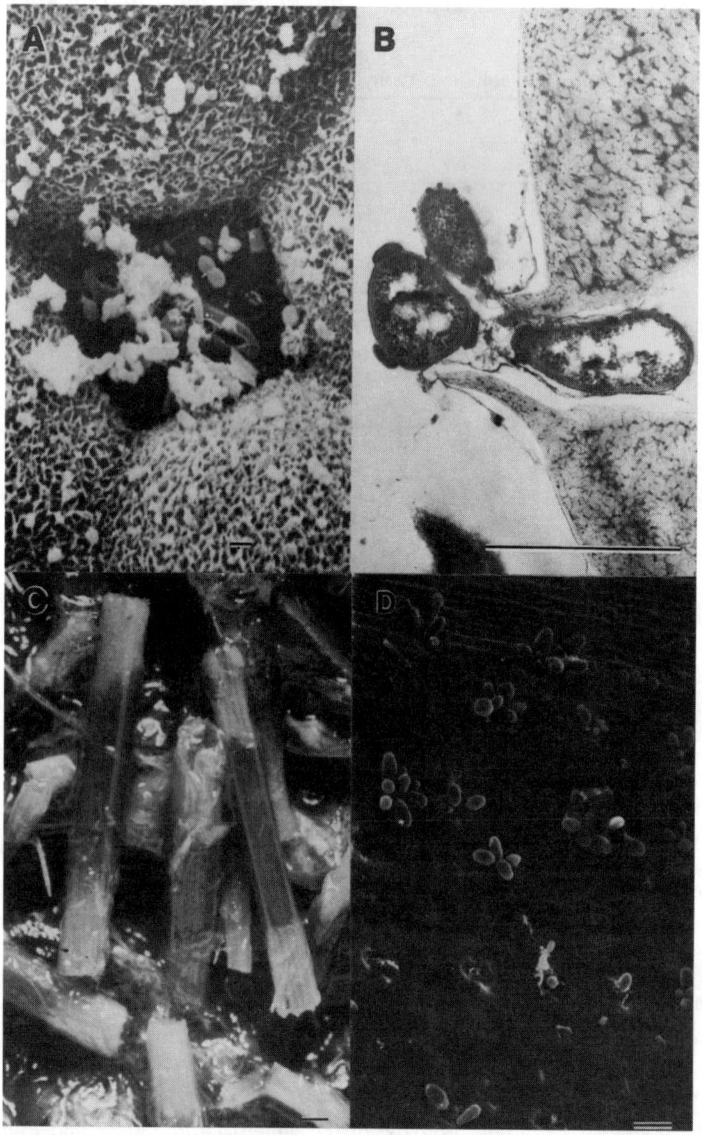

Figure 10.2. Colonization of plant materials. (A) Scanning electron micrograph of an alfalfa leaflet after 2 hours of incubation with ruminal fluid. Note that the bacteria have attached to the stomatal opening. (B) Transmission electron micrograph showing bacterial access to the interior of the leaf cell by penetration of the stomatal opening. (C) Photograph showing that digestion of alfalfa stems has progressed from the cut ends. (D) Scanning electron micrograph showing that rumen fungi have preferentially colonized the stomata of a corn leaf after 12 hours in ruminal fluid. Bars in (A) and (B) = 1 μm; in (C), = 50 μm (Figs. A and B from Cheng et al. (1980) with permission from the publisher; Fig. C, Cheng unpublished data; Fig. D, courtesy of E. Grenet, INRA-Theix, France).

possessing a combination of pectinases and xylanases (Cheng et al. 1991b). Cell walls are subsequently colonized by fibrolytic microbes. These processes will be addressed separately, beginning with the initial event—adhesion to cell walls.

4. Adhesion of Ruminal Microorganisms to Plant Cell Walls and to Plant Polysaccharides

Ruminal bacteria and fungi grow in close association with the cellulose-rich plant cell wall materials and cereal grains that they colonize (Cheng et al. 1991b). Figure 10.3 illustrates the apparent involvement of strands of exopolysaccharide in the adhesion of *R. flavefaciens* to crystalline cellulose. Latham et al. (1978a) observed that *F. succinogenes* and *R. flavefaciens* were bound to rye grass during its digestion. *F. succinogenes* adhered to cut edges of most plant cells except those of xylem. This species also adhered to the uncut surfaces of mesophyll, epidermis, and phloem cell walls. In contrast, *R. flavefaciens* predominated on uncut surfaces of epidermis, phloem, and sclerenchyma cell walls. Thus the two bacteria demonstrated uniquely different specificities for binding, which served to reduce competition. Additional documentation of adhesion specificity was provided when Bhat et al. (1990) also found that *R. flavefaciens* and *F. succinogenes* had separate specific adhesion sites on barley straw. Latham et al. (1978a) also determined with periodate staining that both *F. succinogenes* (Latham et al. 1978a) and *R. flavefaciens* had distinctive carbohydrate coats at the cell surface, that of *R. flavefaciens* being thicker, which suggested a role for these glycosylated structures in adhesion. In *R. albus* strain 8, the two aromatic fatty acids phenylacetate and 3-phenylpropanoate have been suggested to influence both the formation of the large cellulase complex of *R. albus* 8 (Hungate and Stack 1982, Stack et al. 1983) and stimulate the degradation of and affinity for cellulose of *R. albus* (Morrison et al. 1990, Stack and Hungate 1984). It is thought that 3-phenylpropanoate is involved in the formation of the capsule structure that surrounds the cell wall and facilitates formation of the cellulase complex of *R. albus* 8 (Stack and Hungate 1984). Adhesion not only provides close contact for maximal enzyme-substrate interaction, it also confers upon the adherent cells a strategic positional advantage for substrate uptake. The necessity of adhesion for cellulose digestion by ruminal microorganisms was further demonstrated by the observations that a low concentration of methylcellulose which blocked adhesion of bacteria (Kudo et al. 1987) and of fungi (Cheng et al. 1991a) to cellulose also blocked cellulose digestion. Morris and Cole (1987) found that several strains of *R. albus* that did not adhere to cellulose exhibited little cellulolytic activity although the correlation was poor. By comparison, Gong and Forsberg (1989) found that mutants of *F. succinogenes* which exhibited reduced adhesion also either lacked or had reduced

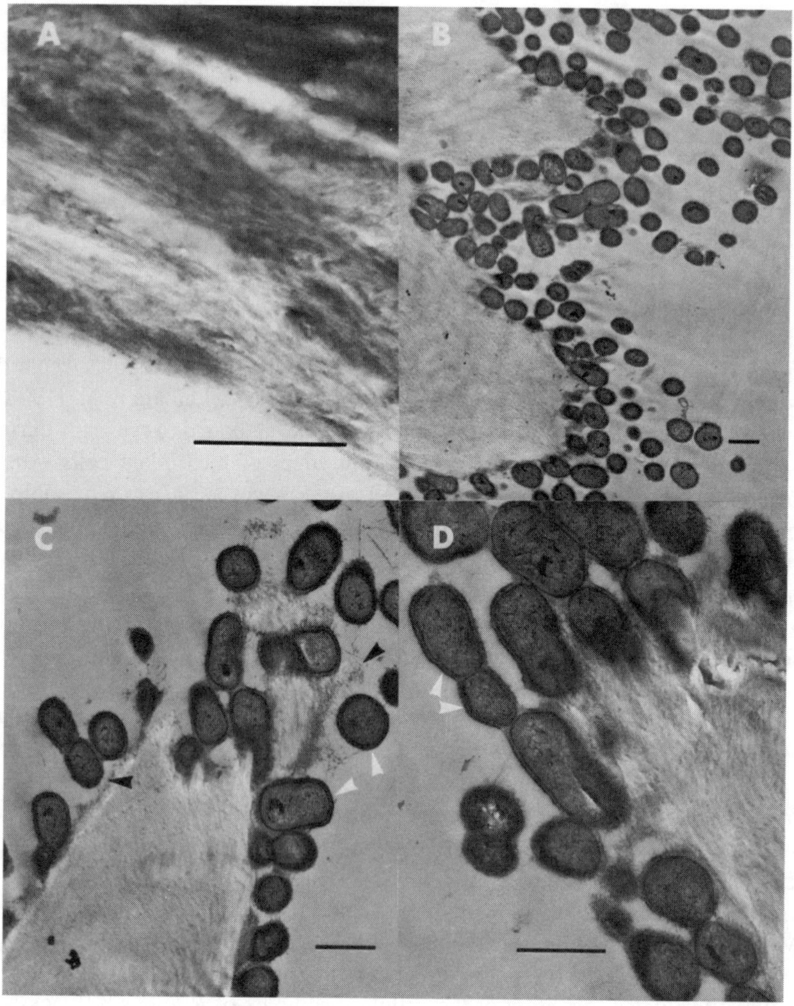

Figure 10.3. Transmission electron micrographs of thin sections of *R. flavefaciens* adhering to cellulose paper. *R. flavefaciens* was grown in MADAM (+cell) and, during primary fixation, stained with ruthenium red. (A) Uninoculated cellulose paper. (B) Low magnification showing the numerous *R. flavefaciens* around the cellulose paper after 48 hours incubation. (C, D) Note the close proximity of *R. flavefaciens* (white arrows) to the cellulose paper and the strands of exopolysaccharide (black arrows) holding the bacterial cells and the biofilm in place (Beaudette 1994). All bars = 1 μm.

cellulolytic activity, again supporting the link between adhesion and cellulose digestion by predominant fibrolytic bacteria.

Turbidity and radioisotope labeling assays have been used to determine the extent of binding of cells to cellulose. The turbidity assay is conducted by mixing cells with cellulose, usually particles of Avicel cellulose, removing the cellulose and any cells bound to it by differential sedimentation and quantifying unbound cells in the supernatant by measuring the turbidity. The radioactive assay involves growing cells in the presence of a radioactively labeled nutrient, harvesting and washing the cells, and then mixing them either with filter paper discs (Rasmussen et al. 1989) or with Avicel cellulose (Gong J and Forsberg CW, unpublished data). The proportion of cells bound to cellulose is determined by assaying the bound radioactivity after washing the cellulose. In the case of the *Ruminococcus* sp. and *F. succinogenes*, greater sensitivity in this assay has been achieved by using carbon-14-labeled isobutyric acid to label cells.

Minato and Suto (1978) demonstrated via a turbidity assay, that cellobiose blocked binding of *F. succinogenes* to cellulose. Cells grown on cellobiose at a concentration of 0.2% (wt/vol) exhibited decreased binding for cellulose. The repressive effect of cellobiose in the culture medium on synthesis of adhesin was confirmed by Roger et al. (1990); however, in their study a much higher concentration of cellobiose (4%, wt/vol) in the medium was necessary to block synthesis of the adhesin. Mitsumori and Minato (1993a,b) followed up their earlier work on *F. succinogenes* with the isolation and purification of putative cellulose-binding adhesins of 120 and 225 kDa that were released from cellulose by washing with cellobiose. Neither of the cellulose binding proteins exhibited catalytic activity.

Gong and Forsberg (1993) developed a method for separating the outer membrane of *F. succinogenes* from other cellular fractions. A key feature of this method was washing cells with 0.5 M NaCl, which released fragments of outer membrane. In addition, the process prompted release of the most abundant cellulose-binding protein in a monomeric form. It was identified as endoglucanase 2 (Gong and Forsberg 1993, and unpublished data). By dissociating the outer membrane in a nondenaturing detergent (CHAPS) and mixing the extract with cellulose, at least a dozen cellulose-binding proteins were detected. From these a 180-kDa noncatalytic cellulose-binding glycoprotein was used as an antigen for antibody production. The binding of cells to cellulose was reduced by the antibodies by approximately 65%, which suggests that the 180-kDa protein has a role in adhesion. Cross-reaction of the antibodies to the 180-kDa glycoprotein with many other glycoproteins, including the chloride-stimulated cellobiosidase (Huang et al. 1990), complicates the interpretation of these findings but strongly suggests that many of the protein(s) involved in the adhesion process are glycoproteins. The role of glycoproteins in adhesion to cellulose is further supported by the observation that the glycoprotein epitopes of wild-type cells grown on cellulose congregated at the cell surface sites of binding to cellulose as shown by immunoelectron microscopy.

There is little direct evidence for the presence of cellulose-binding domains on cellulases from ruminal microbes. In one study, the N-terminal 347 residues of xylanase C from *Pseudomonas fluorescens* subsp. *cellulosa*, containing a noncatalytic, cellulose-binding domain (CBD), was fused to the N terminus of endoglucanase A (EGA) from *Ruminococcus albus* SY3 (Poole et al. 1991). The hybrid enzyme bound to insoluble cellulose and could be eluted such that cellulose-binding capacity and catalytic activity were retained. The catalytic properties of the fusion enzyme was similar to EGA. In contrast to this study, when the cellulose binding domain from the *Thermomonospora fusca* endoglucanase E2 was fused to the catalytic domain of a *Prevotella ruminicola* endoglucanase, the endoglucanase's properties were dramatically altered (Maglione et al. 1992). The hybrid enzyme not only bound to insoluble cellulose, but the catalytic properties of the fusion enzyme were altered such that the degraded cellulose was nearly 10-fold more efficient than the native enzyme. Clearly, the structure of cellulose-binding domains can dramatically affect both binding and catalytic activities of cellulases. The question remains as to whether cellulose-binding domains of enzymes such as endoglucanase 2 of *F. succinogenes*; noncatalytic, cellulose-binding proteins; or both play a role in adhesion to cellulose.

The adhesion of ruminal bacteria to hemicellulose, starch, pectin, and other polymers has received little attention other than the work of Minato and Suto (1979) on the binding of rumen bacteria to starch. They documented the binding of *Ruminobacter* (*Bacteroides*) *amylophilus* and *Prevotella* (*Bacteroides*) species to starch granules. As will be discussed later, Tancula et al. (1992) have identified specific starch binding proteins in the intestinal bacterium *Bacteroides thetaiotaomicron* using elegant immunological techniques. A similar approach could be applied to the major amylolytic ruminal bacteria.

5. General Mechanisms for the Degradation of Plant Cell Walls, Cellulose, and Hemicellulose

As outlined above, the low porosity, the complex matrix of polymers, and the varying crystallinity of materials making up the cell wall together would necessitate that a wide range of hydrolytic enzymes acting simultaneously is probably necessary for cell wall digestion. Figure 10.4 indicates the enzymes that cleave the major bonds in cell wall polymers. The diagram is not inclusive because of the diversity of cell wall components, and as well, it is likely that some enzymes are as yet undiscovered. Research by Williams and co-workers has revealed the broad diversity of glycanases and glycosidases in the ruminal bacteria (Williams and Withers 1982a), fungi (Williams et al. 1994, Williams and Orpin 1987a,b),

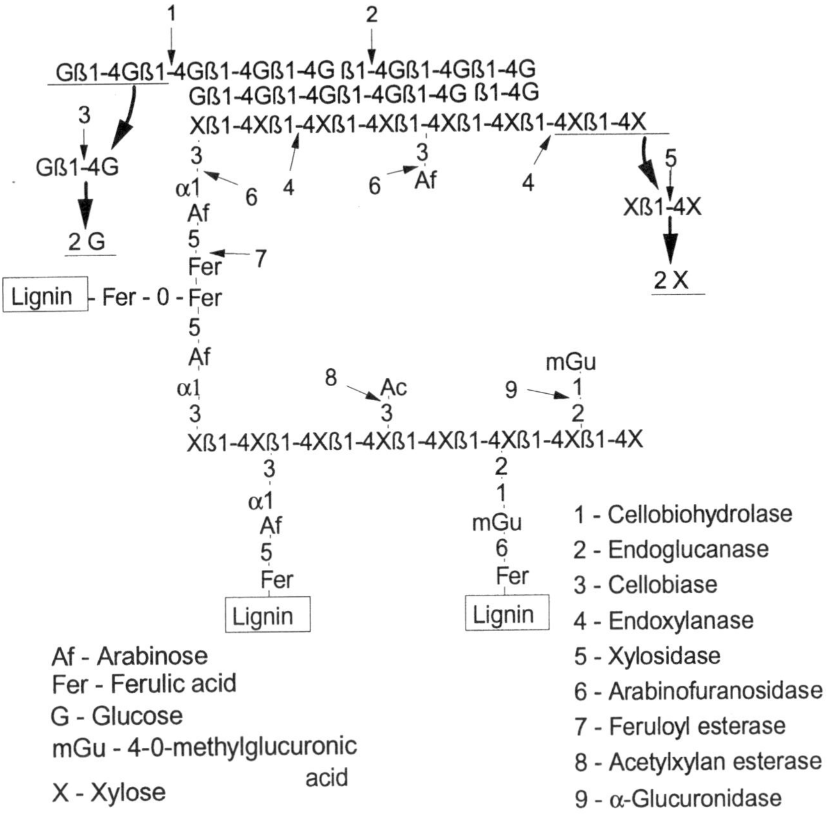

Figure 10.4. Schematic diagram illustrating the major linkages in cell walls of grasses and legumes and various enzyme cleavage sites.

and protozoa (Williams et al. 1984, 1986). The organization of these enzymes within the cell to facilitate the digestion of the mesh of plant cell wall polymers is an important aspect of their function. Since the fibrolytic enzyme systems of ruminal organisms are incompletely understood, hydrolytic systems of several other cellulase systems have been reviewed.

The enzymology of cellulose degradation is complex because it involves depolymerization of a crystalline substrate by multiple enzymes. The basic phenomenon of crystalline cellulose biodegradation was worked out some time ago for several aerobic fungi including *Phanerochaete* (*Sporotrichum*) *pulverulentum* and *Trichoderma koningii* and a model to illustrate their mechanism of action was developed. This model has been reviewed by Béguin and Aubert (1994) and by Wood (1992) and is briefly recounted here. Endoglucanases first hydrolyze amorphous regions of cellulose fibres. The nonreducing ends generated are attacked by cellobiohydrolase(s), which proceed with the degradation of the crystalline regions. β-Glucosidases prevent the accumulation of cellobiose which, if it accumulates, usually inhibits cellobiohydrolases. This model has been extended to a number of aerobic bacteria and to the anaerobic thermophilic bacterium *Clostridium stercorarium* (see Béguin and Aubert 1994). The component cellulase enzymes have been purified from the culture fluid as individual enzymes; however, protein-protein interactions have been shown to occur in the course of hydrolysis of crystalline cellulose, so some similarity exists between this model and systems such as that of *Clostridium thermocellum*, which involved enzyme complexes.

Cellulolytic enzymes from the thermophilic bacterium *C. thermocellum* have been isolated as multienzyme complexes called cellulosomes (Lamed et al. 1983a,b) that have been enormously difficult to dissociate in order to purify individual components (Wu et al. 1988). Lamed et al. (1983a,b) determined that the cellulosomes have a mass of 2 MDa and are composed of at least 14 different polypeptides with M_r values ranging from 48 kDa to 210 kDa. Zymogram analysis has shown that the majority of the polypeptides have cellulase or xylanase activity (Kohring et al. 1990, Lamed et al. 1983a, Morag et al. 1990). Although cellulosomes can be isolated from the culture fluid, they are also present on the surfaces of cells where they form protrusions comprising as many as several hundred cellulosomes (Lamed et al. 1983a,b).

The cellulosome cellulase complex of *C. thermocellum* consists of a scaffolding protein (CipA) which possesses both cellulase binding sites and a cellulose binding domain. Nucleotide sequence analysis of nine endoglucanase genes, one xylanase and one licheninase gene showed that although the N terminus sequences of the translated proteins differed among genes, the C terminus sequences contained two highly similar repeat segments of 22 amino acid residues, each of which bound to complementary sites on CipA, thereby forming a hydrolytic complex (Béguin and Aubert 1994). According to Béguin and Aubert (1994), the story as yet is incomplete; but there is a general understanding of the molecular makeup of the cellulosome.

High-molecular-mass complexes containing numerous cellulase enzymes have been identified in a number of rumen bacteria and fungi including *R. albus* (Lamed et al. 1987, Wood et al. 1982), *F. succinogenes* (Groleau and Forsberg 1981), and *N. frontalis* (Wilson and Wood 1992a,b). A high M_r complex containing several xylanases was also purified from the culture supernatant fluid of *B. fibrisolvens* (Lin and Thomson 1991a). The xylanase complex from *B. fibrisolvens* seems to fit the definition of a cellulosomelike complex, or indeed, a xylanosome as has been suggested (Lin and Thomson 1991a). However, whether the high molecular weight complexes from *R. albus* or *F. succinogenes* can be classified as cellulosomes remain to be determined. Miron and co-workers (Miron et al. 1989, 1990; Miron 1991; Miron and Ben-Ghedalia 1992) observed protuberant structures on the surfaces of *R. albus*, *R. flavefaciens*, and *F. succinogenes* grown on plant cell walls (but not on *F. succinogenes* grown on cellobiose as a carbon source) by staining cells with cationized ferritin prior to observation in the electron microscope. These data were considered to support the polycellulosome theory (Miron and Ben-Ghedalia 1992); however, the mechanism by which cationized ferritin causes the cellulosome appearance is unresolved. At least in the case of *F. succinogenes* the only types of structures we have observed by transmission electron microscopy (TEM) after fixation and staining were blebs arising from the outer membrane, and they had complex protein patterns with more resemblance to outer membranes of gram-negative bacteria than to the typical *C. thermocellum* cellulosome protuberances. As sequence information for these individual proteins and knowledge of the structure of the complex are currently lacking, a useful comparison with cellulosomes from *C. thermocellum* is not possible at this time.

Primary sequence information is available for more than 300 genes coding for cell wall degrading enzymes. The enzymes have been separated into families on the basis of similarities of their primary sequences and secondary structure (Gilkes et al. 1991, Henrissat 1991, Henrissat and Bairoch 1993). The cellulases and xylanases comprise 11 families (Henrissat and Bairoch 1993). Enzymes of bacterial, fungal and even plant origin are found in the same families, suggesting that horizontal gene transfer has been a factor in evolution. Exo- and endoglucanases have been placed in the same families, emphasizing that minor changes in sequence can dramatically alter the catalytic site. The reaction mechanism for cleavage of the β-1,4-linkage occurs via an acid-base mechanism involving only two residues. The details are reviewed by Béguin and Aubert (1994).

Many cellulases and xylanases contain noncatalytic CBD. These are usually located at either the N or C terminus of the enzyme and are separated from the catalytic domain by linker segments which are rich in glycosylated proline, threonine and serine residues. The CBDs have been grouped into four families (Gilkes et al. 1991, Béguin and Aubert 1994).

An important conclusion arising from careful analysis of the biochemical and amino acid sequence data is that domains of the same family appear to share similar biochemical properties (Claeyssens and Henrissat 1992). Therefore from a sequence analysis one can deduce the major properties of the translated protein.

The presence of a CBD presumably enhances the activity of the cellulase toward crystalline cellulose by allowing numerous catalytic cycles while remaining tethered to the substrate. In addition, there is evidence that CBDs open up crystalline regions of cellulose by peeling off cellulose chains (Knowles et al. 1988), and in some cases whole microfibrils (Din et al. 1991), from the top layers of cellulose.

6. Mechanisms of Cellulose and Hemicellulose Degradation by Ruminal Bacteria and Fungi

6.1. Fibrobacter succinogenes

F. succinogenes completely digests all forms of cellulose (Weimer 1993, Groleau and Forsberg 1981). Roger et al. (1993) observed that *F. succinogenes* digested wheat straw and maize stems more rapidly than, and to a similar extent as, *R. flavefaciens* 007, the highly cellulolytic ruminal fungi *N. frontalis* MCH3 and *Orpinomyces joyonii* TP 90–9. All four organisms attacked similar tissues and completely digested the phloem and inner parenchyma within 48 hours. Despite the capability of *F. succinogenes* to degrade cellulose completely, no potent cellulase complex has been isolated, even though a broad range of hydrolytic conditions have been tested (Groleau and Forsberg 1981, Wilson DB and Russell JB, personal communication). Nevertheless, eight different glucanase genes (Malburg and Forsberg 1993) and one cellodextrinase gene (Iyo and Forsberg 1994) have been cloned and several sequenced from *F. succinogenes* S85 (Broussolle et al. 1993, McGavin et al. 1993) (Table 10.5). These did not include endoglucanase 2 (Table 10.6) (McGavin and Forsberg 1988), the chloride-stimulated cellobiosidase (Huang et al. 1988), or the cellobiase (Gong and Forsberg 1993). In addition, one endoglucanase from *F. succinogenes* strain AR1 has been sequenced (Cavicchioli et al. 1991). The sequenced glycanases belong to the families A and E and none of them appear to have CBDs. However, endoglucanase 2 does have a CBD (McGavin and Forsberg 1989), and the chloride-stimulated cellobiosidase binds to cellulose (Huang et al. 1988) and probably has a CBD as well. Several interesting aspects of the *F. succinogenes* cellulases have recently been noted. The gene *endB* (Broussolle et al. 1993, Forano et al. 1994) codes for an enzyme with a mass similar to that of endoglucanase 1 described by McGavin and Forsberg (1988); furthermore, the two enzymes have similar catalytic properties and require divalent cations for activity, which suggests that *endB* codes for endoglucanase 1. Endoglucanase 2 is present both as a peripheral membrane protein and associated with isolated glycogen granules (Gong and Forsberg 1993). Most of the cellobiase was also isolated as a glycogen granule bound enzyme that could be released by treatment of the granules with salt (Gong and Forsberg 1993). Whether the enzymes are present on glycogen granules during growth or have an affinity for glycogen and bind after disruption of the cells has not been resolved, nor is it known how these characteristics relate to regulation of cellulose digestion.

F. succinogenes also has multiple xylanases, four of which have been cloned (Malburg et al. 1993), and these do not include a debranching xylanase, endoxylanase 1 (Table 10.6) (Matte and Forsberg 1992). Thus the total number of xylanases must be at least five. The two xylanase genes sequenced, *xynB* and *xynC*, are from the families F and G, respectively. The XynC xylanase is a unique multido-

Table 10.5. Cellulase and glucosidase genes from rumen bacteria and fungi and their gene products

Bacterium	Gene (aa)[a]	Enzyme	Substrates Cleaved[b]	Protein (kDa)[c]	Reference
B. fibrisolvens A46	*celA* (432), sp, [A2]	Endoglucanase	BBG, CMC, L, pNPC	48.9/47.0	Hazlewood et al. 1990
B. fibrisolvens H17c	*endI* (547), sp, [A4]	Endoglucanase	L, CMC, pNPC, G_4-G_6	61.0/—[d]	Berger et al. 1989
B. fibrisolvens H17c	*cedI* (547), [E9]	Cellodextrinase	CMC, pNCP, G_3-G_6	61.0/61.0	Berger et al. 1990
B. fibrisolvens H17c	*bglA* (830), [A3]	β-Glucosidase	G_2-G_5	91.8/94.0	Lin et al. 1990
F. succinogenes S85	*cel*-3 (658), ss, [A3]	Endoglucanase	CMC, L, BBG, pNPC	73.4/118	McGavin et al. 1989
F. succinogenes S85	*endB* (555) [E9]	Endoglucanase	CMC, BBG, L	58.0/—	Broussolle et al. 1993 L14436
F. succinogenes S85	*cedA* (357) [A5]	Cellodextrinase	pNPC, pNPL, G^3-G^6	41.9/50.0	Iyo and Forsberg 1994
F. succinogenes AR1	*endA*$_{FS}$ (453) [E9]	Endoglucanase	CMC, ASC, Avicel	47.0/46.5	Cavicchioli et al. 1991
N. patriciarum	*celB* (474) [A5]	Endoglucanase	CMC, BBG, XYN, L	53.1/—	Zhou et al. 1994
P. ruminicola B_14	pC3, sp, [26]	Endoglucanase	CMC, XYN	—/88.0	Matsushita et al. 1991
P. ruminicola AR20	*celA* (506), sp, [A4]	Endoglucanase	CMC, ASC, XYN	52.9/52.0	Vercoe and Gregg 1992
P. ruminicola 23	(584) [A5]	Endoglucanase	CMC, XYN		S27500
R. albus F-40	*EgI* (407), sp, [E9]	Endoglucanase	CMC	40.8/50.0	Ohmiya et al. 1989, Deguchi et al. 1991
R. albus SY3	*celA* (365) [A4]	Endoglucanase	CMC, XYN	41.2/44.5	Poole et al. 1990
	celB (409 [A4]	Endoglucanase	CMC	45.5/—	
R. albus F-40	*egII* (407) [A5]	Endoglucanase			P16216
R. albus F-40	*egIV* (936) [A2]	Endoglucanase	CMC	35.8/35.0	Karita et al. 1993
R. albus F-40	pRA201 (947) [A3]	β-Glucosidase	pNPG, G_2-G_6	104/120	Takano et al. 1992 Ohmiya et al. 1985
R. albus 8	*celA* (411) [A5]	Endoglucanase	CMC, ASC	45.7/41.3	Attwood et al. 1996 L04563
R. albus AR67	*celA* (414) [A5]	Endoglucanase			Vercoe and Gregg 1995 L10243
R. flavefaciens 17	*endA* (605 +) [A5]	Glucanase	ASC, CMC, L, G_5, G_6	—	Cunningham et al. 1991
R. flavefaciens FD-1	*celA* (336) [A5]	Cellodextrinase	pNPC, G_4-G_6	39.4/—	Wang and Thomson 1990, 1992
R. flavefaciens FD-1	*celB* (632) [44]	Endoglucanase	CMC	69.4/—	Vercoe et al. 1995b
R. flavefaciens FD-1	*celD* (405) [ ]	Endoglucanase	CMC, XYN	44.61/—	Vercoe et al. 1995a
R. flavefaciens FD-1	*celE* (963), sp, [-]	Endoglucanase	CMC	35.9/35.0	Wang et al. 1993

[a] ( ), Amino acid residues in the protein; sp, signal peptide; [ ], glucosyl hydrolase family.
[b] Abbreviations: CMC, carboxymethylcellulose; ASC, acid swollen cellulose; BBG, barley-β-glucan; L, lichenin; pNPC, p-nitrophenyl-β-D-cellobioside; MUC, 4-methylumbelliferyl-β-D-cellobioside; MUX, 4-methylumbelliferyl-β-D-xyloside; XYN, xylan; AXYN, arabinoxylan.
[c] The molecular mass of a protein derived from the DNA sequence is presented before the slash, and that of enzyme purified from either the *E. coli* host or the original rumen bacterium after the slash.
[d] Not available.

Table 10.6. Xylanase and xylosidase genes from rumen bacteria and fungi and their gene products

Bacterium	Gene (aa)[a]	Enzyme	Substrates cleaved[b]	K_m	V_{max}	Protein (kDa)[c]	Reference
B. fibrisolvens 49	*xynA* (411), sp [F10]	Xylanase	XYN			46.7/—[d]	Mannarelli et al. 1990
B. fibrisolvens H17	*xynB* (635) [F10]	Xylanase	XYN, pNPC			73.2/72.0	Lin and Thomson 1991b
B. fibrisolvens GS113	*xylB* (517) [43]	Xylosidase	pNPX, pNPA, X_2-X_5			62.0/60.0	Sewell et al. 1989; Utt et al. 1991
F. succinogenes S85	pJ110 (349), sp, [A5]	Lichenase	Lichenin, oat β-glucan			35.2/ 32.2–37.2	Teather and Erfle 1990; Erfle et al. 1988
F. succinogenes S85	*xynC* (608), sp, 2cd [G11]	Xylanase	XYN, CMC, AXYN	2.09	6	66.0/63.0	Paradis et al. 1993
	(233) [G11]	XynC-A	XYN	1.83	689	—/30.0	
	(214) [G11]	XynC-B	XYN, AXYN, CMC	2.38	92	39.1/40.0	Zhu et al. 1994b
N. patriciarum	*xynA* (607), sp, 2 cd, [G11]	Xylanase	Xylan			66.2/53.0	Gilbert et al. 1992
N. patriciarum	*xyn B* (860), sp [F10]	Xylanase	Xylanase	XYN, AXYN, pNPC, MUX		88.1/93.0	Black et al. 1994
P. ruminicola 23	(584), sp, [At]	Xylanase	XYN, CMC			65.7/62.0	Whitehead 1993
R. flavefaciens 17	*xynA* (954), sp, 2 cd	Xylanase	XYN			111/90.0	Zhang and Flint 1992
	(248) [G11]	XynA-A	XYN			28.5/40.0	
	(332) [F10]	XynA-C	XYN			—/—	
R. flavefaciens 17	*xynB* Θ, sp [G11]	Xylanase	XYN				Zhang et al. in press
R. flavefaciens 17	*xynD* (802), sp, 2cd	Xylanase/ glucanase	XYN, L, laminarin			89.5/65.0	Flint et al. 1993
	(213) [G11]		XYN			53.0/35.0	
	(249) [16]	XynD-A Xylanase XynD-C, β-1,3-1,4-glucanase	L, laminarin			37.0/43.0	

[a] ( ), Amino acid residues in the protein; sp, signal peptide; [ ], glycosyl hydrolase family.
[b] Abbreviations: CMC, carboxymethylcellulose; ASC, acid swollen cellulose; AXYN, arabinoxylan; BBG, barley-β-glucan; L, lichenin; pNPC, p-nitrophenyl-β-D-cellobiosidase; MUC, 4-methylumbelliferyl-β-D-cellobioside; MUX, 4-methylumbelliferyl-β-D-xyloside; XYN, xylan.
[c] The molecular mass of a protein derived from the DNA sequence is presented before the slash, and that of enzyme purified from either the *E. coli* host or the original microorganism after the slash.

main enzyme that possesses two similar catalytic sites (Table 10.7) (Paradis et al. 1993, Zhu et al. 1994b), whereas XynB typically exhibits glucanase as well as xylanase activity (Malburg and Forsberg, unpublished).

Other enzymes of *F. succinogenes* which have not been cloned include an α-glucuronidase (Smith and Forsberg 1991), a ferulic acid esterase, an acetylxylan esterase and an arabinofuranosidase (McDermid et al. 1990b). The debranching xylanase, endoxylanase 1 (Matte and Forsberg 1992), and the acetylxylan esterase (McDermid et al. 1990a) are unique in that they remove arabinose and acetyl

Table 10.7. Cellulolytic, hemicellulolytic, and amylolytic enzymes from rumen bacteria and fungi that are different from the cloned gene products

Organism	Enzyme	Substrates[a]	K_m[b]	V_{max}[c]	Mass (kDa)	Ref
Cellulases						
F. succinogenes S85	Endoglucanase 1	CMC, BβG, L, AMC	3.6 mg/mL (CMC)	84	65 (m)[d]	McGavin and Forsberg 1988
F. succinogenes S85	Endoglucanase 2	CMC	12.2 mg/mL (CMC)	10.4	118 (m)	McGavin and Forsberg 1988
F. succinogenes S85	Cellobiosidase (anion-stimulated)	CMC, pNPC, pNPL, XYN	0.1 mM (pNPC)	6.0-14.0	75 (m)	Huang et al. 1988
N. frontalis MCH3	Glycoside hydrolase	pNPF, pNPG	0.91 mM (pNPA) 0.27 mM (pNPG)	— —	120 (m)	Hebraud and Fevre 1990b
N. frontalis MC-2	Cellobiase	pNPG, G_2, G_3, G_5	0.053 mM (G_2)	5.9	85	Li and Calza 1991
R. albus F-40	Endoglucanase	CMC	7.2 mg/mL	—	50 (m)	Ohmiya et al. 1993
R. albus F-40	Endoglucanase	CMC, G_5, G_6	0.72 mM (G_6)	13	53 (m)	Watanabe et al. 1992
R. albus F-40	Cellobiosidase	pNPC	—	—	200 (d)	Ohmiya et al. 1982
R. albus F-40	β-glucosidase	pNPC, pNPG	26 mM (Cel)	—	82 (m)	Ohmiya et al. 1985
R. albus AR67	Exo-β-1,4-D-glucosidase[e]	Cel, pNPG, pNPC, Lam, Lamn	12.5 mM (Cel)	5.3	64 (m)	Ware et al. 1990
R. flavefaciens FD-1	Exoglucanase	AMC, CC, pPNC, pPNL	3.1 mM (pNPC)	0.3	230 (d)	Gardner et al. 1987
Hemicellulases						
B. fibrisolvens GS113	α-L-Arabinofuranosidase	pNPA, mubA, arabinobiose	0.7 mM (pNPA)	109	31 (o)	Hespell and O'Bryan 1992
F. succinogenes S85	Endoxylanase 1	XYN, AXYN	2.6 mg/mL (AXYN)	33.6	53.7 (m)	Matte and Forsberg 1992
F. succinogenes S85	Endoxylanase 2	XYN, CMC	1.3 mg/mL (XYN)	118	66.0 (m)	Matte and Forsberg 1992
F. succinogenes S85	Acetylxylan esterase	NAC, AXYN	2.7 mM (NAC)	9,100	65 (m)	McDermid et al. 1990a
N. frontalis MCH3	Xylanase 1	XYN, CMC	1.22 mg/mL (XYN)	37[f]	45 (m)	Gomez de Segura and Fevre 1993
	Xylanase 2	XYN, CMC	2.5 mg/mL (XYN)	317[f]	70 (m)	
R. albus 8	Xylanase 1	XYN	0.23 mg/mL (XYN)	—	720	Greve et al. 1984[a]
R. albus 8	Xylanase 2	XYN	0.28 mg/mL (XYN)	—	260	Greve et al. 1984[a]
R. albus 8	α-L-Arabinofuranosidase	pNPA	1.3 mM (pNPA)	—	75 (t)	Greve et al. 1984[b]
N. patriciarum 27	Xylosidase	pNPX, X_2, Birchwood XYN	0.13 mM (pNPX)	8.9	85 (d)	Zhu et al. 1994a
N. frontalis RK21	Xylosidase	pNPX, pNPA, X_2, X_3	—	0.9[f]	53,83 (d)	Garcia-Campayo and Wood 1993
N. frontalis MCH3	Xylosidase	pNPX, XYN	1.2 mM (pNPX)	—	90 (d)	Hebraud and Fevre 1990a
Neocallimastix MC-2	Feruloyl esterase FAE1	FAXX, PAXX	31.9 μM (FAXX)	2.9	69	Borneman et al. 1992
	FAE2	FAXX, PAXX	9.6 μM (FAXX)	11.4	24	

(continued)

Table 10.7. *(continued)* Cellulolytic, hemicellulolytic, and amylolytic enzymes from rumen bacteria and fungi that are different from the cloned gene products

Organism	Enzyme	Substrates[a]	K_m[b]	V_{max}[c]	Mass (kDa)	Ref
Neocallimastix MC-2	p-Coumaroyl esterase	PAXX	19.4 μM	5.1	5.8 (d)	Borneman et al. 1991
Orpinomyces sp. PC-2	β-Glucosidase	pNPG, G_2, β-1, 2-glucobiose, LAM	0.25 mM (G_2) 0.35 mM (pNPG)	27.1 27.2	85.4 (m)	
Pectinolytic enzymes						
L. multiparus 685	Poly (1,4-α-L) galacturonate lyase	PGA, TGA	—	—	—	Wojciechowicz et al. 1980
S. ruminantium 777	1,4-α-D-Galactosiduronate-hydrolase	NaP, TGA	27.6 mM	28.3	—	
Amylases						
Streptococcus bovis JB1	α-Amylase	amylose, amylopectin	0.88 mg/mL (potato starch)	2510	77 (m)	Freer 1993
Ruminobacter amylophilus 70	α-Amylase	amylose, amylopectin, dextrin	—	—	92 (m)	McWethy and Hartman 1977

[a] Substrates: CMC, carboxymethycellulose; AMC, amorphous cellulose; AXY, acetylxylan; AXYN, arabinoxylan; BβG, barley β-glucan; CC, crystalline cellulose; FAXX, O-{5-O-[(E)-feruloyl]-α-L-arabinofuranosyl}-(1-3)-O-β-D-xylopyranosyl-(1-4)-D-xylopyranose; PAXX, the equivalent p-coumaroyl ester; G_2, cellobiose; G_5, cellopentaose; L, Lichenan; Lam, laminarbiose; Lamn, laminarin; mubA, methylumbelliferyl-α-L-arabinofuranoside; NAC, α-naphthyl acetate; pNPF, p-nitrophenyl-β-D-fucopyranoside; pNPC, p-nitrophenyl-β-D-cellobioside; pNPG, p-nitrophenyl-β-D-glucoside; pNPL, p-nitrophenyl-β-D-lactoside; NaP, Sodium pectate; TGA, trigalacturonate; PGA, polygalacturonate; XYN, xylan.

[b] Substrates used for K_m determination are in parentheses.

[c] Expressed as μmol of substrate hydrolyzed/min/mg protein.

[d] m, monomer; d, dimer; t, tetramer; o, octamer.

[e] Data refer to cloned gene product expressed in *E. coli.*

[f] A specific activity value; V_{max} not available.

residues from intact xylan, which then would allow the unhindered action of the multiplicity of xylanases (Table 10.6). In contrast to these, the debranching α-glucuronidase present in the extracellular culture fluid (Smith and Forsberg 1991) seemed to act only on small fragments of glucuronoxylan and required the concerted action of the XynC xylanase for detectable activity. However, immunoblotting has demonstrated that the XynC enzyme is present at a very low concentration in cultures of the wild-type organism (Zhu H and Forsberg CW, unpublished), which therefore may restrict the role of the debranching α-glucuronidase in vivo. As documented by Dehority (1993) in his review and confirmed by Miron et al. (1994), *F. succinogenes* efficiently digests the hemicellulose fraction of plant cell walls, presumably as a result of the action of hemicellulases described, but lacks the ability to use pentose sugars as a carbon source, which in the case of xylose seems to be due to the absence of a permease, xylose isomerase, and xylulose-5-PO_4 kinase (Matte et al. 1992).

Most of the hydrolytic enzymes of *F. succinogenes*, including endoglucanase 2, the chloride-stimulated cellobiosidase, the cellodextrinase, endoxylanases 1 and 2, and the acetylxylan esterase (McDermid et al. 1990b), appear to be produced constitutively. Endoglucanase 1 was not synthesized in cellobiose-grown cells (Huang et al. 1990, McGavin et al. 1990), documenting that its synthesis is subject to carbon source-dependent regulation.

6.2. Butyrivibrio fibrisolvens

Most *B. fibrisolvens* strains grow poorly on insoluble substrates including both plant cell walls and cellulose (Dehority 1993, Miron et al. 1994, Stewart and Bryant 1988). However, in association with the primary fibrolytic bacteria, it utilizes the solubilized degradation products extensively (Strobel and Dawson 1993, Miron et al. 1994). Attempts to separate the xylanase and glucanase enzymes of *B. fibrisolvens* by chromatography were unsuccessful, which has led to the proposal of a xylanosome consisting of a complex of at least 11 xylanolytic activities and endoglucanase activity, but lacking β-glucosidase and β-xylosidase (Lin and Thomson 1991a). This species has been shown to produce feruloyl and coumaroyl esterases (Akin et al. 1993) and α-L-arabinofuranosidase (Table 10.7) (Hespell and O'Bryan 1992). The xylanolytic enzymes are induced by xylan, xylooligosaccharides, and xylobiose (Williams and Withers 1993). A number of genes coding for fibrolytic enzymes have been characterized from *B. fibrisolvens*, including two endoglucanases and a cellodextrinase from family A, a β-glucosidase from family E, two xylanases from family F, and a bifunctional xylosidase-arabinofuranosidase placed in family 43 (Table 10.6). It is curious that although the bacterium has an array of fibrolytic capabilities it seems to lack one or more proteins/enzymes necessary for digestion of crystalline cellulose and plant cell walls.

6.3. Ruminococcus albus and Ruminococcus flavefaciens

As noted previously, the intact cellulase complex of *R. albus* is present in different molecular forms (Wood et al. 1982, Wood and Wilson 1984, Yu and Hungate 1979). Individual components of the cellulase system of *R. albus* strain F-40 have been purified, including an endo-β-1,4-glucanase, a β-glucosidase, and a cellobiosidase (Ohmiya et al. 1982, 1985, 1987). Two endo-β-1,4-glucanase genes from strain F-40 have been cloned in *E. coli* (Kawaii et al. 1987, Ohmiya et al. 1988, Karita et al. 1993), and the DNA sequences have been determined (Ohmiya et al. 1989, Karita et al. 1993). The β-glucosidase gene from strain F-40 has also been cloned and the DNA sequence has been determined (Takano et al. 1992). Ten distinct β-glucanase clones from *R. albus* strain 8 have been reported; however, characterization of these clones has not been achieved owing to difficulties in molecular analysis of DNAs cloned from *Ruminococcus* (Howard and White 1988). Therefore, another gene library was generated and an endoglucanase gene *celA* from *R. albus* 8 was isolated and characterized at the DNA and protein level (Attwood et al. 1996). The protein, designated CelA, showed extensive homology with *Ruminococcus* endoglucanases by both primary amino acid sequence alignment and hydrophobic cluster analysis (Attwood et al. 1996). Recently, endo-β-1,4-glucanase genes and an exo-β-1,4-D-glucosidase gene from *R. albus* strain AR67 have been cloned into *E. coli* (Ware et al. 1989, 1990). The endo-β-1,4-glucanase genes from strain AR67 have been compared by DNA-DNA hybridization and restriction mapping to cloned endo-β-1,4-glucanases from *R. albus* strain AR68 (Ware et al. 1989). By these techniques, it was concluded that endo-β-1,4-glucanase genes from these two strains were not closely related. Two endo-β-1,4-glucanases from *R. albus* strain SY3 have been cloned in *E. coli* (Romaniec et al. 1989), and the DNA sequences of these genes (*cel*A and *celB*) have been determined (Poole et al. 1990). In contrast to the work of Ware et al. (1989), these two endo-β-1,4-glucanase genes have a high degree of homology with each other and also show significant homology with the endo-β-1,4-glucanase gene from strain F-40, the *End1* gene from *B. fibrisolvens*, and the *celE* gene from *C. thermocellum* (Poole et al. 1990).

The enzyme system of *R. flavefaciens* is comprised of exo-β-1,4-glucanase, endo-β-1,4-glucanase, and cellodextrinase activities (Doerner and White 1990b, Gardner et al. 1987, Pettipher and Latham 1979, Rasmussen et al. 1988). The main end products of cellulolysis by *R. flavefaciens* are cellotriose and cellobiose with only small amounts of glucose being produced (Rasmussen et al. 1988, Russell 1985). A number of endoglucanase activities have been observed which belong to two main complexes termed Endoglucanase A and B (Doerner and White 1990b). At least 13 activities are present in the endo-β-1,4-glucanase A complex and five unique activities in the endo-β-1,4-glucanase B complex, some of which are glycosylated (Doerner and White 1990a). A single exoglucanase (ExoA) has been purified and is a dimer of 114 kDa subunits. The main end product from hydrolysis of insoluble cellulose by ExoA is cellobiose (Gardner et al. 1987).

Five β-glucanase genes from *R. flavefaciens* strain FD-1 have been cloned in *E. coli* (Barros and Thomson 1987, Howard and White 1990, Vercoe et al. 1995a,b, Wang et al. 1993, White et al. 1990). The DNA sequences for four of these genes have been reported, and their amino acid sequences have been derived (Vercoe et al. 1995a,b, Wang and Thomson 1990, Wang et al. 1993). Two of these endoglucanases (*celE*, Wang et al. 1993, *celB*, Vercoe et al. 1995b) are not homologous to any of the current cellulase families. Interestingly, in the native host, there appears to be differential expression of the cellulases that have been cloned. The *celB*, *celD*, and *celE* genes are inducible, whereas *celC* appears to be expressed constitutively (White et al. 1993a). There is some contention about the regulation of expression of *celA*. The original report demonstrated that *celA* was expressed constitutively (Doerner et al. 1992), but a later study suggested that expression was inducible by cellulose (Wang et al. 1993). Genes encoding CMCase activity, β-glucosidase activity, and activity against 4-methylumbelliferyl-β-D-cellobioside (MUC) have been cloned from *R. flavefaciens* 186 (Huang et al. 1989), and these genes have been characterized at the DNA level (Asmundson et al. 1990). Flint et al. (1989) have also cloned xylanases from *R. flavefaciens* 17, some of which also show activity toward cellulose. An endo-β-1,4-glucanase gene from *R. flavefaciens* 17 has also been sequenced (Cunningham et al. 1991). While there is clearly a diversity of cellulase families represented among the ruminococci, it is striking to note that all the currently characterized cellulases lack identifiable cellulose binding domains.

Many *R. albus* and *R. flavefaciens* strains degrade crystalline cellulose (Miron et al. 1994), but some lack this capacity (Morris and Cole 1987). From the cellulolytic *R. albus* strain F-40 a variety of glucanase, cellobiosidase and β-glucosidase enzymes have been purified and characterized while only an exoglucanase has been purified from *R. flavefaciens* (Table 10.7). Difficulties encountered in fractionating the cellulase systems would appear to be due to the presence of the enzymes in high-molecular-weight protein complexes resembling cellulosomes (Doerner and White 1990, Lamed et al. 1987a). Furthermore, extracellular cellulases of *Ruminococcus* strain AR67 retain some activity on crystalline cellulose, but much of the activity is oxygen labile (Gregg et al. 1993). A broad range of cellulase genes have been cloned from both *R. albus* and *R. flavefaciens* (Table 10.5) and many have been sequenced suggesting that both bacteria possess multiple genes, many of which are of the family A subtypes.

Gregg et al. (1993) reported the cloning of an exoglucanase gene from *R. albus* detected by screening *E. coli* clones for cellobiosidase activity under anaerobic conditions. An oxygen-sensitive exoglucanase was found with the capacity to degrade crystalline cellulose into fibers. When this enzyme was mixed with another, less active cellobiohydrolase and three previously cloned endoglucanases (Ware et al. 1989) from the same strain, the mixture of enzymes acted synergistically to degrade cellulose to cellooligosaccharides (Gregg et al. 1993). Thus it would appear that the cellulase system of *R. albus* has been largely reconstructed in vitro.

The cellulase system of *R. flavefaciens* is not as well characterized as that of *R. albus*, although it appears to be composed of endoglucanase, exoglucanase (ExoA), and cellodextrinase activities (Doerner and White 1990, Gardner et al. 1987, Rasmussen et al. 1988). The involvement of ExoA in cellulose hydrolysis was demonstrated by growth inhibition when an ExoA-specific monoclonal antibody was added to the culture (Doerner et al. 1994).

Ruminococcus species actively degrade xylan. Several xylanases which hydrolyze xylan only have been purified from *R. albus* (Table 10.7) (Greve et al. 1984a) as well as an arabinofuranosidase (Greve et al. 1984b) which functions synergistically with the xylanases. No xylanase genes from *R. albus* have been cloned to date, but a variety of xylanase genes have been cloned from *R. flavefaciens* strain 17. These include two coding for multicatalytic domain enzymes, *xynA* and *xynD*, each of which possesses two catalytic domains (Zhang and Flint 1992, Flint et al. 1993), and a third gene *xynB* coding for a xylanase (Zhang et al. 1994), with a single catalytic domain (Table 10.6). The XynA domains exhibited similar substrate specificities, but belonged to different families while the XynD catalytic domains exhibited different specificities, domain A cleaved xylan and domain C exhibiting β-1,3-glucanase activity and cleaving lichenin and laminarin. The functional significance of the multidomain organization is unclear. It could either be a mechanism for co-ordinating the expression of multiple gene products or could provide a catalytic advantage because of the association of various xylans and glucans in the complex matrix of the plant cell wall. No debranching enzymes have been reported in the ruminococci and *R. flavefaciens* FD1 was found to lack ferulic acid esterase (Akin et al. 1993).

The RNA levels of *xynA* and *xynB* and a third region encoding multiple activities of *R. flavefaciens* 17 were increased by growth on xylan as compared with cellobiose, suggesting induction of the enzymes by xylan as a carbon source. Furthermore, total cell-associated xylanase and β-xylosidase activities and extracellular xylanase were similarly induced by xylan (Flint et al. 1991).

6.4. Ruminal Fungi

The rumen fungi possess a wide array of fibrolytic enzymes that hydrolyze a range of glycosidic linkages and esterases to solubilize phenolic compounds. These enzymes enable them to colonize highly lignified plant stem tissues (Akin et al. 1989, Trinci et al. 1994, Wubah et al. 1993); however, they cannot utilize the lignin moiety (Akin and Benner 1988, Gordon and Phillips, 1989a, McSweeney et al. 1994). Fungi are able to penetrate cuticle external to the plant epidermis layer (Akin et al. 1983), whereas bacteria totally lack this attribute, but no cutinase enzymes have been characterized in the fungi. Many of the polysaccharide-hydrolyzing enzymes are localized on rhizoids and rhizomycelia (vegetative stage) and also are secreted into the extracellular culture medium (Lowe et al. 1987).

Neocallimastix spp. produce a cellulase complex with very high activity against crystalline cellulose. The cellulase specific activity of *N. frontalis* RK21 reportedly exceeds that of *Trichoderma reesei* C30, a fungus recognized for its high cellulase activity (Wood et al. 1986). The cellulase complex from *Neocallimastix frontalis* RK21 has a mass of between 750 and 1,000 kDa, and has been fractionated by adsorption to cellulose into a component that retains the cotton solubilizing activity but very little of the endoglucanase or β-glucosidase activity (Wilson and Wood 1992a). This largely glucanase-free fraction exhibits a higher specific activity against cotton cellulose than the *C. thermocellum* cellulosome, making it perhaps the most active cellulase complex yet reported (Wilson and Wood

1992b). Cotton degrading activity is selectively lost when the complex is treated with chitinase, indicating that association of the enzymes with the fungal cell wall is essential for activity (Wilson and Wood 1992b). Cotton degrading activity was greatly enhanced relative to other activities when *N. frontalis* was grown in coculture with a methanogen (Wilson and Wood 1992b, Wood et al. 1986). This indicates the importance of activities other than endoglucanase in the biodegradation of cellulose, but the mechanism of stimulation is not known.

Both the cellulose-binding, cellulose solubilizing fraction and the non-cellulose binding, endoglucanase/β-glucosidase fraction from *N. patriciarum* acted synergistically with cellobiohydrolases from the aerobic fungus *Penicillium pinophilum* confirming that this interaction is possible between components of different fungal cellulase systems (Wood et al. 1994). This may not be as unusual as it would seem, given that gene cloning confirmed lateral transfer of genes among bacteria, fungi and plants. As a result, critical peptide domains of the cellulase systems of the two organisms may be closely related. A unique β-glucosidase was purified from *Orpinomyces* sp. PC-2 (Chen et al. 1994). This enzyme, in addition to hydrolyzing cellobiose, also cleaved β-1,2-glucobiose and β-1,3-glucobiose (laminarobiose) indicating that it could function in the terminal hydrolysis of cereal β-glucans as well as cellulose.

To dissect the complex fungal cellulase system, Xue et al. (1992a,b) screened a cDNA library prepared from *N. patriciarum* in the vector λZapII. They obtained four notable cDNAs which they designated *celA*, *celB*, *celC* (Xue et al. 1992b), and *celD* (Xue et al. 1992a). *celA* encoded a cellobiohydrolase which hydrolyzed crystalline cellulose, producing cellobiose and exhibiting cellulose binding activity. *celB* and *celC* encoded endoglucanase activities. The enzyme encoded by *celD* was a highly active hydrolase with three seemingly identical multifunctional catalytic domains each with endoglucanase, licheninase, cellobiosidase, and xylanase activity, and exhibiting cellulose binding capability. As yet, no results have been reported on the extent of hydrolysis of crystalline cellulose by either the individual enzymes or the possible synergistic interactions of combinations. Since *N. frontalis* and *N. patriciarum* are closely related it will be interesting to determine whether the cloned genes from *N. patriciarum* require complex formation (i.e., as a cellulosome) for maximum catalytic activity, as do those from *N. frontalis*. The *celA, celB*, and *celC* gene products were induced by growth of *N. patriciarum* on cellulose while *celD* was constitutive.

Neocallimastix strains also exhibit high secreted xylanase activity as well as a range of other hemicellulases including xylosidase, arabinofuranosidase, acetylxylan esterase, feruloyl esterase, and coumaroyl esterase (Borneman et al. 1990; Wubah et al. 1993). Many of these enzymes have been purified and characterized (Table 10.7). In the case of xylanases 1 and 2 from *N. frontalis* (Gomez de Segura and Fevre 1993), both exhibited endoglucanase activity while only xylanase 1 exhibited affinity for cellulose. Xylosidases purified from *N. frontalis* and *N. patriciarum* also exhibited low xylanase activity (Zhu et al. 1994a, Hebraud and Fevre 1990a). Low-molecular-mass feruloyl and coumaroyl esterases were isolated from *Neocallimastix* MC-2 (Table 10.7).

To date, three xylanase genes have been cloned from *N. patriciarum* strains (Black et al. 1994, Gilbert et al. 1992, Tamblyn Lee et al. 1993). *XynA* was composed of a signal sequence and two catalytic domains separated from each other and from a tandem 40-residue C-terminal repeat by a threonine/proline linker sequence (Table 10.6) (Gilbert et al. 1992). The catalytic domains exhibited considerable sequence homology, and both degraded xylan to primarily xylose and xylobiose. The *XynB* exhibited both xylanase and cellobiosidase activity, but reportedly was subject to extensive processing in *E. coli*, presumably proteolytic degradation (Black et al. 1994). The xylanase cloned by Tamblyn Lee et al. (1993) was also sensitive to some proteases.

6.5. Ruminal Protozoa

There is a lack of information on the exact role and extent of the contribution of the ruminal protozoa to plant cell wall degradation (Nagaraja et al. 1992). Furthermore, the actual presence of cell wall hydrolases is equivocal because no rumen protozoa have been grown in axenic culture; the possibility therefore exists, as suggested earlier, that the enzyme activities demonstrated by protozoa actually originate from the associated bacteria. However, studies with antibiotics indirectly illustrate that at least a proportion of the enzymatic activities originate with the protozoa. For example, *Epidinium ecaudatum* digests mesophyll tissue when grown in the presence of antibacterial antibiotics. Coleman (1978) also found that antibiotics had no effect on the cellulolytic activity of *Eudiplodinium maggi* and subsequently documented the cellulase content of 15 species of entodiniomorphid protozoa. Recently Coleman (1992) reported that *E. maggi* engulfed cellulose particles more rapidly than starch grains and the *E. maggi* and *Epidinium caudatum* synthesized amylopectin from the cellulose. Williams et al. (1984, 1986) demonstrated that the ciliate protozoa contain a broad array of glycanases and glycosidases which suggests that they possess hydrolytic activities similar to those of the ruminal bacteria and fungi. Experiments on ruminant digestion in the presence and absence of rumen protozoa indicate a positive effect of protozoa on fiber digestion (Ushida et al. 1991).

7. Pectin Degradation by Ruminal Bacteria and Fungi

Pectin is at its highest concentration in actively growing forage. The major enzymes involved in the hydrolysis of pectin include methylesterase, galacturonidase, and galacturonate lyase (Fig. 10.5). As shown in Tables 10.2, 10.3, and

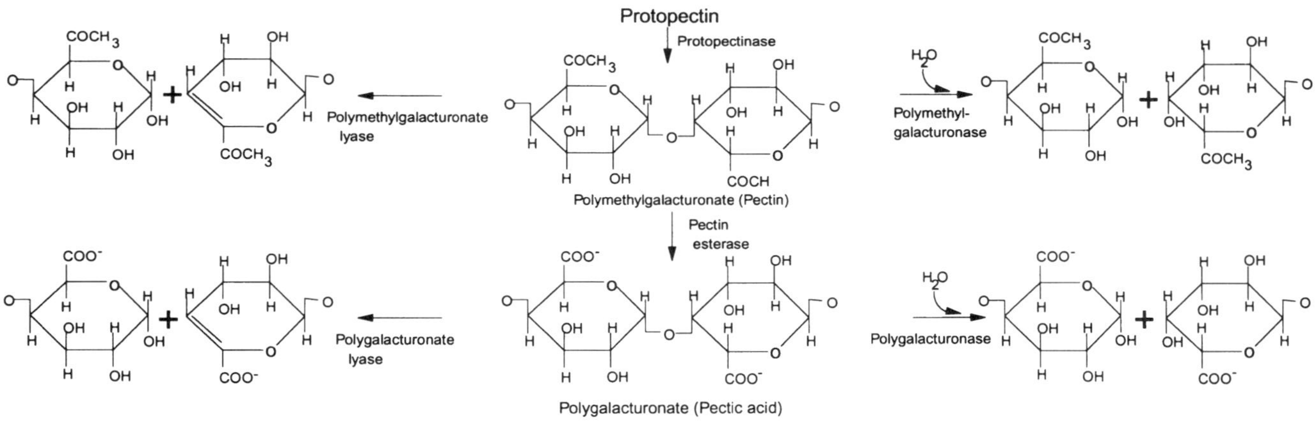

Figure 10.5. Schematic diagram illustrating the enzymes involved in the cleavage of pectin. Redrawn from Sakai (1992).

10.4, the major fibrolytic bacteria and fungi have limited capacity to degrade and use pectin as a carbon source for growth. Instead, *Lachnospira multiparus*, *B. fibrisolvens*, *Succinivibrio dextrinosolvens*, and *Treponema bryantii* seem to be the major species active in this regard. *L. multiparus*, which is very active in the colonization of fresh leaves (Cheng et al. 1991b), possesses polygalacturonate lyase (Table 10.7) and an exopolygalacturonase which cleaves the short chains of polygalacturonate to monomeric galacturonate residues that are used as a carbon source (Wojciechowicz et al. 1980). *B. fibrisolvens* produces a pectin esterase and an exopectate lyase (Wojciechowicz et al. 1982). Treponemes cultured by Wojciechowicz and Ziolecki (1979) possessed a complex of enzymes including pectin esterase, polygalacturonate lyase, polygalacturonate glucanhydrolase, and, in addition, a low level of exogalacturonase which was more active on polygalacturonic acid than on pectin, but did degrade it. *P. (Bacteroides) ruminicola* also possesses a polygalacturonate lyase (Wojciechowicz, 1972). *S. ruminantium* was unable to degrade intact pectin but did degrade polygalacturonic acid via a polygalacturonase (Heinrichova et al. 1989). In contrast to these bacteria, the anaerobic fungi do not use pectin or its degradation products as a carbon source; however, Gordon and Phillips (1992) detected extracellular pectinolytic activity in several fungi and have identified the enzyme to be a pectin lyase, presumably a polymethylgalacturonate lyase, since the extract lacked activity against polygalacturonic acid. They suggested that the enzyme may aid penetration of the fungal rhizoids through the pectin-containing middle lamella located between adjacent plant cells to allow the fungus access to other fermentable carbohydrates.

8. Effect of Lignin and Tannins on Cell Wall Polymer Degradation

The interference of lignin with cell wall digestion is routinely demonstrated by the negative correlation observed between lignin content and digestibility of organic matter in the rumen (Susmel and Stefanon 1993). Even the ruminal fungus *N. patriciarum*, shown by McSweeney et al. (1994) to be able to solubilize 33.6% of lignin in a sorghum stem fraction, was unable to effect digestion of this polymer. The fungus was found not to cleave the ether-linked phenolics. Rather, it hydrolyzed ester linkages, promoting solubilization of lignin-carbohydrate complexes. Phenolic esters, in particular, *p*-coumaric acid esters, have been suggested to be the primary impedance to degradation of some cell wall types, and biodegradation was improved by reducing the level of ester crosslinks (Akin et al. 1991, Hartley et al. 1992). It therefore seems that the presence of feruloyl esterase and coumaroyl esterase produced by the ruminal fungi (Borneman et al. 1991, 1992) and feruloyl esterase of *F. succinogenes* (McDermid et al. 1990b) in conjunction with xylanases may be crucial to the ruminal digestion and solubilization of highly lignified cell walls. It is possible that, with its esterases, *B. fibrisolvens* could also play a role (Akin et al. 1993). The product of fungal and bacterial esterases and hemicellulase enzymes are lignin-carbohydrate complexes such as those described by Ben-Ghedalia et al. (1994) and McSweeney (1994). These results provide support for the model of cell wall degradation proposed by Chesson (1993).

Condensed tannins, hydroxyflavanols formed from polymerization of leucoanthocyanidine and catechin inhibit digestion of plant carbohydrates and reduce protein digestibility (Susmel and Stefanon 1993). Little research has been done on the mechanism of action of condensed tannins; however, recent studies have shown that condensed tannins from birdsfoot trefoil inhibit cellulose digestion by *F. succinogenes* (Bae et al. 1993) and several ruminal fungi (McAllister et al. 1994).

9. Starch Structure and Degradation by Ruminal Microorganisms

The utilization of starch by ruminal microorganisms has been reviewed by Kotarski et al. (1992) and Chesson and Forsberg (1988). For completeness we will present some of the same information and provide an update.

9.1. Structure of Starch Granules

Starch is a heterologous polysaccharide composed of two polymers—amylose and amylopectin. Amylose is a linear molecule of 940 to 3,200 D-glucopyranose residues linked by α-1,4-bonds. Amylopectin is a larger, highly branched polymer averaging 10^4 to 10^5 glucose residues and consisting of α-1,4-linked D-glucose chains of 19 to 27 residues joined at branch points by α-1,6-bonds (Fig. 10.6) (Guilbot and Mercier 1985, Manners 1985). Starches purified from non-waxy and heterowaxy cultivars of various grains contain 14% to 34% amylose while those from waxy cultivars have less than 1% amylose. These polymers are deposited within endosperm cells in semicrystalline granules, 2 to 38 μm in diameter, which are minimally hydrated and stabilized by inter- and intramolecular hydrogen bonding. The proportions of amylose and amylopectin within the granules vary with plant species and cultivar (Guilbot and Mercier 1985). Starch in nature exists in granules in three different X-ray patterns described as A, B, or C (Katz 1934). Type A is found in cereals and is generally open to attack by amylase. Type B is typical of potato and banana and is resistant to amylase. Type C is a combination of A and B and is found in certain pea and bean starches which are also resistant (Fuewa et al. 1980). However, availability of starch in processed feed depends on the treatment process. When starch is cooked it becomes gelati-

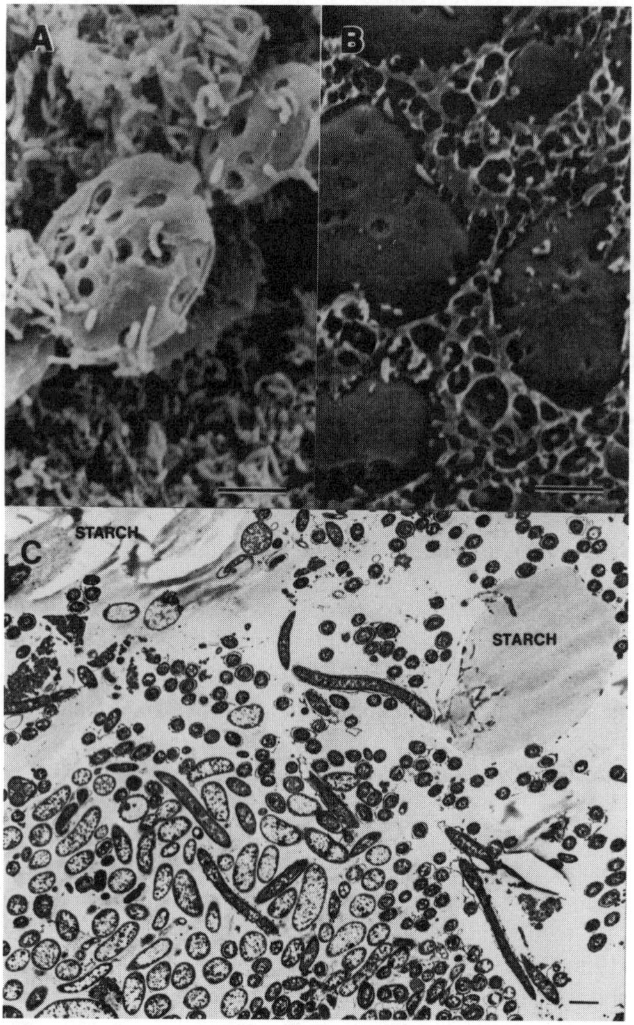

Figure 10.6. Colonization of starch granules by ruminal microorganisms. Scanning electron micrographs of (A) rumen bacteria attached to starch granules in the endosperm of hammer-milled barley. After 24 hours' incubation the protein matrix surrounding starch granules has been almost completely digested. (B) Intact protein matrix of barley that has been treated with formaldehyde. This treatment slows the rate of barley starch digestion because the intact protein matrix restricts access of digestive bacteria to the starch granules. (C) Transmission electron micrograph of the microbial biofilm on the surface of milled barley after 16 hours of exposure in the rumen. Note that regions immediately adjacent to starch granules is consistently dominated by a long thin gram-negative, rod-shaped bacterium. Bars in (A) and (B) = 5 μm; (C) = 1 μm. (Figures A and B are from McAllister et al. (1990a); Fig. C, from McAllister et al. (1993), with permission of the publishers.

nized and accessible to amylases. During other processing steps, it can undergo retrogradation which renders it resistant to pancreatic amylase.

9.2. Digestion of Starch by Ruminal Microorganisms

Fragmentation of the starch granule and uptake of water leads to gelatinization with the release of amylose and amylopectin. Cereal grains processed by rolling, cracking or grinding are readily digested in the rumen with barley the most digestible (87% to 90%), while corn (50% to 90%) and sorghum (42% to 89%) are less digestible (reviewed by Kotarski et al. 1992). Waxy grains are digested more rapidly than the non-waxy grains (MacGregor and Ballance 1980).

Bacteria, protozoa, and all fungi contribute to starch degradation within the rumen. Colonization of starch granules by ruminal microorganisms is illustrated in Figure 10.7. The relative contribution of each group undoubtedly varies; however, under conditions of high grain ration being fed to the ruminant, the ruminal pH is below 6.0, and growth of ruminal fungi (Gordon and Phillips 1989b) and protozoa (Hobson and Jouany 1988) is somewhat inhibited. The major ruminal bacteria reported to ferment starch include *Streptococcus bovis*, *P. ruminicola*, *Butyrivibrio* sp., and *Ruminobacter (Bacteroides) amylophilus* (Cotta 1988). Other amylolytic species reported include *F. succinogenes*, *Succinimonas amylolytica*, and *Eubacterium ruminantium* (Hungate 1966). McAllister et al. (1990b) found that *S. bovis*, *R. amylophilus*, and *B. fibrisolvens* exhibited different digestive capacities for starch in barley, maize, and wheat. *S. bovis* digested wheat starch to a greater extent than either barley or maize starch, while *R. amylophilus* exhibited greatest digestion of barley starch, and *B. fibrisolvens*, which digested

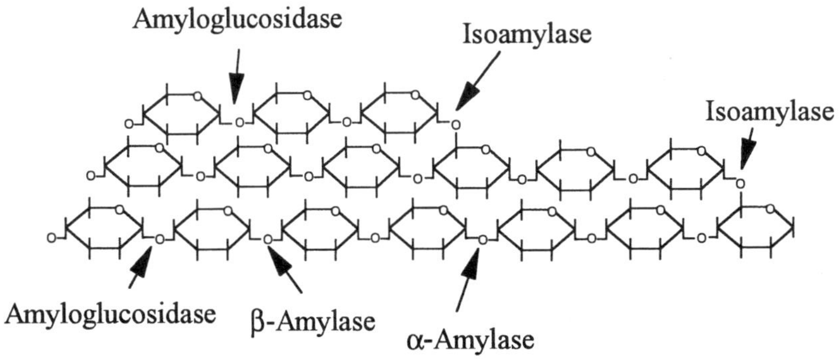

Figure 10.7. Schematic diagram of amylopectin. The polymer is composed of short chains of α-(1–4) linked glucose residues in chains cross-linked by α-(1-6) bonds. The sites of cleavage by various amylolytic enzymes are shown.

starch in barley and maize to a similar extent, was unable to digest wheat starch. Scanning electron microscopy revealed that *B. fibrisolvens* digested mainly the protein coating on the starch granules while *S. bovis* colonized the endosperm randomly and *R. amylophilus* preferentially colonized the starch granules. Thus the mode of attack of amylolytic organisms on endosperm tissue can vary significantly with digestive species. Work by Minato and Suto (1978) indicates that attachment to starch may be important for its hydrolysis by *R. amylophilus* and various *Prevotella* strains but not by either *S. bovis* or *B. fibrisolvens*.

Cereal starch digestion by ruminal fungi including *N. patriciarum*, *O. joyonii*, and *Piromyces communis* was investigated by McAllister et al. (1993). Of these, *O. joyonii* exhibited the greatest digestive capability and the least specificity for cereal type. Although the fungi penetrated the surface protein matrix of the granule and remained in close association with it, they did not penetrate into the granule. Pitting, presumably from extracellular enzymes, was observed instead.

A general concept supported by mounting evidence is that starch digestion by ruminal microorganisms is limited by structural carbohydrates and protein in the endosperm matrix (Kotarski et al. 1992). The role of protein in the endosperm matrix in selection of the microbial species degrading starch and slowing the rate of degradation is supported by studies of the effect of formaldehyde treatment of barley grain, which amplifies the protein effect on digestion of starch granules (McAllister et al. 1990a).

9.3. Enzymology of Starch Hydrolysis by Ruminal Microorganisms

The complete biodegradation of amylodextrin starch by different microorganisms involves the combined action of several enzymes including α-amylase, isoamylase, β-amylase, glucoamylase, and α-glucosidase (Fig. 10.6). α-Amylase randomly cleaves internal α-1,4 linkages of the amylose backbone, isoamylase cleaves the α-1,6 branch points of amylodextrin, and glucoamylase and β-amylase cleave glucose and maltose residues, respectively, from the nonreducing termini of amylose. Glucoamylase is the least specific of the starch-degrading enzymes and enzymes from most sources cleave α-1,3 and α-1,6 as well as α-1,4 linkages, but at different rates of hydrolysis (IUB Enzyme Nomenclature, 1992, Pazur and Kleppe 1962). All amylolytic organisms seem to possess an α-glucosidase. There is limited knowledge of the enzymology and fine structure of the starch-degrading enzymes of the rumen microbiota. Only two α-amylases have been purified (Table 10.6). These are a highly active 77-kDa enzyme from *S. bovis* (Freer 1993) and a lower-activity enzyme from *R. (Bacteroides) amylophilus* (McWethy and Hartman 1977). The *S. bovis* enzyme degraded both potato and corn starches, while the *R. amylophilus* enzyme had relatively low activity against potato starch.

Cotta (1988) found that amylase from *S. bovis*, *B. fibrisolvens*, and *P. ruminicola* were inhibited by glucose, but detailed molecular studies are lacking. The amylases were only partially secreted, except from *S. bovis* and *B. fibrisolvens* grown on maltose where all of the enzyme was secreted.

Three α-amylase genes have been cloned, two from *S. bovis* strains (Cotta and Whitehead 1993, Clark et al. 1992) and one from *B. fibrisolvens* (Rumbak et al. 1991a). The amylase gene from *B. fibrisolvens* codes for a 107-kDa enzyme. Amino acid sequence analysis indicated that it possesses conserved regions for substrate binding, catalysis, and binding of calcium, which are essential for catalysis. The enzyme hydrolyzes amylose, amylopectin, and glycogen, but is neither adsorbed to nor hydrolyzed raw corn starch granules. A glycogen branch-forming enzyme that exhibited starch clearing activity was also cloned from *B. fibrisolvens* (Rumbak et al. 1991b). Apparently the enzyme also inserted branches into the starch substrate present in the surrounding medium. This increased the solubility of the polymer and enabled it to diffuse in the agar, giving rise to a halo on a starch azure plate. Much more work on the enzymes involved in starch hydrolysis is necessary because of complex interaction of amylase with the particulate forms of starch.

10. Microbial Interactions

The impact of microbial interactions within the rumen on plant cell wall digestion may be either positive or negative. Rapid and extensive degradation of forage to monomeric components depends on the presence of a combination of cellulolytic and noncellulolytic microbes (Dehority 1993). For example, *F. succinogenes* lacks the ability to use pentose sugars, but *B. fibrisolvens* has this capacity, and the combination of the two species brings about extensive utilization of pentoses as well as hexoses in plant cell walls (Miron et al. 1994). Kudo et al. (1986) showed that noncellulolytic treponemes stimulate in vitro degradation of straw by *F. succinogenes*. Coculture with methanogens has been shown to stimulate cellulolytic and xylanolytic activities of anaerobic fungi (Joblin et al. 1990, Marvin-Sikkema et al. 1990, Teunissen et al. 1992) and of *R. albus* (Pavlostathis et al. 1990). These interactions are dramatically illustrated in Figure 10.8 for *R. flavefaciens*, *T. bryantii*, and *Methanobacterium smithii*. Presence of either or both of the latter two species in coculture with *R. flavefaciens* substantially increased cellulolytic activity, even though neither of them can degrade this substrate alone.

An example of negative impact of some microbial associations on ruminal forage digestion is the restriction of fungal colonization of plant material by the presence of ruminal bacteria. The production of an extracellular protein by *R. albus* and *R. flavefaciens* inhibits fungal cellulolytic activity by binding either to

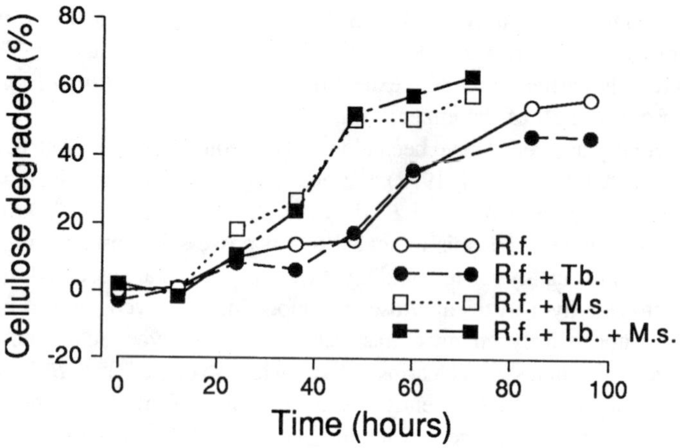

Figure 10.8. Percentage of cellulose digestion in cultures consisting of *Ruminococcus flavefaciens* (*R.f.*), *R. flavefaciens* + *Treponema bryantii* (*T.b.*), *R. flavefaciens* + *Methanobrevibacter smithii* (*M.s.*), *R. flavefaciens* + *M. smithii* + *Treponema bryantii.*

critical fungal cellulases or to the cellulose substrate (Bernalier et al. 1993, Stewart et al. 1992).

Furthermore, recent research using 16S rRNA-targeted, oligonucleotide hybridization probes for *R. albus* and *R. flavefaciens* and *F. succinogenes* to measure relative populations during in vitro competition studies showed that *R. albus* 8 was able to produce a heat-stable protein factor which inhibits the growth of *R. flavefaciens* FD-1 (Odenyo et al. 1994a,b). This is the first demonstration of the production of a bacteriocinlike substance by a rumen bacterium. Presumably this is a mechanism used to compete for nutrients.

11. General Features of the Microbial Fermentation in the Large Intestine

The large intestine consists of the cecum, colon, and rectum. For convenience it will be referred to as the large intestine. Much of the information available is from studies of bacteria which are considered important in the human large intestine, since most research on intestinal fermentations has been done on the human tract. The major polysaccharide-degrading bacteria found in the large intestine include species of *Bacteroides*, *Bifidobacterium*, *Ruminococcus*, *Eubacterium*,

Table 10.8. Approximate composition of substrates thought to be available for fermentation by bacteria in the large intestine in persons consuming Western diets[a]

Substrate	% (Range)	Origin
Resistant starch	28–42	Starchy foods
Nonstarch plant polysaccharides	28–42	All plant foods
Oligosaccharides	7–9	Legumes, root vegetables, artichokes
Unabsorbed sugars	7–9	Milk (lactose deficiency), raffinose, stachyose
Dietary protein	9–10	Whole seeds and grains
Pancreatic enzymes and other gut secretions	6–14	Host
Mucus	3–7	Host
Sloughed epithelial cells	Unknown	Host

[a] The total substrates available per day ranges from 29 to 94 g day. Data recalculated from Cummings and Macfarlane (1991) and some additional information added.

and *Clostridium* (Cummings and Macfarlane 1991, Macfarlane and Macfarlane 1993). They are present at $\log_{10}$ numbers ranging from 9.6 to 11.3/g dry weight of fecal material. Protozoa are not found in the healthy human gut, and yeasts and the fungi are present only in low numbers (Cummings and Macfarlane 1991, Macfarlane and Macfarlane 1993). Bacteria are the major component of feces, accounting for 40% to 55% of the fecal solids in persons consuming Western diets (Cabotaje et al. 1990). The intestinal biota of other monogastric animals may include both fungi (Trinci et al. 1994) and protozoa.

Composition of the substrate materials available to the gut biota of persons consuming Western diets is shown in Table 10.8. The structures of several of the polymers are illustrated in Figures 10.6 and 10.9. Starch that has bypassed the small intestine is the major substrate in intestinal fermentation. The amount and availability of starch in the large intestine depend upon the type of food eaten, the manner in which it was processed for feed, and the extent to which it had been ground and chewed (see also Sect. 9.1). Starch in intact seeds is practically inaccessible, but breaching the seed coat makes it readily available. Nonstarch plant polysaccharides include plant cell walls and their constituents, pectins, inulin, seed mucilage, and plant gums. Guar gum, a polysaccharide from a legume cultivated in India, is commonly added to foods as an emulsifier. It is composed of β(1-4)-linked mannose residues forming the backbone with α(1-6)-linked galactose branches and has a galactose to mannose ratio of approximately 1:2 (Fig. 10.9). The oligosaccharides raffinose and stachyose (Fig. 10.9) from beans, and

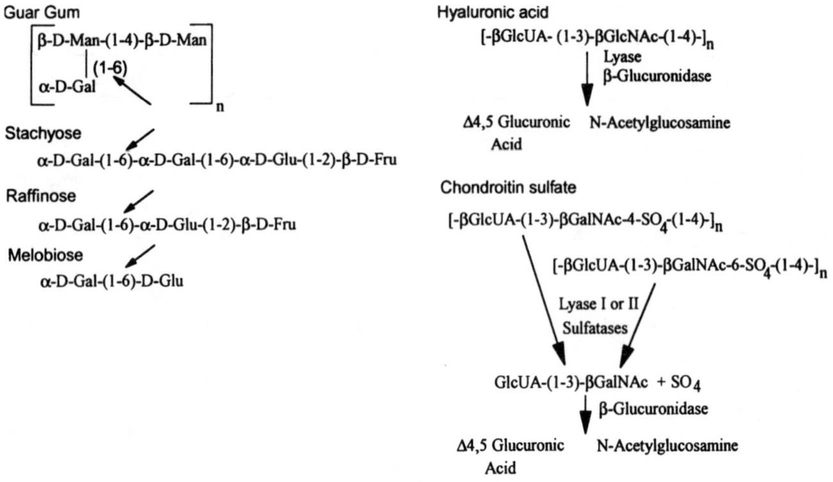

Figure 10.9. Structures of the α-galactosides, melibiose, raffinose, stachyose and guar gum, and the site of cleavage by α-galactosidase (indicated by arrows); and structures of the polysaccharides hyaluronic acid and chondroitin sulfate, and enzymes involved in their hydrolysis.

the fructooligosaccharides from artichokes, onions, and some root crops are the main sources of sugars because they are not absorbed in the small intestine. The oligosaccharides are completely hydrolyzed to sugars and fermented. Up to 50% of cellulose and 80% of noncellulosic polysaccharides are in addition degraded and utilized by intestinal microbes (Cummings 1981) with accompanying gas production.

In addition to carbohydrates of dietary origin, host-produced substances, mainly glycoproteins, such as mucin, chondroitin sulfate (Fig. 10.9), and hyaluronic acid (Fig. 10.9) contribute to substrate for microbial activity. Mucins have a peptide core with carbohydrate side chains that contain mainly galactose and hexosamines such as N-acetylglucosamine and some fucose (Hoskins and Boulding 1981). The carbohydrate moieties occur as linear- and branched-chain oligosaccharides which can constitute up to 85% of the molecular mass (Smith and Podolsky 1986). Hyaluronic acid and chondroitin sulfate are linear mucopolysaccharides consisting of repeating units of glucuronic acid linked to N-acetylgalactosamine (Fig. 10.9). Chondroitin sulfate differs from hyaluronic acid by being substituted with sulfate on the N-acetylgalactosamine residue at either the 4 or the 6 hydroxyl group.

12. Enzymology of Polymer Degradation by Intestinal Microorganisms

12.1. Starch

Practically all major species of bacteria in the intestine use starch as a substrate for growth (Table 10.9). Tancula et al. (1992) have shown that *B. thetaiotaomicron* cells bind starch to their surface in the course of digesting the polysaccharide. Glucan strands are translocated through the outer membrane into the periplasm, where digestion occurs. Three maltose-inducible outer membrane bound proteins and one cytoplasmic membrane protein appear essential for digestion. The genes coding for these proteins are clustered on the *B. thetaiotaomicron* chromosome. In other studies in the same laboratory, amylase and pullulanase were found to be either periplasmic or associated with the membrane of this organism (Anderson and Salyers 1989a,b, Smith and Salyers 1991). Pullulanase cleaved α-(1-6)-D-glucosidic linkages, a neopullulanase cleaved α-(1-4) linkages, and α-glucosidase cleaved α-(1-4)-D-glucosidic linkages of maltooligosaccharides ranging in size from maltose to maltoheptose, and in addition cleaved α-(1-6)-linked oligosaccharides such as panose (6-α-glucosyl-maltose).

Geun Eog et al. (1992) reported the isolation of a largely extracellular 66-kDa amylase from *Bifidobacterium* sp. Int-57 which cleaves starch in an endo fashion, producing maltose and maltotriose. The enzyme exhibited low activity on pullulan and glycogen. The gene was also cloned, but no further information is available.

Table 10.9. Polysaccharides degraded by intestinal bacteria of humans

Bacteria	Polysaccharide	References
Bacteroides ovatus, B. thetaiotaomicron, and other *Bacteroides* spp.	Starch, pullulan, arabinogalactan, pectin, chondroitin sulfate, guar gum, heparin, mucins	Salyers et al. 1977, Macfarlane et al. 1990
Bifidobacterium spp.	Starch, pullulan, xylan, arabinogalactan, gum arabic, pectin, inulin, raffinose, stachyose	Hoskins et al. 1985, Sakai et al. 1987, Gavini et al. 1991; McKellar and Modler 1989, Geun Eog et al. 1992
Clostridium spp.	Starch, pectin	
Eubacterium spp.	Starch, pectin	
Ruminococcus spp.	Guar gum, mucin	

Source: Macfarlane and Macfarlane (1993), with additional documentation listed in the table.

Information is lacking on strains able to use forms of starch resistant to pancreatic amylase.

12.2. Nonstarch Plant Polysaccharides, Sugars, and Miscellaneous Compounds

In rat and pig intestines cellulolytic *F. succinogenes* and *R. flavefaciens* species were found to account for 4% to 6% of the total viable cells (Macy et al. 1982, Varel et al. 1984, Montgomery et al. 1988). These potent cellulolytic bacteria are important hemicellulose digesters as well. Although no work has been done on the fibrolytic enzymes of *F. intestinalis*, the *Fibrobacter* species may be predominant since it is commonly found in the intestine (Montgomery et al. 1988). In contrast to the rat and pig, cellulolytic bacteria have not been reproducibly found in fecal samples from humans (Montgomery 1988).

B. thetaiotaomicron is an important pectin digester and produces both a polygalacturonic acid lyase and a polygalacturonic hydrolase (McCarthy et al. 1985). The hydrolase appears to be on the inner membrane. The main hydrolysis product of the lyase was an unsaturated dissacharide, while the hydrolase produced primarily galacturonic acid. No methyl esterase has been reported, although the bacterium undoubtedly requires the enzyme to carry out extensive digestion of the polymer.

B. ovatus, another predominant species of the gut microbiota, readily degrades galactans and α-galactosides. It produces two cell-associated galactomannanases and two cell-associated α-galactosidases, designated α-galactosidase I and II. These enzymes, with the exception of α-galactosidase II, are induced by growth on guar gum and can completely degrade the polymer to monosaccharides (Gherardini et al. 1985, Gherardini and Salyers 1987, Valentine et al. 1992).

α-Galactosidase II is produced by *B. ovatus* when it is grown on galactose, melibiose, raffinose, or stachyose. Neither α-galactosidase I nor II releases galactose from intact guar gum, but both act after partial hydrolysis of the molecule by galactomannanase. Both α-galactosidases I and II release galactose from melibiose, raffinose, and stachyose (Gherardini et al. 1985, Gherardini and Salyers 1987). The α-galactosidase II exhibited the highest affinity for substrates. Hydrolysis products from raffinose and stachyose are galactose and sucrose, which is subsequently hydrolyzed by invertase. Two-dimensional gel analysis of *B. ovatus* extracts prepared from cells grown on guar gum revealed 20 membrane polypeptides and 12 soluble polypeptides regulated in a manner similar to that of the enzymes, indicating that the hydrolysis system was exquisitely complex (Valentine et al. 1992).

Many *Bifidobacterium* strains utilize raffinose and stachyose and this is attributed to the presence of a 330-kDa α-galactosidase composed of eight 39-kDa subunits (Sakai et al. 1987). Some *Bifidobacterium* species grow on inulin and shorter fructooligosaccharides because of a cell-associated β-fructosidase (McKellar and Modler 1989). Fructooligosaccharides do not support the growth of *Clostridium perfringens*, *E. coli*, or other undesirable bacteria (Hidaka et al. 1986), which is the basis for use of fructooligosaccharides to enhance the population density of *Bifidobacterium* species in the intestine, thereby promoting a "more desirable" intestinal fermentation. *Bifidobacterium* spp. also utilize several pentose sugars (Geun Eog et al. 1992, Gavini et al. 1991) and xylooligosaccharides. Xylooligosaccharides are promoted as an alternative to fructooligosaccharides as species-selective agents (Okazaki et al. 1990). However, in comparison to the *Bacteroides* species, the hydrolytic capabilities of *Bifidobacterium* spp. are more limited.

12.3. Mucin, Chondroitin Sulfate, and Hyaluronic Acid

Bifidobacterium spp., *Ruminococcus* spp., and some *Bacteroides* spp. have major roles in degradation of mucin, chondroitin sulfate, and hyaluronic acid (Hoskins et al. 1985, Roberton and Stanley 1982). The extensive degradation of mucin affected by these species requires a combination of enzymes including a sialidase, a specific α-glycosidase, the requisite β-D-galactosidase, and β-N-acetylglucosaminidase. Hoskins et al. (1985) reported the identification of two *Ruminococcus torques* strains which produced all of these enzymes and released greater than 90% of hexoses from hog gastric mucin. *Bifidobacterium* strains lacked the specific α-galactosidase, but possessed sialidase and β-glycosidases. Some salivary glycoproteins were partially degraded by a *R. albus* strain which possessed the specific α-galactosidase activity, but lacked N-acetylhexosaminidase. Various other fecal *Bacteroides*, *E. coli*, and *S. faecalis* strains produced a similar complement of enzymes.

Mucopolysaccharides such as chondroitin sulfate and hyaluronic acid are degraded by *Bacteroides* sp. (Gibson et al. 1988, Salyers et al. 1982). *B. thetaiotaomicron* produces five separate enzymes—two lyases, two sulfatases, and a β-glucuronidase—which degrade the polymer to glucuronic acid, N-acetylgalactosamine, and sulfate (Hwa and Salyers 1992).

13. Uptake and Metabolism of Monosaccharides and Disaccharides

The oligosaccharides and sugars arising from plant cell wall digestion serve as a carbon and energy source for the attached fibrolytic bacteria. Those degradation products that escape are used by the unattached microorganisms; however, the maximum size of oligosaccharides transported into bacterial cells is unknown. The diversity of uptake systems that exist is evidenced by the number of sugars utilized (Tables 10.2, 10.3). Uptake of glucose, cellobiose, xylose, and arabinose by a variety of rumen bacteria has been examined. *Fibrobacter* spp. apparently possess separate sodium-dependent symport systems for uptake of glucose (Franklund and Glass 1987) and for cellobiose (Maas and Glass 1990). Both transport systems were produced constitutively. In contrast, *F. succinogenes* lacks the ability to take up and utilize xylose, presumably because of the lack of a xylose permease (Matte et al. 1992). *R. flavefaciens* possesses a constitutively synthesized cellobiose phosphorylase for cellobiose uptake (Helaszek and White 1993). However, *R. albus* seems to possess a common ATP driven permease for glucose and xylose uptake (Thurston et al. 1994) though cellobiose is the preferred substrate over xylose and arabinose. Furthermore, cellobiose was observed to decrease pentose metabolism by repression of both transport activity and the catabolic enzymes isomerase and kinase (Thurston et al. 1993, 1994).

The transport mechanisms for uptake of sugars by *B. fibrisolvens* have not been characterized, but the bacterial strains show different discrete preferences for sugar utilization, either co-utilizing glucose or cellobiose and pentoses or with a preference for cellobiose over pentose sugars (Strobel and Dawson 1993).

Prevotella ruminicola metabolizes pentoses and glucose or maltose simultaneously, but preferentially ferments pentoses over cellobiose (Strobel 1993b). A sodium-dependent symport mechanism for xylose and arabinose uptake was apparent in this species (Strobel 1993b). In contrast to cellulolytic and hemicellulolytic bacteria, *S. ruminantium* uses glucose phosphotransferase and a pentose-proton symport system for hexose and pentose uptake, respectively (Strobel 1993a). Grown in the presence of glucose and either xylose or arabinose, *S. ruminatium* used the glucose preferentially. However, utilization of cellobiose and pentoses was simultaneous.

Hexoses taken up are metabolized via glycolysis in all ruminal bacteria (Joyner and Baldwin 1966, Kistner and Kotze 1973), while pentoses are generally isomerized, phosphorylated, and further metabolized via the transketolase pathway (see Matte et al. 1992 for a discussion).

14. Genetic Manipulation of the Fibrolytic Capability of Ruminal and Nonruminal Bacteria

Although numerous reviews have been written on the potential for genetic manipulation of ruminal bacteria to enhance plant cell wall digestion (Forsberg and Cheng 1992, Forsberg et al. 1986, Hespell and Whitehead 1990, Russell and Wilson 1988, Smith and Hespell 1983), only now are we on the threshold of modification of microorganisms. For example, Maglione et al. (1992) have reported that fusion of a gene coding for a cellulose-binding domain from *Thermomonospora fusca* with the catalytic domain from *P. ruminicola* $B_1 4$ gave a 10-

fold increase in activity against crystalline cellulose. Introduction of this construct into *P. ruminicola* may produce a bacterium with enhanced capacity for survival and cellulolytic activity in a low-pH environment, such as that encountered in the rumina of cattle receiving high-energy (grain-based) diets. Establishment of such an organism in the ruminal population would then improve cellulose utilization by the ruminant (Russell and Wilson 1988). In a different approach, Whitehead et al. (1991) introduced an endoxylanase gene from *B. ruminicola* into the chromosome of *B. thetaiotaomicron* and obtained stable amplified endoxylanase production in the absence of a selective antibiotic to maintain the plasmid in the cell. The reported objective (Whitehead et al. 1991) was to introduce the modified intestinal bacterium into the rumen with the expectation that it would dramatically enhance xylanase activity and perhaps plant cell wall digestion. These are interesting test cases, many more of which will follow as transformation systems and suitable vectors are developed for other candidate ruminal bacteria.

15. Future Prospects in Ruminant and Nonruminant Animal Feeding

In the past the emphasis has been on chemical pretreatment of plant materials to improve the extent of digestion (Chesson 1993). This will continue, although the approach will likely shift to enzymic pretreatment of plant materials (e.g., with cellulases and hemicellulases) as evidence mounts of the efficiency of this method in increased digestibility in the rumen (Nakashima et al. 1988) and in the monogastric digestive system (Campbell and Bedford 1992). Therefore, it is possible that genetic modification of ruminal microorganisms to overproduce selected cellulase and hemicellulase enzymes will benefit the ruminal fermentation. In monogastric animals we may see a change from the feeding of microbial enzymes to enhance feed digestion to the use of genetically modified forages and cereals which contain the appropriate hydrolytic enzymes in vacuoles or in the cytosol to be released from these inactive sites by damage to the plant tissues. Plant breeding is now sufficiently developed that this is a feasible approach (Comai 1993). For example, a *Bacillus* amylase has been expressed in seeds of tobacco plants (Pen et al. 1992), and Ni et al. (1994) recently reported the generation of transgenic plants which are blocked in lignin synthesis. Plants lacking lignin will have enhanced digestibility in ruminants and monogastric animals.

In monogastric animals such as the pig and chicken, it will eventually be possible for the animal itself to produce the key fibrolytic enzymes through the generation of transgenic animals which secrete glucanase or a xylanase from the pancreas, as has been demonstrated in the mouse (Hall et al. 1993), or perhaps directly from epithelial cells lining the small intestine. These modifications may prove to have a dual benefit as intestinal secretion of a xylanase by swine would enhance nutrient absorption and the xylooligosaccharide products from xylanase action would enrich for *Bifidobacterium* spp. in the intestine, thus providing a more favorable intestinal environment (Okazaki et al. 1990).

These examples invite contemplation of the exciting possibilities the future holds for improving feed conversion in domestic animals.

References

Akin DE, Benner R (1988) Degradation of polysaccharides and lignin by ruminal bacteria and fungi. Appl Environ Microbiol 54:1117–1125.

Akin DE, Gordon GRL, Hogan JP (1983) Rumen bacterial and fungal degradation of *Digitaria pentzii* grown with or without sulfur. Appl Environ Microbiol 46:738–748 .

Akin DE, Lyon CE, Windham WR, Rigsby LL (1989) Physical degradation of lignified stem tissues by ruminal fungi. Appl Environ Microbiol 55:611–616.

Akin DE, Rigsby LL, Hanna WW, Gates RN (1991) Structure and digestibility of tissues in normal and brown midrib pearl millet (*Pennisetum glaucum*). J Sci Food Agric 56: 523–538.

Akin DE, Borneman WS, Rigsby LL, Martin SA (1993) *p*-Coumaroyl and feruloyl arabinoxylans from plant cell walls as substrates for ruminal bacteria. Appl Environ Microbiol 59:644–647.

Anderson KL, Salyers AA (1989a) Biochemical evidence that starch breakdown by *Bacteroides thetaiotaomicron* involves outer membrane starch-binding sites and periplasmic starch-degrading enzymes. J Bacteriol 171:3192–3198.

Anderson KL, Salyers AA (1989b) Genetic evidence that outer membrane binding of starch is required for starch utilization by *Bacteroides thetaiotaomicron*. J Bacteriol 171:3199–3204.

Asmundson RV, Huang C-M, Kelly WJ, Yu P-L, Curry MM (1990) The cellulase of *Ruminococcus flavefaciens* 186: characterization, cloning and use in ruminant nutrition. In: Akin DE, Ljungdahl LG, Wilson JR, Harris PJ, eds. Microbial and Plant Opportunities to Improve Lignocellulose Utilization by Ruminants, pp. 401–409. New York: Elsevier.

Attwood GT, Davies NJ, Herreras F, White BA (1992) Cloning and partial characterization of a cellulase gene from the rumen anaerobe, *Ruminococcus albus*. Abstr. Ann. Meet. Amer Soc Microbiol p. 281.

Attwood GT, Herreras F, Weissensten L, White BA (1996) An endo-β-1,4-glucanase gene (*cel*A) from the rumen anaerobe *Ruminococcus albus* 8: cloning, sequencing and transcriptional analysis. Can J Microbiol. 42:267–278.

Bacic A, Harris PJ, Stone BA (1988) Structure and function of plant cell walls. In: Stumpf PK, Conn EE, eds. The Biochemistry of Plants, pp. 297–371. New York: Academic Press.

Bae HD, McAllister TA, Yanke J, Cheng KJ, Muir AD (1993) Effects of condensed tannins on endoglucanase activity and filter paper digestion by *Fibrobacter succinogenes* S85. Appl Environ Microbiol 59:2132–2138.

Barros MEC, Thomson JA (1987) Cloning and expression in *Escherichia coli* of a cellulase gene from *Ruminococcus flavefaciens*. J Bacteriol 169:1760–1762.

Beaudette, LA (1994) The effect of methanogenesis on the consortial degradation of cellulose. Ph.D. thesis, University of Calgary, Calgary, Alberta, Canada.

Béguin P, Aubert JP (1994) The biological degradation of cellulose. FEMS Microbiol Rev 13:25–58.

Ben-Ghedalia D, Yosef E, Solomon R, et al. (1994) Size exclusion chromatography of cotton stalk lignins isolated from rumen digesta and feces of sheep. J Agric Food Chem 41:1160–1163.

Berger E, Jones WA, Jones DT, Woods DR (1989) Cloning and sequencing of an endoglucanase (*end1*) gene from *Butyrivibrio fibrisolvens* H17c. Mol Gen Genet 219:183–198.

Berger E, Jones WA, Jones DT, Woods DR (1990) Sequencing and expression of a cellodextrinase (*ced1*) gene from *Butyrivibrio fibrisolvens* H17c cloned in *Escherichia coli*. Mol Gen Genet 223:310–318.

Bernalier A, Fonty G, Bonnemoy F, Gouet P (1993) Inhibition of the cellulolytic activity of *Neocallimastix frontalis* by *Ruminococcus flavefaciens*. J Gen Microbiol 139:873–880.

Bhat S, Wallace RJ, Orskov ER (1990) Adhesion of cellulolytic ruminal bacteria to barley straw. Appl Environ Microbiol 56:2698–2703.

Black GW, Hazlewood GP, Xue GP, Orpin CG, Gilbert HJ (1994) Xylanase B from *Neocallimastix patriciarum* contains a non-catalytic 455-residue linker sequence comprised of 57 repeats of an octapeptide. Biochem J 299:381–387.

Borneman WS, Hartley RD, Morrison WH, Akin DE, Ljungdahl LG (1990) Feruloyl and *p*-coumaroyl esterase from anaerobic fungi in relation to plant cell wall degradation. Appl Microbiol Biotechnol 33:345–351.

Borneman WS, Ljungdahl LG, Hartley RD, Akin DE (1991) Isolation and characterization of p-coumaroyl esterase from the anaerobic fungus *Neocallimastix* strain MC-2. Appl Environ Microbiol 57:2337–2344.

Borneman WS, Ljungdahl LG, Hartley RD, Akin DE (1992) Purification and partial characterization of two feruloyl esterases from the anaerobic fungus *Neocallimastix* strain MC-2. Appl Environ Microbiol 58:3762–3766.

Broussolle V, Forano E, Gaudet G, Ribot Y (1993) Nucleotide sequence of the *endB* gene of *Fibrobacter succinogenes* S85 encoding a novel family E β-1,4-endoglucanase. GenBank L14436.

Cabotaje LM, Lipez-Guisa JM, Shinnick FL, Marlett JA (1990) Neutral sugar composition and gravimetric yield of plant and bacterial fractions of faeces. Appl Environ Microbiol 56:1786–1792.

Campbell GL, Bedford MR (1992) Enzyme applications for mongastric feeds: a review. Can J Anim Sci 72:449–466.

Carpita N, Gibeaut DM (1993) Structural models of primary cell walls of flowering plants: consistency of molecular structure with the physical properties of the walls during growth. Plant J 3:1–30.

Cavicchioli R, East PD, Watson K (1991) *end*A_{FS}, a novel family E endoglucanase gene from *Fibrobacter succinogenes* AR1. J Bacteriol 173:3265–3268.

Chen H, Xinliang L, Ljungdahl LG (1994) Isolation and properties of an extracellular β-glucosidase from the polycentric rumen fungus *Orpinomyces* sp. strain PC-2. Appl Environ Microbiol 60:64–70.

Cheng K-J, Fay JP, Howarth RE, Costerton JW (1980) Sequence of events in the digestion of fresh legume leaves by rumen bacteria. Appl Environ Microbiol 40:613–625.

Cheng K-J, Kudo H, Duncan SH, Mesbah A, Stewart CS, Bernalier A, Fonty G, Costerton JW (1991a) Prevention of fungal colonization and digestion of cellulose by the addition of methylcellulose. Can J Microbiol 37:484–487.

Cheng KJ, Forsberg CW, Minato H, et al (1991b) Microbial ecology and physiology of feed degradation within the rumen. In: Tsuda T, Sasaki H, Kawashima R, eds. Physiological Aspects of Digestion and Metabolism in Ruminants, pp. 595–624. New York: Academic Press.

Chesson A (1993) Mechanistic models of forage cell wall degradation. In: Jung HG, Buxton DR, Hatfield RD, eds. Forage Cell Wall Structure and Digestibility, pp. 347–376. Madison: American Society of Agronomy.

Chesson A, Forsberg CW (1988) Polysaccharide degradation by rumen microorganisms. In: Hobson PN, ed. The Rumen Microbial Ecosystem, pp. 251–284. London: Elsevier Science.

Chesson A, Gordon AH, Lomax JA (1985) Methylation analysis of mesophyll, epidermis, and fibre cell-walls isolated from the leaves of perennial and Italian ryegrass. Carbohydr Res 141:137–147.

Claeyssens M, Henrissat B (1992) Specificity mapping of cellulolytic enzymes—classification into families of structurally related proteins confirmed by biochemical analysis. Protein Sci 1:1293–1297.

Clark, RG, Cheng K-J, Selinger LB, Hynes MF (1994) A conjugative transfer system for rumen bacterium *Butyrivibrio fibrisolvens* based on Tn916-mediated transfer of the *Staphylococcus aureus* plasmid pUB110. Plasmid. 32:295–305.

Clark RG, Hu Y-J, Hynes MF, Salmon RK, Cheng K-J (1992) Cloning and expression of an amylase gene from *Streptococcus bovis* in *Escherichia coli*. Arch Microbiol 157: 201–204.

Coleman GS (1978) The metabolism of cellulose, glucose, and starch by the rumen ciliate protozoa *Eudiplodinium maggii*. J Gen Microbiol 107:359–366.

Coleman GS (1985) The cellulase content of 15 species of entodiniomorphid protozoa, mixed bacteria and plant debris isolated from the ovine rumen. J Agric Sci 107:709–721.

Coleman GS (1986) The amylase activity of 14 species of entodiniomorphid protozoa and the distribution of amylase in digesta fractions of sheep containing no protozoa or one of seven different protozoal populations. J Agric Sci 107:709–721.

Coleman GS (1992) The rate of uptake and metabolism of starch grains and cellulose particles by *Entodinium* species, *Eudiplodinium maggii*, some other entodiniomorphid protozoa and natural protozoal populations taken from the ovine rumen. J Appl Bacteriol 73:507–513.

Comai L (1993) Impact of plant genetic engineering on foods and nutrition. Annu Rev Nutr 13:191–215.

Cotta MA (1988) Amylolytic activity of selected species of ruminal bacteria. Appl Environ Microbiol 54:772–776.

Cotta MA, Whitehead TR (1993) Regulation and cloning of the gene encoding amylase activity of the ruminal bacterium *Streptococcus bovis*. Appl Environ Microbiol 59: 189–196.

Cummings JH (1981) Dietary fibre. Br Med Bull 37:65–70.

Cummings JH, Macfarlane GT (1991) The control and consequences of bacterial fermentation in the human colon. J Appl Bacteriol 70:443–459.

Cunningham C, McPherson CA, Martin J, Harris WJ, Flint HJ (1991) Sequence of a cellulase gene from the rumen anaerobe *Ruminococcus flavefaciens* 17. Mol Gen Genet 228:320–323.

Deguchi H, Watanabe Y, Sasaki T, Matsuda T, Shimizu S, Ohmiya K (1991) Purification and properties of the endo-1,4-β-glucanase from *Ruminococcus albus* and its gene products in *Escherichia coli*. J Ferment Bioeng 71:221–225.

Dehority BA (1993) Microbial ecology of cell wall fermentation. In: Jung HG, Buxton DR, Hatfield RD, Ralph J, eds. Forage Cell Wall Structure and Digestibility. pp. 425–453. Madison: American Society of Agronomy.

Din N, Gilkes NR, Tekant B, Miller RC Jr, Warren RAJ, Kilburn DG (1991) Non-hydrolytic disruption of cellulose fibres by the binding domain of a bacterial cellulase. Biotechnology 9:1096–1099.

Dinwoodie JM (1975) Timber—a review of the structure-mechanical property relationship. J Microsc 104:3–32.

Doerner KC, Gardner RM, Schook LB, Mackie RI, White BA (1994) Inhibition of the exo-β-1,4-glucanase from *Ruminococcus flavefaciens* FD-1 by a specific monoclonal antibody. Enzyme Microb Technol 16:2–9.

Doerner KC, Howard GT, Mackie RI, White BA (1992) β-glucanase expression by *Ruminococcus flavefaciens* FD-1. Microbiol Lett 93:147–154.

Doerner K C, White BA (1990a) Detection of glycoproteins separated by non-denaturing polyacrylamide gel electrophoresis. Anal Biochem 187:147–150.

Doerner KC, White BA (1990b) Assessment of the endo-β-1,4-glucanase components of *Ruminococcus flavefaciens* FD-1. Appl Environ Microbiol 56:1844–1850.

Enzyme nomenclature (1992) Recommendations of the Nomenclature Committee of the International Union of Biochemistry and Molecular Biology on the Nomenclature and Classification of Enzymes. New York: Academic Press.

Erfle JD, Teather RM, Wood PJ, Irvin JE (1988) Purification and properties of a 1,3-1,4-β-D-glucanase (lichenase, 1,3-1,4-β-D-glucan 4-glucanohydrolase, EC-3.2.1.73) from *Bacteroides succinogenes* cloned in *Escherichia coli*. Biochem J 255:833–841.

Flint HJ, Martin J, McPherson CA, Daniel AS, Zhang J-X (1993) A bifunctional enzyme, with separate xylanase and β(1,3-1,4)-glucanase domains, encoded by the *xynD* gene of *Ruminococcus flavefaciens*. J Bacteriol 175:2943–2951.

Flint HJ, McPherson CA, Bisset J (1989) Molecular cloning of genes from *Ruminococcus flavefaciens* encoding xylanase and β(1-3,1-4) glucanase activities. Appl Environ Microbiol 55:1230–1233.

Flint HJ, McPherson CA, Martin J (1991) Expression of two xylanase genes from the rumen cellulolytic bacterium *Ruminococcus flavefaciens* 17 cloned in pUC13. J Gen Microbiol 137:123–129.

Forano E, Broussolle V, Gaudet G, Bryant JA (1994) Molecular cloning, expression, and characterization of a new endoglucanase gene from *Fibrobacter succinogenes* S85. Curr Microbiol 28:7–14.

Forsberg CW, Cheng KJ, Krell PJ, Phillips JP (1993) Establishment of rumen microbial gene pools and their manipulation to benefit fibre digestion by domestic animals. In: Proceedings VII World Conference on Animal Production, pp. 281–316. Edmonton: World Association for Animal Production.

Forsberg CW, Cheng KJ (1992) Molecular strategies to optimize forage and cereal digestion by ruminants. In: Bills DD, Kung S-D, eds. Biotechnology and Nutrition, pp. 109–147. Stoneham: Butterworth-Heinemann.

Forsberg CW, Crosby B, Thomas DY (1986) Potential for manipulation of the rumen fermentation through the use of recombinant DNA techniques. J Anim Sci 63:310–325.

Francisco JA, Stathopoulos C, Warren RAJ, Kilburn DG, Georgiou G (1993) Specific adhesion and hydrolysis of cellulose by intact *Escherichia coli e*xpressing surface anchored cellulase or cellulose binding domains. Biotechnology 11:491–495.

Franklund CV, Glass TL (1987) Glucose uptake by the cellulolytic ruminal anaerobe *Bacteroides succinogenes*. J Bacteriol 169:500–506.

Freer SN (1993) Purification and characterization of the extracellular α-amylase from *Streptococcus bovis* JB1. Appl Environ Microbiol 59:1398–1402.

Fry SC (1988) The Growing Plant Cell Wall: Chemical and Metabolic Analysis. London: Longman Scientific and Technical.

Fuewa H, Takaya T, Sugimoto Y (1980) Degradation of various starch granules by amylases. In: Marshall JJ, ed. Mechanisms of Saccharide Polymerization and Depolymerization, pp. 73–100. New York: Academic Press.

Garcia-Campayo V, Wood TM (1993) Purification and characterisation of a β-D-xylosidase from the anaerobic rumen fungus *Neocallimastix frontalis*. Carbohydr Res 242: 229–245.

Gardner RM, Doerner KC, White BA (1987) Purification and characterization of an exo-β-1,4-glucanase from *Ruminococcus flavefaciens* FD-1. J Bacteriol 169:4581–4588.

Gavini F, Pourcher A-M, Neut C, Monget D, Romond C, Oger C, Izard D (1991) Phenotypic differentiation of bifidobacteria of human and animal origins. Int J Syst Bacteriol 41:548–557.

Geun Eog J, Han H-K, Yun S-W, Rhim SL (1992) Isolation of amylolytic *Bifidobacterium* sp. Int-57 and characterization of amylase. J Microbiol Biotechnol 2:85–91.

Gherardini F, Babcock M, Salyers AA (1985) Purification and characterization of two α-galactosidases associated with catabolism of guar gum and other α-galactosides by *Bacteroides ovatus*. J Bacteriol 161:500–506.

Gherardini F, Salyers AA (1987) Purification and characterization of a cell-associated, soluble mannanase from *Bacteroides ovatus*. J Bacteriol 169:2038–2043.

Gibson GR, Cummings JH, Macfarlane GT (1988) Competition for hydrogen between sulphate-reducing bacteria and methanogenic bacteria from the human large intestine. J Appl Bacteriol 65:241–247.

Gilbert HJ, Hazlewood GP, Laurie JI, Orpin CG, Xue GP (1992) Homologous catalytic domains in a rumen fungal xylanase: evidence for gene duplication and prokaryotic origin. Mol Microbiol 6:2065–2072.

Gilkes NR, Henrissat B, Kilburn DG, Miller J, R.C., Warren RAJ (1991) Domains in microbial β-1,4-glycanases: sequence conservation, function, and enzyme families. Microbiol Rev 55:303–315.

Gold, MH, Alic M (1993) Molecular biology of the lignin-degrading Basidiomycete *Phanaerochaete chrysosporium*. Microbiol Rev 57:605–622.

Gomez de Segura B, Fevre M (1993) Purification and characterization of 2 1,4-β-xylan endohydrolases from the rumen fungus *Neocallimastix frontalis*. Appl Environ Microbiol 59:3654–3660.

Gong J, Forsberg CW (1989) Factors affecting adhesion of *Fibrobacter succinogenes* subsp. *succinogenes* S85 and adherence-defective mutants to cellulose. Appl Environ Microbiol 55:3039–3044.

Gong J, Forsberg CW (1993) Separation of outer and cytoplasmic membranes of *Fibrobacter succinogenes* and membrane and glycogen granule locations of glycanase and cellobiase. J Bacteriol 175:6810–6821.

Gordon GLR, Phillips M (1989a) Degradation and utilization of cellulose and straw by three different anaerobic fungi from the ovine rumen. Appl Environ Microbiol 55:1703.

Gordon GL, Phillips MW (1989b) Comparative fermentation properties of anaerobic fungi from the rumen. In: Nolan JV, Leng RA, Phillips MW, eds. The Roles of Protozoa and Fungi in Ruminant Nutrition, pp. 127–138. Armidale, Australia: Penambul Books.

Gordon GLR, Phillips MW (1992) Extracellular pectin lyase produced by *Neocallimastix* sp: LM1, a rumen anaerobic fungus. Lett Appl Microbiol 15:113–115.

Gregg K, Rowan A, Ware C (1993) Digestion of filter-paper by cellulases cloned from *Ruminococcus albus* AR67. In: Shimada K, Ohmiya K, Kobayashi Y, Hoshino S, Sakka K, Karita S, eds. Genetics, Biochemistry and Ecology of Lignocellulose Degradation, pp. 166–172. Tokyo: Uni Publishers.

Greve LC, Labavitch JM, Hungate RE (1984a) Xylanase action on alfalfa cell walls. In: Dugger WM, Bartnicki-Garcia S, eds. Structure, Function, and Biosynthesis of Plant Cell Walls, pp. 150–166. Riverside: University of California.

Greve LC, Labavitch JM, Hungate RE (1984b) α-L-Arabinofuranosidase from *Ruminococcus albus* 8: purification and possible role in hydrolysis of alfalfa cell wall. Appl Environ Microbiol 47:1135–1140.

Groleau D, Forsberg CW (1981) Cellulolytic activity of the rumen bacterium *Bacteroides succinogenes*. Can J Microbiol 27:517–530.

Guilbot A, Mercier C (1985) Starch. In: Aspinall GO, ed. The Polysaccharides, pp. 209–282. New York: Academic Press.

Hall J, Ali S, Surani MA, et al. (1993) Manipulation of the repertoire of digestive enzymes secreted into the gastrointestinal tract of transgenic mice. Biotechnology 11:376–379.

Harris PJ (1990) Plant cell wall structure and development. In: Akin DE, Ljungdahl LG, Wilson JR, Harris PJ, eds. Microbial and Plant Opportunities to Improve Lignocellulose Utilization by Ruminants, pp. 71–90. New York: Elsevier Science.

Hartley RD, Morrison III WH, Borneman WS, et al. (1992) Phenolic constitutents of cell wall types of normal and brown midrib mutants of pearl millet (*Pennisetum glaucum* (L) R Br) in relation to wall biodegradability. J Sci Food Agric 59:211–216.

Hazlewood GP, Davidson K, Laurie JI, Romaniec MPM, Gilbert HJ (1990) Cloning and sequencing of the *cel*A gene encoding endoglucanase A of *Butyrivibrio fibrisolvens* strain A46. J Gen Microbiol 136:2089–2097.

Hebraud M, Fevre M (1990a) Purification and characterization of an extracellular β-xylosidase from the rumen anaerobic fungus *Neocallimastix frontalis*. FEMS Microbiol Lett 72:11–16.

Hebraud M, Fevre M (1990b) Purification and characterization of an a specific glycoside hydrolase from the anaerobic ruminal fungus *Neocallimastix frontalis*. Appl Environ Microbiol 56:3164–3169.

Heinrichova K, Wojciechowicz M, Ziolecki A (1989) The pectinolytic enzyme of *Selenomonas ruminantium*. J Appl Bacteriol 66:169–174.

Helaszek CT, White BA (1991) Cellobiose uptake and metabolism by *Ruminococcus flavefaciens*. Appl Environ Microbiol 57:64–68.

Henrissat B (1991) A classification of glycosyl hydrolases based on amino acid sequence similarities. Biochem J 280:309–316.

Henrissat B, Bairoch A (1993) New families in the classification of glycosyl hydrolases based on amino acid sequence similarities. Biochem J 293:781–788.

Hespell RB, O'Bryan PJ (1992) Purification and characterization of an α-L-arabinofuranosidase from *Butyrivibrio fibrisolvens* GS113. Appl Environ Microbiol 58:1082–1088.

Hespell RB, Whitehead TR (1990) Physiology and genetics of xylan degradation by gastrointestinal tract bacteria. J Dairy Sci 73:3013–3022.

Hidaka H, Eida T, Takizawa T, Tokunada T, Tashiro Y (1986) Effects of fructooligosaccharides on intestinal flora and human health. Bifidobacteria Microflora 5:37–50.

Hobson PN, Jouany JP (1988) Models, mathematical and biological, of the rumen function. In: Hobson PN, ed. The Rumen Microbial Ecosystem, pp. 461–511. New York: Elsevier Science.

Hoskins LC, Agustines M, McKee WB, Boulding ET, Kriaris M, Niedermeyer G (1985) Mucin degradation in human colon ecosystems. Isolation and properties of fecal strains that derade ABH blood group antigens and oligosaccharides from mucin glycoproteins. J Clin Invest 75:944–953.

Hoskins LC, Boulding ET (1981) Mucin degradation in human colon ecosystems. Evidence for the existence and role of bacterial subpopulations producing glycosidases as extracellular enzymes. J Clin Invest 67:163–172.

Howard GT, White BA (1988) Molecular cloning and expression of cellulase genes from *Ruminococcus albus* 8 in *Escherichia coli* bacteriophage λ. Appl Environ Microbiol 54:1752–1755.

Howard GT, White BA (1990) Cloning in *Escherichia coli* of a bi-functional cellulase xylanase enzyme from *Ruminococcus flavefaciens* FD-1. Anim Biotechnol 1:95–106.

Huang L, Forsberg CW, Thomas DY (1988) Purification and characterization of a chloride-stimulated cellobiosidase from *Bacteroides succinogenes* S85. J Bacteriol 170: 2923–2932.

Huang C-M, Kelly WJ, Asmundson RV, Yu P-L (1989) Molecular cloning and expression of multiple cellulase genes of *Ruminococcus flavefaciens* strain 186 in *Escherichia coli*. Appl Microbiol Biotechnol 31:265–271.

Huang L, McGavin M, Forsberg CW, Lam JS, Cheng KJ (1990) Antigenic nature of the chloride-stimulated cellobiosidase and other cellulases of *Fibrobacter succinogenes* subsp. *succinogenes* S85 and related fresh isolates. Appl Environ Microbiol 56: 1229–1234.

Hungate RE (1966) The Rumen and Its Microbes. New York: Academic Press.

Hungate R E, Stack J (1982) Phenylpropanoic acid: growth factor for *Ruminococcus albus*. Appl Environ Microbiol 44:79–83.

Hwa V, Salyers AA (1992) Evidence for differential regulation of genes in the chondroitin sulfate utilization pathway of *Bacteroides thetaiotaomicron*. J Bacteriol 174:342–344.

Iiyama K, Lam TBT, Stone BA (1994) Covalent cross-links in the cell wall. Plant Physiol 104:315–320.

Ishii S (1984) Cell wall cementing materials of grass leaves. Plant Physiol 76:959–961.

Iyo AH, Forsberg CW (1994) Features of the cellodextrinase gene from *Fibrobacter succinogenes* S85. Can J Microbiol 40:592–596.

Joblin KN, Naylor GE, Williams AG (1990) Effect of *Methanobrevibacter smithii* on xylanolytic activity of anerobic ruminal fungi. Appl Environ Microbiol 56(8): 2287–2295.

Joyner AE, Baldwin RL (1966) Enzymatic studies of pure cultures of rumen bacteria. J Bacteriol 92:1321–1330.

Karita S, Morioka K, Kajino T, Sakka K, Shimada K, Ohmiya K (1993) Cloning and sequencing of a novel endo-1,4-β-glucanase gene from *Ruminococcus albus*. J Ferment Bioeng 76:439–444.

Katz JR (1934) X-ray investigation of gelatinization and retrogradation of starch and its importance for bread research. Bakers Weekly 81:34–37.

Kawai S, Honda H, Tanase T, Taya M, Iijima S, Kobayashi T (1987) Molecular cloning of *Ruminococcus albus* cellulase gene. Agric Biol Chem 51:59.

Kistner A, Kotze JP (1973) Enzymes of intermediary metabolism of *Butyrivibrio fibrisolvens* and *Ruminococcus albus* grown under glucose limitation. Can J Microbiol 19: 1119–1127.

Knowles J, Teeri TT, Lehtovaara P, Penttilä M, Salohemio M (1988) The use of gene technology to investigate fungal cellulolytic enzymes. In: Aubert J-P, Béguin P, Millet J, eds. Biochemistry and Genetics of Cellulose Degradation. FEMS Symposium No. 43, pp. 153–169. London: Academic Press.

Kohring S, Wiegel J, Mayer F (1990) Subunit composition and glycosidic activities of the cellulase complex from *Clostridium thermocellum* JW20. Appl Environ Microbiol 56:3798–3804.

Kotarski SF, Waniska RD, Thurn KK (1992) Starch hydrolysis by the ruminal microflora. J Nutr 122:178–190.

Kudo H, Cheng K-J, Costerton JW (1986) Interactions between *Treponema bryantii* and cellulolytic bacteria in the in vitro degradation of straw cellulose. Can J Microbiol 32: 244–248.

Kudo, Cheng KJ, Costerton JW (1987) Electron microscopic study of the methylcellulose-mediated detachment of cellulolytic rumen bacteria from cellulose fibers. Can J Microbiol 33:267–272.

Lam TBT, Iiyama K, Stone BA (1990) Primary and secondary walls of grasses and other forage plants:Taxonomic and structural considerations. In: Akin DE, Ljungdahl LG, Wilson JR, eds. Microbial and Plant Opportunities to Improve Lignocellulose Utilization by Ruminants, pp. 43–69. New York: Elsevier Science.

Lamed R, Setter E, Bayer EA (1983a) Characterization of a cellulose-binding, cellulase-containing complex in *Clostridium thermocellum.* J Bacteriol 156:828–836.

Lamed R, Setter E, Kenig R, Bayer EA (1983b) The cellulosome: a discrete cell surface organelle of *Clostridium thermocellum* which exhibits separate antigenic, cellulose-binding and various cellulolytic activities. Biotechnol Bioeng Symp 13:163–181.

Lamed R, Naimark J, Morgenstern E, Bayer EA (1987) Specialized cell surface structures in cellulolytic bacteria. J Bacteriol 169:3792–3800.

Latham MJ, Brooker BE, Pettipher GL, Harris PJ (1978a) Adhesion of *Bacteroides succinogenes* in pure culture and in the presence of *Ruminococcus flavefaciens* to cell walls in leaves of perennial ryegrass (*Lolium perenne*). Appl Environ Microbiol 35:1166–1173.

Latham MJ, Brooker BE, Pettipher GL, Harris PJ (1978b) *Ruminococcus flavefaciens* cell coat and adhesion to cotton cellulose and to cell walls in leaves of perennial rye grass (*Lolium perenne*). Appl Environ Microbiol 35:156–165.

Li XL, Calza RE (1991) Kinetic study of a cellobiase purified from *Neocallimastix frontalis* EB188. Biochim Biophys Acta 1080:148–154.

Lin LL, Rumbak E, Zappe H, Thomson JA, Woods DR (1990) Cloning, sequencing and analysis of expression of a *Butyrivibrio fibrisolvens* gene encoding a β-glucosidase. J Gen Microbiol 136:1567–1576.

Lin LL, Thomson JA (1991a) An analysis of the extracellular xylanases and cellulases of *Butyrivibrio fibrisolvens* H17c. FEMS Microbiol Lett 104:65–82.

Lin LL, Thomson JA (1991b) Cloning, sequencing and expression of a gene encoding a 73 kDa xylanase enzyme from the rumen anaerobe *Butyrivibrio fibrisolvens* H17c. Mol Gen Genet 228:55–61.

Lowe SE, Theodorou MK, Trinci APJ (1987) Cellulases and xylanase of an anaerobic rumen fungus grown on wheat straw, wheat straw holocellulose, cellulose, and xylan. Appl Environ Microbiol 53:1216–1223.

Maas LK, Glass TL (1991) Cellobiose uptake by the cellulolytic ruminal anaerobe *Fibrobacter (Bacteroides) succinogenes*. Can J Microbiol 37:141–147.

Macfarlane GT, Hay S, Macfarlane S, Gibson GR (1990) Effect of different carbohydrates on growth, polysaccharidase and glycosidiase production by *Bacteroides ovatus*, in batch and continuous culture. J Appl Bacteriol 68:179–187.

Macfarlane GT, Macfarlane S (1993) Factors affecting fermentation reactions in the large bowel. Nutr Soc Proc 52:367–373.

MacGregor AW, Ballance DL (1980) Hydrolysis of large and small starch granules from normal and waxy barley cultivars by α-amylase from barley malt. Cereal Chem 57: 397–402.

Macy JM, Farrand JR, Mongomery L (1982) Cellulolytic and non-cellulolytic bacteria in rat gastrointestinal tracts. Appl Environ Microbiol 44:1428–1434.

Maglione G, Matsushita O, Russell JB, Wilson DB (1992) Properties of a genetically reconstructed *Prevotella ruminicola* endoglucanase. Appl Environ Microbiol 58: 3593–3597.

Malburg LM, Tamblyn Lee JM, Forsberg CW (1992) Degradation of cellulose and hemicellulose by rumen microorganisms. In: Winkelmann G, ed. Microbial Degradation of Natural Products, pp. 127–159. Weinheim: VCH Verlagsgesellschaft mbH.

Malburg LM, Smith DC, Schellhorn HE, Forsberg CW (1993) *Fibrobacter succinogenes* S85 has multiple xylanase genes. J Appl Bacteriol 75:564–573.

Malburg LM, Forsberg CW (1993) *Fibrobacter succinogenes* possesses at least nine distinct glucanase genes. Can J Microbiol 39:882–891.

Mannarelli M, Evans S, Lee D (1990) Cloning, sequencing, and expression of a xylanase gene from the anaerobic ruminal bacterium *Butyrivibrio fibrisolvens*. J Bacteriol 172: 4247–4254.

Manners DJ (1985) Some aspects of the structure of starch. Cereal Foods World 30: 461–467.

Marvin-Sikkema FD, Richardson AJ, Stewart CS, Gottschal JC, Prins RA (1990) Influence of hydrogen-consuming bacteria on cellulose degradation by anaerobic fungi. Appl Environ Microbiol 56:3793–3797.

Matsushita O, Russell JB, Wilson DB (1991) A *Bacteroides ruminicola* 1,4-β-D-endoglucanase is encoded in two reading frames. J Bacteriol 173:6919–6926.

Matte A, Forsberg CW, Gibbins AMV (1992) Enzymes associated with metabolism of xylose and other pentoses by *Prevotella (Bacteroides) ruminicola* strains, *Selenomonas ruminantium* D, and *Fibrobacter succinogenes* S85. Can J Microbiol 38:370–376.

Matte A, Forsberg CW (1992) Purification, characterization, and mode of action of endoxylanases 1 and 2 from *Fibrobacter succinogenes* S85. Appl Environ Microbiol 58: 157–168.

McAllister TA, Cheng KJ, Rode LM, Buchanan-Smith JG (1990a) Use of formaldehyde to regulate digestion of barley starch. Can J Anim Sci 79:581–589.

McAllister TA, Cheng KJ, Rode LM, Forsberg CW (1990b) Digestion of barley, maize, and wheat by selected species of ruminal bacteria. Appl Environ Microbiol 56:3146–3153.

McAllister TA, Dong Y, Yanke LJ, Bae HD, Cheng KJ, Costerton JW (1993) Cereal grain digestion by selected strains of ruminal fungi. Can J Microbiol 39:367–376.

McAllister TA, Bae HD, Yanke LJ, Cheng KJ, Muir A (1994) Effect of condensed tannins from birdsfoot trefoil on endoglucanase activity and the digestion of cellulose filter paper by ruminal fungi. Can J Microbiol 40:298–305.

McCarthy RE, Kotarski SF, Salyers AA (1985) Location and characteristics of enzymes involved in the breakdown of polygalacturonic acid by *Bacteroides thetaiotaomicron*. J Bacteriol 161:493–499.

McDermid KP, Forsberg CW, MacKenzie CR (1990a) Purification and properties of an acetylxylan esterase from *Fibrobacter succinogenes* S85. Appl Environ Microbiol 56: 3805–3810.

McDermid KP, MacKenzie CR, Forsberg CW (1990b) Esterase activities of *Fibrobacter succinogenes* subsp. *succinogenes* S85. Appl Environ Microbiol 56:127–132.

McGavin M, Forsberg CW (1988) Isolation and characterization of endoglucanases 1 and 2 from *Bacteroides succinogenes* S85. J Bacteriol 170:2914–2922.

McGavin MJ, Forsberg CW (1989) Catalytic and substrate-binding domains of endoglucanase 2 from *Bacteroides succinogenes*. J Bacteriol 171:3310–3315.

McGavin MJ, Forsberg CW, Crosby B, Bell AW, Dignard D, Thomas DY (1989) Structure of the *cel-3* gene from *Fibrobacter succinogenes* S85 and characteristics of the encoded gene product, endoglucanase 3. J Bacteriol 171:5587–5595.

McGavin M, Lam J, Forsberg CW (1990) Regulation and distribution of *Fibrobacter succinogenes* subsp. *succinogenes* S85 endoglucanases. Appl Environ Microbiol 56: 1235–1244.

McKellar RC, Modler HW (1989) Metabolism of fructo-oligosaccharides by *Bifidobacterium* spp. Appl Microbiol Biotechnol 31:537–541.

McSweeney CS, Dulieu A, Katayama Y, Lowry JB (1994) Solubilization of lignin by the ruminal anaerobic fungus *Neocallimastix patriciarum*. Appl Environ Microbiol 60: 2985–2989.

McWethy SJ, Hartman PA (1977) Purification and some properties of an extracellular α-amylase from *Bacteroides amylophilus*. J Bacteriol 129:1537–1544.

Minato H, Suto T (1978) Technique for fractionation of bacteria in rumen microbial ecosystem. II. Attachment of bacteria isolated from bovine rumen to cellulose powder in vitro and elution of bacteria attached therefrom. J Gen Appl Microbiol 24:1–16.

Minato H, Suto T (1979) Technique for fractionation of bacteria in rumen microbial ecosystem. III. Attachment of bacteria isolated from bovine rumen to starch granules in vitro and elution of bacteria attached therefrom. J Gen Appl Microbiol 25:71–93.

Miron J (1991) The hydrolysis of lucerne cell-wall monosaccharide components by monocultures or pair combinations of defined ruminal bacteria. J Appl Bacteriol 70:245–252.

Miron J, Ben-Ghedalia D (1992) The degradation and utilization of wheat-straw cell-wall monosaccharide components by defined ruminal cellulolytic bacteria. Appl Microbiol Biotechnol 38:432–437.

Miron J, Ben-Ghedalia D, Yokoyama MT, Lamed R (1990) Some aspects of cellobiose effect on bacterial cell surface structures involved in lucerne cell walls utilization by fresh isolates of rumen bacteria. Anim Feed Sci Technol 30:107–120.

Miron J, Duncan SH, Stewart CS (1994) Interactions between rumen bacterial strains during the degradation and utilization of the monosaccharides of barley straw cell-walls. J Appl Bacteriol 76:282–287.

Miron J, Yokoyama MT, Lamed J (1989) Bacterial cell surface structures involved in lucerne cell wall degradation by pure cultures of cellulolytic rumen bacteria. Appl Microbiol Biotechnol 32:218–222.

Mitsumori M, Minato H (1993a) Purification of cellulose-binding proteins 1 and 2 from cell lysate of *Fibrobacter succinogenes* S85. J Gen Appl Microbiol (Tokyo) 39:361–369.

Mitsumori M, Minato H (1993b) Presence of several cellulose binding proteins in cell lysate of *Fibrobacter succinogenes* S85. J Gen Appl Microbiol (Tokyo) 39:273–283.

Montgomery L (1988) Isolation of human colonic fibrolytic bacteria. Lett Appl Microbiol 6:55–57.

Montgomery L, Flesher B, Stahl D (1988) Transfer of *Bacteroides succinogenes* (Hungate) to *Fibrobacter* gen. nov. as *Fibrobacter succinogenes* comb. nov. and description of *Fibrobacter intestinalis* sp. nov. Int J Syst Bacteriol 38:430–435.

Monties B (1989) Lignins. Meth Plant Biochem 1:113–157.

Moore PJ, Staehelin PJ (1988) Immunogold localization of the cell-wall-matrix polysaccharides rhamnogalacturonan I and xyloglucan during cell expansion and eytokinesis in *Trifolium pratense* L:, implication for secretory-pathways-Planta 174:433–445.

Morag E, Bayer EA, Lamed R (1990) Relationship of cellulosomal and non-cellulosomal xylanases of *Clostridium thermocellum* to cellulose-degrading enzymes. J Bacteriol 172: 6098–6105.

Morris EJ, Cole OJ (1987) Relationship between cellulolytic activity and adhesion to cellulose in *Ruminococcus albus*. J Gen Microbiol 133:1023–1032.

Morrison M, Mackie RI, Kistner A (1990) 3-Phenylpropanoic acid improves the affinity of *Ruminococcus albus* for cellulose in continuous culture. Appl Environ Microbiol 56: 3220–3222.

Mountfort DO, Asher RA (1985) Production and regulation of cellulase by two strains of the rumen anaerobic fungus *Neocallimastix frontalis*. Appl Environ Microbiol 49: 1314–1322.

Nagaraja TG, Towne G, Beharka AA (1992) Moderation of ruminal fermentation by ciliated protozoa in cattle fed a high-grain diet. Appl Environ Microbiol 58:2410–2414.

Nakashima Y, Orskov ER, Hotten PM, Ambro K, Takase Y (1988) Rumen degradation of straw. 6. Effect of polysaccharidase enzymes on degradation characteristics of ensiled rice straw. Anim Prod 47:412–427.

Ni WT, Paiva NL, Dixon RA (1994) Reduced lignin in transgenic plants containing a caffeic acid O-methyltransferase antisense gene. Transgenic Res 3:120–126.

Odenyo AA, Mackie RI, Stahl DA, White BA (1994a) The use of 16S ribosomal RNA targeted oligonucleotide probes to study competition between ruminal fibrolytic bacteria. I. Development of probes for *Ruminococcus* species and evidence for bacteriocin production. Appl Environ Microbiol 60:3688–3696.

Odenyo A A, Mackie RI, Stahl DA, White BA (1994b) The use of 16S ribosomal RNA targeted oligonucleotide probes to study competition between ruminal fibrolytic bacteria. II. Pure culture studies with cellulose and alkaline peroxide treated wheat straw. Appl Environ Microbiol 60:3697–3703.

Ohmiya K, Shimizu M, Taya M, Shimizu S (1982) Purification and properties of cellobiosidase from *Ruminococcus albus*. J Bacteriol 150:407–409.

Ohmiya K, Shirai M, Kurachi Y, Shimizu S (1985) Isolation and properties of β-glucosidase from *Ruminococcus albus*. J Bacteriol 161:432–434.

Ohmiya K, Maeda K, Shimizu S (1987) Purification and properties of endo-(1→4)-β-D-glucanase from *Ruminococcus albus*. Carbohydr Res 166:145–155.

Ohmiya K, Nagashima K, Kajino T, Goto E, Tsukada A, Shimuzu S (1988) Cloning of the cellulase gene from *Ruminococcus albus* and its expression in *Escherichia coli*. Appl Environ Microbiol 54:1511–1515.

Ohmiya K, Kajino T, Kato A, Shimizu S (1989) Structure of a *Ruminococcus albus* endo-1,4-β-glucanase gene. J Bacteriol 171:6771–6775.

Ohmiya K, Maeda K, Shimizu S (1993) Purification and properties of endo-(1-4)-β-D-glucanase from *Ruminococcus albus*. Carbohyd Res 166:145–155.

Okazaki M, Fujikawa S, Matsumoto N (1990) Effect of xylooligosaccharide on the growth of bifidobacteria. Bifidobacteria Microflora 9:77–86.

Orpin CG (1988) Genetic approaches to the improvement of lignocellulose degradation in the rumen. In: Aubert J-P, Béguin P, Millet J, eds. Biochemistry and Genetics of Cellulose Degradation, pp. 172–179. London: Academic Press.

Paradis FW, Zhu H, Krell PJ, Phillips JP, Forsberg CW (1993) The *xynC* gene from *Fibrobacter succinogenes* S85 codes for a xylanase with two similar catalytic domains. J Bacteriol 175:7666–7672.

Pavlostathis SG, Miller TL, Wolin MJ (1990) Cellulose fermentation by continuous cultures of *Ruminococcus albus* and *Methanobrevibacter smithii*. Appl Microbiol Biotechnol 33:109–116.

Pazur JH, Kleppe K (1962) The hydrolysis of α-D-glucosidases by amyloglucosidase from *Aspergillus niger*. J Biol Chem 237:1002–1006.

Pen J, Molendijk L, Quax WJ, et al. (1992) Production of active *Bacillus licheniformis* α-amylase in tobacco and its application in starch liquefaction. Bio/Technology 10: 292–296.

Pettipher G L, Latham MJ (1979) Characteristics of enzymes produced by *Ruminococcus flavefaciens* that degrade plant cell walls. J Gen Microbiol 110:21–27.

Phillips MW, Gordon GLR (1988) Sugar and polysaccharide fermentation by rumen anaerobic fungi from Australia, Britain and New Zealand. Biosystems 21:377–383.

Poole DH, Hazlewood GP, Laurie JI, Barker PJ, Gilbert HJ (1990) Nucleotide sequence of the *Ruminococcus albus* SY3 endoglucanase genes celA and celB. Mol Gen Genet 223:217–223.

Poole D M, Durrant AJ, Hazlewood GP, Gilbert HJ (1991) Characterization of hybrid proteins consisting of the catalytic domains of *Clostridium* and *Ruminococcus* endoglucanases, fused to *Pseudomonas* non-catalytic cellulose-binding domains. Biochem J 279:787–792.

Rasmussen MA, White BA, Hespell RB (1989) Improved assay for quantitating adherence of ruminal bacteria to cellulose. Appl Environ Microbiol 55:2089–2091.

Rasmussen MA, Hespell RB, White BA, Bothast RJ (1988) Inhibitory effects of methycellulose on cellulose degradation by *Ruminococcus flavefaciens*. Appl Environ Microbiol 54:890–897.

Roberton AM, Stanley RA (1982) In vitro utilization of mucin by *Bacteroides fragilis*. Appl Environ Microbiol 43:325–330.

Roger V, Fonty G, Komisarczuk-Bony S, Gouet P (1990) Effects of physicochemical factors on the adhesion to cellulose Avicel of ruminal bacteria *Ruminococcus flavefaciens* and *Fibrobacter succinogenes* subsp. *succinogenes*. Appl Environ Microbiol 56: 3081–3087.

Roger V, Bernalier A, Grenet E, Fonty G, Jamot J, Gouet P (1993) Degradation of wheat straw and maize stem by a monocentric and a polycentric rumen fungi, alone or in association with rumen cellulolytic bacteria. Anim Feed Sci Technol 42:69–82.

Romaniec M P M, Davidson K, White BA, Hazlewood GP (1989) Cloning of *Ruminococcus albus* endo-β-1,4-glucanase and xylanase genes. Lett Appl Microbiol 9:101–104.

Rumbak E, Rawlings DE, Lindsey GG, Woods DR (1991a) Cloning, nucleotide sequence, and enzymatic characterization of α-amylase from the ruminal bacterium *Butyrivibrio fibrisolvens* H17c. J Bacteriol 173:4203–4211.

Rumbak E, Rawlings DE, Lindsey GG, Woods DR (1991b) Characterization of the *Butyrivibrio fibrisolvens glgB* gene, which encodes a glycogen-branching enzyme with starch-clearing activity. J Bacteriol 173:6732–6741.

Russell JB (1985) Fermentation of cellodextrins by cellulolytic and non-cellulolytic rumen bacteria. Appl Environ Microbiol 49:572–576.

Russell JB, Wilson DB (1988) Potential opportunities and problems for genetically altered rumen microorganisms. J Nutr 118:271–279.

Sakai K, Tachiki T, Kumagai H, Tochikura T (1987) Hydrolysis of α-D-galactosyl oligosaccharides in soymilk by α-D-galactosidase of *Bifidobacterium breve* 203. Agric Biol Chem 51:315–322.

Sakai T (1992) Degradation of pectins. In: Winkelman G, ed. Microbial Degradation of Natural Products, pp. 58–81. New York: VCH.

Salyers AA, O'Brien M, Kotarski SF (1982) Utilization of chondroitin sulfate by *Bacteroides thetaiotaomicron* growing in carbohydrate-limited continuous culture. J Bacteriol 150:1008–1015.

Salyers AA, Vercellotti JR, West SEH, Wilkins TD (1977) Fermentation of mucin and plant polysaccharides by strains of *Bacteroides* from the human colon. Appl Environ Microbiol 33:319–322.

Sewell GW, Utt EA, Hespell RB, Mackenzie KF, Ingram LO (1989) Identification of the *Butyrivibrio fibrisolvens* xylosidase gene (*xylB*) coding region and its expression *Escherichia coli*. Appl Environ Microbiol 55:306–311.

Showalter AM (1993) Structure and function of plant cell wall proteins. Plant Cell 5: 9–23.

Smith AC, Podolsky DK (1986) Colonic mucin glycoproteins in health and disease. In: Mendeloff A, ed. Clinics in Gastroenterology, Vol. 15, pp. 815–837. Philadelphia: W.B. Saunders.

Smith CJ, Hespell RB (1983) Prospects for development and use of recombinant deoxyribonucleic acid techniques with ruminal bacteria. J Dairy Sci 66:1536–1546.

Smith DC, Forsberg CW (1991) α-Glucuronidase and other hemicellulase activities of *Fibrobacter succinogenes* S85 grown on crystalline cellulose or ball-milled barley straw. Appl Environ Microbiol 57:3552–3557.

Smith KA, Salyers AA (1991) Characterization of a neopullulanase and an α-glucosidase from *Bacteroides thetaiotaomicron* 95–1. J Bacteriol 173:2962–2968.

Stack R J, Hungate RE (1984) Effect of 3-phenylpropanoic acid on capsule and cellulases of *Ruminococcus albus* 8. Appl Environ Microbiol 48:218–223.

Stack RJ, Hungate RE, Opsahl WP (1983) Phenylacetic acid stimulation of cellulose digestion by *Ruminococcus albus* 8. Appl Environ Microbiol 46:539–544.

Stanton TB, Canale-Parola E (1980) *Treponema bryantii* sp. nov., a rumen spirochaete that interacts with cellulolytic bacteria. Arch Microbiol 127:145–156.

Stewart CS, Bryant MP (1988) The rumen bacteria. In: Hobson PN, ed. The Rumen Microbial Ecosystem, pp. 21–75. Barking: Elsevier Science.

Stewart CS, Duncan SH, Richardson AJ, Backwell C, Begbie R (1992) The inhibition of fungal cellulolysis by cell-free preparations from ruminococci. FEMS Microbiol Lett 97:83–87.

Strobel HJ (1993a) Evidence for catabolite inhibition in regulation of pentose utilization and transport in the ruminal bacterium *Selenomonas ruminantium*. Appl Environ Microbiol 59:40–46.

Strobel HJ (1993b) Pentose utilization and transport by the ruminal bacterium *Prevotella ruminicola*. Arch Microbiol 159:465–471.

Strobel HJ, Dawson KA (1993) Xylose and arabinose utilization by the rumen bacterium *Butyrivibrio fibrisolvens*. FEMS Microbiol Lett 113:291–296.

Susmel P, Stefanon B (1993) Aspects of lignin degradation by rumen microorganisms. J Biotechnol 30:141–148.

Takano M, Moriyama R, Ohmiya K (1992) Structure of a β-glucosidase gene from *Ruminococcus albus* and properties of the translated product. J Ferment Bioeng 73:79–88.

Talbott LD, Ray PM (1992) Molecular size and separability features of pea cell wall polysaccharides. Plant Physiol 98:357–368.

Tamblyn Lee JM, Hu Y, Zhu H, Cheng KJ, P.J.Krell, Forsberg CW (1993) Cloning of a xylanase gene from the ruminal fungus *Neocallimastix patriciarum* 27 and its expression in *Escherichia coli*. Can J Microbiol 39:134–139.

Tancula E, Feldhaus MJ, Bedzyk LA, Salyers AA (1992) Location and characterization of genes involved in binding of starch to the surface of *Bacteroides thetaiotaomicron*. J Bacteriol 174:5609–5616.

Teather RM, Erfle JD (1990) DNA sequence of a *Fibrobacter succinogenes* mixed-linkage β-glucanase (1,3–1,4-β-D-glucan 4-glucanohydrolase) gene. J Bacteriol 172: 3837–3741.

Teunissen MJ, Kets EPW, Op den Camp HJM, Huis in t Veld JHJ, Vogels GD (1992) Effect of coculture of anaerobic fungi isolated from ruminants and non-ruminants with methanogenic bacteria on cellulolytic and xylanolytic enzyme activities. Arch Microbiol 157:176–182.

Theodorou MK, Lowe SE, Trinci APJ (1988) Fermentative characteristics of anaerobic rumen fungi. Biosystems 21:371.

Thomson JA (1993) Molecular biology of xylan degradation. FEMS Microbiol Rev 104: 65–82.

Thurston B, Dawson KA, Strobel HJ (1993) Cellobiose versus glucose utilization by the ruminal bacterium *Ruminococcus albus*. Appl Environ Microbiol 59:2631–2637.

Thurston B, Dawson KA, Strobel HJ (1994) Pentose utilization by the ruminal bacterium *Ruminococcus albus*. Appl Environ Microbiol 60:1087–1092.

Trinci APJ, Davies DR, Gull K, Lawrence MI, Nielsen BB, Rickers A, Theodorou MK (1994) Anaerobec fungi in herbivorous animals. Mycol Res 98:129–152.

Ushida K, Jouany JP, Demeyer DI (1991) Effect of presence and absence of rumen protozoa on the efficiency of utilization of concentrate and fibrous feeds. In: Tsuda T, Sasaki Y, Kawashima R, eds. Physiological Aspects of Digestion and Metabolism in Ruminants, pp. 625–654. New York: Academic Press.

Utt EA, Eddy CK, Keshav KF, Ingram LO (1991) Sequencing and expression of the *Butyrivibrio fibrisolvens xylB* gene encoding a novel bifunctional protein with β-D-xylosidase and α-L-arabinofuranosidase activities. Appl Environ Microbiol 57: 1227–1234.

Valentine PJ, Arnold P, Salyers AA (1992) Cloning and partial characterization of two chromosomal loci from *Bacteroides ovatus* that contain genes essential for growth on guar gum. Appl Environ Microbiol 58:1541–1548.

Varel VH, Fryda SJ, Robinson IM (1984) Cellulolytic bacteria from pig large intestine. Appl Environ Microbiol 47:219–221.

Vercoe PE, Gregg K (1992) DNA sequence and transcription of an endoglucanase gene from *Prevotella (Bacteroides) ruminicola* AR20. Mol Gen Genet 233:284–292.

Vercoe PE, Gregg K (1995) Sequence and transcriptional analysis of an endoglucanase gene from *Ruminococcus albus* AR67. Anim Biotechnol 6:59–71.

Vercoe PE, Spight DH, White BA (1995a) Nucleotide sequence and transcriptional analysis of the *cel*D β-glucanase gene from *Ruminococcus flavefaciens* FD-1. Can J Microbiol 41:27–34.

Vercoe PE, Finks JL, White BA (1995b) DNA sequence and transcriptional characterization of the β-glucanase gene (*cel*B) from *Ruminococcus flavefaciens* FD-1. Can J Microbiol 41:869–876.

Wang W, Thomson JA (1990) Nucleotide sequence of the *celA* gene encoding a cellodextrinase of *Ruminococcus flavefaciens* FD-1. Mol Gen Genet 222:265–269.

Wang W, Thomson JA (1992) Nucleotide sequence of the *celA* gene encoding a cellodextrinase of *Ruminococcus flavefaciens* FD-1. Mol Gen Genet 233:492.

Wang WY, Reid SJ, Thomson JA (1993) Transcriptional regulation of an endoglucanase and a cellodextrinase gene in *Ruminococcus flavefaciens* FD-1. J Gen Microbiol 139: 1219–1226.

Wardrop AB, Bland DE (1959) The process of lignification in woody plants. In: Kratzl K, Billet G, eds. Biochemistry of Wood, pp. 92–116. New York: Pergamon.

Ware CE, Bauchop T, Gregg K (1989) The isolation and comparison of cellulase genes from two strains of *Ruminococcus albus*. J Gen Microbiol 135:921–930.

Ware CE, Lachke AH, Gregg K (1990) Mode of action and substrate specificity of a purified exo-1,4-β-D-glucosidase cloned from the cellulolytic bacterium *Ruminococcus albus* AR67. Biochem Biophys Res Commun 171:777–786.

Watanabe Y, Moriyama R, Matsuda T, Shimizu S, Ohmiya K (1992) Purification and properties of the endo-1,4-β-glucanase III from *Ruminococcus albus*. J Ferment Bioeng 73:54–57.

Weimer PJ (1993) Effects of dilution rate and pH on the ruminal cellulolytic bacterium *Fibrobacter succinogenes* S85 in cellulose-fed continuous culture. Arch Microbiol 160: 288–294.

White BA, Clarke JH, Doerner KC, et al. (1990) Improving cellulase activity in *Ruminococcus* through genetic modification. In: Akin DE, Ljungdahl LG, Wilson JR, Harris PJ, eds. Microbial and Plant Opportunities to Improve Lignocellulose Utilization by Ruminants, pp. 389–400. New York: Elsevier Scientific.

White BA, Attwood GT, Vercoe PE, Mackie RI (1993a) Regulation of β-glucanase expression in *Ruminococcus*. In: Shimada K, Ohmiya K, Kobayashi Y, Hoshino S, Sakka K, Karita S, eds. Genetics, Biochemistry and Ecology of Lignocellulose Degradation, pp. 155–165. Tokyo: Japan.

White BA, Mackie RI, Doerner KC (1993b) Enzymatic hydrolysis of forage cell walls. In: Jung HG, Buxton DR, Hatfield RD, eds. Forage Cell Wall Structure and Digestibility. Madison: American Society of Agronomy.

Whitehead TR, Cotta MA, Hespell RB (1991) Introduction of *Bacteroides ruminicola* xylanase gene into the *Bacteroides thetaiotaomicron* chromosome for production of xylanase activity. Appl Environ Microbiol 57:277–282.

Whitehead TR (1993) Analyses of the gene and amino acid sequence of the *Prevotella (Bacteroides)-ruminicola* 23 xylanase reveals unexpected homology with endoglucanases from other genera of bacteria. Curr Microbiol 27:27–33.

Williams AG, Orpin CG (1987a) Polysaccharide-degrading enzymes formed by three species of anaerobic rumen fungi grown on a range of carbohydrate substrates. Can J Microbiol 33:418–426.

Williams AG, Orpin CG (1987b) Glycoside hydrolase enzymes present in the zoospore and vegetative growth stages of the rumen fungi *Neocallimastix patriciarum*, *Piromonas communis*, and an unidentified isolate, grown on a range of carbohydrates. Can J Microbiol 33:427–434.

Williams AG, Withers SE (1982a) The effect of carbohydrate growth substrates on the glycosidase activity of hemicellulose-degrading rumen bacterial isolates. J Appl Bacteriol 52:389–401.

Williams AG, Withers SE (1982b) The production of plant cell wall polysaccharide-degrading enzymes by hemicellulolytic rumen bacterial isolates grown on a range of carbohydrate substrates. J Appl Bacteriol 52:377–387.

Williams AG, Withers SE (1993) The regulation of xylanolytic enzyme formation by *Butyrivibrio fibrisolvens* NCFB 2249. Lett Appl Microbiol 14:194–198.

Williams AG, Withers SE, Coleman GS (1984) Glycoside hydrolases of rumen bacteria and protozoa. Curr Microbiol 10:287–294.

Williams AG, Ellis AB, Coleman GS (1986) Subcellular distribution of polysaccharide depolymerase and glycoside hydrolase enzymes in rumen ciliate protozoa. Curr Microbiol 13:139–147.

Williams AG, Withers SE, Orpin CG (1994) Effect of the carbohydrate growth substrate on polysaccharolytic enzyme formation by anaerobic fungi isolated from the foregut and hindgut of nonruminant herbivores and the forestomach of ruminants. Lett Appl Microbiol 18:147–151.

Wilson CA, Wood TM (1992a) The anaerobic fungus *Neocallimastix frontalis*—Isolation and properties of a cellulosome-type enzyme fraction with the capacity to solubilize hydrogen-bond-ordered cellulose. Appl Microbiol Biotechnol 37:125–129.

Wilson CA, Wood TM (1992b) Studies on the cellulase of the rumen anaerobic fungus *Neocallimastix frontalis*, with special reference to the capacity of the enzyme to degrade crystalline cellulose. Enzyme Microb Technol 14:258–264.

Wilson JR (1993) Organization of forage plant tissues. In: Jung HG, Buxton DR, Hatfield RD, Ralph J, eds. Forage Cell Wall Structure and Digestibility, pp. 1–32. Madison: American Society of Agronomy.

Wilson WD, Barwick JM, Lomax JA, Jarvis MC, Duncan HJ (1988) Lignified and non-lignified cell walls from kale. Plant Sci 57:83–90.

Wilson WD, Jarvis MC, Duncan HJ (1989) In vitro digestibility of kale (*Brassica oleracea*) secondary xylem and parenchyma cell walls and their polysaccharide components. J Sci Food Agric 49:9–14.

Wojciechowicz M (1972) Comparison of the action of *Bacteroides ruminicola* polygalacturonic lyase and of pectinase on lower oligogalacturonides. Acta Microbiol Pol Ser A 4:189–195.

Wojciechowicz M, Ziolecki A (1979) Pectinolytic enzymes of large rumen treponemes. Appl Environ Microbiol 37:136–142.

Wojciechowicz M, Heinrichova K, Ziolecki A (1980) A polygalacturonate lyase produced by *Lachnospira multiparus* isolated from the bovine rumen. J Gen Microbiol 117: 193–199.

Wojciechowicz M, Heinrichova K, Ziolecki A (1982) An exopectate lyase of *Butyrivibrio fibrisolvens* from the bovine rumen. J Gen Microbiol 128:2661–2665.

Wood TM (1992) Fungal cellulases. Biochem Soc Trans 20:46–53.

Wood TM, Wilson CA (1984) Some properties of the endo-(1–4)-β-D-glucanase synthesized by the anaerobic cellulolytic bacterium *Ruminococcus albus*. Can J Microbiol 30: 316–321.

Wood TM, Wilson CA, McCrae SI (1994) Synergism between components of the cellulase system of the anaerobic rumen fungus *Neocallimastix frontalis* and those of the aerobic fungi *Penicillium pinophilum* and *Trichoderma koningii* in degrading crystalline cellulose. Appl Microbiol Biotechnol 41:257–261.

Wood TM, Wilson CA, Stewart CS (1982) Preparation of the cellulase from the cellulolytic anaerobic rumen bacterium *Ruminococcus albus* and its release from the bacterial cell wall. Biochem J 205:129–137.

Wood TM, Wilson CA, McCrae SI, Joblin KN (1986) A highly active extracellular cellulase from the anaerobic ruminal fungus *Neocallimastix frontalis*. FEMS Microbiol Lett 34:37–40.

Wu JHD, Orme-Johnson WH, Demain AL (1988) Two components of an extracellular protein aggregate of *Clostridium thermocellum* together degrade crystalline cellulose. Biochemistry 27:1703–1709.

Wubah DA, Akin DE, Borneman WS (1993) Biology, fiber-degradation, and enzymology of anaerobic zoosporic fungi. Crit Rev Microbiol 19:99–115.

Xue GP, Gobius KS, Orpin CG (1992a) A novel polysaccharide hydrolase cDNA (*celD*) from *Neocallimastix patriciarum* encoding 3 multi-functional catalytic domains with high endoglucanase, cellobiohydrolase and xylanase activities. J Gen Microbiol 138: 2397–2403.

Xue GP, Orpin CG, Gobius KS, Aylward JH, Simpson GD (1992b) Cloning and expression of multiple cellulase cDNAs from the anaerobic rumen fungus *Neocallimastix patriciarum* in *Escherichia coli*. J Gen Microbiol 138:1413–1420.

Yu I, Hungate RE (1979) The extracellular cellulases of *Ruminococcus albus*. Ann Rech Vet 10:251–254.

Zhang JX, Flint HJ (1992) A bifunctional xylanase encoded by the *xynA* gene of the rumen cellulolytic bacterium *Ruminococcus flavefaciens* 17 comprises two dissimilar domains linked by an asparagine/glutamine-rich sequence. Mol Microbiol 6:1013–1023.

Zhang JX, Martin J, Flint HJ (1994) Identification of non-catalytic conserved regions in xylanases encoded by the *xyn*B and *xyn*D genes of the cellulolytic rumen anaerobe *Ruminococcus flavefaciens*. Mol Gen Genet 245:260–264.

Zhou LQ, Xue GP, Orpin CG, Black GW, Gilbert HJ, Hazlewood GP (1994) Intronless *celB* from the anaerobic fungus *Neocallimastix patriciarum* encodes a modular family A endoglucanase. Biochem J 297:359–364.

Zhu H, Cheng K-J, Forsberg CW (1994a) A truncated β-xylosdiase from the anaerobic fungus *Neocallimastix patriciarum* 27. Can J Microbiol 40:484–490.

Zhu H, Paradis FW, Krell PJ, Phillips JP, Forsberg CW (1994b) Enzymatic specificities and modes of action of the two catalytic domains of the XynC xylanase from *Fibrobacter succinogenes* S85. J Bacteriol 176:3885–3894.

11

Digestion of Nitrogen in the Rumen: A Model for Metabolism of Nitrogen Compounds in Gastrointestinal Environments

Michael A. Cotta and James B. Russell

1. Introduction

After energy, nitrogen is quantitatively the most important nutritional requirement for growth of gastrointestinal tract microorganisms. In the rumen and other pregastric environments, nitrogen is derived primarily from the plant material of the diet. In cecal and colonic fermentation, the nitrogen sources include undigested (by the host enzymes) feed materials as well as endogenous sources such as digestive secretions and sloughed intestinal epithelia. In all of these environments, some nitrogen can also be derived from urea, either in the form of dietary urea or endogenous urea that is transferred to the gut via secretion (e.g., saliva) or directly across the epithelium. In addition, gut microorganisms can in themselves provide exogenous nitrogen by lysis and cell turnover. Protozoa derive much of their nitrogen from the uptake of other gut microbes (e.g., bacteria and smaller protozoa).

The significance of nitrogen transformations in gastrointestinal environments depends on the animal and anatomical location of digestive organ. In ruminants and other foregut fermenters, fermentation and the synthesis of microbial protein are essential to the nutrition and survival of the animal, but excessive ammonia production can lead to large nitrogen losses. Fermentation also takes place in the hindgut. In cecal fermenters (rabbits, rodents), much of the energy to drive animal metabolism can be derived from gut fermentation, but the microbial protein that arises from this area of the gut is lost unless the animal practices coprophagy. Pigs and humans are normally described as simple-stomached species, but even these animals can derive some energy from intestinal fermentation. In the hindgut, microbial nitrogen metabolism is more apt to be detrimental than beneficial. Ammonia accumulation can create an additional load on urea synthesis, and toxic end products have been implicated in disease.

Of the gastrointestinal environments studied, the rumen is most clearly understood. Perhaps this is due to the comparatively early application of anaerobic microbiological techniques to the study of ruminant digestive physiology or to the economic significance of ruminant livestock species. In any case, the nitrogen metabolism in the rumen has been described in considerable detail. Because microorganisms isolated from other mammalian gut environments are genetically and physiologically similar to those inhabiting the rumen, one would expect them to have essentially the same pattern of nitrogen metabolism. The host species, however, can have an impact on the types of substrates available for fermentation as well as the gut environmental conditions (e.g., transit time, pH, etc.). Both of these factors will influence the composition of the microbial population and the degree to which the microbes can express their potential activities. This chapter will emphasize nitrogen metabolism in the rumen.

2. Degradation of Nitrogen-Containing Compounds in the Rumen

As with all feed materials consumed by ruminants, dietary nitrogen sources are first subjected to ruminal transformation prior to host digestion. In the case of nitrogen, this takes on special significance. Because the rumen microbiota is capable of synthesizing all the protein-containing amino acids from relatively simple sources of nitrogen such as ammonia or urea, the animal is able to exist on diets devoid of protein (Virtanen 1966). In common practice, however, the major source of nitrogen available to ruminal microorganisms is plant protein. Other sources of nitrogen would include nucleic acids, nitrate, and urea.

Most dietary "energy sources" contain a considerable amount of protein, but protein can also be added in the form of supplements (e.g., soybean meal) to raise the total nitrogen content of the diet. Protein supplementation is of practical significance because modern animal production systems have increased the amino acid requirements of domestic animals. In many cases, microbial protein does not meet the nitrogen requirements of the animal, and it is necessary to add protein that passes undegraded from the rumen.

Ruminal protein degradation is often more rapid than the nitrogen utilization for microbial growth. As a result, ammonia accumulates, is absorbed across the rumen wall into the blood, and is transported to the liver, where it is converted to urea. While a portion of the urea can be recycled into the rumen, much is excreted and excess ammonia production represents a significant loss to the system. The challenge to animal nutritionists and rumen microbiologists is to devise feeding regimens that optimize the degradation of nitrogen in the rumen such

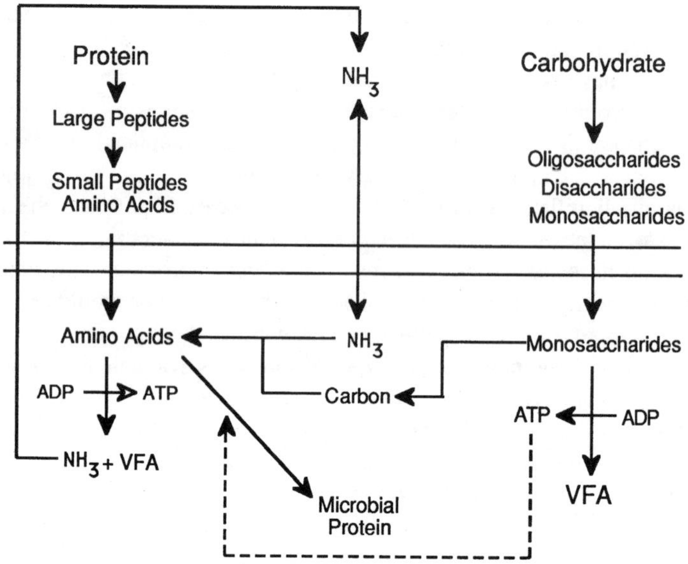

Figure 11.1. A theoretical scheme of carbohydrate and protein utilization by ruminal bacteria.

that maximum output of microbial protein is attained with minimal losses of ammonia. This overall process represents the sum of a number of microbial activities including protein hydrolysis, peptide degradation, amino acid deamination, and various pathways of carbon metabolism (Fig. 11.1). Protein degradation is initially mediated by the action of extracellular or cell-surface-associated enzymes, but the other steps are mediated by intracellular activities.

2.1. *Proteolytic Activity of Mixed Ruminal Bacteria*

Much of the information available on rumen microbial proteases was initially obtained from experiments with mixed ruminal microorganisms. Strained rumen contents were typically fractionated by sedimentation or centrifugation to obtain bacteria and protozoa. Both these populations exhibited proteolytic capabilities, but the specific activity of the bacterial fraction was much greater than that of the protozoal fraction (Brock et al. 1982, Prins et al. 1983). Based on this difference, and the relative ease of cultivation, research on the proteolytic activity of ruminal microorganisms has focused almost completely on the bacterial population.

When Brock et al. (1982) analyzed ruminal protease activity, up to 75% of total ruminal proteolytic activity was associated with the feed particles, and the proportion of the total of activity associated with this fraction varied with the amount of particles present in rumen contents. Similar results have been obtained by other workers (Wallace 1985, Wallace and Cotta 1988). Because rumen microbiologists typically study microorganisms that are present in the fluid phase, such a stratification could potentially introduce a bias. Work with protease inhibitors indicates that the adherent and nonadherent populations are at least qualitatively similar (Wallace 1985). In the case of the ruminal fungi, however, this may not be true. Fungi have distinct growth phases (i.e. vegetative, zoospore), and the distribution of fungal activity in the rumen will vary with the stage of growth (Orpin and Joblin 1988). As with all fungal activities, the importance of fungi to nitrogen metabolism in the rumen is difficult to ascertain.

Proteolytic enzymes of mixed ruminal bacteria appear to be predominantly cell-associated, as little proteolytic activity can be detected in cell-free rumen fluid (Blackburn and Hobson 1960a, Brock et al. 1982). Further examination of the subcellular location of rumen bacteria proteases by Kopecny and Wallace (1982) indicated that most (approximately 80%) of this proteolytic activity could be released from cells by mild blending or shaking treatments. Based on their data, the majority of proteases produced by ruminal bacteria are likely to be associated with capsular or extracellular coat materials. The pH optima for mixed bacterial proteases are generally in the range of pH 5 to 7 (Blackburn and Hobson 1960a, Kopecny and Wallace 1982). Anaerobic conditions are needed for optimal proteolytic activity. However, if anaerobiosis is maintained, the addition of a reducing agent is unnecessary and may actually be inhibitory to some proteases (Brock et al. 1982, Kopecny and Wallace 1982). Based on the effects of various protease inhibitors, ruminal bacteria appear to produce serine proteases, cysteine proteases and metalloproteases, but little aspartate protease activity has ever been observed (Brock et al. 1982, Kopecny and Wallace 1982, Prins et al. 1983). Enzymes with trypsinlike, chymotrypsinlike, carboxypeptidaselike, and aminopeptidaselike activities have been detected using specific synthetic substrates (Brock et al. 1982, Wallace and Kopecny 1983, Prins et al. 1983). Because only a limited number of synthetic substrates were tested, enzymes with additional substrate specificities could not be excluded.

Studies using mixed bacteria demonstrate the combined effects of many enzymes produced by many different organisms, and the results can be difficult to interpret. For example, it would be convenient to equate the quantity of each protease class with the effects of protease inhibitors. Unfortunately, some enzyme inhibitors can affect proteolytic enzymes indirectly (Barrett 1977, Whitaker 1994). Synthetic peptides can also be hydrolyzed at rates that are markedly different from the rates observed with natural substrates. While the proportions of individual classes of proteolytic activities present under a given set of conditions cannot be determined with precision, relative changes can be monitored, and this information can be compared to the activities of individual species.

2.2. Proteases of Bacteria

Proteolytic bacteria comprise from 12% and 43% of the total bacterial population in the rumen, and their numbers vary with diet and animal (Bryant and Burkey 1953, Fulghum and Moore 1963, Prins et al. 1983). Attempts to isolate numerically important ruminal bacteria that could utilize protein as a sole source of energy yielded only facultative anaerobic bacteria, and these organisms were too few in number to explain protein digestion in the rumen (Appleby 1955, Hunt and Moore 1958, Blackburn and Hobson 1960b). Later work indicated that the predominant proteolytic bacteria were saccharolytic bacteria. These bacteria were members of the genera *Prevotella*, *Butyrivibrio*, *Selenomonas*, *Eubacterium*, *Lachnospira*, and *Streptococcus* (Bryant and Small 1956, Bryant et al. 1958, Blackburn and Hobson 1962, Fulghum and Moore 1963, Hazlewood and Nugent 1978, Russell et al. 1981, Hazlewood et al. 1983). Of these, *Prevotella ruminicola*, *Ruminobacter amylophilus*, and *Butyrivibrio fibrisolvens* are thought to be the most important.

Protease production by *R. amylophilus* has been studied is some detail (Blackburn 1968a, Henderson et al. 1969). It produced proteases in simple maltose-ammonium sulfate-salts medium, and protease production was neither induced nor repressed by a number of nutrients (e.g., peptone or casamino acids). In batch culture, protease production paralleled growth and virtually all the activity was located on the cell periphery. When growth ceased and the cells entered a stationary phase, much of the proteolytic activity was released into the culture medium. When *R. amylophilus* was grown in maltose-limited continuous culture, the rate of protease production increased with increasing dilution rate to a peak at $D = 0.24\ h^{-1}$ (Henderson et al. 1969). Further increases in dilution rate resulted in a decrease in the rate of protease production until enzyme production increased again at $D = 0.4\ h^{-1}$. Nitrogen-limited chemostat cultures exhibited the same pattern of enzyme production as carbon-limited cultures, but protease activity was approximately 2.7 times lower. Although these data do not indicate the specific types of metabolic regulation, they do indicate that the physiological state can have an influence on the proteolytic activity of *R. amylophilus*.

Lesk and Blackburn (1971) partially purified a protease released into the culture medium by stationary phase cells of *R. amylophilus*. The purified enzyme was active over a broad pH range (4.5 to 12.0) with optima at pH 6.0 and 11.5. The enzymatic activity exhibited trypsinlike characteristics. N-benzoyl-L-arginine methyl ester (BAME) was hydrolyzed and proteolytic activity was inhibited by tosyl-L-lysine chromomethylketone (TLCK) and phenylmethylsulfonyl fluoride (PMSF). BAME and casein hydrolysis were not affected to the same extent by various enzyme inhibitors and enzyme activity had two optima. These results suggested the presence of more than one protease, but data from heat inactivation, effects of trypsinlike inhibitors, and gelfiltration experiments indicated a single protein with two active molecular weight forms, 30 and 60 kDa. Efforts to disassociate the 60 kDa-molecular-weight form into smaller units were unsuccessful. The cell-bound protease(s) isolated from exponentially growing cultures of *R. amylophilus* were similar to the protease released by stationary phase cultures (Blackburn 1968b, Blackburn and Hullah 1974). These preparations contained considerable quantities of nucleic acids, and efforts to remove nucleic acid contaminants from these samples were either unsuccessful or resulted in the loss of proteolytic activity.

The benefit of protein degradation to *R. amylophilus* is unclear. Although this organism is proteolytic, it is unable to utilize the products of proteolysis—namely, peptides and amino acids (Hobson et al. 1968). One possibility is that cell proteases remove the proteinaceous coat surrounding starch granules, enabling cells to utilize starch, the only polymeric carbohydrate that this species can ferment (Hamlin and Hungate 1956). This hypothesis is consistent with the findings of McAllister et al. (1993), who showed that protease pretreatment enhanced the digestion of corn starch granules by mixed ruminal microorganisms.

In contrast to *R. amylophilus*, *P. ruminicola* can utilize protein as a sole source of nitrogen for growth (Hazlewood and Nugent 1978, Hazlewood et al. 1981). During exponential growth, more than 90% of the protease activity was cell-associated, but as growth rate declined and cells entered the stationary phase, much of this activity was transferred to the surrounding culture medium. It appeared that the released protease was inactivated because there was decrease in total protease activity in stationary phase cultures. The effects of alternate nitrogen sources or other nutrients on the proteolytic activity of *P. ruminicola* have not been extensively examined. Hazlewood et al. (1981) did note that substitution of bovine serum albumin (BSA) for leaf fraction 1 protein diminished production of protease at all stages of growth. The proteolytic activity from washed cell suspensions of *P. ruminicola* exhibited a broad pH optimum with maximal activity occurring between pH 5.9 and 8.2 (Hazlewood et al. 1981). Enzymatic activity was sensitive to oxygen, but inclusion of dithiothreitol as a reducing agent in the assay mixture prevented loss of activity under aerobic assay conditions. The addition of reducing agents (e.g., cysteine or dithiothreitol) stimulated activity even if anaerobiosis was maintained, indicating the possibility of cysteine protease activity. Based on the effects of enzyme inhibitors, *P. ruminicola* produces a mixture of serine, cysteine, and aspartic proteases (Hazelwood et al. 1981, Wallace and Brammall 1985). Synthetic substrates were poorly hydrolyzed. Although *P. ruminicola* produces less protease activity than some other ruminal bacteria (e.g., *R. amylophilus*, *B. fibrisolvens*), the similarity of this activity to that of mixed ruminal bacteria indicates that *P. ruminicola* may be a significant contributor to protein digestion in the rumen (Wallace and Brammall 1985, Wallace and Cotta 1988).

B. fibrisolvens is among the most frequently isolated ruminal bacteria and most strains are proteolytic (Bryant and Small 1956, Dehority 1966, Fulghum and Moore 1963, Hazlewood et al. 1983). *B. fibrisolvens* strains can vary considerably in their proteolytic activity (Cotta and Hespell 1986, Wallace and Brammall 1985). The proteases of the highly proteolytic strains were largely extracellular, while the activity of low protease strains was predominantly cell associated. Production of protease in a highly active strain appeared to be regulated by nitrogen source, and the highest specific activity was observed when low levels (1 g/L) of casamino acids were provided as the nitrogen source (Cotta and Hespell 1986). Increasing the amino acid addition in the form of free amino acids (casamino acids) or peptides (Trypticase) to 20 g/L repressed protease production. The rate of carbohydrate fermentation influenced the level of protease, and rapidly growing cultures produced more protease than slower-growing cultures. Maximal activity occurred over a broad pH range of 5.5 to 7.0. Serine protease(s) appear to be responsible for virtually all the activity, and PMSF was a potent inhibitor of activity. Various thiol inhibitors, metal chelators, soybean trypsin inhibitor, TLCK, or tosylamide phenylethyl chloromethyl ketone (TPCK) were not inhibitory. Similar results have been obtained for other highly proteolytic strains (Wallace and Brammall 1985, Strydom et al. 1986). Analysis of the extracellular protease activities produced by strain H17c by SDS-PAGE activity gels indicated that a number of proteases of different molecular weight may be produced, but that all of these forms had identical characteristics (Strydom et al. 1986). The idea that *B. fibrisolvens* is a dominant organism in protein degradation is consistent with high levels of serine protease activity in mixed ruminal bacteria (Brock et al. 1982), but other workers have indicated that mixed ruminal bacteria have predominantly cell-associated cysteine protease activities (Kopecny and Wallace 1982, Prins et al. 1983, Wallace and Brammall 1985). The low-activity strains produced cell associated protease with characteristics more similar to the latter (Wallace and Brammall 1985).

Aside from its role in acute indigestion, *S. bovis* was originally thought not to be important in the ruminal digestion of feed materials, including protein. However, Russell et al. (1981) noted that extensive digestion of protein in short term incubations of mixed ruminal bacteria was always associated with the proliferation of *S. bovis*. Proteolytic strains of *S. bovis* were isolated from these incubations and their capacity for digestion and utilization of casein examined. Based on these findings, they hypothesized that *S. bovis* might be responsible for the rapid degradation of protein that sometimes occurs shortly after feeding. Subsequently, proteolytic strains of *S. bovis* were isolated from the rumen on a number of occasions (Hazelwood et al. 1983, Wallace and Brammall 1985, Westlake and Mackie 1990). Hazlewood et al. (1983) found that, in cattle fed fresh fodder (alfalfa-cocksfoot), *S. bovis* comprised 84% of the isolates obtained on a medium containing leaf fraction 1 protein as the sole source of nitrogen. The proteolytic activity of *S. bovis* is cell-associated and based on the effects of enzyme inhibitors appears to be a serine protease (Wallace and Brammall 1985, Westlake and Mackie, 1990). The protease activity from *S. bovis* hydrolyzed easily degraded proteins such as leaf fraction 1 protein and casein, but had very poor activity toward more resistant proteins like bovine serum albumin (BSA) (Hazlewood et al. 1983). *S. bovis* was also to produces very high levels of aminopeptidase activity which most other ruminal bacteria lack (Wallace and Brammall 1985, Robinson and Russell 1985).

2.3. Proteolytic Activity of Protozoa

The proteolytic enzyme activity of the protozoal fraction of ruminal contents is almost always considerably less than that of the bacterial fraction (Prins et al. 1983, Brock et al. 1982, Nugent and Mangan 1981). Based on these results, the role of protozoa in the digestion of protein in the rumen is of relatively minor significance. This simple comparison of enzyme activities, however, oversimplifies the role of the protozoal population in the metabolism of protein in the rumen. As with other nutrients, the ruminal ciliate protozoa have a limited capacity for assimilation of soluble nutrients, particularly the entodiniomorphid species. The protozoa do, however, have the capacity to take up particulate materials and internalize these into food vacuoles where they will undergo intracellular digestion (Coleman 1980; Williams 1986; Williams and Coleman 1988, 1992). Ruminal protozoa consumed dyed casein particles (Blackburn and Hobson 1960), and their ability to digest protein was greatly diminished when the protein sources were finely ground (Naga and el-Shazly, 1968). Hino and Russell (1987) also showed that mixed ruminal protozoa had a greater capacity to degrade less soluble, particulate protein than soluble proteins. Because of this, it is thought that protozoa are more active in the degradation of protein associated with feed particles. Protozoa actively engulf and digest chloroplasts which contain ribulose bisphosphate carboxylase (leaf fraction 1 protein), the most abundant protein in plants (Stern et al. 1977, Mangan and West 1977).

The major influence of protozoa on the protein economy of the ruminant is probably related more to their effect on the resident microbial population. Protozoa obtain much of their nitrogen from engulfment and digestion of bacteria and other protozoa (Coleman 1980, Williams and Coleman 1992). The amino acids of digested bacteria are incorporated unmodified into protozoal protein, but peptides and amino acids are also released into the surrounding rumen fluid where they are degraded to ammonia and VFAs by ruminal bacteria (Hino and Russell 1987, Williams and Coleman 1988). Protozoal predation increases the turnover of microbial protein and results in higher ruminal ammonia concentrations (Demeyer and Van Nevel 1979, Leng and Nolan 1984, Wallace and McPherson 1987, Newbold and Hillman 1990). Although the mechanisms involved are unclear, protozoa do not pass out of the rumen in proportion to their presence in the rumen (Leng 1982, Murphy et al. 1985, Leng et al. 1986, Ffoulkes and Leng

1988, Abe and Iriki 1989). As a result, the contribution of protozoal protein to host digestion is usually less than the bacteria. This effect is at least partially offset by the fact that the protozoa are more digestible than the bacteria and have a higher biological value (Onodera and Koga 1987). Positive effects have sometimes been attributed to defaunation of the rumen, but defaunation seems only to be of benefit to animals consuming low-protein diets (Veira 1986, Bird and Leng 1978).

Forsberg et al. (1984) noted that the protease activity of cell extracts from mixed ruminal protozoa had maximal activity at pH 5.8 and was equally effective at hydrolyzing azocasein, casein, and soybean protein, but was less active at degrading hemoglobin and gelatin. Inhibition of activity by thiol reagents (iodoacetic acid, merthiolate) and stimulation by cysteine indicated a predominance of cysteine protease activity. Pepstatin also partially inhibited this activity, suggesting the possible presence of aspartic protease activity. Inhibition by TLCK and TPCK as well as activity patterns against synthetic substrates indicated the presence of enzymes with trypsin and chymotrypsinlike specificity (Forsberg et al. 1984, Prins et al. 1983). Mixed ruminal protozoa produced much higher levels of aminopeptidase activity than mixed ruminal bacteria (Forsberg et al. 1984, Prins et al. 1983). Coleman (1983) further subdivided the protozoal population of the rumen and found that the small protozoa fraction possessed much higher activity toward leaf fraction 1 (per milligram cell protein) than the large protozoa population.

All species of protozoa appear to be proteolytic, but the activities of individual species vary considerably. The highest protease activities were seen with *Diplodinium pentacanthum*, *Entodinium simplex*, and *Entodinium caudatum*, whereas the lowest levels were associated with the cellulolytic species *Eremoplastron bovis* and *Eudiplodinium maggii* (Coleman 1983). In nearly all cases, protease activity was higher when protozoa were cultivated in vitro than when they were reintroduced back into the rumen.

Abou Akkada and Howard (1962) partially characterized the proteolytic activity of *Entodinium caudatum*. The protease (measured with casein) was most active over a pH range of 6.5 to 7.0 and had trypsinlike activity. They also noted peptidase activity with pH optimum between pH 5.8 and 6.0. In contrast, Coleman (1983) found the protease activity from the same species to be most active at pH 3.2. The protease activity described by Coleman (1983) was inhibited by n-ethylmaleimide, leupeptin, and pepstatin, which is similar (except for pH optimum) to what Forsberg et al. (1984) described for mixed ruminal protozoa. Lockwood et al. (1988) conducted a comprehensive examination of the protease activities of a number of protozoal species. In all cases, the pH optimum (azocasein as the substrate) fell between pH 4 and 5. Leupeptin was the most effective inhibitor of protease activity of all six species studied, and the addition of dithiothreitol was stimulatory in all cases. The effects of the other inhibitors were less consistent and suggest some heterogeneity in the types of proteolytic enzymes that may be produced by individual protozoa. SDS-PAGE gels of cell extracts indicated that ruminal protozoa produced a number of different proteases. In general, it appears that ruminal protozoa produce proteases that are primarily of the cysteine class, and leupeptin seems to be the most effective inhibitor of protozoal proteases. Differences in pH optima are difficult to explain. Chemical conditions within the digestive vacuoles of ruminal protozoa are unknown at this time, so the potential effect of pH on enzyme activity cannot be evaluated.

2.4. Proteolytic Activity of Fungi

As noted earlier, the contribution of the fungi to ruminal fermentation is difficult to ascertain. Essentially all of the fungal proteolytic activity probably remains associated with feed particles, the site of fungal colonization. The inability of researchers to separate the fungi from feed particles and other microorganisms makes it impossible to estimate the relative importance of this group of microorganisms.

There have been relatively few in vitro studies of the proteolytic activity of individual fungal species (Wallace and Joblin 1985, Wallace and Munro 1986, Michel et al. 1993, Asao et al. 1993). *Neocallimastix frontalis* strain PNK2 had protease activity as high as the most proteolytic ruminal bacteria (Wallace and Joblin 1985). During the early stages of growth, protease activity was predominantly associated with plant particles, but by the later stages of growth almost all of the activity was found in the extracellular culture fluid. Proteolytic activity was inhibited by ρ-chloromercuribenzoate (pCMB) and the metal chelators EDTA and 1,10 phenanthroline. The EDTA inhibition could be overcome by the addition of the divalent cations, calcium, zinc, or cobalt. TLCK partially inhibited the protease, but other inhibitors were ineffective. The proteolytic activity exhibited maximum activity at pH 7.5. These results are consistent with the production of cysteine and metalloprotease activities. Inhibition by TLCK also suggested the possible production of a trypsinlike protease, but the trypsin substrate, benzoyl-arginine ρ-nitroanilide, was not hydrolyzed. Asao et al. (1993) examined the proteolytic activity of a *Piromyces* sp. in addition to a *Neocallimastix* sp. These strains produced mixtures of proteolytic enzymes. A number of inhibitors partially inhibited activity, but none alone inhibited more than 55% of the original activity. The pattern of inhibition observed was similar for both organisms. The pH optima for azocasein degradation were 7.9 and 8.8 for *Neocallimastix* sp. and *Piromyces* sp., respectively. In contrast to these two studies, Michel et al. (1993) found that none of the seven strains of ruminal fungi in their study produced significant levels of protease activity under any growth condition examined. The strains used included two strains each of *Neocallimastix* sp. and *Piromyces* sp. as well two unidentified polycentric fungal strains and a *Caecomyces* sp. All these strains did, however, produce aminopeptidases. Based on these limited data, it appears that considerable variability exists with regard to the proteolytic capacity of ruminal fungi. Clearly much more research will be required to determine the role of fungi in the metabolism of protein in the rumen.

2.5. Factors Influencing Protein Degradation

The level of proteolytic activity in the rumen can vary with diet. Changes in activity, however, are not always predictable (Blackburn and Hobson 1960a, Prins et al. 1983). The feeding of diets high in cereal grains is associated with increased protease activity of rumen contents (Siddons and Paradine 1981, McAllister et al. 1993). This observation has a certain logical appeal since several of the more active protein degrading bacteria are also amylolytic (e.g., *P. ruminicola*, *R. amylophilus*). However, much of the increase in activity is simply due to increases in the total microbial population that occurs on these diets and the proportion of proteolytic bacteria does not necessarily increase (McAllister et al. 1993). Feeding fresh forage also seems to increase ruminal proteolytic activity (Nugent and Mangan 1981, Hazlewood et al. 1983, Prins et al. 1983). In this case there is an increase in the proportion of proteolytic microorganisms relative to the total microbial population (Prins et al. 1983). Based on the differential hydrolysis of synthetic substrates and effects of protease inhibitors, Prins et al. (1983) showed that the types of proteolytic enzymes present in the rumen could vary, but these differences could not always be explained solely by changes in the diet. Wallace et al. (1987) examined the effect of urea, casein, and albumin supplementation of sheep diets on ruminal proteolytic activity. Their results showed little change in the total activity or the ability to degrade different proteins with changing diet. These results indicate that the rumen population is responsive primarily to changes in carbohydrate rather than nitrogen. This is not surprising since the predominant proteolytic microorganisms of the rumen cannot use protein as a sole energy source and are in fact the major saccharolytic species present in the rumen.

Since more than a single protease-producing organism is likely to be present in the rumen at any given time, the potential for interaction exists. Wallace (1985) combined a number of protease producing species and monitored the effects on growth and proteolytic activity. The synergism was greatest with low-protease-producing strains. When *S. ruminantium* was combined with a number of different species, particularly *S. bovis* and *P. ruminicola*, there was a large increase in protein degradation (Fig. 11.2). In these experiments maximal proteolysis was observed when *S. ruminantium* contributed 50% of the enzyme activity.

Clear induction and repression of protease production does not appear to be the major mechanism of regulation present in ruminal microorganisms. Only high levels of peptides or amino acids not likely to occur in the rumen were able to repress enzyme expression in *B. fibrisolvens* (Cotta and Hespell 1986). Alteration

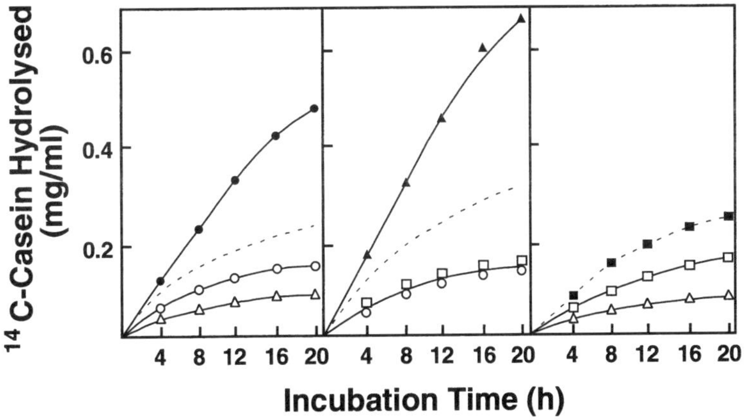

Figure 11.2. Hydrolysis of ^{14}C-labeled casein (2 mg/mL) by suspensions of rumen bacteria: ○, *Selenomonas ruminantium*; △, *Streptococcus bovis*; □, *Prevotella ruminicola*; ●, *S. ruminantium* + *S. bovis*; ▲, *S. ruminantium* + *P. ruminicola*; ■, *P. ruminicola* + *S. bovis*. Broken lines represent the sum of individual activities of the two species in the mixture (redrawn with permission from Wallace 1985).

of growth rate, either in continuous culture or by changing the energy source, significantly affected the production of protease activity by *B. fibrisolvens* and *R. amylophilus* (Cotta and Hespell 1986, Henderson et al. 1969). The effect of growth rate on protease production by other proteolytic organisms is unknown, but even small changes in the concentrations of available energy source can have a profound effect on the growth rate of the microbial population in vivo (Nocek and Russell 1988, Russell et al. 1992). It is not inconceivable that similar changes would also modulate the expression of degradative enzymes as well.

The chemical and physical characteristics of proteins are important determinants of their susceptibility to digestion by rumen microorganisms. Feedstuff proteins vary in their solubility, and protein solubility has in many cases provided convenient means of estimating the degradability of proteins by rumen microorganisms (Hendrickx and Martin 1963, Crooker et al. 1978, Crawford et al. 1978). Because the rapid degradation of protein can lead to excessive ammonia accumulation, a variety of treatments have been employed to reduce the rate of digestion of soluble proteins. Treatments such as heating or reacting with formaldehyde render proteins less soluble and less susceptible to ruminal proteolysis (Ferguson 1975). An interesting technique being explored to reduce ruminal proteolysis involves the feeding of condensed tannins as present in certain plants like Sainfoin. These compounds bind to protein and make them less available for hydrolysis by rumen bacterial proteases (Jones et al. 1994). This binding is reversible, and dissociation occurs under either acidic or alkaline conditions allowing the protein to undergo post ruminal digestion by the host. If provided in the diet at appropriate levels, the presence of condensed tannins can improve the nitrogen economy of the ruminant animal (Waghorn et al. 1987). The addition of condensed tannins may have an added benefit of bloat protection.

Although solubility is an important determinant of the vulnerability of proteins to ruminal digestion, soluble proteins differ in their rate and extent of degradation. BSA and ovalbumin are very soluble but are only slowly degraded by rumen microorganisms (Mangan 1972, Mahadevan et al. 1980, Nugent et al. 1983, Wallace 1983). Disulfide bonds stabilize the structure of proteins and makes them resistant to proteolytic attack. The slowly digested soluble protein, BSA, contains 16 disulfide bonds. Performic acid oxidation or mercaptoethanol reduction of disulfide bonds markedly enhanced the degradation of BSA and a variety of other proteins (Mahadevan et al. 1980, Wallace 1983, Nugent et al. 1983). These treatments had no effect on the degradation of casein (Nugent et al. 1983) or leaf fraction 1 protein (Wallace 1983) because these proteins have no disulfide bonds. Other aspects of the tertiary structure of protein can also influence the degradation of protein (McNabb et al. 1994). The addition of an artificial crosslinking agent (1,6 diisocyanatohexane) to stabilize the tertiary structure of proteins markedly depressed the degradation of proteins by mixed rumen bacterial proteases (Wallace 1983). Ovalbumin has an unusual primary structure in that the N-terminal amino acid is acetylated and the C terminal residue is proline. This renders the protein resistant to exoproteolytic attack by many enzymes with aminopeptidaselike and carboxypeptidaselike specificities.

The use of proteinase inhibitors offers a theoretical method of inhibiting ruminal protein degradation but few researchers have addressed the aspect of kinetics in their assessment of proteolytic activity in the rumen. If the enzymes are already in excess (zero order), subtle changes in protease production (or regulation) will have little impact on the rate of protein degradation. When Chen et al. (1987b) increased the protein content of dairy cattle rations, the peptide concentration of ruminal fluid increased until the crude protein was greater than 17% (Fig. 11.3). This result indicated that the rumen was operating as a zero order system with respect to proteolytic activity. At higher protein values, peptide concentrations did not increase further and the system was clearly first order with respect to proteases. When the soybean meal was autoclaved to decrease its solubility, the relationship between peptides and protein was always zero order with respect to enzyme activity.

3. Peptide Metabolism

Proteolysis is a sequential process that generates amino acid chains of shorter and shorter length, but free amino acids are not generally a significant end product

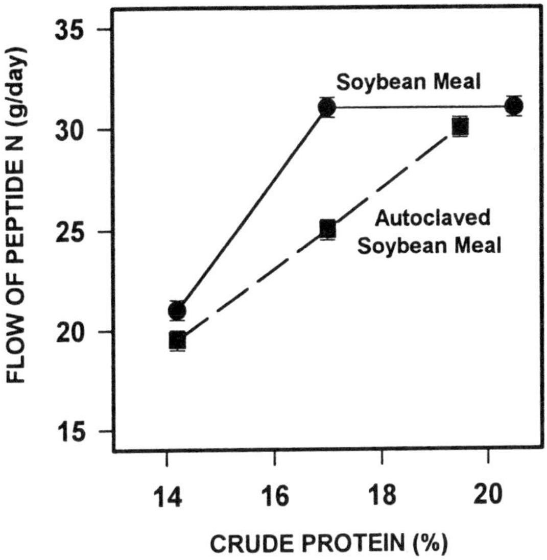

Figure 11.3. Relationship between crude protein content of the ration and the flow of peptide nitrogen from the rumen (from Chen et al. 1987b).

of proteases (Fig. 11.4). Oligopeptides are further hydrolyzed by peptidases. Peptidases have virtually no activity toward proteins and can be produced even by nonproteolytic microorganisms. Bacterial peptidases are found in the extracellular space and intracellularly. Peptides can also be taken up by transport systems. In *Salmonella*, the oligopeptide transport system takes up peptides with as many as five amino acids (Hiles et al. 1987).

Wright and Hungate (1967) noted that ruminal fluid had low concentrations of free amino acids, and subsequent work indicated that mixed ruminal bacteria utilized peptides more efficiently than amino acids (Wright 1967). In vivo (Mangan 1972) and in vitro (Russell et al. 1983) studies, however, indicated that casein catabolism yielded peptides that were fairly resistant to ruminal degradation. Since degradation of leaf fraction 1 protein from alfalfa did not cause a sustained increase in peptides even though it was hydrolyzed rapidly (Nugent and Mangan 1981, Nugent et al. 1983), it appeared that peptide accumulation could not be explained solely by the rate of proteolysis.

Chen et al. (1987a) used a ninhydrin reaction to estimate the amino nitrogen content of ruminal fluid samples that had been previously treated with perchloric acid to remove macromolecules (protein, RNA, and DNA) and brought to a boil under alkaline conditions to remove ammonia. This method of analysis indicated that high-producing dairy cattle fed 12 times a day had peptide flows from the

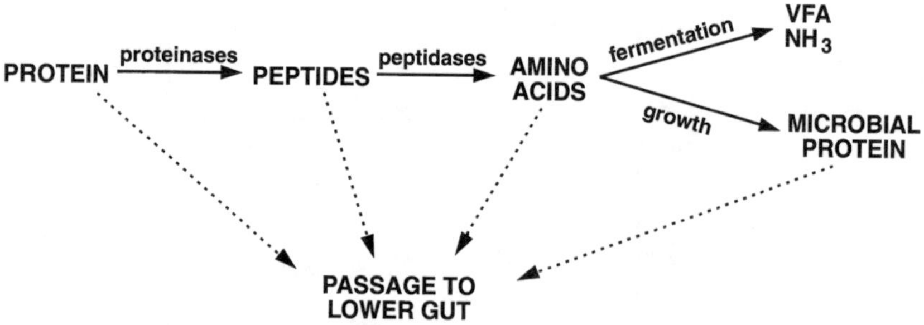

Figure 11.4. Schematic representation of protein degradation, fermentation, and microbial protein synthesis in the rumen and the passage of nitrogenous compounds to the lower gut (from Chen et al. 1987b).

rumen of 120 to 200 g/day. When the animals were fed a 17% crude protein ration once a day, the ruminal peptide concentrations ranged from 1,400 to 340 μg/mL, and peptide flow was 120 g/day. Robinson and McQueen (1994) calculated nonammonia, nonprotein from the difference between total nitrogen and ammonia plus true protein nitrogen and indicated that peptide concentrations in the rumen were even higher than the values reported by Chen et al. (1987b). Williams and Cockburn (1991) found that the accumulation of peptides varied with the protein source in the diet, and peptides (expressed on a nitrogen basis) were always higher than amino acids. Peptide concentrations did not always follow a predictable pattern based on protein solubility or degradability, and peptide concentrations were even high when urea was the primary nitrogen source provided.

Wallace et al. (1990) reported nonammonia, nonprotein nitrogen concentrations that were as high as 192 μg nitrogen/mL, but later indicated that this material could not be recovered as amino acid nitrogen. The difference between nonammonia, nonprotein nitrogen and amino acid nitrogen was explained by ammonia contamination and the presence of hypothetical nonammonia, nonprotein, non-amino-acid nitrogen containing substances in ruminal fluid (Wallace and McKain 1990). Neither of these ideas seems likely. Chen et al. (1987a) indicated that the samples were "brought to a boil" for 20 minutes under alkaline conditions and that this procedure removed "more than 95% of the ammonia." The discrepancy between nonammonia, nonprotein nitrogen and amino acid nitrogen is probably related to amino acid destruction. It is a well-documented fact that amino acids can undergo side reactions when they are exposed to acid hydrolysis in an oxidizing environment (Rodwell 1977). When amino acids were dissolved in an in vitro medium and heated under alkaline conditions (to remove ammonia), the ninhydrin reaction was unaffected, but the reaction with L-amino acid oxidase was greatly diminished (Chen and Russell 1991).

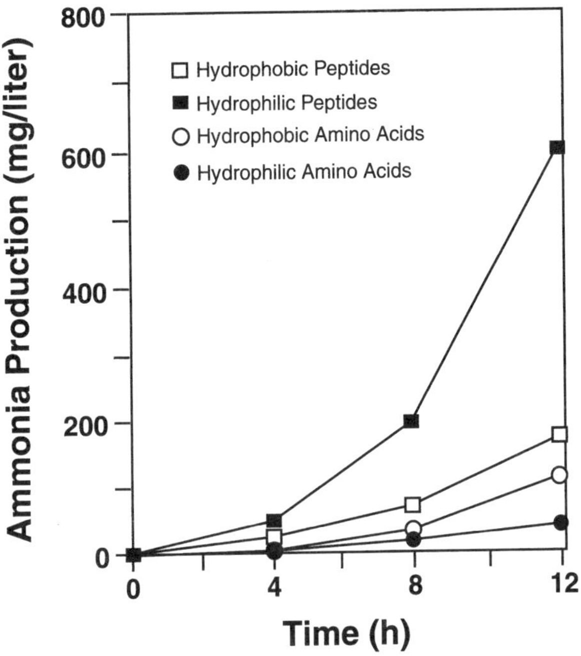

Figure 11.5. Ammonia production from two peptide fractions and free amino acids having the same amino acid composition by mixed ruminal bacteria (270 mg/L) in vitro. The concentration of amino acids or peptides in the culture was 15 g/L (from Chen et al. 1987c).

The structural and compositional characteristics of peptides can have a marked influence on their metabolism. Enzymatically digested casein (Trypticase) can be fractionated by isopropanol (90%) into two peptide fractions of roughly equal mass: a soluble peptide fraction (average chain length 3.3 amino acids) enriched for hydrophobic amino acids, and an insoluble peptide fraction (average chain length 3.2 amino acids) containing a higher proportion of hydrophilic amino acids (Chen and Russell 1987c). When these fractions were incubated with mixed ruminal bacteria, the hydrophilic peptides were degraded more than twice as fast as the hydrophobic peptides (Fig. 11.5). Because free amino acids (similar composition) were degraded at an even slower rate than peptides, it appeared that the arrangement of amino acids within the peptide is a key factor determining the rate of peptide metabolism.

The presence of particular amino acids in peptides can also influence the utilization of peptides. The hydrophobic fraction of Trypticase was highly enriched in proline, and exhaustive digestion of casein and gelatin by mixed ruminal bacteria led to an accumulation of peptides that were enriched in proline and glycine (Yang and Russell 1992). Subsequent experiments with synthetic peptides showed that proline, particularly when it was present in the C terminus, can decrease the degradation rate of peptides. Broderick et al. (1988) noted that mixed ruminal microorganisms took up glycylalanine twice as fast as glycylproline.

The number amino acids in a peptide may also influence the rate of metabolism, but the data are equivocal. Some experiments indicate that certain larger peptides are taken up more rapidly that smaller peptides (Cooper and Ling 1985). However, other work in the same laboratory showed little difference in the uptake and metabolism of large and small peptides (Armstead and Ling 1993). Chemical modifications of peptides can also influence their metabolism. Acetylation of N terminal residues makes peptides resistant to some peptidases and degradation by mixed ruminal bacteria (Wallace 1992).

The role of the protozoal and fungal populations in ruminal metabolism of peptides is less clear. Coleman (1967) showed that *E. caudatum* can assimilate amino acids and incorporate these into protozoal protein, and other protozoal species are able to take up and incorporate amino acids (Coleman 1980). Mixed ruminal protozoa metabolized dipeptides, but it appeared that they only have limited capacity for utilization of larger peptides (Wallace et al. 1990). When Hino and Russell (1987) incubated mixed ruminal protozoa with formalin-treated casein, killed bacteria, or killed yeast, there was an accumulation of peptides in the extracellular fluid. Wallace et al. (1987) observed little difference in the rate peptide degradation when ruminal fluid was obtained from faunated or defaunated sheep. Protozoa have high levels of aminopeptidase and this enzyme may be involved in the degradation of engulfed particulate proteins (Wallace et al. 1987, Prins et al. 1983, Forsberg et al. 1984). Whether the ruminal fungi have the ability to utilize peptides is unknown. It seems likely that the fungi can use peptides. Strains of fungi are proteolytic and peptone is stimulatory to growth (Orpin and Joblin 1988).

Relatively little is known about the mechanisms of peptide uptake and hydrolysis. Mixed ruminal bacteria from sheep hydrolyzed assorted dipeptides, tripeptides, oligopeptides and synthetic substrates (Wallace and McKain 1989, Wallace et al. 1993). The hydrolysis of glycylarginyl-4-methoxy-2-naphthylamide (gly-arg-MNA) and the pattern of products generated during peptide hydrolysis were consistent with the presence of dipeptidyl aminopeptidase activity.

A survey of pure cultures of ruminal bacteria indicated that many of the predominant species of ruminal bacteria have peptidase activity (Wallace and McKain 1991). *Megasphaera elsdenii* exhibited the highest activity toward the alanine containing substrates, while *S. bovis* (as expected) had very high activity for leucine-4-methoxy-2-naphthylamide (leu-MNA), a leucine aminopeptidase substrate. Even the cellulolytic species, *Fibrobacter succinogenes* and *Ruminococcus flavefaciens*, displayed activity toward some substrates. Of the species tested, only *P. ruminicola* rapidly hydrolyzed the synthetic dipeptidase substrate, glycylarginyl-MNA (Wallace and McKain 1991, McKain et al. 1992). Peptide analyses indicated that *P. ruminicola* produces dipeptidyl aminopeptidase, but other peptidase activities could not be ruled out (Wallace et al. 1993).

Early work indicated that *P. ruminicola* utilizes ammonia or oligopeptide nitrogen but not amino acid or short peptide (di-, tripeptide) nitrogen for growth (Pittman and Bryant 1964). Subsequent studies revealed that *P. ruminicola* possesses a general system for the uptake of peptides (Pittman et al. 1967). The rate of peptide incorporation increased with peptide size, but the size exclusion of the peptide transport systems could not be determined. Westlake and Mackie (1990) demonstrated that *S. bovis* possessed the ability to take up a variety of peptides. In their experiments, peptides as large as 750 Da could be detected intracellularly. Many of the peptides, however, underwent hydrolysis prior to or during uptake, and a number of different products were detected in extracellular fluids as well as in cytoplasmic contents. Further work is needed, and our understanding of peptide assimilation is far from clear. A more mechanistic approach of studying peptide metabolism may be achieved through application of molecular biology methods (e.g., generation of specific mutants for uptake of peptides).

4. Deamination

Much of the protein that is degraded in the rumen is deaminated, and the rate of ruminal ammonia production often exceeds the needs of ammonia-utilizing species (Annison 1956). Excess ammonia is absorbed across the ruminal wall, transported to the liver and kidney, converted to urea, and excreted. Ruminants can lose 25% or more of their protein nitrogen as excess ruminal ammonia (Nolan 1975), and this wasteful aspect of ruminal fermentation can have a profoundly detrimental effect on the economics of ruminant production.

Because ammonia accumulation is a balance of ammonia production and utilization, increased ammonia assimilation offers another method of decreasing nitrogen loss. Ammonia assimilation is enhanced by increasing the fermentation rate of ruminally degraded carbohydrate. Bacteria can only utilize ammonia as a nitrogen source when there is sufficient ATP to drive microbial protein synthesis (Fig. 11.1). Some (not all) forms of starch are fermented rapidly, and there is often an inverse relationship between the starch content of a ration and the concentration of ammonia in the rumen (Nocek and Russell 1988).

Most ruminal bacteria can utilize either ammonia or amino acids as a nitrogen source for the synthesis of microbial protein (Allison 1969), but only a few carbohydrate-fermenting bacteria are able to deaminate amino acids and produce significant amounts of ammonia (Bladen et al. 1961). *Selenomonas ruminantium*, *Megasphaera elsdenii*, and *Butyrivibrio fibrisolvens* produced ammonia, but based on its number in the rumen and its capacity to produce ammonia, Bladen et al. (1961) concluded that *Prevotella (Bacteroides) ruminicola* is usually the most important ammonia-producing bacterium in the rumen of mature cattle.

When *P. ruminicola* was grown in a glucose-limited chemostat with an excess of Trypticase, ammonia accumulation was inversely proportional to the dilution rate (Russell 1983) (Fig. 11.6a). When the dilution rate of the chemostat approached the maximum specific growth rate of the bacterium, there was no net accumulation of ammonia. A comparison of chemostats receiving Trypticase and those not, indicated that *P. ruminicola* (1) preferred to use peptides and amino acids as a nitrogen source; (2) had evolved peptide and amino acid transport mechanisms that were sufficient to meet its needs for a nitrogen source, and (3) only diverted amino acids from protein synthesis to deamination when the supply of ATP from carbohydrate fermentation was to insufficient to drive maximal rates of protein synthesis. Because the pathways of amino acid deamination are unable to produce ATP at a rapid rate, *P. ruminicola* cannot utilize peptides and amino acids as a sole energy source for growth (Bladen et al. 1961).

Based on the relationship between ammonia production and the dilution rates of chemostats with an excess of Trypticase, *P. ruminicola* had a specific activity of ammonia production of 13.8 nmol/mg protein/min (Fig. 11.6b). When mixed ruminal bacteria were incubated with a similar amount of Trypticase, the specific activity was 30 nmol/mg protein/min (Table 11.1). Because *S. ruminantium*, *M. elsdenii* and *B. fibrisolvens* also had specific activities that were less than 30 nmol/mg protein/min and most other bacteria could not produce any ammonia, there was a definite contradiction. How could the most active ammonia-producing bacteria have a specific activity that was less than mixed cultures of ruminal bacteria?

Because standard methods of bacterial isolation had always used carbohydrate as an energy source for growth and a low concentration protein hydrolysate (Bryant and Robinson 1961), it is not surprising that all of the ammonia-producing bacteria which were examined by Bladen et al. (1961) required carbohydrates as an energy source. When the carbohydrates were deleted and the concentration of Trypticase was increased 15-fold, several new bacteria were isolated (Russell et al. 1988, Chen and Russell 1989a). rRNA analyses indicated that the large coccus was *Peptostreptococcus anaerobius*, the short motile rod was *Clostridium sticklandii*, and the irregular rod was *Clostridium aminophilum*, sp. nov. (Paster et al. 1993).

These bacteria could not utilize carbohydrates as an energy source for growth, but they grew as fast on peptides and amino acids as most ruminal bacteria can grow on glucose. Pathways of amino acid yield very little ATP per amino acid fermented, but all three of them had very high specific activities of ammonia production (Table 11.1). Because most of the amino acid entering the cell had to be deaminated and fermented as a energy source, only a small fraction ($<$

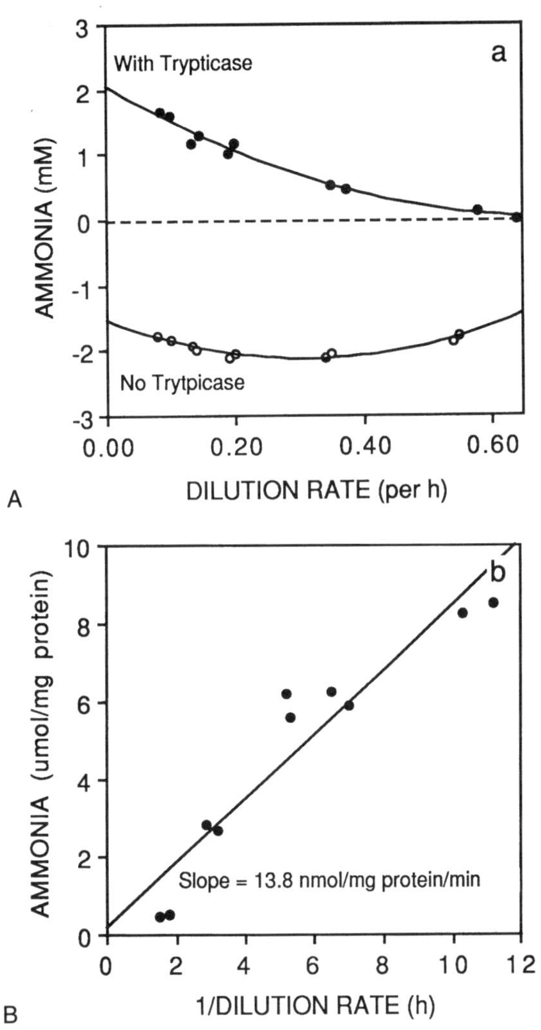

Figure 11.6. The ammonia production and utilization of *P. ruminicola* grown in glucose-limited continuous culture with (●) and without (○) 15 g/L Trypticase (a). The relationship between the reciprocal of dilution rate and ammonia production is shown in (b).

Table 11.1. A comparison of various rumen bacteria and their capacity to produce ammonia, grow in the presence of monensin, and their growth rates on peptides or amino acids

Organism	Ammonia Production (nmol/mg protein/min)	Monensin Sensitive	Growth Rate (L/h)
Mixed rumen bacteria	30		
Megasphaera elsdenii	19	no	< 0.05
Selenomonas ruminantium	15	no	0
Prevotella ruminicola	14	no	0
Peptostreptococcus anaerobius C	346	yes	0.53
Clostridium aminophilum F	318	yes	0.43
Clostridium sticklandii SR	367	yes	0.34

5%) of the amino nitrogen was ever incorporated into microbial protein (Chen and Russell 1988, Chen and Russell 1989a).

The obligate amino acid-fermenting bacteria are not proteolytic and even extracellular peptidase activity is sometimes limiting. When *P. anaerobius* was cocultured with either peptidase-producing strains of *S. bovis* or *P. ruminicola*, there was a large and synergistic increase in the amount of Trypticase that was converted to ammonia (Chen and Russell 1988). *C. sticklandii*, *C. aminophilum*, and *P. ruminicola* had little capacity to utilize gelatin hydrolysate, but rapid ammonia production resulted when *P. ruminicola* was cocultured with either of the clostridia (Chen and Russell 1989a). Each bacterium has a preference for specific amino acids and they probably operate in a cooperative fashion (Table 11.2).

The obligate amino acid-fermenting bacteria cannot grow in the absence of sodium, and sodium plays a dominate role in both amino acid transport and energy derivation from amino acid fermentation (Table 11.2). *C. aminophilum* appears to use a glutaconyl CoA decarboxylase to expel sodium (Chen and Russell 1990), and *C. sticklandii* uses an ornithine efflux system to create a sodium gradient which can then be used to drive arginine or lysine transport (Van Kessel and Russell 1992). *P. anaerobius* has a sodium-dependent transport system for branched chain amino acids, but its mechanism of sodium expulsion has not been defined (Chen and Russell 1989b). In some cases, the bacteria use facilitated diffusion mechanisms to take up amino acids, and these high-capacity systems may be at least partially responsible their high rates of peptide and amino acid transport and deamination.

In the 1930s Stickland studied the growth of several amino acid fermenting bacteria and noted that the bacteria were only able to grow when the amino acids were provided as pairs of highly oxidized and highly reduced amino acids (Nis-

Table 11.2. Some general characteristics of *Peptostreptococcus anaerobius* strain C, *Clostridium aminophilum* strain F, and *Clostridium sticklandii* strain SR

Characteristic	C	F	SR
Gram stain	+	+	variable
Growth in presence of O_2	no	no	no
Ferments carbohydrates	no	no	no
Proteolytic	no	no	no
Requires sodium	yes	yes	yes
Fermentation products			
acetate	+	+ + +	+ + +
propionate	—	—	+
butyrate	—	+ +	+
isobutyrate	+	—	+
isovalerate	+	—	+
valerate	+	—	—
isocaproate	+ +	—	—
hydrogen	+	—	—
Amino acids most rapidly fermented	leu	glu	arg
	ser	gln	ser
	thr	his	lys
	gln	ser	gln
	phe	thr	met

man 1954). *P. anaerobius* was able to deaminate proline, glycine, alanine, valine, and isoleucine faster when they were provided as Stickland pairs, but leucine was deaminated at a rapid rate even if another electron sink was not provided (Chen and Russell 1988). In this latter case, reducing equivalents from isovalerate production were directly shuttled to isocaproate production.

In the rumen methanogenesis is a dominant means of reducing equivalent disposal (Wolin 1975), and the hydrogenase inhibitor, carbon monoxide, caused a large decrease in methane production and a modest decrease in ammonia production (Russell and Jeraci 1984, Hino and Russell 1985). Since virtually all of the decrease in ammonia production could be explained by a decrease in branched-chain volatile fatty acids, it appeared that a decrease in interspecies hydrogen transfer was causing a selective decrease in the fermentation of highly reduced branched-chain amino acids (Fig. 11.7). *P. anaerobius* produces hydrogen when it ferments branched-chain amino acids (Chen and Russell 1988).

Defaunation often causes a decrease in ruminal ammonia (Abou Akkada and el-Shazly 1964, Eadie and Gill 1971), and this observation lead some researchers to speculate that protozoa might have a high specific activity of amino acid deamination (Hungate 1966). In vitro results, however, indicated that protozoa

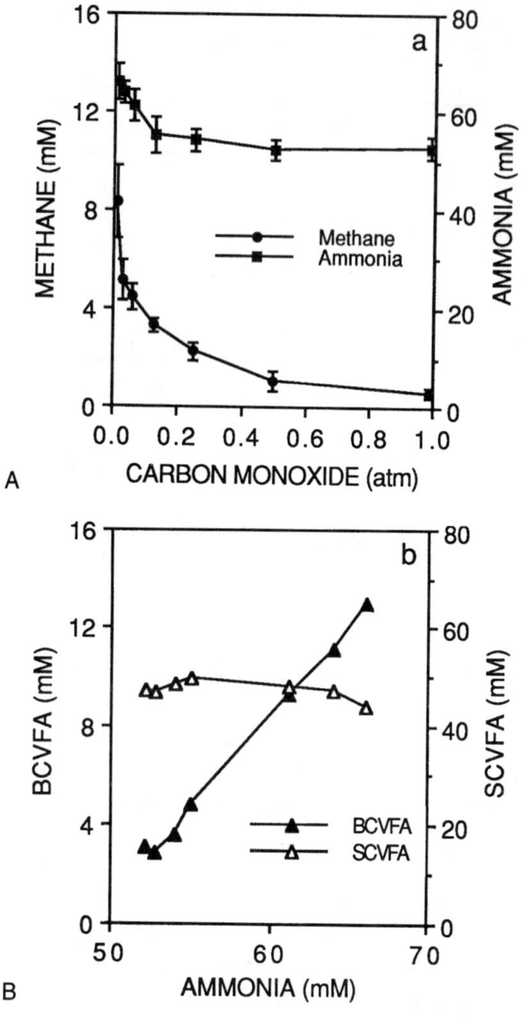

Figure 11.7. The effect of carbon monoxide on methane and ammonia production of mixed ruminal bacteria that were provided with 15 g/L Trypticase (a). The relationship between ammonia production and either branched-chain volatile fatty acid (BCVFA) or straight-chain volatile fatty acid production (SCVFA) is shown in (b).

were only able to produce significant amounts of ammonia from particulate protein, and even then the specific activity of the protozoa was less than half of the bacteria (Hino and Russell 1987). When provided with particulate protein sources, there was a synergistic increase in ammonia production when the protozoa were cocultured with mixed ruminal bacteria. Because the protozoa released large amounts of non-ammonia, non-protein nitrogen it appeared that the protozoa were primarily involved in proteolysis rather than deamination per se (Hino and Russell 1987).

The ionophore monensin has been primarily marketed as product that enhances animal production by shifting the pathways of carbohydrate fermentation from acetate to propionate production (Richardson et al. 1976). However, even early work indicated that monensin could also decrease ruminal ammonia production (Dinius et al. 1976, Van Nevel and Demeyer 1977). Because most of the carbohydrate-fermenting bacteria which deaminate amino acids are monensin-resistant (Chen and Wolin 1979, Nagaraja and Taylor 1987), the effect of monensin was not readily apparent.

All of the obligate amino acid-fermenting bacteria are monensin-sensitive, and recent work showed that monensin could decrease the numbers of these bacteria 10-fold (Yang and Russell 1993a,b). This 10-fold reduction caused a 50% reduction in the specific activity of ruminal ammonia production and the steady state concentration of ammonia in the rumen. Monensin addition caused an increase in peptide and amino acid flow from the rumen when protein hydrolysates were pulse-dosed into the rumen (Yang and Russell 1993a), but there was no increase in the flow of peptides and free amino acids when dietary protein was the amino acid source (Yang and Russell 1993b). Amino nitrogen which was protected from deamination was taken up by carbohydrate-fermenting bacteria, and the flow of unattached bacteria from the rumen increased 30%.

5. Urea Metabolism

Because most ruminal bacteria are able to utilize ammonia as a nitrogen source (Allison 1969), nonprotein nitrogen can be added to ruminant rations as an inexpensive protein substitute (e.g., Briggs 1967). Endogenous urea is also recycled into the rumen by way of salivary secretions and by diffusion across the rumen epithelium (Houpt 1959, Mathison and Milligan 1971, Nolan and Leng 1972, Kennedy and Milligan 1980). The contribution of endogenous urea varies with the diet consumed (Kennedy and Milligan 1980). Endogenous urea contributed as much as 25% of the total nitrogen entering the rumen when animals were fed poor-quality diets (Kennedy 1980).

Urea entering the rumen is rapidly hydrolyzed by the enzyme urease to ammonia and carbon dioxide. All the urease activity in the rumen is microbial in origin, and the bacterial population appears to be solely responsible for this activity (Jones et al. 1964). Early reports of protozoal urease were later shown to be due to bacteria associated with protozoal cells (Onodera et al. 1977). In fact, urease has been used as a method to evaluate bacterial contamination of isolated ruminal protozoa (Forsberg et al. 1984). Some nonruminal fungi produce urease, but the ureolytic activity of ruminal fungi is unknown (Mobley and Hausinger 1989).

Early efforts to isolate urease-producing anaerobic bacteria from the rumen were thwarted by cultural techniques. Gibbons and Doetsch (1959) isolated an anaerobic *Lactobacillus* sp. that hydrolyzed urea, but this organism was present only at 10^5/g rumen contents. Facultative anaerobic species were much more readily isolated than strict anaerobes, but these bacteria were present in low numbers (10^4 to 10^6) and did not produce enough urease to explain the activity of ruminal fluid (Cook 1976, Jones et al. 1964). Slyter et al. (1968) isolated predominant species of ruminal bacteria in high numbers (10^8) from animals fed purified diets containing urea, but the urease activity of these isolates was not determined. Because facultative anaerobes generally produced high levels of urease (Van Wyk and Steyn 1975, Wallace et al. 1979), Cheng and Costerton (1980) proposed that facultative anaerobes that were attached to the rumen wall played a significant role in ureolysis and the transfer of urea across the rumen epithelium.

Varel et al. (1974) demonstrated that the high concentrations of ammonia and complex nitrogen sources in standard media can repress urease production. As a result, attempts to identify ureolytic species using standard media generally fail. Using a modified detection medium containing 50 mM urea and 0.05% each of Trypticase and yeast extract, Wozny et al. (1977) detected ureolytic anaerobes with numbers greater than 10^7/mL rumen fluid. Based on these results, 10% of the total population is ureolytic and includes strains of *S. dextrinosolvens*, *Ruminococcus bromii*, *B. fibrisolvens*, *Peptostreptococcus productus*, *P. ruminicola*, *Treponema* sp., and *Coprococcus* sp.

Of the urease-producing ruminal bacteria, *S. ruminantium* has been studied in the greatest detail. John et al. (1974) isolated a highly ureolytic *S. ruminantium* using a selective medium containing urea as the major nitrogen source. Smith et al. (1981) later showed that this strain only synthesized urease under nitrogen-limited growth conditions. Urease production was repressed by ammonia, amino acids, and even high concentrations of urea. Because the regulation of urease paralleled that of glutamine synthetase, it appeared that urease and glutamine synthetase might be coordinately regulated. In *S. dextrinosolvens* nitrogen limitation resulted in high levels of enzyme activity, while nitrogen excess cultures repressed in urease production (Patterson and Hespell 1985). Control of urease synthesis in nonruminal microorganisms involves more complicated mechanisms than simple induction and repression (Mobley and Hausinger 1989).

The urease of *S. ruminantium* has been purified (Hausinger 1986, Mobley and Hausinger 1989). It has a native molecular weight of 360 kDa and is composed of three subunit species (α, 70 kDa; β, 8 kDa; and γ, 8 kDa) in an undetermined stoichiometry. The enzyme contains two nickel atoms per α subunit. This enzyme, however, differs from partially purified urease from mixed rumen microorganisms (Mahadevan et al. 1977). The mixed microbial urease had a molecular weight of 120 kDa to 130 kDa and did not appear to require nickel for activity. Other workers, however, indicated that the urease activity of rumen contents did require nickel for activity (Spears et al. 1977, Spears and Hatfield 1978).

As previously noted, endogenous urea can enter the rumen in saliva or by diffusion across the rumen epithelium. Because the concentration of urea in saliva parallels plasma, the quantity of urea entering the rumen should be dependent upon blood urea concentration and saliva flow (Norton et al. 1982). The diffusion of urea into the rumen wall is somewhat more complex and appears to be influenced by conditions within the rumen. Kennedy (1980) noted that urea diffusion into the rumen was poorly correlated with blood urea concentrations and more closely associated (negatively) with ruminal ammonia concentrations. Based on this work, it appears that high ruminal ammonia concentrations may inhibit urea diffusion into the rumen. When a fermentable carbohydrate source is added to the diet, ammonia concentration is lowered, and urea transfer is enhanced (Kennedy et al. 1981). These and similar observations have led to the proposal that the urease activity of microorganisms attached to or in close proximity to the rumen wall regulates the diffusion of urea into the rumen (Cheng and Wallace 1979). According to this theory, when ruminal ammonia is low, high urease activity maintains a urea concentration gradient across the rumen wall and enhances the diffusion of urea. In contrast, high ammonia concentrations would repress urease activity, resulting in depressed urea transfer. The identification of ureolytic bacteria that colonize the rumen wall appears to support this idea (Wallace et al. 1979, Cheng and Costerton 1980). This theory, however, fails to address urea toxicity. If urease was repressed, limiting the rate of urea hydrolysis, excess urea should not generate the toxic levels of rumen and blood ammonia that can arise (Austin 1967, Chalupa 1968).

6. Nucleic Acid Metabolism

Although not as abundant as proteins, plants contain significant amounts of RNA and DNA. The nucleic acid content of feedstuffs varies considerably with source, ranging from about 5% (Coelho da Silva et al. 1972) in legumes such as alfalfa to 0.24% in flaked corn (Smith and McAllan 1970). Endogenous sources of DNA and RNA such as sloughed rumen epithelium probably add little to the total nitrogen in the rumen (McAllan 1982), but the contribution of microbial cells to nucleic acids in the rumen is considerable. Microbial cells contain approximately 10% RNA and 1% DNA (Leedle et al. 1982), and the turnover of microorganisms in the rumen is great (as much as 30%/day; McAllan 1982).

Nucleic acids are degraded extensively in the rumen (McAllan and Smith 1973, Smith and McAllan 1970, Russell and Wilson 1988). In vivo and in vitro experiments showed that purified high molecular weight RNA was rapidly (less than 1 hour) converted to oligo- and mononucleotides (McAllan and Smith 1973a). Adenine, guanine, and cytosine were not detected, but their deamination products hypoxanthine, xanthine, and uracil did accumulate (McAllan and Smith 1973a,b). Similar patterns of degradation were observed for purified DNA and nucleic acids from hay, but the rates were slower. Endonucleases are present in cell free rumen fluid, and these enzymes are responsible for the initial rapid degradation of nucleic acids (McAllan and Smith 1973, Russell and Wilson 1988). Several predominant ruminal bacteria produce extracellular nuclease activities that could be involved in DNA digestion (Flint and Thomson 1990). These include strains of *P. ruminicola*, *F. succinogenes*, *S. ruminantium*, and *L. multiparus*.

Little information is available on the metabolism of nucleic acids by individual species of ruminal bacteria. In a few cases, nucleic acids, nucleotides, or bases have been examined for growth-stimulating activity or as possible required nutrients (Gill and King 1958, John et al. 1974, Pittman and Bryant 1964, Roche et al. 1973). Some ruminal bacteria can use adenine or guanine as a nitrogen source (Pittman and Bryant 1964, John et al. 1974), but this ability is probably linked to the deamination these bases and use of the released ammonia. RNA fermenting strains of *S. ruminantium* used ribonucleosides as sole sources of nitrogen and energy (Cotta 1990). The purine nucleosides adenosine and guanosine supported much higher growth than pyrimidine nucleosides. Because bases alone did not support growth, it appeared that only the ribose moiety of the nucleosides was serving as an energy source. In these studies, DNA was not fermented, but deoxyribose-fermenting strains of *S. ruminantium* have been isolated (Rasmussen 1993).

Entodiniomorphid protozoa cannot synthesize purines and pyrimidines and must rely on exogenous sources of these nucleic acid monomers for growth (Coleman 1979, 1980; Williams and Coleman 1988, 1992). Ingested bacterial cells are the most important sources of nucleic acids for protozoa (Coleman 1980). Most information on nucleic acid utilization by protozoa is derived from studies of *E. caudatum*, a protozoon that can be cultivated in vitro for long periods of time (Williams and Coleman 1992). *E. caudatum* assimilates nucleic acids both from ingested bacteria and in the form of free monomers (Coleman 1964, 1968). When *E. caudatum* was incubated with free ^{14}C-labeled purines and pyrimidines, much of the label was incorporated into acid soluble intracellular pools, but a significant portion was integrated into cellular nucleic acids. Distribution of the label in the protozoon demonstrated that bases underwent interconversion: adenine to guanine and cytosine to uracil, and vice versa. Thymine was taken very rapidly and reduced to dihydrothymidine, but then not metabolized further or incorporated into cellular nucleic acids. After prolonged incubation, free bases were metabolized and excreted by the protozoon. The purines were deaminated to xanthine and hypoxanthine, and the pyrimidines were reduced to dihydrouracil and dihyrothymidine. Nucleic acids from *E. coli* cells underwent similar transformations, but a larger portion of the incorporated label was distributed in cellular nucleic acids than in soluble pools. Nucleosides and nucleotides were preferentially used over free bases. Because label from ribose (or glucose or starch) was not incorporated into nucleic acids, *E. caudatum* apparently cannot incorporate ribose into nucleic acids. Information on the metabolism of nucleic acids by other entodiniomorphid protozoa is more limited, but generally confirms the work with *E. caudatum* (Coleman 1972; Coleman and Laurie 1974, 1977; Coleman and Sandford 1979). Nothing is known about the metabolism of nucleic acids by rumen holotrich protozoa, but they may also derive benefit from ingested bacterial nucleic acids (Williams and Coleman 1992, Williams 1986).

7. Coordination of Nitrogen and Carbohydrate Metabolism

The metabolism of nitrogen and carbohydrates in the rumen must not be considered as a separate phenomenon, but rather as an integrated and highly interrelated process. When energy is limiting and nitrogen is in excess, much of the nitrogen in the diet is converted to ammonia and lost. If energy is in excess and nitrogen is limiting (even ammonia nitrogen), the efficiency of microbial growth can decrease.

When *S. bovis* JB1 was incubated in a nitrogen-free medium with an excess of glucose, the cells fermented glucose at a rapid rate in the absence of growth (Russell and Strobel 1990). Because the nongrowth glucose consumption (energy spilling) was eight times greater than the maintenance rate of growing cells, it appeared that this phenomenon could not be categorized as ''maintenance'' per se. Based on the observation that the nongrowth energy transduction was enhanced by TCS and inhibited by DCCD, it appeared that the ATP turnover was due to the membrane-bound F_1F_0 ATPase and a cycle of protons thorough the cell membrane. Since *S. bovis* uses glucose transport mechanisms that are not dependent on Δp and ATP did not decrease, the effect of DCCD could not be explained by an inhibition of glucose uptake or catabolism per se.

Ohm's law indicates that current flow (amperage) should be proportional to voltage, but the relationship between the rate of energy spilling and the protonmotive force across the cell membrane was clearly nonohmic. Because there was little change in proton motive force over wide ranges of energy spilling, it appeared that the resistance of the cell membrane to proton conductance was changing, and

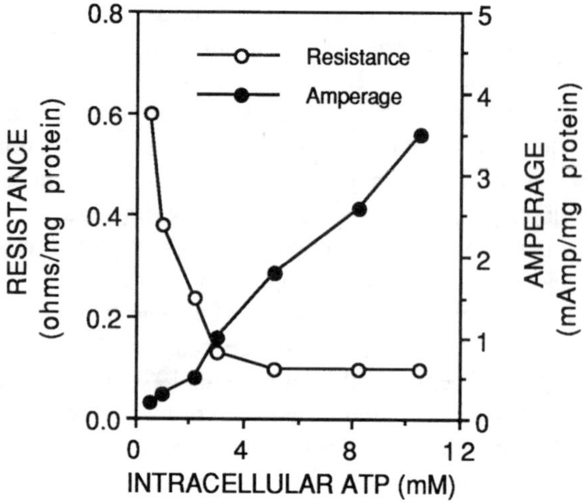

Figure 11.8. The relationship of intracellular ATP, cell membrane resistance, and proton flux (amperage) for nongrowing *S. bovis* cells.

this change was correlated with the intracellular ATP content of the cells (Fig. 11.8). When the rate of ATP formation was greater than the amount that was needed by biosynthetic or maintenance functions, the cells decreased their membrane resistance, allowed a greater influx of protons, and utilized the membrane-bound proton F_1F_0 ATPase as a mechanism to hydrolyze ATP. This ATP hydrolysis allowed the cells to continue fermenting glucose in the absence of growth. Information on other ruminal bacteria is not available, but work with *S. bovis* provides a mechanistic model of energy spilling.

Most ruminal bacteria can utilize ammonia as a nitrogen source, but the growth of many species is stimulated by amino acids (Allison, 1969). Pure cultures of ruminal bacteria grew as much as 40% more efficiently when they were supplemented with protein hydrolysate (Cotta and Russell 1982), and the flow of microbial protein from the rumen was 30% greater when a urea-based diet was supplemented with casein (Hume 1970). The effect of amino acids on rumen microbial growth efficiency was a curiosity (Hespell and Bryant 1979). Stouthamer's calculations (1973) indicated that the amino acid availability should have little impact on bacterial growth efficiency as the cost of transport would offset most of the saving in biosynthesis. Some pathways of amino acid biosynthesis are actually energy-yielding schemes (Fig. 11.9), and polymerization reactions require far more energy than monomer synthesis (Table 11.3).

Streptococcus bovis can utilize ammonia as a sole source of nitrogen (Wolin et al. 1959) but grows nearly twice as fast when provided with amino acids as

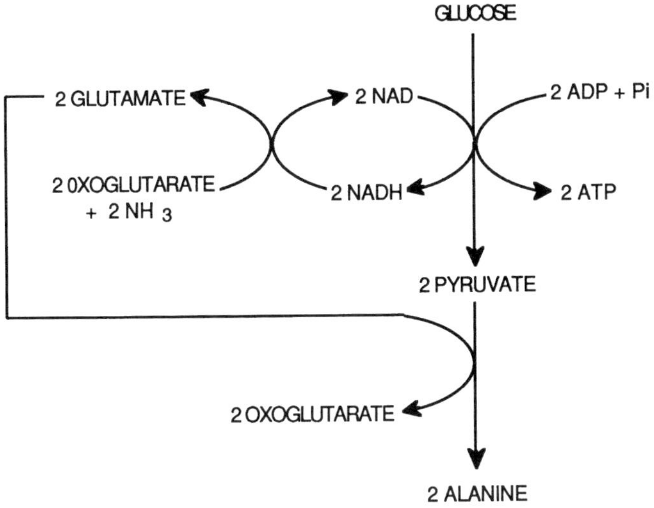

Figure 11.9. Pathway of alanine biosynthesis from glucose and ammonia via pyruvate-oxoglutarate transaminase and glutamate dehydrogenase.

a nitrogen source (Russell 1993a). Because amino acid availability had virtually no effect on rate at which *S. bovis* fermented glucose, it appeared that amino acids were affecting the balance of anabolic and catabolic rates (Fig. 11.10a). When amino acids are available, most of the ATP can be utilized in biosynthetic reactions, little ATP is dissipated in energy spilling reactions, and the growth yield is high (Fig. 11.10b). When amino acids are not available, however, catabolic pathways produce more ATP than the biosynthetic pathways (e.g., amino acid biosynthesis) can utilize, ATP is dissipated in energy spilling reactions, and the growth yield is low.

When *P. ruminicola* B_14 was grown in an ammonia-limited medium with an excess of glucose, little energy was spilled, but the viable cell number decreased 10,000-fold (Russell 1992). The cultures initially stored the excess glucose as polysaccharide. When the polysaccharide stores were saturated, intracellular ATP increased, succinate and acetate production declined, and the cells produced methylglyoxal, a toxic end product (Russell 1993b). Glucose excess or methylglyoxal-treated cells could not maintain a membrane potential and lost their intracellular potassium. Glucose-resistant mutants had a threefold lower rate of glucose uptake than the wild-type cells, but even the mutants were eventually killed by methylglyoxal accumulation. Other genetically distinct strains of *P. ruminicola* produced methylglyoxal when glucose was in excess and nitrogen was limiting, but methylglyoxal production was not a ubiquitous feature of all *P. ruminicola* strains.

Table 11.3. The ATP requirement for the formation of bacterial cells from glucose with or with amino acids

Macromolecule	% Dry Weight	With Ammonia	With Amino Acids
		(mmol ATP/g Cells)	
Polysaccharide	16.6	2.1	2.1
Protein	52.2		
amino acid formation		1.4	1.4
polymerization		19.0	19.0
Lipid	9.4	0.2	0.2
RNA	15.7		
nucleoside formation		3.5	3.5
polymerization		0.9	0.9
DNA	3.2		
nucleoside formation		0.9	0.9
polymerization		0.2	0.2
mRNA turnover		1.4	1.4
Transport functions			
ammonium ions		4.2	0.0
amino acids		0.0	0.0
potassium		0.2	0.2
phosphate		0.8	0.8
Total		34.7	30.5
Y_{ATP} (g cell/mol ATP)		28.8	32.8

Source: Stouthamer (1973).

The experiments with *S. bovis* and *P. ruminicola* indicate that bacteria have not always evolved mechanisms of regulating carbohydrate uptake. If the potential rate of ATP formation from carbohydrate fermentation exceeds the demands of biosynthesis, the cells must either store the glucose as glycogen, shift their metabolism to pathways that produce less ATP (e.g., methylglyoxal), or use other mechanisms of ATP dissipation (Fig. 11.11). Because there is obviously a limit to the amount of polysaccharide that a cell can store and end product like methylglyoxal are toxic, energy spilling may be a ''protective mechanism.''

Ruminal microorganisms have developed the capacity to use an array of different nitrogen sources, but these nitrogen sources are not always equivalent. The efficiency of microbial growth in the rumen is dependent on the rate and release of both nitrogen and energy, but this fact has often been overlooked by ruminant nutritionists. If ruminant nutrition is to progress past the point of random diet evaluation, the kinetics of the system must be considered (Russell et al. 1992, Sniffen et al. 1992, Fox et al. 1992).

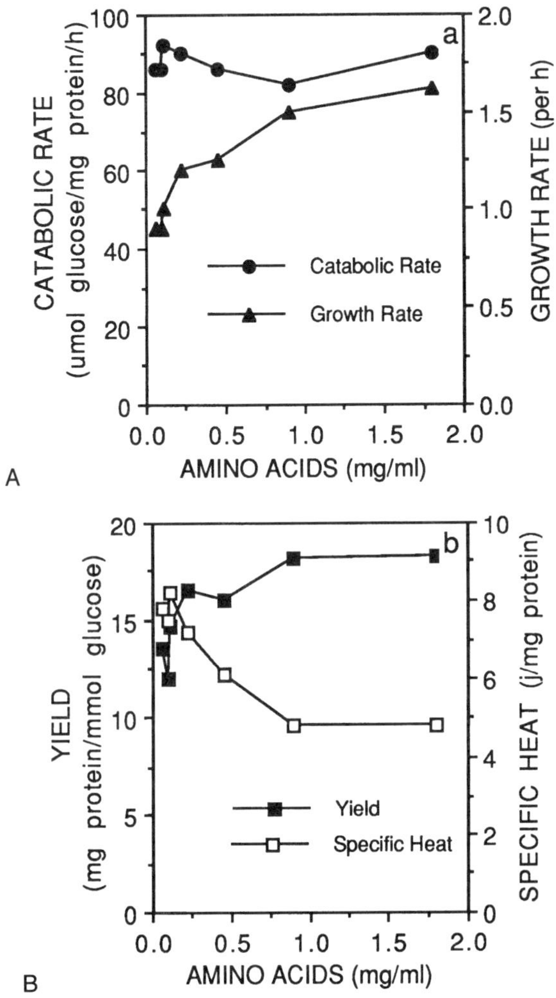

Figure 11.10. The effect of amino acids on the catabolic and anabolic rates of *S. bovis* (a). The relationship between yield and specific heat is shown in (b).

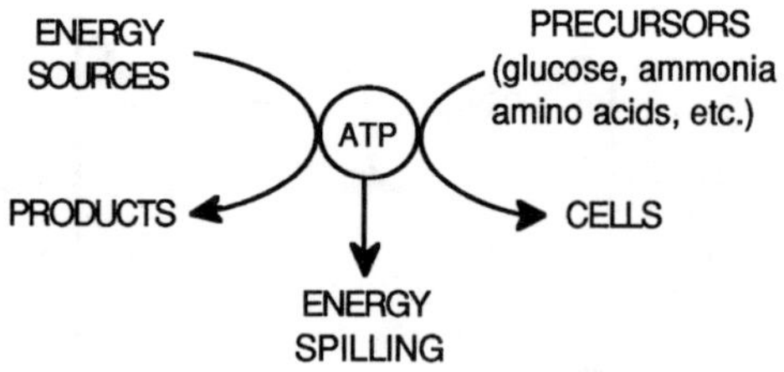

Figure 11.11. A theoretical scheme showing the interrelationship of anabolism, catabolism, and energy spilling.

8. Conclusions

Using the rumen as a model, this chapter has attempted to describe the nitrogen metabolism of gut environments. The rumen ecosystem is clearly one of the best-studied microbial habitats. The rumen environment has been described with respect to motility, turnover, pH, osmolality, etc. Materials entering the rumen have been analyzed and categorized by a variety of procedures. Rumen microorganisms capable of performing all of the major transformations of feed have been isolated. The nutrient requirements of ruminal microorganisms, including nitrogen, have been defined, and at least some patterns of metabolic regulation have been identified. But as Hungate (1960) noted: "Analysis that falls short of quantitative formulation can hardly be considered complete." In this light, the rumen is still poorly understood. The rates of ammonia generation and microbial protein synthesis, the turnover of microbial protein, and the fluxes of nitrogen across the rumen wall are still a matter of gross approximation rather than precise estimation. If rumen microbiology is to progress to a point of meaningful quantitation, microbiologists must be prepared to express themselves mathematically and integrate their concepts in dynamic models. This aspect of ecological analysis is still in its infancy.

References

Abe M, Iriki T (1989) Mechanism whereby holotrich ciliates are retained in the reticulorumen of cattle. Br J Nutr 62:579–587.

Abou Akkada AR, Howard BH (1962) The biochemistry of rumen protozoa. 5. The nitrogen metabolism of *Entodinium*. Biochem J 82:313–320.

Abou Akkada AR, el-Shazly K (1964) Effects of the absence of ciliate protozoa from the rumen on microbial activity and grow of lambs. Appl Microbiol 12:384–390.

Allison MJ (1969) Biosynthesis of amino acids by ruminal microorganisms. J Anim Sci 29:797–807.

Annison EF (1956) Nitrogen metabolism in the sheep. Biochem J 64:705–714.

Appleby JC (1955) The isolation and classification of proteolytic bacteria from the rumen of the sheep. J Gen Microbiol 12:526–533.

Armstead IP, Ling JR (1993) Variations in the uptake and metabolism of peptides and amino acids by mixed ruminal bacteria in vitro. Appl Environ Microbiol 59:3360–3366.

Asao N, Ushida K, Kojima Y (1993) Proteolytic activity of rumen fungi belonging to the genera *Neocallimastix* and *Piromyces*. Lett Appl Microbiol 16:247–250.

Austin J (1967) Urea toxicity and its prevention. In: Briggs MH, ed. Urea as a Protein Supplement, pp. 173–184. London: Pergamon Press.

Barrett AJ (1977) Introduction to the history and classification of tissue proteinases. In: Barrett AJ, ed. Proteinases in Mammalian Cells and Tissues, pp. 1–55. New York: Elsevier/North Holland.

Bird SH, Leng RA (1978) The effects of defaunation of the rumen on the growth of cattle on low-protein high-energy diets. Br J Nutr 40:163–167.

Blackburn TH (1968a) Protease production by *Bacteroides amylophilus* strain H 18. J Gen Microbiol 53:27–36.

Blackburn TH (1968b) The protease liberated from *Bacteroides amylophilus* strain H 18 by mechanical disintegration. J Gen Microbiol 53:37–51.

Blackburn TH, Hobson PN (1960a) Proteolysis in sheep rumen by whole and fractionated rumen contents. J Gen Microbiol 22:272–281.

Blackburn TH, Hobson PN (1960b) Isolation of proteolytic bacteria from the sheep rumen. J Gen Microbiol 22:282–289.

Blackburn TH, Hobson PN (1962) Further studies on the isolation of proteolytic bacteria from the sheep rumen. J Gen Microbiol 29:69–81.

Blackburn TH, Hullah WA (1974) The cell-bound protease of *Bacteroides amylophilus* H18. Can J Microbiol 20:435–441.

Bladen HA, Bryant MP, Doetsch RN (1961) A study of bacterial species from the rumen which produce ammonia from protein hydrolyzate. Appl Microbiol 9:175–180.

Briggs MH (1967) Urea as a Protein Supplement. London: Pergamon Press.

Brock FM, Forsberg CW, Buchanan-Smith JG (1982) Proteolytic activity of rumen microorganisms and effects of proteinase inhibitors. Appl Environ Microbiol 44:561–569.

Broderick GA, Wallace RJ, McKain NJ (1988) Uptake of small peptides by mixed rumen microorganisms in vitro. J Sci Food Agric 42:109–118.

Bryant MP, Burkey LA (1953) Numbers and some predominant groups of bacteria in the rumen of cows fed different rations. J Dairy Sci 36:218–224.

Bryant MP, Robinson IM (1961) An improved nonselective culture media for ruminal bacteria and its use in determining diurnal variation in numbers of bacteria in the rumen. J Dairy Sci 44:1446–1456.

Bryant MP, Small N (1956) The anaerobic monotrichous butyric acid-producing curved rod-shaped bacteria of the rumen. J Bacteriol 72:16–21.

Bryant MP, Small N, Bouma C, Robinson IM (1958) Characteristics of ruminal anaerobic cellulolytic cocci and *Cillobacterium cellulosolvens* n. sp. J Bacteriol 76:529–537.

Chalupa W (1968) Problems in feeding urea to ruminants. J Anim Sci 27:207–219.

Chen M, Wolin MJ (1979) Effect of monensin and lasalocid-sodium on the growth of methanogenic and rumen saccharolytic bacteria. Appl Environ Microbiol 38:72–77.

Chen G, Russell JB (1988) Fermentation of peptides and amino acids by a monensin sensitive ruminal peptostreptococcus. Appl Environ Microbiol 54:2742–2749.

Chen G, Russell JB (1989a) More monensin-sensitive, ammonia-producing bacteria from the rumen. Appl Environ Microbiol 55:1052–1057.

Chen G, Russell JB (1989b) Sodium-dependent transport of branched-chain amino acids by a monensin-sensitive ruminal peptostreptococcus. Appl Environ Microbiol 55: 2658–2663.

Chen G, Russell JB (1990) Transport and deamination of amino acids by a gram-positive, monensin-sensitive ruminal bacterium. Appl Environ Microbiol 56:2186–2192.

Chen G, Russell JB (1991) Effect of monensin and a protonophore on protein degradation, peptide accumulation, and deamination by mixed ruminal microorganisms in vitro. J Anim Sci 69:2196–2203.

Chen G, Russell JB, Sniffen CJ (1987a) A procedure for measuring peptides in rumen fluid and evidence that peptide uptake can be a rate-limiting step in ruminal protein degradation. J Dairy Sci 70:1211–1219.

Chen G, Sniffen CJ, Russell JB (1987b) Concentration and estimated flow of peptides from the rumen of dairy cattle: effects of protein quantity, protein solubility, and feeding frequency. J Dairy Sci 70:983–992.

Chen G, Strobel HJ, Russell JB, Sniffen CJ (1987c) Effect of hydrophobicity on utilization of peptides by ruminal bacteria in vitro. Appl Environ Microbiol 53:2021–2025.

Cheng KJ, Costerton JW (1980) Adherent rumen bacteria—their role in digestion of plant material, urea, and epithelial cells. In: Ruckebusch Y, Thivend P, eds. Digestive Physiology and Metabolism in Ruminants, pp. 227–232. Westport, Conn: AVI Publishing Co.

Cheng KJ, Wallace RJ (1979) The mechanism of passage of endogenous urea through the rumen wall and the role of ureolytic epithelial bacteria in the urea flux. Br J Nutr 42:553–557.

Coelho da Silva JF, Seeley RC, Thomson DJ, Beever DE, Armstrong DG (1972) The effect in sheep of physical form on the sites of digestion of a dried lucerne diet. Br J Nutr 28:43–61.

Coleman GS (1964) The metabolism of *Escherichia coli* and other bacteria by *Entodinium caudatum*. J Gen Microbiol 37:209–223.

Coleman GS (1967) The metabolism of free amino acids by washed suspensions of the rumen ciliate *Entodinium caudatum*. J Gen Microbiol 47:433–447.

Coleman GS (1968) The metabolism of bacterial nucleic acid and free components of nucleic acid by the rumen ciliate *Entodinium caudatum*. J Gen Microbiol 54:83–96.

Coleman GS (1972) The metabolism of starch, glucose, amino acids, purines, pyrimidines and bacteria by the rumen ciliate *Entodinium simplex*. J Gen Microbiol 71:117–131.

Coleman GS (1979) Rumen ciliate protozoa. In: Biochemistry and Physiology of Protozoa, pp. 381–408. New York: Academic Press.

Coleman GS (1980) Rumen ciliate protozoa. In: Lumsden WHR, Muller R, Baker JR, eds. Advances in Parasitology, pp. 121–172. London: Academic Press.

Coleman GS (1983) Hydrolysis of fraction 1 leaf protein and casein by rumen entodiniomorphid protozoa. J Appl Bacteriol 55:111–118.

Coleman GS, Laurie JI (1974) The metabolism of starch, glucose, amino acids, purines, pyrimidines and bacteria by three *Epidinium* spp. isolated from the rumen. J Gen Microbiol 85:244–256.

Coleman GS, Laurie JI (1977) The metabolism of starch, glucose, amino acids, purines, pyrimidines and bacteria by the rumen ciliate *Polyplastron multivesiculatum*. J Gen Microbiol 98:29–37.

Coleman GS, Sandford DC (1979) The uptake and utilization of bacteria, amino acids, and nucleic acid components by the rumen ciliate *Eudiplodinium maggii*. J Appl Bacteriol 47:409–419.

Cook AR (1976) Urease activity in the rumen of sheep and the isolation of ureolytic bacteria. J Gen Microbiol 92:32–48.

Cooper PB, Ling JR (1985) The uptake of peptides and amino acids by rumen bacteria. Proc Nutr Soc 44:144.

Cotta MA (1990) Utilization of nucleic acids by *Selenomonas ruminantium* and other ruminal bacteria. Appl Environ Microbiol 56:3867–3870.

Cotta MA, Hespell RB (1986) Proteolytic activity of the ruminal bacterium *Butyrivibrio fibrisolvens*. Appl Environ Microbiol 52:51–58.

Cotta MA, Russell JB (1982) Effect of peptides and amino acids on efficiency of rumen bacterial protein synthesis in continuous culture. J Dairy Sci 65:226–234.

Crawford RJ, Hoover WH, Sniffen CJ, Crooker BA (1978) Degradation of feedstuff nitrogen in the rumen vs. nitrogen solubility in three solvents. J Anim Sci 46:1768–1775.

Crooker BA, Sniffen CJ, Hoover WH, Johnson LL (1978) Solvents for soluble nitrogen measurements in feedstuffs. J Dairy Sci 61:437–447.

Dehority BA (1966) Characterization of several bovine rumen bacteria isolated with a xylan medium. J Bacteriol 91:1724–1729.

Demeyer DI, Van Nevel CJ (1979) Effect of defaunation on the metabolism of rumen microorganisms. Br J Nutr 42:515–524.

Dinius DA, Simpson ME, Marsh PB (1976) Effect of monensin fed with forage on digestion and the ruminal ecosystem of steers. J Anim Sci 42:229–234.

Eadie JM, Gill JC (1971) The effect of the presence and absence of rumen ciliate protozoa on the total rumen bacteria counts in lambs. Appl Environ Microbiol 47:101.

Ferguson KA (1975) The protection of dietary proteins and amino acids against microbial fermentation in the rumen. In: McDonald IW, Warner ACI, eds. Digestion and metabolism in the ruminant, pp. 448–464. Armidale, Australia: The University of New England Publishing Unit.

Ffoulkes D, Leng RA (1988) Dynamics of protozoa in the rumen of cattle. Br J Nutr 59: 429–436.

Flint HJ, Thomson AM (1990) Deoxyribonuclease activity in rumen bacteria. Lett Appl Microbiol 11:18–21.

Forsberg CW, Lovelock LKA, Krumholz L, Buchanan-Smith JG (1984) Protease activities of rumen protozoa. Appl Environ Microbiol 47:101–110.

Fox DG, Sniffen CJ, O'Conner JD, Russell JB, Van Soest PJ (1992) A net carbohydrate and protein system for evaluating cattle diets. III. Cattle requirements and diet adequacy. J Anim Sci 70:3578–3596.

Fulghum RS, Moore WEC (1963) Isolation, enumeration, and characteristics of proteolytic ruminal bacteria. J Bacteriol 85:808–815.

Gibbons RJ, Doetsch RN (1959) Physiological study of an obligately anaerobic ureolytic bacterium. J Bacteriol 77:417–428.

Gill JW, King KW (1958) Nutritional characteristics of a *Butyrivibrio*. J Bacteriol 75: 666–673.

Hamlin LJ, Hungate RE (1956) Culture and physiology of a starch-digesting bacterium (*Bacteroides amylophilus* n. sp.) from the bovine rumen. J Bacteriol 72:548–554.

Hausinger RP (1986) Purification of a nickel-containing urease from the rumen anaerobe *Selenomonas ruminantium*. J Biol Chem 261:7866–7870.

Hazlewood GP, Edwards R (1981) Proteolytic activities of a rumen bacterium, *Bacteroides ruminicola* R8/4. J Gen Microbiol 125:11–15.

Hazlewood GP, Nugent JHA (1978) Leaf fraction 1 protein as a nitrogen source for the growth of a proteolytic rumen bacterium. J Gen Microbiol 106:369–371.

Hazlewood GP, Orpin CG, Greenwood Y, Black ME (1983) Isolation of proteolytic rumen bacteria by use of selective medium containing leaf fraction 1 protein (ribulosebisphosphate carboxylase). Appl Environ Microbiol 45:1780–1784.

Henderson C, Hobson PN, Summers R (1969) The production of amylase, protease and lipolytic enzymes by two species of anaerobic rumen bacteria. In: Malek I, ed. Continuous Cultivation of Microorganisms, pp. 189–204. New York: Academic Press.

Hendrickx H, Martin J (1963) In vitro study of the nitrogen metabolism in the rumen. C R Rech IRSIA 31:9–66.

Hespell RB, Bryant MP (1979) Efficiency of rumen microbial growth:influence of some theoretical and experimental factors on Y_{ATP}. J Anim Sci 49:1640–1659.

Hiles ID, Gallagher MP, Jamieson DJ, Higgins CF (1987) Molecular characterization of the oligopeptide permease of *Salmonella typhimurium*. J Mol Biol 195:125–142.

Hino T, Russell JB (1985) The effect of reducing equivalent disposal and NADH/NAD on the deamination of amino acids by intact and cell-free extracts of rumen microorganisms. Appl Environ Microbiol 50:1368–1374.

Hino T, Russell JB (1987) Relative contributions of ruminal bacteria and protozoa to the degradation of protein in vitro. J Anim Sci 64:261–270.

Hobson PN, McDougall EI, Summers R (1968) The nitrogen sources of *Bacteroides amylophilus*. J Gen Microbiol 50:i. Abstract.

Houpt TR (1959) Utilization of blood urea in ruminants. Am J Physiol 197:115–120.

Hume ID (1970) Synthesis of microbial protein in the rumen. III. The effect of dietary protein. Aust J Agric Res 21:305–314.

Hungate RE (1960) Microbial ecology of the rumen. Bacteriol Rev 24:353–364.

Hungate RE (1966) The Rumen and Its Microbes. New York: Academic Press.

Hunt WG, Moore RO (1958) The proteolytic system of a gram negative rod isolated from the bovine rumen. Appl Microbiol 6:36–39.

John A, Isaacson HR, Bryant MP (1974) Isolation and characteristics of a ureolytic strain of *Selenomonas ruminantium*. J Dairy Sci 57:1003–1014.

Jones GA, MacLeod RA, Blackwood AC (1964) Ureolytic rumen bacteria. I. Characteristics of the microflora from a urea-fed sheep. Can J Microbiol 10:371–378.

Jones GA, McAllister TA, Muir AD, Cheng K-J (1994) Effects of sainfoin (*Onobrychis viciifolia* Scop.) condensed tannins on growth and proteolysis by four strains of ruminal bacteria. Appl Environ Microbiol 60:1374–1378.

Kennedy PM (1980) The effects of dietary sucrose and the concentration of plasma urea and rumen ammonia on the degradation of urea in the gastrointestinal tract of cattle. Br J Nutr 43:125–140.

Kennedy PM, Milligan LP (1980) The degradation and utilization of endogenous urea in the gastrointestinal tract of ruminants:a review. Can J Anim Sci 60:205–221.

Kennedy PM, Clark RTJ, Milligan LP (1981) Influences of dietary sucrose and urea on transfer of endogenous urea to the rumen of sheep and numbers of epithelial bacteria. Br J Nutr 46:533–541.

Kopecny J, Wallace RJ (1982) Cellular location and some properties of proteolytic enzymes of rumen bacteria. Appl Environ Microbiol 43:1026–1033.

Leedle JAZ, Bryant MP, Hespell RB (1982) Diurnal variations in bacterial numbers and fluid parameters in ruminal contents of animals fed low- or high-forage diets. Appl Environ Microbiol 44:402–412.

Leng RA (1982) Dynamics of protozoa in the rumen of sheep. Br J Nutr 48:399–415.

Leng RA, Nolan JV (1984) Nitrogen metabolism in the rumen. J Dairy Sci 67:1072–1089.

Leng RA, Dellow D, Waghorn G (1986) Dynamics of large ciliate protozoa in the rumen of cattle fed on diets of freshly cut grass. Br J Nutr 56:455–462.

Lesk EM, Blackburn TH (1971) Purification of *Bacteroides amylophilus* protease. J Bacteriol 106:394–402.

Lockwood BC, Coombs GH, Williams AG (1988) Proteinase activity in rumen ciliate protozoa. J Gen Microbiol 134:2605–2614.

Mahadevan S, Sauer FD, Erfle JD (1977) Purification and properties of urease from bovine rumen. Biochem J 163:495–501.

Mahadevan S, Erfle JD, Sauer FD (1980) Degradation of soluble and insoluble proteins by *Bacteroides amylophilus* protease and by rumen microorganisms. J Anim Sci 50: 723–728.

Mangan (1972) Quantitative studies on nitrogen metabolism in the bovine rumen. The rate of proteolysis of casein and ovalbumin and the release and metabolism of free amino acids. Br J Nutr 27:261–293.

Mangan JL, West J (1977) Ruminal digestion of chloroplasts and the protection of protein by glutaraldehyde treatment. J Agric Sci (Camb) 89:3–15.

Mathison GW, Milligan LP (1971) Nitrogen metabolism in sheep. Br J Nutr 25:351–366.

McAllan AB (1982) The fate of nucleic acids in ruminants. Proc Nutr Soc 41:309–317.

McAllan AB, Smith RH (1973a) Degradation of nucleic acids in the rumen. Br J Nutr 29:331–345.

McAllan AB, Smith RH (1973b) Degradation of nucleic acid derivatives by rumen bacteria in vitro. Br J Nutr 29:467–474.

McAllister TA, Phillippe RC, Rode LM, Cheng K-J (1993) Effect of the protein matrix on the digestion of cereal grains by ruminal microorganisms. J Anim Sci 71:205–212.

McKain N, Wallace RJ, Watt ND (1992) Selective isolation of bacteria with dipeptidyl aminopeptidase type I activity from the sheep rumen. FEMS Microbiol Lett 95:169–174.

McNabb WC, Spencer D, Higgins TJ, Barry TN (1994) In-vitro rates of rumen proteolysis of ribulose-1,5-bisphosphate carboxylase (Rubisco) from lucerne leaves, and of ovalbumin, vicilin and sunflower albumin 8 storage proteins. J Sci Food Agric 64:53–61.

Michel V, Fonty G, Millet L, Bonnemoy F, Gouet P (1993) In vitro study of the proteolytic activity of rumen anaerobic fungi. FEMS Microbiol Lett 110:5–10.

Mobley HLT, Hausinger RP (1989) Microbial ureases:significance, regulation and molecular characterization. Microbiol Rev 53:85–108.

Murphy MR, Drone PEJ, Woodford ST (1985) Factors stimulating migration of holotrich protozoa into the rumen. Appl Environ Microbiol 49:1329–1331.

Naga MA, el-Shazly K (1968) The metabolic characterization of the ciliate protozoon *Eudiplodinium medium* from the rumen of buffalo. J Gen Microbiol 53:305–315.

Nagaraja TG, Taylor MB (1987) Susceptibility of ruminal bacteria to antimicrobial feed additives. Appl Environ Microbiol 53:1620–1625.

Newbold CJ, Hillman K (1990) The effect of ciliate protozoa on the turnover of bacterial and fungal protein in the rumen of sheep. Lett Appl Microbiol 11:100–102.

Nisman B (1954) The Stickland reaction. Bacteriol Rev 18:16–42.

Nocek JE, Russell JB (1988) Protein and energy as an integrated system. Relationship of ruminal protein and carbohydrate availability to microbial synthesis and milk production. J Dairy Sci 71:2070–2107.

Nolan JV (1975) Quantitative models of nitrogen metabolism in sheep. In: McDonald IW, Warner ACI, eds. Digestion and Metabolism in the Ruminant, pp. 416–431. Armidale, Australia: University of New England Publishing Unit.

Nolan JV, Leng RA (1972) Dynamic aspects of ammonia and urea metabolism in sheep. Br J Nutr 27:177–194.

Norton BW, MacKintosh JB, Armstrong DG (1982) Urea synthesis and degradation in sheep given pelleted-grass diets containing flaked barley. Br J Nutr 48:249–264.

Nugent JHA, Mangan JL (1981) Characteristics of the rumen proteolysis of fraction I (18S) leaf protein from lucerne (*Medicago sativa* L.). Br J Nutr 46:39–58.

Nugent JHA, Jones WT, Jordan DJ, Mangan JL (1983) Rates of proteolysis in the rumen of the soluble proteins casein, fraction 1 (18S) leaf protein, bovine serum albumin and bovine submaxillary mucoprotein. Br J Nutr 50:357–368.

Onodera R, Koga K (1987) Effect of inhabitation by rumen protozoa on the nutritive value of protein in rumen contents. Agric Biol Chem 51:1417–1424.

Onodera R, Nakagawa Y, Kandatsu M (1977) Ureolytic activity of the washed cell suspension of rumen ciliate protozoa. Agric Biol Chem 41:2177–2182.

Orpin CG, Joblin KN (1988) The rumen anaerobic fungi. In: Hobson PN, ed. The Rumen Microbial Ecosystem, pp. 129–150. London: Elsevier Applied Science.

Paster B, Russell JB, Yang CMJ, Chow JM, Woese CR, Tanner R (1993) Phylogeny of ammonia-producing ruminal bacteria, *Peptostreptococcus anaerobius, Clostridium sticklandii* and *Clostridium aminophilum* sp. nov. Int J Sys Bacteriol 43:107–110.

Patterson JA, Hespell RB (1985) Glutamine synthetase activity in the ruminal bacterium *Succinivibrio dextrinosolvens*. Appl Environ Microbiol 50:1014–1020.

Pittman KA, Bryant MP (1964) Peptides and other nitrogen sources for growth of *Bacteroides ruminicola*. J Bacteriol 88:401–410.

Pittman KA, Lakshmanan S, Bryant MP (1967) Oligopeptide uptake by *Bacteroides ruminicola*. J Bacteriol 93:1499–1508.

Prins RA, van Rheenen DL, van't Klooster AT (1983) Characterization of microbial proteolytic enzymes in the rumen. Ant Van Leeuwenhoek 49:585–595.

Rasmussen MA (1993) Isolation and characterization of *Selenomonas ruminantium* capable of 2-deoxyribose utilization. Appl Environ Microbiol 59:2077–2081.

Richardson LF, Raun AP, Potter EL, Cooley CO, Rathmacher RP (1976) Effect of monensin on rumen fermentation in vitro and in vivo. J Anim Sci 43:657–664.

Robinson PH, McQueen RE (1994) Influence of supplemental protein source and feeding frequency on rumen fermentation and performance. J Dairy Sci 77:1340–1353.

Roche C, Albertyn H, van Gylswyk NO, Kistner A (1973) The growth response of cellulolytic acetate-utilizing and acetate-producing butyrivibrios to volatile fatty acids and other nutrients. J Gen Microbiol 78:253–260.

Rodwell VW (1977) Amino acids and peptides. In: Harper HA, Rodwell VW, Mayes PA, eds. Review of physiological chemistry, 16th Ed., pp. 18–35. Los Altos, Calif: Lange Medical Publications.

Russell JB (1983) Fermentation of peptides by *Bacteroides ruminicola* B_14. Appl Environ Microbiol 45:1566–1574.

Russell JB (1992) Glucose toxicity and the inability of *Bacteroides ruminicola* to regulate glucose transport and utilization. Appl Environ Microbiol 58:2040–2045.

Russell JB (1993a) Effect of amino acids on the heat production and growth efficiency of *Streptococcus bovis*:balance of anabolic and catabolic rates. Appl Environ Microbiol 59:1747–1747.

Russell JB (1993b) The glucose toxicity of *Prevotella ruminicola*: methylglyoxal accumulation and its effect on membrane physiology. Appl Environ Microbiol 59:2844–2850.

Russell JB, Jeraci JL (1984) Effect of carbon monoxide on fermentation of fiber, starch, and amino acids by mixed rumen microorganisms in vitro. Appl Environ Microbiol 48: 211–217.

Russell JB, Robinson PH (1984) Composition and characteristics of strains of *Streptococcus bovis*. J Dairy Sci 67:1525–1531.

Russell JB, Strobel HJ (1990) ATPase-dependent energy spilling by the ruminal bacterium, *Streptococcus bovis*. Arch Microbiol 153:378–383.

Russell JB, Wilson DB (1988) Potential opportunities and problems for genetically altered rumen microorganisms. J Nutr 118:271–279.

Russell JB, Bottje WG, Cotta MA (1981) Degradation of protein by mixed cultures of rumen bacteria: identification of *Streptococcus bovis* as an actively proteolytic rumen bacterium. J Anim Sci 53:242–252.

Russell JB, Sniffen CJ, Van Soest PJ (1983) Effect of carbohydrate limitation on degradation and utilization of casein by mixed rumen bacteria. J Dairy Sci 66:763–775.

Russell JB, Strobel HJ, Chen G (1988) The enrichment and isolation of a ruminal bacterium with a very high specific activity of ammonia production. Appl Environ Microbiol 54: 872–877.

Russell JB, O'Conner JD, Fox DG, Van Soest PJ, Sniffen CJ (1992) A net carbohydrate and protein system for evaluating cattle diets: I. ruminal fermentation. J Anim Sci 70: 3551–3561.

Siddons RC, Paradine J (1981) Effect of diet on protein degrading activity in the sheep. J Sci Food Agric 32:973–981.

Slyter LL, Oltjen RR, Kern DL, Weaver JM (1968) Microbial species including ureolytic bacteria from the rumen of cattle fed purified diets. J Nutr 94:185–192.

Smith RH, McAllan AB (1970) Formation of microbial nucleic acids in the rumen in relation to the digestion of food nitrogen and the fate of dietary nucleic acids. Br J Nutr 24:545–556.

Smith CJ, Hespell RB, Bryant MP (1981) Regulation of urease and ammonia assimilatory enzymes in *Selenomonas ruminantium*. Appl Environ Microbiol 42:89–96.

Sniffen CJ, O'Conner JD, Van Soest PJ, Fox DG, Russell JB (1992) A net carbohydrate and protein system for evaluating cattle diets. II. Carbohydrate and protein availability. J Anim Sci 70:3562–3577.

Spears JW, Hatfield EE (1978) Nickel for ruminants. I. Influence of dietary nickel on ruminal urease activity. J Anim Sci 47:1345–1350.

Spears JW, Smith CJ, Hatfield EE (1977) Rumen bacterial urease requirement for nickel. J Dairy Sci 60:1073–1076.

Stern, MD, Hoover WH, Leonard JB (1977) Ultrastructure of rumen holotrichs by electron microscopy. J Dairy Sci 60:911–918.

Stouthamer AH (1973) A theoretical study on the amount of ATP required for synthesis of microbial cell material. Ant van Leeuwenhoek 39:545–565.

Strydom E, Mackie RI, Woods DR (1986) Detection and characterization of extracellular proteases in *Butyrivibrio fibrisolvens* H17c. Appl Microbiol Biotechnol 24:214–217.

Van Kessel JS, Russell JB (1992) The energetics of arginine and lysine transport by whole cells and membrane vesicles of strain SR, a monensin-sensitive ruminal bacterium. Appl Environ Microbiol 58:969–975.

Van Wyk L, Steyn PL (1975) Ureolytic bacteria in sheep rumen. J Gen Microbiol 91: 225–232.

Van Nevel CJ, Demeyer DI (1977) Effect of monensin on rumen metabolism in vitro. Appl Environ Microbiol 34:251–257.

Varel VA, Bryant MP, Holdeman LV, Moore WEC (1974) Isolation of ureolytic *Peptostreptococcus productus* from feces using defined medium; failure of common urease tests. Appl Microbiol 28:594–599.

Veira DM (1986) The role of ciliate protozoa in nutrition of the ruminant. J Anim Sci 63:1547–1560.

Virtanen AI (1966) Milk production of cows on protein free feed. Science 153:1603–1614.

Waghorn GC, Ulyatt MJ, John A, Fisher MT (1987) The effect of condensed tannins on the site of digestion of amino acids and other nutrients in sheep fed on *Lotus corniculatus* L. Br J Nutr 57:115–126.

Wallace RJ (1983) Hydrolysis of ^{14}C-labelled proteins by rumen micro-organisms and by proteolytic enzymes prepared from rumen bacteria. Br J Nutr 50:345–355.

Wallace RJ (1985) Synergism between different species of proteolytic rumen bacteria. Curr Microbiol 12:59–64.

Wallace RJ (1985) Proteolytic activity of large particulate material in the rumen. Abstracts of XIII Internat Congr Nutr, Brighton, U.K., p. 10.

Wallace RJ (1993) Acetylation of peptides inhibits their degradation by rumen micro-organisms. Br J Nutr 68:365–372.

Wallace RJ, Brammall ML (1985) The role of different species of bacteria in the hydrolysis of protein in the rumen. J Gen Microbiol 131:821–832.

Wallace RJ, Cotta MA (1988) Metabolism of nitrogen-containing compounds. In: Hobson PN, ed. The Rumen Microbial Ecosystem, pp. 217–250. London: Elsevier Applied Science.

Wallace RJ, Joblin KN (1985) Proteolytic activity of a rumen anaerobic fungus. FEMS Microbiol Lett 29:19–25.

Wallace RJ, Kopecny J (1983) Breakdown of diazotized proteins and synthetic substrates by rumen bacterial proteases. Appl Environ Microbiol 45:212–217.

Wallace RJ, McKain N (1989) Analysis of peptide metabolism by ruminal microorganisms. Appl Environ Microbiol 55:2372–2376.

Wallace RJ, McKain N (1990) A comparison of methods for determining the concentration of extracellular peptides in rumen fluid of sheep. J Agric Sci (Camb) 114:101–105.

Wallace RJ, McKain N (1991) A survey of peptidase activity in rumen bacteria. J Gen Microbiol 137:2259–2264.

Wallace RJ, Munro CA (1986) Influence of the rumen anaerobic fungus *Neocallimastix frontalis* on the proteolytic activity of a defined mixture of rumen bacteria growing on solid substrate. Lett Appl Microbiol 3:23–26.

Wallace RJ, McPherson CA (1987) Factors affecting the rate of breakdown of bacteria protein in rumen fluid. Br J Nutr 58:313–323.

Wallace RJ, Cheng K-J, Dinsdale D, Orskov ER (1979) An independent microbial flora of the epithelium and its role in the ecomicrobiology of the rumen. Nature 279:424–426.

Wallace RJ, Broderick GA, Brammall ML (1987a) Protein degradation by ruminal microorganisms from sheep fed dietary supplements of urea, casein, or albumin. Appl Environ Microbiol 53:751–753.

Wallace RJ, Broderick GA, Brammall ML (1987b) Microbial protein and peptide metabolism in rumen fluid from faunated and ciliate-free sheep. Br J Nutr 58:87–93.

Wallace RJ, Newbold CJ, McKain N (1990a) Patterns of peptide metabolism by rumen microorganisms. In: Hoshino S, Onodera R, Minato H, Itabashi H. eds. The Rumen Ecosystem: The Microbial Metabolism and Its Regulation, pp. 43–50. Berlin: Springer-Verlag.

Wallace RJ, McKain N, Newbold CJ (1990b) Metabolism of small peptides in rumen fluid. Accumulation of intermediates during hydrolysis of alanine oligomers, and comparison of peptidolytic activities of bacteria and protozoa. J Sci Food Agric 50:191–199.

Wallace RJ, McKain N, Broderick GA (1993) Breakdown of different peptides by *Prevotella (Bacteroides) ruminicola* and mixed microorganisms from the sheep rumen. Curr Microbiol 26:333–336. Westlake K, Mackie RI (1990) Peptide and amino acid transport in *Streptococcus bovis*. Appl Microbiol Biotechnol 34:97–102.

Whitaker JR (1994) Principles of Enzymology for the Food Sciences. New York: Marcel Dekker.

Williams AG (1986) Rumen holotrich ciliate protozoa. Microbiol Rev 50:25–49.

Williams AG, Coleman GS (1988) The rumen protozoa. In: Hobson PN. ed. The Rumen Microbial Ecosystem. pp. 72–128. London: Elsevier Applied Science.

Williams AG, Coleman GS (1992) The Rumen Protozoa. New York: Springer-Verlag.

Williams AP, Cockburn JE (1991) Effect of slowly and rapidly degraded protein sources on the concentrations of amino acids and peptides in the rumen of steers. J Sci Food Agric 56:303–314.

Wright DE (1967) Metabolism of peptides by rumen microorganisms. Appl Microbiol 15: 547–550.

Wright DE, Hungate RE (1967) Amino acid concentrations in rumen fluid. Appl Microbiol 15:148–151.

Wolin MJ, Manning GB, Nelson WO (1959) Ammonium salts as a sole source of nitrogen for the growth of *Streptococcus bovis*. J Bacteriol 78:147–149.

Wolin MJ (1975) Interactions between the bacterial species in the rumen. In: Warner ACI, McDonald IW, eds. Digestion and Metabolism in the Ruminant, pp. 134–148. Armidale, Australia: University of New England Publishing Unit.

Wozny MA, Bryant MP, Holdeman LV, Moore WEC (1977) Urease assay and urease-producing species of anaerobes in the bovine rumen and human feces. Appl Environ Microbiol 33:1097–1104.

Yang C-MJ, Russell JB (1992) Resistance of proline-containing peptides to ruminal degradation in vitro. Appl Environ Microbiol 58:3954–3958.

Yang CMJ, Russell JB (1993a) The effect of monensin on the specific activity of ammonia production by ruminal bacteria and disappearance of amino nitrogen from the rumen. Appl Environ Microbiol 59:3250–3254.

Yang CMJ, Russell JB (1993b) The effect monensin supplementation on ruminal ammonia accumulation in vivo and the numbers of amino-acid fermenting bacteria. J Anim Sci 71:3470–3476.

12

Biosynthesis of Nitrogen-Containing Compounds

Mark Morrison and Roderick I. Mackie

1. Introduction

Nitrogen is essential for growth in all biological systems, and its assimilation into a variety of life-sustaining compounds has been the topic of study for many microbiologists. This chapter focuses on ammonia assimilation and the biosynthesis of amino acids, polyamines, pyrimidines, and purines. Wherever possible, emphasis will be directed toward findings obtained from ruminal and colonic bacteria, although the knowledge base developed for these bacteria is relatively superficial. To overcome these limitations, some discussion pertaining to gram-negative enteric bacteria (*Escherichia coli*, *Salmonella typhimurium* [official designation, *Salmonella enterica*, serovar *typhimurium*], and *Klebsiella* spp.) and gram-positive bacteria (*Bacillus subtilis* and *Clostridium* spp.) has been included for the sake of clarity and reference. Readers interested in detailed information concerning the topics covered in this chapter, as well as the biosynthesis of nitrogen-containing vitamins and coenzymes, should refer to the volumes edited by Neidhardt et al. (1996) and Sonenshein et al. (1993), as well as the recent review of nitrogen control in bacteria by Merrick and Edwards (1995). The goals of this chapter are to provide a cohesive overview that complements the well-chronicled field of knowledge developed from these intensively studied species, and to highlight opportunities where further studies of ruminal and colonic bacteria may expand our understanding of these processes.

2. Ammonium Transport

It is logical to perceive that under conditions of high substrate availability, passive diffusion of ammonia across the cell membrane(s) will occur and satisfy metabolic requirements. However, since the pK_a for the dissociation of the ammonium ion is 9.25 at 24°C, the majority of the substrate is present in the ionized form at physiological pH, and nitrogen-limited cells will probably require some form of ammonium translocation mechanism. Indeed, ammonium transport systems are ubiquitous amongst bacteria isolated from a variety of habitats. In general, the affinity of the ammonium carriers is high with K_m values between 5 and 50 μM (Kleiner 1985). Most of the ammonium transport systems have been studied using [^{14}C]methylammonium as a substrate. With the exception of some cyanobacteria, the ammonium and methylammonium gradients are similar in magnitude (when both are measured), and intracellular concentrations range from 50-fold to 200-fold higher than the external concentration. Russell and Strobel (1987) reported that concentration gradients of ammonium across the cell membranes of mixed ruminal bacteria ranged from 1.8-fold to 15-fold, far less than bacteria from other habitats. Similar studies with colonic bacteria do not appear to have been conducted.

Ammonium transport systems in *E. coli* were recently described by Barnes and Jayakumar (1993). It appears that this bacterium possesses a K^+/NH_4^+ antiporter and as such, a case can be made for a facilitated diffusion mechanism with the membrane potential ($\Delta\Psi$) as the driving force behind transport (Kleiner 1985). The first indication that ammonium transport system(s) in enteric bacteria might be nitrogen-regulated was derived from a mutational analysis of a *K. pneumoniae* strain which was unable to transport methylammonium (Kleiner 1982). This mutant was also found to lack glutamine synthetase (GS) and nitrogenase activities, as well as being incapable of growth with histidine as a nitrogen source. A more detailed study carried out with *E. coli* strains containing mutations in different *ntr* genes showed that both the RpoN (NtrA) and NR_I (NtrC) proteins were required to activate expression of ammonium transporter(s), while the NR_{II} (NtrB) protein played a role in their repression (Servin-Gonzalez and Bastarrachea 1984, Castroph and Kleiner 1984, Jayakumar et al. 1987). Glutamine and glutamine analogs also appear to allosterically regulate ammonium transport (Jayakumar et al. 1986). The *ntr*-regulated *amtA* gene of *E. coli*, which is thought to encode a 27-kDa peripheral membrane protein, has been cloned, sequenced, and mapped on the *E. coli* chromosome. However, at the time of writing, the role of the AmtA protein in ammonium transport still remains uncertain.

Under conditions of K^+ limitation, some bacteria use ammonium as a counter ion (Buurman et al. 1989) and in *E. coli*, ammonium uptake under these conditions is mediated by the high-affinity, ATP-dependent K^+ transporter (Kdp; Buurman et al. 1989, Barnes and Jayakumar 1993). The rumen is a moderately halophilic environment, and many bacteria maintain intracellular K^+ concentrations in excess of 200 mM (e.g., Russell 1994), much higher than the external environment. Kinetic models of rumen microbial growth indicate that bacteria can direct as much as 40% of their ATP toward satisfying maintenance functions (Russell et al. 1992), but whether ammonium transport can serve as alternative counter ion in ruminal bacteria is uncertain. Irrespective of whether ammonium transport is related to nitrogen or K^+ limitation, influx is counteracted when the bacterial cytoplasm is more alkaline than the external environment. Kleiner (1985) estimated that up to 6 mol of ammonium may be transported in *E. coli* before 1 mol is trapped in glutamate by the concerted action of glutamine synthetase and glutamate synthase enzymes. Ammonia efflux results in the establishment of futile cycles, ATP consumption is elevated, and bacterial growth efficiency is adversely affected (see Russell and Cook 1995). Considering that Smith and Hespell (1980) could explain less than 50% of the decrease in the growth efficiency of *Selenomonas ruminantium* under nitrogen-limited conditions by ATP-expenditure associated with glutamine synthetase activity, it is tempting to speculate that futile cycle(s) similar to those described by Russell and Cook (1995) might be responsible for the remainder of the change.

Evidence is also available that ruminal pH can have an impact on the ammonia requirements necessary to maximize microbial protein synthesis. Requirements when animals are fed forage-based rations (which favor ruminal pH values close to neutral) often have resulted in values much lower than those determined from similar experiments where concentrate-based diets are fed, and which favor pH values of 6.0 and below (Orskov et al. 1979, Boniface et al. 1988). Thus, while ruminal ammonium concentrations are often considered to be nonlimiting for growth, their impact upon growth efficiency, and the role of ruminal pH, seem worthy of further investigation. Little is known about the mechanism(s) and possible regulation of ammonium (or potassium) transport in ruminal bacteria; or the extent to which futile cycles involving ammonia negatively impact growth efficiency in ruminal microorganisms.

3. Enzymes of Ammonia Assimilation

In the enteric bacteria, all cellular nitrogen required for the synthesis of macromolecules can be derived from the amido group of glutamine, the amino group of glutamate, or directly from incorporation of ammonia. Glutamate serves as an amino group donor, whereas glutamine is an essential precursory molecule and/or amino group donor for the synthesis of purines, pyrimidines, amino sugars, histidine, tryptophan, asparagine, NAD, and *p*-aminobenzoate. Of the 11- to 12-g atoms of nitrogen incorporated into a kilogram (dry weight basis) of *E. coli* cells, as much as 90% is utilized for the biosynthesis of glutamate and its products, with the remainder consumed during the biosynthesis of glutamine and related nitrogen-containing compounds (Reitzer 1996). The route of ammonia assimilation depends on its extracellular concentration: at concentrations in excess of 1 mM, ammonia is assimilated directly into glutamate, glutamine, and asparagine, while at concentrations less than 0.1 mM, ammonia is assimilated solely into glutamine (Reitzer and Magasanik 1987). Although there are exceptions to this generality, the observations outlined above probably are applicable to many of the prokaryotes which prefer to utilize ammonia as a nitrogen source. As such, emphasis in this section will be directed toward those enzymes responsible for the assimilation of ammonia into glutamate, glutamine, and asparagine. Glutamate can also be formed directly as a product of the catabolism of some amino acids and other nitrogen sources or, through the transamination of 2-ketoglutarate. Knowledge of these mechanisms in enteric bacteria was reviewed by Reitzer (1996) and will not be reiterated here, since they do not contribute to assimilation of ammonia.

3.1. Glutamate Dehydrogenase

Biosynthetic glutamate dehydrogenase (GDH) catalyzes the reductive amination of 2-ketoglutarate with ammonia (Fig. 12.1). In general, GDHs that are involved in the biosynthesis of glutamate specifically use reducing equivalents donated by NADPH, although there are notable exceptions (e.g., GDH enzymes of *Porphyromonas gingivalis*, *Clostridium symbiosum*, and *Clostridium difficile)*. The biosynthetic GDHs are characterized by their 3 × 2 hexameric configuration of a subunit protein of 48 to 50-kDa molecular weight (Smith et al. 1975). Kinetic studies have shown the NADPH-dependent enzymes possess relatively high affinities for substrates and low affinities for products in the amination reaction, confirming their presumed biosynthetic roles. The K_ms for ammonia have been reported to be of the order of 1 mM (Coulton and Kapoor 1973, Sakamoto et al. 1975), although Lin and Reeves (1991a) reported a K_m for ammonia of 36 mM for *E. coli* $D_5H_3G_7$.

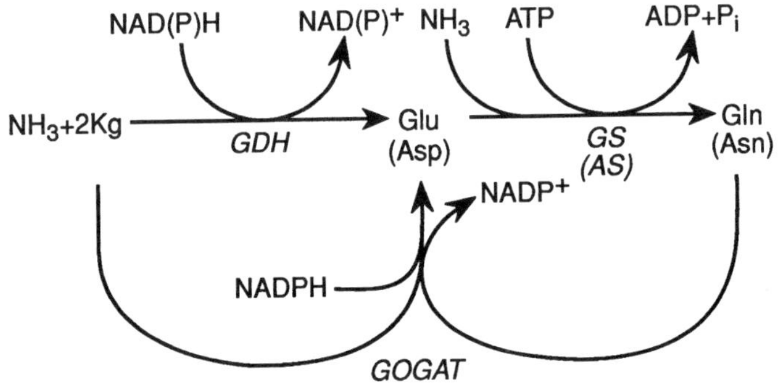

Figure 12.1. Main pathways of ammonia assimilation in bacteria. Enzymes: *GDH* (glutamate dehydrogenase), *GS* (glutamine synthase), *AS* (asparagine synthetase), and *GOGAT* (glutamate synthase). Substrates: 2Kg (2-Ketoglutarate), Glu (Glutamate), Asp (Aspartate), Gln (Glutamine), and Asn (Asparagine).

High levels of both NADH- and NADPH-linked GDH activity have been observed in ruminal contents (Chalupa et al. 1970) and in continuous culture of mixed ruminal bacteria (Erfle et al. 1977). In studies with pure cultures of ruminal bacteria, NAD(H)-dependent GDH activity was present in *Ruminococcus albus* and *Megasphaera elsdenii*, whereas most other ruminal bacteria possess NADPH-dependent activity (Joyner and Baldwin 1966). *Selenomonas ruminantium* was shown to possess both types of activity in a relatively constant activity ratio (0.25 to 0.35) under both glucose- and ammonia-limited growth. Apparent K_m values for ammonia and 2-ketoglutarate were estimated as 6.7 and 23 mM, respectively (Smith et al. 1980). Glutamate dehydrogenases with subunit molecular weights of approximately 48.5 kDa have been purified from *Bacteroides thetaiotaomicron* (Glass and Hylemon 1980) and *B. fragilis* (Yamamoto et al. 1987a), and the *gdhA* genes of *B. thetaiotaomicron* and *Prevotella ruminicola* strain B_14 have recently been cloned by *E. coli* mutant complementation (Baggio and Morrison 1996, Wen and Morrison 1996). Where determined, the substrate affinity values were similar to those calculated for *S. ruminantium.* A distinctive feature of all these GDHs relative to other eubacteria is their ability to catalyze reactions with either NADP(H) or NAD(H) as cofactors. It is also of interest that the GDH-specific activities are also influenced by monovalent salt concentration, first observed in ruminal bacteria by Smith et al. (1980). Using GDH enzyme purified from *B. thetaiotaomicron*, Glass and Hylemon (1980) also found that NADP-dependent activity increased by 73% and NAD-dependent activity decreased by 50% in the presence of 0.1 M NaCl. In *P. ruminicola* cell extracts the effects upon GDH activity are similar, with 0.2 M KCl increasing NADPH-dependent activity two-fold while eliminating NADH-dependent activity. The NADP(H)-dependent GDH of the gram-positive ruminal bacterium *R. flavefaciens* FD-1 has also been purified and characterized (Duncan et al. 1992), and, like the enzymes from the gram-negative bacteria, 0.5 M KCl was required for optimal enzyme activity. However, relative to the gram-negative ruminal and colonic bacteria described above, the K_m for ammonia is relatively high (19 mM).

What makes further investigation of GDH enzymes in ruminal and colonic bacteria potentially interesting to the rest of the microbiology community? First, in human colonic *Bacteroides* and *P. ruminicola*, growth under nitrogen-limited conditions results in as much as a 10-fold increase in GDH specific activity (Yamamoto et al. 1984, Baggio and Morrison 1996, Wen and Morrison 1996). It has also been proposed that not only do specific proteases play a role in pathogenicity (Macfarlane et al. 1992) but that their expression and secretion is nitrogen-regulated (Gibson and MacFarlane 1988). Considering that the modulation of GDH-specific activity in response to ammonia concentration in some ruminal and colonic bacteria is contrary to those observations made in enteric bacteria (see Reitzer 1996; also Sect. 4.3) the molecular mechanisms underpinning this type of regulation seem worthy of further investigation. Second, GDHs are thought to be a good model for a molecular-based study of evolutionary relationships, because of their ubiquitous nature, plus the fact that their mutation rate is slow (1.8 amino acid substitutions 100 residues^{-1} 10^8 years^{-1}; Wilson 1988). Previous multiple sequence alignments of hexameric *gdh* sequences has led to the hypothesis that two closely related, but already different, *gdh* genes were present in the last common ancestor of eubacteria, archaebacteria, and eukaryotes (Benanchenhou-Lahfa et al. 1993), and that each of these enzyme families is distinguishable by the presence of highly conserved amino acid motifs. However, GDH sequences from the Bacteriodaceae were not available for inclusion in these studies, and while the dual cofactor specificity has been observed for other eubacterial dehydrogenases (see Glass and Hylemon 1980), it is still relatively unusual. As part of the *Cytophaga-Flavobacter-Bacteroides* (CFB) group, the colonic *Bacteroides* spp. probably also represent an ancient branch in the eubacterial line of descent (Weisburg et al. 1987), and the gene encoding GS enzyme activity from *B. fragilis* established a new precedent in relation to this enzyme's molecular structure. For these reasons, the presumptive *B. thetaiotaomicron* GdhA sequence, along with the sequence encoding an NAD(H)-dependent GDH isolated from *Porphyromonas (Bacteroides) gingivalis* (McBride 1988) and other hexameric GDH sequences in the database, were examined using the FastA and ''Growtrees'' programs available through the Wisconsin Genetics Program (version 8.0). The predicted GdhA sequence from *B. thetaiotaomicron* possesses motifs typical of the family I hexameric GDH proteins, although the predicted ORF possesses a novel amino acid substitution (P to A at position 155) within one of these motifs. The *B. thetaiotaomicron* GdhA sequence and *P. gingivalis* were also found to be divergent from all other family I sequences, with the exception of the *Clostridium symbiosum* Gdh sequence (Fig. 12.2). Moreover, the analysis indicated that these sequences were more closely related to, but preceded all of the family II type sequences. Considering that *B. thetaiotaomicron* appears to possess two distinct genes encoding GDH activity (Chen 1995), further phylogenetic analysis of the glutamate dehydrogenases from the Bacteriodaceae may provide useful information in regards to the biodiversity and evolution of these enzymes.

3.2. Glutamate Synthase

The studies by Meers et al. (1970) and Tempest et al. (1970) were the first to provide evidence that an alternative pathway of glutamate biosynthesis existed in prokaryotes. When *K. aerogenes* was cultured under nitrogen-limited conditions, GDH activity was repressed, but the intracellular pool of glutamate was largely unaffected. The enzyme responsible for glutamate biosynthesis was identified as glutamate synthase (GOGAT), which catalyzes the transfer of the amide group of glutamine, generated via glutamine synthetase (GS), to 2-ketoglutarate and

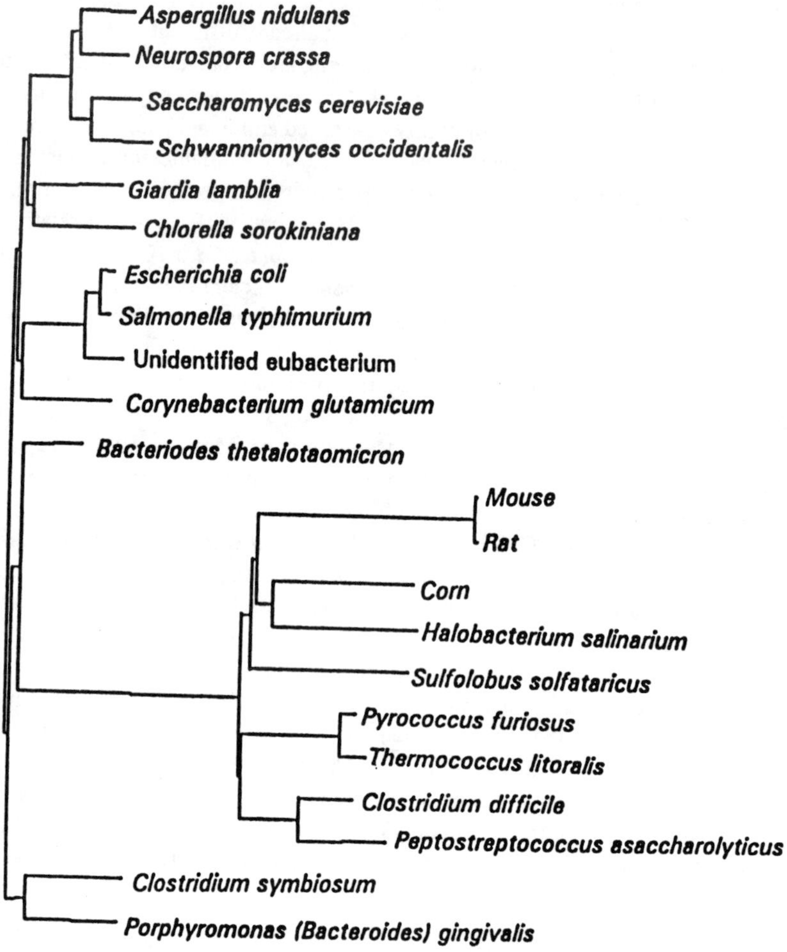

Figure 12.2. Comparison of the predicted amino acid sequences from the *gdhA* gene of *B. thetaiotaomicron* and the *gdh* sequence of *Porphyromonas gingivalis*, by multiple alignment with 22 other glutamate dehydrogenase sequences, using the neighbor joining method of the Growtrees program (Wisconsin Genetics Package, version 8.0). From Baggio and Morrison (in press).

yielding 2 molecules of glutamate (Fig. 12.1). The discovery also explained how *B. subtilis*, which lacks assimilatory GDH, could grow and hence meet its needs for glutamate in ammonia-containing medium (Fisher and Sonenshein 1991).

The GOGAT enzyme has since been identified in many bacteria, and the enzyme has been purified from *E. coli* and *K. aerogenes* cell extracts. The active enzyme of enteric (and probably other) bacteria is comprised of equimolar amounts of a small (~53 kDa) and a large (135 to 175 kDa) subunit (Miller and Stadtman 1972, Trotta et al. 1974). The *E. coli* enzyme, at least, may form aggregates of 4 dimers. The structural genes for the two subunits of *E. coli* GOGAT have been mapped to 69 minutes on the *E. coli* chromosome by studying mutations resulting in loss of GOGAT activity (see Reitzer and Magasanik 1987, Reitzer 1996). The genes *glt*B (large subunit) and *gltD* (small subunit) are linked and form an operon, *gltBD* (Garciarrubio et al. 1983, Lozoya et al. 1980, Madonna et al. 1985). Antibodies raised against the *E. coli* enzyme subunits are cross reactive with two similarly sized polypeptides from extracts of *S. typhimurium* (Madonna et al. 1985). A third gene, *gltF*, has also been identified as part of the *glt* operon in *K. pneumoniae* (Kuczins et al. 1991) as well as *E. coli* (Castano et al. 1992). The GltF protein is thought to be a regulatory protein affecting glutamate biosynthesis (see Reitzer 1996; Sect. 4.4).

Several studies have addressed the involvement of the subunits in the GOGAT reaction. The large subunit contains flavin, iron and sulfide, and can also catalyze glutaminase activity resulting in the deamidation of glutamine (Mantsala and Zalkin 1976b), although such activity might be an artifact of damage to the subunit (see Reitzer 1996). The small subunit is known to catalyze an ammonia-dependent (rather than glutamine-dependent) amination of 2-ketoglutarate when ammonia levels are sufficiently high (Mantsala and Zalkin 1976c), although mutational loss of the GltB protein is sufficient to result in a GOGAT-deficient phenotype. Moreover, a *gltB gdh* mutant strain of *S. typhimurium* is also a glutamate auxotroph (Madonna et al. 1985), which supports the contention that the small subunit of GOGAT is not capable of catalyzing GDH-like activity in vivo (Reitzer 1996).

Since *B. subtilis* has no assimilatory GDH activity, it relies on GS to incorporate ammonia into glutamine, and GOGAT to provide glutamate (Fisher and Sonenshein 1991). The *B. subtilis* GOGAT is also composed of equimolar amounts of a large and small protein subunit, which are encoded by the genes *gltA* and *gltB*, respectively (Deshpande and Kane 1980). These two structural genes appear to be closely linked on the *B. subtilis* chromosome but are not transcribed contiguously (Bohannon et al. 1985, Merrick and Edwards 1995). A gene upstream of *gltA*, which results in a *glt*$^-$ phenotype (lacking GOGAT activity, or glutamate auxotrophy) when insertionally inactivated, was identified by Tn*917* insertion mutagenesis (Bohannon et al. 1985). The protein encoded by this gene, GltC, has subsequently being shown to be a DNA-binding protein involved with transcriptional regulation of *gltA* (see Sect. 4.4).

The physiological requirement(s) and characteristics of GOGAT activity have been more difficult to demonstrate in ruminal and colonic bacteria. No GOGAT activity was measurable in either *S. bovis* (Griffith and Carlsson 1974) or *P. ruminicola* strain $B_1$4 (Wen and Morrison, unpublished observations), and its detection in *Ruminobacter (Bacteroides) amylophilus* has been variable (Jenkinson et al. 1979, Hespell 1984). Yamamoto et al. (1984) observed only low levels of NADP(H)-dependent GOGAT activity in *B. fragilis* cell-free extracts, and in studies with *S. ruminantium*, neither NADPH or NADH were suitable electron donors (Smith et al. 1980). Enzyme activity was measurable only when dithionite-reduced methyl viologen was used as the electron donor, and similar findings have been reported with the methane-producing archaebacteria *Methanobacterium thermoautotrophicum* and *Methanosarcina barkeri* (Kenealy et al. 1982). Thus, some unidentified low-potential electron carrier is required and may help to explain problems involved in measuring GOGAT activity in obligate anaerobes. The fact that ferrodoxin-dependent GOGAT enzymes exist in algae (Lee and Miflin 1975) and plants (Miflin and Lee 1975), suggests that any future studies of the GOGAT enzymes of ruminal and colonic bacteria should evaluate this electron donor as the preferred cofactor.

3.3. Glutamine Synthetase

The glutamine synthetase (GS) enzyme of enteric bacteria exists as a 600-kDa dodecamer of identical subunits (Ginsburg and Stadtman 1973). The ability of GS to couple ATP hydrolysis to the amidation of glutamate shifts the equilibrium far in favor of ammonia incorporation, rather than release. Since the affinity of GS for ammonia is typically much higher than that of GDH, Umbarger (1969) suggested that GS may act as an ammonia scavenger under N-limited conditions, a role confirmed by intensive investigation of the enteric bacteria (see Reitzer 1996). In enteric bacteria only one GS enzyme (encoded by the *glnA* gene) is present, and its inactivation results in an absolute requirement for glutamine.

The enzyme present in the enteric bacteria is the most common form of GS found in eubacteria (GS-I), although the GS-I enzymes can be further subdivided according to whether and how the mature enzyme is modified in response to nitrogen availability. While the enteric enzymes are all subject to inactivation by adenylylation, similar enzymes in *Bacillus* and *Clostridium* spp. are not (see Sect. 4.1). The enzymes also differ significantly from their counterpart enzymes in the enteric bacteria in terms of amino acid composition and posttranslational control mechanisms, but share some similarities in relation to the effects of divalent cations upon specific activity (see Schreier 1993). Until several years ago, the DNA sequence data of all known GS structural genes could be divided into two classes: the GS-I-type enzymes described above; and the octameric, 288-kDa GS-II enzymes present in *Rhizobium*, *Bradyrhizobium*, *Agrobacterium*, *Streptomyces*, and *Frankia* spp. (see Merrick and Edwards 1995). The cloning and isolation of the GS structural gene (*glnA*) of *B. fragilis* established a new precedent in the current knowledge base. Nucleotide sequence analysis by Hill et al. (1989) of *B. fragilis* DNA which complement an *E. coli glnA* deletion mutant identified an open reading frame of 2187 base pairs which would encode a protein of 75 kDa, confirming previous observations by Southern et al. (1986), that the GS subunit in *B. fragilis* was nearly twice as large as the known GS-I and GS-II enzymes and would exist as a hexamer. Further DNA sequence alignments of the *B. fragilis glnA* gene with other *gln* structural genes illustrated that the *B. fragilis* sequence contains only four of the five highly conserved motifs present in all other known GS sequences, and thought to be involved with active site formation (Almassy et al. 1986, Janson et al. 1986). The spatial distribution of these highly conserved motifs relative to each other was similar in all the *gln* sequences analyzed, including *B. fragilis* (Hill et al. 1989). Thus, the *B. fragilis* GS appears to be comprised of a central, catalytic domain similar to all other GS enzymes, but flanked by an amino terminal region of 149 amino acids and a carboxy-terminal region of 391 amino acids. The *glnA* gene from the anaerobe *Butyrivibrio fibrisolvens* possesses strong DNA sequence homology with the *B. fragilis glnA* gene (Woods and Santangelo 1993), as does a DNA fragment isolated from a library of *Ruminococcus flavefaciens* FD-1 DNA, by low stringency colony blot hybridizations (Mackie, unpublished data). The GS-III type enzymes have also now been identified in several cyanobacteria (Merrick and Edwards 1995).

Measurement of GS enzyme activity can be accomplished by either measuring glutamine biosynthesis or alternatively, using ''forward'' and ''transferase'' assays with hydroxylamine hydrochloride as a substrate (see Bender et al. 1977 as an example). Studies with mixed ruminal bacteria in continuous culture showed that GS activity increased approximately 10-fold in response to ammonia-limitation (Erfle et al. 1977), and similar findings have been reported for several ruminal bacteria grown in pure cultures. In *S. ruminantium*, GS activity could measured using the forward assay described by Bender et al. (1997), but not by the γ-glutamyltransferase (GGT) procedure (Smith et al. 1980). In contrast, *R. amylophilus* (Jenkinson et al. 1979), *Succinivibrio dextrinosolvens* (Patterson and Hespell 1985), and *R. flavefaciens* strain FD-1 (Duncan 1992) possess GGT activity, but only the *R. amylophilus* and *S. dextrinosolvens* enzymes appear to be responsive to inhibitors in the GGT assay (indicative of the adenylylation-denadenylylation regulatory mechanism characteristic of the enteric bacteria; see Bender et al. 1977). The biosynthetic assay appears to be the only means of measuring GS activity in *B. fragilis* cell extracts (Yamamoto et al. 1984, Southern et al. 1986), but heterologous expression of the *B. fragilis glnA* in *E. coli* can be measured using the GGT assay. The GGT activity of the heterologous protein was lost if extracts from *B. fragilis* cells were added to the assay buffer, suggesting the presence of an inhibitory compound(s) that may be involved with regulating enzyme activity in vivo. Yamamoto et al. (1984) used the GS inhibitor methionine sulfoxamine to effectively eliminate measurable GS enzyme activity in *B. fragilis* but, despite enzyme inhibition, saw no effect upon growth rate or yield of ammonia-limited cultures. While the authors concluded that the GS enzyme was not essential for ammonia assimilation during nitrogen-limited growth, there was no explanation as to why growth of the bacterium was not glutamine-dependent.

Probably, the salient interests in the future studies of GS enzymes from ruminal and colonic bacteria will be the extent to which GS-III-type enzymes are distributed throughout members of these microbial habitats, enzyme structure-function relationships, and mechanism(s) of enzyme regulation. From a more practical standpoint, the relevance and dependence upon GS enzyme activity as a means of ammonia assimilation in the ruminal environment is still unclear. Concentrations of ammonia in ruminal fluid, combined with the ammonia saturation constants calculated for predominant ruminal bacteria (~50 μM; Schaefer et al. 1980), would suggest that GS activity is not required. Several reasons have been put forward to explain why GS activity could be a quantitatively important route of ammonia assimilation in the rumen (e.g., Wallace and Cotta, 1988), but equally important is the potential role of the enzyme in meeting the bacterium's requirement for glutamine, and its subsequent use as a high-energy nitrogen donor. Such a role for GS could be especially relevant in those ruminal bacteria with an obligate growth requirement for ammonia (such as the cellulolytic ruminal bacteria).

3.4. Asparagine Synthetase

The genetics, biochemistry, and regulation of asparagine synthetases (AS) in enteric bacteria were reviewed recently by Merrick (1988) and Reitzer (1996). The enteric bacteria possess two ASs: one that is glutamine-dependent (encoded by *asnB*) and expressed primarily during nitrogen-limited growth; and a second, ammonia-dependent enzyme (encoded by *asnA*). The AsnA protein appears to be restricted to prokaryotes (see Reitzer and Magasanik 1982, Reitzer 1983) and the substrate specificities of the respective enzymes are sufficient to explain the phenotypes of the strains that lack either AS: *asnA* mutants can still utilize glutamine as a nitrogen donor and so have no discernable phenotype; but *asnB* mutants are incapable of growth in nitrogen-limited media, presumably because of a limited supply of ammonia. In *E. coli,* expression of *asnA* appears to be mediated via the *asnC* gene product, which is located adjacent to, but transcribed divergently, from *asnA*. There is additional evidence which relates the *ntr* system of regulation (see Sect. 4.1) with the control of synthesis of ASs, but the mechanism(s) underpinning this regulation are not understood (see Reitzer 1996).

Even less is known about ASs in gram-positive bacteria (Schrier 1993), although ammonia-dependent enzymes have been purified from *Streptococcus bovis* (Burchall et al. 1964) and *Lactobacillus arabinosus* (Ravel 1970). Enzyme activity was inhibited by L-asparagine, but through different mechanisms: in *L. arabinosus*, formation of the enzyme appears to be terminated; while in *S. bovis*, enzyme formation is not affected. A mutant *S. bovis* strain defective in AS activity was incapable of growth in complete amino acid medium minus L-asparagine. Although both AS and GS activity have been identified in different strains of *S. bovis*, the latter enzyme has not always been demonstrated (Hespell 1984, Wallace and Cotta 1988), which raises the possibility that some strains depend on AS activity as a primary route of ammonia assimilation. The K_m (ammonia) for the *S. bovis* enzyme was calculated to be 4 mM (Burchall et al. 1964), consistent with the enzyme possessing a role in ammonia assimilation.

4. Control of Assimilatory Pathways

4.1. Glutamine Synthetase and the Ntr *System*

The pathway catalyzed by GS can be considered to be a major branch point in bacterial metabolism. The control of GS catalytic activity and the transcriptional control of *glnA* by nitrogen availability have been intensively investigated in the enteric bacteria, and are the foundation of much of our current understanding of the control of bacterial nitrogen metabolism. Moreover, the studies of the enteric *ntr* system were instrumental in the development of our understanding of two-component regulatory systems; a theme that has fundamental relevance in both prokaryotic and eukaryotic cells. The following model of regulation of GS enzyme activity was first described by Ginsburg and Stadtman (1973) and updated most recently by Reitzer (1996). Understanding of the broader nitrogen regulation (Ntr) model has been periodically updated (e.g., Magasanik 1982, Kustu et al. 1986, Reitzer and Magasanik 1987). However, as noted in the Introduction, it is not our objective to duplicate the most detailed descriptions and recent reviews of the Ntr system (see Merrick and Edwards 1995, Reitzer 1996), but to outline the paradigm to which all other prokaryotes are compared.

The GS enzyme of enteric bacteria is subject to feedback inhibition and is most sensitive to such regulation during growth in nitrogen-rich medium. Inhibitory compounds can be subdivided into two major classes. The first class of inhibitors are alanine, glycine, and serine, all of which obtain their nitrogen from glutamate. The second class of inhibitors include those products that depend absolutely on the amide group of glutamine for their biosynthesis, and include histidine, tryptophan, CTP, AMP, carbamoyl-phosphate, and glucosamine-6-phosphate. The mechanism of feedback inhibition for each class of inhibitors is also thought to be different (see Reitzer 1996).

The degree of feedback inhibition appears to be reduced when enteric bacteria are grown in nitrogen-limited media, and the difference is thought to be a reflection of the adenylylation state of the GS enzyme, which is altered in response to the nitrogen status of the cell. The cyclic adenylylation of GS is illustrated in Figure 12.3. In simplistic terms, when the intracellular ratio of glutamine to 2-ketoglutarate is high, the bacterium senses nitrogen sufficiency, and a cascade of events leads to the adenylylation (and inactivation) of GS. Once the ratio of glutamine to 2-ketoglutarate favors the latter, the bacterium senses nitrogen deficiency and the enzymes utilized previously in the adenylylation cascade are reversed (although the stoichiometry of the deadenylylation reaction is not the same as that for adenylylation; see Reitzer 1996).

The (de)adenylylation of GS is mediated by an adenylyltransferase enzyme (encoded by the *glnE* gene), and the direction of the transferase activity is influenced by the P_{II} protein (encoded by the *glnB* gene); which itself can be modified

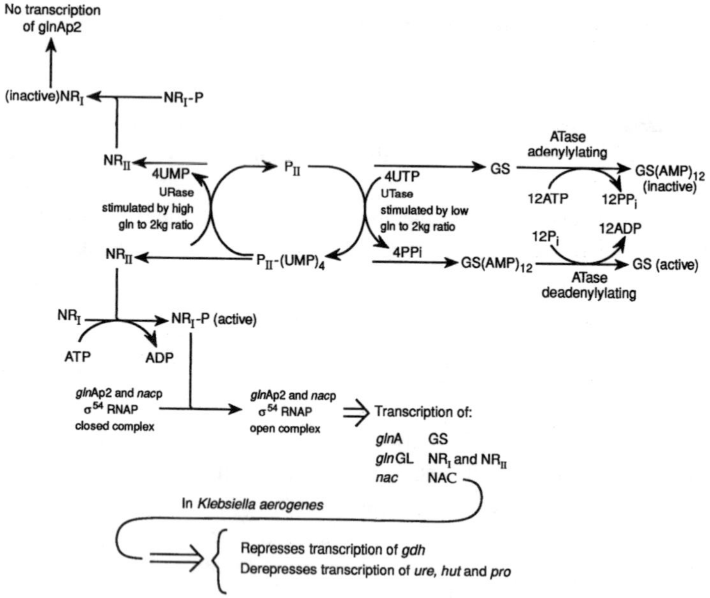

Figure 12.3. Three levels of nitrogen regulation: (1) adenylylation and deadenylylation control of GS; (2) regulation of transcription of *glnALG* from σ^{54}-dependent promoter; and (3) regulation of transcription of NAC and σ^{70}-dependent operons which allow use of less preferred nitrogen sources. See text for abbreviations.

by the bifunctional uridylyltransferase (UTase)/uridylyl removing (UR) enzyme (encoded by the *glnD* gene). Thus, the UTase/UR enzyme serves as the actual sensor of nitrogen availability and mediates its effects first via the P_{II} protein: when P_{II} is uridylylated, its interaction with the adenylyltransferase enzyme favors GS adenylylation (nitrogen excess and GS inhibition); when the uridylyl groups are removed from P_{II}, its interaction with the adenyltransferase enzyme favors GS deadenylylation (nitrogen limitation and GS activation). It should be noted that partial adenylylation of GS does not result in a loss of catalytic activity by the entire enzyme, although the sensitivity to feedback inhibition is altered (see Reitzer 1996). The differences in feedback inhibition of GS relative to its adenylylation state is also a reflection of its altered roles in nitrogen-deficient and nitrogen-sufficient cultures. In the former instance the enzyme's role is primarily for ammonia assimilation, while in the latter, the primary role for GS activity is to provide glutamine for biosynthetic reactions, not to serve as the primary route of ammonia assimilation.

The P_{II} protein also is the transducer of nitrogen status, sensed by the UR/UTase protein, into transcriptional regulation of the *glnL* and *glnG* genes (sometimes referred to as *ntrB* and *ntrC*, respectively) which are a part of the *glnALG* (or *glnAntrBC*) operon (see Fig. 12.3). Two promoters precede the *glnA* structural gene. The first, *glnAp*$_1$, is located 190 basepairs upstream of the *glnA* translation start site and requires RNA polymerase (RNAP) complexed with the major σ subunit of the cell, σ^{70}. Transcription is initiated from this promoter during carbon-limited growth, requires cAMP complexed with cAMP receptor protein, and is terminated completely at the end of *glnA*. A basal level of *glnL* and *glnG* mRNAs is maintained under these conditions by transcription initiated at a relatively weak promoter located within the *glnA-glnL* intercistronic region *(glnLp*; see Reitzer 1996). The second promoter element preceding *glnA, glnAp*$_2$, is located only 70 bp upstream of the translation start site, but only recognizes RNAP complexed with a minor σ subunit, σ^{54} (encoded by the *rpoN* [also *ntrA*] gene; detailed reviews of this promoter-RNAP-σ^{54} interaction have recently been published: Kustu et al. 1991, Weiss et al. 1992). As depicted in Figure 12.3, during periods of nitrogen-excess P_{II} is not uridylylated, and its interaction with the protein kinase NR_{II} (NtrB) prevents the transfer of a phosphate group from NR_{II} to NR_I (NtrC), and instead stimulates the desphosphorylation of NR_I. Dephosphorylated NR_I can still bind to DNA and interferes with transcription from both *glnAp and glnLp,* which ensures that transcription of all three genes is maintained at a low, basal level. During periods of nitrogen-limitation, the P_{II} is uridylylated (by the GlnD protein) and P_{II}-UMP is no longer capable of interaction with NR_{II}. In this case NR_{II} phosphorylates itself (see Stock et al. 1989, Merrick and Edwards 1995, for a description of the possible phosphate donors) and in turn phosphorylates NR_I. The phosphorylated form of NR_I acts as a positive control element (enhancer) for the initiation of transcription from *glnAp*$_2$; and results in elevated level of *glnALG* transcription, as well as transcription at other σ^{54}-dependent promoters. These include genes encoding for glutamine, arginine, histidine, and ammonium transport; nitrite and nitrate assimilation; nitrogen fixation genes; and the nitrogen assimilation control (Nac) protein.

Currently, there is no genetic evidence for the existence of an Ntr-like system in the major gram-negative species of either ruminal or colonic bacteria. Earlier results of GS enzyme assays with *S. dextrinosolvens* and *R. amylophilus* are consistent with both these bacteria possessing an Ntr-like regulatory system; but findings with *S. ruminantium* are claimed to be the first example of a gram-negative bacterium being devoid of this regulatory activity (Hespell 1984). Indeed, eventhough P_{II} and other Ntr-homologs are being identified in a diverse range of microorganisms (see Merrick and Edwards 1995), there is clearly an opportunity to characterize what appear to be novel mechanisms of nitrogen regulation in the predominant species of ruminal and colonic bacteria (see Sect. 4.5).

4.2. Regulation of GS in Gram-Positive Bacteria

Bacillus subtilis depends entirely on GS-GOGAT for ammonia assimilation, even when ammonia is not limiting (Pan and Coote 1979), and Merrick (1988) listed several other species which also lack GDH and thus also depend on GS-GOGAT for ammonia assimilation. The activity of GS in the bacilli appears to be sensitive to feedback inhibition, although not to the same extent as the enzymes from enteric bacteria. The primary effectors appear to be alanine and glutamine; and a coincident decrease in GS activity with the changes in amino acid pool size perhaps reflects a need to regulate enzyme activity during stationary phase and onset of sporulation (Schreir 1993). Expression of the *glnA* structural gene, which is part of the *glnRA* operon, has also been examined. The protein encoded by the *glnR* inhibits the transcription of the operon, *glnRA*, but is not a sensor of intracellular glutamine levels. Glutamine synthetase itself may sense critical nitrogen levels and transmit the message through GlnR. Indeed, GS was reported to increase the efficiency of GlnR binding to the -20 to -60 promoter region (Sonenshein 1992). In their review, Fisher and Sonenshein (1991) also suggested that GS may interact with other genes through their regulatory proteins. Furthermore, they suggested that the expression of *glnRA* operon may be regulated by factors other than just GlnR and GS. *Bacillus cereus* also has a *glnRA* operon that is similarly regulated (Nakano et al. 1989, Nakano and Kimura 1991).

The emergence of potential elements of a Ntr system in the bacilli, and some understanding of GS regulation in *Clostridium acetobutylicum,* by the production of an antisense mRNA which hybridizes across the translational start sites of the *glnA* transcript (Merrick and Edwards 1995), have provided additional evidence that nitrogen regulation in gram positive bacteria should be explored further. As pointed out by Merrick and Edwards (1995), there is still enormous scope for advances to be made with gram-positive bacteria.

4.3. The nac *Extension of the* Ntr *System*

Exogenous sources of aspartate and(or) glutamate have been shown to regulate GDH synthesis in enteric bacteria, and although the exact mechanisms have not been clarified, *gdhA* transcription is reduced in response to exogenous sources of glutamate (Reitzer 1996). Posttranscriptional regulation has also been suggested as a regulatory mechanism (Riba et al. 1988), and there is some preliminary evidence that the GDH enzyme of *E. coli* strain $D_5H_3G_7$ may be phosphorylated by an endogenous protein kinase at a histidine residue(s) (Lin and Reeves 1991, 1992).

Nitrogen limitation strongly represses GDH synthesis in *K. aerogenes*, but has moderate to no effect in *E. coli* (strain-dependent) and no effect at all in *S. typhimurium.* Bender et al. (1983) described a certain glutamate auxotroph, which lacked GOGAT activity but constitutively expressed the *glnALG* operon (and consequently repressed GDH synthesis), that reverted to glutamate-independent growth. Approximately one-half of the revertant strains grew in glucose-ammonia minimal medium, but had lost the ability to use histidine and proline as N sources. The mutated allele was subsequently mapped to be closely linked to the *his* gene, and designated as *nac-1* (for nitrogen assimilation control). The Nac protein has since been shown to be a LysR-like transcriptional regulator that strongly represses GDH synthesis, moderately represses GOGAT synthesis, and activates a subset of genes previously thought to be under the direct control of the *ntr* system (histidine, proline, and urea utilization; see Bender 1992, Schwaca and Bender 1993a,b, Goss and Bender 1995). The Nac protein is synthesized as part of the Ntr response, and effectively serves as an extension of this regulatory network, to include genes which are transcribed by RNAP-σ^{70}. While the *nac* gene appears to be present in *E. coli*, it has not been identified in *S. typhimurium*, leading to the speculation that both these bacteria either have lost or are in the process of losing Nac (Bender 1992). Of the ruminal and colonic bacteria studied to date, only *R. amylophilus* appears to demonstrate a strong repression of GDH activity (and increased GS activity) in response to ammonia-limited growth (Jenkinson et al. 1979), consistent with the existence of an *ntr* system with *nac* extension.

Those ruminal and colonic bacteria which are capable of growth with organic nitrogen sources also demonstrate moderate to strong repression of GDH synthesis in response to the availabilty of peptide nitrogen, most probably at the level of transcription (Morrison and Mackie 1996, Wen and Morrison 1996). Studies with pure cultures of ruminal bacteria have shown that only small amounts of peptide or amino acids stimulate bacterial growth yield (e.g., Pittman and Bryant 1964, Cotta and Russell 1982), attributed to peptide fermentation partially satisfying the bacterium's maintenance energy requirements (Russell 1983). Theoretically, amino acid biosynthesis is considered to represent a relatively small metabolic burden for bacteria (Stouthammer 1979). However, the ''opportunity cost'' associated with the synthesis of amino acid precursors, such as 2-ketoglutarate, might impact growth efficiency in obligate anaerobes. For instance in energy-limited continuous cultures, the cell yield of *P. ruminicola* is impacted by the fate of succinate. Not only does 2-ketoglutarate biosynthesis via succinyl-CoA carboxylation appear to require ATP, but Strobel (1992) has shown that propionate production from succinate depends on coenzyme B_{12}, and inadequate concentrations result in a decreased cell yield per mol glucose fermented. Stimulation of bacterial cell yield by peptide nitrogen might arise from the repression of GDH synthesis and the ''sparing'' of succinate (and ATP) from 2-ketoglutarate (and glutamate) biosynthesis, permitting additional ATP to be formed via succinate decarboxylation and propionate production.

4.4. GOGAT, **gltF** *in Enteric Bacteria, and* **gltC** *in* **Bacillus subtilis**

In enteric bacteria, GOGAT activity is high in ammonia-containing minimal medium, but is repressed in nitrogen-limited cultures which are supplemented with either glutamate or nitrogen sources which give rise to glutamate (see Reitzer 1996). The repression requires a functional *ntr* system (to facilitate amino acid uptake and catabolism). No fewer than four other mechanisms for regulating GOGAT synthesis have been identified in some or all members of the enteric bacteria (see Reitzer 1996 for detailed information). First, the product of the third gene of the *gltBDF* operon, GltF, may be a transmembrane protein kinase and participate glutamate-dependent repression (Castano et al. 1992). An alternative hypothesis is that GltF coordinates an Ntr-independent regulation of genes encoding for proline, arginine, and glycine catabolism (Kuczius et al. 1991). Second, the leucine-responsive protein (Lrp) appears to bind upstream of the *gltBDF* transcription start site and is thought to be a leucine-insensitive activator, although it can slightly repress transcription in the presence of leucine (Ernsting et al. 1993). Third, carbon starvation appears to moderately repress GOGAT (and GDH) synthesis and a cAMP-CRP binding site overlapping the −35 promoter region has been identified (Oliver et al. 1987). Fourth, the stringent response has been implicated as having a role in decreasing energy-consuming nitrogen assimilation by decreasing GS and GOGAT activity (Reitzer 1996).

In contrast to other aspects of ammonia assimilation and nitrogen regulation, regulation of GOGAT synthesis in *B. subtilis* seems to be better characterized than in the enteric bacteria. In many bacilli, growth with both glutamine and ammonia are associated with high levels of GOGAT activity, while growth with glutamate results in low GOGAT activity. Among the glutamate auxotrophs obtained by Tn*917* mutagenesis were insertions in the *gltC* gene. The *gltC* gene is oriented opposite to *gltAB*, but their promoter regions overlap. The GltC protein is a positive regulator of *gltA* and presumably *glt*B, and an autoregulator of its own expression. Bohannon and Sonenshein (1989) noted that the sequence of *gltC* shares considerable homology with the LysR family of bacterial activator proteins described by Henikoff et al. (1988). Furthermore, the amino-terminal protein sequence bears considerable similarity with the corresponding sequences in the λcI repressor and Cro, and also with the Nac protein of the enteric bacteria (see Schrier 1993 for detailed information).

4.5. Other Potential Mechanisms

Perhaps the most interesting feature of nitrogen control in ruminal and colonic bacteria is the general observation that GDH activity increases as much as ten-fold in response to ammonia limitation (see Wallace and Cotta 1988; Yamamoto et al. 1984, 1987a,b; Morrison and Mackie 1996). Some knowledge of the potential regulatory mechanisms has been obtained from the study of the human colonic *Bacteroides.* It has been proposed that GDH enzyme activity is controlled by a reversible inactivation/activation mechanism. This hypothesis has been developed from studies which showed that when cells were subjected to ammonia shock, there was little change in the amount of extractable GDH protein, but enzyme activity was decreased. No phosphorylation or adenylylation of the enzyme has been detected (Yamamoto et al. 1987a,b). Evidence has since been obtained with *B. thetaiotaomicron* that suggests GDH activity is modulated by genetic material located downstream of *gdhA*, which can function in *trans* (Baggio and Morrison, 1996). The *gdhA* gene from *B. thetaiotaomicron* was cloned and isolated within a 7.0-kb DNA fragment by *E. coli* mutant complementation. Northern blot analysis using a *gdh*A-specific DNA probe indicated that a single transcript of approximately 1.5 kb was expressed in both *E. coli* transformants and *B. thetaiotaomicron.* Although *gdhA* transcription was unaffected, no GdhA enzyme activity could be detected in *E. coli* transformants when smaller DNA fragments, which contained the entire *gdhA* gene, were analyzed. Enzyme activity was restored if these *E. coli* strains were cotransformed with a second plasmid which contained a 3-kb segment of DNA *(gdhX)* located downstream of the *gdhA* coding region. Frameshift mutagenesis within the DNA downstream of *gdhA* in the original clone also resulted in the loss of GdhA enzyme activity. These results indicate that additional gene product(s) are involved in modulating GDH activity in *B. thetaiotaomicron.* The *gdhX* locus is currently being characterized in greater detail, and given the similarities between *B. fragilis* and *B. thetaiotaomicron* in nitrogen control of GDH activity, it is likely that studies with *B. thetaiotaomicron* should be applicable to other species of colonic *Bacteroides.*

Regulation of GS synthesis in human colonic *Bacteroides* also seems worthy of further investigation. Abratt et al. (1993) constructed fusions between the *glnA* promoter and a promoterless *eglA* (β-1,4-endoglucanase) gene from *Clostridium acetobutylicum,* in the plasmid shuttle vector pVAL-1. The constructs were conjugatively transferred to *B. fragilis* and *B. thetaiotaomicron*. The results confirmed that *glnA* expression in *B. fragilis* is transcriptionally regulated and that a 584-bp region, which contains two near-perfect direct repeats, was required for regulation. Interestingly, the greatest and most consistent affect upon reporter gene expression was found in *B. fragilis* cells grown with nonlimiting concentrations of ammonia. Reporter gene activity decreased more than 10-fold between early and late exponential phase of growth in ammonia-rich medium, but remained at relatively high levels through all stages of growth when cells were ammonia limited. Therefore, negative regulation of *gln*A transcription in response to excess ammonia (and/or products requiring glutamine for their biosynthesis) appears to be a reasonable hypothesis; and in this respect bears some similarity with observations made with *Bacillus subtilis* (Schreier 1993). However, there is currently no published evidence for a *glnR* homologue in *B. fragilis*, and the regulatory mechanisms modulating GS activity remain undefined in the human colonic *Bacteroides* spp. The advent of molecular genetics protocols applicable to human colonic *Bacteroides* now facilitates a more in-depth characterization of ammonia assimilation and nitrogen regulation in these interesting bacteria.

5. Biosynthesis of Other L-Amino Acids

Much of the research defining the pathways of amino acid biosynthesis has utilized bacterial cultures grown with glucose as a carbon source. The important precursors and intermediates in amino acid biosynthesis with this carbon source are illustrated in Figure 12.4, and the monograph by Gottschalk (1986) provides a more detailed inspection of the metabolic routes and intermediates in *Escherichia coli*. The "carbon skeleton" precursors of L-amino acids invariably include pyruvate, (alanine, valine and leucine); oxaloacetate (aspartate, asparagine, methionine, lysine, threonine, and isoleucine); 2-ketoglutarate (glutamate, glutamine, arginine, proline); 3-phosphoglycerate (serine, glycine, and cysteine); phosphoenolpyruvate plus erythrose-4-phosphate (phenylalanine, tyrosine, and tryptophan); and 5-phosphoribosyl-1-pyrophosphate (PRPP, histidine).

5.1. Biosynthetic Pathways

Collectively, oxaloacetate and pyruvate can be utilized for the biosynthesis of at least nine L-amino acids. L-aspartate is produced from the transamination of

Figure 12.4. Amino acid biosynthetic pathways for bacteria growing on glucose (adapted from Takanami and Oishi 1986). Amino acids in boxes.

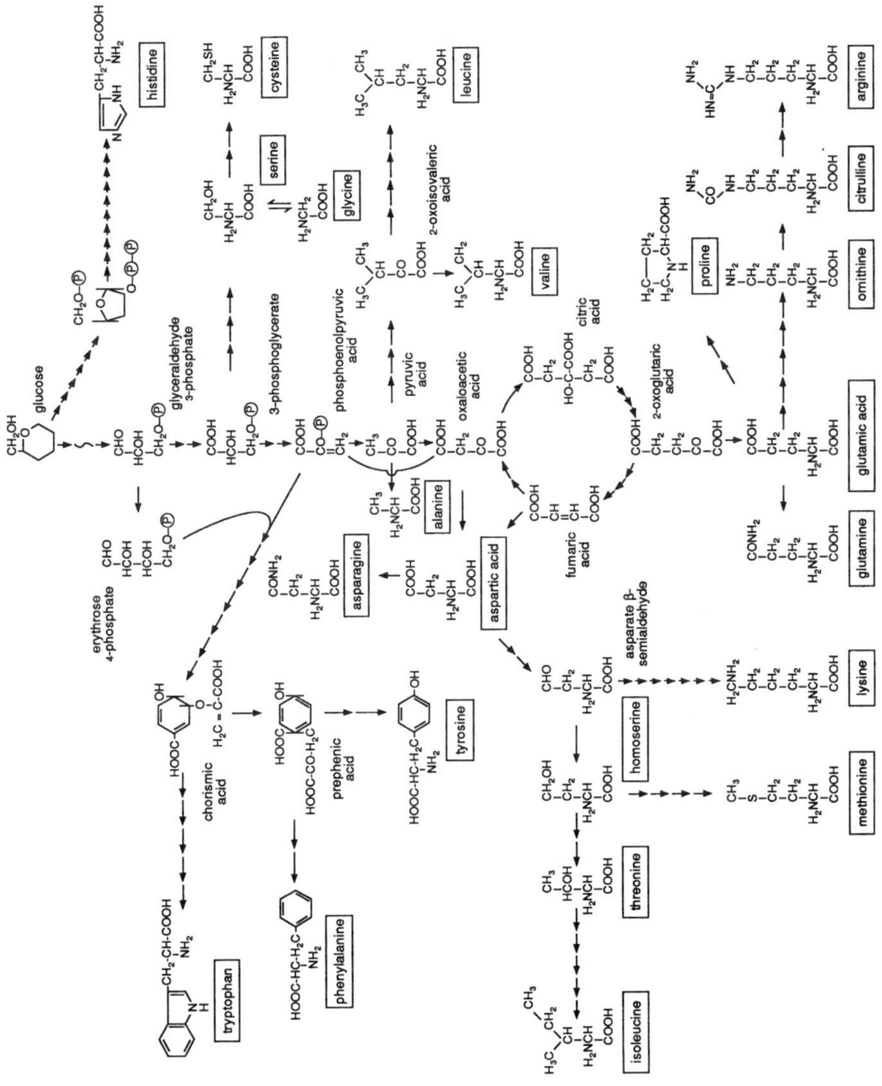
glucose
glyceraldehyde 3-phosphate
3-phosphoglycerate
phosphoenolpyruvic acid
pyruvic acid
oxaloacetic acid
citric acid
2-oxoglutaric acid
fumaric acid
erythrose 4-phosphate
2-oxoisovaleric acid
histidine
cysteine
serine
glycine
leucine
valine
alanine
asparagine
aspartic acid
proline
ornithine
citrulline
arginine
glutamic acid
glutamine
chorismic acid
prephenic acid
tyrosine
tryptophan
phenylalanine
asparate β-semialdehyde
homoserine
lysine
methionine
threonine
isoleucine

oxaloacetate, and glutamate is used as the nitrogen donor. Asparagine biosynthesis in the enteric bacteria can be catalyzed by either of two asparagine synthetases, which are regulated by the nitrogen status of the growth medium (Merrick 1988). Lysine, threonine, and methionine biosyntheses also proceed from aspartate. In both the enteric bacteria and *B. subtilis*, the biosynthesis of these three amino acids begins with the phosphorylation of aspartate by either one of three aspartokinases. Each one of these aspartokinases is different in terms of the patterns regulating gene expression and enzyme activity (Cohen and Saint-Girons 1987, Paulus 1993, Patte 1996, Greene 1996). Diaminopimelate, which is required in many bacteria for the biosynthesis of cell wall and(or) spore cortex peptidoglycans, is an important intermediate of the lysine biosynthetic pathway. An additional ''branch'' in the lysine biosynthetic pathway, for the production of dipicolinate, is present in bacteria which produce heat resistant endospores (e.g., clostridia). Further phylogenetic differences in the biosynthesis of the ''aspartate family'' of amino acids are addressed by Paulus (1993).

Alanine biosynthesis results from the amination of pyruvate and can be catalyzed by transaminase and(or) alanine dehydrogenase. In fact, studies by Blake et al. (1983) suggested that alanine dehydrogenase might be an important pathway of ammonia assimilation in the rumen. Although alanine has sometimes been identified as the predominant amino acid, the ammonia concentrations associated with these findings tended to be very high, consistent with the high K_m (ammonia) often found for alanine dehydrogenases (e.g., Kenealy et al. 1982). Pyruvate is also used for the biosynthesis of the branched-chain amino acids (isoleucine, leucine, and valine). The biosynthesis of the branched chain amino acids has been very well characterized in the enteric bacteria, and the most detailed information concerning the molecular biology and physiology of their formation can be found in reviews by Umbarger (1987, 1990, 1996). The eubacterial paradigm for the biosynthesis of branched chain amino acids begins with threonine and pyruvate, which are ultimately converted to isoleucine and valine, respectively, by a common set of (iso)enzymes. An intermediate in valine biosynthesis, α-ketoisovalerate, serves as the precursory molecule for leucine biosynthesis. The enteric bacteria possess multiple acetohydroxy acid synthase (AHAS) isozymes, while *B. subtilis* and other gram-positive bacteria are dependent on a single AHAS enzyme. Studies with *E. coli* K-12 and *S. typhimurium* LT2, by Dailey and Cronan (1986) and Dailey et al. (1987), respectively, have demonstrated a requirement for a specific AHAS isozyme, dependent upon the carbon source provided for growth. AHAS I-deficient mutants of both bacteria required isoleucine and valine for growth on acetate or oleate minimal media, whereas AHAS II-deficient mutants were able to grow on these media without isoleucine supplementation. Barak et al. (1987) concluded that the AHAS I isozyme enables enteric bacteria to synthesize branched-chain amino acids during growth upon poor carbon sources, and the concomitant low endogenous concentrations of pyruvate. Under similar growth conditions, the AHAS II and AHAS III isozymes probably cannot catalyze the amounts of acetolactate required for valine and leucine biosynthesis, and function effectively only when glucose is the available carbon source. Detailed information of the genetic and biochemical characteristics of AHAS enzymes (including a Type IV isoenzyme from *E. coli* K-12), and the need for multiple enzymes for acetohydroxy acid synthesis, can be found in the review by Umbarger (1996).

Research conducted with ruminal and colonic bacteria by Allison and co-workers in the 1960s, and reviewed by Allison (1969), clearly outlined how many of these bacteria could also synthesize branched chain amino acids from branched chain, volatile fatty acids. Isovalerate, 2-methylbutyrate, and isobutyrate are reductively carboxylated, then aminated to produce leucine, isoleucine, and valine, respectively. Subsequently, similar pathway(s) were described for *Clostridium sporogenes* by Monticello and Costilow (1982) and Monticello et al. (1984). With these exceptions, the biosynthesis of isoleucine invariably requires 2-ketobutyrate however, the origin of this ketoacid can be either threonine (Umbarger 1987, Fink 1993), homoserine (Cohen and Sallach 1961), propionate (Buchanan 1969, Sauer et al. 1975, Eikmanns et al. 1983a), glutamate (Phillips et al. 1972), or pyruvate and acetyl-CoA (Westfall et al. 1983, Eikmanns et al. 1983b). It should be noted that evidence for these other routes of branched-chain amino acid biosynthesis have been derived from radiotracer experiments (e.g., Sauer et al. 1975, Eikmanns et al. 1983b) and/or using cell-free, partially purified extracts (e.g., Allison and Peel 1971). More in-depth biochemical and genetic studies do not appear to have been undertaken.

Strictly anaerobic bacteria of the rumen and colon are capable of utilizing reductive carboxylations to yield the appropriate carbon skeletons for most amino acids. For example, succinate was shown to be reductively carboxylated to produce 2-ketoglutarate in *P. ruminicola* (Allison and Robinson 1970) as well as *Bacteroides fragilis, Veillonella*, and *Selenomonas* spp. (Allison et al. 1979). It is also clear that phenylacetate, hydroxyphenylacetate, and indoleacetate are suitable substrates for reductive carboxylations and can be utilized for the biosynthesis of phenylalanine, tyrosine, and tryptophan, respectively (Allison 1965, 1969; Allison and Robinson 1967). The reductive carboxylation pathways present in some rumen bacteria appear to be in addition to the biosynthesis of these aromatic amino acids from a common precursor, chorismate. Chorismate biosynthesis requires phosphoenolpyruvate and erythrose-4-phosphate as substrates and proceeds via the shikimate pathway. Sauer et al. (1975) provided evidence to support the presence of both routes of aromatic amino acid biosyntheses with mixed rumen contents, but the quantitative importance of the two pathways remains unclear.

In addition to the reductive carboxylation of hydroxyphenylacetate, there appear to be at least two other pathways for tyrosine biosynthesis present in gram-positive bacteria, although both require prephenate as the precursor (Henner and

Yanofsky 1993). Despite the extensive research on the biosynthesis of aromatic amino acids (see Pittard 1987, 1996; Henner and Yanofsky 1993, for detailed reviews), it appears that in all other bacteria studied to date, the enzymatic reactions resulting in phenylalanine and tryptophan biosyntheses are identical in nature and involve the shikimate pathway. The conservation of the shikimate pathway is not surprising, considering that chorismate is also utilized for the biosynthesis of folate, ubiquinone, menaquinone, and enterochelin. However, the regulatory strategies are more divergent but can be ''classified'' relative to 16S rRNA phylogenetic groupings of bacteria (Byng et al. 1982, Jensen 1992).

Allison (1969) had earlier speculated that imidazole acetate might serve as a precursor for histidine biosynthesis in rumen and gastrointestinal bacteria, but no conclusive data appear to be available. The results of Sauer et al. (1975), however, support the contention that histidine biosynthesis proceeds via the same pathway described for the enteric bacteria (Winkler 1987, 1996).

In ruminal bacteria, the availability of branched-chain and phenylsubstituted fatty acids has also been demonstrated to modulate the flux of glucose carbon into amino acids. Specifically, Allison et al. (1984) found that glucose carbon was incorporated into leucine when isovaleric acid was not present in the growth medium. Supplementation of the growth medium with this fatty acid effectively blocked glucose carbon flow into leucine. Similarly, the provision of phenylacetate and 2-methylbutyrate reduced the utilization of glucose carbon for phenylalanine and isoleucine biosynthesis, respectively. Little more than this is known concerning the regulation of amino acid biosynthetic pathways in gastrointestinal bacteria.

Sauer et al. (1975) concluded from radiotracer studies with ruminal bacteria that only 3-hydroxypyruvate, the precursor of serine, appeared to be synthesized by an oxidative step. The serine-glycine pathway of *E. coli* and *S. typhimurium*, described in detail by Stauffer (1987, 1996), appears to be the same pathway used in ruminal bacteria. The initial reaction is the oxidation of 3-phosphoglycerate to 3-hydroxypyruvate, followed by transamination (to 3-phosphoserine) and dephosphorylation to serine. A single enzyme, serine hydroxymethyltransferase, catalyzes the interconversion between serine and glycine and also produces 5,10-methylenetetrahydrofolate (Stauffer 1987,1996). The metabolic relationships between pyridoxine and serine biosynthesis were recently described by Lam and Winkler (1990), and serine is also used for both cysteine and tryptophan biosynthesis (see Fig. 12.1). In anaerobic environments containing sulfide, cysteine biosynthesis is thought to be fairly simple. Serine is enzymatically converted to O-acetylserine, which then reacts with sulfide to produce cysteine (Kredich 1987). Serine can also serve as the amino group donor in the terminal reaction of tryptophan biosynthesis (Pittard 1987,1996). An alternate pathway for serine biosynthesis in *E. coli,* which utilizes threonine as a precursor, was described by Ravnikar and Somerville (1987). This pathway was defined as the threonine utilization (Tut) cycle. Serine auxotrophs utilized the Tut cycle when the medium was supplemented with leucine, arginine, lysine, threonine, or methionine. It was also postulated that during growth on acetate, the Tut cycle would spare the 3-phosphoglycerate pool for gluconeogenesis. Members of the clostridia have also been shown to utilize threonine for serine biosynthesis and constitutively produce a threonine aldolase which converts threonine to acetaldehyde and glycine. Glycine and a C_1 unit are then converted to serine (Dainty and Peel 1970).

The glutamate family of amino acids include proline and arginine, and there is little information to suggest that the biosynthesis of these amino acids in ruminal and colonic bacteria deviates greatly from the pathways described for the enteric bacteria (see Neiderman and Wolin 1967; Leisinger 1987, 1996; Glansdorff 1987, 1996). The biosynthesis of both arginine and proline requires that the 5-carboxyl group of glutamate be activated, then reduced, to yield glutamate-5-semialdehyde. Acetylation of glutamate-5-semialdehyde prevents its spontaneous cyclization to the precursor of proline and commits the molecule to the biosynthesis of ornithine. Ornithine carbamoyltransferase catalyzes the transfer of the carbamoyl moiety from carbamoyl phosphate to the 5-amino group of ornithine to yield citrulline. Citrulline is then converted to argininosuccinate by an ATP-dependent reaction which also requires aspartate. Finally, fumarate is cleaved from argininosuccinate to yield arginine (Cunin et al. 1986). Both ornithine and arginine can also be utilized for the biosynthesis of polyamines (see Sect. 6).

5.2. *Amino Acid Biosynthesis and Osmoregulation*

Most microbial habitats are likely to experience at least some variation in osmolality, and the mammalian gastrointestinal tract is probably no exception. The most extensive studies of the physiological and genetic responses of bacteria to variations in osmolality have been conducted with enteric bacteria and were most recently reviewed by Csonka (1996). In addition to potassium ions, these bacteria obtain compatible solutes by de novo synthesis (e.g., glutamate, glutamine) or exogenous sources (e.g., proline, betaine, and glycine-betaine) are transported into the bacterial cell.

Tempest et al. (1970) were among the first to observe that glutamate levels in gram-negative bacteria increased dramatically in response to high osmolality. The gram-positive bacteria studied by Tempest et al. (1970) maintained much higher cytoplasmic concentrations of glutamate but still accumulated more glutamate in response to high osmolality. In other gram-positive bacteria such as *Staphylococcus* spp., glutamine and(or) alanine will accumulate in the cytoplasm in response to osmotic stress (Anderson and Witter 1982). The regulation of glutamate accumulation in response to increased osmolality has become more clearly defined in the enteric bacteria. Glutamate accumulation appears to be dependent upon K^+ uptake (McLaggan et al. 1994), and mutational loss of either GDH or GOGAT does not alter the kinetics of glutamate accumulation (Botsford et al. 1994, McLaggan et al. 1994). More detailed information relating to putrescine, neutral solutes, and hierarchy of osmoprotectants in the enteric bacteria can be found in the review by Csonka (1996).

Osmolality in the rumen fluctuates within a fairly narrow range (200 to 280 mOsmol kg^{-1}), although feeding rapidly fermentable carbohydrates will increase the solute concentration (principally as VFA), which may increase osmotic pressure substantially (Slyter 1976). The water-absorbing capacity of the human colon and the fermentative capacity of its resident microflora are also likely to result in a hypertonic environment during stool formation (Davenport 1982). Mackie and Therion (1984) presented preliminary results with *S. ruminantium* that showed the specific growth rate of this bacterium was severely affected by osmolality in excess of 500 mOsmol kg^{-1}. However, the physiological and genetic responses to this stress have not been further examined. Likewise, virtually no studies of osmoregulation have been conducted with colonic bacteria, such as members of the *B. fragilis* group. The relative importance of amino acid biosynthesis and/or transport of compatible solutes has been unexplored with this predominant group of gastrointestinal bacteria.

6. Biosynthesis of Polyamines

Polyamines are present in almost all living organisms and appear to play a role in a wide variety of biological reactions including nucleic acid and protein biosyntheses (see Tabor and Tabor 1985). The polyamines include aliphatic linear molecules, from diamines to hexamines, tertiary branched tetramines and quartenary branched pentamines. No fewer than 24 linear and four branched types of polyamines have been isolated from bacteria by acid extraction and chromatographic procedures. The structural features of these polyamines have recently been examined, and were used as a chemotaxonomic marker in bacterial systematics (Hamana and Matsuzaki 1992).

The major precursory molecules and the biosynthesis of bacterial polyamines are illustrated in Figure 12.5. A detailed description of the polyamine metabolic enzymes was presented in volume 94 of *Methods of Enzymology*, and for microorganisms by Tabor and Tabor (1985). Polyamine biosynthesis in the enteric bacteria has also been recently reviewed (Glansdorff 1996). Hamana and Matsuzaki (1992) concluded that the polyamine distribution patterns in bacteria were primarily dependent upon the presence (or absence) of four amino acid decarboxylases: arginine decarboxylase, diaminobutyric acid decarboxylase, lysine decarboxylase, and ornithine decarboxylase. Three diamines—1,3-diaminopropane, putres-

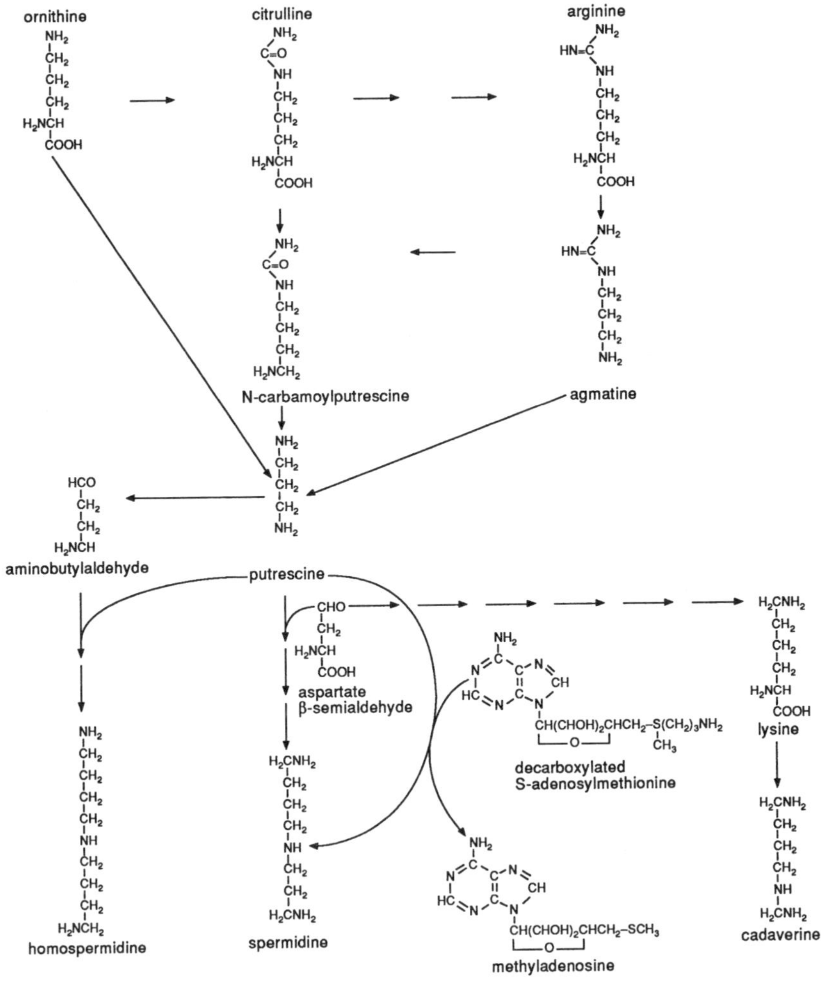

Figure 12.5. Pathways for the biosynthesis of putrescine, spermidine, homospermidine, and cadaverine in bacteria (adapted from Tabor and Tabor 1985, Hamana and Mitzuoka 1992).

cine, and cadaverine—result from the decarboxylation of L-1,2-diaminobutyric acid, L-ornithine, and L-lysine, respectively. Arginine may also serve as a precursory molecule, being converted first to citrulline and(or) agmatine, and ultimately to putrescine. Putrescine biosynthesis in bacteria may therefore proceed via three different pathways: (1) decarboxylation of ornithine; (2) hydrolysis of agmatine by "agmatinase," giving rise to putrescine and urea; and (3) synthesis from N-carbamoylputrescine generated from either the decarboxylation of citrulline or, deamidation of agmatine (Tabor et al. 1983, 1985; see Fig. 12.2). Aminopropyl and aminobutyl transferases are also important biosynthetic enzymes which affect the polyamine profiles of bacteria. For instance, spermidine biosynthesis can proceed via putrescine and L-aspartic β-semialdehyde, which is catalyzed by an aminobutyltransferase. Spermidine and spermine synthases are both aminopropyl transferases, which utilize decarboxylated S-adenosylmethionine for the synthesis of spermidine and spermine, respectively (see Fig. 12.4). In general terms, spermine is limited to eukaryotes (Tabor and Tabor 1985) and thermophilic bacteria of various genera (Hamana and Matsuzaki 1992), and will not be discussed here in further detail. The N-methylation and N-acetylation of polyamine terminal amino groups is also observed in bacteria and contributes to the diversity of bacterial polyamines (Hamana and Matsuzaki 1992).

Polyamines are required by many bacteria for optimal growth, but the underlying physiological principles for the requirement(s) are unclear (see Tabor and Tabor 1985). The basic nature of bacterial polyamines results in ionic interactions with nucleic acids, acidic proteins, and phospholipids. Moreover, in the rumen bacterium *Selenomonas ruminantium*, the polyamine cadaverine is covalently linked to the peptidoglycan layer (Kamio et al. 1981). Putrescine and cadaverine are also constituents of the peptidoglycan in both *Veillonella alcalescens* and *V. parvula*, with more than 40% saturation of the α-carboxyl group of D-glutamate observed (Kamio and Nakamura 1987). However, these bacteria appear reliant upon exogenous sources of polyamines, because their omission from the growth medium results in abnormal cell surface and septum development. The polyamine biosynthetic enzymes also appear to be absent in other gastrointestinal bacteria. Within the *Bacteroides fragilis* cluster, *B. fragilis*, *B. ovatus*, and *B. thetaiotaomicron* fall into this category, while *B. distasonis*, *B. uniformis*, and *B. vulgatus* possess spermidine as their primary polyamine. A number of oral isolates representing the genus *Prevotella* were also found devoid of known polyamines, as was *Mitsuokella (Bacteroides) multiacidus* (Hamana and Matsuzaki 1992).

In terms of gram-positive gastrointestinal bacteria, *Eubacterium limosum* is devoid of polyamines, but putrescine and spermidine are commonly found in the Clostridia, particularly in mesophilic species such as *C. butyricum* (Kneifel et al. 1986). Members of the genera *Lactobacillus*, *Streptococcus*, and *Enterococcus* also possess putrescine and spermidine, albeit in relatively low concentrations (Hamana et al. 1989). The three orders of methanogenic archaebacteria each have a characteristic polyamine profile. Species of *Methanobacterium* and *Methanobrevibacter*, which belong to the Methanobacteriales, have no detectable polyamines (Kneifel et al. 1986). Methanogens belonging to the Methanococcales possess spermidine and homospermidine as major polyamines, while members of the Methanosarcinacea possess homospermidine as the major polyamine (Hamana and Matsuzaki 1992).

7. Biosynthesis of Pyrimidines and Purines

Pyrimidines and purines are constituents of ribonucleotides, which participate in nearly all biochemical processes. Nucleotides are precursors of DNA and RNA, and can be metabolic regulators (cAMP) as well as part of activated intermediates in anabolic processes, such as glycogen biosynthesis. Adenine nucleotides, especially ATP, have an obvious role in microbial physiology. Adenine nucleotides are also important components of the coenzymes NAD(P), FAD, and CoA. The de novo synthesis of pyrimidine and purine nucleotides only will be covered here, although "salvage reactions," to reutilize or interconvert purine and pyrimidine bases, also exist. These groups of enzymes also make the ribose moiety of nucleosidase and the amino groups of cytosine and adenosine nucleotides available as carbon, energy, and nitrogen sources. Readers interested in detailed descriptions of these salvage reactions are directed to reviews by Nygaard (1993), Zalkin and Nygaard (1996) and Neuhard and Kelln (1996).

7.1. *PRPP, an Essential Substrate in* De Novo *Pyrimidine and Purine Biosynthesis*

The compound PRPP is required as a substrate in no fewer than 10 different enzymatic reactions in *E. coli*, including purine and pyrimidine nucleotide biosyntheses. When *E. coli* is growing exponentially in glucose minimal medium, approximately 80% of the available PRPP pool is utilized for the biosynthesis of nucleotides. A further 10% to 15% of the PRPP pool appears to be utilized for tryptophan and histidine biosynthesis, while little more than 1% was incorporated into nicotinamide coenzymes (Jensen 1983). The synthesis of PRPP is catalyzed by PRPP synthetase, and ADP is an allosteric inhibitor of enzyme activity (Switzer and Sogin 1973). In both *E. coli* and *S. typhimurium*, a large intracellular pool of adenine nucleotides result in a depletion of the PRPP pool, as will an addition of purine bases to exponentially growing cultures. Expression of the *prs* gene, which encodes PRPP synthetase, is also regulated and uracil starvation results in the greatest levels of derepression (Neuhard and Nygaard 1987, Zalkin and Nygaard, 1996).

7.2. De Novo *Pyrimidine Biosynthesis*

The enzymatic steps involved with de novo pyrimidine biosynthesis are illustrated in Figure 12.6 and the primary precursors are carbamoyl phosphate, aspartate and PRPP. Pyrimidine biosynthesis begins with the condensation of aspartate with carbamoyl phosphate, catalyzed by the enzyme aspartate transcarbomylase. The resulting carbamoyl aspartate moiety undergoes cyclization yielding 4,5-dihydroorotate. Dehydrogenation then yields orotate, which is linked to PRPP. A final decarboxylation step yields uridine monophosphate. This pathway is universal in all eukaryotic and prokaryotic organisms studied to date. However, derivation of the carbamoyl phosphate used for pyrimidine biosynthesis, regulation of the biosynthetic pathway, and the arrangement of the genes encoding the biosynthetic enzymes are varied.

The genetics and biochemistry of carbamoyl phosphate biosynthesis were reviewed by Makoff and Radford (1978). Three main classes of enzyme organization were proposed. Class I organisms, which include *E. coli* and *S. typhimurium*, were defined as those possessing a single carbamoyl phosphate synthetase (CPS) enzyme, responsible for providing a substrate pool utilized for both pyrimidine and arginine biosynthesis. The second (class II) organization includes *B. subtilis,* and separate CPS enzymes are utilized for pyrimidine and arginine biosynthesis. The third class of organisms include gram-positive bacteria such as *Lactobacillus* and *Enterococcus* spp., and there is a total of lack of CPS activity. Accordingly, the class III bacteria possess an absolute requirement for arginine, and carbamoyl phosphate is produced by the degradation of either arginine or citrulline.

The structure and function, allosteric, and genetic regulation of the other enzymes for de novo pyrimidine biosynthesis have most recently been reviewed in detail for *E. coli* and *S. typhimurium* by Neuhard and Nygaard (1987) and Neuhard and Kelln (1996), and for *B. subtilis* and other gram-positive bacteria by Switzer and Quinn (1993). Allosteric regulation of pyrimidine biosynthesis in *E. coli* and *S. typhimurium* occurs at three points: (1) CPS is inhibited by UMP and stimulated by ornithine; (2) aspartate carbamoyl transferase is feedback inhibited by the endproduct of the pathway, CTP; and (3) CTP synthetase activity in vivo appears to be affected positively by UTP and negatively by high concentrations of CTP (see Neuhard and Nygaard 1987, Neuhard and Kelln 1996). In gram-positive bacteria, CPS is the only enzyme in the pathway clearly demonstrated to be subject to allosteric regulation. Allosteric activators are GTP, PRPP, GDP, and GMP; inhibitors include uridine nucleotides, with UMP being the most effective (see Switzer and Quinn 1993). In *E. coli* and *S. typhimurium*, the genes encoding pyrimidine biosynthetic enzymes are ubiquitous throughout the chromosome and are regulated noncoordinately. Conversely, the *pyr* genes map within a tight cluster spanning 12 kb of the *B. subtilis* chromosome, and transcription of *pyr* genes appears to be as a single polycistronic message. Only *pyrG (ctrA)*, which encodes

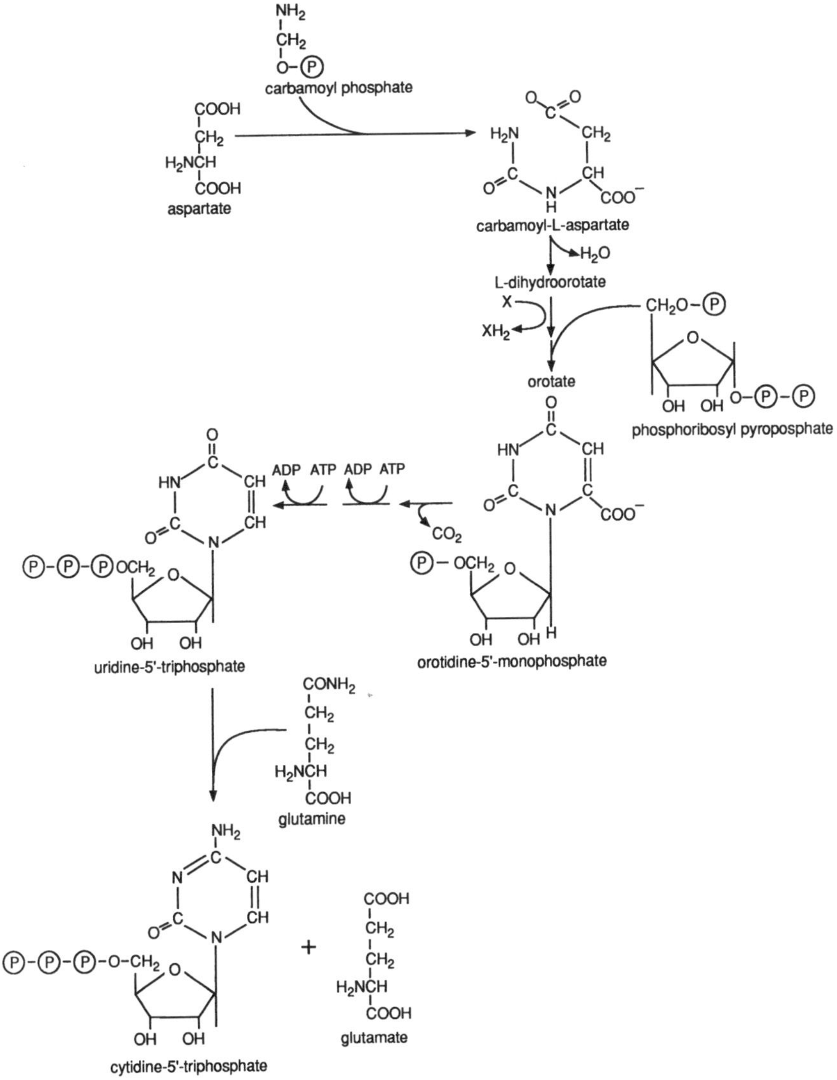

Figure 12.6. Pathways for de novo biosynthesis of pyrimidines indicating entry points for aspartate, carbamoyl phosphate, and phosphoribosyl pyrophosphate. Each arrow represents an enzymatic step. Two ATP molecules are utilized to produce uridine-5′-triphosphate, and glutamine serves as amine group donor for conversion to cytidine-5′-triphosphate.

the CTP synthetase enzyme in *B. subtilis,* is not a part of this gene cluster (Switzer and Quinn 1993). The genetic organization of *pyr* genes in *E. faecalis* appears to be identical to that found in *B. subtilis* (Li et al. 1994). Virtually nothing appears to be known of the genes or enzymes of pyrimidine biosynthesis for the archaebacteria.

7.3. De Novo *Purine Biosynthesis*

Relative to pyrimidine nucleotides, the de novo synthesis of purine nucleotides requires a great many more enzymatic steps. The 10 enzymatic steps required for the synthesis of inosinic acid (IMP) are illustrated in Figure 12.7. Two further enzymatic steps result in the production of adenosine monophosphate (AMP): (1) adenylsuccinate is produced by the GTP-mediated condensation of aspartate with IMP, and (2) cleavage of the fumarate moiety from adenylsuccinate gives rise to AMP. Alternatively, the IMP moiety is dehydrogenated to produce xanthylic acid, which in turn is the substrate for the ATP-mediated amidation that produces guanosine monophosphate (GMP).

The de novo pathway of purine biosynthesis is conserved in all organisms investigated to date. However, like the genes encoding the enzymes for pyrimidine biosynthesis, their chromosomal organization, the structural features of the gene products and their regulation, varies. The genes encoding for the synthesis of IMP are found clustered in *B. subtilis*, and at least some of these same genes are similarly arranged in *L. casei* (Gu et al. 1992). Conversely, in *E. coli* and *S. typhimurium*, the genes encoding IMP biosynthesis are scattered throughout the chromosome (He et al. 1990). Watanabe et al. (1989) proposed that the REP sequences in these bacteria, thought to play role in chromosomal rearrangement (Stern et al. 1984), might be partly responsible for the scattering of *pur* genes throughout the chromosome. In this context, Watanabe et al. (1989) noted that REP sequences are found in the upstream and downstream regions of the *purF* and *purL* operons, respectively.

Rolfes and Zalkin (1988b) cloned *purR* from *E. coli*, encoding a repressor protein for purine nucleotide biosynthesis. Both Rolfes and Zalkin (1988a) and Kilstrup et al. (1989) presented genetic evidence for the role of the PurR repressor protein in the regulation of at least some of the *pur* (genes *D* and *F*) genes in *E. coli*. A highly conserved, 16-nucleotide motif (the PUR box) positioned to overlap the −35 region of many *pur* promoters has since been reported (see Rolfes and Zalkin 1990); detailed information concerning gene regulation in enteric bacteria by the DNA-binding protein PurR can be found in the review by Zalkin and Nygaard (1996). Expression of *purR* is autoregulated, and involves two operator sites, located downstream of the promoter.

In *B. subtilis*, transcriptional regulation of the *pur* operon appears to involve both development of secondary structure in *pur* transcripts and direct repression

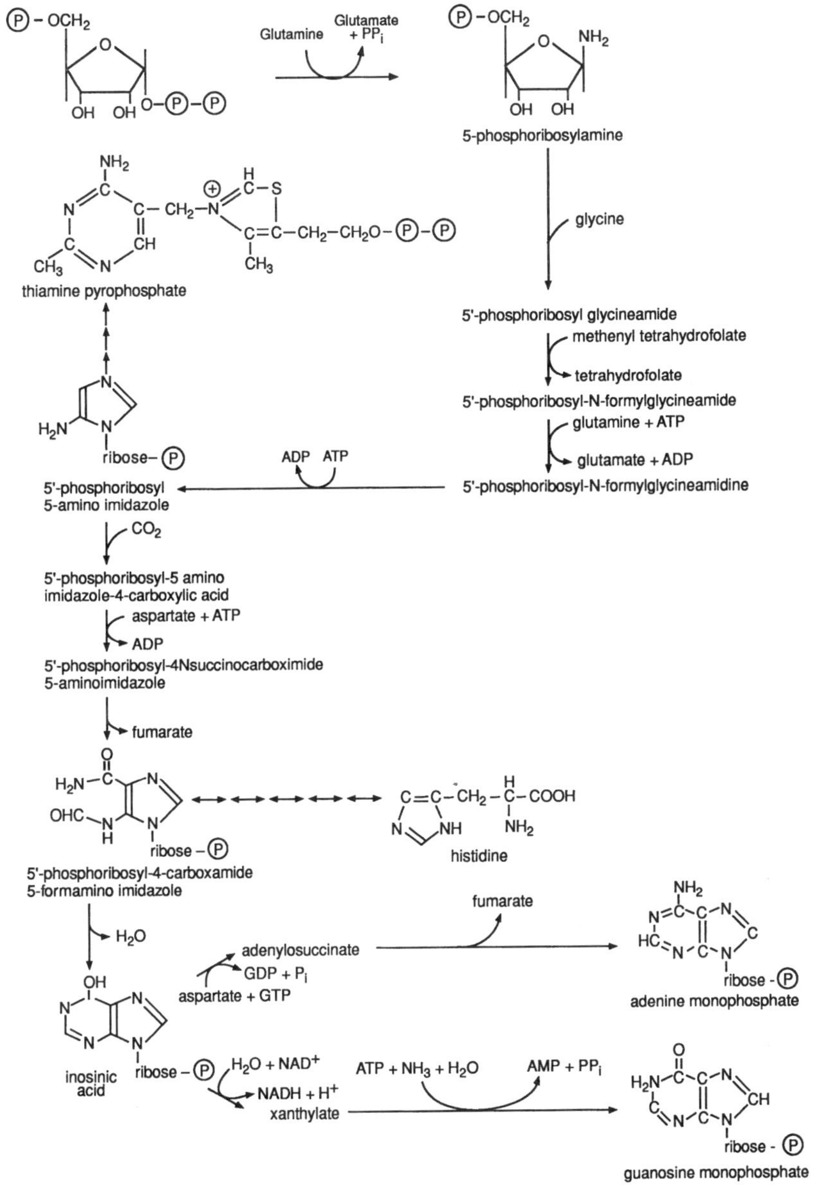

Figure 12.7. De novo pathways(s) of purine nucleotide biosynthesis illustrating branch points for thiamine pyrophosphate and histidine biosynthesis (adapted from Gottschalk 1986, Neuhard and Nygaard 1987, Zalkin 1993).

of *pur* transcription. Both mechanisms are described in some detail by Zalkin (1993). The first model involves the activation of a *trans*-acting regulatory protein by guanine. The activated protein binds with leader mRNA, allowing a stem-loop structure to form which prematurely terminates transcription. High adenosine levels represses transcription initiation, and a *cis*-acting region extending from −145 to −29 relative to the *pur* transcription initiation site is required for repression (Ebbole and Zalkin 1989). The protein-DNA interaction does not appear to share any similarity with the PurR protein-PUR site interaction described for *E. coli.*

There are also some distinctive features between the glutamine PRPP amidotransferase and phosphoribosyl formylglcineamidine (FGAM) synthetase enzymes of *B. subtilis* and *E. coli.* The catalytic properties and effect of purines upon glutamine PRPP amidotransferase activity are similar in both bacteria, and were reviewed in detail by Switzer (1989). However, the *B. subtilis* enzyme possesses an 11 amino-acid N-terminal "propeptide," which must first be cleaved to produce a mature enzyme and, a 4Fe-4S cluster, which is absent from the *E. coli* enzyme. The Fe-S cluster appears to be involved with oxidative inactivation of the *B. subtilis* enzyme (Grandoni et al. 1989; the *E. coli* enzyme is not subject to such inactivation), and the four cysteine residues involved as ligands in the Fe-S cluster of *B. subtilis* are not conserved in the *E. coli* sequence (Zalkin 1993).

The FGAM synthetase enzyme of *E. coli* is a monomer of approximately 135,000 M_r, and possesses separate domains within the polypeptide which catalyze either: (1) the *ammonia-dependent* synthesis of FGAM or (2) the *glutamine-dependent* synthesis of FGAM (Schendel et al. 1989, Sampei and Mizobuchi 1989). These two catalytic functions are encoded within separate genes, *purQ* and *purL,* in *B. subtilis* (Ebbole and Zalkin 1987), and the mature FGAM synthetase enzyme exists as a heterodimer. Based upon DNA sequence homology, it would appear that *L. casei* also possesses a heterodimeric FGAM synthetase. The *L. casei purL* gene has been completely sequenced (Gu et al. 1992) and was found to overlap regions with strong amino acid identity to the *purQ* and *purF* gene products of *B. subtilis.* These gene overlaps also exist in *B. subtilis* and are similar in length to those found in *L. casei* (Ebbole and Zalkin 1987).

The study of purine biosynthesis in other bacteria is minimal. Hamilton and Reeve (1985a,b) determined the nucleotide sequences of chromosomal DNA from *Methobrevibacter smithii* and *Methanobacterium thermoautotrophicum* which complemented *purE1* and *purE2 (purK)* mutations of *E. coli.* The *purEK* genes encode phosphoribosyl-5-aminoimidazole (AIR) carboxylase activity in both *E. coli* and *B. subtilis.* However, the DNA sequence data indicated that the archaebacterial DNA encoded a single polypeptide chain, comprised of two domains. These domains were highly homologous, and could represent a tandem duplication of a single gene (probably *purE*), which has ultimately been fused into a single open reading frame. Thus, the AIR carboxylases of the methanogenic archaebacteria are encoded by a single gene, whereas two genes are required for AIR carboxylase activity in *E. coli* and *B. subtilis.* Interestingly, the *purK* gene product is thought to facilitate the catalysis of AIR carboxylation, when CO_2 is present in low concentrations (see Gots et al. 1977). Whether other strict anaerobes, capable of autotrophic growth with CO_2 possess greater similarity with either the eubacterial or archaebacterial species studied to date, seems an attractive and informative topic of molecular evolution.

8. Conclusions

The study of ammonia assimilation, amino acid biosynthesis, and nitrogen regulation in ruminal and colonic bacteria has been quite descriptive, relative to the fundamental studies conducted primarily with the enteric bacteria. Although many of the findings appear to be similar and(or) applicable to ruminal and colonic bacteria, there are some noteworthy exceptions. Specific examples include regulation of GDH synthesis, GS structure-function relationships, nitrogen regulation in human colonic *Bacteroides*, and the molecular aspects of reductive carboxylations of volatile fatty acids by obligate anaerobes for amino acid biosynthesis. With the continuing development and application of molecular genetics in the study of ruminal and colonic bacteria, it should be possible to better characterize the genetics and biochemistry of these processes. The attainment of such knowledge will not only be of fundamental interest to microbial physiologists, but should also find some application in the continuing efforts to optimize ruminal function and animal productivity. Indeed, future improvements in the more practical areas of ruminal microbiology are largely dependent upon a more fundamental understanding of the physiology of ruminal (and colonic) bacteria.

Acknowledgments

M. Morrison would like to thank Dr. K.W. Nickerson for helpful discussions. Research performed in M. Morrison's laboratory has been funded by the University of Nebraska's Agricultural Research Division, and USDA NRICGP #944081.

References

Abratt VR, Zappe H, Woods DR (1993) A reporter gene to investigate the regulation of glutamine synthetase in *Bacteroides fragilis* Bf1. J Gen Microbiol 139:59–65.

Allison MJ (1965) Phenylalanine biosynthesis from phenylacetic acid by anaerobic bacteria from the rumen. Biochem Biophys Res Commun 18:30–35.

Allison MJ (1969) Biosynthesis of amino acids by ruminal microorganisms. J Anim Sci 29:797–807.

Allison MJ, Peel JL (1971) The biosynthesis of valine from isobutyrate by *Peptostreptococcus elsdenii* and *Bacteroides ruminicola.* Biochem J 121:431–437.

Allison MJ, Robinson IM, Baetz AL (1979) Synthesis of α-ketoglutarate by reductive carboxylation of succinate in *Veillonella*, *Selenomonas*, and *Bacteroides* species. J Bacteriol 140:980–986.

Allison MJ, Baetz AL, Wiegel J (1984) Alternative pathways for the biosynthesis of leucine and other amino acids in *Bacteroides ruminicola* and *Bacteroides fragilis.* Appl Environ Microbiol 48:1111–1117.

Allison MJ, Robinson IM (1967) Tryptophan biosynthesis from indole-3-acetic acid by anaerobic bacteria from the rumen. Biochem J 102:36–37.

Allison MJ, Robinson IM (1970) Biosynthesis of α-ketoglutarate by the reductive carboxylation of succinate in *Bacteroides ruminicola.* J Bacteriol 104:50–56.

Almassy RJ, Janson CA, Hamlin R, Xuong NH, Eisenberg D (1986) Novel subunit-subunit interactions in the structure of glutamine synthetase. Nature 323:304–309.

Anderson DB, Winter LD (1982) Glutamine and proline accumulation by *Staphylococcus aureus* with reduction in water activity. J Bacteriol 143:1501–1503.

Baggio L, Morrison M (1996) The NAD(P)H-dependent glutamate dehydrogenase of *Bacteroides thetaiotaomicron* belongs to enzyme family I, and its activity is affected by *trans*-acting gene(s) positioned downstream of *gdhA*. J Bacteriol. 178:in press.

Barak Z, Chipman DM, Gollop N (1987) Physiological implications of the specificity of acetohydroxyacid synthase isozymes of enteric bacteria. J Bacteriol 169:3750–3756.

Barnes EM Jr, Jayakumar A (1993) NH_4^+ transport systems in *Escherichia coli.* In: Bakker EP, ed. Alkali Cation Transport Systems in Prokaryotes, pp. 397–409. Boca Raton, Fla: CRC Press.

Benachenhou-Lahfa N, Forterre P, Labedan B (1993) Evolution of glutamate dehydrogenase genes: evidence for two paralogous protein families and unusual branching patterns of the archaebacteria in the universal tree of life. J Mol Evol 36:335–346.

Bender RA, Janssen KA, Resnick AD, Blumberg M, Foor F, Magasanik B (1977) Biochemical parameters of the glutamine synthetase from *Klebsiella aerogenes*. J Bacteriol 129: 1001–1009.

Bender RA, Snyder PM, Bueno R, Quinto M, Magasanik B (1983) Nitrogen regulation system of *Klebsiella aerogenes:* the *nac* gene. J Bacteriol 156:444–446.

Bender RA (1991) The role of the NAC protein in the nitrogen regulation of *Klebsiella aerogenes*. Mol Microbiol 5:2575–2580.

Blake JS, Salter DN, Smith RH (1983) Incorporation of nitrogen into rumen bacterial fractions of steers given protein- and urea-containing diets. Ammonia assimilation into intracellular bacterial amino acids. Br J Nutr 50:769–782.

Bohannon D, Rosenkrantz MS, Sonenshein AL (1985) Regulation of *Bacillus subtilis* glutamate synthase genes by the nitrogen source. J Bacteriol 163:957–964.

Boniface AN, Murray RM, Hogan JP (1986) Optimum level of ammonia in the rumen liquor of cattle fed a tropical grass hay. Proc Aust Soc Anim Prod 16:151–154.

Botsford JL, Alvarez M, Hernandez R, Nichols R (1994) Accumulation of glutamate by *Samonella typhimurium* in response to osmotic stress. Appl Environ Microbiol 60: 2568–2574.

Buchanan BB (1969) Role of ferredoxin in the synthesis of α-ketobutyrate from propionyl coenzyme A and carbon dioxide by enzymes from photosynthetic and nonphotosynthetic bacteria. J Biol Chem 244:4218–4223.

Burchall JJ, Reichelt EC, Wolin MJ (1964) Purification and properties of the asparagine synthetase of *Streptococcus bovis.* J Biol Chem 239:1794–1798.

Byng GS, Kane JF, Jensen RA (1982) Diversity in the routing and regulation of complex biochemical pathways as indicators of microbial relatedness. Crit Rev Microbiol 9: 227–252.

Castano I, Flores N, Valle F, Covarrubias AA, Bolivar F (1992) *glt*F, a member of the *glt*BDF operon of *Escherichia coli,* is involved in nitrogen-regulated gene expression. Mol Microbiol 6:2733–2741.

Castroph H, Kleiner D (1984) Some properties of a *Klebsiella pneumoniae* ammonium transport negative mutant (Amt). Arch Microbiol 139:245–247.

Chalupa W, Clark J, Opliger P, Larker R (1970) Ammonia metabolism in rumen bacteria and mucosa from sheep fed soy protein or urea. J Nutr 100:161–169.

Chen T (1995) Molecular cloning, nucleotide sequence analysis, and expression of a gene encoding NADH-dependent glutamate dehydrogenase activity from *Bacteroides thetaiotaomicrom.* MS Thesis, University of Nebraska-Lincoln.

Cohen PP, Sallach HJ (1961) Nitrogen metabolism of amino acid. In: Greenberg DM, ed. Metabolic Pathways, Vol. II, pp. 1–66. New York: Academic Press.

Cotta MJ, Russell JB (1982) Effect of peptides and amino acids on efficiency of rumen bacterial protein synthesis in continuous culture. J Dairy Sci 65:226–234.

Coulton JW, Kapoor M (1973a) Studies on the kinetics and regulation of glutamate dehydrogenase of *Salmonella typhimurium.* Can J Microbiol 19:427–438.

Coulton JW, Kapoor M (1973b) Studies on the kinetics and regulation of glutamate dehydrogenase of *Salmonella typhimurium.* Can J Microbiol 19:439–450.

Csonka LN, Epstein W (1996) Osmoregulation. In: Neidhardt FC, Curtiss R III, Ingraham JL, et al., eds. *Escherichia coli* and *Salmonella typhimurium*: Cellular and Molecular Biology, pp. 1210–1224. Washington, DC: American Society for Microbiolology.

Dailey FE, Cronan JE Jr (1986) Acetohydroxy acid synthase I, a required enzyme for isoleucine and valine biosynthesis in *Escherichia coli* K-12 during growth on acetate as the sole carbon source. J Bacteriol 165:453–460.

Dailey FE, Cronan JE Jr, Maloy SR (1987) Acetohydroxy acid synthase I is required for isoleucine and valine biosynthesis by *Salmonella typhimurium* LT2 during growth on acetate or long-chain fatty acids. J Bacteriol 169:917–919.

Dainty RH, Peel JL (1970) Biosynthesis of amino acids in *Clostridium pasteurianum.* J Biochem 117:573–584.

Davenport HW (1982) Physiology of the Digestive Tract, 5th Ed. Chicago: Yearbook.

Deshpande KL, Katze JR, Kane JF (1980) Regulation of glutamate synthase from *Bacillus subtilis* by glutamine. Biochem Biophys Res Commun 95:55–60.

Duncan PA, White BA, Mackie RI (1992) Purification and properties of NADP-dependent glutamate dehydrogenase from *Ruminococcus flavefaciens* FD-1. Appl Environ Microbiol 58:4032–4037.

Ebbole DJ, Zalkin H (1987) Cloning and characterization of a 12-gene cluster from *Bacillus subtilis* encoding nine enzymes for de novo purine nucleotide synthesis. J Biol Chem 262:8274–8287.

Ebbole DJ, Zalkin H (1989) Interaction of a putative repressor protein with an extended control region of the *Bacillus subtilis pur* operon. J Biol Chem 264:3553–3561.

Eikmanns B, Jaenchen R, Thauer RK (1983a) Propionate assimilation by methanogenic bacteria. Arch Microbiol 136:106–110.

Eikmanns B, Linder D, Thauer RK (1983b) Unusual pathway of isoleucine biosynthesis in *Methanobacterium thermoautotrophicum*. Arch Microbiol 136:111–113.

Erfle JD, Sauer FD, Mahadevan S (1977) Effect of ammonia concentration on activity of enzymes of ammonia assimilation and on synthesis of amino acids by mixed rumen bacteria in continuous culture. J Dairy Sci 60:1064–1072.

Ernsting, BR, Denninger JW, Blumenthal RM, Matthews RG (1993) Regulation of *gltBDF* operon of *Escherichia coli*: how is a leucine-insenstive operon regulated by the leucine-responsive regulatory protein? J Bacteriol 175:7160–7169.

Fink PS (1993) Biosynthesis of branch chained amino acids. In: Sonenshein AL, Hoch JA, Losick R, eds. *Bacillus subtilis* and Other Gram-Positive Bacteria, pp. 307–318. Washington DC: American Society for Microbiology.

Fisher SH, Sonenshein AL (1991) Control of carbon and nitrogen metabolism in *Bacillus subtilis*. Annu Rev Microbiol 45:107–135.

Garciarubbio A, Lozoya E, Covarubbias A, Bolivar F (1983) Structural organization of genes that encode two glutamate synthase subunits of *Escherichia coli*. Gene 26: 165–170.

Gibson SAW, Macfarlane GT (1988) Studies on the proteolytic activity of *Bacteroides fragilis*. J Gen Microbiol 134:19–27.

Ginsburg A, Stadtman ER (1973) Regulation of glutamine synthetase in *Escherichia coli*. In: Prusiner S, Stadtman ER, eds. The Enzymes of Glutamine Metabolism, pp. 9–44. New York: Academic Press.

Glansdorff N (1987) Biosynthesis of arginine and polyamines. In: Neidhardt FC, Ingraham JL, Low KB, Magasanik B, Schaechter M, Umbarger HE, eds. *Escherichia coli* and *Salmonella typhimurium*: Cellular and Molecular Biology, pp. 321–344. Washington, DC: American Society for Microbiology.

Glansdorff N (1996) Biosynthesis of arginine and polyamines. In: Neidhardt FC, Curtiss R III, Ingraham JL, et al., eds. *Escherichia coli* and *Salmonella typhimurium*: Cellular and Molecular Biology, pp. 408–433. Washington, DC: American Society for Microbiology.

Glass TL, Hylemon PB (1980) Characterization of a pyridine nucleotide-nonspecific glutamate dehydrogenase from *Bacteroides thetaiotaomicron*. J Bacteriol 141:1320–1330.

Gots JS, Benson CE, Jochimsen B, Koduri KR (1977) Microbial models and regulatory elements in the control of purine metabolism. CIBA Found Symp 48:23–41.

Gottschalk G (1986) Biosynthesis of amino acids. In: Bacterial Metabolism, 2nd Ed., pp. 43–55. Berlin: Springer-Verlemyer Press.

Goss TJ, Bender RA (1995) The nitrogen assimilation control protein NAC is a DNA binding transcription activator in *Klebsiella aerogenes*. J Bacteriol 177:3546–3555.

Grandoni AA, Switzer RL, Makaroff CA, Zalkin H (1989) Evidence that the iron sulfate cluster of *Bacillus subtilis* glutamine phosphoribosylpyrophosphate amidotransferase determines the stability of the enzyme to inactivation in vivo. J Biol Chem 264: 6058–6064.

Greene RC (1996) Biosynthesis of methioinine. In: Neidhardt FC, Curtiss R III, Ingraham JL, et al., eds. *Escherichia coli* and *Salmonella typhimurium*: Cellular and Molecular Biology, pp. 542–560. Washington DC: American Society for Microbiology.

Griffith CJ, Carlsson J (1974) Mechanism of ammonia assimilation in *Strepococci*. J Gen Microbiol 82:253–260.

Gu Z, Martindale DW, Lee BH (1992) Isolation and complete sequence of the *purL* gene encoding FGAM synthase II in *Lactobacillus casei*. Gene 119:123–126.

Hamana K, Matsuzaki S (1992) Polyamines as a chemotaxonomic marker in bacterial systematics. Microbiol 18:261–283.

Hamana K, Akiba T, Uchino F, Matsuzaki S (1989) Distribution of spermine in bacilli and lactic acid bacteria. Can J Microbiol 35:450–455.

Hamilton PT, Reeve JN (1985a) Structure of genes and an insertion element in the methane producing archaebacterium *Methanobrevibacter smithii*. Mol Gen Genet 200:47–59.

Hamilton PT, Reeve JN (1985b) Sequence divergence of an archaebacterial gene cloned from a mesophilic and a thermophilic methanogen. J Mol Evol 22:351–360.

Henner D, Yanofsky C (1993) Biosynthesis of aromatic amino acids. In: Sonenshein AL, Hoch JA, Losick R, eds. *Bacillus subtilis* and Other Gram-Positive Bacteria, pp. 269–280. Washington DC: American Society for Microbiology.

He B, Shiau A, Choi, KY, Zalkin H, Smith JM (1990) Genes of the *Escherichia coli purR* negatively controlled by a repressor operator interaction. J Bacteriol 172:4555–4562.

Henikoff S, Haughn GW, Calvo JM, Wallace JC (1988) A large family of bacterial activator proteins. Proc Natl Acad Sci USA 85:6602–6606.

Hespell RB (1984) Influence of ammonia assimilation pathways and survival strategy on rumen microbial growth. In: Herbivore Nutrition in the Subtropics and Tropics, pp. 346–358. Craighall, South Africa: Science Press.

Hill RT, Parker JR, Goodman HJK, Jones DT, Woods DR (1989) Molecular analysis of a novel glutamine synthetase of the anaerobe *Bacteroides fragilis*. J Gen Microbiol 135: 3271–3279.

Janson CA, Kayne PS, Almassy RJ, Grunstein M, Eisenberg D (1986) Sequence of glutamine synthetase from *Salmonella typhimurium* and implications for the protein structure. Gene 46:297–300.

Jayakumar A, Schulamn I, MacNeil D, Barnes EM (1986) Role of the *Eschericia coli glnALG* operon in regulation of ammonium transport. J Bacteriol 166:281–284.

Jayakumar A, Hong J-S, Barnes EM (1987) Feedback inhibition of ammonium (methylammonium) ion transport in *Escherichia coli* by glutamine and glutamine analogs. J Bacteriol 169:553–557.

Jenkinson HF, Buttery PJ, Lewis D (1979) Assimilation of ammonia by *Bacteroides amylophilus* in chemostat cultures. J Gen Microbiol 113:305–313.

Jensen KF (1983) Metabolism of five phosphoribosyl-pyrophosphate (PRPP) in *Escherichia coli* and *Salmonella typhimurium*. In: Munch-Petersen A, ed. Metabolism of Nucleotides, Nucleosidase, and Nucleobases in Microorganisms, pp. 1–25. London: Academic Press.

Jensen RA (1992) An emerging outline of the evolutionary history of aromatic amino acid biosynthesis. In: Mortlock RP, ed. The Evolution of Microbial Function, pp. 205–236. West Coldwell NJ: Tulsa Press.

Joyner AE, Baldwin RL (1966) Enzymatic studies of pure cultures of rumen microorganisms. J Bacteriol 92:1321–1330.

Kamio Y, Nakamura K (1987) Putrescine and cadaverine are constituents of peptidoglycan in *Veillonella alcalescens* and *Veillonella parvula*. J Bacteriol 169:2881–2884.

Kamio Y, Itoh Y, Terawaki Y, Kusano T (1981) Cadaverine is covalently linked to peptidoglycan in *Selenomonas ruminantium*. J Bacteriol 145:122–128.

Kenealy WR, Thompson TE, Schebt KR, Zeikus JG (1982) Ammonia assimilation and biosynthesis of alanine, aspartate, and glutamate in *Methanosarcina barkeri* and *Methanobacterium thermoautotrophicum*. J Bacteriol 150:1357–1365.

Kilstrup M, Meng LM, Neuhard J, Nygaard P (1989) Genetic evidence for a repressor of synthesis of cytosine deaminase and purine biosynthesis enzymes in *Escherichia coli*. J Bacteriol 171:2124–2127.

Kleiner D (1982) Ammonium (methylammonium) transport by *Klebsiella pneumoniae*. Biochim Biophys Acta 688:702–708.

Kleiner D (1985) Bacterial ammonium transport. FEMS Microbiol Rev 32:87–100.

Kneifel H, Stetter, KO, Andreesen JR, Wiegel J, Konig H, Schoberth SM (1986) Distribution of polyamines in representative species of archaebacteria. Syst Appl Microbiol 7: 241–245.

Kredich NM (1987) Biosynthesis of cysteine. In: Neidhardt FC, Ingraham JL, Low KB, Magasanik B, Schaechter M, Umbarger HE, eds. *Escherichia coli* and *Salmonella typhimurium*: Cellular and Molecular Biology, pp. 419–428. Washington, DC: American Society for Microbiology.

Kuczius T, Eilinger T, D'Ari R, Castroph H, Kleiner D (1991) The *gltF* gene of *Klebsiella pneumoniae*: cloning and initial characterization. Mol Gen Genet 229:479–482.

Kustu S, Sei K, Keener J (1986) Nitrogen regulation in enteric bacteria. In: Booth IR,

Higgins CR, eds. Regulation of Gene Expression, pp. 139–154. Cambridge: Cambridge University Press.

Kustu S, North AK, Weiss DS (1991) Prokaryotic transcriptional enhancers and enhancer-binding proteins. Trends Biochem Sci 16:397–402.

Lam H-M, Winkler ME (1990) Metabolic relationships between pyridoxine (vitamin B_6) and serine biosynthesis in *Escherichia coli* K-12. J Bacteriol 172:6518–6528.

Leisinger T (1987) Biosynthesis of proline. In: Neidhardt FC, Ingraham JL, Low, KB, Magasanik B, Schaechter M, Umbarger HE, eds. *Escherichia coli* and *Salmonella typhimurium*: Cellular and Molecular Biology, pp. 345–351. Washington, DC: American Society for Microbiology.

Li X, Singh KV, Weinstock, EN, Murray DE (1994) Organization of the *Enterococcus faecalis* purine biosynthesis gene cluster. Abst of 94th Meeting Am Soc Microbiol Paper, p. 132.

Lin H-P, Reeves HC (1991) Phosphorylation of *Escherichia coli* $NADP^+$ specific glutamate dehydrogenase. Curr Microbiol 22:181–184.

Lozoya E, Sanchez-Pescador R, Covarubbias A, Vichido I, Bolivar F (1980) Tight linkage of genes that encode the two glutamate synthase subunits of *Escherichia coli* K-12. J Bacteriol 144:616–621.

MacFarlane GT, MacFarlane S, Gibson GR (1993) Synthesis and release of proteases by *Bacteroides fragilis*. Curr Microbiol 24:55–59.

Mackie RI, Therion JJ (1984) Influence of mineral interactions on growth efficiency of rumen bacteria. In: Gilchrist FMC, Mackie RI, eds. Herbivore Nutrition in the Subtropics and Tropics, pp. 455–477. Craighall, South Africa: Science Press.

Madonna MJ, Fuchs RL, Brenchley JE (1985) Fine structure analysis of *Salmonella typhimurium* glutamate synthase genes. J Bacteriol 161:353–360.

Magasanik B (1982) Genetic control of nitrogen assimilation in bacteria. Annu Rev Genet 16:135–168.

Magasanik B (1988) Reversible phosphorylation of an enhancer binding protein regulates the transcription of bacterial nitrogen utilization genes. Trends Biochem Sci 13: 475–479.

Makoff AJ, Radford A (1978) Genetics and biochemistry of carbamoyl phosphate biosynthesis and its utilization in the pyrimidine biosynthetic pathway. Microbiol Rev 42: 307–328.

McBride BC, Joe A, Singh U (1990) Cloning of *Bacteroides gingivalis* surface antigens involved with adherence. Arch Oral Biol 55:59–68.

McLaggan D, Naprstek J, Buurman ET, Epstein W (1994) Interdependence of K^+ and glutamate accumulation during osmotic adaptation of *Escherichia coli*. J Biol Chem 261:1911–1917.

Meers JL, Tempest DW, Brown GM (1970) Glutamine (amide): oxoglutarate amino transferase (NADP), an enzyme involved in the synthesis of glutamate by some bacteria. J Gen Microbiol 64:187–194.

Mehrez AZ, Orskov ER, MacDonald I (1977) Rates of ruminal fermentation in relation to ammonia concentration. Br J Nutr 38:437–443.

Merrick MJ (1988) Regulation of nitrogen assimilation by bacteria. In: Cole JK, Ferguson SK, eds. The Nitrogen and Sulfur Cycles, pp. 331–361. Symposium of the Society of General Microbiology. Cambridge: Cambridge University Press.

Merrick MJ, Edwards RA (1995) Nitrogen control in bacteria. Microbiol Rev 59:604–622.

Miller RE, Stadtman ER (1972) Glutamate synthase from *Escherichia coli*: an iron-sulfide flavoprotein. J Biol Chem 247:7407–7419.

Montellico DJ, Hadioetomo RS, Costilow RN (1984) Isoleucine synthesis by *Clostridium sporogenes* from propionate or α-methylbutyrate. J Gen Microbiol 130:309–318.

Montellico DJ, Costilow RN (1982) Interconversion of valine and leucine by *Clostridium sporogenes*. J Bacteriol 152:946–949.

Morrison M, Mackie RI (1996) Nitrogen metabolism by ruminal microorganisms: current understanding and future perspectives. Aust J Agric Res 47:227–246.

Nakano Y, Kimura K (1991) Purification and characterization of a repressor for the *Bacillus cereus glnRA* operon. J Biochem 109:223–228.

Nakano Y, Tanaka E, Kato C, Kimura K, Horikoshi K (1989) The complete nucleotide sequence of the glutamine synthetase gene *(glnA)* of *Bacillus subtilis*. FEMS Microbiol Lett 57:81–86.

Neidhardt FC, Curtiss R III, Ingraham JL, et al., eds. *Escherichia coli* and *Salmonella typhimurium*: Cellular and Molecular Biology. Washington, DC: American Society for Microbiology.

Neuhard J (1983) Utilization of preformed pyrimidine basis and nucleosides. In: Munch-Petersen A, ed. Metabolism of Nucleotides, Nucleosides, and Nucleobases in Micoorganisms, pp. 95–148. London: Academic Press.

Neuhard J, Kelln RA (1996) Biosynthesis and conversion of pyrimidines. In: Neidhardt FC, Curtiss R III, Ingraham JL, et al., eds. *Escherichia coli* and *Salmonella typhimurium*: Cellular and Molecular Biology, pp. 580–599. Washington, DC: American Society for Microbiology.

Neuhard J, Nygaard P (1987) Purines and pyrimidines. In: Neidhardt FC, Ingraham JL, Low KB, Magasanik B , Schaechter M, Umbarger HE, eds. *Escherichia coli* and *Salmonella typhimurium*: Cellular and Molecular Biology, pp. 445–473. Washington, DC: American Society for Microbiology.

Niederman RA, Wolin MJ (1967) Arginine biosynthesis by *Streptococcus bovis*. J Bacteriol 94:100–102.

Nygaard P (1983) Utilization of preformed purine bases and nucleosides. In: Munch-Petersen A, ed. Metabolism in Nucleotides, Nucleosides, and Nucleobases in Microorganisms, pp. 27–93. London: Academic Press.

Nygaard P (1993) Purine and pyrimidine salvage pathways. In: Sonenshein AL, Hoch JA, Losick R, eds. *Bacillus subtilis* and other gram-positive bacteria, pp. 359–380. Washington, DC: American Society of Microbiology.

Oliver G, Gossett G, Sanchez-Pescador E, et al. (1987) Determination of the nucleotide sequence for the glutamate synthase structural genes of *Escherichia coli* K-12. Gene 60:1–11.

Pan FL, Coote JG (1979) Glutamine synthetase and glutamate synthase activities during growth and sporulation of *Bacillus subtilis.* J Gen Microbiol 131:1903–1910.

Patte J-C (1996) Biosynthesis of threonine and lysine. In: Neidhardt FC, Curtiss R III, Ingraham JL, et al., eds. *Escherichia coli* and *Salmonella typhimurium*: Cellular and Molecular Biology, pp. 528–541. Washington, DC: American Society for Microbiology.

Patterson JA, Hespell RB (1985) Glutamine synthetase activity in the ruminal bacterium *Succinivibrio dextrinosolvens*. Appl Environ Microbiol 50:1014–1020.

Paulus H (1993) Biosynthesis of the aspartate family of amino acids. In: Sonenshein AL, Hoch JA, Losick R, eds. *Bacillus subtilis* and Other Gram-Positive Bacteria, pp. 237–268. Washington, DC: American Society of Microbiology.

Pegg AE (1986) Recent advances in the biochemistry of polyamines in eukaryotes. Biochem J 234:249–262.

Phillips AT, Nuss JI, Moosic J, Foshay C (1972) Alternate pathway for isoleucine biosynthesis in *Escherichia coli.* J Bacteriol 109:714–719.

Pittard AJ (1987) Biosynthesis of the aromatic amino acids. In: Neidhardt FC, Ingraham JL, Low KB, Magasanik B, Schaechter M, Umbarger HE, eds. *Escherichia coli* and *Salmonella typhimurium*: Cellular and Molecular Biology, pp. 368–394. Washington, DC: American Society for Microbiology.

Pittard AJ (1996) Biosynthesis of aromatic amino acids. In: Neidhardt FC, Curtiss R III, Ingraham JL, et al., eds. *Escherichia coli* and *Salmonella typhimurium*: Cellular and Molecular Biology, pp. 458–484. Washington, DC: American Society for Microbiology.

Pittman KA, Bryant MP (1964) Peptides and other nitrogen sources for growth of *Bacteroides rumincola.* J Bacteriol 88:401–410.

Ravnikar PD, Somerville RL (1987) Genetic characterization of a highly efficient alternate pathway of serine biosynthesis in *Escherichia coli.* J Bacteriol 169:2611–2617.

Reitzer LJ (1983) Aspartate and asparagine biosynthesis. In: Herrmann KM, Somerville RL, eds. Amino Acids: Biosynthesis and Genetic Regulation, pp. 133–145. Reading, Mass: Addison-Wesley.

Reitzer LJ (1996) Ammonia assimilation and the biosynthesis of glutamine, glutamate, aspartate, asparagine, L-alanine, and D-alanine. In: Neidhardt FC, Curtiss R III, Ingraham JL, et al., eds. *Escherichia coli* and *Salmonella typhimurium*: Cellular and Molecular Biology, pp. 391–407. Washington, DC: American Society for Microbiology.

Reitzer LJ, Magasanik B (1982) Asparagine synthetases of *Klebsiella aerogenes:* properties and regulation of synthesis. J Bacteriol 151:1299–1313.

Reitzer LJ, Magasanik B (1987) Ammonia assimilation and the biosynthesis of glutamine, glutamate, aspartate, asparagine, L-alanine, and D-alanine. In: Neidhardt FC, Ingraham JL, Low KB, Magasanik B, Schaechter M, Umbarger HE, eds. *Escherichia coli* and *Salmonella typhimurium*: Cellular and Molecular Biology, pp. 302–320. Washington, DC: American Society for Microbiology.

Riba L, Becerril B, Servin-Gonsalez L, Valle F, Bolivar F (1988) Identification of a functional promoter for the *Escherichia coli gdhA* gene and its regulation. Gene 71: 233–246.

Robinson, IM, Allison MJ (1969) Isoleucine biosynthesis from 2-methyl-butyric acid by anaerobic bacteria from the rumen. J Bacteriol 97:1220–1226.

Rolfes R, Zalkin H (1988a) Regulation of *Escherichia coli purF*. Mutations that define the promoter, operator, and purine repressor gene. J Biol Chem 263:19649–19652.

Rolfes RJ, Zalkin H (1988b) *Escherichia coli* gene *purR* encoding a repressor protein for purine nucleotide synthesis. Cloning, nucleotide sequence, and interaction with the *purF* operator. J Biol Chem 263:19653–19661.

Rolfes RJ, Zalkin H (1990) Autoregulation of *Escherichia coli purR* requires two control sites downstream of the promoter. J Bacteriol 172:5758–5766.

Russell JB (1983) Fermentation of peptides by *Bacteroides ruminicola* $B_1$4. Appl Environ Microbiol 45:1566–1574.

Russell (1992) Glucose toxicity and the inability of *Bacteroides ruminicola* to regulate glucose transport and utilization. Appl Environ Microbiol 58:2040–2045.

Russell JB, Strobel HJ (1987) Concentration of ammonia across cell membranes of mixed rumen bacteria. J Dairy Sci 70:970–976.

Russell JB, Cook GM (1995) Energetics of bacterial growth: balance of anabolic and catabolic rates. Microbiol Rev 59:48–62.

Sakamoto N, Kotre AM, Savageau MA (1975) Glutamate dehydrogenase from *Escherichia coli:* purification and properties. J Bacteriol 124:775–783.

Sampei GI, Mizobuchi K (1989) The organization of the *purL* gene encoding 5′-phosphoribosylformyl-glycine-amide transferase of *Escherichia coli*. J Biol Chem 264: 21230–21238.

Sauer FD, Erfle JD, Mahadevan S (1975) Amino acid biosynthesis in mixed rumen cultures. Biochem J 150:357–372.

Schaefer DM, Davis CL, Bryant MP (1980) Ammonia saturation constants for predominant species of rumen bacteria. J Dairy Sci 63:1248–1263.

Schendel FJ, Mueller E, Subbe J, Shiau A, Smith JM (1989) Formylglycinamide ribonucleotide synthetase from *Escherichia coli*; cloning, sequencing, overproduction, isolation and characterization. Biochem J 28:2459–2471.

Schwaca A, Bender RA (1993a) The *nac* (nitrogen assimilation control) gene from *Klebsiella aerogenes.* J Bacteriol 175:2107–2115.

Schwaca A, Bender RA (1993b) the product of the *Klebsiella aerogenes nac* (nitrogen assimilation control) gene is sufficient for activation of the *hut* operons and the *gdh* operon. J Bacteriol 175:2116–2124.

Servin-Gonzalez L, Bastarrachea F (1984) Nitrogen regulation of synthesis of the high affinity methylammonium transport system of *Escherichia coli*. J Gen Microbiol 130: 3071–3077.

Schreier HJ (1993) Biosynthesis of glutamine and glutamate and the assimilation of ammonia In: Sonenshein AL, Hoch JA, Losick R, eds. *Bacillus subtilis* and other gram-positive bacteria, pp. 281–298. Washington, DC: American Society for Microbiology.

Schreier HJ, Sonenshein AL (1986) Altered regulation of the *gln*A gene in the glutamine synthetase mutants of *Bacillus subtilis*. J Bacteriol 167:35–43.

Slyter LL (1976) Influence of acidosis on rumen function. J Anim Sci 43:910–929.

Smith CJ, Hespell RB, Bryant MP (1980) Ammonia assimilation and glutamate formation in the anaerobe *Selenomonas ruminantium*. J Bacteriol 141:593–602.

Smith EL, Austin BM Blumenthal KM, Nyc JF (1975) Glutamate dehydrogenases. In: Boyer PD, ed. The Enzymes, Vol. XI, pp. 293–367. New York: Academic Press.

Sonenshein AL, Hoch JA, Losick R (1993) *Bacillus subtilis* and Other Gram-Positive Bacteria. Washington, DC: American Society for Microbiology.

Southern JA, Parker JA, Woods DR (1986) Expression and purification of glutamine synthetase cloned from *Bacteroides fragilis*. J Gen Microbiol 132:2827–2835.

Stauffer GV (1987) Biosynthesis of serine and glycine. In: Neidhardt FC, Ingraham JL, Low KB, Magasanik B, Schaechter M, Umbarger HE, ed. *Escherichia coli* and *Salmonella typhimurium*: Cellular and Molecular Biology, pp. 412–418. Washington, DC: American Society for Microbiology.

Stauffer GV (1996) Biosynthesis of serine, glycine, and one-carbon units. In: Neidhardt FC, Curtiss R III, Ingraham JL, et al., eds. *Escherichia coli* and *Salmonella typhimurium*: Cellular and Molecular Biology, pp. 506–513. Washington, DC: American Society for Microbiology.

Stock JB, Ninfa AJ, Stock AM (1989) Protein phosphorylation and regulation of adaptive responses in bacteria. Microbiol Rev 53:450–490.

Stouthammer AH (1979) The search for correlation between theoretical and experimental growth yields. In: Quayle JR, ed. International Review of Biochemistry. Microbial Biochemistry, Vol. 21, pp. 28–47. Baltimore: University Park Press.

Strobel HJ (1992) Vitamin B_{12} dependent propionate production by the ruminal bacterium *Prevotella ruminicola* 23. Appl Environ Microbiol 58:2331–2333.

Switzer L (1989) Regulation of bacterial glutamine phosphoribosylpyrophosphate amidotransferase. In: Herve G, ed. Allosteric Enzymes, pp. 129–151. Boca Raton, Fla: CRC Press.

Switzer L, Sogin DC (1973) Regulation and mechanism of phosphoribosylpyrophosphate synthetase. J Biol Chem 248:1063–1073.

Switzer RL, Quinn CL (1993) De novo pyrimidine nucleotide synthesis. In: Sonenshein AL, Hoch JA, Losick R, eds. *Bacillus subtilis* and other gram-positive bacteria, pp. 343–358. Washington, DC: American Society for Microbiology.

Tabor CW, Tabor H (1985) Polyamines in microorganisms. Microbiol Rev 49:81–99.

Tabor CW, Tabor H, Hafner EW, Markham GD, Boyle SM (1983) Cloning of *Escherichia coli* genes for the biosynthetic enzymes for polyamines. Methods Enzymol 94:117–124.

Tempest DW, Meers JL, Brown CM (1970) Influence of environment on the content and composition of microbial free amino acid pools. J Microbiol 64:171–185.

Trotta PP, Platzer KEB, Haschmeyer RH, Meister A (1974) Glutamine binding subunit of glutamate synthase and partial reactions catalyzed by this amidotransferase. Proc Natl Acad Sci USA 71:4607–4611.

Umbarger HE (1987) Biosynthesis of the branched-chain amino acids. In: Neidhardt FC, Ingraham JL, Low KB, Magasnik B, Schaechter M, Umbarger HE, eds. *Escherichia coli* and *Salmonella typhimurium*: Cellular and Molecular Biology, pp. 352–367. Washington, DC: American Society for Microbiology.

Umbarger HE (1990) The study of branched chain amino acid biosynthesis—its roots and its fruits. In: Barak Z, Chipman DM, Schloss JV, eds. Biosynthesis of Branched Chain Amino Acids, pp. 1–24. New York: VCH Publishers.

Umbarger HE (1996) Biosynthesis of branched chain amino acids. In: Neidhardt FC, Curtiss R III, Ingraham JL, et al., eds. *Escherichia coli* and *Salmonella typhimurium*: Cellular and Molecular Biology, pp. 442–457. Washington, DC: American Society for Microbiology.

Wallace RJ, Cotta MA (1988) Metabolism of nitrogen-containing compounds. In: Hobson PN, ed. The Rumen Microbial Ecosystem, pp. 217–249. New York: Elsevier.

Watanabe W, Sampel G-I, Aiba A, Mizobuchi K (1989) Identification and sequence analysis of *Escherichia coli purE* and *purK* genes encoding 5′-phosphoribosyl-5-amino-4-imidazole carboxylase for de novo purine biosynthesis. J Bacteriol 171:198–204.

Weisburg WG, Oyaizu H, Woese CR (1985) Natural relationship between *Bacteroides* and flavobacteria. J Bacteriol 164:230.

Weiss DS, Klose KE, Hoover TR, et al. (1992) Prokaryotic transcriptional activators. In: McKnight SL, Yamamoto KR, eds. Transcriptional Regulation, pp. 667–694. Cold Spring Harbor, NY: Cold Spring Harbor Laboratory Press.

Wen Z, Morrison M (1996) The NAD(P)H-dependent glutamate dehydrogenase activities of *Prevotella ruminicola* $B_1$4 can be attributed to one enzyme (Gdh A) and gdhA expression is regulated in response to nitrogen source available for growth. Appl Environ Microbiol. 62:3826–3833.

Westfall NJH, Charon NW, Peterson DE (1983) Multiple pathways for isoleucine biosynthesis in the spirochete *Leptospira*. J Bacteriol 154:846–853.

Winkler ME (1987) Biosynthesis of histidine. In: Neidhardt FC, Ingraham JL, Low KB, Magasnik B, Schaechter M, Umbarger HE, eds. *Escherichia coli* and *Salmonella typhimurium*: Cellular and Molecular Biology, pp. 395–411. Washington, DC: American Society for Microbiology.

Wilson AC, Carlson SS, White TJ (1977) Biochemical evolution. Annu Rev Biochem 46: 573–639.

Winkler ME (1996) Biosynthesis of histidine. In: Neidhardt FC, Curtiss R III, Ingraham JL, et al., eds. *Escherichia coli* and *Salmonella typhimurium*: Cellular and Molecular Biology, pp. 485–505. Washington, DC: American Society for Microbiology.

Woods DR, Santangelo J (1993) Molecular analysis and expression of nitrogen metabolism and electron transport genes of *Clostridium*. In: Woods DR, ed. The Clostridia and Biotechnology, pp. 201–225. Boston: Butterworth-Heinemann.

Yamamoto I, Abe A, Saito H, Ishimoto M (1984) The pathway of ammonia assimilation in *Bacteroides fragilis*. J Gen Appl Microbiol 30:499–508.

Yamamoto I, Saito H, Ishimoto M (1987a) Properties of glutamate dehydrogenase purified from *Bacteroides fragilis*. J Biochem 101:1391–1397.

Yamamoto I, Saito H, Ishimoto M (1987b) Regulation of synthesis and reversible inactivation in vivo of dual coenzyme-specific glutamate dehydrogenase in *Bacteroides fragilis*. J Gen Microbiol 133:2773–2780.

Zalkin H (1993) De novo purine nucleotide synthesis. In: Sonenshein AL, Hoch JA, Losick R, eds. *Bacillus subtilis* and other gram-positive bacteria, pp. 335–341. Washington, DC: American Society for Microbiology.

Zalkin H, Nygaard P (1996) Biosynthesis of purine nucleotides. In: Neidhardt FC, Curtiss R III, Ingraham JL, et al., eds. *Escherichia coli* and *Salmonella typhimurium*: Cellular and Molecular Biology, pp. 561–579. Washington, DC: American Society for Microbiology.

13

Biotransformation of Bile Acids, Cholesterol, and Steroid Hormones

Stephen F. Baron and Phillip B. Hylemon

1. Introduction

1.1. Nomenclature

Steroids are a family of organic compounds that have a five-ring perhydrocyclopentanophenanthrene nucleus. The numbering system and skeletal structures for steroids are shown in Figure 13.1. Steroids vary in the number and location of double bonds; in the type, number, and position of functional groups; and in the stereochemical configuration of substituents below (α) or above (β) the plane of the nucleus. Most steroids have 18β- and 19β-methyl groups bonded to C-10 and C-13. The hydrogen atom in steroids saturated at at C-5 can be β-oriented or α-oriented (allo). Alkyl side chains with various functional groups can be present at C-17, usually in the 17β orientation.

Estrogens (C_{18} or phenolic steroids) lack both a C-17 side chain and the 19β-methyl group and contain a 3-hydroxylated, aromatic A ring. Androgens (C_{19} steroids) lack a C-17 side chain and have a 3-oxo group and Δ^4 double bond. Corticosteroids and pregnane hormones (C_{21} steroids) contain a two-carbon side chain at C-17, a 3-oxo group, and a Δ^4 double bond. The C_{19} and C_{21} steroids are collectively termed neutral steroids. Cholesterol and other C_{27} sterols contain an eight-carbon, branched, aliphatic side chain, a 3β-hydroxy group, and a Δ^5 double bond. Bile acids possess a side chain substituted with a terminal carboxyl group. Although the side chain lengths of bile acids can vary, the most common class has a branched, five-carbon side chain with a carboxyl group at C-24 (C_{24} bile acids). The carboxyl group can be conjugated to glycine or taurine via an

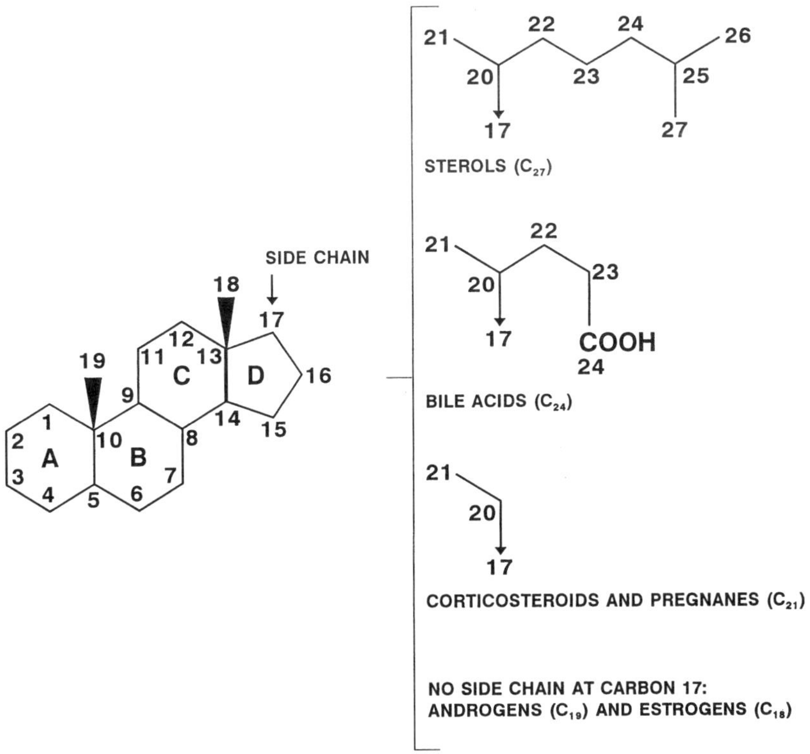

Figure 13.1. Simplified structures of steroids found in the human intestinal tract. Refer to the other figures for specific structures.

amide linkage or to other groups. Under physiological conditions, conjugated and free bile acids exist as their sodium and potassium salts. However, the term "bile acid" will be used throughout most of the text.

1.2. Systematic and Trivial Names

The systematic names of steroids mentioned in the text by their trivial names are listed in Table 13.1. We followed the guidelines proposed by Hofmann et al. (1992) for nomenclature of bile acids.

1.3. Enterohepatic Circulation

The gastrointestinal tract of humans contains about 10^{14} bacteria, represented by 300 to 400 species (Moore et al. 1988, Savage 1977). At least 99% of these

Table 13.1. Trivial and systematic name of steroids

Trivial Name	Systematic Name
Bile Acids	
5β-Cholanoic acid	5β-Cholan-24-oic acid
2-Cholenoic acid	2-Cholen-24-oic acid
3-Cholenoic acid	3-Cholen-24-oic acid
Lithocholic acid	3α-Hydroxy-5β-cholan-24-oic acid
Isolithocholic acid	3β-Hydroxy-5β-cholan-24-oic acid
Sulfolithocholic acid	3α-Sulfo-5β-cholan-24-oic acid
Deoxycholic acid	3α,12α-Dihydroxy-5β-cholan-24-oic acid
Allodeoxycholic acid	3α,12α-Dihydroxy-5α-cholan-24-oic acid
3-Dehydrocholic acid	12α-Hydroxy-3-oxo-5β-cholan-24-oic acid
3-Dehydroallodeoxycholic acid	12α-Hydroxy-3-oxo-5α-cholan-24-oic acid
3-Dehydro-4-deoxycholenoic acid	12α-Hydroxy-3-oxo-cholen-24-oic acid
3-Dehydro-4,6-deoxycholdienoic acid	12α-Hydroxy-3-oxo-6-choldien-24-oic acid
7-oxolithocholic acid	12α-Hydroxy-7-oxo-5β-cholan-24-oic acid
Ursodeoxycholic acid	3α,7β-Dihydroxy-5β-cholan-24-oic acid
Chenodeoxycholic acid	3α,7α-Dihydroxy-5β-cholan-24-oic acid
Cholic acid	3α,7α,12α-Trihydroxy-5β-cholan-24-oic acid
Isocholic acid	3β,7α,12α-Trihydroxy-5β-cholan-24-oic acid
7-Epicholic acid	3α,7β,12α-Trihydroxy-5β-cholan-24-oic acid
12-Epicholic acid	3α,7α,12β-Trihydroxy-5β-cholan-24-oic acid
Cholyl-Coenzyme A	3α,7α,12α-Trihydroxy-5β-cholan-24-oyl-coenzyme A
Cholylglycine	3α,7α,12α-Trihydroxy-5β-cholan-24-oylglycine
Cholyltaurine	3α,7α,12α-Trihydroxy-5β-cholan-24-oyltaurine
3-Dehydrocholic acid	7α,12α-Dihydroxy-3-oxo-5β-cholan-24-oic acid
3-Dehydro-4-cholenoic acid	7α,12α-Dihydroxy-3-oxo-4-cholen-24-oic acid
7-Dehydrocholic acid	3α,12α-Dihydroxy-7-oxo-4-cholen-24-oic acid
12-Dehydrocholic acid	3α,7α-Dihydroxy-12-oxo-4-cholen-24-oic acid
β-Muricholic acid	3α,6β,7β-Trihydroxy-5β-cholan-24-oic acid
β-Hyocholic acid	3α,6α,7β-Trihydroxy-5β-cholan-24-oic acid
C_{27} Sterols	
Cholesterol	5-Cholesten-3β-ol
Coprostanone	5β-Cholestan-3-one
Coprostanol	5β-Cholestan-3β-ol
Cholestanol	5α-Cholestan-3β-ol
Campestrol	24α-Methyl-5-cholesten-3β-ol;24[R]-ergost-5-en-3β-ol
β-Sitosterol	24β-Ethyl-5-cholesten-3β-ol
Stigmasterol	3β-Hydroxy-24-ethyl-5,22-cholestadiene

(continued)

Table 13.1. *(continued).*

Trivial Name	Systematic Name
C_{19} Steroids	
11β-Hydroxy-androstenedione	11β-Hydroxy-4-androstene-3,17-dione
Dehydroepiandrosterone sulfate	3β-Hydroxy-5-androsten-17-one
Testosterone	17β-Hydroxy-4-androsten-3-one
Epitesterone	17α-Hydroxy-4-androsten-3-one
C_{21} Steroids	
Cortisol	11β,17α,21-Trihydroxy-4-pregnene-3,20-dione
3α-hydroxy-5β-tetrahydrodcortisol	3α,11β,17α,21-Tetrahydroxy-5β-pregnane-20-one
3β-hydroxy-5β-tetrahydrodcortisol	3β,11β,17α,21-Tetrahydroxy-5β-pregnane-20-one
5α-Dihydrocortisol	11β,17α,21-Trihydroxy-5β-pregnane-3,20-dione
5β-Dihydrocortisol	11β,17α,21-Trihydroxy-5α-pregnane-3,20-dione
20α-Dihydrocortisol	11β,17α,20α,21-Tetrahydroxy-4-pregnen-3-one
20β-Dihydrocortisol	11β,17α,20β,21-Tetrahydroxy-4-pregnene-3-one
21-Deoxycortisol	11β,17α-Dihydroxy-4-pregnene-3,20-dione
Cortisone	17α,21-Dihydroxy-4-pregnene-3,11,20-trione
11-Desoxycortisol	17α,21-Dihydroxy-4-pregnene-3,20-dione
16αHydroxyprogesterone	16α-Hydroxy-4-pregnene-3,20-dione
Δ^{16}-Progesterone	4,16-Pregnadience-3,20-dione
17α-Progesterone	17α-pregn-4-en-3,20-dione
17α-Hydroxyprogesterone	17α-Hydroxy-4-pregnene-3,20-dione

are obligately anaerobic bacteria; the other 1% are facultatively anaerobic bacteria. The bacterial count increases from 10^4 to 10^8/g dry weight in the jejunum to 10^{11}/g dry weight in the colon (Moore et al. 1988). During passage through the lower intestinal tract, exogenous and endogenous compounds are exposed to the indigenous microbiota, which can biotransform them to various metabolites. These metabolites can often be absorbed by the host and exert beneficial or harmful effects. The steroid components of bile are notable examples of endogenous compounds that are extensively modified by the intestinal microbiota.

Human gallbladder bile typically contains (g/g dry weight) 67% conjugated bile acids, 22% phospholipids, 4.5% protein, 4% cholesterol, 0.3% bilirubin, electrolytes, and small amounts of steroid hormones (Carey and Cahalane 1988). Cholesterol is obtained from dietary sources or synthesized de novo from acetate in the liver and other tissues. Primary bile acids are synthesized from cholesterol in the liver and conjugated to either glycine or taurine by an amide linkage at the C-24 carboxyl. The types and proportions of primary bile acids synthesized in the liver vary widely among different species of animals; the primary bile acids in humans are cholic and chenodeoxycholic acids. Neutral and phenolic steroids are synthesized from cholesterol in the adrenal cortex or gonads, carried by globulins in the bloodstream, and eventually taken up by the liver. They are then conjugated at C-3, C-17, and C-21 hydroxy groups with glucuronic acid or sulfate before secretion in bile (Macdonald et al. 1983a). The lipid components of bile associate to form mixed micelles, which are physiologically active in fat emulsification. The gallbladder concentrates, stores, and secretes bile into the intestinal lumen during digestion.

In the terminal ileum and proximal cecum, bile acids are actively absorbed into the hepatic portal venous system and returned to the liver. This process is termed enterohepatic circulation (EHC), which has been described in detail (Carey and Cahalane 1988; Hofmann 1976, 1977, 1979; Vlahcevic et al. 1990). Greater than 95% of the bile acids secreted in bile are returned to the liver during each of the four to 12 daily cycles of the EHC. Therefore, most of the bile acid pool in humans (about 3 g) is retained within the EHC. The remaining 5% of biliary bile acids which escape active absorption, about 0.2 to 0.6 g daily, pass through the colon and are biotransformed by bacterial enzymes. The biotransformation products are largely excreted in feces, but some are passively absorbed, returned to the liver, conjugated, and secreted in bile. The bile acids lost through fecal excretion are replaced by de novo synthesis in the liver.

Biliary cholesterol is reabsorbed in the intestine and enters the mesenteric lymphatic system as chylomicrons (Cooper 1990, Stange and Dietschy 1985). These are large lipoproteins consisting of free and esterified cholesterol, apolipoproteins, dietary triglycerides, and phospholipids. The chylomicrons are then modified by lipases located on endothelial cell surfaces. Chylomicron remnants are carried in the bloodstream to the liver, where they are taken up by receptor-mediated endocytosis. About 1 g of cholesterol per day escapes intestinal absorption, passes into the colon, and is modified by bacterial enzymes. The biotransformation products of cholesterol are not absorbed by the colon but are excreted in the feces.

After secretion in bile (up to 13 mg/day), conjugated steroid hormones are deconjugated by the lower small intestinal microbiota. Most of the deconjugated steroids are passively absorbed and returned to the liver *via* EHC (Adlercreutz et al. 1979, Bokkenheuser and Winter 1983, Taylor 1971). However, some (about 2 mg/day) enter the colon, are biotransformed by the colonic microbiota, passively absorbed, and transported to the liver for conjugation. Biliary steroids may undergo several cycles of conjugation and deconjugation, but are eventually excreted in urine. Only minor amounts of steroid hormones are excreted in feces, primarily as unconjugated forms.

2. Metabolism of Bile Acids by the Intestinal Microbiota

Normal human feces contain over 20 different secondary bile acid metabolites formed by bacterial modification of the primary bile acids, cholic acids, and chenodeoxycholic acids (Ali et al. 1966, Hayakawa 1973, Midvedt 1974). Known bacterial transformations of primary bile acids (Fig. 13.2) include (1) hydrolysis of glycine- and taurine-conjugated bile acids, yielding free bile acids (reaction 1); (2)stereospecific oxidation of hydroxy bile acids (reactions 2, 4, and 6) and reduction of the resulting oxo-bile acids to hydroxy epimers (reactions 3, 5, and 7); (3) 7-dehydroxylation—i.e., removal of 7α-hydroxy groups (reactions 8 and 9) or 7β-hydroxy groups (reaction 10); (4) monoester and polyester formation (reaction 11); and (5) removal of 3-sulfo groups (not shown).

Secondary bile acids which undergo EHC enter the bile acid pool and can thus influence the physiology of the host. Secondary bile acids are generally more hydrophobic and less effective in solubilizing lipids than primary bile acids (Carey 1985, Hofmann and Roda, 1984). Certain secondary bile acids are also cytotoxic (Vlahcevic et al. 1990), decrease de novo synthesis of cholesterol and primary bile acids (Heuman et al. 1988a,b), and possibly promote colon carcinogenesis (Cohen et al. 1980, Mower et al. 1979).

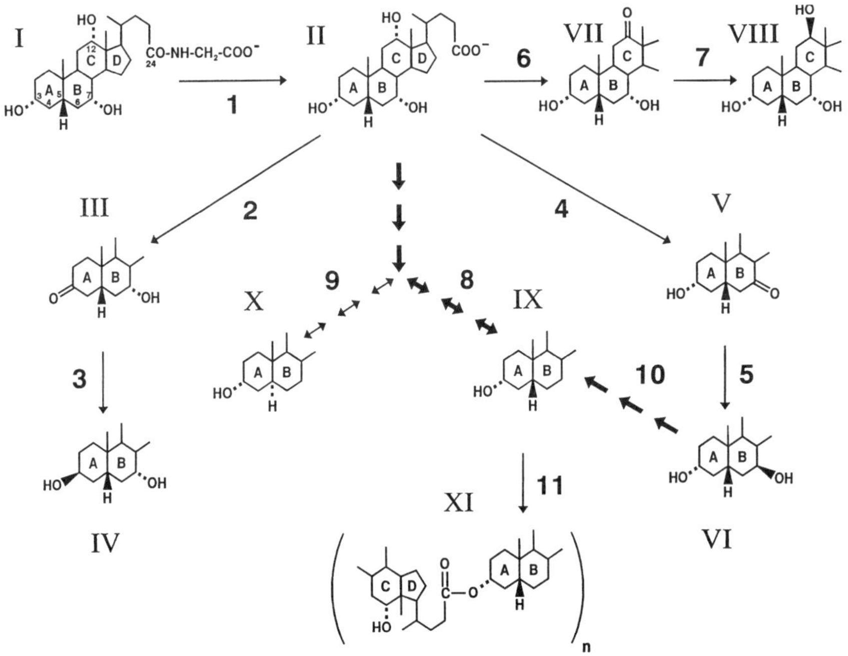

Figure 13.2. Bacterial transformations of cholic acid. Large arabic numerals refer to reactions cited in section 2. Structures: I, cholylglycine; II, cholic acid; III, 3-dehydrocholic acid; IV, isocholic acid; V, 7-dehydrocholic acid; VI, 7-epicholic acid; VII, 12-dehydrocholic acid; VIII, 12-epicholic acid; IX, deoxycholic acid; X, allodeoxycholic acid; XI, polyester of deoxycholic acid. 12-Dehydroxy analogs of intermediates I, III-VI, IX, and X can also be formed from chenodeoxycholic acid.

2.1. Deconjugation

The hydrolysis of the amide bond of glycine- and taurine-conjugated bile acids (Fig. 13.2, reaction 1) is catalyzed by conjugated bile acid hydrolase (CBH). This reaction is so rapid and complete that conjugated bile acids are usually not detected in intestinal contents (Macdonald et al. 1983a). Unconjugated bile acids are considerably less soluble than conjugated bile acids, particularly at low pH, and are less effective detergents for fat solubilization. Therefore, they must be reconjugated by the liver after EHC. CBH activity depresses growth in poultry (Feighner and Dashkevicz 1988) and contributes to the small-bowel syndrome in humans (Gorbach and Tabaqchali 1969, Northfield et al. 1973). CBH activity has been detected in intestinal strains of *Bacteroides*, *Bifidobacterium*, *Fusobacterium*, *Clostridium*, *Lactobacillus*, *Peptostreptococcus*, *and Streptococcus* (Hylemon and Glass 1983, Kobashi et al. 1978, Macdonald et al. 1983a, Masuda 1981, Midvedt and Norman 1967). The contribution of these genera to overall CBH activity depends on the host. For example, studies with germ-free and conventional mice indicate that lactobacilli contribute at least 74% of the total intestinal CBH activity (Tannock et al. 1989).

CBHs have been purified and characterized from *Bacteroides fragilis*, *Bacteroides vulgatus*, *Clostridium perfringens*, and *Lactobacillus* sp. strain 100–100 (Table 13.2). The optimum pH range for the activity of all CBHs is acidic, ranging from 4.2 to 6.4. Most of these enzymes are composed of one type of subunit.

Table 13.2. Characteristics of conjugated bile acid hydrolases purified from intestinal bacteria

	M_r (10^3)			Apparent K_m (mM)[a]							
Organism	Subunit	Native	Subunit Composition	CG	DCG	CDCG	CT	DCT	CDCT	pH Optimum	Reference[b]
Bacteroides fragilis	32.5	250	α_8?	0.35	0.20	0.26	0.45	0.17	0.29	4.2–4.5	1
Bacteroides vulgatus	36	140	α_4	—	—	—	+	+	+	5.6–6.4	2
Clostridium perfringens	ND[c]	ND	ND	3.6	1.2	14	37	3.5	3.0	5.6–5.8	3
Lactobacillus sp. strain 100–100	42 (α) 38 (β)										5,6
isozyme A	42		α_3	+	+	ND	0.76	+	+	4.2–4.5	
isozyme B	42, 38	115	$\alpha_2\beta_1$	+	+	ND	0.95	+	+	"	
isozyme C	42, 38	105	$\alpha_1\beta_2$	ND	ND	ND	0.45	ND	ND	"	
isozyme D	38	95 80	β_3	ND	ND	ND	0.37	ND	ND	"	
Lactobacillus plantarum 80	37.1[d]	ND	ND	+	+	+	TR	TR	TR	4.7–5.5	7

[a] CG, cholylglycine; DCG, deoxylcholylglycine; CDCG, chenodeoxycholylglycine; CT, cholyltaurine; DCT, deoxycholyltaurine; CDCT, chenodeoxycholyltaurine; +, activity detected; TR, trace of activity; —, no activity detected.

[b] References: 1. Stellway and Hylemon (1976); 2. Kawamoto et al. (1989); 3. Nair et al. (1967); 4. Gopal-Srivastava and Hylemon (1988); 5. Lundeen and Savage (1990); 6. Lundeen and Savage (1992b); 7. Christiaens et al. (1992).

[c] ND, not determined.

[d] Calculated from the deduced amino acid sequence of the conjugated bile acid hydrolase gene (GenBank accession number S51638).

However, *Lactobacillus* sp. strain 100–100 produces four CBH isozymes composed of different trimeric combinations of immunologically distinct α and β subunits. These isozymes have similar kinetic properties, although the V_{max} of isozyme D with cholyltaurine is 10-fold lower than that of isozymes A, B, and C. Most CBHs hydrolyze both glycine- and taurine-conjugated dihydroxy and trihydroxy bile acids to some extent. However, the CBH from *B. vulgatus* hydrolyzes only taurine-conjugated bile acids. The CBH gene from the silage isolate, *Lactobacillus plantarum* 80, has been cloned and expressed in *Escherichia coli* MC1061 (Christiaens et al. 1992) using a direct plate screening technique (Dashkevicz and Feighner 1989). The CBH expressed in *E. coli* MC1061 extracts preferentially hydrolyzes glycine-conjugated bile acids. The deduced amino acid sequence of the CBH gene shares 52% similarity with that of penicillin V amidase from *Bacillus sphaericus*. CBH is synthesized constitutively in most intestinal bacteria. However, CBH activity in *B. fragilis* (Hylemon and Stellwag 1976) and *Lactobacillus* sp. strain 100–100 (Lundeen and Savage 1990) increases 300- and 70-fold, respectively, when the cells enter stationary phase. Furthermore, Lundeen and Savage (1990, 1992a) detected an extracellular factor in *Lactobacillus* sp. strain 100–100 which is induced by conjugated bile acids and stimulates intracellular CBH activity. The factor has an apparent M_r of 12,000 to 25,000; is stable to air, acid, heat, and pronase treatment; and can be partially extracted into organic solvents.

2.2. Hydroxy Group Oxidation and Epimerization

Bile acid hydroxysteroid dehydrogenases (HSDH) catalyze the reversible oxidation of hydroxy groups to oxo groups at various positions in the bile acid molecule. These enzymes are NAD(P)-dependent and are stereospecific for the α or β orientation of the hydroxy group. HSDHs are widely distributed among intestinal bacteria, including enzymes specific for α- or β-hydroxy groups at C-3, C-6, C-7, and C-12 (Hylemon and Glass 1983).

Through the concerted action of α- and β-HSDHs, the intestinal microbiota can epimerize hydroxy groups of bile acids by stereospecific oxidation (Fig. 13.2, reactions 2, 4, and 6), followed by stereospecific reduction of the resulting oxo group (reactions 3, 5, and 7). Epimerization can be performed by a single species containing both α- and β-HSDHs (intraspecies) or by cooperation between one species having an α-HSDH and another having a β-HSDH (interspecies). Epimerization of α-hydroxy groups to β-hydroxy groups is more common than the opposite conversion.

3α- AND 3β-HSDH

The 3α-hydroxy group of primary bile acids can be epimerized in vivo (Fig. 13.2, reactions 2 and 3), since the feces of humans and laboratory animals contain low amounts of the 3β epimers (Hylemon and Glass 1983). Intraspecies 3-hydroxy epimerization has been demonstrated in fecal isolates of *C. perfringens* (Hirano et al. 1981b, Macdonald et al. 1983b), *Eubacterium lentum* (Hirano and Masuda 1981b), and *Peptostreptococcus productus* (Edenharder et al. 1989a), suggesting that these bacteria have both 3α- and 3β-HSDHs. However, to date both enzymes have been detected only in *P. productus* (Edenharder et al. 1989a). Anaerobic conditions favor epimerization of 3α-hydroxy bile acids in growing cultures or cell suspensions of *C. perfringens* or *E. lentum*, while aeration favors oxidation of these compounds to 3-oxo bile acids (Hirano et al. 1981b, Macdonald et al. 1983b).

3α- and 3β-HSDHs in various stages of purity have been characterized in *E. lentum*, *P. productus*, *Ruminococcus* sp. PO1–3, and three species of *Clostridium* (Table 13.3). These 3-HSDHs differ in pyridine nucleotide cofactor specificity, molecular weight, and substrate range. However, most have alkaline pH optima and exhibit lower K_ms for dihydroxy or dioxo bile acids than trisubstituted forms. The 3β-HSDHs from *Clostridium innocuum* and *Clostridium* sp. 25.11.c reduce 3-oxobile acids but do not oxidize 3β-hydroxy bile acids. The 3-HSDHs from *C. perfringens*, *E. lentum*, and *Ruminococcus* recognize certain 3-hydroxy- and 3-oxo-C_{19} steroids as well as bile acid substrates. All of the 3-HSDHs in Table 13.3 are synthesized constitutively. However, the synthesis of all but the

Table 13.3. Characteristics of bile acid 3α- and 3β-hydroxysteroid dehydrogenases (HSDH) from intestinal bacteria

HSDH	Organism	Purity[a]	Cofactor	Native M_r (10^3)	pH Optimum	Substrates Used[b]	Apparent K_m (mM)	Ref.[c]
3α	*Clostridium perfringens*	CE	NADP	ND[d]	11.3	3α-HBA	0.020–0.050	1
						3α-HAN	< 0.001	
3α	*Eubacterium lentum*	CE	NAD	ND	11.3	3α-HBA (CON)	0.008–0.020	2
						3α-HAN	ND[d]	
3α	*Peptostreptococcus productus*	CE	NAD	95	8.5	3α-HBA	ND	3
						3-OBA		
3β	*Peptostreptococcus productus*	CE	NAD	132	9.5	3-OBA	ND	3
3β	*Clostridium innocuum*	PP	NADH	56	10.0–10.2	3-OBA (F)	0.024–0.146	4
						3-OBA (CON)	0.0334	
3β	*Clostridium* sp. 25.11.c	CE	NADPH	104	7.3	3-OBA (F)	0.015–0.134	5
						3-OBA (CON)	0.115	
3β	*Ruminococcus* sp. PO1–3	HP	NADP	90	ND	3β-HBA (F)	0.003–0.015	6
						3-OBA (F)	0.030–0.050	
						3β-HAN	ND	
						3-OAN	ND	

[a] CE, cell extract; PP, partially purified; HP, highly purified.

[b] HBA, hydroxybile acids; OBA, oxobile acids; F, free bile acids; CON, glycine or taurine conjugates; HAN, hydroxyandrostanes; OAN, oxoandrostanes. Reduced NAD(P) was used with 3-oxo substrates.

[c] References: 1. Macdonald et al. (1976); 2. Macdonald et al. (1977); 3. Edenharder et al. (1989a); 4. Edenharder and Pfützner (1989); 5. Edenharder et al. (1989b); 6. Akao et al. (1986).

[d] ND, not determined.

3α-HSDH of *C. perfringens* is repressed by addition of their bile acid substrates to the growth medium.

6α- AND 6β-HSDH

The primary bile acid β-muricholic acid (3α,6β,7β-trihydroxy-5β-cholanoic acid) is found in the feces of germ-free rats, mice, and pigs (Madsen et al. 1976, Wostmann et al. 1979), but its 6α-epimer, β-hyocholic acid (3α,6α,7β-trihydroxy-5β-cholanoic acid) is present in the feces of conventional animals. This disparity suggests that the intestinal microbiota epimerize the 6β-hydroxy group of muricholic acid.

There is evidence for both intraspecies and interspecies epimerization of the 6β-hydroxy group. A *Clostridium* sp. (R6 × 76) isolated from rat feces converted β-muricholic acid to β-hyocholic acid (Sacquet et al. 1979). Eyssen et al. (1983) isolated strains of *E. lentum* and *Fusobacterium* sp. from rats. The *E. lentum* isolate oxidized the 6β-hydroxy group of β-muricholic acid to the 6-oxo derivative, while the *Fusobacterium* strains reduced the 6-oxo group to the 6α-hydroxy derivative. When cocultured, these two species epimerized β-muricholic acid to β-hyocholic acid. No 6α- or 6β-HSDHs have yet been purified or characterized.

7α- AND 7β-HSDH

The 7β-epimers of cholic and chenodeoxycholic acids (7-epicholic and ursodeoxycholic acids, respectively) are often detected in human bile and feces, indicating that epimerization of the 7α-hydroxy group occurs in vivo (Fig. 13.2, reactions 4 and 5). The intestinal microbiota can also epimerize 7β-hydroxy bile acids to 7α-hydroxy bile acids. However, the extent of 7-epimerization in vivo is difficult to quantitate, since it competes with the irreversible 7-dehydroxylation of bile acids (Hylemon and Glass 1983).

The epimerization of 7α-hydroxy bile acids has been demonstrated in vitro with mixed fecal cultures (Fedorowski et al. 1979, Higashi et al. 1979, Hirano et al. 1981a) and by coculturing 7α-HSDH-producing fecal isolates with 7β-HSDH-producing isolates (Hirano and Masuda 1981a, 1982b; Macdonald et al. 1982). Bacteria capable of intraspecies 7-epimerization include *Clostridium absonum* (Macdonald and Roach 1981), *Clostridium limosum* (Sutherland and Williams 1985), and lecithinase-lipase-negative strains of *Clostridium* (Edenharder and Knaflic 1981).

7α-HSDHs from members of the genera *Bacteroides*, *Clostridium*, *Escherichia*, and *Eubacterium* have been characterized at various levels of purity (Table

13.4). Although these enzymes are diverse in pyridine nucleotide requirement, molecular weight, and bile acid inducibility, all have particularly alkaline pH optima and use free and conjugated bile acid substrates efficiently. *B. fragilis* and *C. absonum* have distinct NAD- and NADP-dependent 7α-HSDHs. These enzymes in *B. fragilis* also differ in thermal stability, requirement for divalent metal cations, and gel filtration elution profiles (Hylemon and Sherrod 1975). To date, two 7α-HSDHs have been purified and characterized: an NAD-linked enzyme from *E. coli* HB101 (Yoshimoto et al. 1991), and an NADP-linked enzyme from *Eubacterium* sp. VPI 12708 (Franklund et al. 1990). Their associated genes have also been cloned and sequenced (Baron et al. 1991, Yoshimoto et al. 1991). The two enzymes are both homotetramers of comparable M_r, are synthesized constitutively, and share 36% amino acid sequence identity. However, the NAD-linked enzyme has 100-fold higher K_ms for its bile acid substrates than the NADP-linked enzyme. Both 7α-HSDH's share significant amino acid sequence homology with short chain, non-metal-containing alcohol and polyol dehydrogenases, which all contain a putative pyridine nucleotide binding domain. Sequence alignments of the two 7α-HSDHs with these enzymes reveals seven perfectly conserved amino acid residues (Baron et al. 1991). Two of these, a tyrosine and a lysine located in the middle of the amino acid sequence, are essential for catalytic activity as demonstrated by site-directed mutagenesis of *Drosophila* alcohol dehydrogenase (Chen et al. 1993).

7β-HSDHs have been characterized in *C. absonum*, *Clostridium* sp. 25.11.c

Table 13.4. Characteristics of bile acid 7α-hydroxysteroid dehydrogenases from intestinal bacteria

				M_r (10^3)		pH	Apparent K_m (mM)[c]				
Organism	Purity[a]	Inducibility[b]	Cofactor	Native	Subunit	Optimum	C	CG	CDC	CDCG	Ref.[d]
Bacteroides fragilis strains	CE	G	NAD	ND[e]	ND	9.5–10	0.34	0.33	0.10	0.10	1,2
	CE	G	NADP	ND	ND	7.0–9.0	ND	ND	ND	ND	1,2
Bacteroides thetaiotaomicron	PP	G?	NAD	320	ND	8.5–9	0.22	0.32	0.048	0.083	3
Clostridium absonum	PP	I	NAD	ND	ND	9.5–11.5	0.25	ND	ND	ND	4,5
	PP	I	NADP	ND	ND	9.5–11.5	0.090	ND	0.0065	ND	4,5
Clostridium perfringens	CE	C	NADP	ND	ND	ND	ND	ND	ND	ND	6
Clostridium sp. 25.11.c	CE	S	NADP	82	ND	8.5–8.7	ND	ND	ND	ND	7
Escherichia coli	HP	C	NAD	120	28	8.5	1.2	1.25	0.43	ND	8
Eubacterium sp. VPI 12708	HP	C	NADP	124	32	8.5–10.5	0.011	0.0083	0.0056	0.0055	9

[a] CE, cell extract; PP, partially purified; HP, highly purified.

[b] G, synthesized during transition from log to stationary phse; I, bile acid inducible; S, constitutive but synthesis stimulated by bile acids; C, constitutive.

[c] C, cholic acid; CG, cholylglycine; CDC, chenodeoxycholic acid; CDCG, chenodeoxycholylglycine.

[d] References: 1. Macdonald et al. (1975); 2. Hylemon and Sherrod (1975); 3. Sherrod and Hylemon (1977); 4. Macdonald et al. (1983c); 5. Macdonald and Roach (1981); 6. Macdonald et al. (1976); 7. Edenharder et al. (1989b); 8. Yoshimoto et al. (1991); 9. Franklund et al. (1990).

[e] ND, not determined.

Table 13.5. Characteristics of bile acid 7β-hydroxysteroid dehydrogenases from intestinal bacteria

Organism	Purity[a]	Indicibility[b]	Cofactor	M_r (10^3) Native	Subunit	pH Optimum	Apparent K_m (mM)[c] UD	UDG	7OL	Ref.[d]
Clostridium absonum	CE, PP	I	NADP	200	ND[e]	9.0–10.0	0.072	ND	ND	1,2
Clostridium sp. 25.11.c	CE	S	NADP	115	ND	8.5–8.7	ND	ND	ND	3
Eubacterium aureofaciens	CE	C	NADP	45	ND	10.5	0.108	0.909	ND	4
Peptostreptococcus productus	CE	C	NADP	53	ND	9.8	0.022	0.238	ND	4
	CE	S	NADP	82	ND	10.0	TR[e]	ND	0.0550	5
Ruminococcus sp. PO1–3	HP	ND	NADP	60	30	8.0–9.0	0.005	ND	0.0085	6

[a] CE, cell extract; PP, partially purified; HP, highly purified.
[b] I, bile acid inducible; S, constitutive but synthesis stimulated by bile acids; C, constitutive.
[c] UD, ursodeoxycholic acid; UDG, Ursodeoxycholyl glycine; 7OL, 7-oxolithocholic acid (with NADPH as electron donor).
[d] References: 1. Macdonald et al. (1983c); 2. Macdonald and Roach (1981); 3. Edenharder et al. (1989b); 4. Hirano and Masuda (1982a); 5. Edenharder et al. (1989a); 6. Akao et al. (1987).
[e] ND, not determined.
[f] TR, trace of activity.

(a lecithinase-lipase-negative strain), *Eubacterium aureofaciens*, *P. productus*, and *Ruminococcus* sp. PO1–3 (Table 13.5). All of these are NADP-dependent and have alkaline pH optima but vary widely in native M_r and bile acid inducibility. Only one, the 7β-HSDH of *Ruminococcus* sp. PO1–3, has been purified to homogeneity. It is a homodimer of a 30,000-M_r subunit. Activity of the enzyme is inhibited by sulfhydryl reagents and stimulated by dithiothreitol, suggesting the presence of a reactive sulfhydryl group in the active site.

12α- AND 12β-HSDH

Low amounts of 12-oxo and 12β-hydroxy bile acids occur in human feces (Ali et al. 1966, Eneroth et al. 1966a,b), suggesting that the intestinal microbiota carry out the 12-epimerization of primary bile acids (Fig. 13.2, reactions 6 and 7). Edenharder and Schneider (1985) demonstrated 12-epimerization of deoxycholic acid by mixed human fecal cultures and by cocultures of *Clostridium paraputrificum* and *E. lentum*, which produce 12α- and 12β-HSDH, respectively. To date, no intestinal isolate has been found to possess both 12α- and 12β-HSDH.

12α-HSDHs have been characterized in *E. lentum* and four species of *Clostridium* (Table 13.6). Most of these are NADP-dependent, have native M_rs near 100,000, and have alkaline pH optima. The 12α-HSDHs can all use cholic and deoxycholic acids as substrates but have lower K_m's for the latter. Most of these enzymes prefer unconjugated to conjugated substrates. The NADP-dependent 12α-HSDHs of *Clostridium* group P, strain C 48–50 (Braun et al. 1991) and *Clostridium leptum* (P. de Prada, Ph.D. dissertation, Virginia Commonwealth

Table 13.6. Characteristics of bile acid 12α- and 12β-hydroxysteroid dehydrogenases (HSDH) from intestinal bacteria

				M_r (10^3)			Apparent K_m (mM)[b]						
HSDH	Organism	Purity[a]	Cofactor	Native	Subunit	pH Optimum	C	DC	CG	DCG	12β-HBA	12-OBA (F,CON)	Ref.[c]
12α	*Clostridium* group P, strain C 48–50	HP	NADP	104	26	8.5–9.5	0.072	0.045	ND[d]	ND			1
12α	*Clostridium leptum*	HP	NADP	110	27	8.5	+	0.067	+	+			2
12α	*Clostridium perfringens*	CE	NAD	ND	ND	10.5	+	0.80	+	1.0			3
12α	*Eubacterium lentum*	PP	NAD	125	ND	8.0–10.5	0.059	0.028	0.25	0.17			4
12β	*Clostridium paraputrificum*	CE	NADP	126	ND	7.8, 10.0[e]					+	0.09–0.27	5

[a] HP, highly purified; PP, partially purified; CE, cell extract.

[b] C, cholic acid; CG, cholylglycine; DC, deoxycholic acid; DCG, deoxycholylglycine; 12β-HBA, 12β-hydroxybile acids; 12-oxoBA, 12-oxobile acids; F and CON, free and conjugated bile acids, respectively; +, K_m not determined, but activity detected.

[c] References: 1. Braun et al. (1991); 2. P. de Prada (1983, doctoral dissertation, Virginia Commonwealth University, Richmond; 3. Macdonald et al. (1976); 4. Macdonald et al. (1979) 5. Edenharder and Pfützner (1988).

[d] ND, not determined.

[e] for 12β-HBA oxidation with NADP and 12-OBA reduction with NADPH, respectively.

University, Richmond, Virginia, 1993) have been purified to homogeneity. Both are homotetramers of a ca. 27,000-M_r subunit. Product inhibition studies indicate that 12α-hydroxy oxidation by these enzymes proceeds by an ordered bi bi mechanism, with NADP binding first and leaving last with respect to the bile acid substrate. The N-terminal amino acid sequence of the *C. leptum* enzyme shows homology to those of short chain, non-metal-containing alcohol and polyol dehydrogenases.

NADP-dependent 12β-HSDH has been detected in *Clostridium tertium* and *Clostridium difficile* (Edenharder and Schneider 1985) and partially characterized in *C. paraputrificum* (Edenharder and Pfützner 1988) (Table 13.6). The *C. paraputrificum* enzyme oxidized 12β-hydroxy bile acids and reduced 12-oxo bile acids, although the oxidative activity was 75% lower than the reductive activity. Other properties of the enzyme are similar to those of the 12α-HSDHs. The synthesis of all 12α- and 12β-HSDHs is constitutive, although addition of 12-oxo or 12β-hydroxy bile acids to cultures of *C. paraputrificum* stimulates synthesis of its 12β-HSDH about threefold (Edenharder and Pfützner 1988).

2.3. 7-Dehydroxylation

The most physiologically significant biotransformation of bile acids in humans is the 7α-dehydroxylation of the primary bile acids: cholic and chenodeoxycholic acids, generating deoxycholic and lithocholic acids, respectively (Fig. 13.1, reaction 8). Deoxycholic acid is passively absorbed from the colon and consequently comprises up to 25% of the circulating bile acid pool in humans (Sjövall 1960) and up to 95% in rabbits (Lindstedt and Sjövall 1957). The deoxycholic acid content of human bile remains relatively constant since the liver cannot 7α-hydroxylate this compound to regenerate cholic acid. However, this reaction is thought to occur in rats (Norman and Sjövall 1958). Lithocholic acid is so insoluble that most adsorbs to and is excreted in feces; only minor amounts undergo EHC.

The number of 7α-dehydroxylating bacteria in human feces is relatively low, ranging from 10^3 to 10^5/g wet weight of feces (Ferrari et al. 1977, Stellwag and Hylemon 1979). Most of the intestinal bacteria which carry out 7α-dehydroxylation are members of the genera *Clostridium* (Hayakawa and Hattori 1970, Hirano et al. 1981b,c, Stellwag and Hylemon 1979) and *Eubacterium* (Gustafsson et al. 1966, Hirano et al. 1981c, White et al. 1980). These bacteria use unconjugated but not C-24 conjugated bile acids as substrates for 7α-dehydroxylation, yet lack CBH (Batta et al. 1990, Stellwag and Hylemon 1979, White et al. 1983). Therefore, conjugated bile acids must first be deconjugated by nondehydroxylating bacteria before 7α-dehydroxylation. Moreover, bile acids conjugated at C-24 with N-methylated amino acids (Schmassmann et al. 1990a,b) or with sulfate (Kihira et al. 1991) resist both deconjugation and 7-dehydroxylation by the gut microbiota in rodents. Steric hindrance of the 7-hydroxy group also inhibits 7-dehydroxylation, as shown with 7-hydroxy bile acid analogs containing 7α- or 7β-methyl groups (Hylemon et al. 1984, Kuroki et al. 1987).

The human intestinal isolate, *Eubacterium* sp. VPI 12708, has a bile acid-inducible enzyme system which dehydroxylates 7α-hydroxy bile acids (White et al. 1980, 1981, 1983). A mechanism for bile acid 7α-dehydroxylation in this organism has been proposed, based on the isolation of bile acid intermediates and conversion of these intermediates to dehydroxylated products (Björkhem et al. 1989, Coleman et al. 1987a, Hylemon et al. 1991, Mallonee et al. 1992). This mechanism is illustrated in Figure 13.3, with cholic acid as the dehydroxylation substrate; however, this pathway might also function in the 7α-dehydroxylation of chenodeoxycholic acid to lithocholic acid. Cholic acid (I) is first actively transported into the bacterial cell by a specific carrier protein. Inside the cell, cholic acid is conjugated at the C-24 carboxyl either to ADP-3′-phosphate (Coleman et al. 1987a) or to coenzyme A (CoA) by an ATP-dependent bile acid CoA ligase (Mallonee et al. 1992). The cholic acid conjugate (II) is oxidized twice and dehydrated, yielding 3-dehydro-4,6-deoxycholdienoic acid (V). This intermediate is reduced to 3-dehydro-4-deoxycholenoic acid (VI), which is reduced twice to yield deoxycholic acid (VIII). Deoxycholic acid is transported out of the cell by a specific carrier. Most of the steps in the pathway appear to be pyridine nucleotide-dependent. However, reduction of 3-dehydro-6-deoxycholenoic acid, an analog of the 3-dehydro-4,6-deoxycholdienoic acid intermediate (V), is stimu-

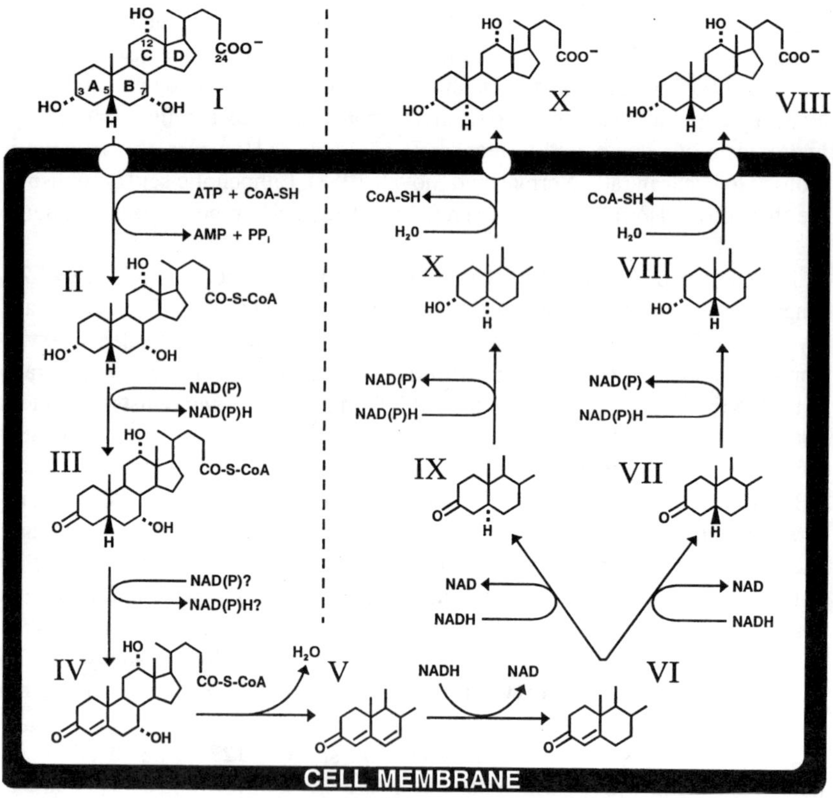

Figure 13.3. Proposed pathway for the 7α-dehydroxylation of cholic acid by *Eubacterium* sp. VPI 12708. Intermediates: I, cholic acid; II, cholyl-coenzyme A; III, 3-dehydrocholyl-coenzyme A; IV, 3-dehydro-4-cholenoyl-coenzyme A; V, 3-dehydro-4,6-deoxycholdienoic acid; VI, 3-dehydro-4-deoxycholenoic acid; VII, 3-dehydrodeoxycholic acid; VIII, deoxycholic acid; IX, 3-dehydroallodeoxycholic acid; X, allodeoxycholic acid. CoA-SH, coenzyme A; PP$_i$, pyrophosphate. Open circles represent bile acid transport proteins. Dashed line separates the oxidative reactions from the reductive reactions.

lated by addition of reduced flavins (White et al. 1983). Although the ADP conjugate of the 3-dehydro-4-cholenoic acid intermediate (IV) has been detected (Coleman et al. 1987a), it is unclear whether the subsequent intermediates remain conjugated at C-24.

Cell extracts of *Eubacterium* sp. VPI 12708 produce appreciable amounts of 3-dehydroallodeoxycholic and allodeoxycholic acids (Fig. 13.3, X and XI) in addition to 3-dehydrodeoxycholic and deoxycholic acids (VII and VIII) during the 7α-dehydroxylation of cholic acid (Hylemon et al. 1991). These results suggest that reduction of the 3-dehydro-4-deoxycholdienoic intermediate (VI) generates both the 5β and 5α configurations of the C-5 hydrogen. Allobile acids are the major bile acid class in lower animals and certain fishes (Elliott 1971), and small amounts occur in humans and other mammals (Eneroth et al. 1966b, Tammar 1966).

Bile acid 7α-dehydroxylation activity in *Eubacterium* sp. VPI 12708 is induced by culturing the organism in the presence of unconjugated C_{24} bile acids which have a 7α-hydroxy group (White et al. 1980). This treatment also induces the biosynthesis of several new polypeptides presumably involved in 7α-dehydroxylation (Paone and Hylemon 1984, White et al. 1981). Most of the genes encoding these polypeptides are clustered on a large, cholic acid-inducible operon (>10 kb) containing nine or more open reading frames (*baiA2-baiI*) (Franklund et al. 1993, Mallonee et al. 1990, White et al. 1988a,b) (Fig. 13.4). This operon encodes gene products with M_rs of 9,000, 19,500, 27,000, 47,500, 50,000, 58,000, 59,000, and 72,000. Two additional cholic acid-inducible genes, *baiA1* and *baiA3*, are located on separate monocistronic transcripts about 1 kb long (Coleman et al. 1987b, 1988; Gopal-Srivastava et al. 1990; White 1988b) (Fig. 13.4). These genes encode identical 27,000-M_r polypeptides which share 92% amino acid sequence identity with the *baiA2* gene product. Enzymatic functions have tentatively been assigned to several of the *bai* gene products based on biochemical activities of the gene products expressed in *E. coli* or on amino acid sequence homology to

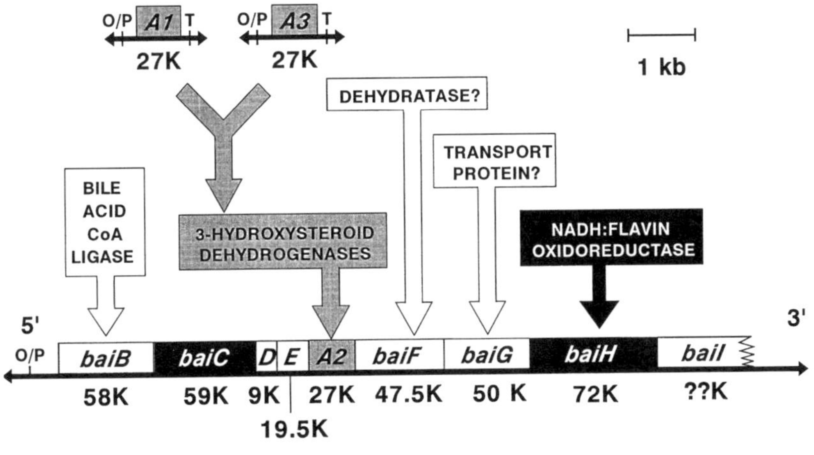

Figure 13.4. Bile acid inducible (*bai*) genes in *Eubacterium* sp. VPI 12708. The predicted M_rs in thousands (K) of the gene products are shown below each gene, and their proposed functions in bile acid 7α-dehydroxylation above the genes. Genes that share homology are shaded or blackened. Symbols: O/P, operator/promoter region; T, stem-loop terminator structure; *A1*, *A2*, *A3*, *D*, and *E*–genes *baiA1*, *baiA2*, *baiA3*, *baiD*, and *baiE*, respectively.

proteins with known functions. The 58,000-M_r protein has ATP-dependent bile acid-CoA ligase activity (Mallonee et al. 1992). The three 27,000-M_r proteins share homology with short-chain NAD(P)-dependent alcohol/polyol dehydrogenases and are thus probably 3α-HSDHs which catalyze 3α-hydroxy oxidation or 3-oxo reduction of intermediates II, VIII, and X (Fig. 13.3) (Coleman et al. 1988). The 50,000-M_r protein has homology to bacterial tetracycline export proteins and may thus function in bile acid transport (D.H. Mallonee and P.B. Hylemon, unpublished data). The 47,500-M_r protein has homology to carnitine dehydratase of *E. coli* (D.H. Mallonee and P.B. Hylemon, unpublished data). The 72,000-M_r protein has NADH:flavin oxidoreductase activity and has been purified from cell extracts of *Eubacterium* sp. VPI 12708 (Franklund et al. 1993, Lipsky and Hylemon 1980). It has amino acid sequence homology with certain soluble NAD(P)H dehydrogenases, trimethylamine dehydrogenase, and the 59,000-M_r protein encoded by the above operon, but its physiological role in 7α-dehydroxylation is unclear (Franklund et al. 1993).

The 7β-hydroxy bile acid ursodeoxycholic acid is used clinically to dissolve cholesterol gallstones and to treat certain cholestatic liver diseases. Washed fecal suspensions convert this compound to lithocholic acid (Fedorowski et al. 1979). *Eubacterium* sp. strains VPI 12708 (White et al. 1982) and C-25 (Takamine and Imamura 1985) can 7β-dehydroxylate ursodeoxycholic acid, forming lithocholic acid. This reaction proceeds directly in these strains, and not by 7α-dehydroxylation of chenodeoxycholic acid formed by 7β-epimerization of ursodeoxycholic acid. The 7β-dehydroxylation activity in *Eubacterium* sp. strain C-25 is greatly stimulated by coincubation with *Bacteroides distasonis* strain K-5 whole cells or cell extracts; the latter strain had no 7-HSDH or 7-dehydroxylation activity alone. The 7α-dehydroxylation and 7β-dehydroxylation activities in *Eubacterium* sp. VPI 12708 have similar bile acid inducibility, cofactor requirements, reaction kinetics, and chromatographic behavior, suggesting that they are catalyzed by the same enzyme system (White et al. 1982, 1983).

2.4. Esterification

Saponifiable derivatives (esters) of bile acids have been detected in human and rodent feces from untreated subjects or from subjects given oral doses of 24-^{14}C-labeled bile acids (Benson et al. 1993, Korpela et al. 1986, 1988, Norman 1964, Norman and Palmer 1964). The proportion of saponifiable bile acids in human feces is estimated to be 10% to 37% of the total bile acid content and consists mainly of deoxycholic acid esters (Korpela et al. 1986, 1988). Recently, Benson et al. (1993) demonstrated that human and hamster feces contain considerable amounts of deoxycholic acid oligomers, formed by esterification of the C-24 carboxyl group of one molecule with the 3α-hydroxy group of the next (Fig. 13.2, reaction 11). The chain lengths of these polyesters range from two to 22 deoxycholic acid subunits. Since these compounds are not present in bile, they are probably produced by intestinal bacteria.

Kelsey and Sexton (1976) showed that mixed cultures from human feces incubated anaerobically with 24-^{14}C-lithocholic acid and ethanol produced the C-24 ethyl esters of lithocholic acid and isolithocholic acid. However, under similar conditions, fecal isolates of *B. thetaiotaomicron*, *Citrobacter* sp., and *P. productus* produced only the lithocholic acid ester (Kelsey and Thompson 1976). Edenharder and Hammann (1985) showed that strains of *Bacteroides*, *Eubacterium*, and *Lactobacillus* isolated from human feces converted cholic acid to deoxycholic acid and the C-24 methyl ester of deoxycholic acid. However, this activity was only rarely detected in mixed fecal cultures. The esterification activity in the pure cultures did not require addition of exogenous methanol and was lost after serial transfers of the cultures. To date, enzymes catalyzing bile acid esterification by intestinal bacteria have not been purified.

2.5. Desulfation

Bacterially produced lithocholic acid which escapes fecal excretion is returned to the liver via EHC. There, it is conjugated to glycine or taurine at the C-24 carboxyl group and esterified with sulfate at the 3α-hydroxy group before secretion in bile. After deconjugation of the glycine or taurine moiety by bacterial CBH, the sulfate group can also be removed by bacterial sulfatases. Sulfobile acids are poorly absorbed from the gastrointestinal tract (De Witt and Lack 1980) but are readily excreted in feces. Studies with germ-free and conventional rats indicate that bacterial desulfation of sulfobile acids decreases their excretion and promotes EHC of the desulfated products (Eyssen et al. 1985, Robben et al. 1988).

Bile acid desulfation has been studied in mixed fecal cultures (Kelsey et al. 1980, 1981; Pacini et al. 1987; Palmer 1972) and fecal isolates. The bacterial genera capable of bile acid desulfation include *Clostridium* (Borriello and Owen 1982, Huijghebaert and Eyssen 1982, Robben et al. 1986), *Peptococcus* (Van Eldere et al. 1988), and *Fusobacterium* (Robben et al. 1989). The desulfation products of 3-sulfolithocholic acid generated by mixed and pure cultures include lithocholic acid, isolithocholic acid and certain of its 3β-fatty acyl esters, 2-cholenoic acid, 3-cholenoic acid, and 5β-cholanoic acid.

Bile acid sulfatase activity in pure cultures of fecal bacteria requires a 3α- or 3β-sulfo group and a free C-24 or C-26 carboxyl group. Bile acids sulfated at positions other than C-3 are not desulfated. Desulfation by these bacteria is also stereospecific, depending on the configuration of both the 3-sulfo group (3α or 3β) and the C-5 hydrogen atom (5α or 5β). These configurations also determine the desulfated products formed. For example, *Fusobacterium* sp. strains H35 and H83 convert 3α- or 3β-sulfo-5β-bile acids to 5β-cholen-3-oic acids but convert

3β-sulfo-5α-bile acids to 5α-cholen-2-oic acids and 3α-sulfo-5α-bile acids; no products are formed from 3α-sulfo-5α-bile acids (Robben et al. 1989). Bile acid desulfation could hypothetically occur by elimination or hydrolysis. However, the mechanism has not been determined, and enzymes catalyzing the reaction have not been purified.

2.6. Physiological Function of Bile Acid Metabolism in Intestinal Bacteria

Bile acid deconjugation and desulfation may provide growth substrates for the bacteria which perform these reactions, or perhaps inhibit the growth of competing bacteria due to the toxicity of the free bile acids produced. Dehydroxylating bacteria may gain a similar competitive advantage, since dehydroxylated bile acids are considerably more hydrophobic and toxic than their parent compounds (Binder et al. 1975).

Esterification and epimerization of hydroxy bile acids may reduce their toxicity and thus protect bile acid-sensitive bacteria. For example, ursodeoxycholic acid (7β-hydroxy) is less hydrophobic than chenodeoxycholic acid (7α-hydroxy) and presumably less deleterious to cell membranes (Armstrong and Carey 1982). Similarly, bile acid esters are sparingly soluble and may form harmless aggregates.

Oxidation of hydroxy bile acids by HSDHs may generate reducing equivalents for electron transport phosphorylation or biosynthetic purposes. In fact, addition of exogenous electron acceptors such as menadione or fumarate to cultures of *B. thetaiotaomicron* greatly increases the oxidation of cholic acid to 7-dehydrocholic acid (Sherrod and Hylemon 1977). Oxo- and ring unsaturated intermediates formed during epimerization and 7-dehydroxylation reactions may serve as electron sinks or as terminal electron acceptors for electron transport chains. Because 7α-dehydroxylation is a net reductive process, it may serve as a key electron accepting reaction in the energy metabolism of dehydroxylating bacteria. These bacteria may occupy a unique niche in the colon ecosystem since their numbers are relatively low (Sect. 2.3) while the amounts of their bile acid substrates are considerable (up to 600 mg).

3. Metabolism of Cholesterol by the Intestinal Microbiota

3.1. Reduction to Coprostanone and Coprostanol

The colon receives up to 1 g of cholesterol per day originating from bile, the diet, and sloughing of intestinal mucosa. Cholesterol is reduced to coprostanol and minor amounts of coprostanone by the colonic microbiota (Fig. 13.5). Together, cholesterol and its transformation products comprise 95% of the total neutral steroid detected in rat and human feces (McNamara et al. 1981). The role of the intestinal microbiota in the formation of coprostanol and coprostanone was confirmed by comparing the fecal sterols of germ-free and conventional rats (Kellogg and Wostmann 1969); the former excrete unmodified cholesterol, but the latter also excrete coprostanol and coprostanone in amounts up to 55% of the total fecal sterols. The cecum is the site where cholesterol is most extensively transformed in rats (Eyssen et al. 1972, Kellogg 1973). The proportion of cholesterol converted in vivo varies within populations. Wilkins and Hackman (1974) found that 23 out of 31 in a group of North American Caucasians excreted high amounts of coprostanol and coprostanone in their feces (88% of the total sterols), while the other eight excreted only low amounts (10%).

Both mixed fecal biota (Snog-Kjaer et al. 1956) and fecal isolates have been shown to convert cholesterol to its transformation products. Crowther et al. (1977) reported that strains of *Bifidobacterium*, *Clostridium*, and *Bacteroides* reduce cholesterol to coprostanol, although this has not been confirmed. Most cholesterol-reducing bacteria that have been isolated and characterized are members of the genus *Eubacterium*.

A cholesterol-reducing bacterium, *Eubacterium* ATCC 21408, was isolated

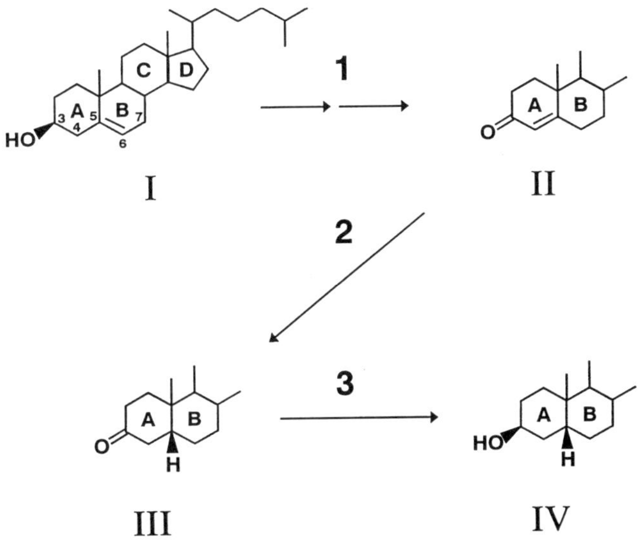

Figure 13.5. Bacterial transformations of cholesterol. Large Arabic numerals refer to reactions cited in section 3. Structures: I, cholesterol; II, 4-cholesten-3-one; III, coprostanone; IV, coprostanol.

from rat cecal contents, using a cholesterol-rich calf brain powder medium (Eyssen et al. 1973). This organism required 3β-hydroxy-Δ^5 animal or plant sterols for growth, including cholesterol, stigmasterol, campesterol, and β-sitosterol, and reduced them to 3β-hydroxy-5β-saturated derivatives. In addition, the organism reduced 4-cholesten-3-one, coprostanone, and its 5α-epimer (cholestanone) to coprostanol (see Fig. 13.5 for structures). Reduction of the Δ^5 double bond required a 3β-hydroxy group, since sterols with a 3α-hydroxy group, with other C-3 substituents, or without a 3-hydroxy group were not reduced. Double bonds at positions other than Δ^4 or Δ^5 were not reduced. Eyssen et al. (1973) suggested that this bacterium used cholesterol as an external electron acceptor, since it required up to 2.0 mg/mL cholesterol for optimal growth. Subsequently, using brain-supplemented media, Sadzikowski et al. (1977) isolated cholesterol-reducing gram-positive diplobacilli from rat cecal contents and human feces, and Mott and Brinkley (1979) isolated the cholesterol-reducing *Eubacterium* strain 403 from baboon feces. In all of these studies (Eyssen et al. 1973, Mott and Brinkley 1979, Sadzikowski et al. 1977), the bacteria would not form viable colonies on solid media, and were purified in broth culture by serial dilution, filtration, and addition of selective inhibitors.

Mott and Brinkley (1979) discovered that the cholesterol-reducing *Eubacterium* strains ATCC 21408 and 403 required both cholesterol and a plasmalogen component of calf brain lipids, plasmenylethanolamine, for growth. Plasmalogens are phosphoglyceride analogs in which the C-1 substituent of glycerol is a long chain 1-alkenyl ether. These membrane phospholipids are particularly abundant in muscle cells, nerve cells, and certain anaerobic bacteria. Although the role of the plasmalogen was unclear, its alkenyl ether group was metabolized during growth. Moreover, plasmalogens with a 1-alkyl ether rather than a 1-alkenyl ether would not support growth of these *Eubacterium* strains (Mott et al. 1980). Ultimately, Brinkley et al. (1980) developed a solid medium containing lecithin, brain powder, and 5% added cholesterol, which adequately supported colony formation by cholesterol-reducing *Eubacterium* strains. With this medium, nine new strains of cholesterol-reducing bacteria were isolated from baboon feces and intestinal contents (Brinkley et al. 1982). Unlike previous cholesterol reducers, none of these strains required cholesterol plus plasmalogen for growth, although seven required plasmalogen for cholesterol reduction. The above studies suggest that cholesterol-reducing intestinal bacteria can be classified into three categories: (1) those that require plasmalogen (and cholesterol) for growth and cholesterol reduction; (2) those that require plasmalogen for cholesterol reduction but not growth; and (3) those that require plasmalogen for neither cholesterol reduction nor growth.

The mechanism of cholesterol reduction was studied in cultures of *Eubacterium* ATCC 21408 fed [4β-^{3}H,4-^{14}C]-cholesterol (Eyssen and Parmentier 1974, Parmentier and Eyssen 1974) (Fig. 13.5). The reaction most likely proceeds by a sequential oxidation of the 3β-hydroxy group and isomerization of the Δ^5 to a Δ^4 double bond, forming 4-cholesten-3-one (II), reduction of the Δ^4 double bond, yielding coprostanone (III), and reduction of the 3-oxo group, yielding coprostanol (IV). This hypothesis is further supported by the detection of coprostanone in feces and by the reduction of 4-cholesten-3-one and coprostanone to coprostanol by fecal bacteria (Eyssen et al. 1973, Mott et al. 1980). To date, the enzymology of cholesterol reduction has not been studied.

4. Metabolism of Steroid Hormones by the Intestinal Microbiota

Steroid hormones that undergo EHC are conjugated to either glucuronide or sulfate at C-3, C-17, and C-21 by the liver and secreted in biliary bile (Laatikaninen 1970b). Total concentrations of steroids in human bile show gender differences, with adult men and women secreting approximately 13 mg/day and 6 mg/day, respectively (Laatikaninen 1970a). Pregnant women are reported to have an increased biliary secretion of steroid hormones (Eriksson et al. 1970). Four 17α-hydroxylated C_{21} steroids have been identified in biliary bile, including 5β- (and 5α-)pregnan-3α,17α,20α-triol and 5-pregnen-3α (and 3β),17α,20α-triol (Laatinkanien 1970c).

Studies of neutral steroid hormone patterns in feces of germ-free and conventional animals indicate that extensive biotransformations of steroid hormones are carried out by the intestinal microbiota (Eriksson and Gustafsson 1970, Eriksson et al. 1969a, Gustafsson 1968). Known microbial biotransformations include (Table 13.7, Fig. 13.6) (1) hydrolysis of glucuronide and sulfate conjugates by microbial glucuronidases and sulfatases (not shown); (2) 21-dehydroxylation (reaction 10); (3) side chain cleavage (reaction 5); (4) saturation of the Δ^4 double bond (reactions 1 and 4); (5) oxidation/reduction of hydroxy/oxo groups at C-3, C-17, and C-20 (reactions 2, 3, 6–8, and 9); and (6) 16α-dehydroxylation, yielding 17β-side chains (Fig. 13.7). Steroid hormone metabolites generated by the intestinal microbiota can be absorbed and secreted in the urine (Wade et al. 1959). However, the significance of the intestinal biotransformations of steroid hormones to the host is not clear.

4.1. Hydrolysis of Glucuronides and Sulfates

Steroid hormones that are secreted into bile are deconjugated in the intestine by bacterial sulfatases and glucuronidases (Eriksson and Gustafsson 1970). Glucuronidase activity may be derived from intestinal bacteria or mucosal cells. However, the sulfatases are strictly bacterial, and the intestinal microbiota can hydrolyze sulfate conjugates in the 3α-, 3β-, 17β-, and 21-hydroxy positions (Macdonald et al. 1983a). Bacterial desulfation has been demonstrated to increase

Table 13.7. Steroid hormone modifications by intestinal microbiota

Biotransformation(s)	Organism(s)	Enzyme(s)	Cofactor(s)	M_r (10^3)	Comments	Ref[a]
Hydrolysis of sulfate and glucuronide conjugates	*Clostridium* sp. *Peptococcus* sp. *Eubacterium* sp. *Lactobacillus* sp. *Bacteroides* sp. *Escherichia coli*	sulfatase(s) and β-glucuronidase(s)	— —	? ?	β-glucuronidase activity found both in bacteria and intestinal mucosa	1–3
21-CH_3OH → 21-CH_3	*Eubacterium lentum*	21-dehydroxylase	$FMNH_2$	582[b]	Requires α-ketol steroids as substrates	4
C_{17} side chain → 17-oxo	*C. scindens, Eubacterium desmolans*	desmolase	AD, B_{12}, divalent cations	?	Steroid inducible	5,6
4-ene → 5α-H	*Clostridum* sp. J-1, *Eubacterium* sp. strain 144	5α-reductase	?	?	Soluble and membrane bound forms; H_2 or pyruvate stimulatory	7,8
4-ene → 5β-H	*C. paraputrificum C. inocuum*	5β-reductase	NADH	?		9–11
16α-OH → Δ^{16}	*Eubacterium* sp. strain 144	16α-dehydratase	—	42.4[c]	Δ^{16} steroid chemically reactive	12,13
Δ^{16} → 17α-Pregnanes	*Eubacterium* sp. strain 144	Δ^{16}-steroid reductase	?	?	H_2 or pyruvate stimulatory	14
3-oxo → 3β-OH	*C. innocuum*	3β-hydroxysteroid dehyrogenase	NADH	80.0[b]		11
3-oxo → 3α-OH	*E. lentum C. paraputrificum*	3α-hydroxysteroid dehydrogenase	NAD(P)H	?		10,15
17-oxo → 17β-OH	*E. lentum C. innocuum B. fragilis*	17β-hydroxysteroid dehydrogenase	?	?		16
17-oxo → 17α-OH	*Eubacterium* sp. VPI 12708	17α-hydroxysteroid dehydrogenase	NADPH	42.0[c]	O_2-sensitive	17
20-oxo → 20β-OH	*B. fragilis Bifidobacterium adolescentis*	20β-hydroxysteroid dehydrogenase	?	?		18
20-oxo → 20α-OH	*C. scindens*	20α-hydroxysteroid dehydrogenase	NADH	40.0[c]	Uses only steroids with 17α, 21-dihydroxy group	19

[a] Selected references: 1. Graves et al. (1977); 2. Van Eldere et al. (1988); 3. Van Eldere et al. (1991); 4. Feighner and Hylemon (1980); 5. Winter et al. (1984a); 6. Krafft et al. (1987); 7. Bokkenheuser et al. (1983); 8. Glass et al. (1991); 9. Schubert et al. (1967); 10. Glass et al. (1979); 11. Stokes and Hylemon (1985); 12. Glass and Lamppa (1985); 13. Glass et al. (1982); 14. Watkins and Glass (1991); 15. Bokkenheuser et al. (1979); 16. Winter et al. (1984b); 17. dePrada et al. (1994); 18. Winter et al. (1982a); 19. Krafft and Hylemon (1989)

[b] Native M_r.

[c] Subunit M_r.

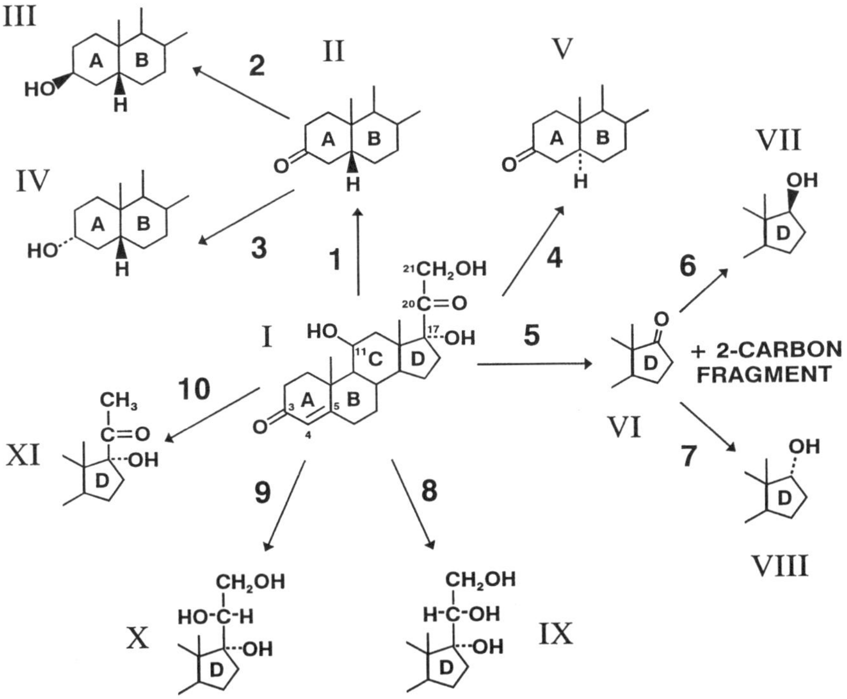

Figure 13.6. Bacterial transformations of cortisol. Large arabic numerals refer to reactions cited in section 4. Structures: I, cortisol; II, 5β-dihydrocortisol; III, 3α-hydroxy-5β-tetrahydrocortisol; IV, 3β-hydroxy-5β-tetrahydrocortisol; V, 5α-dihydrocortisol; VI, 11β-hydroxy-androstenedione; VII, 11β,17β-Dihydroxy-4-androsten-3-one; VIII, 11β,17α-Dihydroxy-4-androsten-3-one; IX, 20α-dihydrocortisol; X, 20β-dihydrocortisol; XI, 21-deoxycortisol. Intermediates VII and VIII have not been detected in vivo. However, their 11-dehydroxy analogs, testosterone and epitestosterone, are formed by the action of 17α- and 17β-hydroxysteroid dehydrogenases on androstenedione (Section 4).

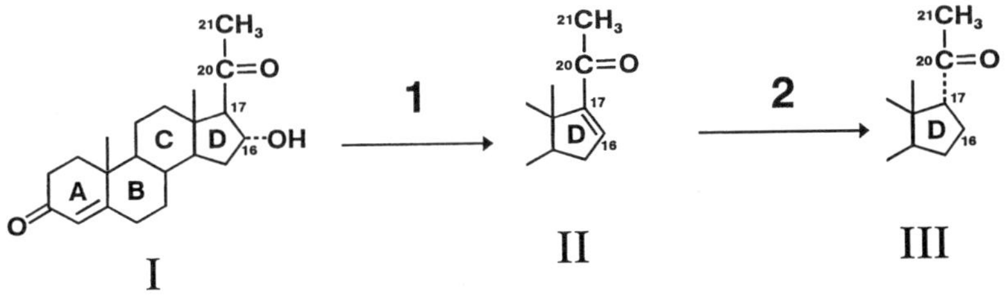

Figure 13.7. Bacterial metabolism of 16α-hydroxyprogesterone. Large arabic numerals refer to reactions cited in Section 4. Structures: I, 16α-hydroxyprogesterone; II, Δ^{16}-progesterone; III, 17α-progesterone.

the elimination of dehydroepiandrosteone sulfate from rats by increasing its EHC and urinary excretion (Van Eldere et al. 1990). Steroid sulfatase activity has been detected in certain species of the genera *Clostridium*, *Lactobacillus*, *Eubacterium*, *Peptococcus*, and *Bacteroides* (Van Eldere et al. 1988). However, the substrate specificity for steroid sulfates varies considerably among these species, suggesting that there are many different steroid sulfatases among the intestinal microbiota. Indeed, *Peptococcus niger* synthesizes at least three different steroid sulfatases (Van Eldere et al. 1991).

4.2. 21-Dehydroxylation of Corticosteroids

21-Dehydroxylation of corticosteroids is restricted to those that undergo EHC (Fig. 13.6, reaction 10). The role of intestinal bacteria in 21-dehydroxylation was first shown in the late 1960s by comparing the steroid hormone patterns in germ-free and conventional animals (Eriksson et al. 1969b, Gustafsson 1968, Gustafsson and Sjövall 1968). Studies by Eriksson et al. (1969) showed that anaerobic incubation of fecal suspensions with 3β,21-dihydroxy-5α-pregnane-20-one resulted in the formation of 3,20-dihydroxypregnane metabolites. Moreover, it has also been demonstrated that the removal of the 21-hydroxy group of corticosteroids by intestinal bacterial shifts the secretion from biliary to renal.

To date, all bacteria capable of 21-dehydroxylation are members of the genus *Eubacterium*. Bokkenheuser et al. (1977) successfully isolated a bacterium from human fecal suspension, which was capable of the 21-dehydroxylation of corticosteroids with an α-ketol side chain. The bacterium was later identified as a strain of *Eubacterium lentum* (Bokkenheuser et al. 1979). Feighner et al. (1979) demonstrated 21-dehydroxylase activity in cell extracts of *E. lentum* and showed that the enzyme required reduced free flavins for activity. The M_r of 21-dehydroxylase was estimated to be 582,000 by gel filtration chromatography. 21-Dehydroxylation increased sevenfold when whole cells of *E. lentum* were sparged with hydrogen gas (Feighner and Hylemon 1980).

21-Dehydroxylation of corticosteroids requires a 20-oxo substrate and reduced flavins. Holland and Riemland (1984) studied the mechanism of 21-dehydroxylation using substrates labelled with deuterium at C-17 and C-21. Their proposed mechanism is consistent with an initial formation of an enediol followed by reduction, dehydration, and enol-oxo tautomerism, yielding the 21-dehydroxylated steroid product. Such a mechanism suggests that 21-dehydroxylation requires more than one enzymatic activity. However, 21-dehydroxylase has not yet been purified to homogeneity or characterized.

4.3. Side-Chain Cleavage

Evidence for the side-chain cleavage of glucocorticoids by members of the intestinal microbiota was first reported by Nabarro et al. (1957) (Fig. 13.6, reaction 5). Studies by Eriksson and Gustafsson (1971) demonstrated that mixed fecal biota from healthy human subjects removed the side-chain of cortisol. Bokkenheuser et al. (1984) isolated from human feces a species of *Clostridium* that had desmolase (C17-C20 lyase) activity. The bacterium was classified as a new species, *Clostridium scindens*. It was demonstrated using whole cells of this bacterium that side-chain cleavage required a hydroxy group at C-17 and preferred an α-ketol group at C-20 or C-21 for optimal activity (Winter et al. 1984a). It was shown later that desmolase activity in *C. scindens* was induced by cortisol, cortisone, 11-desoxycortisol, and 17α-hydroxyprogesterone. Desmolase activity in cell extracts of *C. scindens* prepared from induced cultures required NAD^+ and a divalent metal cation—i.e., Mg^{2+}, Mn^{2+}, or Fe^{2+} (Krafft et al. 1987). Additional studies demonstrated that desmolase activity was further stimulated by the addition of vitamin B_{12} to cell extracts of *C. scindens* (A.E. Krafft and P.B. Hylemon, unpublished data). Finally, Morris et al. (1986) isolated from cat feces a new species, *Eubacterium desmolans*, which had desmolase activity.

4.4. Reduction of Ring A

Reduction of ring A in steroid hormones markedly decreases their hormonal activity. The reduction of ring A in steroid hormones is normally thought to be carried out in the liver. However, the intestinal microbiota rapidly reduce ring A of steroid hormones containing the Δ^4-3-oxo-steroid structure. Bacterial reduction of such hormones occurs by a sequential two step reaction (Glass et al. 1979). The Δ^4 double bond is initially reduced to either to the 5α- or 5β configuration (Fig. 13.6, reactions 1 and 4) by stereospecific 5α- or 5β-steroid reductases. The 3-oxo group is then reduced to either a 3α- or 3β-hydroxy group by stereospecific 3α- or 3β-HSDHs (Fig. 13.6, reactions 2 and 3). Species of intestinal clostridia have been shown to carry out the reduction of Δ^4-3-oxo-steroids. *Clostridium innocuum* and *Clostridium paraputrificum* reduce Δ^4-3-oxosteroids to 3β-5β and 3α-5β derivatives, respectively (Bokkenheuser et al. 1979, Glass et al. 1979, Stokes and Hylemon 1985), while *Clostridium* sp. J-1 produces a 3β-5α derivative (Bokkenheuser et al. 1983). Ring A in synthetic progestins, which are used in oral contraceptives, is reduced three to 10 times more slowly than in naturally occurring progestins (Bokkenheuser et al. 1983). This observation may explain the pharmacological superiority of synthetic progestins over natural progestins. In cell extracts prepared from *C. innocuum* or *C. paraputrificium*, NADH markedly stimulated the 5β-reductase and 3α- and 3β-HSDH activities. The relative molecular weight of 5β-reductase and 3β-HSDH from *C. innocuum* was estimated to be 80,000 by gel filtration chromatography. *Eubacterium lentum* has also been reported to have both 3α- and 3β-HSDH activities and can epimerize the 3α-hydroxy group to the 3β-orientation (Bokkenheuser et al. 1979).

4.5. Oxidation/Reduction of 17- and 20-Hydroxy/Oxo Groups

The reduction of the 17-oxo group to a 17β-hydroxy derivative can be carried out by a number of species of intestinal bacteria including: *Eubacterium lentum, Clostridium paraputrificum, Clostridium* J-1, *Clostridium innocuum*, and *Bacteroides fragilis*. The 17-oxo function of phenolic steroids is reduced specifically by certain species of the genera *Eubacterium* and *Clostridium*, while the 17-oxo group of androstenedione is reduced solely by *B. fragilis* (Winter et al. 1984b). The reduction of the 17-oxo group of androstenedione to the 17α-hydroxy derivative (yielding epitestosterone) was recently discovered in the human intestinal isolate, *Eubacterium* sp. VPI 12708 (de Prada et al. 1994). A steroid inducible 17α-HSDH was purified and characterized from this organism. The enzyme was oxygen-sensitive and was highly specific for NADPH and the 17-oxo group of C_{19} steroids. The N-terminal amino acid sequence suggested that this enzyme belongs to a disulfide reductase gene family (de Prada et al. 1994). The presence of both 17α- and 17β-HSDHs suggests that the intestinal microbiota can epimerize the 17β-hydroxy group of testosterone to the 17α-hydroxy group of epitestosterone. The pathways of epitestosterone biosynthesis in the body are controversial. However, there has been renewed interest in the biosynthesis of this compound because of its potential as an antiandrogen (Bicikova et al. 1992). In addition, the International Olympic Committee uses the urinary ratio of testosterone to epitestosterone as a marker for testosterone doping in athletes (Kicman et al. 1990).

Bacteroides fragilis and *Bifidobacterium adolescentis* have been reported to have 20β-HSDH activity (Bokkenheuser et al. 1975, Winter et al. 1982a). Reduction of the 20-oxo group makes the steroid resistant to further modification of the steroid side chain. Steroid substrate specificity studies using whole cells of *B. fragilis* and *B. adolescentis* suggest that there may be more than one form of 20β-HSDH among the intestinal microbiota. However, 20β-HSDH has not been purified or characterized from members of the intestinal biota.

20α-HSDH has been detected in a variety of mammalian tissues (Nancarrow et al. 1981) and, in microorganisms, isolated from soil and water. Studies by Winter et al. (1984a) showed that the intestinal microbiota contain 20α-HSDH activity. Bokkenheuser et al. (1984) isolated from human feces a species of *Clostridium* that had 20α-HSDH activity. This bacterium was later determined to be a new species, *Clostridium scindens* (Morris et al. 1985). The 20α-HSDH from *C. scindens* has been purified and characterized (Krafft and Hylemon 1989). The enzyme was NAD^+/NADH dependent and was highly specific for adrenocorticosteroids having 17α- and 21-hydroxy groups. The subunit M_r of the enzyme was approximately 40,000 Da. The N-terminal amino acid sequence (first 11 residues) suggested that this enzyme may belong to the glyceraldehyde-3-phosphate dehydrogenase gene family (Krafft and Hylemon 1989).

4.6. 16α-Dehydroxylation

Steroid hormones with a 17α side chain have been detected in feces and urine of rats and humans (Calvin and Lieberman 1962, Eriksson et al. 1969a). These steroid hormones have been shown to originate from the 16α-dehydroxylation of steroid hormones by the intestinal microbiota. 16α-Dehydroxylation activity has been detected in human feces and rat intestinal contents (Eriksson and Gustafsson 1971). Two intestinal strains of *Eubacterium* with 16α-dehydroxylation activity were isolated by Bokkenheuser et al. (1980). Winter et al. (1982b) showed that the 16α-dehydroxylation of corticoids by isolates of intestinal bacteria was a two-step biotransformation (Fig. 13.7). The 16α-hydroxysteroid substrate was dehydrated to a Δ^{16}-steroid intermediate (reaction 1), which accumulated in culture media and was subsequently reduced to a 17α side chain steroid product (reaction 2). Glass et al. (1982) showed that the Δ^{16} steroid intermediate would react nonenzymatically with L-cysteine in the culture medium, forming a 16-thioether bond. The high reactivity of Δ^{16} steroids is believe to be due to the formation of a highly reactive α,β-unsaturated ketone associated with the D-ring of the steroid. It is not known if cysteine conjugates of steroids are formed in vivo; however, Δ^{16} steroids have been detected in the urine of female subjects (Calvin and Lieberman 1962). A 16α-dehydratase has been purified from *Eubacterium* sp. strain 144 and was found to have a subunit M_r of 42,400 and native M_rs of 181,000 and 326,000 by gel filtration chromatography (Glass and Lamppa 1985). The reduction of Δ^{16} steroids by whole cell suspensions of *Eubacterium* sp. strain 144 required growth in the presence of hemin and an inducing steroid. In addition, it was necessary to incubate the cell suspensions in the presence of either pyruvate or molecular hydrogen. It was proposed that hemin was used for the synthesis of a cytochrome-containing electron transport chain which then uses Δ^{16} steroids as an electron acceptor (Glass et al. 1991, Watkins and Glass 1991). It was discovered that these same growth conditions induced a membrane-associated steroid 5α-reductase which also required either pyruvate or molecular hydrogen as an electron donors. These results suggest that these steroid biotransformations might be linked to energy metabolism in this bacterium.

5. Summary

In the 1960's and early 1970's, the various intestinal biotransformations of bile acids, cholesterol, and steroid hormones were shown to occur in intact animals and humans. In the late 1970's and 1980's, intestinal bacteria catalyzing these steroid biotransformations were isolated and identified. Enzymes catalyzing some of the steroid modifications have been purified and characterized, and the corresponding genes cloned and sequenced. Some steroid modifying enzymes have been assigned to gene families. In the late 1980's, a multi-step biochemical pathway for bile acid 7-dehydroxylation was discovered. However, the physiological significance of the steroid biotransformations to intestinal bacteria has not been determined. Finally, a thorough understanding of the importance of these biotransformations to the host in health and disease awaits further research.

Acknowledgments

We thank Drs. Darrell Mallonee and Paloma de Prada for permission to cite unpublished data. We are pleased to acknowledge support from the National Institutes of Health, Public Health Service USA, through grants RO1-DK 40986 and PO1-DK 38030.

References

Adlercreutz H, Martin F, Jarvenpaa R, Fotsis T (1979) Steroid absorption and enterohepatic recycling. Contraception 20:201–223.

Akao T, Akao T, Hattori M, Namba T, Kobashi K (1986) 3β-Hydroxysteroid dehydrogenase of *Ruminococcus* sp. from human intestinal bacteria. J Biochem 99:1425–1431.

Akao T, Akao T, Kobashi K (1987) Purification and characterization of 7β-hydroxysteroid dehydrogenase from *Ruminococcus* sp. of human intestine. J Biochem 102:613–619.

Ali SS, Kuksis A, Beveridge JMR (1966) Excretion of bile acids by three men on a fat-free diet. Can J Biochem 44:957–969.

Armstrong MJ, Carey MC (1982) The hydrophobic-hydrophilic balance of bile salts. Inverse correlation between reverse-phase high performance liquid chromatographic mobilities and micellar cholesterol-solubilizing capacities. J Lipid Res 23:70–80.

Baron SF, Franklund CV, Hylemon PB (1991) Cloning, sequencing, and expression of the gene coding for bile acid 7α-hydroxysteroid dehydrogenase from *Eubacterium* sp. VPI 12708. J Bacteriol 173:4558–4569.

Batta AK, Salen F, Arora R, Shefer S, Batta M, Person A (1990) Side chain conjugation prevents bacterial 7-dehydroxylation of bile acids. J Biol Chem 265:10925–10928.

Benson GM, Haskins NJ, Eckers C, Moore PJ, Reid DG, Mitchell RC, Waghmare S, Suckling KE (1993) Polydeoxycholate in human and hamster feces: a major product of cholate metabolism. J Lipid Res 34:2121–2134.

Bicikova M, Hampl R, Starka L (1992) Epitestosterone- a potent competitive inhibitor of C-21 steroid side chain cleavage in the testes. J Steroid Biochem Molec Biol 43: 721–724.

Binder HJ, Filburn B, Floch M (1975) Bile acid inhibition of intestinal anaerobic organisms. Amer J Clin Nutrit 28:119–125.

Björkhem I, Einarsson K, Melone P, Hylemon PB (1989) Mechanism of intestinal formation of deoxycholic acid from cholic acid in humans: evidence for a 3-oxo-Δ^4-steroid intermediate. J Lipid Res 30:1033–1039.

Bokkenheuser VD, Winter J (1983) Biotransformation of steroids. In: Hentges DJ, ed.

Human Intestinal Microflora in Health and Disease, pp. 215–239. New York: Acad Press.

Bokkenheuser VD, Suzuki JB, Polowsky SD, Winter J, Kelly WG (1975) Metabolism of deoxycorticosterone by human fecal flora. Appl Microbiol 30:82–90.

Bokkenheuser VD, Winter J, Dehazya P, Kelly WG (1977) Isolation and characterization of human fecal bacteria capable of 21-dehydroxylating corticoids. Appl Environ Microbiol 34:571–575.

Bokkenheuser VD, Winter J, Finegold SM, Sutter VL, Ritchie AE, Moore WEC, Holdeman LV (1979) New markers for *Eubacterium lentum*. Appl Environ Microbiol 37: 1001–1006.

Bokkenheuser VD, Winter J, O'Rourke-Locascio S, Ritchie AE (1980) Isolation and characterization of fecal bacteria capable of 16α-dehydroxylating corticoids. Appl Environ Microbiol 40:803–808.

Bokkenheuser VD, Winter J, Cohen BI, O'Rourke-Locascio S, Mosbach EH (1983) Inactivation of contraceptive steroid hormones by human intestinal clostridia. J Clin. Microbiol 18:500–504.

Bokkenheuser VD, Morris GN, Richie AE, Holdeman LV, Winter J (1984) Biosynthesis of androgen from cortisol by a species of *Clostridium* recovered from human fecal flora. J Infect Dis 149:489–494.

Borriello SP, Owen RW (1982) The metabolism of lithocholic acid and lithocholic acid-3α-sulfate by human fecal bacteria. Lipids 17:477–482.

Braun M, Lünsdorf H, Bückmann AF (1991) 12α-Hydroxysteroid dehydrogenase from *Clostridium* group P, strain C 48–50. Production, purification and characterization. Eur J Biochem 196:439–450.

Brinkley AW, Gottesman AR, Mott GE (1980) Growth of cholesterol-reducing *Eubacterium* on cholesterol-brain agar. Appl Environ Microbiol 40:1130–1132.

Brinkley AW, Gottesman AR, Mott GE (1982) Isolation and characterization of new strains of cholesterol-reducing bacteria from baboons. Appl Environ Microbiol 43: 86–89.

Calvin HI, Lieberman S (1962) Studies on the metabolism of 16α-hydroxyprogesterone in humans: conversion to urinary 17-isoprogesterone in humans: conversion to urinary 17-progesterone. Biochemistry 1:639–645.

Carey MC (1985) Physical-chemical properties of bile acids and their salts. In: Danielsson H, Sjövall J, eds. Sterols and Bile Acids, pp. 345–403. Amsterdam: Elsevier Science.

Carey MC, Cahalane MJ (1988) The enterohepatic circulation. In: Arias I, Jakoby WB, Popper H, Schachter D, Shafritz DA. eds. The Liver: Biology and Pathobiology, 2nd Ed., pp. 573–616. New York: Raven Press.

Chen Z, Jiang JC, Lin Z-G, Lee WR, Baker ME, Chang SH (1993) Site-specific mutagenesis of *Drosophila* alcohol dehydrogenase: evidence for involvement of tyrosine-152 and lysine-156 in catalysis. Biochemistry 32:3342–3346.

Christiaens H, Leer RJ, Pouwels PH, Verstraete W (1992) Cloning and expression of a conjugated bile acid hydrolase gene from *Lactobacillus plantarum* by using a direct plate assay. Appl Environ Microbiol 53:331–336.

Cohen BI, Raicht RF, Deschener EE, Takahashi M, Sarwal AM, Fazzini E (1980) The effect of cholic acid feeding on N-methyl-N-nitrosourea colon tumors and cell kinetics in rats. J Natl Cancer Inst 64:573–578.

Coleman JP, White WB, Egestad B, Sjövall J, Hylemon PB (1987a) Biosynthesis of a novel bile acid nucleotide and mechanism of 7α-dehydroxylation by an intestinal *Eubacterium* species. J Biol Chem 262:4701–4707.

Coleman JP, White WB, Hylemon PB (1987b) Molecular cloning of bile acid 7-dehydroxylase from *Eubacterium* sp. strain VPI 12708. J Bacteriol 169:1516–1521.

Coleman JP, White WB, Lijewski M, Hylemon PB (1988) Nucleotide sequence and regulation of a gene involved in bile acid 7α-dehydroxylation in *Eubacterium* sp. strain VPI 12708. J Bacteriol 170:2070–2077.

Cooper AD (1990) Hepatic lipoprotein and cholesterol metabolism. In: Zakim D, Boyer TD, eds. Hepatology. A Textbook of Liver Disease, pp. 96–123. Philadelphia: W.B. Saunders.

Crowther JS, Drasar BS, Goddard P, Hill MJ, Johnson K (1977) The effect of chemically defined diet on the faecal flora and a faecal steroid concentration. Gut 14:490–493.

Dashkevicz MP, Feighner SD (1989) Development of a differential medium for bile salt hydrolase-active *Lactobacillus* spp. Appl Environ Microbiol 55:11–16.

de Prada P, Setchell KDR, Hylemon PB (1994) Purification and characterization of a novel 17α-hydroxysteroid dehydrogenase from an intestinal *Eubacterium* sp. VPI 12708. J Lipid Res 35:922–929.

De Witt EH, Lack L (1980) Effects of sulfation patterns on intestinal transport of bile salt sulfate esters. Am J Physiol 238:G34-G39.

Edenharder R, Hammann R (1985) Deoxycholic acid methyl ester-a novel bacterial metabolite of cholic acid. Sys Appl Microbiol 6:18–22.

Edenharder R, Knaflic T (1981) Epimerization of chenodeoxycholic acid to ursodeoxycholic acid by human intestinal lecithinase-lipase-negative *Clostridia.* J Lipid Res 22: 652–658.

Edenharder R, Pfützner A (1988) Characterization of NADP-dependent 12β-hydroxysteroid dehydrogenase from *Clostridium paraputrificum.* Biochim Biophys Acta 962: 362–370.

Edenharder R, Pfützner M (1989) Partial purification and characterization of an NAD-dependent 3β-hydroxysteroid dehydrogenase from *Clostridium innocuum.* Appl Environ Microbiol 55:1656–1659.

Edenharder R, Schneider J (1985) 12β-dehydrogenation of bile acids by *Clostridium paraputrificum, C. tertium,* and *C. difficile* and epimerization at carbon-12 of deoxycholic acid by cocultivation with 12α-dehydrogenating *Eubacterium lentum.* Appl Environ Microbiol 49:964–968.

Edenharder R, Pfützner A, Hammann R (1989a) Characterization of NAD-dependent 3α-

and 3β-hydroxysteroid dehydrogenase and of NADP-dependent 7β-hydroxysteroid dehydrogenase from *Peptostreptococcus productus*. Biochim Biophys Acta 1004: 230–238.

Edenharder R, Pfützner M, Hammann R (1989b) NADP-Dependent 3β-, 7α, and 7β-hydroxysteroid dehydrogenase activities from a lecithinase-lipase-negative *Clostridium* species 25.11.c. Biochim Biophys Acta 1002:37–44.

Elliott WH (1971) Allo bile acids. In: Nair PP, Kritchevsky D, eds. The Bile Acids: Chemistry, Physiology and Metabolism, Vol. 1, pp. 47–92. New York: Plenum Press.

Eneroth P, Gordon B, Ryhage R, Sjövall J (1966a) Identification of mono- and dihydroxy bile acids in human feces by gas-liquid chromatography and mass spectrometry. J Lipid Res 7:511–523.

Eneroth P, Gordon B, Sjövall J (1966b) Characterization of trisubstituted cholanoic acids in human feces. J Lipid Res 7:524–530.

Eriksson H, Gustafsson J-A (1970) Steroids in germfree and conventional rats. Sulfo- and glucuronohydratase activities of fecal contents from conventional rats. Eur J Biochem 13:198–202.

Eriksson H, Gustafsson J-A (1971) Excretion of steroid hormones in adults. Eur J Biochem 18:146–150.

Eriksson H, Gustafsson J-A, Sjövall J (1968) Steroids in germ free and conventional rats: 4. Identification and bacterial formation of 17α-pregnane derivatives. Eur J Biochem 6:219–226.

Eriksson H, Gustafsson J-A, Sjövall J (1969a) Steroids in germ free and conventional rats: free steroids in faeces from conventional rats. Eur J Biochem 9:286–290.

Eriksson H, Gustafsson J-A, Sjövall J (1969b) Steroids in germ free and conventional rats: 21-dehydroxylation by intestinal microorganisms. Eur J Biochem 9:350–354.

Eriksson H, Gustafsson J-A, Sjövall J (1970) Excretion of steroid-hormones in adults: C_{19} and C_{21} steroids in feces from pregnant women. Eur J Biochem 12:520–626.

Eyssen H, Parmentier G (1974) Biohydrogenation of sterols and fatty acids by the intestinal microflora. Am J Clin Nutr 27:1329–1340.

Eyssen H, Piessens-Denef M, Parmentier G (1972) Role of the cecum in maintaining Δ^5-steroid and fatty acid reducing activity of the rat intestinal microflora. J Nutr 102: 1501–1512.

Eyssen HJ, Parmentier GG, Compernolle FC, De Pauw G, Piessens-Denef M (1973) Biohydrogenation of sterols by *Eubacterium* ATCC 21,408-*nova* species. Eur J Biochem 36:411–421.

Eyssen H, DePauw G, Stragier J, Verhulst A (1983) Cooperative formation of ω-muricholic acid by intestinal microorganisms. Appl Environ Microbiol 45:141–147.

Eyssen H, Van Eldere J, Parmentier G, Huijghebaert S, Mertens J (1985) Influence of microbial bile salt desulfation upon the fecal excretion of bile salts in gnotobiotic rats. J Steroid Biochem 22:547–554.

Fedorowski T, Salen G, Tint GS, Mosbach EH (1979) Transformation of chenodeoxycholic acid and ursodeoxycholic acid by human intestinal bacteria. Gastroenterology 77: 1068–1073.

Feighner SD, Dashkevicz MP (1988) Effect of dietary carbohydrates on bacterial cholyltaurine hydrolase in poultry intestinal homogenates. Appl Environ Microbiol 54:337–342.

Feighner SD, Hylemon PB (1980) Characterization of a corticosteroid 21-dehydroxylase from the intestinal anaerobic bacterium, *Eubacterium lentum*. J Lipid Res 21:585–593.

Feighner SD, Bokkenheuser VD, Winter J, Hylemon PB (1979) Characterization of C_{21} and neutral steroid hormone transforming enzyme, 21-dehydroxylase in crude cell extracts of *Eubacterium lentum*. Biochim Biophys Acta 571:154–163.

Ferrari A, Pacini N, Canzi E, Brano R (1980) Prevalence of O_2-intolerant microorganisms in primary bile acid dehydroxylating mouse intestinal microflora. Curr Microbiol 4: 257–260.

Franklund CV, de Prada P, Hylemon PB (1990) Purification and characterization of a microbial, NADP-dependent bile acid 7α-hydroxysteroid dehydrogenase. J Biol Chem 265:9842–9849.

Franklund CV, Baron SF, Hylemon PB (1993) Characterization of the *baiH* gene encoding a bile acid-inducible NADH:flavin oxidoreductase from *Eubacterium* sp. VPI 12708. J Bacteriol 175:3002–3012.

Glass TL, Lamppa RS (1985) Purification and properties of 16α-hydroxyprogesterone dehydroxylase from *Eubacterium* sp. strain 144. Biochim Biophys Acta 837:103–110.

Glass TL, Wheeler LA, Sutter VL, Finegold SM (1979) Transformation of 4-androsten-3,17-dione by growing cultures and cell extracts of *Clostridium paraputrifium*. Biochim Biophys Acta 573:332–342.

Glass TL, Winter J, Bokkenheuser VD, Hylemon PB (1982) Biotransformation of 16α-hydroxyprogesterone by *Eubacterium* sp. 144: non-enzymatic addition of L-cysteine to Δ^{16}-progesterone. J Lipid Res 23:352–356.

Glass TL, Saxerud MH, Casper HH (1991) Properties of a 4-ene-3-ketosteroid-5α-reductase in cell extracts of the intestinal anaerobe, *Eubacterium* sp. strain 144. J Steroid Biochem Molec Biol 39:367–374.

Gopal-Srivastava R, Hylemon PB (1988) Purification and characterization of bile salt hydrolase from *Clostridium perfringens*. J Lipid Res 29:1079–1085.

Gopal-Srivastava R, Mallonee DH, White WB, Hylemon PB (1990) Multiple copies of a bile acid-inducible gene in *Eubacterium* sp. strain VPI 12708. J Bacteriol 172: 4420–4426.

Gorbach SL, Tabaqchali S (1969) Bacteria, bile and the small bowel. Gut 10:963–972.

Graves V, Suraya E, Nishikiza, O (1977) Hydrolysis of steroid glucuronides with β-glucuronidase from bovine liver, *Helix pomatia* and *Escherichia coli*. Clin Chem 23: 532–535.

Gustafsson BE, Midvedt T, Norman A (1966) Isolated fecal microorganisms capable of 7α-dehydroxylating bile acids. J Exp Med 123:413–432.

Gustafsson J-A (1968) Steroids in germ free and conventional rats. Identification of C_{19} and C_{21} steroids in faeces from conventional rats. Eur J Biochem 6:248–255.

Gustafsson J-A, Sjövall J (1968) Steroids in germ free and conventional rats. 6. Identification of 15- and 21-hydroxylated C_{21} steroids in faeces from germfree rats. Eur J Biochem 6:236–247.

Hayakawa S (1973) Microbiological transformation of bile acids. Adv Lipid Res 11: 143–192.

Hayakawa S, Hattori T (1970) 7α-dehydroxylation of cholic acid by *Clostridium bifermentans* strain ATCC 9714 and *Clostridium sordelii* strain NCIB. FEBS Lett 6:131–133.

Heuman DM, Hernandez CR, Hylemon PB, Kubaska WM, Hartman C, Vlahcevic ZR (1988a) Regulation of bile acid synthesis: I. Effects of conjugated ursodeoxycholate and cholate on bile acid synthesis in chronic bile fistula rat. Hepatology 8:358–365.

Heuman DM, Vlahcevic ZR, Bailey ML, Hylemon PB (1988b) Regulation of bile acid synthesis: II. Effect of bile acid feeding on enzymes regulating hepatic cholesterol and bile acid synthesis in the rat. Hepatology 8:892–897.

Higashi S, Setoguschi T, Katsuki T (1979) Conversion of 7-ketolithocholic acid to ursodeoxycholic acid by human intestinal anaerobic microorganisms: interchangeability of chenodeoxycholic acid and ursodeoxycholic acid. Gastroenterol Jpn 14:417–424.

Hirano S, Masuda N (1981a) Epimerization of the 7-hydroxyl group of bile acids by the combination of two kinds of microorganisms with 7α- and 7β-hydroxysteroid dehydrogenase activity, respectively. J Lipid Res 22:1060–1068.

Hirano S, Masuda N (1981b) Transformation of bile acids by *Eubacterium lentum.* Appl Environ Microbiol 42:912–915.

Hirano S, Masuda N (1982a) Characterization of the NADP-dependent 7β-hydroxysteroid dehydrogenase from *Peptostreptococcus productus* and *Eubacterium aureofaciens.* Appl Environ Microbiol 43:1057–1063.

Hirano S, Masuda N (1982b) Enhancement of the 7α-dehydroxylase activity of a gram-positive intestinal anaerobe by *Bacteroides* and its significance in the 7-dehydroxylation of ursodeoxycholic acid. J Lipid Res 23:1152–1158.

Hirano S, Masuda N, Oda H (1981a) In vitro transformation of chenodeoxycholic acid and ursodeoxycholic acid by human intestinal flora, with particular reference to the mutual conversion between the two bile acids. J Lipid Res 22:735–743.

Hirano S, Masuda N, Oda H, Mukai H (1981b) Transformation of bile acids by *Clostridium perfringens.* Appl Environ Microbiol 42:394–399.

Hirano S, Nakama R, Tamaki M, Masuda N, Oda H (1981c) Isolation and characterization of thirteen intestinal microorganisms capable of 7-dehydroxylating bile acids. Appl Environ Microbiol 41:737–745.

Hofmann AF (1976) The enterohepatic circulation of bile acids in man. Adv Intern Med 21:501–534.

Hofmann AF (1977) The enterohepatic circulation of conjugated bile acids in healthy man: quantitative description and function. In: Polonovsky J, ed. Exposes Annuels de Biochimie Medicale (33e serie), pp. 69–86. New York: Masson Publishing.

Hofmann AF (1979) The enterohepatic circulation of bile acids in man. In: Paumgartner

G, ed. Bile Acids Clinics in Gastroenterology, Vol. 6, pp. 3–24. Philadelphia: W.B. Saunders.

Hofmann AF, Roda A (1984) Physiological properties of bile acids and their relationship to biological properties: an overview of the problem. J Lipid Res 25:1477–1489.

Hofmann AF, Sjövall J, Kurz G, Radominska A, Schteingart CD, Tint GS, Vlahcevic ZR, Setchell KDR (1992) A proposed nomenclature for bile acids. J Lipid Res 33:599–604.

Holland HL, Riemland E (1984) The mechanism of C-21 dehydroxylation of tetrahydrodeoxycorticosterone by *Eubacterium lentum*. J Chem Soc Chem Commun 22:1481–1482.

Huijghebaert SM, Mertens JA, Eyssen HJ (1982) Isolation of a bile salt sulfatase-producing *Clostridium* sp. strain from rat intestinal microflora. Appl Environ Microbiol 44: 1030–1034.

Hylemon PB, Glass TL (1983) Biotransformation of bile acids and cholesterol by the intestinal microflora. In: Hentges DJ, ed. Human Intestinal Microflora in Health and Disease, pp. 189–213. New York: Academic Press.

Hylemon PB, Sherrod JA (1975) Multiple forms of 7α-hydroxysteroid dehydrogenase in selected strains of *Bacteroides fragilis*. J Bacteriol 122:418–424.

Hylemon PB, Stellwag EJ (1976) Bile acid biotransformation rates of selected gram-positive and gram-negative intestinal anaerobic bacteria. Biochem Biophys Res Commun 69:1088–1094.

Hylemon PB, Moody DP, Cohen BI, Une M, Mosbach EH (1984) Effect of bile acid analogs on 7α-dehydroxylase activity in *Eubacterium* sp. VPI 12708. Steroids 44: 329–336.

Hylemon PB, Melone PD, Franklund CV, Lund E, Björkhem I (1991) Mechanism of intestinal 7α-dehydroxylation of cholic acid: evidence that allo-deoxycholic acid is an inducible side-product. J Lipid Res 32:89–96.

Kawamoto K, Horibe I, Uchida K (1989) Purification and characterization of a new hydrolase for conjugated bile acids, chenodeoxycholyltaurine hydrolase, from *Bacteroides vulgatus*. J Biochem (Tokyo) 106:1049–1053.

Kellogg TF (1973) On the Site of the microbiological reduction of cholesterol to coprostanol in the rat. Lipids 8:658–659.

Kellogg TF, Wostmann BS (1969) Fecal neutral steroids and bile acids from germfree rats. J Lipid Res 10:495–503.

Kelsey MI, Sexton SA (1976) The biosynthesis of ethyl esters of lithocholic acid and isolithocholic acid by rat intestinal microflora. J Steroid Biochem 7:641–647.

Kelsey MI, Thompson RJ (1976) The biosynthesis of ethyl lithocholate by fecal microorganisms. J Steroid Biochem 7:117–124.

Kelsey MI, Molina JE, Huang S-KS, Hwang K-K (1980) The identification of microbial metabolites of sulfolithocholic acid J Lipid Res 21:751–759.

Kelsey MI, Hwang K-K, Huang S-KS, Shaikh B (1981) Characterization of microbial metabolites of sulfolithocholic acid by high-performance liquid chromatography. J Steroid Biochem 14:205–211.

Kicman AT, Brooks RV, Collyer SC, Cowan DA, Nanjee MN, Southan GJ, Wheeler MJ (1990) Criteria to indicate testosterone administration. Br J Sports Med 17:354–359.

Kihira K, Okamoto A, Ikawa S, Mikami T, Yoshii M, Mosbach EH, Hoshita T (1991) Metabolism of sodium $3\alpha,7\alpha$-dihydroxy-5β-cholane-24-sulfonate in hamsters. J Biochem (Tokyo) 109:879–881.

Kobashi K, Nishizawa I, Yamada T, Hase J (1978) A new hydrolase specific for taurine conjugates of bile acids. J Biochem (Tokyo) 84:495–497.

Korpela JT, Fotsis T, Adlercreutz H (1986) Multicomponent analysis of bile acids in faeces by anion exchange and capillary column gas-liquid chromatography: application in oxytetracycline treated subjects. J Steroid Biochem 25:277–284.

Korpela JT, Adlercreutz H, Turunen MJ (1988) Fecal free and conjugated bile acids and neutral sterols in vegetarians, omnivores, and patients with colorectal cancer. Scand J Gastroenterol 23:277–283.

Krafft AE, Hylemon PB (1989) Purification and characterization of a novel form of 20α-hydroxsteroid dehydrogenase from *Clostridium scindens*. J Bacteriol 171:2925–2932.

Krafft AE, Winter J, Bokkenheuser VD, Hylemon PB (1987) Cofactor requirements of steroid-17,20-desmolase and 20A-hydroxysteroid dehydrogenase activities in cell extracts of *Clostridium scindens*. J Steroid Biochem 28:49–54.

Kuroki S, Mosbach EH, Cohen BI, Stenger RJ, McSherry CK (1987) 7-Methyl bile acids: 7β-methyl-cholic acid inhibits bacterial 7-dehydroxylation of cholic acid and chenodeoxycholic acid in the hamster. J Lipid Res 28:856–863.

Laatikainen T (1970a) Excretion of neutral steroid hormones in human bile. Ann Clin Res 2:3–38.

Laatikainen T (1970b) Identification of $C_{21}O_3$ and $C_{21}O_4$ steroids in human bile. Eur J Biochem 14:372–378.

Laatikainen T (1970c) Quantitative studies on the excretion of glucuronide and mono- and disulfate conjugates of neutral steroids in human bile. Ann Clin Res 2:338–349.

Lindstedt S, Sjövall J (1957) On the formation of deoxycholic acid from cholic acid in the rabbit. Acta Chem Scand 11:421–426.

Lipsky RH, Hylemon PB (1980) Characterization of a NADH:flavin oxidoreductase induced by cholic acid in a 7α-dehydroxylating intestinal *Eubacterium* species. Biochim Biophys Acta 612:328–336.

Lundeen SG, Savage DC (1990) Characterization and purification of bile salt hydrolase from *Lactobacillus* sp. strain 100–100. J Bacteriol 172:4171–4177.

Lundeen SG, Savage DC (1992a) Characterization of an extracellular factor that stimulates bile salt hydrolase activity in *Lactobacillus* sp. strain 100–100. FEMS Microbiol Lett 94:121–126.

Lundeen SG, Savage DC (1992b) Multiple forms of bile salt hydrolase from *Lactobacillus* sp. strain 100–100. J Bacteriol 174:7217–7220.

Macdonald IA, Roach PB (1981) Bile salt induction of 7α- and 7β-hydroxysteroid dehydrogenases in *Clostridium absonum*. Biochim Biophys Acta 665:262–269.

Macdonald IA, Williams CN, Mahony DE, Christie WM (1975) NAD- and NADP-Dependent 7α-hydroxysteroid dehydrogenases from *Bacteroides fragilis*. Biochim Biophys Acta 384:12–24.

Macdonald IA, Meier EC, Mahony DE, Costain GA (1976) 3α-, 7α-, and 12α-Hydroxysteroid dehydrogenase activities from *Clostridium perfringens*. Biochim Biophys Acta 450: 142–153.

Macdonald IA, Mahony DE, Jellet JF, Meier CE (1977) NAD-Dependent 3α- and 12α-hydroxysteroid dehydrogenase activities from *Eubacterium lentum* ATCC No. 25559 Biochim Biophys Acta 489:466–476.

Macdonald IA, Jellett JF, Mahony DE, Holdeman LV (1979) Bile salt 3α- and 12α-hydroxysteroid dehydrogenases from *Eubacterium lentum* and related organisms. Appl Environ Microbiol 37:992–1000.

Macdonald IA, Rochon YP, Hutchison DM, Holdeman LV (1982) Formation of ursodeoxycholic acid from chenodeoxycholic acid by a 7β-hydroxysteroid dehydrogenase-elaborating *Eubacterium aureofaciens* strain co-cultured with 7α-hydroxysteroid dehydrogenase-elaborating organisms. Appl Environ Microbiol 44:1187–1195.

Macdonald IA, Bokkenheuser VD, Winter J, McLeron AM, Mosbach EH (1983a) Degradation of steroids in the human gut. J Lipid Res 24:675–700.

Macdonald IA, Hutchison DM, Forrest TP, Bokkenheuser VD, Winter J, Holdeman LV (1983b) Metabolism of primary bile acids by *Clostridium perfringens*. J Steroid Biochem 18:97–104.

Macdonald IA, White BA, Hylemon PB (1983c) Separation of 7α- and 7β-hydroxysteroid dehydrogenase activities from *Clostridium absonum* ATCC# 27555 and cellular response of this organism to bile acid inducers. J Lipid Res 24:1119–1126.

Madsen D, Beaver M, Chang L, Gruchner-Kardoss E, Wostmann B (1976) Analysis of bile acids in conventional and germfree rats. J Lipid Res 17:107–112.

Mallonee DH, White WB, Hylemon PB (1990) Cloning and sequencing of a bile acid-inducible operon from *Eubacterium* sp strain VPI 12708. J Bacteriol 172:7011–7019.

Mallonee DH, Adams JL, Hylemon PB (1992) The bile acid-inducible *baiB* gene from *Eubacterium* sp. strain VPI 12708 encodes a bile acid-coenzyme A ligase. J Bacteriol 174:2065–2071.

Masuda N (1981) Deconjugation of bile salts by *Bacteroides* and *Clostridium*. Microbiol Immunol 25:1–11.

McNamara DJ, Prosa A, Miettinen TA (1981) Thin-layer and gas-liquid chromatographic identification of neutral steroids in human and rat feces. J Lipid Res 22:474–484.

Midvedt T (1974) Microbial bile acid transformation. Am J Clin Nutr 27:1341–1347.

Midvedt T, Norman A (1967) Bile acid transformations by microbial strains belonging to genera found in intestinal contents. Acta Pathol Microbiol Scand 71:629–638.

Moore WEC, Moore LVH, Cato EP (1988) You and your flora. US Fed Culture Collect 18:7–22.

Morris GN, Winter J, Cato EP, Ritchie AE, Bokkenheuser VD (1985) *Clostridium scindens* sp. nov. a human intestinal bacterium with desmolitic activity on corticoids. Int J Syst Bacteriol 35:478–481.

Morris GN, Winter J, Cato EP, Ritchie AE, Bokkenheuser VD (1986) *Eubacterium desmolans* sp. nov. a steroid desmolase-producing species from cat fecal flora. Int J Syst Bacteriol 36:183–186.

Mott GE, Brinkley AW (1979) Plasmenylethanolamine: growth factor for cholesterol-reducing *Eubacterium.* J Bacteriol 139:755–760.

Mott GE, Brinkley AW, Mersinger (1980) Biochemical characterization of cholesterol-reducing *Eubacterium.* Appl Environ Microbiol 40:1017–1022.

Mower HF, Ray RM, Shoff R, Stemmerman GM, Monura A, Glober G, Kamiyama S, Shimada A, Yamakawa H (1979) Fecal bile acids in two Japanese populations with different colon cancer risks. Cancer Res 39:328–331.

Nabarro JD, Moxham N, Walker G, Slater JDH (1957) Rectal hydrocortisone. Br Med J 2:272–274.

Nair PP, Gordon M, Reback J (1976) The enzymatic cleavage of the carbon-nitrogen bond in $3\alpha,7\alpha,12\alpha$-trihydroxy-5β-cholan-24-oylglycine. J Biol Chem 242:7–11.

Nancarrow Cd, Sharof MA, Sweet F (1981) Purification of 20α-hydroxysteroid oxidoreductase from bovine fetal erythrocytes. Steroids 37:539–553.

Norman A (1964) Faecal excretion products of cholic acid in man. Br J Nutr 18:173–186.

Norman A, Palmer RH (1964) Metabolites of lithocholic acid-24-C^{14} in human bile and feces. J Lab Clin Med 63:986–1001.

Norman A, Sjövall J (1958) On the transformation and enterohepatic circulation of cholic acid in the rat. Bile acids and steroids, 68. J Biol Chem 233:872–885.

Northfield TC, Drassar BS, Wright JT (1973) Value of small intestinal bile acid analysis in the diagnosis of the stagnant loop syndrome. Gut 14:341–347.

Pacini N, Albini E, Ferrari A, Zanchi R, Marca G, Bandiera T (1987) Transformation of sulfated bile acids by human intestinal microflora. Arzneimittelforschung 37:983–987.

Palmer RH (1972) Metabolism of lithocholate in humans. In: Back P, Gerock W, eds. Bile Acids in Human Disease, pp. 65–69. Stuttgart: F.K. Schattauer Verlag.

Paone DAM, Hylemon PB (1984) HPLC purification and preparation of antibodies to cholic acid-inducible polypeptides from *Eubacterium* sp. VPI 12708. J Lipid Res 25: 1343–1349.

Parmentier G, Eyssen H (1974) Mechanism of biohydrogenation of cholesterol into coprostanol. Eur J Biochem 348:279–284.

Robben J, Parmentier G, Eyssen H (1986) Isolation of a rat intestinal *Clostridium* strain producing 5α- and 5β-bile salt 3α-sulfatase activity. Appl Environ Microbiol 51:32–38.

Robben J, Caenepeel P, Van Eldere J, Eyssen H (1988) Effects of intestinal microbial bile salt sulfatase activity on bile salt kinetics in gnotobiotic rats. Gastroenterology 94: 494–502.

Robben J, Janssen G, Merckx R, Eyssen H (1989) Formation of Δ^2- and Δ^3-cholenoic acids from bile acid 3-sulfates by a human intestinal *Fusobacterium* strain. Appl Environ Microbiol 55:2954–2959.

Sacquet EC, Raibard PM, Mejean C, Riottot MJ, Leprince C, Leglise PC (1979) Bacterial formation of ω-muricholic acid in rats. Appl Environ Microbiol 37:1127–1131.

Sadzikowski MR, Sperry JF, Wilkins TD (1977) Cholesterol-reducing bacterium from human feces. Appl Environ Microbiol 34:355–362.

Savage DC (1977) Microbial ecology of the gastrointestinal tract. Annu Rev Microbiol 31:107–133.

Schmassmann A, Angellotti MA, Ton-Nu HT, et al. (1990a) Transport, metabolism, and effect of chronic feeding of cholylsarcosine, a conjugated bile acid resistant to deconjugation and dehydroxylation. Gastroenterology 98:163–174.

Schmassmann A, Hofmann AF, Angellotti MA, et al. (1990b) Prevention of ursodeoxycholate hepatotoxicity in the rabbit by conjugation with N-methyl amino acids. Hepatology 11:989–996.

Schubert K, Schlegal J, Bohme KH, Horhold C (1967) Mikrobielle hydrierungs-und dehydrierungs-Reaktionen bei Δ^4-3-ketosteroiden mit einer 5-hydroxy-gruppe. Biochim Biophys Acta 144:132–138.

Sherrod JA, Hylemon PB (1977) Partial purification and characterization of NAD-dependent 7α-hydroxysteroid dehydrogenase from *Bacteroides thetaiotaomicron*. Biochim Biophys Acta 486:351–358.

Sjövall J (1960) Bile acids in man under normal and pathological conditions: bile acids and steroids. Clin Chim Acta 5:33–41.

Snog-Kjaer A, Prange J, Dam H (1956) Conversion of cholesterol to coprostanol by bacteria. J Gen Microbiol 14:256–260.

Stange EF, Dietschy JM (1985) Cholesterol absorption and metabolism by the intestinal epithelium. In: Danielsson H, Sjövall J, eds. Sterols and Bile Acids, pp. 121–149. Amsterdam: Elsevier.

Stellwag EJ, Hylemon PB (1976) Purification and characterization of bile salt hydrolase from *Bacteroides fragilis* subsp. *fragilis*. Biochim Biophys Acta 452:165–176.

Stellwag EJ, Hylemon PB (1979) 7α-dehydroxylation of cholic acid and chenodeoxycholic acid by *Clostridium leptum*. J Lipid Res 20:325–333.

Stokes NA, Hylemon PB (1985) Characterization of Δ^4-3-ketosteroid-5β-reductase and 3β-hydroxysteroid dehydrogenase in cell extracts of *Clostridium innocuum*. Biochim Biophys Acta 836:255–261.

Sutherland JD, Williams CN (1985) Bile acid induction of 7α- and 7β-hydroxysteroid dehydrogenases in *Clostridium limosum*. J Lipid Res 26:344–350.

Takamine F, Imamura T (1985) 7β-Dehydroxylation of 3,7-dihydroxy bile acids by a *Eubacterium* species strain C-25 and stimulation of 7β-dehydroxylation by *Bacteroides distasonis* strain K-5. Microbiol Immunol 29:1247–1252.

Tammar AR (1966) Presence of allocholic acid in mammalian (including human) bile. Biochem J 98:25p-26p.

Tannock GW, Dashkevicz MP, Feighner SD (1989) Lactobacilli and bile salt hydrolase in the murine intestinal tract. Appl Environ Microbiol 55:1848–1851.

Taylor W (1971) The excretion of steroid hormone metabolites in bile and feces. Vitam Horm 29:201–285.

Van Eldere J, Robben J, De Pauw G, Merckx R, Eyssen H (1988) Isolation and identification of intestinal steroid-desulfating bacteria from rats and humans. Appl Environ Microbiol 54:2112–2117.

Van Eldere J, Mertens J, Eyssen H (1990) Influence of intestinal bacterial desulfation on the enterohepatic circulation of dehydroepiandrosterone sulfate. J Steroid Biochem 36: 451–456.

Van Eldere J, Parmentier G, Asselberghs, Eyssen H (1991) Partial characterization of the steroid sulfatases in *Peptococous niger* H4. App Environ Microbiol 57:69–76.

Vlahcevic ZR, Heuman DM, Hylemon PB (1990) Physiology and pathophysiology of enterohepatic circulation of bile acids. In: Zakim D, Boyer TD, eds. Hepatology. A Textbook of Liver Disease, pp. 341–377. Philadelphia: W.B. Saunders.

Wade AP, Slater JDH, Kellie AE, Holliday ME (1959) Urinary excretion of 17-ketosteroids following rectal infusion of cortisol. J Clin Endocrinol Metab 19:444–453.

Watkins WE, Glass TL (1991) Characteristics of 16-dehydroprogesterone reductase in cell extracts of the intestinal anaerobe, *Eubacterium* sp. strain 144. 38:257–263.

White BA, Lipsky RL, Fricke RJ, Hylemon PB (1980) Bile acid induction specificity of 7α-dehydroxylase activity in an intestinal *Eubacterium* species. Steroids 35:103–109.

White BA, Cacciapuoti AF, Fricke RJ, Whitehead TR, Mosbach EH, Hylemon PB (1981) Cofactor requirements for 7α-dehydroxylation of cholic and chenodeoxycholic acid in cell extracts of the intestinal anaerobic bacterium, *Eubacterium* species VPI 12708. J Lipid Res 22:891–898.

White BA, Fricke RJ, Hylemon PB (1982) 7β-dehydroxylation of ursodeoxycholic acid by whole cells and cell extracts of the intestinal anaerobic bacterium, *Eubacterium* species VPI 12708. J Lipid Res 23:145–153.

White BA, Paone DAM, Cacciapuoti AF, Fricke RJ, Mosbach EH, Hylemon PB (1983) Regulation of bile acid 7-dehydroxylase activity by NAD+ and NADH in cell extracts of *Eubacterium* species VPI 12708. J Lipid Res 24:20–27.

White WB, Coleman JP, Hylemon PB (1988a) Molecular cloning of a gene encoding a 45,000-dalton polypeptide associated with bile acid 7-dehydroxylation in *Eubacterium* sp. strain VPI 12708. J Bacteriol 170:611–616.

White WB, Franklund CV, Coleman JP, Hylemon PB (1988b) Evidence for a multigene family involved in bile acid 7-dehydroxylation in *Eubacterium* sp. strain VPI 12708. J Bacteriol 170:4555–4561.

Wilkins TD, Hackman AS (1974) Two patterns of neutral steroid conversion in the feces of normal North Americans. Cancer Res 34:2250–2254.

Winter J, Cerone-Mclernon A, O'Rourke-Locascio S, Ponticervo L, Bokkenheuser VD (1982a) Formation of 20β-dihydrosteroids by anaerobic bacteria. J Steroid Biochem 17:661–667.

Winter J, O'Rourke-Locascio S, Bokkenheuser VD, Hylemon PB, Glass TL (1982b) 16α-Dehydroxylation of corticoids by bacteria isolated from rat fecal flora. J Steroid Biochem 16:231–237.

Winter J, Norris GN, O'Rourke-Locascio S, Bokkenheuser VD, Mosbach EH, Cohen BI and Hylemon PB (1984a) Mode of action of steroid desmolase and reductases synthetsized by *Clostridium scindens* (formerly *Clostridium* Strain 19). J Lipid Res 25: 1124–1131.

Winter J, O'Rourke-Locascio S, Bokkenheuser VD, Mosbach EH, Cohen BI (1984b) Reduction of 17-keto steroids by anaerobic microorganisms isolated from human fecal flora. Biochim Biophys Acta 795:208–211.

Wostmann BS, Beaver M, Madsen D (1979) Bile acids in germfree piglets. In: Fliedner TM, Heit H, Niethammer D, Pflieger H, eds. Clinical and Experimental Gnotobiotics. Zentrabl Bakteriol, Suppl. 7, pp. 121–123. Stuttgart: Fischer.

Yoshimoto T, Higashi H, Kanatani A, et al. (1991) Cloning and sequencing of the 7α-hydroxysteroid dehydrogenase gene from *Escherichia coli* HB101 and characterization of the expressed enzyme. J Bacteriol 173:2173–2179.

14

Gastrointestinal Toxicology of Monogastrics

King-Thom Chung

1. Introduction

The gastrointestinal tract is one of the most active sites in the body where metabolism takes place. A large number of microorganisms are found in the gastrointestinal tract. The number of microorganisms per gram colon contents exceeds 10^{11}/g dry weight, constituting more than 40% of the total fecal mass (Stephen and Cummings 1980). These microorganisms represent a sources of great metabolic power and may play a role in many disease processes including cancer (Goldin 1986a,b, 1990, Gorbach and Goldin 1990, Hill 1987, Mallett and Rowland 1990). The microbiota produce a wide range of substances de novo and/or from ingested sources including the diet. These toxic substances may be mutagens, tumor promoters, initiators or carcinogens in addition to many types of toxins. Toxins are defined as chemicals which have adverse effects on living systems. Toxicology today involves studies of the effects of specific substances on any biological system, whether a whole organism, organ or tissue, through the interaction of toxin with target.

There are many types of natural toxins present in food plants and animals. Some toxins are formed through microbiological and intestinal enyzymatic conversion of food constituents, or by the processing of food. Some are added as additives, but converted into toxins after metabolism. In this chapter, the types of toxins that are formed in, or likely to enter, the gastrointestinal tract will be reviewed briefly. In addition, the roles played by the intestinal microbiota and intestinal enzymes in transforming the food toxicants or promoting potential hazardous compounds will also be discussed.

2. Toxins from Animal Sources

2.1. Bile Acids

Bile acids occurring in livers of bear, cattle, sheep, goat, and rabbit are toxic. The bile acid acts as a suppressant on the central nervous system. Cholic acid, deoxycholic acid and taurolithocholic acid are some of the bile acids (Fig. 14.1). Animal studies have shown that bile acids can promote tumor formation in the bowel (Narisawa et al. 1974, Campbell et al. 1975, Mastromarino et al. 1976, 1978) and liver

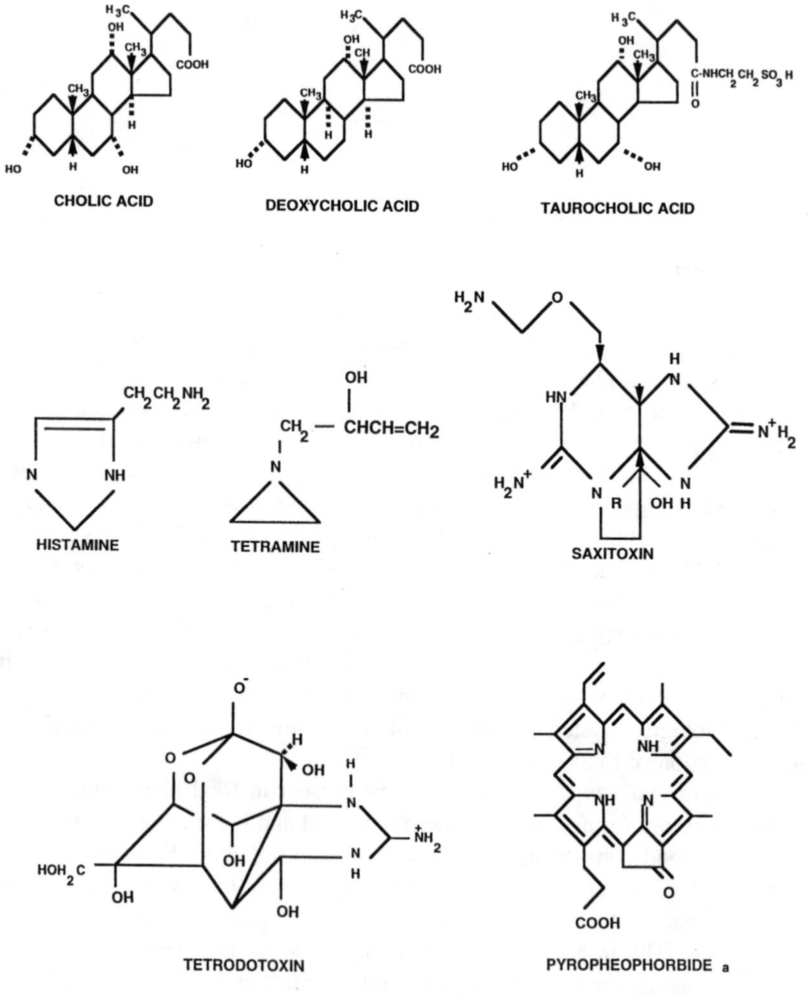

Figure 14.1. Chemical structures of toxins from animal sources.

(Hiasa et al. 1971). Generally, the livers commonly eaten in the western world do not contain sufficient quantities of bile acid to produce toxic effects.

2.2. Scombroid Poisoning

Many seafoods, particularly *Scombroides* species, including mackerel, tuna, bonito, skipjack, Spanish mackerel, and Scomberesocidae (saury), as well as some nonscombroid species such as mahi-mahi, blue fish, jack mackerel, yellow tail, amberjack, herring, sardines, and anchovies contain high concentrations of the amino acid histidine. Histidine can be converted to histamine through the action of histidine decarboxylase from contaminating microorganisms mostly *Morganella morganii, Klebsiella pneumoniae, Lactobacillus* species and a few other enteric bacteria (Taylor 1986, 1988). The consumption of these fish species with high histamine can cause the onset of symptoms lasting a few minutes to a few hours, although the duration of the illness is rather short. The symptoms resemble allergic reactions caused by histamine. The gastrointestinal tract is affected in most cases with associated pain, nausea, vomiting, diarrhea and cramping. Poisoning may also be accompanied by neurological symptoms such as headache, palpitations, flushing, tingling, burning and itching. In serious cases, hypotension, rash, urica-ria, edema, and localized inflammation may occur. A histamine hazard level of 50 mg/100 g fish has been established by the FDA although it can reach toxic concentrations of up to 500 mg/100 g in fish without the development of off-flavors (Taylor 1988).

2.3. Saxitoxin

A few species of dinoflagellates, notably *Gonyaulax catenella* (on the U.S. West Coast) and *G. tamarensis* (east coast) produce a series of closely related toxins that accumulates in shellfish such as clams, crabs, mussels, cockles, scallops and oyster which feed on these algae. This often occurs during the so called red tides or blooms on the sea coast. The toxin is called saxitoxin. The basic chemical structure of saxitoxin is shown in Figure 14.1 (Hall and Reichardt 1984). Saxitoxin toxicity is also commonly called paralytic shellfish poisoning (PSP). The estimated minimum lethal dose in humans is 4 mg or less (Taylor 1988). Symptoms of total toxic reaction from saxitoxin poisoning are numbness of lips, hands and feet, 15 min to 2 or 3 hour after ingestion of poisoning shellfish. This is followed by difficulty in walking, vomiting, coma, and death. In the United States, the maximum allowable amount of saxitoxin in seafood products is 80 μg/100 g (Taylor 1988).

2.4. Tetramine

The marine gastropods of the family *Buccinidae* (whelk, seasnail), notably *Neptunea arthritica* and *N. antiqua*, contain tetramine, which is toxic. The symptoms of tetramine poisoning are headache localized in the back of the head, dizziness, loss of balance, eye pain, vomiting, and hives (Taylor 1988).

2.5. Tetrodotoxin

Many species of puffer fish (*Sphaeroides*), ocean sunfish and porcupine fish contain tetrodotoxin which is also called puffer or fugu poison. The highest concentrations of the toxin are found in the ovaries, roe and liver, with lesser amounts in the skin and intestines. Small amounts can be found in the muscles and blood. The symptoms of tetrodotoxin intoxication are numbness of the lips, tongue; and fingers; nausea; vomiting; and anxiety. There could also be muscular paralysis of extremities without loss of tendon reflexes, followed by lack of muscular coordination, loss of consciousness, respiratory paralysis, and death. The fatal oral dose for humans is about 1 to 2 mg crystalline tetrodotoxin. Tetrodotoxin is an amino perhydroquinazoline compound (Fig. 14.1) having a molecular formula of $C_{11}H_{17}N_3$. There are about seven different derivatives of this compound with different degrees of toxicity (Taylor 1988).

2.6. Ciguatoxin

The ciguatoxin or ciguatera poisoning results from the ingestion of fish that have become toxic by feeding on toxic dinoflagellates such as *Gambierdiscus toxicus, Prococentrum concavum*, and others. The commonly implicated fish are grouper, barracuda, red snapper, and sea bass. The internal organs of the fish, the liver, intestine, roe and gonads, are more toxic than muscle (Hokama and Miyahara 1986). The symptoms are tingling on the lips, tongue, and throat; numbness in these areas follows. It may also cause nausea, vomiting, a metallic taste, dryness of the mouth, abdominal cramps, diarrhea, headache, prostration, chill, fever, profuse sweating, dilated eyes and blurred vision, temporary blindness, paralysis, and death. Thus far the toxic principle of ciguatera poisoning is not clear, but several neurotoxins have been isolated from ciguatoxic fishes. The major neurotoxin is ciguatoxin, which is a lipid soluble polyether compound of uncertain structure with a molecular weight of approximately 1,100 Da (Miller et al. 1984).

2.7. *Pyropheophorbide* a

Abalone may contain pyropheophorbide *a* (Fig. 14.1), a derivative of chlorophyll which comes from seaweed. Pyropheophorbide *a* is photoactive and can cause red edema on face and hands (Shibamoto and Bjeldanes 1993a).

2.8. Other Shellfish Poisonings

The edible tissues of filter-feeding mollusks occasionally contain neurotoxic shellfish poison called breves toxin B and C which are polyether-type compounds.

These toxins are produced by the marine dinoflagellate *Ptychodiscus brevis*. These toxins probably bind to nerve cells, but the mechanism of action remains to be investigated. The symptoms of these neurotoxic shellfish poisonings include tingling and numbness of the lips, tongue, throat, and area around the mouth; muscular aches; gastrointestinal discomfort; and dizziness (Baden et al. 1984; Taylor 1988).

Diarrhetic shellfish poisoning also occurs from mussels, scallops, and short-necked clams fed on *Dinophysis fortii* and *D. acuminata*. These toxins are polyether compounds and at least five of them are known: okadaic acid, two dinophysistoxins, and two pectenotoxins. The symptoms of these diarrhetic shellfish poisonings include gastrointestinal complaints, diarrhea, nausea, vomiting, and abdominal pain. This type of poisoning has been identified primarily in Japan, the Netherlands, and Chile (Kat 1983, Yasumoto et al. 1984).

2.9. Others

There are several other toxins associated with foods of animal origin. Rancid or oxidized fat is known to be toxic, especially when vitamin E is deficient (Pryor 1985; Gurr 1988). Common amino acids can be toxic at high levels, and are promoters of bladder cancer in rats (Hegarty 1986). Leucine, isoleucine, and valine are tumor promoters (Nishio 1986). Tryptophan is a tumor promoter (NAS/NRC 1989), and many tryptophan metabolites are carcinogens in animals (Bryan 1969, 1971, Chung et al. 1975, Bowden et al. 1976).

3. Toxins from Plant Sources

Plants provide energy, protein, fiber, and vitamins important for human health, but also produce compounds toxic to animals and humans. A few of these will be discussed in this section.

3.1. Goitrogens

Many food plants such as soybeans, pine nuts, peanuts, millet, fruits, and vegetables contain goitrogens, substances that cause goiters. *Brassica* such as broccoli, Brussels sprouts, cabbage, cauliflower, horseradish, kale, kohlrabi, mustard, and mustard seed are the primary dietary sources of goitrogens (Jones 1992a). Turnips, rutabaga, carrot, peach, pear, radish, and strawberry may also contain goitrogens. Together with the lack of adequate iodine, goitrogenic foods are primarily responsible for the world's goiter cases (Gaitan 1990). The goitrogens have been identified as glucosinolate (Fenwick et al. 1989) (Fig. 14.2A). The glucosinolate can be converted to several products by thioglucosidase which is present in plants and gastrointestinal microbiota. Products of this reaction are nitriles, thiocyanates, and oxazolidines. The oxazolidine goitrin is a thyroid-suppressing agent. It can cause mild thyroid enlargement (Shibamoto and Bjeldanes 1993b). Thiocyanate can also cause thyroid enlargement. Nitrile is a well-known toxin, which can cause acute toxicity.

3.2. Cyanogenic Glycosides

Cyanogenic glycosides are present in many food plants. For example, amygdalin is found in bamboo shoots, sorghum, choke cherries, pin cherries, and in the pits of apple, apricot, cherry, plum, quince, and peach (Conn 1981), while linamarin is found in cassava root, flax, and lima bean (Jones 1992a), and dhurrin is found in sorghum. These cyanogenic glycosides can be cleaved by β-glucosidase to release the sugar and cyanohydrin moieties. The cyanohydrin will further decompose spontaneously or by hydroxynitrile lyase to the corresponding ketone or aldehyde and hydrogen cyanide (Shibamoto and Bjeldane 1993b). Hydrogen cyanide is a potent respiratory inhibitor. β-Glucosidase is present in plant tissues and in many intestinal bacteria including *Streptococcus faecium* and other *Streptococcus* species (MacDonald et al. 1984), *Bacteroides uniformis*, and *Bacteroides ovatus* (Bokkenheuser et al.1987). Brassicas contain both cyanogens and goitrogens. Allyl isothiocyanate present in these vegetables can cause chromosomal aberration in hamster cells and cancer in rats (Ames 1983). Another cyanogen, benzyl cyanide, is present in cress, which is toxic to mice (Jones 1992a).

3.3. Favism

Fava bean, commonly known as broad bean, contains vicine and convicin. Vicine and convicin can be converted to divicine and isouramil by β-glycosidase (Fig. 14.2). Divicine and isouramil are toxic to humans, producing favism, which is a syndrome of acute hemolytic anemia. The clinical symptoms include pallor, fatigue, shortness of breath, nausea, abdominal pain, fever, and chills. In more severe cases, renal failure may occur. Fatalities have been reported in infants and children (Marquardt 1989).

Fava beans also contain high amounts of monoamine oxidase inhibitor which can cause headaches, palpitations and sharp increases in blood pressure (Jones 1992a). Fava bean has also been reported to be associated with a certain type of cancer in Latin America (Ames 1983, Nagao et al. 1986).

3.4. Lathyrism

Certain peas of the genus *Lathyrus*, known as vetch peas, chick-pea or garbanzo, and sweet pea (*L. sativus*) contain β-L-glutamylaminopropionitrile (BAPN). BAPN can inhibit the activity of lysyl oxidase which is an important enzyme for the cross-linking of collagen, the primary structural protein of connec-

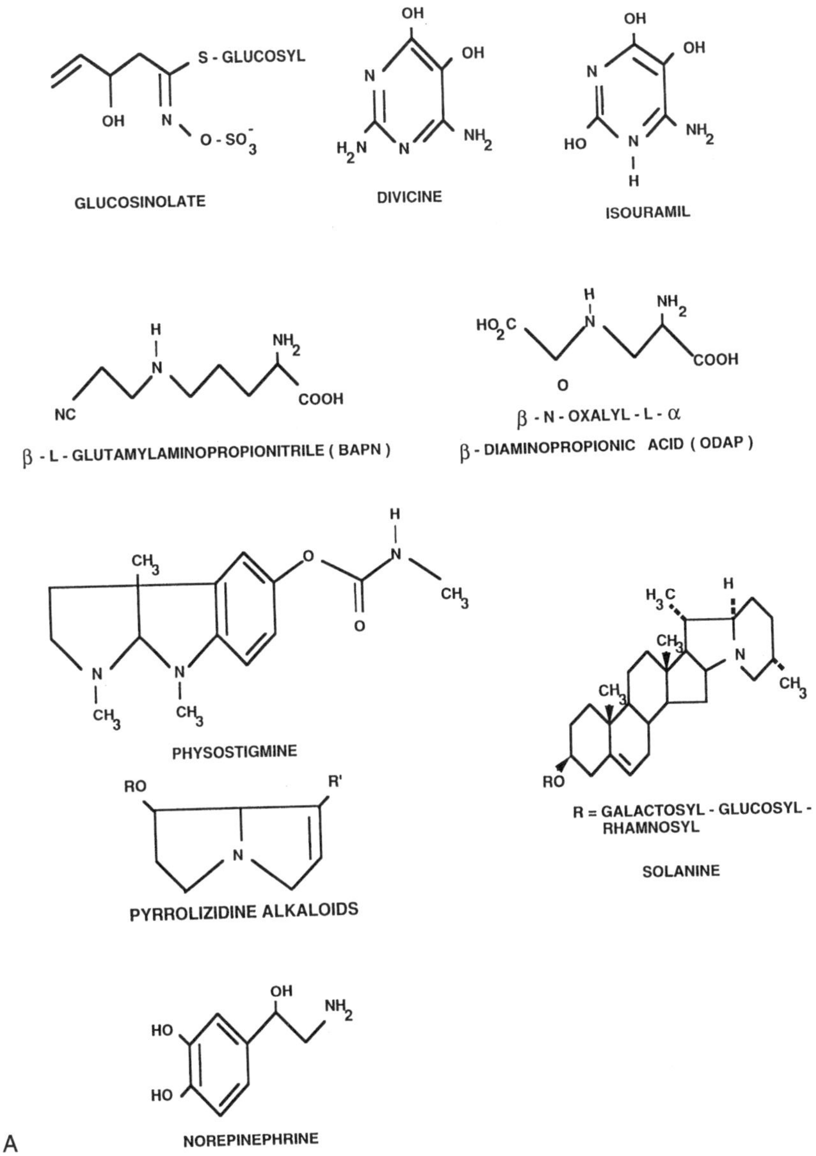

Figure 14.2. (A) Chemical structures of toxins from plant sources. (B) Chemical structures of some mutagens in food plants (following pages).

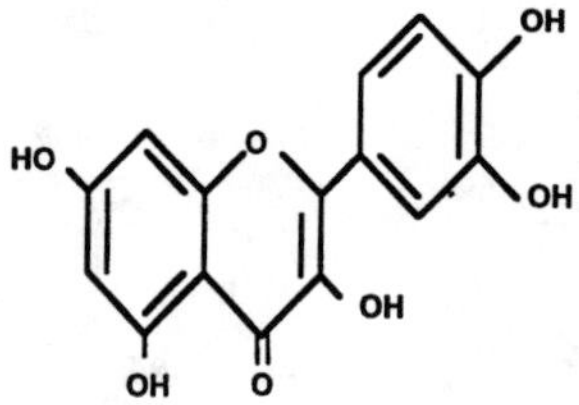

QUERCETIN

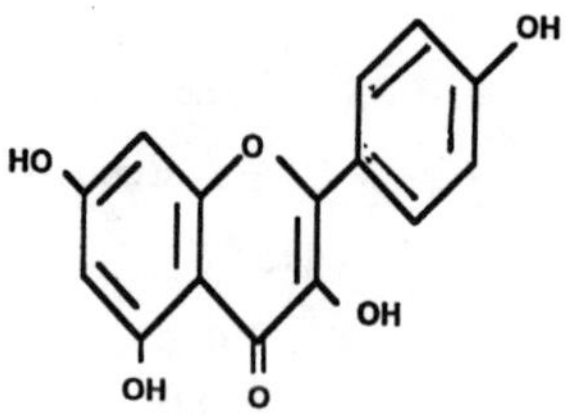

KAEMPFEROL

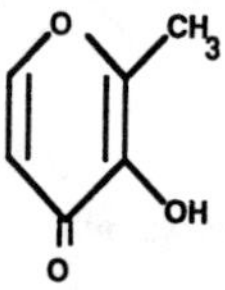

MALTOL

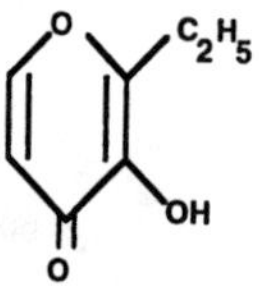

ETHYL MALTOL

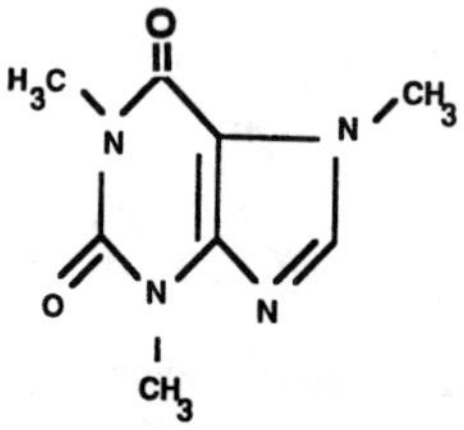

CAFFEINE

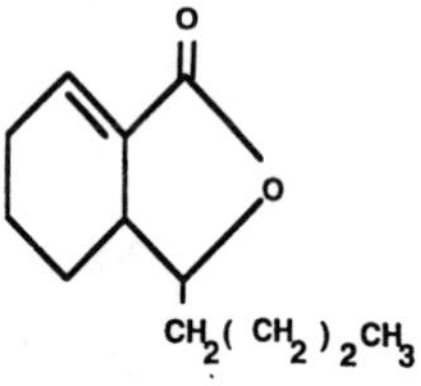

SEDANOLIDE

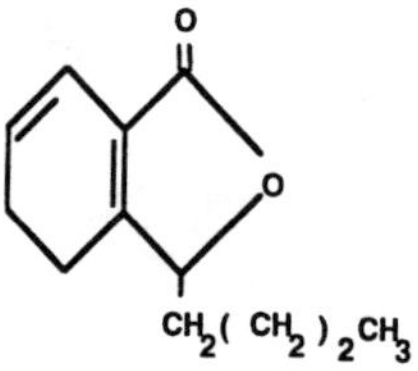

SEDANENOLIDE

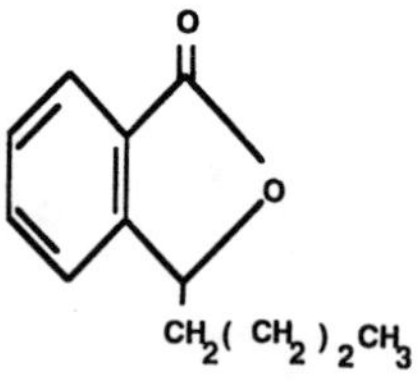

B

3 - N - BUTYLPHTHALIDE

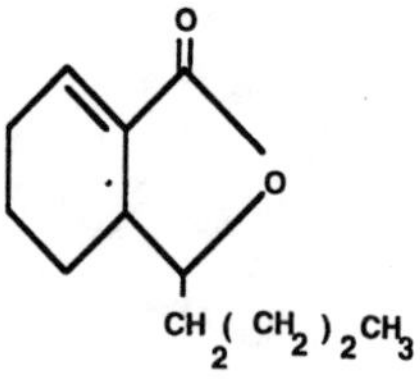

SEDANONIC ANHYDRIDE

Figure 14.2. *(continued).*

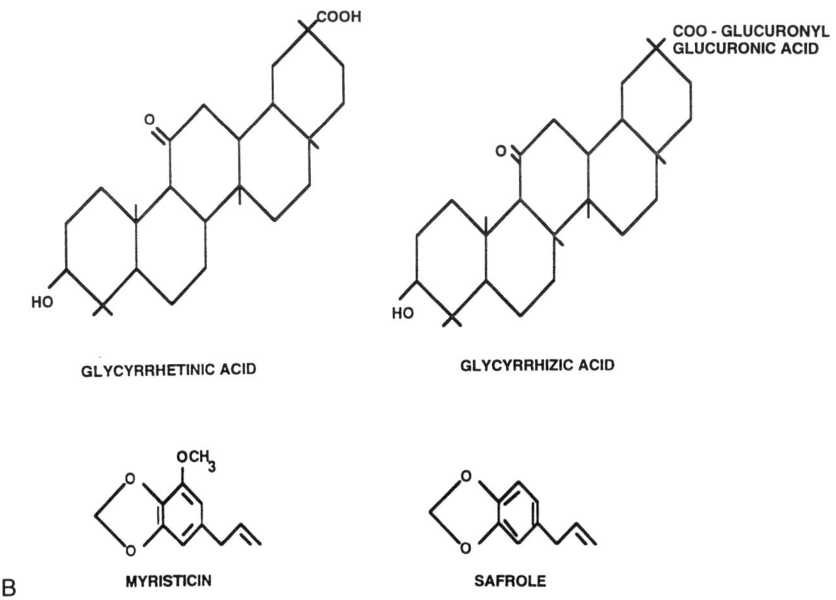

Figure 14.2. *(continued).*

tive tissue and bone (Shibamoto and Bjeldanes 1993b). The consequence of consumption of BAPN is osteolathyrism, characterized by bone deformation and weakness in aortic and connective tissue (Jones 1992a). Another substance in *L. sativus* is β-N-oxalyl-L-α,β-diaminopropionic acid (ODAP). ODAP was believed to be responsible for neurlathyrism, characterized by increasing paralysis of the legs, followed by general weakness and muscular rigidity. Neurological damage is seen only in humans, particularly in males (Roy and Spencer 1989). The chemical structures of BAPN and ODAP are shown in Figure 14.2A.

3.5. Lectins (Hemagglutinins)

Various legumes used as feed or food sources, such as black beans, soybeans, lima beans, kidney beans, peas, and lentils contain lectins—proteins and glycoproteins that possess the ability to bind certain carbohydrates. Lectin will cause the agglutination of the cells when these carbohydrates are present as components of cell walls. Lectin can cause blood cell agglutination or bind to the carbohydrate components of intestinal epithelial cells, resulting in decreased absorption of nutrients from the gastrointestinal tract. Lectins are therefore considered to be antinutritional agents.

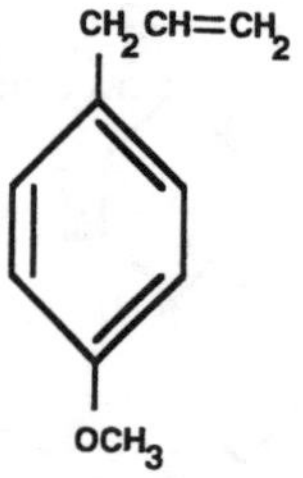

ESTRAGOLE

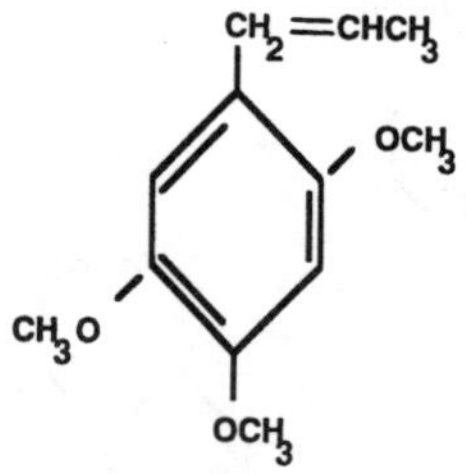

β-asarone

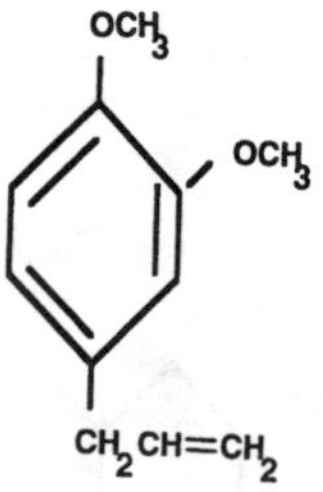

METHYL EUGENOL

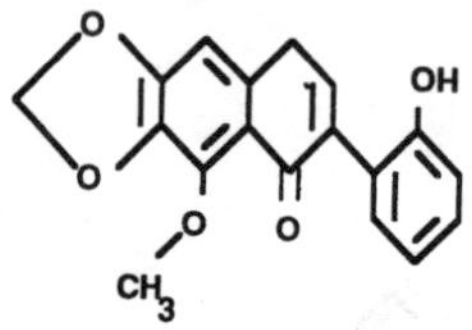

β - VULGARIN

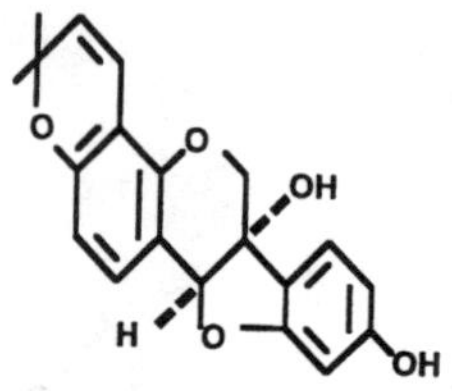

GLYCEOLIN

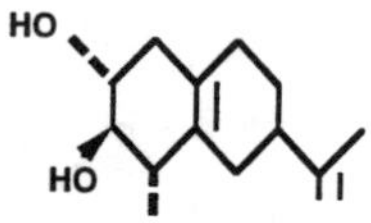

RISHITIN

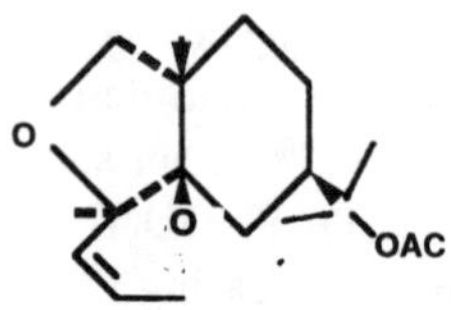

PHYTUBERIN

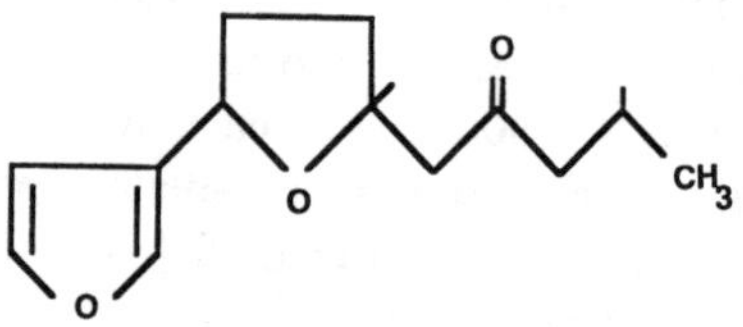

IPOMEAMARONE

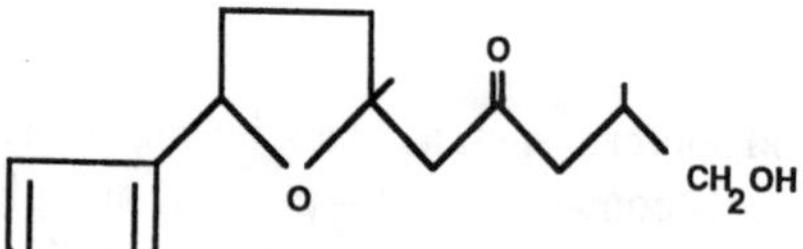

B IPOMEAMARONOL

Figure 14.2. *(continued)*.

Lectins from black beans or soybeans retard the growth of rats. Lectins from kidney beans produce death in rats fed on this antinutritional agent at 0.5% of the diet for 2 weeks. Ricin, a kind of lectin from nonlegume castor beans, is extremely toxic, with an LD_{50} of 0.05 mg/kg in rats by injection. The mechanism of toxicity of lectins requires further study, but the intestinal microbiota may play an important role in legume and lectin-induced toxicity (Shibamoto and Bjeldane 1993b). Lectins may impair the body's defense system against bacterial infection. On the other hand, some lectins are poorly absorbed. Lectin can reach the colon in a biologically intact form, which may have some beneficial effect. Lectin has been reported to protect the human body against colon cancer (Leiner 1981).

3.6. Enzyme Inhibitors

There are many enzyme inhibitors in food plants. Two groups of enzyme inhibitors are well known—protease inhibitors, and cholinesterase inhibitors.

PROTEASE INHIBITORS

Most varieties of legumes and grains, as well as other food plants, such as potatoes, eggplant, onion, beets, lettuce, peas, peanuts, sweet potatoes, and turnips, contain protease inhibitors, which inhibit trypsin, chymotrypsin, and other protein digesting enzymes (Leiner 1980, Rackie and Gumbmann 1981). Trypsin inhibitors play an antinutritional role and cause growth retardation in test animals and cause pancreatic hypertrophy (Bender 1987). In most cases, cooking destroys these inhibitors (Jones 1992). But one of the potato enzyme inhibitors, carboxypeptidase inhibitor, is stable to microwave cooking, boiling, and baking (Ryan and Hass 1981). Certain protease inhibitors have been reported to inhibit tumor formation (Merz 1983).

CHOLINESTERASE INHIBITORS

Cholinesterase is an enzyme that mediates hydrolysis of acetylcholine to acetate and choline. This is an important enzyme in neurotransmission. A variety of food plants contain cholinesterase inhibitor, including the edible parts of raw asparagus, broccoli, carrot, cabbage, celery, radish, pumpkin, raspberry, vegetable marrow, orange, pepper, strawberry, tomatoes, turnip, apple, eggplant, and potatoes (Silverstone 1985). The most active inhibitor is in potatoes (Concon 1988). One of the cholinesterase inhibitors is identified as physostigmine. Another anticholinesterase substance is solanine (a glucosidal alkaloid), which is found primarily in eggplant, potatoes, and tomatoes. In the United States, healthy potatoes contain 1 to 5 mg solanine per small potato (100 g); if the potato is sunburned or blighted, the solanine level increases (Jones 1992). The symptoms of solanine poisoning include gastric pain, nausea, and vomiting. Respiration becomes difficult, combined with weakness and prostration. Death has been reported from consumption of blighted potatoes (Morgan and Coxon 1987, Sharma and Salunke 1989). The chemical structures of physostigmine and solanine are shown in Figure 14.2A.

3.7. Pyrrolizidine Alkaloids

Many food plants, including range plants such as *Senecio*, *Crotalaria*, and *Heliotrophium*, contain pyrrolizidine alkaloids with common chemical structures shown in Figure 14.2A. Some of these—for example, monocrotaline and lasiocarpine—are potent carcinogens (Shibamoto and Bjeldanes 1993b). The carcinogenicity of these compounds is dependent upon metabolism to an ultimate reactive form. Intestinal metabolism of this group of compounds should be further studied.

Solanine can also be classified as an alkaloid. Another alkaloid, tomatine, is present in high concentrations in tomato and is an effective pesticide (Silverstone 1985). Another common alkaloid is quinine, which is a drug. Quinine is found in beverage tonic water, and is toxic to humans, particularly to children (Fletcher 1976).

3.8. Vasoactive Amines

Many food plants and animals contain putrescine and cadaverine. Food plants such as banana, tomato, and avocado contains serotonin, tyramine, and dopamine. Cheese also contains a very high concentration of tyramine. Banana pulp contains about 2 μg per gram norepinephrine (Shibamoto and Bjeldanes 1993b). These vasoactive amines affect the vascular system. Vasoactive amines that constrict blood vessels are known as pressor amines. Dopamine and norepinephrine are pressor amines, important neurotransmitters released from adrenergic nerve cells. Their chemical structures are shown in Figure 14.2A.

The circulating levels of vasoactive amines are controlled by monoamine oxidase (MAO). The ingestion of foods such as aged cheese containing high concentrations of vasoactive amines, in someone treated with MAO inhibitors would cause severe complications (McCabe 1986). The symptoms caused by vasoactive amines included hypertensive crisis, migraine headaches, and, in some cases, intracranial bleeding leading to death.

Large amounts of the amine 5-indoleacetic acid in plantain are related to an increased incidence of heart disease (Silverstone 1985). Phenylethylamine from chocolate has also been reported to have an vasoactive effect (Joint Committee of the Royal College of Physicians and the British Nutrition Foundation 1984).

3.9. Mutagens in Food Plants

There are many mutagens in food plants. Flavonoids are a group of compounds widely occurring in food plants, some of which are used for flavoring or as colorants. However, many of them are mutagens such as quercetin, kaempferol, and rutin (Fig. 14.2B).

Quercetin occurs in many fruits, vegetables, and teas, and has been shown to be mutagenic (Jones 1992). Close relatives of quercetin, kaempferol and its conjugated form astragadalin, were widespread in food and are shown to be mutagenic in the Ames *Salmonella* mutagenicity assay (MacGregor and Jurd 1978; MacGregor 1986a,b). Rutin is a conjugated form of quercetin, in which the 3-hydroxy group of quercetin is linked with carbohydrates by a glycosidic linkage. Many intestinal microorganisms have glycosidases to cleave the glycosidic bond. Rutin is present in red wine and has been shown to become mutagenic through activation by liver and colonic enzymes (Yu et al. 1986). Both rutin and quercetin have been shown to cause bladder tumors in rats (Pamuku et al. 1980), while other studies indicated that both quercetin and rutin might have anticancer properties (Carr 1985, Birt 1989).

Quercetin is considered a kind of tannin. Tannins are found in almost every food plant, particularly banana, grapes, raisins, sorghum, spinach, red wine, and beer. Unripe persimmons have an unusually high tannin content. Coffee, chocolate, and tea also have some tannins. Vegetable tannins, commonly referred to as tannic acid, are a group of phenolic compounds. They can be divided into either hydrolyzable tannins or condensed tannins. The hydrolyzable tannins are usually esterified with sugars. Tannins in food plants are usually hydrolyzable tannins. Tannins can bind protein, penetrate through the superficial cells, and cause liver damage (Mehansho et al. 1987). Tannins inhibit digestive enzymes (Butler 1989) and reduce the bioavailability of iron and vitamin B_{12} (Leiner 1980). Tannins are believed to be responsible for the high incidences of cheek and esophageal cancers in regions where people chew betel nut or consume large amounts of herbal teas, which contain considerable quantities of tannins (Stolz 1982; Jone 1992a).

Maltole, ethyl maltole, and diacetyl are maltoles present in the diet in relatively large amounts. The levels of maltole added to baked goods, ice creams, and candy are approximately 110 mg/100 g; they are also added to certain beverages (Shibamoto and Bjeldanes 1993b). Although there is no conclusive evidence showing their carcinogenic potential, they are mutagenic (Zeiger and Dunkel 1991).

Caffeine is an analogue of xanthine and is present in coffee, tea, cola, and cocoa products. Caffeine causes central nervous system stimulation, diuresis, relaxation of stomach muscles, cardiac muscle stimulation, and increased gastric secretion. Excessive consumption of caffeine can result in nervousness, irritability, and cardiac arrhythmias (Shibamoto and Bjeldanes 1993b). Caffeine has been implicated with cancer of the ovary, bladder, pancreas, and large bowel (Ames 1986). Other xanthine analogs are theobromine and theophylline. No strong biological activity of theobromine or theophylline have been reported (Vanos and Hofstaetter 1987). Caffeine, theobromine, and theophylline are plant alkaloids.

Many spices have been reported to be mutagenic. Onion and garlic essential oils contain high levels of organosulfur compounds that are reported to have some anticancer effects (Belman 1983, Hayes et al. 1987, Wargovich 1987, Sparnins et al. 1988). The phthalides (i.e., sedanolide, sedanenolide, sedanonic anhydride, and 3-n-butylphthalide) present in celery seed oil are reported to be mutagens and have sedative effects on humans (Shibamoto and Bjeldanes 1993b). Glycyrrhizins (glycyrrhetinic acid and glycyrrhizic acid) are present in the root of the licorice plant (*Glycyrrhize glabra* L) and are currently used in candies, root beer, liquors, chocolate, and vanilla (IFT Panel 1986b, Sela and Steinberg 1989). Consumption of large amounts of licorice candy over an extended period of time may lead to hypertension, sodium retention and heart enlargement (Shibamoto and Bjeldanes 1993b).

Myristicin is present in nutmeg and mace and also in black pepper, parsley, celery, dill, and members of the carrot family. It can cause headache, cramps, nausea, liver damage, and even death (Shibamoto and Bjeldanes 1993b).

Safrole (Fig. 14.2B) is the main component of the oil of sassafras which is used to flavor root beer. Safrole is a component of essential oils such as star, anise and camphor oil. It is also present in small quantities in mace, nutmeg, Japanese wild ginger, California bay laurel and cinnamon leaf oil. Besides being a hepatocarcinogen, safrole has been shown to damage the liver (Kapadia et al. 1978, Miller et al. 1982, Alvares 1984). β-Asarone (Fig. 14.2B) is chemically related to safrole and is found in calamus oil. β-Asarone induced malignant tumors in the small intestine of rats fed high doses (Shibamoto and Bjeldanes 1993b). Another substance, estragole (Fig. 14.2B), present in tarragon oil also caused liver cancer in mice (Miller and Miller 1986). Methyl eugenol, structurally similar to estragole, is present in bay leaves, cloves, and lemon grass. Methyl eugenol (Fig. 14.2B) has carcinogenic activity in rodents (Miller and Miller 1986).

Phytoalexins are compounds produced by plants in response to environmental stresses such as microbial invasion, cold, ultraviolet light irradiation, physical damage, and certain chemical stimuli. Chemicals capable of stimulating phytoalexin production include metal salts, polyamines, and pecticides. Beuchat and Golden (1989) reported that many phytoalexins have antimicrobial properties. Some phytoalexins are mutagenic. Examples of these are β-vulgarin produced from beets after fungal infections; glyceolin produced from soybeans when soybeans were infected with the fungus *Phytophthera megasperma*; rishitin and phytuberin produced from potatoes after fungal infections; and ipomeamarone and ipomeamaronol produced by sweet potatoes. Some phytoalexins are very toxic. For example, ipomeamarone and ipomeamaronol can cause respiratory distress, pulmonary edema, congestion, and death in cattle. They also cause liver degeneration in experimental animals (Grisebach and Ebel 1978, Bailey and Mansfield

1982, Shibamoto and Bjeldanes 1993b). Further studies are needed to demonstrate the toxicology of phytoalexins.

4. Toxicants from Food Processing

Many types of food processing techniques, such as frying, roasting, toasting, evaporation, smoking, sterilization, pasteurization, irradiation, pickling, freezing, canning, and home cooking are employed to prepare food. Some toxic compounds can be formed during these processes (Shibamoto and Bjeldanes 1993c).

4.1. Polycyclic Aromatic Hydrocarbons (PAHs)

PAHs are found in water, soil, dust, and many foods. Beef, cheese, herring, dried herring, salmon, sturgeon, frankfurters, and ham have been found to contain PAHs. Cooked sausage, smoked hot sausage, smoked turkey fat, charcoal-broiled steak, and barbecued ribs may contain benzo(a)pyrene, a major kind of PAH. PAHs are formed from carbohydrates at high temperature in the absence of oxygen. Amino acids and fat also produce PAHs upon high-temperature treatment. Many cooking processes will cause the production of PAHs. For example, broiling meat over hot ceramic or charcoal briquetts will allow the dripped fat to come in contact with a very hot surface, and PAHs are formed in the ensuing reaction. The surface temperature of baking bread can reach 400°C. Deep frying, reaching 400 to 600°C, facilitates the production of PAHs.

On the other hand, the canning of proteinaceous foods, the caramelizing of sugar, or the roasting of coffee will also produce PAHs (Krone et al. 1986, Powrie et al. 1986). Pickling and fermentation are also involved with the production of PAHs (Sugimura 1986a).

The most commonly known carcinogenic PAH is benzo(a)pyrene (BP), which is present in various foods at concentrations ranging from 0.31 to 18.8 mg/100 g. BP is a potent carcinogen, causing leukemia, lung adenoma, stomach and skin tumors. BP can cross the placenta and produce tumors in fetuses of animals treated during pregnancy (Shibamoto and Bjeldanes 1993c).

Metabolic activation is required for the mutagenicity and carcinogenicity of BP. The metabolic activation involves initially cytochrome P450 and produces the 7,8-epoxide derivative. The 7,8-epoxide derivative is further metabolized to 7,8-diol by epoxide hydrolase. The 7,8-diol is then converted to diol-epoxide, which in turn interacts with DNA and produces mutation. The metabolic activation of many PAHs has been extensively studied during the last decade (Combes 1989, Thies and Siegers 1989).

Other PAHs such as benz(a)anthracene, benzo(b)fluoranthene, benzo(c)phenathrene, chrysene, 5-methylchrysene, dibenz(a,h)anthracene, and dibenzo(a,i)-

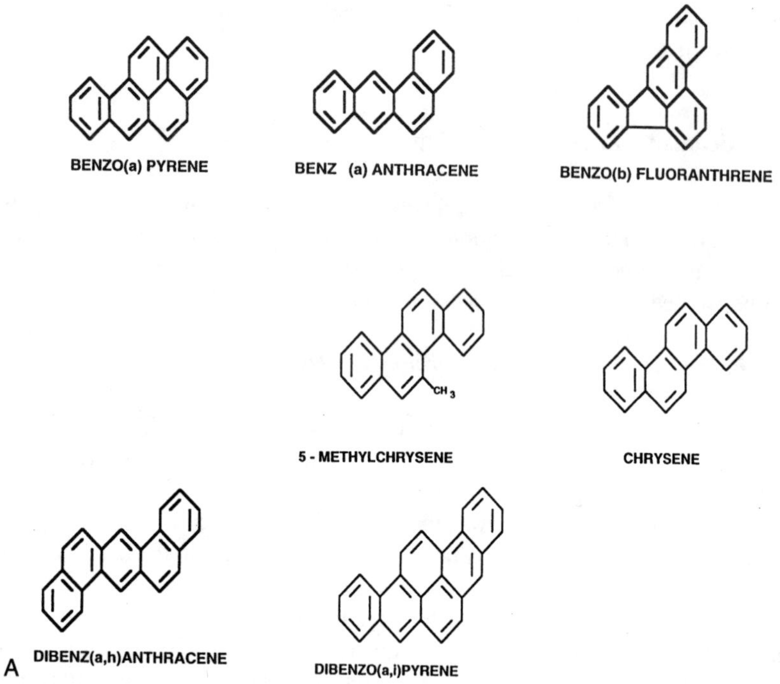

Figure 14.3. (A) Chemical structures of some polycyclic aromatic hydrocarbons (PAHs). (B) Chemical structures of some amino acid pyrolysates (following pages).

pyrene have also been found in food. The chemical structures of these PAHs are shown in Figure 14.3A.

4.2. Amino Acid Pyrolysates

Nagao et al. (1977a) reported the high mutagenicity of smoked condensates and the charred surface of fish and meat. Later, it was found that proteins and amino acids could generate strong mutagenic compounds upon heating. Among the amino acids tested, tryptophan was the most active (Nagao et al. 1977b). The mutagenic pyrolysis products were identified as Trp-P-1 and Trp-P-2 (Sugimura et al. 1977). Since then, Sugimura et al. (1985, 1986b, 1988, 1989) reported many heterocyclic aromatic amines from pyrolysates of food, proteins, and amino acids. Others also reported the formation of potent mutagens from pyrolysates of amino acids or proteinaceous foods (Yoshida et al. 1978, Commoner et al. 1978, Yamamoto et al. 1978, Kasai et al. 1980a,b, 1981, Yokoda et al. 1981, Nigishi et al. 1984, 1985; Felton et al. 1986, Becher et al. 1988). There are at least 18 different

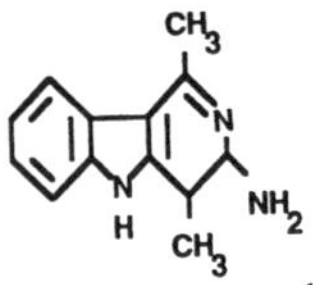

3 - AMINO - 1,4 - DIMETHYL - 5H - PYRIDO[4,3 - b] INDOLE (TRP - P - 1)

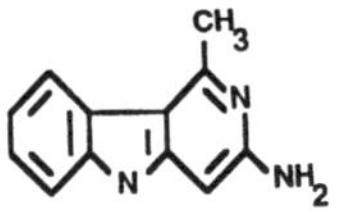

3 - AMINO - 1 - METHYL - 5H - PYRIDO [4,3 -b] INDOLE [TRP - P - 2]

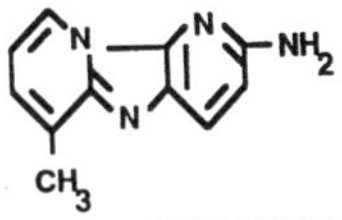

2 - AMINO - 6 - METHYLDIPHRIDO [1,2 - a : 3',2' - d] IMIDAZOLE [GLU - P -1]

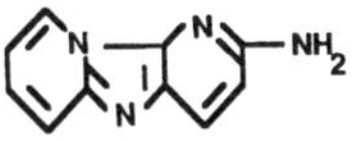

2 - AMINODIPYRIDOL[1, 2 -a :3', 2' - d] IMIDAZOLE [GLU - P - 2]

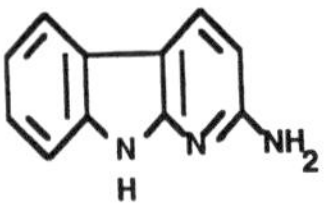

2 - AMINO - 9H - PYRIDO [2,3 - b] INDOLE (A α C)

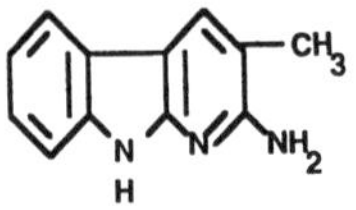

2 - AMINO - 3 - METHYL - 9H - PYRIDOL [2,3 -b] INDOLE [MEA α C]

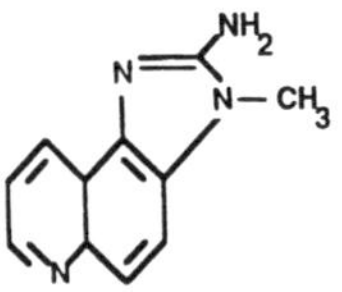

2 - AMINO - 3 - METHYLIMIDAZO[4,5 -f] QUINOLINE (IQ)

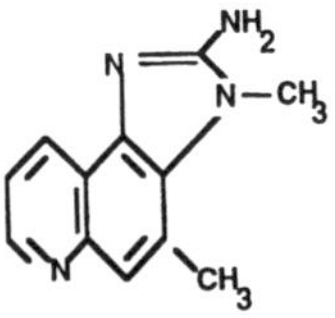

2 - AMINO - 3,4 - DIMETHYLIMIDAZO[4,5 -f] GUINOLINE(MeIQ)

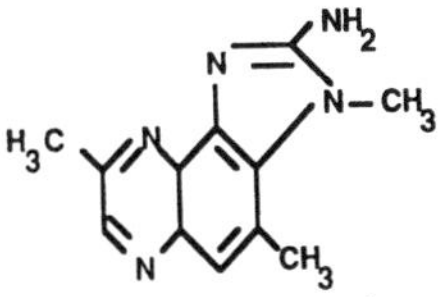

2 - AMINO - 3,4,8 - TRIMETHYLIMIDAZO[4,5 - f] QUINOXALINE(4,8 - DiMeIQx)

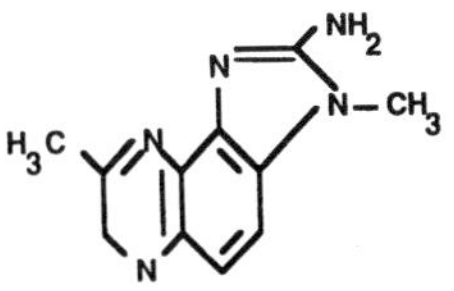

2 - AMINO - 3,8 - DIMETHYLIMIDAZO[4,5 - f] QUINOXALINE (MeIQx)

B

Figure 14.3. *(continued)*.

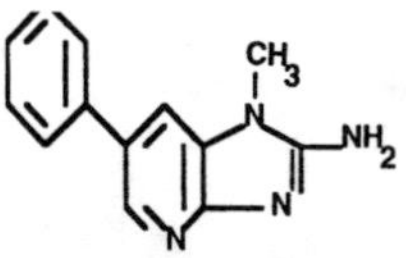

2 - AMINO - METHYL - 6 - PHENYLIMIDAZO[4,5 -b] PYRIDINE(PhIP)

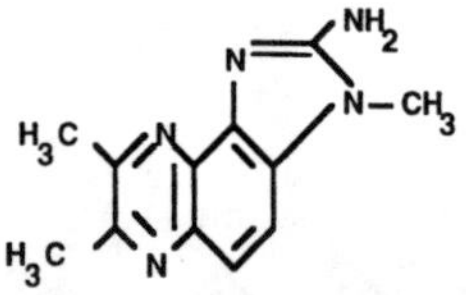

2 - AMINO - 3,7,8 - TRIMETHYLIMIDAZO[4,5 - f] QUINOXALINE(7,8 - DiMeIQx)

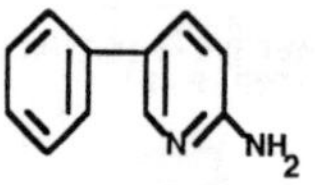

2 - AMINO - 5 - PHENYLPYRIDINE(PHE - P - 1)

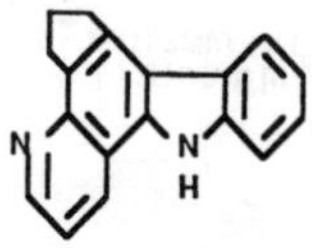

3,4 - CYCLOPENTENOPYRIDO[3,2 -a] CARBAZOLE(Lys - P - 1)

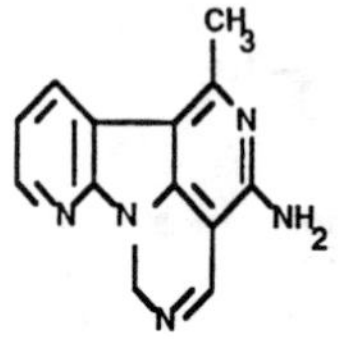

4 - AMINO - 6 - METHYL - IH - 2,5,10,10b - TETRAAZAFLUORANTHENE(ORN -P - 1)

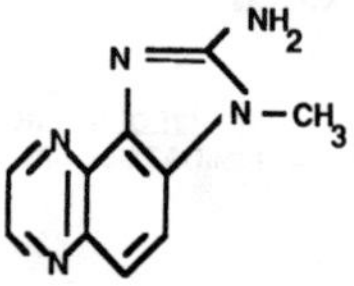

2 - AMINO - 3 - METHYLIMIDAZO[4,5 -f] QUINOXALINE(IQx)

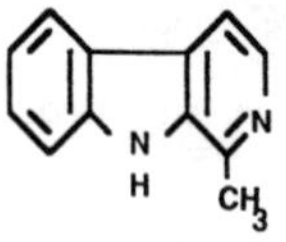

1 METHYL - 9H - PYRIDO[3,4 - b]INDOLE (HARMAN)

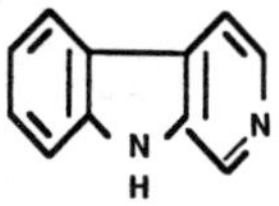

9H - PYRIDO[3,4 - b]INDOLE (NORHARMAN)

B

Figure 14.3. *(continued).*

heterocyclic amines from protein pyrolysates reported; their chemical structures are shown in Figure 14.3B (Negish et al. 1991a). Extensive reviews have been published concerning these mutagens (Hatch et al. 1984, 1988, de Meester 1989, Jagerstad et al. 1991). DNA modification in vitro and in vivo with these heterocyclic amines has also been demonstrated (Negishi et al. 1991b). Formation of heterocyclic aromatic amines is a subject of importance and interest. Major precursors are single amino acids or amino acids together with creatine or creatinine (Felton and Knize 1991). The Maillard reaction, which generally acounts for the browning of food products, may be involved (Wei et al. 1981). This reaction occurs when carbonyl groups of aldose sugars which interact with amino compounds (alkylamine, amino acid or protein) to form a type of N-substituted glycosylamine, which is then converted to 1-amino-1-deoxy-2-ketoses and then further transformed into heterocyclic aromatic amines. Many compounds such as D-glucose, L-rhamnose, lactose, maltole, starch, cysteamine, cyclotenes, glycine, diacetyl, and ammonia are substrates for this reaction (Shibamoto and Bjeldane 1993c). Jagerstad et al. (1991) studied the formation of heterocyclic aromatic amines using model systems and found that many factors such as time, temperature, concentrations of precursors, type of amino acids, certain divalent ions, and moisture content influenced the formation of heterocyclic aromatic amines. However, the mechanisms involved in the formation of mutagenic heterocyclic aromatic amines are still unclear. There may be routes other than the Maillard reaction (Jagerstad et al. 1991).

4.3. N-nitrosamines

Nitrates are present in foods, particularly in the leaf and root vegetables, cured meat, beer, and many other food items (Ames 1986). The dietary intake for adult Americans has been estimated at 100 mg/day (Shibamoto and Bjeldanes 1993c). An interaction between nitrate and other components of the diet, such as soy sauce, may occur. Nagao et al. (1986) demonstrated that soy sauce mixed with nitrate forms a variety of potent mutagens. In the Orient, over 10 L of soy sauce is consumed annually per person.

The nitrate is reduced to nitrite by oral and gastrointestinal microorganisms. Nitrite is also present in many foods or added to cured meat or other products as a food additive because of its antimicrobial, flavor, and color properties. Nitrite will interact with amines, which are also present in the gastrointestinal tract, in a reaction called nitrosation. Nitrosation of secondary and tertiary amines produces stable nitrosamines which are potent carcinogens. Halogens and thiocyanates promote the nitrosation process; antioxidants such as ascorbate and vitamin E inhibit the nitrosation. Diethylnitrosamine (DEN) and dimethylnitrosamine are major nitrosamines in the gastric juices of humans. Nitrosation also occurs during high-temperature heating of foods such as bacon, which contain nitrite and amines. Spice premixes also contain high concentration of nitrosamines.

Nitrosamines are known carcinogens in experimental animals (Wasserman and Wolff 1981), producing various types of tumors. The nitrosamines exhibit a great deal of organ specificity in their carcinogenic effect. The important nitrosamines producing gastrointestinal cancers are ethylbutylnitrosamine for stomach cancer, methylnitrosourea for duodenal cancer, and dimethylnitrosamine for liver cancer (Shibamoto and Bjeldanes 1993c). The mechanism of carcinogenicity remains to be studied, but nitrosamines require metabolic activation. In some cases, hydroxylation of the α-carbon is required, leading to the formation of alkylating agents which exert their genotoxic effect.

4.4. Food Irradiation

The use of ionizing radiation such as gamma rays is a general trend for the preservation of foods. Gamma-ray irradiation is very effective in killing a wide range of microorganisms. This process does not produce radioactive compounds in foods. However, the energies used for food irradiation are sufficient to produce free radicals, which can combine with each other or form new bonds with other compounds in foods. The toxicity of the newly formed compounds is not known at present. Therefore, the biological safety of the process is still being evaluated.

5. Toxins from Food Additives

A food additive is a substance other than food component that is added intentionally to food in a controlled amount to achieve certain functions important for food preparation or preservation. Food additives include preservatives, antimicrobial agents, bleaching agents, antioxidants, sweeteners, coloring agents, flavoring agents, and nutrient supplements. There are more than 300 substances recognized as food additives. Many of them are safe, but a few of them were discovered to be toxic or metabolically converted to toxicants by enzymes of the intestinal microbiota or digestive systems.

5.1. Food Preservatives

A chemical used as a food preservative should be nontoxic and suitable for application; must not impart off-flavors when used at levels effective in controlling microbial growth; must exhibit antimicrobial properties over the pH range of each particular food; and should be economical and practical to use (Shibamoto and Bjeldanes 1993d). However, some have been found to be toxic.

Benzoic acid or sodium benzoate is used as an antimicrobial additive in carbonated and still beverages, syrups, fruit salads, icings, jams, jellies, preserves, salted margarine, mince meat, pickles, and relishes, pie, pastry fillings, prepared salads, fruit cocktail, soy sauce and caviar. The general level of use is about 0.1% (Jay 1992). Benzoic acid was found to be toxic to rats, rabbits, dogs, and mice. The subacute toxicity symptoms of benzoic acid or its sodium salt in mice include weight loss, diarrhea, irritation of internal membranes, internal bleeding, enlargement of liver and kidney, hypersensitivity, paralysis, and death (Shibamoto and Bjeldanes 1993d). There were no chronic toxicities found. Benzoic acid has not been shown to be toxic to humans (Chipley 1993). The compound is excreted in the urine in the form of hippuric acid (conjugated with glycine) or benzyl glucuronides (White et al. 1964, Hall and James 1980).

2-(2-Furyl)-3-(5-nitro-2-furyl) acrylamide (AF-2) is a nitrofuran compound that has been widely used in clinical and veterinary medicines and as an antimicrobial for animal feed. In Japan, AF-2 was added to soybean (tofu), ham, sausage, fish ham, fish sausage, and fish paste in 1965. But AF-2 was found to be mutagenic in the Ames mutagenicity test, later identified as a carcinogenic (Hayashi 1991), and banned from use. The chemical structure of AF-2 is shown in Figure 14.4A.

5.2. Food Antioxidants

Antioxidants are added to prevent oxidative deterioration of foods and are therefore widely used in the food industry. L-ascorbic acid (vitamin C) is widely present in plants. Its physiological functions include stimulation of collagen formation, prevention of scurvy, hydrolysis of proline, oxidation of tyrosine, and the stimulation of the formation of adrenal steroid hormones. Vitamin C has also been shown to prolong the life of cancer patients (Cameron and Pauling 1976, 1978) and to inhibit tumor cell growth and carcinogen-induced DNA damage (Chen et al. 1989). Lathia et al. (1988) demonstrated that vitamin C is effective in preventing nitrosation of amines under physiological conditions. Reed et al. (1989) also revealed that high doses of vitamin C might reduce the incidence of gastric cancer. However, overdosage of vitamin C induces perspiration and nervous tension, and it lowers pulse rate. Ascorbic acid is readily oxidized to dehydroascorbic acid, which is reduced by glutathione. The dehydroascorbic acid was diabetogenic in rats by intravenous injections (Shibamoto and Bjeldanes 1993d). The chemical structures of ascorbic acid and dehydroascorbic acid are shown in Figure 14.4A.

α-Tocopherol (vitamin E) is present in many plant products, particularly lettuce. It is an effective antioxidant. High doses of vitamin E cause headache, nausea, fatigue, dizziness, and blurred vision (Shibamoto and Bjeldane 1993d). The chronic toxicity of vitamin E has not been thoroughly studied.

Propyl gallate (PG) is an antioxidant used in vegetable oil. Chung et al. (1993) demonstrated that PG inhibited the growth of many food-borne bacteria, with the inhibition related to the ester linkage between gallic acid and propanol. Gallic acid has been shown to reduce mutagen formation under specific conditions (Stich et al 1982, Ohshima and Bartsch 1983, Yamamoto et al. 1988). Propyl gallate has been shown to be toxic to rats, although no other animal studies showed serious problems (Shibamoto and Bjeldanes 1993d).

Other commonly used antioxidants are butylated hydroxyanisol (BHA) and butylated hydroxytoluene (BHT). BHA has been shown to produce mild diarrhea in dogs, chronic allergic reactions, and damage to the metabolic system. Williams et al. (1990a,b) performed a detailed study on the genotoxicity and chronic toxicity of BHA and BHT. No evidence of genotoxicity was found for either BHA or BHT using several tests. No prenatal toxicity for BHA in studies with mice, rats, rhesus monkeys, hamsters, pigs, or rabbits was found (Sherwin 1990). However,

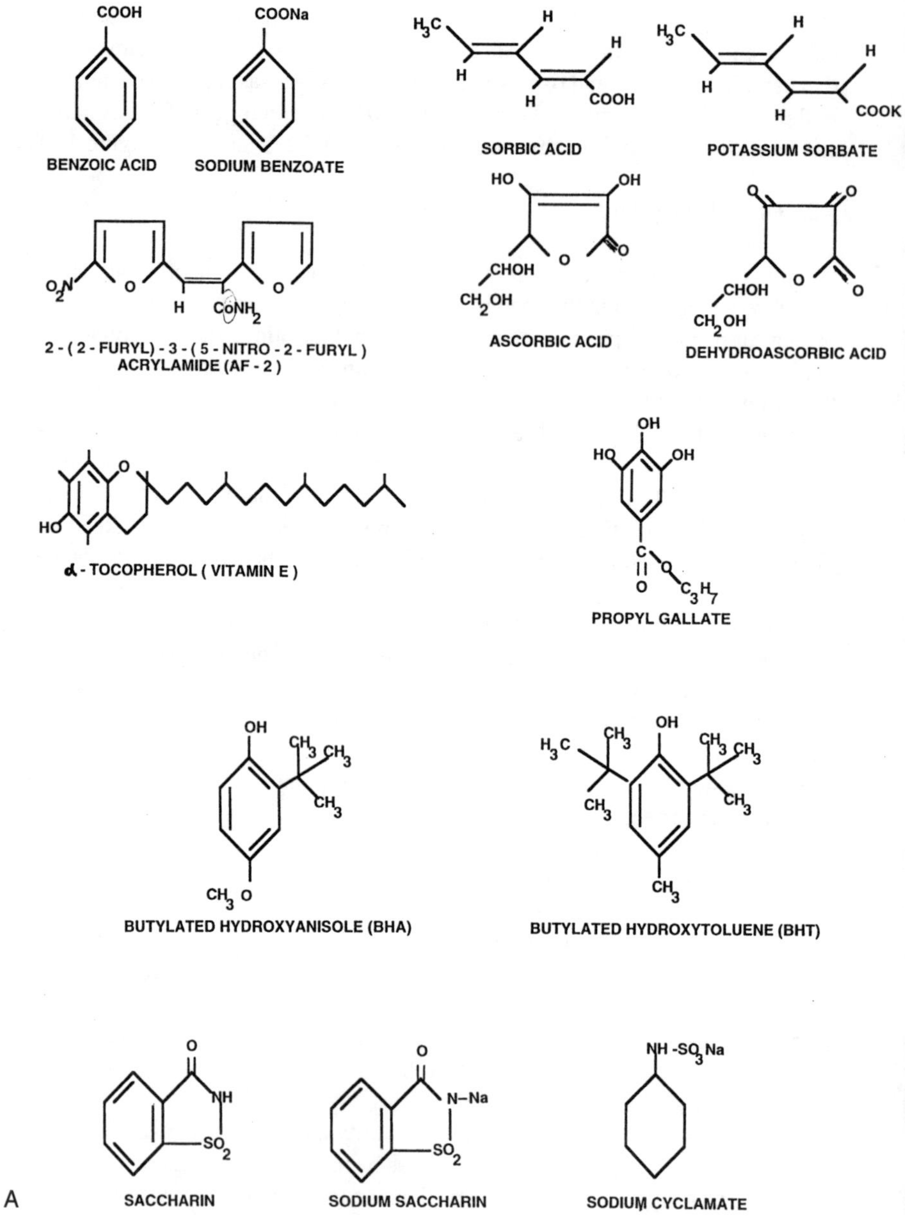

Figure 14.4. (A) Chemical structures of some potential toxins from food additives. (B) Chemical structures of food colorants (following pages).

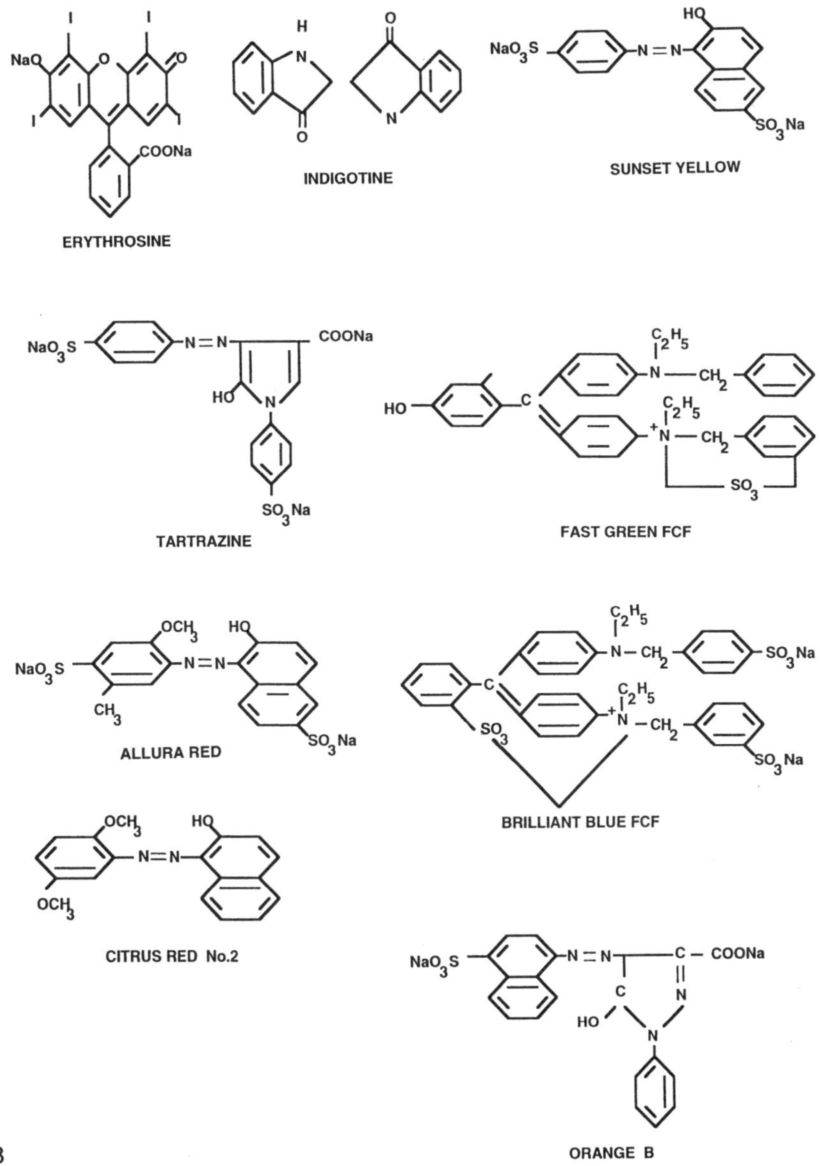

B

Figure 14.4. *(continued).*

in chronic toxicity testing with male rats, BHA at 12,000 mg/100 g in the diet caused a small increase in the number of squamous cell papillomas in the stomach (Williams et al. 1990b). Other toxic effects of BHA and BHT also have been reported (Davidson 1993). There are many reports indicating that both BHA and BHT have some anticarcinogenic activities (Wattenberg 1975, 1978, Milner 1985).

Antioxidants (α-tocopherol and PG) have been shown to inhibit the formation of nitrosamines (Gray and Dugan 1975). α-Tocopherol, BHA, BHT, and PG also were shown to inhibit the hepatotoxicity that occurs after feeding sodium nitrite and secondary amines in rats (Astill and Mulligan 1977, Kamm et al. 1977)

5.3. *Food Sweeteners*

Saccharin and sodium cyclamate are two major sweeteners added to foods. In 1972, it was found that 7.5% saccharin in feed produced bladder cancer in the second generation of rats (Jones 1992b). However some reports showed contradictory results (IFT Expert Panel 1986). Saccharin is approved for use in 80 countries and has been determined as safe by both the Food and Agriculture Organization (FAO)/World Health Organization (WHO) Joint Expert Committee on Food Additives (FECFA) and the Scientific Committee for Foods of the European Economic Community (Arnold and Clayson 1985, Arnold and Munro 1983).

Sodium cyclamate showed some acute toxicity to experimental animals. Capillary transitional cell tumors were found in the urinary bladder of rats that received a mixture of sodium cyclamate and sodium saccharin (Price et al. 1970). Bladder cancer also appeared when sodium cyclamate alone was fed in drinking water (Rudali et al. 1969). It was considered that sodium cyclamate was carcinogenic. Sodium cyclamate can be converted into cyclohexylamine, which is very toxic. The use of sodium cyclamate as a food additive was therefore prohibited in 1969 in the U.S. (Jones 1992b). Later studies showed that cyclamate did not cause cancer (Calorie Control Council 1985, Morris and Przybyla 1985). But cyclamate is still not allowed as a food additive in the U.S., although it is considered as a safe additive by WHO and the European Community (Malaspina 1987).

5.4. *Food Colorants*

Food colorants are added to food purely for improvement of appearance, not for nutritional value. There were about 80 synthetic dyes used in the United States for coloring food in 1900, when no regulations regarding their safety were established. Many of these were later banned from use because of their toxicities. For example, butter yellow (dimethylaminoazobenzene [DAB]), which was used to enhance the yellow color of butter, induced maligant tumors in rats (Miller and Baumann 1945, Miller and Miller 1948). Now there are 29 food colorants in use. Nine of them are synthetic colorants including Erythrosine (FD & C Red No. 3), Indigotine (FD & C Blue No. 2), Tartrazine (FD & C Yellow No. 5), Sunset Yellow (FD & C Yellow No. 6), Fast Green FCF (FD & C Green No. 3), Allura Red (FD & C Red No. 40), Brilliant Blue FCF (FD & C Blue No. 1), Citrus Red No. 2 (FD & C Citrus Red. No. 2) and Orange B. Chemical structures of these food colorants are shown in Figure 14.4B. Five of them—Tartrazine, Sunset Yellow, Allura Red, Citrus Red No. 2, and Orange B—are azo dyes. No significant toxicities were found among the non-azo dye food colorants (Jones 1992b). Citrus Red No. 2 is allowed at a level not exceeding 2 mg/100 g (by weight) to color the skins of oranges not intended or used for further processing. Orange B is permitted at a level not exceeding 150 mg/100 g (by weight) for use only for coloring casings and surface of frankfurters and sausages, but it has been withdrawn from sale by its only manufacturer (IFT Expert Panel on Food Safety and Nutrition and the Committe on Public Information 1980). More than 2 million kg/year of Tartrazine, Sunset Yellow, and Allura Red together are certified for use in food by the U.S. Food and Drug Administration (Prival et al. 1988).

Tartrazine has been used as a food colorant since 1916. The medium acute oral lethal dose in mice is 12.17 g/kg. Cases of individual sensitivity to tartrazine with symptoms of urticaria (hives), asthma, and eczema, have been reported (Millstone 1985, Zlotlow and Settipane 1977, Murdoch et al. 1987).

Amaranth was banned by the FDA in 1976 as a food colorant. It had been used in almost every processed food with reddish or brownish color including soft drinks, ice cream, salad dressing, gelatin dessert, cereals, cake mixes, wines, jams, gums, and chocolate, as well as drugs and cosmetics. However, it was suspected of being a carcinogen and a teratogen, although interpretations of experimental data were controversial (FDA 1980). Nevertheless, amaranth is still legally used as a food colorant in Canada and other Western countries. People in this country are certainly ingesting some amaranth in food imported from countries that did not ban amaranth.

There are at least 3,000 azo dyes used in different industries. Although the commonly used azo dyes may not produce toxic effects, they may get into the gastrointestinal tract through ingestion of contaminated foods or water. There is unequivocal evidence that azoreductase from the intestinal microbiota and, to a lesser extent from mammals, catalyzes cleavage of the azo bond to produce aromatic amines. Aromatic amines, such as benzidine, are genotoxic (Walker 1970, Combes and Haveland-Smith 1982, Chung 1983) and are also regarded as environmental pollutants (Chung and Stevens 1993). Chung et al. (1992) reviewed the kinds of intestinal bacteria which catalyze azo reduction. It was also observed that mutagenic azo dyes are those containing either a *p*-phenylenediamine or benzidine moiety (Chung and Cerniglia 1992). Introduction of a methyl or nitro group to the *p*-phenylenediamine increases the mutagenicity significantly (Chung et al. 1995). Benzidine is a potent carcinogen but is only weakly mutagenic in the Ames mutagenicity assay (Chung et al. unpublished results). Azo dye mutagenesis and carcinogenesis merits further study.

Because of the toxicity problems associated with synthetic food colorants, industry has turned attention to natural coloring agents. However, natural coloring agents are not necessarily safe. Caramel, for example, which is used for the light brown color in foods, is found to contain benzo(a)pyrene in small quantities. Curcumin used for the yellow color in curry is very toxic (Shibamoto and Bjeldanes 1993d). Natural colors are usually not as stable as synthetic colorants.

5.5. Food Flavoring Agents

There are over 3,000 synthetic chemicals used as flavoring ingredients. Most of them are generally recognized as safe (GRAS) because of their natural occurrence in foods and are widely ingested with no apparent ill effects. However, there are many toxic chemicals in natural products, and their chronic toxicity should be carefully examined.

Methyl anthranilate is used for grapelike flavor in many products. It is present in the essential oils of orange, lemon, and jasmine. But it promotes an allergiclike reaction on human skin and is toxic to mice, rats, and guinea pigs (Shibamoto and Bjeldanes 1993d).

Safrole (3,4-methylene dioxallyl benzene) (Fig. 14.2B) has been used as a flavoring agent for more than 60 years. Its toxicity has been discussed above.

Monosodium glutamate (MSG) is widely used as a flavor enhancer. MSG has been associated with the Chinese restaurant syndrome which includes headache, nausea, flushing, paresthesia, chest pain, facial pressure, dizziness, sweating, and drowsiness (Kwok 1968; Schaumburg et al. 1969; Settipane 1986). Although FDA affirmed the GRAS status of MSG as not posing any risk of lasting injury, some individuals are susceptible to transient discomfort following ingestion of MSG (JECFA 1988).

There are many other food flavoring agents and flavor enhancers whose toxicities should be further studied.

6. Toxins from Pharmaceutical Agents

Intestinal microbiota can metabolize a variety of pharmacological agents resulting in the production of metabolites required for the physiological activity of these agents or, in contrast, for the inactivation of these agents. A few examples of this type of biotransformation are discussed here (Fig. 14.5).

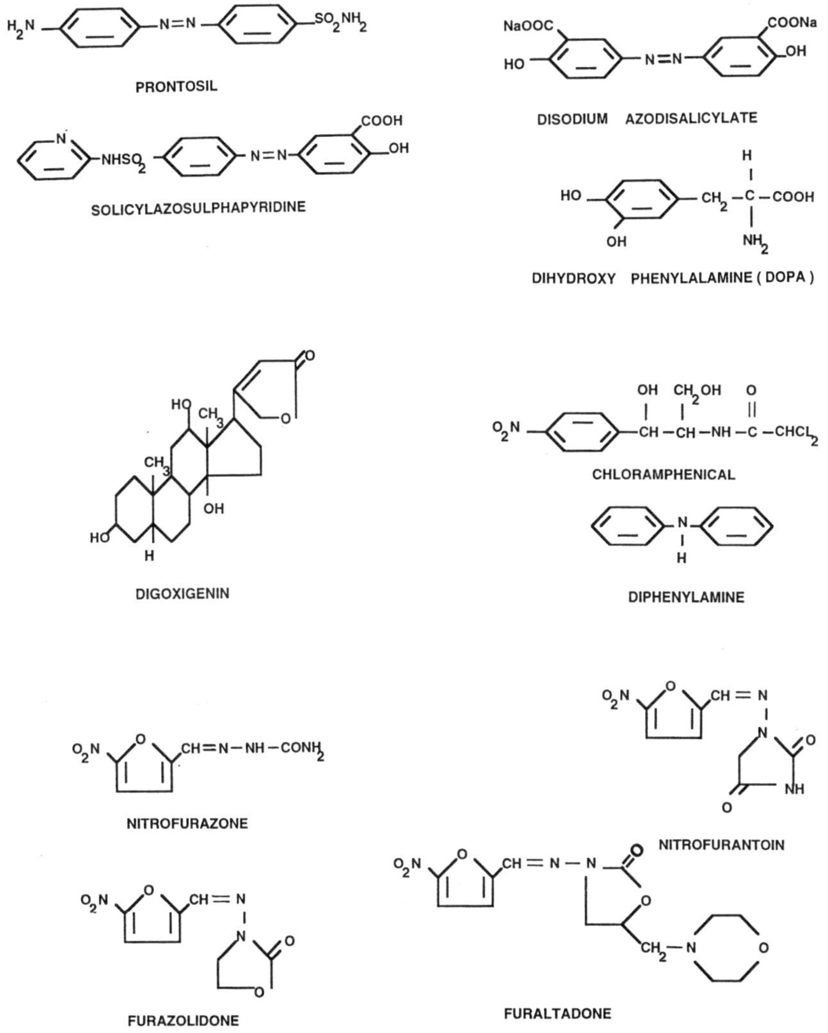

Figure 14.5. Chemical structures of some pharmaceutical agents.

6.1. Prontosil

Prontosil is a red dye, discovered by Dogmak (1935), which has strong antimicrobial properties. It was found that the active component of this drug is sulfanilamide which results from the azo reduction of Prontosil by intestinal microbiota (Fuller 1937; Fouts et al. 1957) Sulfanilamide was the first sulfa drug effective against a variety of pathogens (Trefouel et al. 1935).

6.2. Salicylazosulfapyridine (salfasalazine)

Salfasalazine is a chemotherapeutic drug for the treatment of ulcerative colitis in human patients (Smith and Tong 1975). It is quantitatively metabolized by intestinal microbiota to produce 5-aminosalicylic acid and sulfapyridine, which are active metabolites (Peppercorn and Goldman 1972, 1973, Schroder and Gustafsson 1973). Azodisalicylate, a second-generation salfasalazine, is also metabolized by intestinal microbiota (Azad Khan et al. 1983, Lauritsen et al. 1984).

6.3. Digoxin

Digoxin, a cardiac drug, is hydrolyzed in the gastrointestinal tract by intestinal bacteria to release digoxigenin, which is pharmacologically effective. However, about 10% of the patients who received digoxin did not achieve the predicted serum level of effective digoxin, because some intestinal bacteria further metabolized digoxigenin to dihydrodigoxigenin, which is pharamacologically inactive. The microorganism *Eubacterium lentum* was reported to be exclusively responsible for this metabolism (Lindenbaum et al. 1981).

6.4. 3,4-dihydroxyphenylalanine (DOPA)

DOPA is a drug for the treatment of Parkinson's disease. Intestinal microbiota can modify DOPA by dehydroxylation at the para-position, leading to the formation of metahydroxyphenylacetic acid, which is pharmacologically inactive (Goldin et al. 1973). DOPA can also be decarboxylated by *Streptococccus faecalis*, resulting in an increase in urinary excretion of amines. Therefore, a high dose of DOPA is often needed in order to deliver a small quantity to the brain of patients (Goldin 1990).

6.5. Antibiotics

Aminoglycoside antibiotics are used for the treatment of various diseases. However, intestinal bacteria can inactivate the aminoglycoside by esterification. Intestinal microbiota can also deesterify the antibiotic chloramphenicol with release of the dichloracetyl group resulting in the inactivation of the drug (Goldin 1990).

6.6. *Diphenylamine and Other Secondary Amines*

Diphenylamine is a secondary amine. Intestinal microbiota can add a nitroso group to secondary amines, producing N-nitrosamines, many of which are potent carcinogens. For example, *Streptococcus faecalis* and the common anaerobes *Clostridium, Bacteroides* and *Bifidobacterium* can catalyze the N-nitrosation of diphenylamine. Intestinal microbiota can also remove alkyl groups from secondary amines. For example, methyl amphetamine can be demethylated by intestinal bacteria to produce amphetamine, a commonly used drug (Goldin 1990).

6.7. *Nitro-Aromatic Drugs*

Nitro-aromatic drugs such as nitrofurans which include nitrofurazone, nitrofurantoin, furazolidone, furaltadone, and N-[4-(5-nitro-2-furyl)-2-thiazoly]formamide (FANFT) are antibacterial agents (Olive and McCalla 1977; Bryan 1978). Work with intestinal *Escherichia coli* and *Salmonella typhimurium* showed that these nitrofurans can be reduced to the corresponding aminofurans with the intermediate N-hydroxy derivative being genotoxic (Olive and McCalla 1977; McCalla 1983; Ni et al. 1987; Goldin 1990).

The intestinal microbiota can reduce a variety of nitro-polycyclic aromatic hydrocarbons (nitro-PAH) which are important environmental toxicants (Fu 1990). A detailed review of nitro-arenes regarding their occurrence, metabolism, and biological impact is available (Howard et al. 1990). The role of intestinal microbiota in their genotoxicity will be discussed in the following section.

7. Bacterial Toxins Produced in the Gastrointestinal Tract

Some bacteria produce toxins in the gastrointestinal tract of animals following colonization. Most of these are enterotoxins, proteins can cause diarrhea and vomiting. They are generally divided into two groups: (1) cytotoxic enterotoxins which kill eukaryotic cells by inhibiting protein synthesis or by other mechanisms, and (2) cytotonic enterotoxins which damage cell metabolism in specific ways such as elevating cyclic nucleotide levels (Betley et al. 1986).

Over the years, extensive research has been conducted to study the mechanism(s) of interaction between bacterial toxins and their receptors in the gastrointestinal tract (Middlebrook and Dorland 1984, Field et al. 1989). The receptors are either glycolipids or glycoproteins on the apical surface of enterocytes. The core sequences of some membrane receptors for bacterial toxins have been well

Table 14.1. Major bacterial toxins encountered in the gastrointestinal tract

Producing Organism	Name of Toxin	Cytotoxic or Cytotonic	Heat Stability	Molecular Weight (kDa)	Gene Location
Vibrio cholera	Cholera toxin (CT)	Cytotonic	Labile	84	Chromosome
	Zonula occuluden toxin (ZOT)	Cytotoxic	Labile	10–30	Chromosome
	Accessory cholera enterotoxin (ACE)	Cytotoxic	Labile	—	Chromosome
Escherichia coli	Enterotoxin (LT)	Cytotonic	Labile	91	Plasmid
	Enterotoxin (ST)	Cytotonic	Stable	2	Plasmid
Yersinia enterocolitica	Enterotoxin	Cytotonic	Stable	71	Chromosome
Enterobacter cloacae	Enterotoxin	N/A	Stable	1–10	—
Klebsiella pneumonia	Enterotoxin	N/A	Stable	1–10	—
Shigella flexneri	Enterotoxin	N/A	Stable	1–10	—
Aeromonas hydrophila	Enterotoxin	Cytotonic	N/A	25	—
Salmonella typhimurium	Enterotoxin	Cytotonic	Labile	25	—
Pseudomonas aeruginosa	Enterotoxin(s)	N/A	Mix of Labile and Stable	—	Plasmid
Campylobacter jejuni	Enterotoxin	Cytotoxic	Labile	68	—
Shigella dysenteriae	Shiga toxin	Cytotoxic	N/A	70	Phage
Esherichia coli	Shiga-like toxin	Cytotoxic	N/A	70	Phage
Vibrio cholerae	Shiga-like toxin	Cytotoxic	N/A	70	Phage
Clostridium perfringens	Enterotoxin	Cytotoxic	Labile	35	—
Clostridium difficile	Enterotoxin	Mix of cytotoxic and cytotonic	N/A	250	—
Staphylococcus aureus	Enterotoxin	Cytotoxic	Stable	26–30	Chromosome
Bacillus cereus	Enterotoxin	Cytotoxic	Stable	> 30	—

characterized (Chu and Walker 1993). It appears that toxin-intestine interactions require three major steps: (1) toxin binding to its microvillus membrane receptor; (2) an intracellular response to toxin-receptor interaction (signal transduction); and (3) an enterocyte-effector response or an increased release of neurotransmitter and a secondary enterocyte-effector response. It has also been shown that receptor binding and effector responses are subject to regulation which relates to the development of microvillus membrane in the intestine (Chu and Walker 1991; 1993).

This review will briefly describe the bacterial toxins which are often encountered in the intestine. The major types of bacterial toxins and their properties are outlined in Table 14.1.

7.1. Cholera Toxins

Cholera toxins (CT) are produced by gram-negative, curved *Vibrio cholera*. *V. cholera* attaches to the mucosal surface of the upper intestine and excretes heat-labile enterotoxin (Field 1980, Mekalanos 1985). The enterotoxin is composed of two types of subunits—type A, with a molecular weight of 27,215 Da, and type B (molecular weight 11,677 Da), encoded by genes *ctxA* and *ctxB*, respectively (Mekalanos et al. 1983). Each toxin molecule is composed of five B subunits and one A subunit (Gill 1976). The B subunit is responsible for binding of this toxin to the eucaryotic cell-surface receptor, monosialoglanglioside (GM_1), and facilitates entry of the A subunit into the target cell (Cutrecasas 1973). The A subunit must be nicked by proteolytic enzymes in order to be toxic (Pearson et al. 1982, Mekalanos et al. 1985). The amino terminal peptide of A subunit, A_1, catalyzes the transfer of the ADP-ribose from NAD to the Gs component of the mammalian adenylate cyclase regulatory system (Gill and Meren 1978). This causes an increase in adenylate cyclase activity and an elevated cAMP level. The elevated cAMP causes ion flux changes in the bowel and produces the watery diarrhea (Field 1980). Cholera toxin has been reported to be associated with immunomodulating effects in stimulating mucosal IgA and other immune response to admixed unrelated antigen after oral immunization. These effects include enhanced antigen presentation by a variety of cell types, promotion of isotype differentiation in B cells leading to increased IgA formation, and complex stimulatory as well as inhibitory effects on T-cell proliferation and lymphokine production (Holmgren et al. 1993).

Vibrio cholera also produces a second enterotoxin called Zonula Occluden Toxin (ZOT), which affects the intestinal intercellular tight junctions. The activity of the toxin is reversible. The toxin is heat-labile and is sensitive to protease digestion. The molecular weight of this toxin is between 10 and 30 kDa (Fasano et al. 1991). A third toxin called Accessory Cholera Enterotoxin (ACE) has also been recently reported. ACE increases short-circuit current in Ussing chambers and causes fluid secretion in ligated rabbit ileal loops (Trucksis et al. 1993).

Genes *ctxA* and *ctxB* map between the *nal* and *his* markers on the El Tor strain of *V. cholerae* chromosome, and are part of a large conserved genetic element (Mekalanos 1983). The gene *ace* is located immediately upstream of the genes encoding ZOT and CTX. The *ctx, zot* and *ace* genes comprise a *V. cholerae* ''virulence cassette'' (Trucksis et al. 1993).

Laboratory production of cholera toxins is affected by many environmental factors. Generally, low temperature (30°C) and good aeration with rich amino acid sources favor its production. Depletion of serine and asparagine from the medium or supplements of phosphate, magnesium, and iron to the medium stimulate production (Callahan et al. 1971, 1973). Regulation of cholera toxin production is a complicated matter (Betley et al. 1986).

7.2. Heat-labile Enterotoxin of *Escherichia coli*

Heat-labile enterotoxin (LT) is produced by some strains of *Escherichia coli* isolated from humans and porcines. These toxins are designated as LT_h and LT_p, respectively (Yamamoto et al. 1982). The gene for LT_h is called elt_h or *tor* and for LT_p; it is designated as elt_p. These genes are carried on a family of plasmids called Ent plasmids (Betley et al. 1986). LT_h and LT_p are closely related immunologically (Yamamoto et al. 1984).

The LT also consists of a holotoxin of one A subunit and five B subunits (AB_5). The 14-amino acids of the C-terminal of the A subunit comprise two functional domains that differentially affect oligomerization and holotoxin stability (Streatfield et al. 1992). The LT and CT toxins share similar subunit structure, activities and antigenic determinants (Betley et al. 1986). The nucleotide sequences of the genes for LT have considerable homology with those of CT (Spicer and Noble 1982; Mekalanos et al. 1983).

The production of LT is stimulated by methionine and lysine with either asparatic or glutamic acid (Betley et al. 1986). Addition of zinc to the medium also increases LT production (Sugarman and Epps 1984). Lincomycin and tetracycline at sublethal concentrations are reported to stimulate LT synthesis (Yoh et al. 1983).

7.3. Heat-stable Enterotoxins of *E. coli*

Heat-stable enterotoxins (STs) are a heterogenous group of low molecular weight proteins produced by some strains of *E. coli*. The STs are classified into two groups, methanol-extractable and methanol-insoluble.

Methanol-extractable enterotoxin STI can cause diarrhea in newborn mice or newborn pigs by increasing mucosal levels of cyclic guanosine-3′,5′-monophosphate (cGMP). The heat-stable, enterotoxin-induced diarrheal disease results from a decreased absorptive capacity in the colon in the face of increased small intestinal fluid secretion (Mezoff et al. 1992). There are three types of ST1, designated STI_a (or STI_p), STI_b (or STI_h), and STI_c (or $STIA_2$) (Betley et al. 1986). The genes encoding for ST also are on plasmids (Elwell and Shipley 1980, Smith 1984). The STI gene products have molecular weights of 10 and 7 kDa, which may represent the unprocessed and processed forms of STI (Lathe et al. 1980). STI is an extracellular protein. Transcription of the STI gene is regulated by cAMP. D-Glucose, S-gluconate, and L-arabinose repressed the synthesis of STI (Betley et al. 1986).

Methanol-insoluble enterotoxin STII is composed of 48 amino acid residues and has an isoelectric point of 9.7 (Fujii et al. 1994). It causes diarrhea in weaned piglets and rabbits but not in infant mice. The mechanism does not involve the increase in intracellular levels of cAMP or cGMP, but it may involve prostaglandin E2 (Hitotsubashi et al. 1992). The plasmidborne genes for STII have been cloned and sequenced (Pickens et al. 1983; Lee et al. 1983).

7.4. Enterotoxins Related to ST and LT

Many diverse bacteria in the gastrointestinal tract produce toxins that are related biochemically or genetically to *E. coli* ST or LT.

YERSINIA ENTEROCOLITICA ENTEROTOXIN

Yersinia enterocolitica produces a heat-stable enterotoxin (Yst) that can cause diarrhea by activation of guanyl cyclase (Rao et al. 1981). Yst has been implicated as an important factor in causing diarrhea in young children infected with *Y. enterocolitica* (Delor and Cornelis 1992). The Yst gene is chromosomal and encodes a 71-amino acid polypeptide. The C terminal domain has 30 amino acids; the N-terminal has 18 amino acids and has the properties of a signal sequence. The central domain of 22-amino acid is removed during or after the secretion process. This organization in three domains resemble that of the enterotoxin ST of *E. coli* (Delor et al. 1990).

Nutritional requirements for the synthesis of Yst have been studied. Ca^{2+} or Mn^{2+} had no effect on the biosynthesis of Yst, but Fe^{2+} at above 10 μM inhibited its synthesis. Addition of a mixture of pyrimidines containing thymine, cytosine, and uracil, each at 2 mM, stimulated the synthesis of this toxin, whereas a mixture of adenine and guanine at 2 mM inhibited synthesis (Amirmozafari and Robertson 1993).

ENTEROTOXINS FROM *ENTEROBACTER CLOACAE, KLEBSIELLA PNEUMONIAE,* AND *SHIGELLA FLEXNERI*

Enterotoxins from *Enterobacter cloacae, Klebsiella pneumoniae, Shigella flexneri*, have been reported to produce heat-stable enterotoxins. The enterotoxins from *E. cloacae* and *K. pneumoniae* have molecular weights of 1 to 10 kDa. These bacteria have also been reported to produce LT-like protein (Klipstein and Engert 1977; Betley et al. 1986).

AEROMONAS HYDROPHILA TOXINS

It has been found that clinical isolates of *Aeromonas hydrophila* produce four exotoxins: a hemolysin active against rabbit erythrocytes, a cytotoxin and enterotoxin detectable with Vero cell cultures, and a cholera-toxin-like factor (Vadivelu et al. 1991). The *A. hydrophila* enterotoxin is different from CT, LT, and *E. coli* ST. It is a cytotonic toxin, which elevates intracellular cAMP and prostaglandin E2 levels in cultured Chinese hamster ovary (CHO) cells (Chopra et al. 1992). *Aeromonas sobria* was also reported to produce cytotonic enterotoxin and hemolysin. The purified cytotonic enterotoxin had a molecular weight of 15 kDa, caused fluid accumulation in infant mice, but was nonhemolytic to rabbit erythrocytes. It caused an increase in cAMP activity in tissue culture cells. This cytotonic enterotoxin did not crossreact immunologically with components of cholera toxin. The purified hemolysin had a molecular weight of 55 kDa. It lysed rabbit erythrocytes, but did not cause fluid accumulation in the infant mouse test (Gosling et al. 1993). Both *A. hydrophila* and *A. sobria* produce these toxins at 5°C (Majeed et al. 1990).

SALMONELLA ENTEROTOXINS

Salmonella is the leading etiological agent of bacterial foodborne disease. Human salmonellosis is generally a self-limiting episode of enterocolitis; however, the disease can degenerate into chronic and debilitating conditions. *Salmonella typhimurium* produces a heat-labile enterotoxin that precipitates a net efflux of water and electrolytes into the intestinal lumen (D'Aoust 1991). The enterotoxin, a virulence factor in *S. typhimurium*, caused fluid secretion in ligated intestinal loops of rabbits and elongation of CHO cells (Chopra et al. 1994). The enterotoxin gene (*stn*) is confined to an 800-bp *ClaI-EcoRI* genomic DNA fragment that codes for two polypeptides (25 kDa and 12 kDa). The enterotoxin is a 25-kDa peptide, not the 12-kDa peptide. Formation of the heat-labile enterotoxin was blocked by GM_1 gangliosides and neutralized by antibody made against cholera toxin (Chary et al. 1993). The mechanism of enterotoxin-induced diarrhea has been studied; it appears to involve protein kinase C and arachidonic acid metabolites and may not involve the extracellular calcium pools (Khurana et al. 1991).

Other *Salmonella* strains such as *S. newport* (Harne et al. 1994), *S. stanley* (Sharma et al. 1992), *S. infantis*, and *S. haardt* (Aguero et al. 1991) were also reported to produce enterotoxins.

PSEUDOMONAS AERUGINOSA ENTEROTOXIN

Pseudomonas aeruginosa produces a heat-labile toxin which is similar to the vascular permeability factor (Shiriniwas and Bhujwala 1979).

Production of a heat-stable toxin by *P. aeruginosa* has also been reported (Timmis et al. 1984). Crude enterotoxin obtained from a recent isolate of *P. aeruginosa* CTM3 from milk was shown to be fairly stable at pH 4.5 to 7.5. Formalin and hydrogen peroxide had no significant effect on enterotoxin activity. The enterotoxin was probably encoded by a plasmid (Grover et al. 1990).

CAMPYLOBACTER JEJUNI ENTEROTOXINS

Campylobacter jejuni often colonizes the upper part of the gastrointestinal tracts of both humans and other animals. It produces a cytotoxic enterotoxin which has a closer immunological relationship to *E. coli* LT than to CT (Fernandez et al. 1983; Klipstein and Engert 1985; Hariharan and Panigrahi 1990). The cytotoxic toxin has a molecular weight of 68 kDa and is heat-labile. This toxin is inactivated at pH 3.0 or 9.0, sensitive to trypsin, and lethal to fertilized chicken eggs. It also has toxic effects on chicken embryo fibroblasts and CHO cells. In enzyme-linked immunosorbent assays, this toxin does not bind to GM_1 ganglioside. Adherence of this toxin to host cell membranes is reduced by prior treatment of the cells with proteolytic enzymes, neuraminidase, or glutaraldehyde, but not by treatment with β-galactosidase, lipase, Nonidet P-40, or sodium metaperiodate (Mahajan and Rodgers 1990). In addition to this enterotoxin, *C. jejuni* also produces hepatotoxic factors which cause hepatitis in mice (Kita et al. 1990).

7.5. Shigella dysenteriae Type I Toxin and Related Toxins

Shigella dysenteriae type I produces a cytotoxin called Shiga toxin. Certain strains of *E. coli* and *V. cholerae* produce protein toxins that are structurally and immunologically related to Shiga toxin and are referred to as Shiga-like toxins. Both Shiga toxin and Shiga-like toxins cause fluid accumulation in ligated rabbit intestine (O'Brian and LaVeck 1983; Smith et al. 1983).

Shiga-like toxins resemble cholera toxin in having a single A subunit (MW 315 kDa), which can be nicked by trypsin to A_1 (MW 27 kDa) and A_2 (MW 4.5 kDa) fragments, and several copies of the B subunit (MW 7.5 kDa). The Shiga-like toxins inhibit protein synthesis by catalytic inactivation of the 60S ribosomal subunit of target cells. These toxins are produced in the intestine and absorbed into the blood stream, affecting vascular endothelial cells in the target organs. They may also have a direct toxic effect on enterocytes. They are therefore cytotoxic (Gyles 1992). Several Shiga-like toxins have been reported to be encoded by a family of related bacteriophages (Smith et al. 1983; Newland et al. 1985). The genes encoding A and B subunits have been demonstrated to localize on one of these phages using molecular cloning (Newland et al. 1985).

7.6. Clostridial Toxins

CLOSTRIDIUM PERFRINGENS ENTEROTOXINS

Clostridium perfringens type A produces an enterotoxin (CPE), which is a spore specific protein with a molecular weight of 35 kDa and an isoelectric point of 4.3. It is heat sensitive and pronase sensitive, but is resistant to trypsin, chymotrypsin and papain (Stark and Duncan 1971). Raffinose (Labbe and Rey 1979) and theobromine (Labbe and Nolan 1981) stimulate toxin production. The 319-amino acid protein is encoded by the gene *cpe*, which was cloned, sequenced and expressed in *E. coli* (Czeczulin et al. 1993). The CPE utilizes a unique mechanism to directly affect the membrane permeability of mammalian cells. The CPE binds to a protein receptor on the mammalian plasma membrane and progressively makes the toxin more resistant to release by protease resulting in the formation of a large complex on plasma membranes which contains CPE.

The large complex eventually disrupts the cell and causes cell death (McClane 1994).

CLOSTRIDIUM DIFFICILE TOXINS

Clostridium difficile is an important nosocomial pathogen associated with antibiotic use (Anand et al. 1994). It produces two toxins: an enterotoxin (toxin A), and a cytotoxin (toxin B) (Depitre et al. 1993). The two share several functional properties with other bacterial toxins, like LT enterotoxin of *E. coli* and CT. These toxins cause disruption of microfilaments in the cell. The functional and binding properties of *C. difficile* toxins are confined within one large polypeptide with a molecular weight of 250 kDa. This is the largest bacterial toxin found to date (Wolfhagen et al. 1994).

7.7. Staphylococcal Enterotoxins

Staphylococcus aureus produces enterotoxins which are responsible for staphylococcal food poisoning syndrome characterized by vomiting, retching, abdominal cramping and diarrhea. There are several serological groups of staphylococcal enterotoxins (SE): A, B, C_1, C_2, C_3, D, and E. Each of these toxins is a heat-stable, single-chain protein (MW 26 to 30 kDa). The major ones are A, B, C, and D (Betley et al. 1986).

The SEs are highly hydrophilic and exhibit low α helix and high β pleated sheet content. These enterotoxins are potent activators of T lymphocytes. The receptors for SE on antigen-presenting cells are major histocompatibility complex (MHC) class II molecules (Johnson et al. 1991).

Recent work of Hoffmann et al. (1994) indicates that SEC_1 bonds to an α helix of MHC class II through a central groove in the toxin and thereby promotes or stabilizes the interaction between antigen-presenting cells and T cells. Staphylococcal enterotoxin A has also been shown to induce interferon (IFN) gamma product in spleen cells from BCG-immunized mice (Kato et al. 1990). Staphylococcal enterotoxin B (SEB) has also been recently shown to selectively upregulate the interleukin-2 receptor (IL-2R) β chain, without upregulating the IL-2R α chain, in a human tonsillar B cell population depleted of T cells. The action of SEB, mediated by binding to MHC class II, renders B cells sensitive to T cell-derived IL-2, and is sufficient for induction of vigorous DNA synthesis with low concentration of IL-2 (Franz et al. 1993). Staphylococcal enterotoxin B and C are associated with toxic-shock syndrome (TSS) which is an acute-onset, multiorgan illness resembling severe scarlet fever (Bohach et al. 1990). Staphylococcal enterotoxin D has been reported to be involved in modulating an autoimmune encephalomyelitis (Matsumoto and Fujiwara 1993).

The gene for SEA *(entA)* is located on the chromosome, maps between purine (*pur*) and isoleucine-valine (*ilv*) genes, and has been cloned (Betley et al. 1984). The gene for SEB is chromosomal and also may be located on a phage or a transposon (Ranelli et al. 1985). The genetics of SE genes were reviewed by Betley et al. (1986).

7.8. **Bacillus cereus** *Toxins*

Bacillus cereus is a gram-positive aerobic or facultatively anaerobic spore-forming rod. It is a cause of food poisoning frequently associated with the consumption of rice-based dishes. *Bacillus cereus* produces an emetic or diarrheal syndrome induced by an emetic toxin or enterotoxin, respectively. Other toxins produced during growth include phospholipases, proteases, and hemolysin (Drobniewski 1993). These toxins contribute to the pathogenicity of *B. cereus* in various diseases (Granum et al. 1993). The molecular mass of the enterotoxin is > 30 kDa. It can cause vascular permeability in rabbits, lethality in mice and cytoxicity in Vera and Hela cells (Guaycurus et al. 1993). The emetic toxin is a cytostatic toxin and is stable upon exposure to heat, pH 2 and pH 11, and trypsin (Mikami et al. 1994).

7.9. Other Bacterial Toxins

There are other bacterial toxins. *Streptococcus pyogenes* produces a superantigen which is a 260-amino acid protein. This toxin causes toxic-shock-like syndrome (TSLS) and is very similar to staphylococcal enterotoxin B (Reda et al. 1994). *Bacteroides fragilis* has also been reported to produce a soluble enterotoxin (Duimstra et al. 1991). There are also other bacterial toxins such as botulinal toxin, which produced outside the body when *C. botulinum* grows in food.

8. Biotransformation Mechanisms of Toxins in the Monogastrics

When toxicants reach the gastrointestinal tract, many of them are biotransformed through enzymes of the liver, intestinal microbiota, or epithelial cells. The diversity and metabolic activities of intestinal microorganisms are reviewed elsewhere in these books. Since transformation does not occur exclusively in the gastrointestinal tract, the present discussion focuses on the biochemical mechanisms which take place both in the liver and gastrointestinal tract.

Biotransformation reactions are generally categorized as phase I or phase II reactions. Phase I reactions involve alteration of the chemical structure of the incoming chemicals (toxicants). Oxidation, reduction or hydrolysis are the major types of phase I reactions. During phase I, a polar group is added to the molecule. Usually this will increase the molecule's water solubility, but the most important effect is to render the entering compound a suitable substrate for phase II reactions (Hodgson and Levi 1994). Phase II reactions involve the biosynthetic union or coupling of foreign compounds or phase I metabolites of foreign compounds with endogenous intermediates. The endogenous intermediates are conjugating agents which may be derived from a carbohydrate, a protein or amino acid source, a sulfur or phosphate component, or an intermediate of lipid metabolism. Examples of phase II reactions are glucuronidation, glucosidation, sulfation, acetylation, methylation, amino acid conjugation, glutathione conjugation, and lipophilic conjugation. These conjugations may be viewed as normal biochemical processes serving a dual role in the intermediary metabolism of endogenous compounds as well as in the metabolism (intoxication or detoxification) of foreign compounds including toxicants (Dauterman 1994).

Many reactions associated with conjugation were considered to result in the formation of only hydrophilic conjugation products. These products were generally less lipid soluble, more polar, more easily excreted and were assumed to be less toxic. However, recent findings have demonstrated examples of the lipophilic conjugation of xenobiotics. The lipophilicity of the conjugation results in storage in the organism and delay in excretion. So phase II reactions can no longer be considered synonymous with detoxification (Dauterman 1994).

Although both phase I and phase II reactions have been studied in detail in the liver and other sites, the intestinal tract possesses all of the major phase I and phase II biotransformation enzymes (Hoensch et al. 1976, Dawson and Bridges 1981, Chhabra and Eastin 1984, Combes 1989, Dauterman 1994, Hodgson and Levi 1994, Levi 1994). Interaction of phase I and phase II enzymes is illustrated in Figure 14.6.

8.1. Phase I Reactions

The most important phase I reaction is oxidation. This is mediated by a cytochrome P450-containing enzyme system, known as a ''monooxygenase,'' formerly called ''mixed-function oxidase system'' (MFO). This is a group of hemeproteins whose Fe^{2+}-CO complex show an absorption spectrum with a maximum near 450 nm (in some instances 448 nm), a characteristic attributed to the unique ligation of the heme by a cysteinyl thiolate. The cytochrome P450 enzymes are distributed throughout nature. There are many bacterial forms of cytochrome P450 (Guengerich 1992a). In the higher forms, however, the oxidation system is present in many tissues, including intestinal mucosa, but the principal site of oxidation of most toxic compounds is the smooth endoplasmic reticulum of the liver. The membrane topology of mammalian cytochrome P450 has been extensively studied (Black 1992). Small intestinal mucosa of both the human and rat account for about 5% to 20% of the specific activity as compared to the liver (Thies and Siegers 1989). Along the length of the small intestine a continuous

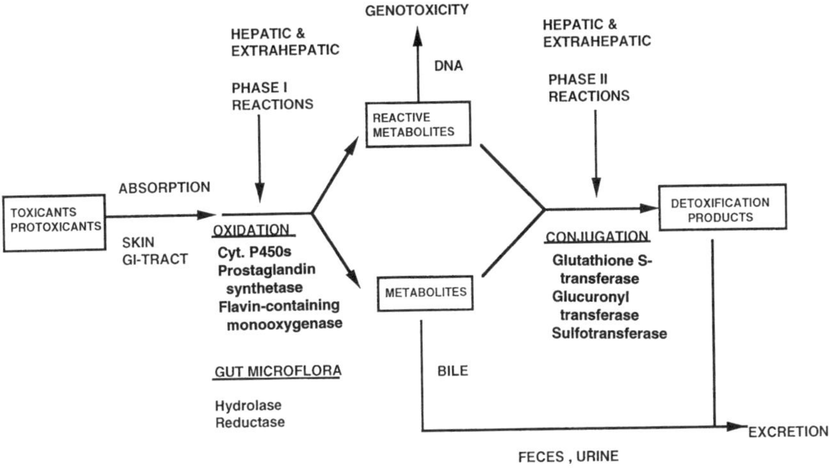

Figure 14.6. Interaction of phase I and phase II enzymes.

decrease in P450 enzyme activity was found in the cells of the upper villous from proximal to distal segments. The mucosa of the large bowel demonstrates a high activity of cytochrome P450-dependent enzymes with decreasing activity towards distal parts of the colon (Newaz et al. 1983). The mixed function oxidase system can be isolated in the form of microsomes upon homogenization and centrifugation.

The nomenclature for P450 genes has been a concern. A Committee on Standardized Nomenclature of the P450 genes has been formed, and recommended nomenclature has been established. The nomenclature for the P450 gene superfamily is based on evolution; recomendations include Roman numerals for distinct gene family, capitals letters for subfamily, and Arabic numerals for individual genes (Nebert et al. 1987). The P450 gene superfamily encodes numerous enzymes, of which more than 150 have been characterized. There are now 28 families, of which 11 are predicted to exist in all mammals (Coon et al. 1992). At least 18 human hepatic cytochrome P450 genes have been sequenced (Crotty 1994). Different substrates, inhibitors, and inducers for each products have been noted. The complement of the various cytochrome P450 enzymes in a given individual varies markedly (Guengerich 1992b). The structure, function, evolution, and regulation of P450 genes have been reviewed (Nebert and Gonzalez 1987; Hodgson and Levi 1994).

The substrates for cytochrome P450 metabolism include endogenous compounds such as steroids and prostaglandins as well as many xenobiotics such as carcinogens, mutagens, and drugs. There are multiple cytochrome P450s with various specificities for different substrates. For example, polycyclic aromatic hydrocarbons are catalyzed by P450 1A1; aflatoxin B_1 by 2C11, 1A1, and 1A2; acetaminophen by 2C11, 1A1, and 2E1; and cyclophosphamide, fluoroxene, benzene, and dimethylnitrosamine by 2E1 (Levi 1994). Cytochrome P450 metabolism of compounds may result in either activation or detoxification. The metabolism by P450 enzymes of many toxic compounds such as polycyclic aromatic hydrocarbons, aromatic amines, nitroaromatic hydrocarbons, nitrosamines, hydrazine, and halogenated hydrocarbons has been reviewed (Kadlubar and Hammons 1986, Combes 1989, Koop 1992, Yang et al. 1992, Hodgson and Levi 1994).

Synthesis of P450 protein in liver is induced by many chemicals (Hodgson 1994). Regulatory mechanisms of the induction process have been investigated (Fujii-Furiyama et al. 1992). Many factors including diet, age, and species differences affect the activities of these enzymes (Conney 1982, Yang et al. 1992; Chadwick et al. 1993, Hodgson 1994, Levi 1994).

Other important enzymes in the phase I reactions of toxic substances include flavin-containing monooxygenase (FMO), prostaglandin synthetase (PGS), molybdenum hydroxylases (aldehyde oxidase and xanthine oxidase), alcohol dehydrogenase, aldehyde dehydrogenase, esterases and amidases, epoxide hydrolase, DDT-dehydrochlorinase, and glutathione reductase (Hodgson and Levi 1994).

The flavin-containing monooxygenase (FMO), like P450, is located in the endoplasmic reticulum. It is mainly responsible for the oxidation of many organic compounds containing nitrogen, sulfur or phosphorus heteroatoms as well as of some inorganic ions. In addition to monooxygenase, there are other enzymes that are involved with the oxidation of xenobiotics. These oxidoreductases are located in either the mitochondrial fraction or the 100,000g supernatants of tissue homogenates. Examples of these enzymes are alcohol dehydrogenases, aldehyde dehydrogenase, amine oxidases (monoamine oxidases and diamine oxidases) and molybdenum hydroxylases (Hodgson and Levi 1994).

Prostaglandin synthetase (PGS), a membrane bound glycoprotein with a subunit molecular mass of about 70 kDa, is present in virtually all mammalin tissues. PGS catalyzes the oxygenation of polysaturated fatty acid to hydroxy endoperoxides. The preferential substrate in vivo is arachidonic acid. PGS has two activities: one is the fatty acid cyclooxygenase activity, which brings about the oxygenation of polyunsaturated fatty acid, such as arachidonic acid, to form a cyclic endoperoxide, prostaglandin G (PGG); the other is prostaglandin hydroperoxidase activity, which reduces the cycloperoxide to a hydroperoxy endoperoxide termed prostaglandin H (PGH). Many xenobiotics may be cooxidized during this second hydroperoxidase reaction (Hodgson and Levi 1994). Xenobiotic cooxidation involving PGS includes: dehydrogenation of acetaminophen, benzidine, diethylstilbestrol, and epinephrine; demethylation of dimethylaniline, benzphetamine, and aminocarb; hydroxylation of benzo(a)pyrene, 2-aminofluorene, and phenylbutazone; epoxidation of 7,8-dihydroxybenzo(a)pyrene; sulfoxidation of methylphenylsulfide; and oxidation of 2-amino-4-[5-nitro-2-furyl]-thiazole (ANFT), n-[4-(5-nitro-2-furyl)-2-thiazoyl]-formide (FANET), and bilirubin (Hodgson and Levi 1994).

Phase I reactions also include hydrolysis by hydrolases which are found in many tissues including intestinal mucosa. Hydrolytic reactions are the only phase I reactions that do not utilize energy. Many xenobiotics, such as esters, amides, or substituted phosphates that contain ester bonds, are susceptible to hydrolysis. Esterase is a kind of hydrolase which mediates the addition of water to an ester bond with the formation of the corresponding alcohol; likewise amidase mediates the addition of water to an amide bond with the production of the corresponding amines. A variety of xenobiotics, such as phthalic acid esters (plasticizers), phenoxyacetic acid and picolinic acid esters (herbicides), pyrethroids and their derivatives (insecticides), and a variety of amide derivatives of drugs are detoxified by hydrolases found in the the gastrointestinal tracts of animals (Hodgson and Levi 1994). Epoxide hydrolase mediates the addition of water to epoxide to form the trans-dihydrodiol which is more water soluble. Epoxide hydrolase is found in human, rat and rabbit intestinal mucosal microsomes (Thies and Siegers 1989). There are also cytosolic epoxide hydrolases which are broadly distributed in different organs. Both microsomal and cytosolic epoxide hydrolases are broadly non-specific, but many substrates hydrolyzed well by cytosolic epoxide hydrolase are hydrolyzed poorly by microsomal epoxide hydrolase and vice versa (Hodgson and Levi 1994).

DDT-dehydrochlorinase, a reduced glutathione (GSH)-dependent enzyme, catalyzes the degradation of pp-DDT to pp-DDE or the degradation of 2,2,-bis(p-chlorophenyl)-1,1-dichloroethane (pp-DDD) to the corresponding DDT ethylene, 2,2,-bis(p-chlorophenyl)-1-chloroethylene (TDEE). Glutathione reductase reduces disulfiram (Antabuse), a drug used for the treatment of alcoholism (Hodgson and Levi 1994). The involvement of these enzymes has been illustrated in the metabolism of many types of toxic compounds (Butler et al. 1989, Combes 1989, Hodgson and Levi 1994).

There are other phase I enzymes. For example, quinone reductase [NAD(p): quinone oxidoreductase] (DT-diaphorase) converts toxic quinones into stable hydroquinones. This is an effective detoxification process (Thies and Siegers 1989). This enzyme is strongly induced by (-)-1-isothiocyanato-4R-(methylsufinyl)-butane, which is present in broccoli (Proschaska et al. 1992, Zhang et al 1992).

Phase I enzymes are more likely involved in the activation of toxicants. There are several enzymes present in many intestinal microbes which are phase I enzymes. These include glycosidase, azo reductase, nitroreductase, hydroxylase, and sulfatase (Crotty 1994). The intestinal microbiota also contain P450 and other enzymes, but this field has not been extensively explored.

GLYCOSIDASE

As mentioned previously, many toxicants are glycosides, particularly flavonoids such as cyclamate, rutin, and quercetin, which consist of a nonsugar moiety (aglycone) and sugar linked by either α- or β-glycosidic linkages. Some intestinal bacteria hydrolyze the glycosidic bond and release biologically active aglycones, many of which are toxic, mutagenic, or carcinogenic.

There are three types of glycosidase: β-glucosidase, β-galactosidase, and β-glucuronidase (Perez et al. 1986, Bokkenheuser et al 1987, Chang et al. 1989). Chadwick et al. (1992) recently reviewed intestinal microbiota with glycosidase activity.

AZOREDUCTASE

The azo dye family includes those used in the manufacturing of textiles, food and drinks, pharmaceuticals, cosmetics, plastics, paper and leather, printing ink, paint, varnish, lacquer, and wood stains. When the water-soluble azo compounds reach the intestinal tract, these dyes are reduced to phenyl and napthyl substituted amines by azoreductase of intestinal microbiota (Goldin 1990, Mallet and Rowland 1990, Chung et al. 1992). Many of these amines are toxic, mutagenic, or carcinogenic (Chung 1983; Levine 1991). For example, Chung et al. (1978b, 1981) showed that Methyl Orange and Methyl Yellow are reduced by intestinal anaerobes to the mutagen N,N-dimethyl-*p*-phenylenediamine. Direct Black was reduced by human intestinal microbiota to the metabolites 4-aminobiphenyl, 4-acetylaminobiphenyl, benzidine, and monacetylbenzidine (Cerniglia et al. 1986). Chung et al. (1992) reviewed the reduction of azo dyes by intestinal microbiota and found that a number of dietary factors affected the azo reductase activity in vivo. Bacteria with azo reductase activity have been identified (Chung et al. 1978a, 1992, Rafii et al. 1990).

NITROREDUCTASE

Zacharin and Juchau (1974) showed that gastrointestinal bacteria play an important role in the reduction of aromatic nitro-groups in vitro and in vivo. Rosenkranz et al. (1981) also noticed in the Ames *Salmonella* tester strains the presence of nitroreductase which reduced nitroarenes (2-nitrofluorene, 1-nitropyrene, 5-nitroacenaphthene and mononitrobenz[a]-pyrenes) to the corresponding amines. McCoy et al. (1981) demonstrated the existence of a family of bacterial nitroreductases capable of activating nitrated polycyclics to mutagens. It is now known that many human (El-Bayoumy et al. 1983, Cerniglia et al. 1984, Manning et al. 1986, Fu et al. 1988, Delclos et al. 1990, Rafii et al. 1991) and rodent (Howard et al. 1983, Cerniglia 1985, Kinouchi and Ohnishi 1986, King et al.

1990) intestinal bacteria have nitroreductase activity. This nitroreductase carries out the reaction as a two-electron transfer (nitro → nitroso → hydroxylamine → amino), forming an anion free-radical intermediate, a nitroso group, and a hydroxyl group. All three of these functional groups have been associated with potentially deleterious genotoxic events (Chadwick et al. 1992). Nitrocompounds that have been found to be reduced by intestinal microbiota include 1-nitropyrene, 4-nitropyrene, 1,3-dinitropyrene, 1,6-dinitropyrene, 1 to 8, dinitropyrene, 6-nitrobenzo-(a)pyrene, 4-nitrobiphenyl, 1-nitronaphthalene, 2-nitronaphthalene, 5-nitroacenaphthalene, 2-nitrofluorene, 3-nitrofluoranthene, 2-nitroanthracene, 9-nitroanthracene, 6-nitrochrysene, 2,4-dinitrotoluene, 2,6-dinitrotoluene, 2-nitrotoluene, 3-nitrotoluene, 4-nitrotoluene, 1-nitrobenzene, and 2-nitrobenzene (Rickert 1987; Fu 1990). It has been found that diet and antibiotics have a large influence on intestinal nitroreductase activity (Goldin and Gorbach 1984).

DEHYDROGENASE

Bile acids, implicated to be associated with colon cancer, are synthesized in the liver, are conjugated with taurine and glycine in the bile, and then enter the gastrointestinal tract. The intestinal microbiota deconjugate the bile acids and release the free acids, primarily cholic and chenodeoxycholic acids. The cholic and chenodeoxycholic acids are further dehydroxylated to deoxycholic and lithocholic acids by 7-α-hydroxysteroid dehydrogenase (Sjovall 1960, Carey 1973, Goldin 1986a). This topic is extensively reviewed in Chapter 13 of the present volume, by Baron and Hylemon.

SULFATASE

There are intestinal bacterial sulfatases which hydrolyze C-sulfonates, O-sulfates and N-sulfonates in the gastrointestinal tract (Scheline 1973).

A typical example is the hydrolysis of cyclamate, which was used as an artificial sweetener. The cyclamate can be converted by sulfatase to produce cyclohexylamine, which is a carcinogen producing bladder tumors (Wallace et al. 1970, Renwick and Williams 1972).

8.2. Phase II Reactions

The primary phase II reaction is the addition of an endogenous compound to a previously oxidized or reduced product (from a phase I reaction). The most important phase II reactions are conjugations that involve UDP-glucuronyl transferase, UDP-glucosyltransferase, sulfotransferase, methyltransferase, acetyltransferase, acyltransferase, and glutathione S-transferase (GST).

UDP-GLUCURONYL TRANSFERASE

UDP-Glucuronyl transferase mediates the addition of glucuronic acid to various substrates. Four general catagories of substrates of O-, N-, S-, and C-containing glycosides have been reported (Dauterman 1994). One typical example is the addition of UDP-glucuronic acid to diethylstilbestrol to form o-glucuronyl diethylstilbestrol. The UDP-glucuronyl transferase is present in nearly all tissues, located in the membrane of the endoplasmic reticulum (Chowdbury et al. 1985). In the intestine, a longitudinal distribution of glucuronyl transferase was noted with high activities in the duodenum and low activities in the distal end of colon (Bock 1977; Koster et al. 1985). Intestinal glucuronyl transferase is inducible by phenobarbital and methylcholanthrene (Laitinen and Watkins 1986). In the stomach, glucuronyl transferase in rats is inducible by dimethylhydrazine, but no induction of this enzyme in the liver, duodenum, or colon was observed (Sandforth and Gutschmidt 1984). Many factors—e.g., age, diet, gender, species and strain differences, genetic factors, and disease—affect glucuronidation (Dauterman 1994).

Normally, the glycuronide conjugates are excreted in the urine and bile, unless thay are involved in the enterohepatic circulation. The conjugates may be hydrolyzed by β-glucuronidase present in various tissues or organs, or upon biliary excretion, by intestinal bacterial β-glucuronidase.

UDP-GLUCOSYLTRANSFERASE

Both UDP-glucuronyl transferase and UDP-glucosyltransferase carry out glycosidation. Glucuronyl transferase mediates the addition of glucuronic acid, while glucosyl transferase mediates the addition of glucoside to various substrates. Glucosidation of certain xenobiotics such as *p*-nitrophenol has been reported in mammals (Dauterman 1994).

SULFOTRANSFERASE

Sulfotransferase mediates the addition of sulfate to various substrates. One example is the formation of N-sulfoacetylaminofluorene from 2-hydroxyacetylaminofluorene by sulfotransferase. With many xenobiotics, sulfation appears to be a detoxification reaction. However, a number of compounds are converted into highly labile sulfate conjugates that form reactive intermediates which have been implicated in carcinogenesis and tissue damage. For example, mono- and dinitrotoluenes, N-hydroxyphenacetin, and other N-hydroxyarylamides are in part activated by sulfation (Dauterman 1994).

METHYL TRANSFERASE

Methylation involves the transfer of methyl groups from one of two methyl donor substrates—S-adenosylmethionine, or N^5-methyltetrahydrofolic acid. The transfer occurs with many xenobiotics. In most cases involving methylation, products are less water-soluble than the parent compound, except for the tertiary amines. However, it is generally considered a detoxification reaction.

ACETYL TRANSFERASE

The xenobiotics containing amino or hydroxyl groups can easily interact with acetyl CoA, an activated conjugating intermediate. This reaction referred to as acetylation, is catalyzed by acetyl transferase. Acetyl transferase activity has been demonstrated in liver, intestinal mucosa, colon, kidney, thymus, lung, pancreas, brain, erythrocytes, and bone marrow. There are also different acetyltransferases in different species. N-ethylmaleimide, iodoacetate, *p*-chloromercuribenzoate, Cu^{2+}, Zn^{2+}, Mn^{2+}, and Ni^{2+} have been reported to inhibit acetyltransferase (Dauterman 1994).

ACYLTRANSFERASE

Acyltransferase carries out the conjugation of carboxylic acid-containing xenobiotics with amino acids. This conjugation requires initial activation of the xenobiotics to a CoA derivative in a reaction catalyzed by acid: CoA ligase. The acyl CoA subsequently reacts with an amino acid to produce acylated amino acid conjugate and CoA. Amino acid conjugation has been reported to occur with glycine, glutamine, arginine and taurine in mammals. The acyltransferase occurs in both liver and kidney and is located in the mitochondria. There is more than one type of N-acyltransferase present in mammalian tissues (Dauterman 1994).

GLUTATHIONE-S-TRANSFERASE (GST)

Perhaps the best-known phase II reaction is GST, which mediates the formal addition of glutathione to various substances by displacement reactions. There are three main groups of GST: α (basic), μ (neutral), and π (acidic), mainly present in the liver, but also present in various extrahepatic tissues including intestinal epithelium and stomach (McKay et al. 1993). A new class of glutathione S transferase, θ, has recently been proposed, since the amino acid sequences showed little similarity to the α, μ, and π classes. They are active against 1,2-epoxy-3-(*p*-nitrophenoxy)-propane, *p*-nitrobenzyl chloride, *p*-nitrophenethyl bromide, cumene hydroperoxide, and dichloromethane (Dauterman 1994). In the stomach, GST activity is generally high in the glandular stomach and low in the forestomach (Aponte et al. 1977, Boyd et al. 1979). GST is generally considered a detoxification process. It inhibits the macromolecular binding of the polycyclic aromatic hydrocarbon metabolites—in particular, epoxides generated from the P450 reactions. Examples are benzo[a]pyrene and aflatoxin B_{12} epoxides from P450 enzyme metabolism (Thies and Siegers 1989, Chen et al. 1995). GST not only produces a more water-soluble substance, but also is involved in the direct detoxification of these epoxides. However, GST may also stimulate macromolecular alkylation by an agent such as N-methyl-N′-nitro-N-nitrosoguanidine, which is an activation, not detoxification (Wiestler et al. 1983). The GST are involved in the metabolism of a wide variety of electrophilic substrates that include antibiotics, vasodilators, herbicides, insecticides, analgesics, anticancer agents and carcinogens.

OTHERS

There are other enzymes that carry phase II reaction. For example β-lyases present in mammalian intestinal bacteria, liver, and kidney are capable of converting a number of cysteine S-conjugates into pyruvate, ammonia, and thiols. Premercapturic acids, but not mercapturic acids, serve as substrate for β-lyase.

8.3. Factors Affecting Biotransformation

Many factors, especially those present in the diet, can affect phase I and phase II biotransformations, increasing or decreasing the toxicity of potential toxicants in the gastrointestinal tract.

Riboflavin is an essential nutrient. Short-term deficiency leads to an increase in the levels of cytochrome P450 and an increase in the oxidation of certain toxicants; long-term deficiency produces a substantial decrease in P450 activity (Shibamoto and Bjeldanes 1993e).

Vitamin E regulates the synthesis of heme, an essential component of cytochrome P450. Vitamin E deficiency will thus decrease the activity of P450. Vitamin C deficiency can also result in the decreased metabolism of many substances, with decreases in cytochrome P450 and NADPH cytochrome P450 reductase activity (Shibamoto and Bjeldanes 1993e).

Protein deficiency affects metabolism of many substances, decreasing or increasing the toxicity depending on the specific substrates. Generally, protein deficiency leads to reduction in levels of NADPH-cytochrome P450 reductase and cytochrome P450 enzymes.

Dietary fat also plays a role in microsomal metabolism. High animal fat and protein, with low fiber intake, increased intestinal cytochrome P450 content (Smith-Babaron et al. 1981a,b). Increased levels of saturated fatty acids in the diets produced increased levels of aniline oxidation, but not hexobarbital oxidation. Supplementation with unsaturated fat resulted in an increase in concentration of cytochrome P450 and a decrease in glucose-6-phosphate dehydrogenase activity (Shibamoto and Bjeldanes 1993e).

Iron deficiency had little effect on the level of cytochrome P450 or cytochrome P450 reductase in liver tissues; however, iron deficiency caused a rapid decrease in the content of P450 and benzo[a]pyrene hydroxylase activity in the small intestine in rats (Hoensch et al. 1976, 1981, 1982).

Magnesium deficiency caused a lower level of P450 and cytochrome P450 reductase, which would significantly decrease the metabolism of many toxicants in the liver. Decreased selenium uptake also affects glutathione peroxidase activity, resulting in failure to remove peroxides and thus unable to avoid oxidative damage to macromolecules. In rats, supplementation with selenium reduced dimethylhydrazine-induced colon cancer (Jacob 1983). Human colon cancer may also be correlated with selenium deficiency (Nelson 1984). A large variety of compounds including drugs, carcinogens or foreign compounds that are widespread in the environment will induce metabolic enzymes in the liver or intestines. If an animal is treated with the carcinogen 3-methylcholanthrene, the rate of hydroxylation of polycyclic aromatic hydrocarbons increases and cytochrome P4501A1 is induced. Phenobarbital treatment, on the other hand, leads to an increase of P4502B1 and a corresponding increase in the rate of metabolism of hexobarbital and aminopyrene. Hodgson (1994) reviewed the induction of P450 and concluded that P4501A1 and 1A2 are induced by PAHs; 2B1 and 2B2 are induced by phenobarbital; 3A1 is induced by PCN/glucocorticoid; 2E1 is induced by ethanol; and 4A1 is induced by peroxisome proliferator such as clofibrate, tridiphane, phthalates, 2,4-D, and 2,4,5-T.

Some cruciferous vegetables including Brussels sprouts, cabbage, broccoli, and cauliflower contain indoles (indole-3-acetonitrile, indole-3-carbinoles, 3,3′-diindolylmethane) that significantly increase cytochrome P4501A enzymes in the intestine and liver. These indoles account for the majority of the inducing effect of these vegetables (Loub et al. 1975). Administration of indoles inhibits benzo[a]-pyrene-induced neoplasia in the forestomach of mice and also inhibits 7,12-dimethyl-benzo[a]-anthracene-induced mammary tumors in rats (Shibamoto and Bjeldanes 1993e). GST activity also is induced by substances from cruciferous vegetables.

Chronic intake of alcohol increased the cellular concentration of cytochrome P450 and the activity of NADPH-cytochrome P450 reductase and benzo[a]pyrene hydroxylase in the small intestine and liver, but not in the large intestine of rats (Seitz et al. 1982). This induction increased intestinal biotransformation of certain procarcinogens such as benzo[a]pyrene and aminofluorene. Parallel to the induction of these enzymes, an increased tumor rate was demonstrated in dimethylhydrazine treated rats (Seitz 1985). However, there was also evidence that did not support this observation (Siegers et al. 1986, Thies and Siegers 1989).

Many drugs such as phenobarbitone, hydrocortisone, benz[a]anthracene, or benz[a]anthracene-7,12 -dione have been shown to enhance mucosal hydroxylation in human colon (Strobel et. al. 1983). Aroclor appears to induce a wide spectrum of enzymatic activity in the liver. Gastric hormones such as pentagastrin, secretin, and cholecystokinin have been shown to induce several enzymatic activities in colon mucosa (Strobel et al. 1981). In rats, the onset of the sexually dimorphic pattern of growth hormone (GH) secretion and increased hepatic GH-binding capacity at puberty are correlated with the expression of some hepatic P450 enzymes involved in steroid metabolism (Legraverend et al. 1992). Many factors such as fat, protein content, plant cell constituents, cholesterol, micronutrients (vitamins and trace elements), and minor components such as flavones, indoles, and miscellaneous phenolics have been shown to substantially modify many intestinal enzyme activities (Wollenberg and Ullrich 1980, Wattenberg 1983). Yang et al. (1992) recently reviewed dietary effects on cytochrome P450 and xenobiotic metabolism and toxicity.

8.4. Interactions

The environment of the intestinal tract can influence the biotransformation capacity of intestinal microbiota (Cole et al. 1985). On the other hand, the microbiota may also affect susceptibility of the mucosa and the expression of its metabolizing enzymes (Rolls et al. 1978). The presence of intestinal microbes can also influence the metabolism of toxic compounds in the gastrointestinal tract and at other sites such as the liver (Wostmann 1984). The intimate metabolic interactions between the enzymes of the intestinal microbes and those of host are an important aspect of the biotransformation of toxins.

Bioactivation of water-soluble azo dyes is an example. The majority of azo dyes undergo reduction by enzymes of intestinal microorganisms and/or the hepatic system (Levine 1991; Chung et al. 1992a). Their bioactivation may depend upon their lipid solubility. Water-soluble azo dyes are anaerobically reduced by intestinal bacteria; then liver enzymes activate reducing products (primary amines) to mutagens. On the other hand, azoreduction of a lipophilic dye (e.g., 4-dimethylamino-azobenzene) reduced its carcinogenicity, and oxidative metabolism of the primary amines yielded mutagenic products (Levine 1991).

Bioactivation of nitropolycyclic aromatic hydrocarbons (Nitro-PAHs) is another example of the interaction between microbial and liver enzymes. Fu (1990) reviewed this class of chemicals and showed that combined metabolism of both intestinal bacterial and liver enzymes contributed to the ring oxidation of nitro-PAHs followed by nitroreduction and/or esterification.

Metabolism of the hepatocarcinogen 2,4-dinitrotoluene (DNT) also involves interaction between microbial and liver enzymes. DNT is oxidized in the liver by P450 to 2,4-dinitrobenzyl alcohol, conjugated with glucuronic acid and excreted in the bile (Rickert et al. 1984, Leonard et al. 1987). The glucuronide is transported to the intestinal tract and hydrolyzed by the β-glucuronidase of intestinal bacteria. The nitro group is also reduced by microbial nitroreductase. The resultant compound is reabsorbed and returned to the liver, where it is further activated to a compound capable of forming adducts with hepatic DNA (Chadwick et al. 1991a). This is a sequential interaction of both microbial and liver enzymes. Any factors which affect enzyme activities of either the liver or intestinal microbia would affect the genotoxicity of DNT. For example, pretreatment of rodents with pentachlorophenol (PCP) has been reported to potentiate the toxicity of DNT. PCP elevates β-glucuronidase activity and reduces nitroreductase activity in the small intestine (Chadwick et al. 1991a, George et al. 1991). PCP also induces monooxygenase activity (Vizethum and Goerz 1979). Likewise, Aroclor 1254 significantly potentiates the genotoxicity of DNT (Chadwick et al. 1991b). However, Aroclor1254 accelerates the activation of DNT after only 1 week of treatment, whereas 4 to 5 weeks of treatment was required with PCP (Chadwick et al. 1993).

Rowland et al. (1987) discovered that intestinal microbiota decreased the activation capacity of the liver toward 2-amino-3,4-dimethylimidazole[4,5-f]quinoline (MIQ) and 3-amino-1-methyl-5H-pyrido [4,3-b]indole (Trp-P-2). In contrast, intestinal microbiota increased the activation capacity of the liver toward the hepatocarcinogen aflatoxin B_1.

Intestinal microbiota can affect the metabolism of known hepatocarcinogens in the liver and the induction of tumors. Spontaneous incidences of mouse hepatoma are significantly influenced by the presence of intestinal microbiota. Conventional animals had a higher number of hepatomas than found in germ-free mice. When germ-free C3H mice were exposed to specific intestinal bacteria isolated from humans, the incidence of tumor increased significantly (Mizutani and Mitsuoka 1981).

A synergistic bioactivation of 2-aminoanthracene and aminofluorene by a mixture of microsomes from rat colon mucosa and cell-free extracts from a human strain of *Bacteriodes fragilis* has been reported (McCoy et al. 1979). It was suggested that bacteria biotransformed a colon specific carcinogen to a penultimate metabolite, then mucosal enzymes further activated it. Tasich and Piper (1983) showed that the colonic microsomal fraction or the cell-free extract of *Bacteroides fragilis* alone demonstrated little bioactivation of 2-aminoanthracene, but together bioactivated the chemical to a genotoxic metabolite.

There are numerous other examples showing that the interaction of both intestinal bacterial and mucosal enzymes is required for the bioactivations of chemicals to ultimate toxicants (Vernet and Seiss 1986, Rickert 1987). The gastrointestinal tract provides the ideal environment for these interactions. These interactions affect not only the toxicity of toxicants, but extend to nutrition, natural immunity and other aspects of health as well (Drasar and Barrow 1985, Balish 1986).

9. Conclusion

Many toxicants occur naturally in animal and plant foodstuffs, are produced during food processing, or are included as additives. Many bacteria produce toxins in the gastrointestinal tract. Some pharmaceutical agents can also be toxicants. Others such as mycotoxins (Hsieh 1989; Bresinsky and Besl 1990), pesticides (Shibamoto and Bjeldanes 1993), and contaminants from industrial wastes are not reviewed in this chapter. Over 53,000 chemicals used by society can enter the gastrointestinal tract (Laitinen and Watkins 1986) and can be considered as toxicants. The intestinal tract is the largest receptive organ of the body to encounter these toxicants. A few toxicants are absorbed through the circulatory system. A majority of toxicants are biotransformed. Mucosal enzymes from the small and large intestines and colon, as well as enzymes from the huge amount of microorganisms, are responsible for these biotransformations. Although the liver is usually considered to be the drug metabolizing organ, ingested chemicals must pass through the intestine, where intestinal microbiota and mucosal enzymes can metabolize them via various pathways (Hanninen et al. 1987, Ilett et al. 1990). The role of the intestinal microbiota in the biotransformation of these chemicals has been more throughly studied in recent decades. Enzymes of intestinal bacteria and mucosa serve as the first defense mechanism against these toxicants. These metabolizing enzymes work sequentially and/or cooperatively in biotransforming these toxicants. Various types of interactions occur between them and are influenced by dietary and other environmental factors. Many dietary factors can either induce or suppress certain enzyme activities, thus greatly affecting the biotransformation and toxicity of toxicants. Whether a chemical is toxic or not is largely dependent upon how it is metabolized (bioactivated). The understanding of these biotransformation reactions is important in interpreting many disease processes, particularly chronic diseases such as cancer. Although a number of examples are provided in this chapter, the understanding of gastrointestinal toxicology is just beginning. Other aspects such as nutrition and immunity are also greatly affected, but our present knowledge is still limited. A clear illustration of interactions between chemicals-microbes-mucosal enzymes is crucial for the total understanding of health and merits further studies using modern research techniques.

Acknowledgments

The author thanks Dr. Barbara J. Taller of the Department of Biology, University of Memphis, and Dr. Cheng-I. Wei of the Food Science and Human Nutrition Department, University of Florida, Gainesville, for their kind editing of this paper.

References

Aguero J, Faundez G, Nunez M, Wormser GP, Lieberman JP, Gabello FC (1991) Choleriform syndrome and production of labile enterotoxin (CT/LT1)-like antigen by species of *Salmonella infantis* and *Salmonella haardt* isolated from the same patients. Rev Infect Diseases 13:420–423.

Alvares AP (1984) Environmental influences on drug bio-transformations in humans. World Rev Nutr Diet 43:45–49.

Ames BN (1983) Dietary carcinogens and anticarcinogens. Science 221:1256–1264.

Ames BN (1986) Food constituents as a source of mutagens, carcinogens, and anticarcinogens. In: Genetic Toxicology of the Diet, pp. 3–32. New York: Alan R. Liss.

Amirmozafari N, Robertson DC (1993) Nutritional requirements for synthesis of heat-labile enterotoxin by *Yersinia enterocoliticca*. Appl Environ Microbiol 59:3314–3320.

Anand A, Bashey B, Mir T, Glatt AE (1994) Epidemiology, clinical manifestations, and outcome of *Clostridium difficile* associated diarrhea. Am J Gastroenterol 89:519–523.

Aponte GE, Triolo HJ, Herr DL (1977) Induction of aryl hydrocarbon hydroxylase and forestomach tumors by benzo(a)pyrene. Cancer 37:3018–3021.

Arnold DL, Munro IC (1983) Artificial sweeteners: their toxicological etiology is an interesting mix. In: Wynder EL, Leveille GA, Weisburger JH, Livingston GE, eds., Environmental Aspects of Cancer: The Role of Macro and Micro Components of Foods, pp. 211–229. Westport, Conn: Food and Nutrition Press.

Arnold DL, Clayson DB (1985) Saccharin—a bitter-sweet case. In: Clayson DB, Krewski D, Munro I, eds., Toxicological Risk Assessment, pp. 227–244. Boca Raton, Fla: CRC Press.

Astill BD, Mulligan LT (1977) Phenolic antioxidants and the inhibition of hepatotoxicity from N-dimethylnitrosamine formed in situ in the rat stomach. Food Cosmet Toxicol 15:167–171.

Azad Khan AKA, Guthrie G, Johnson HH, Truelove SC, Williamson DH (1983) Tissue and bacterial splitting of sulphasalazine. Clin Sci 64:349–354.

Baden DG, Mende TJ, Poli MA, Block RE (1984) Toxins from Florida's red tide dinoflagellate, *Ptychodiscus brevis*. In: Ragelis EP, ed. Seafood Toxins. Washington, DC: American Chemical Society.

Bailey JA, Mansfield JW (1982) Phytoalexins. New York: Wiley.

Balish E (1986) Intestinal flora and natural immunity. Microecol Ther 16:157–167.

Becher G, Knize MG, Nes IF, Felton JS (1988) Isolation and identification of mutagen from a fried Norwegian meat product. Carcinogenesis 9:247–253.

Belman S (1983) Onion and garlic oils inhibit tumor promotion. Carcinogenesis 4: 1063–1065.

Bender AE (1987) Effect on nutritional balance: antinutrients. In: Watson DH, ed., Natural Toxicants in Foods, Progress and Prospects, pp. 110–124, New York: VCH Publishers.

Betley MJ, Lofdahl S, Kreiswirth BN, Bergdoll MS, Novick RP (1984) Staphylococcal enterotoxin A gene is associated with a variable genetic element. Proc Natl Acad Sci USA 81:5179–5183.

Betley MJ, Virginia LM, Mekalanos JJ (1986) Genetics of bacterial enterotoxins. Annu Rev Microbiol 40:577–605.

Beuchat LR, Golden DA (1989) Antimicrobials occurring naturally in foods. Food Technol 43:134–142.

Birt DF (1989) Dietary inhibition of cancer. In: Ragsdale NN, Menzer RE, eds. Carcinogenicity and Pesticides: Principles, Issues, and Relationships, pp. 107–121. Washington, DC: American Chemical Society.

Black SD (1992) Membrane topology of the mammalian P450 cytochromes. FASEB J 6: 680–685.

Bock KW (1977) Dual role of glucuronyl and sulfotransferases converting xenobiotics into reactive or biologically inactive and easily excretable compounds. Arch Toxicol 39:77–85.

Bohach GA, Fast DJ, Nelson RD, Schlievert PM (1990) Staphylococcal and streptococcal pyrogenic toxins involved in toxic shock syndrome and related diseases. Crit Rev Microbiol 17:251–272.

Bokkenheuser VD, Shackleton CHL, Winter J (1987) Hydrolysis of dietary flavonoid glycosides by strains of intestinal anaerobes from humans. Biochem J 248:953–956.

Bowden JP, Chung KT, Andrews AW (1976) Mutagenic activity of tryptophan metabolites produced by rat intestinal microflora. J Natl Cancer Inst 57:921–924.

Boyd SC, Sasame HA, Boyd MR (1979) High concentrations of glutathione in glandular stomach: possible implications for carcinogenesis. Science 205:1010–1012.

Bresinsky A, Besl H (1990) A Colour Atlas of Poisonous Fungi: A Handbook for Pharmacists, Doctors and Biologists. London: Woolfe.

Bryan GT (1969) Role of tryptophan metabolites in urinary bladder cancer. Am Indust Hyg Assoc J 30:27–34.

Bryan GT (1971) The role of tryptophan metabolites in the etiology of bladder cancer. Am J Clin Nutr 24:841–847.

Bryan GT (1978) Nitrofurans: Chemistry, Metabolism, Mutagenesis and Carcinogenesis. New York: Raven.

Butler L (1989) Polyphenols. In: Cheeke PR, ed., Toxicants of Plant Origin, Vol. 4. Phenolics. pp. 95–122. Boca Raton, Fla: CRC Press.

Butler MA, Guengerich FP, Kadlubar FF (1989) Metabolic oxidation of the carcinogens 4-aminobiphenyl and 4,4′-methylenebis(2-chloroaniline) by human hepatic microsomes and by purified rat hepatic cytochrome P-450 monooxygenase. Cancer Res 49:25–31.

Callahan LT, III, Ryder RC, Richardson SH (1971) Biochemistry of *Vibrio cholerae* virulence. II. skin permeability factor cholera enterotoxin production in a chemically defined medium. Infect Immun 4:611–618.

Callahan LT, Richardson SH (1973) Biochemistry of *Vibrio cholerae* virulence. III. Nutritional requirements for toxin production and the effects of pH on toxin elaboration in chemically defined media. Infect Immun 7:767–774.

Calorie Control Council (CCC) (1985) Alternative Sweeteners. Atlanta, Ga: Council.

Cameron E, Pauling L (1976) Supplementary ascorbate in the supportive treatment of cancer: prolongation of survival times in terminal human cancer. Proc Natl Acad Sci USA 73:3685–3689.

Cameron E, Pauling L (1978) Supplementary ascorbate in the supportive treatment of cancer: prolongation of survival times in terminal cancer. Proc Natl Acad Sci USA 75: 4538–4542.

Campbell RL, Singh DV, Nigis ND (1975) Importance of the fecal stream on the induction of colon tumors by azoxymethane in rats. Cancer Res 35:1369–1371.

Carr CJ (1985) Food and drug interactions. In: Fenley JW, Schwass DE, eds., Xenobiotics in Food and Feeds, pp. 221–228. ACS Symposium 234. Washington, DC: American Chemical Society.

Carey JB Jr (1973) Bile salt metabolism in man. In: Nair PP, Kritchevsky D, eds. Bile Acid,Vol. 2, pp. 62–63. New York: Plenum.

Cerniglia CE, Howard PC, Fu PP, Franklin W (1984) Metabolism of nitropolycyclic aromatic hydrocarbons by human intestinal microflora. Biochem Biophys Res Commun 123:262–270.

Cerniglia CE (1985) Metabolism of 1-nitropyrene and 6-nitrobenzo(a)pyrene by intestinal microflora. Prog Clin Biol Res 181:133–137.

Cerniglia CE, Zhuo Z, Manning BW, Federle TW, Heflich RH (1986) Mutagenic activation of the benzidine-based dye Direct Black by human intestinal microflora. Mutat Res 175:11–16.

Chadwick RW, George SE, Chang J, et al. (1991a) Potentiation of 2,6-dinitrotoluene in Fisher 344 rats by pretreatment with pentachlorophenol. Pest Biochem Physiol 39: 168–181.

Chadwick RW, George SE, Chang J, et al. (1991b) Aroclor 1254 alters intestinal enzyme activity and biotransformation of 2,6-dinitrotoluene in rats. Toxicologist 11:255.

Chadwick RW, George SE, Claxton LD (1992) Role of the gastrointestinal mucosa and microflora in the bioactivation of dietary and environmental mutagens or carcinogens. Drug Metab Rev 24(4):425–492.

Chadwick RW, George SE, Chang J, et al. (1993) Effects of age, species differences, antibiotics and toxicants on intestinal enzyme activity and genotoxicity. Environ Toxicol Chem 12:1339–1352.

Chang GW, Bill J, Lum R (1989) Proportion of β-D-glucuronidase-negative *Escherichia coli* in human fecal samples. Appl Environ Microbiol 55:335–339.

Chary P, Prasad R, Chopra AK, Peterson JW (1993) Location of the enterotoxin gene from *Salmonella typhimurium* and characterization of the gene products. FEMS Microbiol Lett 111:87–92.

Chen LH, Boissonneault GA, Glauert HP (1989) Vitamin C, vitamin K and cancer (review). Anticancer Res 8:739–748.

Chen W, Nichols J, Zhou Y, et al. (1995) Effect of dietary restriction on glutathione-S-transferase activity specific toward Aflatoxin B_1-8,9-epoxide. Toxicol Lett. 78:235–243.

Chhabra RS, Eastin WC (1984) Intestinal absorption and metabolism of xenobiotics in laboratory animals: In: Schiller CM, ed., Intestinal Toxicology, pp. 145–160. New York: Raven Press.

Chipley JR (1993) Sodium benzoate and benzoic acid. In: Davidson PM, Branen AL, eds. Antimicrobials in Foods, 2nd Ed., pp. 11–48. New York: Marcel Dekker.

Chopra AK, Vo TN, Houston CW (1992) Mechanism of action of a cytotonic enterotoxin produced by *Aeromonas hydrophila*. FEMS Microbiol Lett 70(1):15–19.

Chopra AK, Brasier AR, Das M, Xu XJ, Peterson JW (1994) Improved synthesis of *Salmonella typhimurium* enterotoxin using gene fusion expression systems. Gene 144: 81–85.

Chowdhury JR, Novikoff PM, Chowdhury NR, Novikoff AB (1985) Distribution of UDP-glucuronyltransferase in rat tissue. Proc Natl Acad Sci USA 82:2990–2994.

Chu SW, Walker WA (1991). Developmental control of bacterial receptors in the gastrointestinal tract. Microecol Ther 20:7–21.

Chu SW, Walker WA (1993) Bacterial toxin interaction with the developing intestine. Gastroenterology 104:916–925.

Chung KT (1983) The significance of azo reduction in the mutagenesis and carcinogenesis of azo dyes. Mutat Res 114:269–181.

Chung, KT, Cerniglia CE (1992) Mutagenicity of azo dyes: structure-activity relationships. Mutat Res 277:201–220.

Chung KT, Stevens SE Jr (1993) Degradation of azo dyes by environmental microorganisms and the helminths. Environ Toxicol Chem 12:2121–2132.

Chung KT, Fulk GE, Slein MW (1975) Tryptophanase of fecal flora as a possible factor in the etiology of colon cancer. J Natl Cancer Inst 54:1073–1078.

Chung, KT, Fulk GE, Egan M (1978a) Reduction of azo dyes by the intestinal anaerobes. Appl Environ Microbiol 35:558–562.

Chung KT, Fulk GE, Andrews AW (1978b) The mutagenicity of methyl orange and metabolites produced by intestinal anaerobes. Mutat Res 58:375–379.

Chung KT, Fulk GE, Andrews AW (1981) Mutagenicity testing of some commonly used dyes. Appl Environ Microbiol 42:641–648.

Chung, KT, Stevens SE Jr, Cerniglia CE (1992) The reduction of azo dyes by the intestinal microflora. Crit Rev Microbiol 18:175–190.

Chung KT, Stevens SE Jr, Lin WF, Wei CI (1993) Growth inhibition of selected food-borne bacteria by tannic acid, propyl gallate and related compounds. Lett Appl Microbiol 17:29–33.

Chung KT, Murdock CA, Stevens SE Jr, et al. (1995) Mutagenicity and toxicity studies of *p*-phenylenediamine and it derivatives. Toxicol Lett. 81:23–32.

Cole CB, Fuller R, Mallet AK, Rowland IR (1985) The influence of the host on expression of intestinal microbial enzyme activities involved in metabolism of foreign compounds. J Appl Bacteriol 59:549–553.

Combes RD, Haveland-Smith RB (1982) A review of the genotoxicity of food, drug and cosmetic colours and other azo, triphenylmethane and xanthene dyes. Mutat Res 98: 101–248.

Combes RD (1989). Cytochrome p-450, mixed function oxidases and formation of genotoxic metabolites by the intestinal tract. In: Koster AS, Richter E, Lauterbach F, Hartmann F., eds. Intestinal Metabolism of Xenobiotics, pp. 119–145. Stuttgart: Gustav Fisher Verlag.

Commoner B, Vithayathil AJ, Dolara P, Nair S, Madyastha P, Cuca GC (1978) Formation of mutagens in beef and beef extract during cooking. Science 201:913–916.

Concon JM (1988) Food Toxicology, Part A & B. New York: Mercel Dekker.

Conn EE (1981) Unwanted biological substances in foods. In: Ayres J, Kirschman J, eds., Impact of Toxicology on Food Processing, pp. 105–121. Westport, Conn: AVI.

Conney AH (1982) Induction of microsomal enzymes by foreign chemicals and carcinogenesis by polycyclic aromatic hydrocarbons: G.H.A. Clowes Memorial Lecture. Cancer Res 42:4875–4917.

Coon MJ, Ding X, Pernecky SJ, Vaz ADN (1992) Cytochrome P450: progress and predictions. FASEB 6:669–673.

Crotty B (1994) Ulcerative colitis and xenobiotic metabolism. Lancet 343:35–38.

Cutrecasas P (1973) Gangliosides and membrane receptors for cholera toxin. Biochem 12:3558–3566.

Czeczulin JR, Hanna PC, McClane BA (1993) Cloning, nucleotide sequencing, and expression of *Clostridium perfringens* enterotoxin gene in *Escherichia coli*. Infect Immun 61: 3429–3439.

D'Aoust JY (1991) Pathogenicity of foodborne *Salmonella*. Int J Food Microbiol 12: 17–40.

Dauterman WC (1994) Metabolism of toxicants: phase II reactions. In: Hodgson E, Levi PE, eds., Introduction to Biochemical Toxicology, 2nd ed., pp. 113–132. Norwalk, Conn: Appleton & Lange.

Davidson PM (1993) Parabens and phenolic compounds. In: Davidson PM, Branen AL, eds. Antimicrobials in Foods, 2nd ed., pp. 263–306. New York: Marcel Dekker.

Dawson J, Bridges JW (1981) Intestinal microsomal drug metabolism: A comparison of rat and guinea-pig enzymes and of rat crypt and villous tip cell enzymes. Biochem Pharmacol 30:2415–2420.

De Meester C (1989) Bacterial mutagenicity of heterocyclic amines found in heat-processed food. Mutat Res 221:235–262.

Delclos KB, Cerniglia CE, Dooley KL, Campbell WL, Franklin W, Walker RP (1990) The role of intestinal microflora in the metabolic activation of 6-nitrochrysene to DNA-binding derivatives in mice. Toxicol 60:137–150.

Delor I, Kaeckenbeeck A, Wauters G, Cornelis GR (1990) Nucleotide sequence of *yst*, the *Yersinia enterocoliticca* gene encoding the heat-stable enterotoxin, and prevalence of the gene among pathogenic and nonpathogenic yersiniae. Infect Immun 58:2983–2988.

Delor I, Cornelis GR (1992) Role of *Yersinia enterocolitica* Yst toxin in experimental infection of young rabbits. Infect Immun 60:4269–4277.

Depitre C, Delmee M, Avesani V, et al. (1993) Serogroup F strains of *Clostridium difficile* produce toxin B but not toxin A. J Med Microbiol 38:434–441.

Domagk G (1935) Ein Beitrag zur Chemotherapie der bakteriellen infektionen. Dtsch Med Wochenschr 61:250–253.

Drasar BS, Barrow PA (1985) Intestinal Microbiology. Washington DC: American Society for Microbiology.

Drobniewski FA (1993) *Bacillus cereus* and related species. Clin Microbiol Rev 6: 324–338.

Duimstra JR, Myers LL, Collins JE, Benfield DA, Shoop DS, Bradbury WC (1991) Enterovirulence of enterotoxingenic *Bacteroides fragilis* in gnotobiotic pigs. Vet Pathol 28: 514–518.

El-Bayoumy K, Sharma C, Louis YM, Reddy B, Hecht SS(1983) The role of intestinal microflora in the metabolic reduction of 1-nitropyrene to 1-aminopyrene in conventional and germfree rats and in humans. Cancer Lett 19:311–316.

Elwell L, Shipley PL (1980) Plasmid-mediated factors associated with virulence of bacteria to animals. Annu Rev Microbiol 34:465–496.

Fasano A, Baudry B, Pumplin DW, et al. (1991) *Vibrio cholerae* produces a second enterotoxin, which affects intestinal tight junction. Proc Natl Acad Sci USA 88: 5242–5246.

Felton JS, Knize MG, Shen NH, et al. (1986) The isolation and identification of a new mutagen from fried ground beef:2-amino-1-methyl-6-phenylimidazo(4,5-b)pyridine (PhIP). Carcinogenesis 7:1081–1086.

Felton JS, Knize MG (1991) Mutagen formation in muscle meats and model heating systems. In: Hayatsu H, ed. Mutagens in Food: Detection and Prevention, pp. 57–66. Boca Raton, Fla: CRC Press.

Fenwick GR, Heaney RK, Mawson R (1989) Glucosinolates. In: Cheeke PR ed., Toxicants of Plant Origin, Vol. II. Glycosides, pp. 2–42. Boca Raton, Fla: CRC Press.

Fernandez H, Neto UF, Fernandes F, de Almeida Pedra M, Trabulsi LR (1983) Culture supernatants of *Campylobacter jejuni* segments of adult rats. Infect Immun 40:429–431.

Field M (1980) Intestinal secretion and its stimulation by enterotoxins. In: Ouchterlony O, Holmgren J, eds. Cholera and Related Diarrheas, pp. 46–52. Basel: Karger.

Field M, Rad MC, Chang EB (1989) Intestinal electrolyte transport and diarrheal disease. N Engl J Med 321:800–806, 879–883.

Fletcher DC (1976) Can the quinine in tonic water be hazardous? JAMA 236:305.

Food and Drug Administration (FDA) (1980) FD & C Red No. 2. Denial of petition for permanent listing: final decision. Food and Drug Administration. Federal Register Jan. 25, p. 6252.

Fouts JR, Kamm JJ, Brodie BB (1957) Enzymatic reduction of prontosil and other azo dyes. J Pharmacol Exp Ther 120:291–300.

Franz A, Bryant A, Farrant J (1993) Staphylococcal enterotoxin B up-regulates interleukin-2 receptor beta chain expression on tonsillar B cells. Europ J Immunol 23:2696–2699.

Fu PP, Cerniglia CE, Richardson KE, Heflich RH (1988) Nitroreduction of 6-nitrobenzo(a-)pyrene: a potential activation pathway in humans. Mutat Res 209:123–129.

Fu PP (1990) Metabolism of nitro-polycyclic aromatic hydrocarbons. Drug Metabol Rev 22:209–268.

Fujii Y, Okamuro Y, Hitotsubashi S, Saito A, Akashi N, Okamoto K (1994) Effect of alterations of basic amino acid residues of *Escherichia coli* heat-stable enterotoxin II on enterotoxicity. Infect Immun 62:2295–2301.

Fujii-Kuriyama Y, Imataka H, Sogawa K, Yasumoto K-I, Kikuchi Y (1992) Regulation of CYP1A1. FASEB J 6:706–710.

Fuller AT (1937) Is p-aminoazobenzenesulfonamide the active agent in prontosil. Lancet i:194–198.

Gaitan E (1990) Goitrogens in food and water. Annu Rev Nutr 10:21–39.

George SE, Chadwick RW, Creason JP, Kohan, Dekker JP (1991) Effect of pentachlorophenol on the activation of 2,6-dintrotoluene to genotoxic urinary metabolites in CD-1 mice: a comparison of GI enzyme activities and urine mutagenicity. Environ Molec Mutagen 18:92–101.

Gill DM (1976) The arrangement of subunits in cholera toxin. Biochem J 15:1242–1248.

Gill DM, Meren R (1978) ADP-ribosylation of membrane proteins catalyzed by cholera toxin: basis of the activation of adenylate cyclase. Proc Natl Acad Sci USA 75: 3050–3054.

Goldin BR (1986a) The metabolism of the intestinal microflora and its relationship to dietary fat, colon and breast cancer. Prog Clin Biol Res 222:655–685.

Goldin BR (1986b) In situ bacterial metabolism and colon mutagens. Annu Rev Microbiol 40:367–393.

Goldin BR (1990) Intestinal microflora: metabolism of drugs and carcinogens. Ann Med 22:43048.

Goldin BR, Gorbach SL (1984) Alterations of the intestinal microflora by diet, oral antibiotics and lactobacillus: decreased production of free amines from aromatic nitro compounds, azo dyes and glucuronides. J Natl Cancer Inst 73:689–693.

Goldin BR, Peppercorn MA, Goldman P (1973) Contribution of host and intestinal microflora in the metabolism of L-Dopa by the rat. J Pharmacol Exp Ther 186:160–166.

Gorbach SL, Goldin BR (1990) The intestinal microflora and the colon cancer connection. Rev Infect Dis 12:S252–256.

Gosling PJ, Turnbull PC, Lightfoot NF, Pether JV, Lewis RJ (1993) Isolation and purification of *Aeromonas sobria* cytotonic enterotoxin and β-haemolysin. J Med Microbiol 38:227–234.

Granum PE, Brynestad S, Kramer JM (1993) Analysis of enterotoxin production by *Bacillus cereus* from diary products, food poisoning incidents and non-gastrointestinal infections. Int J Food Microbiol 17:269–279.

Gray JI, Dugan LR, Jr., (1975) Inhibition of N-nitrosamine formation in model food systems. J Food Sci 40:981–984.

Grisebach H, Ebel J (1978) Phytoalexins, chemical defense substances of higher plants? Angew Chem 17:635–647.

Grover S, Batish VK, Srinivasan RA (1990) Production and properties of crude enterotoxin of *Pseudomonas aeruginosa*. Int J Food Microbiol 10:201–208.

Guaycurus TV, Vicente AC, de Simone SG, Rabinovitch L (1993) Partial isolation and some properties of enterotoxin produced by *Bacillus cereus* strains. Memorias do Instituto Oswald Cruz 88:131–134.

Guengerich, FP (1992a) Cytochrome P450: advances and prospects. FASEB J 6:667–668.

Guengerich, FP (1992b) Characterization of human cytochrome P450 enzymes. FASEB J 6:745–748.

Gurr MI (1988) Lipids: products of industrial hydrogenation, oxidation and heating. In: Walker R, Quattrucci E, eds. Nutritional and Toxicological Aspects of Food Processing, pp. 139–158. New York: Taylor and Francis.

Gyles CL (1992) *Escherichia coli* cytotoxins and enterotoxins. Can J Microbiol 38: 734–746.

Hall BE, James SP (1980) Some pathways of xenobiotic metabolism in the adult and neonatal marmoset (*Callithrix jacchus*). Xenobiotics 10:421–434.

Hall S, Reichardt PB (1984) Cryptic paralytic shellfish toxins. In Ragelis EP: Seafood Toxins, p. 113 Washington, DC: American Chemical Society.

Hanninen O, Lindstrom-Seppa P, Pelkonen K (1987) Role of gut in xenobiotic metabolism. Arch Toxicol 60:34–36.

Hariharan H, Panigrahi D (1990) Cholera-like enterotoxin in certain *Campylobacter jejuni* strains: some observations. Microbiologica 13:7–9.

Harne SD, Sharma VD, Rahman H (1994) Purification and antigenicity of *Salmonella newport* enterotoxin. Ind J Medical Research Section A-Infectious Diseases 99:13–17.

Hatch FT, Felton JS, Stuermer DH, Bjeldanes LF (1984) Identification of mutagens from the cooking of food. In: de Serres, FJ., ed. Chemical Mutagens, Principles and Methods for Their Detection, Vol. 9, p. 111. New York: Plenum Press.

Hatch FT, Knize MG, Healy SK, Slezak T, Felton JS (1988) Cooked-food mutagen reference list and index. Environ Mol Mutagen 12(suppl 14):1–85.

Hayashi Y (1991) Cancer risk assessment of food additives and food contaminants. In: Hayatsu H, ed. Mutagens in Food: Detection and Prevention, pp. 243–257. Boca Raton, Fla: CRC Press.

Hayes MA, Rushmore TH, Goldberg MT (1987) Inhibition of hepatocarcinogenic responses to 1,2-dimethylhydrazine by diallyl sulfide, a component of garlic oil. Carcinogenesis 8:1155–1157.

Hegarty MP (1986) Toxic amino acids in foods of animals and man. Proc Nutr Soc Aust 11:73–81.

Hiasa Y, Konishi Y, Kamatomo Y, Watanabe T, Ito N (1971) Effect of lithocholic acid by intestinal microflora. J Steroid Biochem 7:641–647.

Hill M (1987) The role of bacteria in human carcinogenesis. Anticancer Res 7:1079–1084.

Hitotsubashi S, Fujii Y, Yamanak H, Okamoto K (1992) Some properties of purified *Escherichia coli* heat-stable enterotoxin II. Infect Immun 60:4468–4474.

Hodgson E, Levi PE (1994) Metabolism of toxicants phase I reactions. In: Hodgson E, Levi PE, eds. Introduction to Biochemical Toxicology, 2nd Ed., pp. 75–112. Norwalk, Conn: Appleton & Lange.

Hodgson E (1994) Chemical and environmental factors affecting metabolism of xenobiotics. In: Hodgson E, Levi PE, eds. Introduction to Biochemical Toxicology, 2nd ed., pp. 153–175. Norwalk, Conn: Appleton & Lange.

Hoensch HP, Woo CH, Raffin SB, Schmid R (1976) Oxidative metabolism of foreign compounds in rat small intestine: cellular localization and dependence on dietary iron. Gastroenterology 70:1063–1070.

Hoensch HP, Hartmann F (1981) The intestinal enzymatic biotransformation system: potential role in protection from colon cancer. Hepato-gastroenterol 28:221–228.

Hoensch HP, Steinhardt HJ, Malchow H (1982) Metabolism of xenobiotica in human small intestinal mucosa, relationship to, carcinogenic factors. In: Malt RA, Williamson RCN, eds. Colonic Carcinogenesis, pp. 83–90. Boston: MTP Press.

Hoffmann ML, Jablonski LM, Crum KK, et al. (1994). Predictions of T-cell receptor- and major histocompatibility complex-binding sites on staphylococcal enterotoxin C_1. Infect Immun 62:3396–3407.

Hokama Y, Miyahara JT (1986) Ciguatera poisoning: clinical and immunological aspects. J Toxicol Toxin Rev 5:25–53.

Holmgren J, Lycke N, Czerkinsky C (1993) Cholera toxin and cholera B subunit as oral-mucosal adjuvant and antigen vector systems. Vaccine 11:1179–1184.

Howard PC, Beland FA, Cerniglia CE (1983) Reduction of the carcinogen 1-nitropyrene to 1-aminopyrene by rat intestinal bacteria. Carcinogenesis 4:985–990.

Howard PC, Hecht SS, Beland FA (1990) Nitroarenes: Occurrence, Metabolism, and Biological Impact. New York: Plenum Press.

Hsieh DPH (1989) Carcinogenic potentials of mycotoxin in foods. In: Taylor SL, Scanlan RA, eds. Food Toxicology: A Perspective on the Relative Risk, pp. 11–30. New York: Marcel Dekker.

IFT Expert Panel on Food Safety and Nutrition and the Committee on Public Information (1980) Food Colors. Food Technol 34:77–84.

IFT Expert Panel on Food Safety and Nutrition (1986) Sweeteners: nutritive and nonnutritive. Food Technol 40:195–206.

Ilett KF, Tee LBG, Reeves PT, Minchin RF (1990) Metabolism of drugs and other xenobiotics in the gut lumen and wall. Pharmacol Ther 46:67–93.

Jacobs MM (1983) Selenium inhibition of 1,2-dimethylhydrazine-induced colon carcinogenesis. Cancer Res 43:1646–1649.

Jagerstad M, Skog K, Grivas S, Olsson K (1991) Formation of heterocyclic amines using model systems. Mutat Res 259:219–233.

Jay JM. (1992) Modern Food Microbiology, 4th ed., pp. 251–289. New York: Van Nostrand Reinhold.

Johnson HM, Russell JK, Pontzer CH (1991) Staphylococcal enterotoxin microbial superantigens. FASEB J 5:2706–2712.

Joint Committee of the Royal College of Physicians and the British Nutrition Foundation (1984) Food intolerance and food aversion. J R Coll Physicians Lond 18:83–123.

Joint Expert Committee on Food Additives (JECFA) (1988) L-Glutamic acid and its ammonium, calcium, monosodium and potassium salts. In: Toxicological Evaluation of Certain Food Additives and Contaminants, WHO Series 22, pp. 97–161. New York: Cambridge University Press.

Jones JM (1992a) Naturally occurring food toxicants. In: Food Safety, pp. 69–105. St. Paul: Egan Press.

Jones JM (1992b) Food additives. In: Food Safety, pp. 203–257. St. Paul: Eagan Press.

Kadlubar FF, Hammons GJ (1986) The role of cytochrome P-450 in the metabolism of chemical carcinogens. In: Guengerich FP, ed. Mammalian Cytochrome P-450, Vol. II, pp. 81–130. Boca Raton, Fla: CRC Press.

Kamm JJ, Dashman T, Newmark H, Mergens WJ (1977) Inhibition of amine-nitrite hepatoxicity by α-tocopherol. Toxicol Appl Pharmacol 41:575–583.

Kapadia GJ, Chung EB, Ghosh B, et al. (1978) Carcinogenicity of some folk medicinal herbs in rats. J Natl Cancer Inst 60:683–686.

Kasai H, Yamaizumi Z, Wakabayashi K, et al. (1980a) Potent novel mutagens produced by broiling fish under normal conditions. Proc Jpn Acad 56B:278–283.

Kasai H, Yamaizumi Z, Wakabayashi K, et al. (1980b) Structure and chemical synthesis of Me-IQ, a potent mutagen isolated from broiled fish. Chem Lett 1391–1394.

Kasai J, Yamaizumi S, Shiomi T, et al. (1981) Structure of a potent mutagen isolated from fried beef. Chem Lett 485–488.

Kat M (1983) *Dinophysis acuminata* blooms in the Dutch coastal area related to diarrhetic mussel poisoning in the Dutch Waddensea. Sarsia 68:81–84.

Kato K, Shirosita K, Kurosawa S, et al. (1990) Staphylococcal enterotoxin A induced interferon (IFN) gamma production in spleen cells from BCG-immunized mice: the IFN production is dependent on leukotriene C4 but not dependent on interleukin 2. Immunology 18:40–50.

Khurana S, Ganguly NK, Khullar M, Panigrahi D, Walia BN (1991) Studies on the mechanism of *Salmonella typhimurium* enterotoxin-induced diarrhea. Biochim Biophys Acta 1097:171–176.

King LC, Kohan MJ, George SE, Lewtas J, Claxton LD (1990) Metabolism of 1-nitropyrene by human, rat and mouse intestinal flora: mutagenicity of isolated metabolites by direct analysis of HPLC fractions with a micro-suspension reverse mutation assay. J Toxicol Environ Health 31:179–192.

Kinouchi T, Ohnishi Y (1986) Metabolic activation of 1-nitropyrene and 1,6-dinitropyrene by nitroreductases from *Bacteroides fragilis* and distribution of nitroreductase activity in rats. Microbiol Immunol 30:979–992.

Kita E, Oku D, Hamuro A, et al. (1990) Hepatotoxic activity of *Campylobacter jejuni*. J Med Microbiol 33:171–182.

Klipstein FA, Engert RF (1977) Immunological interrelationships between cholera toxin and the heat-labile and heat-stable enterotoxins of coliform bacteria. Infect Immun 18: 110–117.

Klipstein FA, Engert RF (1985) Immunological relationship of the B subunits of *Campylobacter jejuni* and *Escherichia coli* heat-labile enterotoxins. Infect Immun 48:629–633.

Koop DR (1992) Oxidative and reductive metabolism by cytochrome P450 2E1. FASEB J 6:724–730.

Koster A Sj, Frankhuizen-Sierevogel AC, Noordhoek (1985) Distribution of glucuronidation capacity (1-naphthol and morphine) along the rat intestine. Biochem Pharmacol 34:3527–3532.

Krone CA, Yeh SMJ, Iwaoka WT (1986) Mutagen formation during commercial processing of foods. Environ Health Perspect 67:75–88.

Kwok RHM (1968) Chinese restaurant syndrome. N Engl J Med 278:796.

Labbe RG, Rey DK (1979) Raffinose increases sporulation and enterotoxin production by *Clostridium perfringens* type A. Appl Environ Microbiol 37:1196–1200.

Labbe RG, Nolan LL (1981) Stimulation of *Clostridium perfringens* enterotoxin formation by caffeine and theobromine. Infect Immun 34:50–54.

Laitinen M, Watkins JB (1986) Mucosal biotransformations. In: Rozman K, Hanninen O, eds. Gastrointestinal Toxicology, pp. 412–434. Amsterdam: Elsevier Science.

Lathe R, Hirth P, Dewide M, Harford N, Lecocq J-P (1980) Cell-free synthesis of enterotoxin of *E. coli* from a cloned gene. Nature 284:473–474.

Lathia D, Braasch A, Theissen U (1988) Inhibitory effects of vitamin C and E on in-vitro formation of N-nitrosamine under physiological conditions. Front Gastrointest Res 14: 151–156.

Lauritsen K, Hansen J, Ryde M, Rask-Madsen J (1984) Colonic azo disalicylate metabolism determined by in vivo dialysis in healthy volunteers and patients with ulcerative colitis. Gastroenterology 86:1496–1500.

Lee CH, Moseley SL, Moon HW, Whipp SC, Gyles CL, So M (1983) Characterization of the gene encoding heat stable toxin 2 and preliminary studies of enterotoxigenic *Escherichia coli* heat stable toxin 2 producers. Infect Immun 42:264–268.

Legraverend C, Mode A, Wells T. Robinson I, Gustafsson J-A (1992) Hepatic steroid hydroxylating enzymes are controlled by the sexually dimorphic pattern of growth hormone secretion in normal and dwarf rats. FASEB J 6:711–718.

Leiner IE (1980) Toxic Constituents of Plant Foodstuffs. New York: Academic Press.

Leiner IE (1981) The nutritional significance of the plant lectins. In: Ory RL, ed. Antinutrients, Natural Toxicants in Foods, pp. 143–158. Westport, Conn: F & N Press.

Leonard TB, Graichen ME, Popp JA (1987) Dinitrotoluene isomer-specific hepatocarcinogenesis in F344 rats. J Natl Cancer Inst 79:1313–1319.

Levi PE (1994) Reactive metabolites and toxicity. In: Hodgson E, Levi PE eds. Introduction to Biochemical Toxicology, 2nd ed., pp. 219–239. Norwalk, Conn: Appleton & Lange.

Levine WG (1991) Metabolism of azo dyes: implication for detoxication and activation. Drug Metab Rev 23:253–309.

Lindenbaum J, Rund DL, Butler VP, Tse-Eng D, Saka J (1981) Inactivation of digoxin by the gut flora: reversed by antibiotic therapy. N Engl J Med 305:789–794.

Loub WD, Wattenberg LW, Davis DW (1975) Aryl hydrocarbon hydroxylase induction in rat tissues by naturally occuring indoles of cruciferous plants. J Natl Cancer Inst 54: 985–987.

MacDonald IA, Bussard RG, Hutchison DM, Holdeman LV (1984) Rutin-induced β-glucosidase activity in *Streptococcus faecium* VGH-1 and *Streptococcus* sp. strain FRP-17 isolated from human feces: formation of the mutagen, quercetin, from rutin. Appl Environ Microbiol 47:350–355.

MacGregor JT, Jurd JL (1978) Mutagenicity of plant flavonoids: structural requirements for mutagenic activity in *Salmonella typhimurium*. Mutat Res 54:297–309.

MacGregor JT (1986a) Genetic toxicology of dietary flavonoids. In: Genetic Toxicology of the Diet, pp. 33–43. New York: Alan R. Liss.

MacGregor JT (1986b) Mutagenic and carcinogenic effects of flavonoids. In: Plant Flavonoids in Biology and Medicine, pp. 411–424, New York: Alan R. Liss.

Mahajan S, Rodgers FG (1990) Isolation, characterization, and host-cell-binding properties of a cytotoxin from *Campylobacter jejuni*. J Clin Microbiol 28:1313–1320.

Majeed KN, Egan AF, Mac Rae IC (1990) Production of exotoxins by *Aeromonas* spp. at 5 degrees C. J Appl Bacteriol 69:332–337.

Malaspina A (1987) Regulatory aspects of food additives. In: Miller K, ed. Toxicological Aspects of Food, pp. 17–58. New York: Elsevier.

Mallett AK, Rowland IR (1990) Bacterial enzymes: their role in the formation of mutagens and carcinogens in the intestine. Dig Dis 8:71–79.

Manning BW, Cerniglia CE, Federle TW (1986) Biotransformation of 1-nitropyrene to 1-aminopyrene and 1-formyl-1-aminopyrene by the human intestinal microbiota. J Toxicol Environ Health 18:339–346.

Marquardt RR (1989) Vicine, convicine and their aglycones—derivatives and isouramil. In: Cheeke PR, ed. Toxicants of Plant Origin, Vol. 2, Glycosides, pp. 161–200. Boca Raton, Fla: CRC Press.

Mastromarino A, Reddy BS, Wynder EL (1976) Metabolic epidemiology of colon cancer: enzymatic activity of fecal flora. Am J Clin Nutr 29:1455–1460.

Mastromarino A, Reddy BS, Wynder EL (1978) Fecal profiles of anaerobic microflora of large bowel cancer patients with nonhereditary large bowel polyps. Cancer Res 38: 4458–4462.

Matsumoto Y, Fujiwara M (1993) Immunomodulation of experimental autoimmune encephalomyelitis by staphylococcal enterotoxin D. Cell Immunol 149:268–278.

McCabe BJ (1986) Dietary tyramine and other pressor amines in MAOI regimens: a review. J Am Diet Assoc 86:1059–1064.

McCalla DR (1983) Mutagenicity of nitrofuran derivatives. Environ Mutagen 5:745–765.

McClane BA (1994) *Clostridium perfringens* enterotoxin acts by producing small molecular permeability alterations in plasma membranes. Toxicology 87:43–67.

McCoy EC, Petrillo LA, Rosenkranz HS (1979) The demonstration of cooperative action and intestinal mucosa enzymes in the activation of mutagens. Biochem Biophys Res Commun 89:859–862.

McCoy EC, Rosenkranz HS, Mermelstein (1981) Evidence for the existence of a family of bacterial nitroreductases capable of activating nitrated polycyclics to mutagens. Environ Mutagen 3:421–427.

McKay JA, Murray GI, Weaver RJ, Ewen SWB, Melvin WT, Burke MD (1993) Xenobiotic metabolizing enzyme expression in colonic neoplasia. Gut 34:1234–1239.

Mehansho H, Butler LG, Carlson DM (1987) Dietary tannins and salivary proline-rich proteins. Annu Rev Nutr 7:423–440.

Mekalanos JJ (1983) Duplication and amplification of toxin genes in *Vibrio cholerae*. Cell 35:253–263.

Mekalanos JJ (1985) Cholera toxin: genetic analysis, regulation, and role in pathogenesis. Curr Top Microbiol Immunol 118:97–118.

Mekalanos JJ, Swartz DJ, Pearson GDN, Hartford N, Groyne F, de Wilde M (1983) Cholera toxin genes: nucleotide sequence, deletion analysis and vaccine development. Nature 306:551–557.

Mekalanos J, Goldberg I, Miller V, et al. (1985) Genetic construction of cholera vaccine prototypes. In: Lerner RA, Chanock RM, Brown F, eds. Vaccines 85, pp. 101–105. Cold Spring Harbor, NY: Cold Spring Harbor Press.

Merz B (1983) Adding seeds to the diet may keep cancer at bay. JAMA 249:2746.

Mezoff AG, Giannella RA, Eade MN, Cohen MB (1992) *Escherichia coli* enterotoxin (STa) binds to receptors, stimulates guanyl cyclase, and impairs absorption in rat colon. Gastroenterology 102:816–822.

Middlebrook JL, Dorland RB (1984) Bacterial toxins: cellular mechanisms of action. Microbiol Rev 47:596–620.

Mikami T, Horikawa T, Murakami T, et al. (1994) An improved method for detecting cytostatic toxin (emetic toxin) of *Bacillus cereus* and its application to food samples. FEMS Microbiol Lett 119:53–57.

Miller DM, Dickey RW, Tindall DR (1984) Lipid-extracted toxins from a dinoflagellate, *Gambierdiscus toxicus*. In: Ragelis EP, ed. Seafood Toxins. Washington, DC: American Chemical Society.

Miller EC, Miller JA (1986) Carcinogens and mutagens that occur in foods. Cancer 58: 1795–1803.

Miller JA, Baumann CA (1945) The carcinogenicity of certain azo dyes related to *p*-dimethylaminoazobenzene. Cancer Res 5:227–234.

Miller JA, Miller EA (1948) Carcinogenicity of certain derivatives of *p*-dimethylaminoazobenzene in rat. J Exp Med 87:139–156.

Miller JA, Miller EA, Phillips DH (1982) The metabolic activation and carcinogenicity of alkenylbenzenes that occur naturally in many spices. In: Stich HF, ed. Carcinogens and Mutagens in the Environment, Vol. 3, pp. 83–96. Boca Raton, Fla: CRC Press.

Millstone E (1985) Food additives: the balance of risks and benefits. Chem Indust 21: 421–442.

Milner JA (1985) Dietary antioxidants and cancer. Contemp Nutr 10:1–2.

Mizutani T, Mitsuoka T (1981) Relationship between liver tumorigenesis and intestinal bacteria in gnotobiotic C3H/He male mice. In: Sasaki S, ed. Recent Advances in Germ-free Research, pp. 639–644. Tokyo: Tokai University Press.

Morgan MRA, Coxon DT (1987) Tolerances: Glycoalkaloids in Potatoes. In: Watson DH, ed. Natural Toxicants in Foods: Progress and Prospects, pp. 221–230. New York: DCH Publishers.

Morris CE, Przybyla A (1985) Cyclamate. Special report. Chilton's Food Eng Int 57: 67–75.

Murdoch RD, Pollock I, Naeem S (1987) Tartrazine induced histamine release in vivo in normal subjects. J R Coll Phys Lond 21:257–261.

Nagao M, Honda M, Seino Y, Yahagi T, Sugimura T (1977a) Mutagenicity of smoke condensates and the charred surface of fish and meat. Cancer Lett 2:221–226.

Nagao M, Honda M, Seino Y, Yahagi T, Kawachi T, Sugimura T (1977b) Mutagenicities of protein pyrolysates. Cancer Lett 2:335–340.

Nagao M, Wakabayaski F, Fujita Y, Tahura A, Ochrai T, Sugimura T (1986) Mutagenic compounds in soy sauce, Chinese cabbage, coffee and herbal teas. In: Genetic Toxicology of the Diet, pp. 55–62. New York: Alan R. Liss.

Narisawa T, Magadia NE, Weisburger JH, Wynder EL (1974) Promoting effect of bile acids on colon carcinogenesis after intrarectal instillation of N-methyl-N′-nitro-N-nitrosoguanidine in rats. J Natl Cancer Inst 53:1093–1097.

NAS/NRC (1989) Diet and Health: Implications for Reducing Chronic Disease Risk. Washington, DC: National Academy Press.

Nebert DW, Gonzalez FJ (1987) P450 genes: structure, evolution, and regulation. Annu Rev Biochem 56:945–993.

Nebert DW, Adesnik M, Coon MJ, et al. (1987) The p450 gene superfamily: recommended nomenclature. DNA 6:1–11.

Negishi C, Wakabayashi K, Tsuda M, et al. (1984) Formation of 2-amino-3,7,8-trimethylimidazo[4,5-f]quinoxaline, a new mutagen, by heating a mixture of creatine, glucose and glycine. Mutat Res 140:55–59.

Negishi C, Wakabayashi K, Yamaizumi Z, et al. (1985) Identification of 4,8-DimeIQx, a new mutagen. Selected abstracts of papers presented at the 13th annual meeting of the

Environmental Mutagen Society of Japan, Oct. 12–13, 1984. Tokyo. Mutat Res 147: 267–268.

Negishi T, Nagao M, Hiramoto K, Hayatsu H (1991a) A list of mutagenic heterocyclic amines. In: Hayatsu H, ed. Mutagens in Food: Detection and Prevention, pp. 7–19. Boca Raton, Fla: CRC Press.

Negishi, K, Nagao M, Hiramoto K, Hayatsu H (1991b). DNA modification in vitro and in vivo with heterocyclic amines. In: Hayatsu H, ed. Mutagens in Food: Detection and Prevention, pp. 21–27. Boca Raton, Fla: CRC Press.

Nelson RL (1984) Is the changing pattern of colorectal cancer caused by selenium deficiency? Dis Col Rect 27:459–461.

Newaz SN, Fang WF, Strobel HW (1983) Metabolism of the carcinogens 1,2-dimethylhydrazine by isolated human colon microsomes and human colon tumors cells in culture. Cancer 52:794–798.

Newland JW, Strockbine NA, Miller SF, O'Brian AD, Holmes RK (1985) Cloning of Shiga-like toxin structural genes from a toxin converting phage of *Escherichia coli*. Science 230:179–181.

Ni YC, Heflich RH, Kadlubar FF, Fu PP (1987) Mutagenicity of nitrofurans in *Salmonella typhimurium* TA98, TA98NR and TA98/1,8-DNP$_6$. Mutat Res 192:15–22.

Nishio Y, Kakizoe T, Ohtani M, Sato S, Sugimura T, Fukushima. S (1986) Leucine and isoleucine promote bladder cancer. Science 231:843–845.

O'Brien AD, LaVeck GD (1983) Purification and characterization of *Shigella dysenteriae* 1-like toxin produced by *Escherichia coli*. Infect Immun 40:675–683.

Ohshima H, Bartsch H (1983) A new approach to quantitate endogenous nitrosation in humans. In: Stich HF, ed. Carcinogens and Mutagens in the Environment, Vol. 2, pp. 3–15. Boca Raton, Fla: CRC Press.

Olive PL, McCalla DR (1977) Cytotoxicity and DNA damage to mammalian cells by nitrofurans. Chem Biol Interact 16:223–233.

Pamukcu AM, Yaliner S, Hatcher JF, Bryan GT (1980) Quercetin, a rat intestinal and bladder carcinogen present in bracken fern *Pteridium equilinum*. Cancer Res 40: 3468–3472.

Pearson GDN, Mekalanos JJ (1982) Molecular cloning of *Vibrio cholerae* enterotoxin genes in *Escherichia coli* K-12. Proc Natl Acad Sci USA 79:2976–2980.

Peppercorn MA, Goldman P (1972) The role of intestinal bacteria in the metabolism of salicylazosulfapyridine. J Pharmacol Exp Ther 181:555–562.

Peppercorn MA, Goldman P (1973) Distribution studies of salicylazosulfapyridine and its metabolites. Gastroenterology 64:240–245.

Perez JL, Berrocal CI, Berrocal L (1986) Evaluation of a commercial β-glucuronidase test for the rapid and economical identification of *Escherichia coli*. J Appl Bacteriol 61:541–545.

Pickens RN, Mazaitis AJ, Maas WK, Rey M, Heyneker H (1983) Nucleotide sequence of the gene for heat-stable enterotoxin II of *Escherichia coli*. Infect Immun 42:269–275.

Powrie WD, Wu CH, Molund VP (1986) Browning reaction systems as sources of mutagens and antimutagens. Environ Health Perspect 67:47–54.

Price JM, Biava CG, Oser BL, Vogin EE, Seinfeld J, Ley HL (1970) Bladder tumors in rats fed cyclohexamine or high doses of a mixture of cyclamate and saccharin. Science 167:1131–1132.

Prival MJ, Davis VM, Peiperl MD, Bell SJ (1988) Evaluation of azo dyes for mutagenicity and inhibition of mutagenicity by methods using *Salmonella typhimurium*. Mutat Res 206:247–256.

Prosaska HJ, Santamaria AB, Talalay P (1992) Rapid detection of inducers of enzymes that protect against carcinogens. Proc Natl Acad Sci USA 89:2394–2398.

Pryor WA (1985) Free radical involvement in chronic diseases and aging. In: Fenley JW, Schwass DE, eds. Xenobiotics in Foods and Feeds, ACS Symposium 234, pp. 77–96. Washington, DC: American Chemical Society.

Rackie JJ, Gumbmann MR (1981) Protease inhibitors: physiological properties and nutritional significance. In: Ory RL, ed. Antinutrients and Natural Toxicants in Foods, pp. 203–237. Westport, Conn: F & N Press.

Rafii F, Franklin W, Cerniglia CE (1990) Azoreductase activity of anaerobic bacteria isolated from human intestinal microflora. Appl Environ Microbiol 56:2146–2151.

Rafii F, Franklin W, Heflich RH, Cerniglia CE (1991) Reduction of nitroaromatic compounds by anaerobic bacteria isolated from the human gastrointestinal tract. Appl Environ Microbiol 57:962–968.

Ranelli DM, Jones CL, Johns MB, Mussey GJ, Khan SA (1985) Molecular cloning of staphylococcal enterotoxin B gene in *Escherichia coli* and *Staphylococcus aureus*. Proc Natl Acad Sci USA 82:5850–5854.

Rao MC, Orellana SA, Field M, Robertson DC, Giannella RA (1981) Comparison of the biological actions of three purified heat-stable enterotoxins: effect on ion transport and guanylate cyclase activity in rabbit ileum in vitro. Infect Immun 33:167–170.

Reda KB, Kapur V, Mollick JA, Lamphear JG, Musser JM, Rich RR (1994) Molecular characterization and phylogenetic distribution of the streptococcal superantigen gene (ssa) from *Streptococcus pyogenes*. Infect Immun 62:1867–1874.

Reed PI, Johnston BJ, Walters CL, Hill MJ (1989) Effect of ascorbic acid on the intragastric environment in patients at increased risk of developing gastric cancer. Gastroenterology 96: A411.

Renwick AG, Williams RT (1972) The fate of cyclamate in man and other species. Biochem J 129:869–879.

Rickert DE, Butterworth BE, Popp JA (1984) Dinitrotoluene: acute toxicity, oncogenicity, genotoxicity, and metabolism. CRC Crit Rev Toxicol 13:217–234.

Rickert DE (1987) Metabolism of nitroaromatic compounds. Drug Metab Rev 18:23–53.

Rolls BA, Turvey A, Coates ME (1978) The influence of the gut microflora and dietary fiber on epithelial cell migration in the chick intestine. Br J Nutr 39:91–98.

Rosenkranz HS, McCoy EC, Mermelstein R, Speck WT (1981) A cautionary note on the use of nitro-reductase-deficient strains of *Salmonella typhimurium* for the detection of nitroarenes as mutagens in complex mixture including diesel exhaust. Mutat Res 91: 103–105.

Rowland IR, Mallett AK, Cole CB, Fuller R (1987) Mutagen activation by hepatic fractions from conventional germfree and monoassociated rats. Arch Toxicol (suppl)11:261–263.

Roy MP, Spencer PS (1989) Lathrogen. In: Cheeke PR, ed. Toxicants of Plant Origin, Vol. 3. Protein and Amino Acids, pp. 169–202. Boca Raton, Fla: CRC Press.

Rudali G, Coezy E, Muranyi-Kovacs I (1969) Research on the carcinogenic effect of sodium cyclamate in the mouse. C R Hebd Seances Acad Sci Ser D 269:1910–1912.

Ryan CA, Hass GM (1981) Structural, evolutionary and nutritional properties of proteinase inhibitors from potatoes. In: Ory RL, ed. Antinutrients and Natural Toxicants in Foods, pp. 169–185. Westport, Conn: F & N Press.

Sandforth F, Gutschmidt (1984) The intestinal monoxygenase enzyme in DMH-induced rat colonic carcinogenesis. Z Gastroenterol 22:586–591.

Schaumburg HH, Byck R, Gerstl R, Mashman JH (1969) Monosodium glutamate: its pharmacology and role in the Chinese restaurant syndrome. Science 163:826–828.

Scheline RR (1973) Metabolism of foreign compounds by gastrointestinal microorganisms. Pharmacol Rev 25:451–523.

Schroder H, Gustafsson BH (1973) Azo reduction of salicylazosulphapyridine in germ-free and conventional rats. Xenobiotics 3:225–231.

Seitz HK (1985) Ethanol and carcinogenesis. In: Seitz HK, Kommerell B, eds. Alcohol Related Diseases in Gastroenterology, pp. 195–212. Berlin: Springer Verlag.

Seitz HK, Bosche J, Czygan P, Veith S, Kommerell B (1982) Microsomal ethanol oxidation in the colonic mucosa of the rat: effect of chronic ethanol ingestion. Naunyn-Schmiedebergs Arch Pharmacol 320:81–84.

Sela MN, Steinberg D (1989) Glycyrrhizin: the basic facts plus medical and dental benefits. In: Grenby TH, ed. Progess in Sweeteners, pp. 71–79. New York: Elsevier.

Settipane GA (1986) The restaurant syndromes. Arch Intern Med 146:2129–2130.

Sharma RP, Salunke DK (1989) Solanum glycoalkaloids. In: Cheeke PR, ed. Toxicants of Plant Origins, Vol. 1. Alkaloids, pp. 179–236. Boca Raton, Fla: CRC Press.

Sharma VD, Singh SP, Taku A (1992) Purification of *Salmonella stanley* enterotoxin and its immunology and dermatology. Ind J Exp Biol 30:23–25.

Sherwin ER (1990) Antioxidants. In: Branen AL, Davidson PM, Salminen S, eds. Food Additives, pp. 139–193. New York: Marcel Dekker.

Shibamoto T, Bjeldanes LF (1993a) Natural toxins in animal foodstuffs. In: Introduction to Food Toxicology, pp. 43–66. San Diego: Academic Press.

Shibamoto T, Bjeldanes LF (1993b) Natural toxins in plantstuffs. In: Introduction to Food Toxicology, pp. 67–96. San Diego: Academic Press.

Shibamoto T, Bjeldanes LF (1993c). Toxicants formed during food processing. In: Introduction to Food Toxicology, pp. 184–198. San Diego: Academic Press.

Shibamoto T, Bjeldanes LF. (1993d) Food additives. In: Introduction to Food Toxicology, pp. 157–182. San Diego: Academic Press.

Shibamoto T, Bjeldanes LF (1993e) Biotransformation. In: Introduction to Food Toxicology, pp. 35–48. San Diego: Academic Press.

Shibamoto T, Bjeldanes LF (1993f) Pesticides residues in foods. In: Introduction to Food Toxicology, pp. 141–156. San Diego: Academic Press.

Shriniwas UM, Bhujwala RA (1979) Production of permeability factor and enterotoxin by *Pseudomonas aeruginosa.* Ind J Med Res 70:380–383.

Siegers CP, Bubinger I, Lemoine R, Herbst EW, Younes M (1986) Is alcohol a cocarcinogen in dimethylhydrazine-induced colon tumours in rats? Naunyn-Schmiedebergs Arch Pharmacol 334(suppl):R21.

Silverstone GA (1985) Possible sources of food toxicants. In: Seely S, Freed DL, Silverstone GA, Rippere V, eds. Diet-Related Diseases: The Modern Epidemic, pp. 44–60. Westport, Conn: AVI.

Sjovall J (1960) Bile acids in man under normal and pathological conditions: bile acids and steroids. Clin Chim Acta 5:33–41.

Smith GH, Tong TG (1975) Ulcerative colitis. J Am Pharm Assoc NS15: 202–212.

Smith HR (1984) Genetics of enterotoxin production in *Escherichia coli.* Biochem Soc Trans 12:905–915.

Smith HW, Green P, Parsell Z (1983) Vero cell toxins in *Escherichia coli* and related bacteria: transfer by phage and conjugation and toxic action in laboratory animals, chickens and pigs. J Gen Microbiol 129:3121–3127.

Smith-Barbaro PA, Hanson D, Reddy BS (1981a) Effect of fat and microflora on hepatic, small intestinal and colonic HMG CoA reductase, cytochrome P450 and cytochrome b_5. Lipids 16:183–188.

Smith-Barbaro PA, Hanson D, Reddy BS (1981b) Effect of bran and citrus pulp on hepatic, small intestinal and colonic HMG CoA reductase, cytochrome P-450 and cytochrome b_5. J Nutr 111:789–797.

Sparnins VL, Barany G, Wattenberg LW (1988) Effects of organosulfur compounds from garlic and onion on benzo(a)pyrene-induced neoplasia and glutathione S-transferase activity in the mouse. Carcinogenesis 9:131–134.

Spicer EK, Noble JA (1982) *Escherichia coli* heat-labile enterotoxin nucleotide sequence of the A subunit gene. J Biol Chem 257:5716–5721.

Stark RL, Duncan CL (1971) Biological characteristics of *Clostridium perfringens* type A enterotoxin. Infect Immun 4:89–96.

Stephen AM, Cummings JH (1980) The microbial contribution to human faecal mass. J Med Microbiol 13:45–56.

Stich HF, Rosin MP, Bryan L (1982) Inhibition of mutagenicity of a model nitrosation reaction by naturally occurring phenolics, coffee and tea. Mutat Res 95:119–128.

Stoltz DR (1982) The health significance of mutagens in foods. In: Stich HF, ed. Carcinogens and Mutagens in the Environment, Vol. 3, pp. 75–82. Boca Raton, Fla: CRC Press.

Streatfield SJ, Sandkvist M, Sixma TK, Bagdasarian M, Hol WG, Hirst TR (1992) Intermolecular interactions between the A and B subunits of heat-labile enterotoxin from *Escherichia coli* promote holotoxin assembly and stability in vivo. Proc Natl Acad Sci USA 89:12140–12149.

Strobel HW, Fang WF (1981) Role of cytochrome P-450 in the response of the colon to xenobiotics. In: Bruce WR, et al. eds. Gastrointestinal Cancer-Endogenous Factors, Banbury Report No. 7, pp. 141–149. Cold Spring Harbor, NY: Cold Spring Harbor Laboratory.

Strobel HW, Newaz SN, Fang HF, Lau PP, Oshinsky RJ, Stralka DJ, and Salley FF (1983) Evidence for the presence and reactivity of multiple forms of cytochrome P450 in colonic microsomes from rats and humans. In: Rydstrom J, Montelius J, Bengtsson M, eds. Extrahepatic Metabolism and Chemical Carcinogenesis, pp. 57–66. Amsterdam: Elsevier.

Sugarman B, Epps LR (1984) Zinc and the heat-labile enterotoxin of *Escherichia coli*. J Med Microbiol 18:393–398.

Sugimura T (1985) Carcinogenicity of mutagenic heterocyclic amines formed during the cooking process. Mutat Res 150:33–44.

Sugimura T (1986a) Past, present and future of mutagens in cooked food. Environ Health Perspect 67:5–10.

Sugimura T (1986b) Studies on environmental chemical carcinogenesis in Japan. Science 233:312–318.

Sugimura T, Kawachi T, Nagao M, et al. (1977) Mutagenic principle(s) in tryptophan and phenylalanine pyrolysis products. Proc Jpn Acad 52:58–61.

Sugimura T, Sato S, Wakabayashi K (1988) Mutagens/carcinogens in pyrolysates of amino acids and proteins and in cooked foods: heterocyclic aromatic amines. In: Woo Y-T, Lai DY, Arcos JC, Argus MF, eds. Chemical Induction of Cancer, Vol. IIIC, pp. 681–711. San Diego: Academic Press.

Sugimura T, Wakabayashi K, Nagao M, Ohgaki H (1989) Heterocyclic amines in cooked food. In: Taylor SL, Scanlan RA, eds. Food Toxicology: A Perspective on the Relative Risk, pp. 31–55. New York: Marcel Dekker.

Tasich M, Piper DW (1983) Effect of human colonic microsomes and cell-free extracts of *Bacteroides fragilis* on the mutagenicity of 2-aminoanthracene. Gastroenterology 85: 30–34.

Taylor SL (1986) Histamine food poisoning: toxicology and clinical aspects. CRC Crit Rev Toxicol 17:91–128.

Taylor SL (1988) Marine toxins of microbial origin. Food Technol 42:94–98.

Thies E, Siegers C-P (1989) Metabolic activation and tumorigenesis. In: Koster AS, Richter E, Lauterbach F, Hartmann F, eds. Intestinal Metabolism of Xenobiotics, pp. 199–214. Stuttgart: Gustav Fisher Verlag.

Timmis KN, Montenegro MA, Bulling E, Chakraborty T, Sanyal SC (1984) Genetics of toxin synthesis in gram negative enteric bacteria. In: Alouf JE, Freer JH, Fehrenbach FJ, Jeljaszewicz, ed. Bacterial Protein Toxins, pp. 13–27. London: Academic.

Trefouel J, Trefouel J, Nitti F, Bovet D (1935) Activite du *p*-aminophenylsulfamide sur les infections streptococciques. Experimentales de la souris et due lapin. C R Seanc Soc Biol 120:756–762.

Trucksis M, Galen JE, Michalski J, Fasano A, Kaper JB. (1993). Accessory cholera enterotoxin (Ace), the third toxin of a *Vibrio cholerae* virulence cassette. Proc Natl Acad Sci USA 90:5267–5271.

Vadivelu J, Puthucheary SD, Navaratnam P (1991) Exotoxin profiles of clinical isolates of *Aeromonas hydrophila.* J Med Microbiol 34:363–367.

Vanos V, Hofstaetter S, Cox L (1987) The microbiology of instant tea. Food Microbiol 2:187–197.

Vernet A, Seiss MH (1986) Comparison of the effects of various flavonoids on ethoxycoumarin deethylase activity of rat intestinal and hepatic microsomes. Food Chem Toxicol 24:857–861.

Vizethum W, Goerz G (1979) Induction of the hepatic microsomal and nuclear cytochrome P-450 system by hexachlorobenzene, pentachlorophenol, and trichlorophenol. Chem Biol Interact 28:291–299.

Walker R (1970). The metabolism of azo compounds: a review of the literature. Food Cosmet Toxicol 8:659–676.

Wallace WC, Letho ET, Brouwer EA (1970) The metabolism of cyclamate in rats. J Pharmacol Exp Ther 175:325–330.

Wargovich MJ (1987) Diallyl sulfide, a flavor component of garlic (*Allium sativum*), inhibits dimethylhydrazine-induced colon cancer. Carcinogenesis 8:487–489.

Wasserman AE, Wolff IA (1981) Nitrates and nitrosamines in our environment. In: Ory RL, ed. Antinutrients and Natural Toxicants in Foods, pp. 35–52. Westport, Conn: F & N Press.

Wattenberg LW (1975) Effect of dietary constitutents on the metabolism of chemical carcinogenesis. Cancer Res 35:3326–3321.

Wattenberg LW (1978) Inhibition of chemical carcinogenesis. J Natl Cancer Inst 60: 11–18.

Wattenberg LW (1983) Inhibition of neoplasia by minor dietary constituents. Cancer Res S43:2448–2453.

Wei CI, Kitamura K, Shibamoto T (1981) Mutagenicity of Maillard browning products obtained from a starch-glycine model system. Food Chem Toxicol 19:749–751.

White A, Handler P, Smith EL (1964) Principles of Biochemistry, 3rd ed. New York: McGraw-Hill.

Wiestler O, von Deimling A, Kobori O, Kleihues P (1983) Location of N-methyl-N′-nitro-N-nitrosoguanidine-induced gastrointestinal tumors correlates with thiol distribution. Carcinogenesis 4:879–883.

Williams GM, McQueen CA, Tong C (1990a) Toxicity studies of butylated hydroxyanisol and butylated hydroxytoluene. 1. Genetic and cellular effect. Food Chem Toxicol 28: 793–798.

Williams GM, Wang CX, Iatropoulos MG (1990b) Toxicity studies of butylated hydroxyanisole and butylated hydroxytoluene. 2. Chronic feeding studies. Food Chem Toxicol 28:799–806.

Wolfhagen MJ, Torensma R, Fluit AC, Verhoef J (1994) Toxins A and B of *Clostridium difficile*. FEMS Microbiol Rev 13:59–64.

Wollenberg P, Ullrich V (1980) The drug monooxygenase system in the small intestine. In: Gram TE, ed. Extrahepatic Metabolism of Drugs and Other Foreign Compounds, pp. 267–276. New York: S.P. Medical and Scientific Books.

Wostmann BS (1984) Other organs. In: Coates ME, Gusstafsson BE, eds. The Germ-Free Animal in Biomedical Research, pp. 215–231. London: Laboratory Animals.

Yamamoto T, Tsuji K, Kosuge T, et al. (1978). Isolation and structure determination of mutagenic substances in L-glutamic acid pyrolysates. Proc Jpn Acad 54B:248–250.

Yamamoto T, Tamura T, Yokota T, Takano T (1982) Overlapping genes in the heat-labile enterotoxin operon originating from *Escherichia coli* human strain. Mol Gen Genet 188:356–359.

Yamamoto T, Tamura T, Yokota T (1984) Primary structure of heat-labile enterotoxin produced by *Escherichia coli* pathogenic for humans. J Biol Chem 259:5037–5044.

Yamamoto M, Yamada T, Yoshihira K, Tanimura A, Tomita I (1988) Effects of food components and additives on the formation of nitrosamides. Food Addit Contam 5: 289–298.

Yang CS, Brady JF, Hong J-Y (1992) Dietary effects on cytochromes P450, xenobiotic metabolism, and toxicity. FASEB J 6:737–744.

Yasumoto T, Murata M, Oshima Y, Matsumoto GK, Clardy J (1984) Diarrhetic shellfish poisoning. In: Ragelis EP, ed. Seafood Poisoning. Washington, DC: American Chemical Society.

Yoh M, Yamamoto K, Honda T, Takeda Y, Miwatani T (1983) Effect of lincomycin and tetracycline on production and properties of enterotoxigenic *Escherichia coli*. Infect Immun 42:778–782.

Yokoda M, Narita K, Kosuge T, et al. (1981) A potent mutagen isolated from a pyrolysate of L-ornithine. Chem Pharm Bull 29:1473–1475.

Yoshida D, Matsumoto T, Yoshimura R, Mizusaki T. (1978) Mutagenicity of amino-α-carbolines in pyrolysis products of soybean globulin. Biochem Biophys Res Commun 83:915–920.

Yu C, Swaminathan B, Butler L (1986) Isolation and identification of rutin as the major mutagen in red wine. Mutat Res 170:103–113.

Zachariah PK, Juchau MR (1974) The role of gut flora in the reduction of aromatic nitrogroups. Drug Metab Dispos 2:74–78.

Zeiger E, Dunkel VC (1991) Mutagenicity of chemicals added to foods. In: Hayatsu H, ed. Mutagens in Food: Detection and Prevention, pp. 51–55. Boca Raton, Fla: CRC Press.

Zhang Y, Talalay P, Cho C-G, Posner GH (1992) A major inducer of anticarcinogenic protective enzymes from broccoli and elucidation of structure. Proc Natl Acad Sci USA 89:2399–2403.

Zlotlow MJ, Settipane GA (1977) Allergic potential of food additives: a report of a case of tartrazine sensitivity without aspirin intolerance. Am J Clin Nutr 30:1023–1025.

15

Gastrointestinal Detoxification and Digestive Disorders in Ruminant Animals

Christopher S. McSweeney
Roderick I. Mackie

1. Introduction

Nutrition and toxicology are closely intertwined. Toxic substances can interfere with vital functions of the gastrointestinal tract such as digestion, absorption, excretion, and their regulatory control. From a toxicology viewpoint animals are continually challenged by compounds that are without nutritive value—material that they nevertheless ingest, inhale, or absorb. Such exposure to toxic compounds has been going on as long as life has and is not merely a consequence of the modern era of industrialization. In fact, toxin production may be considered a successful evolutionary strategy or adaptation to predators or a form of chemical warfare practiced by species as varied as insects, frogs, and plants (Rosenthal and Janzen 1979).

Recycling of compounds, formed either naturally or synthetically, takes place on a global scale by animals and plants but particularly by the enzymes of bacteria and fungi. Microorganisms are capable of an enormous range of catabolic activity, and their enzymes are generally specific for individual compounds. Their effectiveness in this process is related to their large number, their short generation time, and their ability to exchange genetic material (Clark 1984). The problem of dealing with nutritionally useless compounds is entirely different for higher forms of life. Instead of a vast array of highly specific enzymes, animals have evolved systems for elimination (transport and excretion) rather than utilization of toxic compounds. Each animal has a group of about 30 inducible enzymes of detoxification that function to make xenobiotic or foreign compounds more readily excretable or less pharmacologically active (Jakoby and Ziegler 1990). The different strategies utilized by microbes and animals for dealing with toxic or nutritionally useless compounds are summarized in Table 15.1.

The ability of ruminant animals to survive on fibrous, low-protein feeds that are indigestible by most nonruminant animals is well documented. In this mutually beneficial relationship between the ruminant animal and foregut (pregastric) fermentation by bacteria, protozoa, and fungi, ecophysiological advantage to the host animal is provided by means of food storage capacity and supply of nutrients (energy from VFAs, protein from microbial cells, and B vitamins synthesized by bacteria). However, another important teleological explanation for the evolution of foregut fermentation is the detoxification of phytotoxins (of plant origin) and mycotoxins (of fungal origin) in the feed. Thus, it is suggested that ruminants have greater flexibility in diet choice, and also a better chance of survival when diet choice is limited, due to pregastric microbial detoxification of plants that are poisonous to their nonruminant herbivore competitors. Furthermore, microbial fermentation in ruminant animals can be modified and deliberately managed as a system to detoxify feedstuffs both naturally by adaptation and through modern genetic engineering technology.

Phytotoxins occur in many common feeds including grain, protein supplements, and forages. They range from tannins, alkaloids, goitrogens, gossypol, saponins, glucosinylates, mimosine, and cyanogens to nitrate and oxalate. In many instances, the ruminant forestomach provides a protective function and the ruminal microbiota effectively degrade a wide variety of toxic compounds. In some cases, the opposite can occur with the production of toxic metabolites from innocuous compounds. Some well-documented examples of the formation of toxic compounds as a result of rumen microbial activity are (1) the rapid hydrolysis of cyanogenic glycosides resulting in cyanide poisoning (James et al. 1975, Majak and Cheng 1984); (2) metabolism of miserotoxin and release of 3-nitropropanol or 3-nitropropionic acids, which are toxic (Majak and Clark 1980); and (3) acute

Table 15.1. Different strategies utilized by animals and microbes for dealing with toxic compounds

Animal	Microbial
Enzymes catalyze conversion of functional groups	High degree of enzyme specificity
Narrow range of inducible enzymes making compounds more readily excretable and less pharmacologically active	Enormous range of catalytic activity
Concentrated at points of entry (intestinal mucosa, liver, and lungs)	Spread over vast array of organism
System evolved for elimination (transport and excretion) of toxic compounds	System evolved for utilization of toxic compounds

Table 15.2. Common secondary compounds affecting the nutritive value of forages and grains

Class	Example	Ruminal Action[a]
Alkaloids	Pyrrolizidine	+
	Fescue	
Glycosides	Cyanogenic	—
	Coumarin	
	Isoflavones and coumestans	
Steroids and terpenes	Saponins	+
Simple acids (and their salts)	Fluoroacetate	+
	Oxalates	
	Nitrates	
Proteins and amino acids	Bloat-producing protein	—
	Mimosine	
	Indospicine	
Polyphenols	Tannins	+
Mycotoxins	Zearalenone	+/—
	Trichothecenes	
	Sporidesmin	
	Phomopsin	
	Swainsonine	
	Slaframine	
	Ergot alkaloids	

[a] Positive (+) or negative (−) effect of ruminal activity on nutritive value.

bovine pulmonary edema and emphysema due to production of 3-methylindole from tryptophan with resultant specific toxic effects on the lungs (Carlson and Breeze 1984). However, it should also be noted that prior exposure of rumen bacteria to many of the plant toxicants increases the rate of subsequent detoxification and that the ''adaptation'' phenomenon is an important factor to consider. A list of common secondary compounds or toxins that affect the nutritive value of forages and grains is presented in Table 15.2.

2. Coevolution of Plants and Animals

Plants contain a wide range of nonnutrient compounds, or ''allelochemicals.'' The continuum of evolutionary changes in synthesis of secondary compounds by plants, followed by the evolution of comparable detoxification mechanisms in animals, has been termed coevolution and continues to be the subject of much discussion (Berenbaum and Feeny 1981). Evolutionary adaptation to toxicants appears to have occurred in wild species of animals that have been exposed to particular plants for many generations. In contrast, domestic animals have been removed from their native habitat and therefore have had less opportunity to adapt to specific plants (Fowler 1983).

There are a number of behavioral and other adaptations that may account for the paucity of toxicological problems with poisonous plants in wild animals. Many wild animals have fastidious grazing habits and tend to nibble small quantities of feed from a variety of different plants, minimizing the likelihood of consuming an acute dose of toxin. Large wild herbivores tend to range over extended areas and are not forced to consume poisonous plants because of lack of other feed as sometimes occurs with confined domestic animals. However, overgrazing, range deterioration and fencing may enhance the risk of poisoning of both domestic and wild species. The extent of problems associated with natural toxicants is extremely difficult to assess, but it appears that their grazing and browsing behavior could be an important factor limiting poisoning (Cheeke and Shull 1985).

With domestic animals there are examples of species differences in susceptibility to plant toxins. These factors may include simple differences in management or feeding programs, differences in the gastrointestinal tract, palatability variations or metabolic adaptations. Grazing behavior and palatability differences may be important in cattle and sheep. Breed differences in susceptibility to plant poisoning have also been reported. Thus animal management is critical in controlling losses associated with toxicity due to ingestion of poisonous plants (Cheeke and Shull 1985).

Our understanding of the challenges herbivores face in minimizing phytotoxicity and the mechanisms herbivores use to minimize ingestion of phytotoxins is limited. Variations in environment such as soil fertility, soil moisture, and sunlight affect phytotoxin concentrations. Also the type and amount of toxins vary among plant species and parts as a function of environment. This raises the question of how herbivores determine which foods are nutritious and which are toxic. One way herbivores learn which foods to eat and which foods to avoid is through postingestive feedback from nutrients and toxins (Provenza and Pfister 1991). Research on conditioned food preferences and aversions shows that voluntary intake depends on postingestive feedback. If the food is nutritious (positive postingestive feedback), intake of the food increases. If toxicity ensues (aversive postingestive feedback), intake of the food is limited. Herbivores apparently learn to associate the taste of food with positive postingestive feedback. They also regulate their food intake by associating the taste of food with aversive postingestive consequences. Thus herbivores have physiological mechanisms to counter phytotoxins and if these are exceeded, they become ill and may die. However, mammals usually adjust intake to avoid intoxication. This adjustment is conditioned by sampling foods to determine when concentrations of toxins change as a result of growth or previous herbivore. Sheep, goats, and cattle sample foods and regulate their intake of nutritious plants that contain toxins. If toxicity decreases, the taste of the plant is no longer paired with aversive ingestive consequences, and any nutritional value the plant provides will cause intake of the plant to increase. In contrast, intake decreases as the toxicity of the plant increases.

The inefficiency and risk associated with trial-and-error learning could provide additional selective pressure for herbivores that feed in mixed-generation groups to rely on social learning. Through this process, foraging information is passed from experienced to inexperienced foragers—initially from mother to offspring. Learning from social models decreases the risk inherent in trial and error if young animals remember foods and sample novel foods cautiously. Thus the importance of learning in diet selection, although important, has been overlooked (Provenza and Pfister 1991). Several research areas require further elaboration and clarification. The mechanisms that enable mammalian herbivores to sense the adverse effects of toxins and to limit intake have not been studied. Other mechanisms may also be involved in the control of food selection and intake. Also the role of inhibition of digestion or reduced digestibility caused by toxins in forage selection should be critically examined.

3. Microbial Adaptation to Toxic Compounds

Different biochemical, growth, and molecular mechanisms or processes may be responsible for an adaptive response to toxic compounds (Clarke 1984, Van der Meer et al 1992):

1. The first is induction of specific enzymes in members of a microbial community, resulting in an increase of the observed degradative capacity of the total community (induction). This was originally termed enzymatic adaptation and referred to substrate-activated biochemical variations not involving changes in genotype (Stanier 1951). With the work of Monod and colleagues it was possible to discuss induced enzymes rather than adaptive enzymes and the mechanisms involved in enzyme induction.
2. Another process is growth of a specific subpopulation of a microbial community able to take up and metabolize the substrate (growth). These first two processes are most commonly advanced in explaining the adaptive response in foregut fermenters.
3. Adaptation can also involve the selection of mutants which acquired altered enzymatic specificities or novel metabolic activities which were not present at the onset of exposure of the community to the introduced (toxic) compounds. Such a selective process may require longer adaptation (selection) than the other two processes (induction and growth).

The ability of the ruminal ecosystem to adapt and increase its capacity to detoxify a plant toxin in response to the amount of toxin consumed is a major factor determining the pathogenesis of plant toxicity in these forestomach fermenters. The main pathways of microbial metabolism in the rumen are associated with the degradation and utilization of polysaccharides, nitrogenous compounds, lipids, and nucleic acids. The enzymes involved in these pathways are not usually considered as playing a role in the metabolism of the many plant toxins that are ingested sporadically by foregut fermenters. However, it is likely that the catalytic enzymes produced in the unadapted rumen, have broad substrate specificities and recognize particular chemical bonds which are common to both toxic and nontoxic plant compounds. In many cases, the degradative pathway for a toxin involves a consortium of microorganisms since the enzymes involved may not be present in one organism. Even when a single species of ruminal bacteria is capable of degrading a toxin, there are probably several distinct strains of the species present in the rumen (see Allison et al. 1992).

The initial rate of metabolism of a particular toxin in the rumen is usually a function of the level of expression of enzymes that degrade or transform the toxin and the number of organisms producing these enzymes. The size of the population of toxin-degrading microorganisms in the unadapted rumen is determined by its ability to derive energy for growth from the normal feed constituents and other less obvious traits which enable it to compete with other organisms. The population is likely to increase in size when a toxic substrate is available and can be exploited as an additional source of energy which the remainder of the rumen microbial ecosystem cannot use. The adaptive response of rumen microorganisms to the presence of a plant toxin may also involve the induction of an enzyme(s) involved in the detoxification process. Utilization of the toxin as a source of energy is usually the most important factor driving adaptation in the rumen. However, the toxin-degrading population can also be selected for indirectly and enriched by manipulating the diet to provide other energy yielding substrates, preferred sources of nitrogen, growth factors, and substrates that can act as electron donors or acceptors in the metabolism of the toxin.

The ability of the rumen to adapt does not imply that all plant toxins can be degraded in the rumen. In fact, the anaerobic nature of the rumen and turnover rate of digesta imposes significant limitations on the metabolic capacity of the resident microorganisms. Many plant toxins that have heterocyclic ring structures are readily degraded by oxidative process, but the energy generated from reductive metabolism is insufficient to sustain a growth rate compatible with survival in the rumen. This does not mean that the compounds cannot be degraded in anaerobic ecosystems such as in sludge and sediments where the growth rate of microorganisms can be very slow. Specific examples of adaptation of the ruminal ecosystem to plant toxins will be described in the following section on degradation of phytotoxins and mycotoxins by foregut fermenters.

Genetic mechanisms that possibly contribute to the adaptive response of bacteria to toxic or xenobiotic compounds are presented briefly. The different mechanisms can be divided into three groups: gene transfer, mutational drift, and genetic recombination and transposition (reviewed by Van der Meer et al. 1992). Some of the mechanisms are difficult to prove experimentally.

3.1. Gene Transfer

Genetic interactions in microbial communities are effected by several mechanisms such as conjugation via plasmid replicons, transduction, and transformation (see review by Van der Meer et al. 1992 for references). The occurrence of plasmids in bacteria in the natural environment is a general phenomenon and an important source of genetic information residing on plasmid vehicles may spread among indigenous bacteria. Since many examples of self-transmissible plasmids that carry genes for degradation of aromatic or other organic compounds are known, their role in spreading these genes to other bacteria is well established (Sayler et al. 1990, Mohn and Tiedje 1992). It is possible that a few common self-transmissible ancestor replicons may have been involved in the acquisition and spread of different catabolic modules.

The importance of gene transfer for adaptation of bacteria to new compounds is well documented in many studies on experimental evolution of novel metabolic activities (reviewed by Van der Meer et al. 1992). Such studies identify biochemical barriers in natural pathways which prevent the degradation of novel substrates and overcome these barriers by transferring appropriate genes. Pathways could be expanded by replacing narrow-specificity enzymes by broader ones (horizontal expansion) or by providing peripheral enzymes which could direct substrates into existing degradative pathways (vertical expansion) (Ramos and Timmis 1987). It should be noted that transfer of catabolic plasmids can lead to regulatory and/or metabolic problems for the cell, and therefore additional mutations in the primary transconjugants are often necessary to construct strains with desired metabolic activities.

3.2. Point Mutations

Several examples have demonstrated that single-site mutations can alter substrate specificities of enzymes or effector specificities (Clark 1984). Single-site mutations are believed to arise continuously and at random as a result of errors in DNA replication or error. However, some authors have suggested that directed mutations in response to selective pressure are possible, but whether environmental factors can control the direction of mutations is debatable. It is possible that a diversity of stress factors including toxic compounds stimulate error-prone DNA replication and accelerate DNA evolution.

Despite the important effects of a single basepair mutation on the adaptive process, the accumulation of single base pair changes may not be the sole mechanism for the divergence of properties in contemporary catabolic enzymes. Other mechanisms that would allow faster divergence of DNA sequences include gene conversion or slipped-strand mispairing (Van der Meer et al. 1992).

3.3. *Recombination and Transposition*

DNA REARRANGEMENTS

The order of genes encoding cleavage pathways can differ from one species to another suggesting that various DNA rearrangements occur. As yet, there are no clear indications of which mechanisms may direct these rearrangements, such as longer direct or inverted repeats which would allow general recombination, or the activity of insertion elements (Van der Meer et al. 1992).

GENE DUPLICATIONS

Gene duplications have been considered an important mechanism for the evolution of microbes (Beacham 1987). Once duplicated, the extra gene copy could be essentially free of selective constraints and thus diverge much faster by accumulating mutations. These mutations could eventually lead to full inactivation, rendering this gene copy silent. Reactivation of the silent gene copy could then occur through the action of insertion elements.

TRANSPOSITION

Insertion elements have been shown to play an important role in rearrangements of DNA fragments, in gene transfer, and in activation or inactivation of silent genes. For catabolic pathways several examples of insertion (IS) elements are known (reviewed by Van der Meer et al. 1992).

INSERTIONAL ACTIVATION

Another important role of IS elements is the activation or inactivation of (silent) genes. One end of an IS element often contains promoterlike sequences that can activate the expression of genes outside the IS element.

Although isolation and characterization of individual bacteria from the natural environment have been important in our understanding of mechanisms that are involved in adaptational processes, they have limitations for proper assessment of the different genetic events in the adaptive response in the natural environment. Thus for a full understanding of adaptational processes in nature, in situ genetic interactions among microorganisms and the influence of environmental parameters on the selection and dissemination of catabolic genes requires study (Van der Meer et al. 1992). This requires the utilization of modern technology in studying detoxification mechanisms in the gastrointestinal tract of different animal models.

4. Ruminal Detoxification of Phytotoxins and Mycotoxins

In many countries, grazing herbivores are exposed to toxic forages. Animals that are foregut fermenters can often detoxify or reduce the toxicity of these plants by microbial metabolism, although microbial biotransformation of certain compounds in the gut can also enhance the toxicity (Dawson and Allison 1988, Mackie and McSweeney 1991). Foregut metabolism of toxic compounds has been characterized mainly in the ruminant animal, but in most cases the microorganisms involved have not been isolated and identified. This section will concentrate on the metabolic pathways in the rumen which are known to detoxify phytotoxins or generate intermediates that are toxic to the animal.

4.1. Non-Protein Amino Acids

The family Fabaceae comprises many plants that contain nonprotein amino acids and their derivatives (Bell 1973). Studies of the ruminal metabolism of some of these important amino acids include mimosine, lathrogens, and canavanine.

MIMOSINE

The tropical leguminous shrub *Leucaena leucocephala* is used widely as a nitrogen supplement but is toxic to ruminants in some regions of the world. *Leucaena* contains the amino acid mimosine {β-[N-(3-hydroxy-4-oxopyridyl)]-α-amino-propionic acid}, which is hydrolyzed mainly by rumen bacteria to the goitrogenic toxin pyridinediol 3-hydroxy-4–1(H)-pyridone (3,4-DHP) (Fig. 15.1), but the amino acid also undergoes autolysis by leaf enzymes during inges-

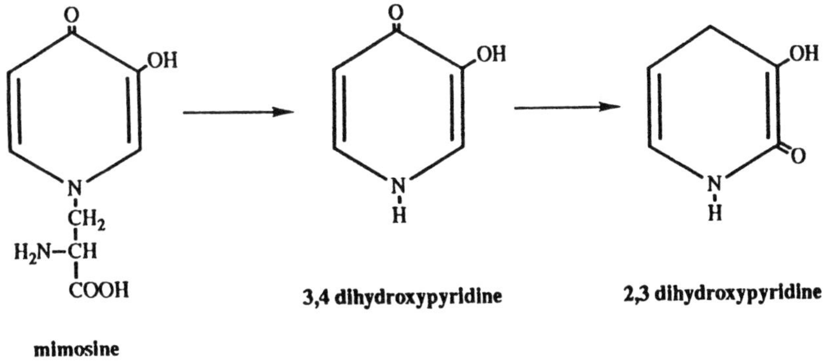

Figure 15.1. Ruminal metabolism of mimosine.

tion and mastication (Hegarty et al. 1964, Lowry et al. 1983). Some ruminants harbor rumen bacteria that are able to degrade 3,4-DHP (Jones 1981, Jones and Megarrity 1983, Jones and Lowry 1984, Dominguez-Bello and Stewart 1990). Ruminants that lack bacteria capable of degrading 3,4-DHP have been protected from intoxication by inoculation with either a mixed population of rumen microorganisms or a bacterium (*Synergistes jonesii* 78–1), both of which came from an Hawaiian goat that was not susceptible to *Leucaena* toxicity (Jones and Megarrity 1986, Hammond et al. 1989).

Bacteria with the ability to degrade 3,4-DHP are not ubiquitous, and rumen microbial populations in some parts of the world are unable to degrade the toxin. Four distinct strains of *Synergistes jonesii* including strain 78–1 were isolated from goat rumen contents. These are all gram-negative, rod-shaped, obligate anaerobic bacteria that degrade 3,4-DHP. The isolates do not ferment carbohydrates but use both 3,4 DHP and its isomer 3-hydroxy-2(1H)-pyridone (2,3 DHP) (Fig. 15.1), as well as arginine and histidine as substrates for growth (Allison et al. 1992). Arginine is catabolized by the arginine deiminase pathway into ornithine, ammonia, and carbon dioxide with the formation of 1 mol ATP/mol arginine consumed (McSweeney et al. 1993a).

The pathway of metabolism of 3,4-DHP has not been determined but it appears that 3,4-DHP is converted to 2,3-DHP before ring cleavage occurs (Jones et al. 1985, Allison et al. 1992). Rumen bacteria that degrade 2,3-DHP but not 3,4-DHP have been isolated from Senepol cattle which were derived from St. Croix, U.S. Virgin Islands (Hammond et al. 1989). These bacteria are probably not very important in the prevention of *Leucaena* toxicity because 2,3-DHP degradation alone will not prevent toxicity and the metabolism of this DHP isomer is not rate-limiting in the degradative pathway.

An oligonucleotide probe which targets a unique region of the 16 rRNA sequence of *Synergistes jonesii* has been developed which will assist in ecological studies of the transmission, colonization, and population of the organism once it is released into the environment (McSweeney et al. 1993b).

NEUROLATHROGENS

The perennial legume crop flatpea (*Lathyrus sylvestris*) is a high-protein forage, but its value as a nitrogen supplement has been limited by reports of toxicity (see Rasmussen et al. 1992). Several nonprotein amino acids have been detected in *L. sylvestris* including 4-N-oxalyl-2,4-diaminobutyric acid, DABA; 2,3-diaminopropionic acid; and 3-N-oxalyl-2,3-diaminopropionic acid. Flatpea toxicity is considered to be due to the neurotoxin, DABA (Van Etten and Miller 1963, Ressler 1964) which causes symptoms of muscle tremors, incoordination, and seizures.

Susceptibility to flatpea toxicity is variable and it has been suggested that ruminants are able to adapt and develop a tolerance to the toxins (Hodgson and Knott 1936, Long et al. 1977). Rasmussen et al. (1993) showed by exchanging rumen contents between alfalfa-fed and flatpea-fed sheep that adaptation to flatpea was probably associated with alterations in rumen metabolism. However the rate of degradation of DABA in the rumen (0.2 to 0.3 μmol/mL/h) was similar for both adapted and unadapted animals. These findings indicate that the ruminal metabolism of other lathrogens in flatpea may be more important than DABA degradation in the pathogenesis of the disease.

4.2. Phytoestrogens

Infertility occurs in sheep grazing pastures of *Trifolium* spp. (clovers) and *Medicago* spp. (medics) owing to the presence of estrogenic substances in these plants. The incidence of infertility is higher with clovers, and accordingly the phytoestrogens in these plants have been studied more extensively. Cattle grazing the same pastures are unaffected. The principal reason for infertility is the impaired transport of spermatozoa to the ova due to excessive secretions of cervical and uterine mucus (Turnbull et al. 1966).

The substances that produce the estrogenic effects are the isoflavones formononetin, daidzein, genistein, and biochanin A (Figs. 15.2, 15.3) in the clovers, and coumestrol (Fig. 15.4) in the *Medicago* species. The coumestans are relatively more estrogenic than the isoflavones based on their biological activity in monogastric laboratory animals. However, rumen metabolism plays a central role in determining the estrogenic activity of these compounds in the ruminant. Phytoestrogens are present in the plant as glycosides which are rapidly hydrolyzed by glucosidases in the rumen. Ewes fed coumesterol and 4′-methoxy-coumesterol became tolerant to the oestrogenic effects of the compounds within 2 weeks (Kelly et al. 1975). The role of rumen metabolism in the development of tolerance to coumestans has not been determined but it has been demonstrated that rumen microorganisms can demethylate 4′-methoxycoumestrol to coumestrol (Adler and Weitzkin-Neiman 1970). Rumen metabolism of the isoflavones results in both the inactivation of compounds that are intrinsically estrogenic and the bioactivation of others. Biochanin A is demethylated to genistein, which is degraded to the nonestrogenic compound *p*-ethylphenol (Fig. 15.3) (Batterham et al. 1965, Braden et al. 1967). The *p*-ethylphenol is probably formed from ring B and part of ring C of genistein, but the phenolic metabolite from ring A has not been identified. Formononetin, which has low or negligible estrogenic activity, is demethylated to daidzein (Fig. 15.2) in both the rumen and liver (Nilsson 1961, 1962). Daidzein is further metabolized by the rumen microorganisms to equol, which is estrogenic (Nilsson et al. 1967). Formononetin can also be reduced to equol in the rumen without prior demethylation, by metabolism to O-methyl equol which is less

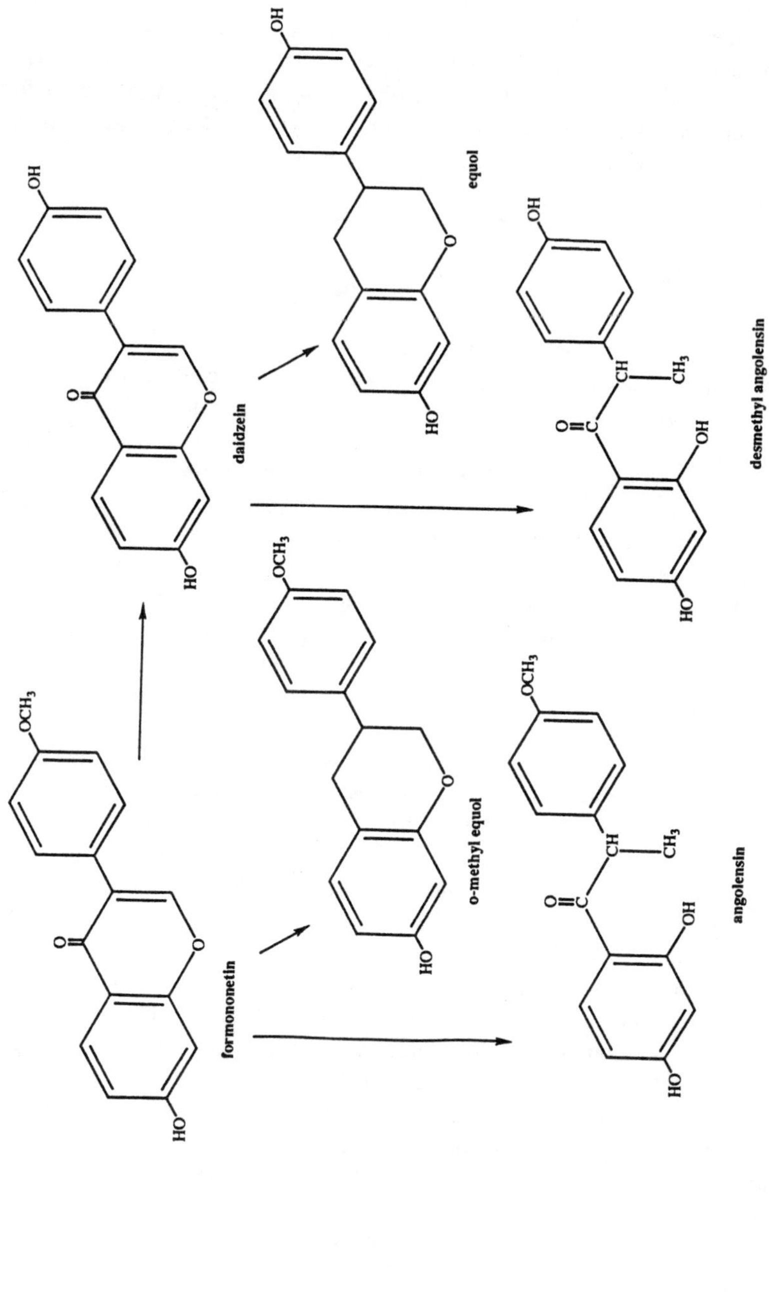

Figure 15.2. Ruminal metabolism of the isoflavones, formononetin, and daidzein.

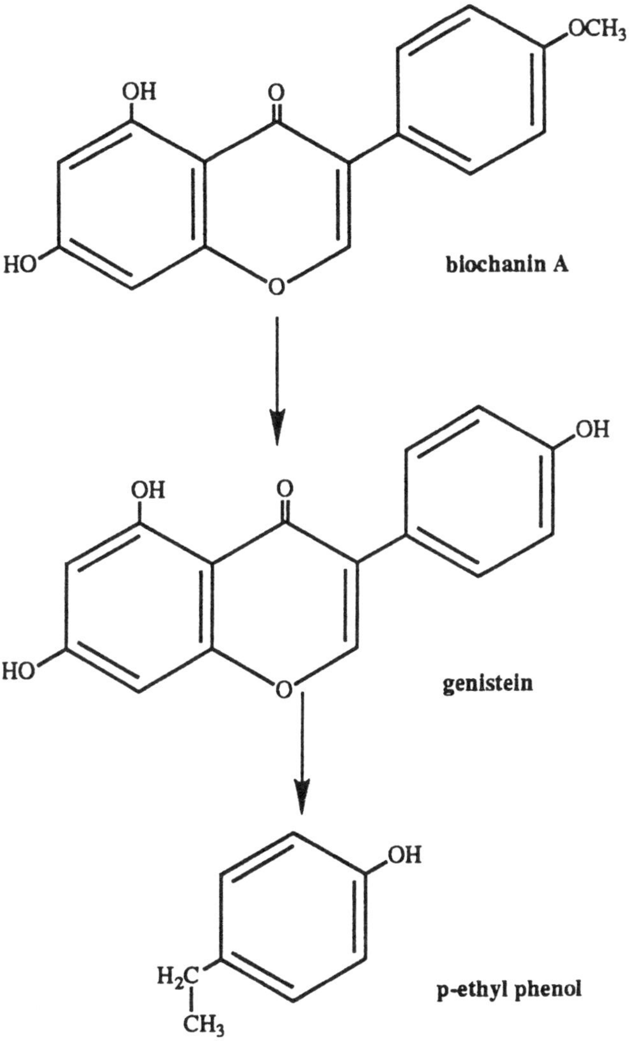

Figure 15.3. Ruminal metabolism of the isoflavones biochanin A and genistein.

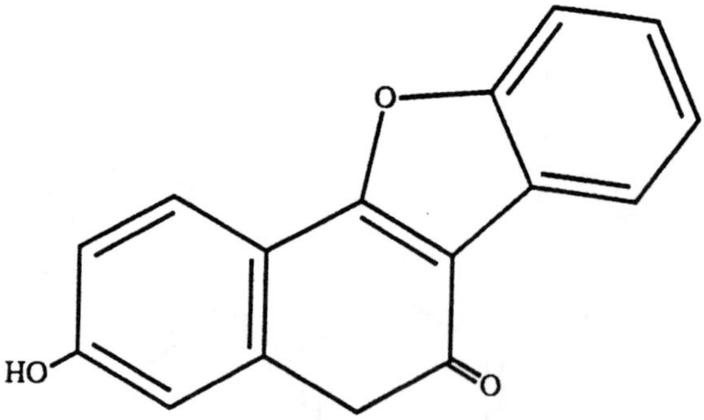

coumesterol

Figure 15.4. Structure of the isoflavone coumesterol.

estrogenic than equol (Cox and Braden 1974). Methylation of the hydroxyl groups of estrogens appears to reduce their estrogenic activity.

The rate of metabolism of these isoflavones and the metabolites produced in the rumen depend on the diet fed (Nilsson et al. 1967, Davies 1987). The same metabolism of isoflavones occurs in the rumen of cattle but the absorbed equol is excreted more rapidly. Thus cattle are less susceptible to the estrogenic effects of the metabolites.

4.3. Aliphatic Nitro Compounds

Leguminous plants containing simple aliphatic compounds with a nitrite moiety are a major cause of poisoning in sheep and cattle (Williams and Barneby 1977). The toxic compounds are 3-nitro-1-propionic acid (NPA) and 3-nitro-1-propanol (NPOH), and occur in the plant as glycosides. These nitrotoxins are found mainly in species of the genus *Astragalas* as well as plants belonging to the genera *Coronilla*, *Indigofera*, and *Lotus*. Some species of *Astragalas* such as cicer milk vetch (*A. cicer*) do not contain nitrotoxins and are valuable forage plants, while timber milk vetch (*A. miser*) contains miserotoxin, which is the glycoside of β-D-glucose and 3-NPOH. Toxicity is associated with the development of methaemoglobinemia which occurs when hemoglobin is oxidized by nitrite that is released from the absorbed nitrotoxin.

Ruminants can tolerate higher concentrations of nitrotoxins than monogastrics owing to detoxification of the compounds by ruminal microorganisms. The degree of protection from toxicity depends on the level of intake of the toxins and their rate of metabolism in the rumen. The glucose conjugates of the nitrotoxins are rapidly hydrolyzed by ruminal microorganisms to NPOH or NPA (Majak and Pass 1989). However, NPA is less toxic than NPOH because it is degraded more rapidly (0.4 vs. 0.1 μmol/mL/min) in the rumen (Gustine et al. 1977, Majak and Cheng, 1981, Majak and Clark 1980, Anderson et al. 1993). It was suggested that detoxification of these nitrotoxins involved cleavage of nitrite from the carbon skeleton and reduction to ammonia since nitrite accumulated in pure cultures of rumen microorganisms that metabolize NPA and NPOH (Majak and Cheng 1981, 1983). However, Anderson and co-workers (1993) demonstrated that the primary pathway of metabolism of NPA and NPOH by ruminal microorganisms from cattle and sheep was the reduction of the nitro groups in the toxins and conversion to β-alanine and 3-amino-1-propanol, respectively.

Practical strategies are being developed to enhance the rate of detoxification of NPOH by ruminal microorganisms. Manipulating the nutritional value of the diet appears to affect the rate of NPOH metabolism, but the factors involved are not fully understood (Majak et al. 1982). The rate of metabolism of NPOH can be increased by supplementing cattle with nitroethane, which is a nontoxic analog of NPOH-(Majak et al. 1986). Presumably the nitroethane acts as a substrate for NPOH-metabolizing microorganisms and increases their population in the rumen. Studies of the microbial metabolism of NPOH and NPA have shown that ferrous and sulfide ions stimulate, while carbon monoxide gas reduces the rate of reduction of NPOH but not NPA (Anderson et al. 1993). This effect of ferrous and sulfide ions on NPOH metabolism was further enhanced when the incubations were done under H_2. These results suggest that NPOH-metabolizing rumen microorganisms may use a nonspecific hydrogenase-ferredoxin system to reduce NPOH (Angermaier and Simon 1983). However, formate, glucose, lactate, pyruvate, and succinate, which can act as hydrogen donors for nitrate, nitrite, and hydroxylamine, did not mimic the effects of hydrogen gas.

Although both NPA and NPOH are metabolized in the rumen, they appear to be inhibitory to growth of some ruminal organisms at concentrations (> 4.2 mM) occurring in the rumen of animals grazing pastures that contain these toxins (Anderson et al. 1993).

4.4. Nitrate-Nitrite

Grain and sugar crop plants and certain grasses and weeds can all accumulate nitrates in concentrations that are toxic to foregut fermenters. Metabolism of nitrate (NO_3^-) to nitrite (NO_2^-) in the rumen by microorganisms predisposes animals to nitrite poisoning although nitrite can be further reduced to ammonia (Sapiro et al. 1949, Lewis 1951a). Nitrite that is absorbed into the blood rapidly oxidizes oxyhemoglobin to methemaglobin, thus reducing the oxygen-carrying capacity of red blood cells. Monogastric animals tend to be tolerant of high levels of NO_3^- in food because they do not readily convert nitrate to nitrite in the gut.

The development of toxicity therefore depends on the relative rates of nitrate and nitrite reduction in the rumen.

Factors that affect the rates of nitrate and nitrite metabolism in the rumen include the number of NO_3^- and NO_2^- reducing organisms, enzymatic activities of these organisms, availability of substrates that act as H^+ donors for reduction of NO_3^- and NO_2^-, and the physiological conditions in the rumen. Adaptation of the rumen to dietary NO_3^- results in a rapid induction of NO_3^- and NO_2^- reductase activities in the mixed rumen population as well as a 10-fold increase in the number of organisms with these enzymatic activities within a 6-day period (Allison and Reddy 1984). However NO_2^- is toxic to some rumen bacteria and animals could be susceptible to toxicity as the rumen ecosystem adapts to the presence of NO_2^- and enriches with bacteria that tolerate low concentrations of NO_2^- (Allison and Reddy 1984). Substrates including formate, pyruvate, succinate, lactate, citrate, malate, and glycine can be used by anaerobic microorganisms to reduce NO_3^- and NO_2^-, and most of these compounds are produced in the rumen from the fermentation of carbohydrates (Holtenius 1957, Jones 1972, Lewis 1951b). Accordingly, rates of reduction of NO_3^- and NO_2^- in the rumen tend to be faster when animals are fed diets rich in readily available carbohydrate (Nakamura et al. 1979, Takahashi et al. 1980). The pH of the rumen may also affect the rates of these reactions since the pH optima for NO_3^- and NO_2^- reduction by rumen microorganisms were 6.5 and 5.7, respectively (Lewis 1951b, Takahashi et al. 1978).

Nitrate reductase activity is membrane bound in ruminal bacteria and appears to be associated with cytochrome-linked electron transport systems (Allison and Reddy 1984, De Vries et al. 1974). Addition of either NO_3^- or NO_2^- to an in vitro fermentation with rumen fluid resulted in a reduction or inhibition of methane production (Allison et al. 1981). Stoichiometric equations indicate that the free energy change ($\Delta G^{o\prime}$) associated with reduction of NO_3^- to ammonia is -598 kJ compared with -131 kJ for the reduction of CO_2 to methane. This indicates that NO_3^--NO_2^--reducing bacteria may outcompete the methanogens for hydrogen. Alternatively, methanogens could decrease in number if NO_2^- accumulated in the rumen and inhibited bacteria that had syntrophic relationships with methanogens. Nitrate can also change the fermentation pattern in the rumen toward acetate production rather than propionate and butyrate due to competition for electrons between NO_3^- and NO_2^- reduction and other hydrogen-utilizing reactions. The predominant bacteria in the rumen that are responsible for NO_3^- and NO_2^- reduction have not been identified, but five major groups of bacteria were isolated from a sheep adapted to NO_3^- (Allison and Reddy 1984). Two of the groups were identified presumptively as selenomonads which could reduce NO_3^- and metabolize formate. An unclassified third group could reduce NO_3^-, metabolize formate, and produce mainly lactate. A fourth group identified as *Anaerovibrio* spp. could not reduce NO_3^- and produced mainly propionate from glucose. The final group were lactate-fermenting short rods that produced mainly acetate but were unable to use glucose as an energy source and varied in their ability to reduce nitrate.

4.5. Oxalate

Oxalate is found in plants of the Oxalidaceae and Chenopodiaceae families. It is present as either a salt of acid oxalate or as soluble and insoluble salts of oxalate (Fig. 15.5). Levels are lower in grasses than forbs, and toxicity is rarely seen in ruminants grazing grass pastures due to the oxalate degrading ability of rumen microorganisms (Dawson et al. 1980a, Allison et al. 1981).

Toxicity occurs when oxalate ions are absorbed in substantial concentrations and damage capillaries in the lungs and the tubular epithelial cells of the kidney. Characteristic crystals of calcium oxalate precipitate in the lumina of the kidney tubules. Tetany can also occur in acute intoxications due to complexing in the circulation of oxalate and ionized calcium. Ruminants that are adapted to diets containing oxalate can tolerate quantities of oxalate that would normally be toxic to nonadapted animals (Allison and Reddy 1984). Acquired tolerance to oxalate toxicity is due to an increased population of oxalate-degrading bacteria in the

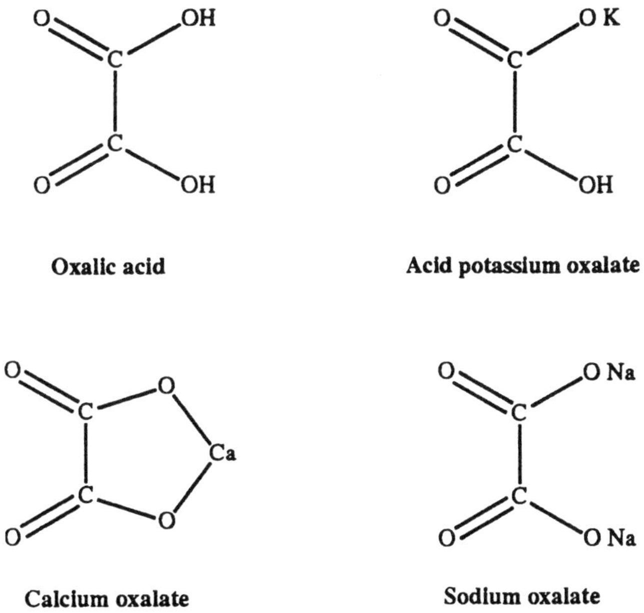

Figure 15.5. Plant oxalates.

rumen (Morris and Garcia-Rivera 1955, Watts 1957). Conversely, fasting appears to predispose animals to toxicity, presumably from a reduction in this specific microbial activity.

The anaerobic, gram-negative, nonmotile rod *Oxalobacter formigenes*, which inhabits the rumen and the large intestine of herbivores at about 10^8 cells/g digesta, appears to be a major oxalate degrader in these animals (Dawson et al. 1980a, Allison et al. 1985). Oxalate is used as a source of energy for growth and is metabolized to CO_2 and formate by this organism. Oxalate was degraded with a K_m of 5.7 mM and a V_{max} of 4.9 nmol oxalate/min/g dry cells, and required activated coenzyme A (CoA) which was supplied as succinyl-CoA (Allison et al. 1985). Formate produced from oxalate degradation is metabolized by methanogenic bacteria to methane (Dawson et al. 1980b). However, the relationship between oxalate degraders and the methanogenic population is not well understood. Oxalate degradation is inhibited by H_2, formate, and benzyl viologen when the fermentation proceeds to methane production, but these agents are not inhibitory in mixed cultures that are not producing methane. These observations have led to the conclusion that unidentified oxalate-degrading microorganisms exist in these cultures in a syntrophic relationship with methanogens while growth of *O. formigenes* is not dependent on a methanogenic population (Allison and Reddy 1984).

Oxalyl-CoA decarboxylase and formyl-CoA transferase are both involved in metabolism of oxalate to formate by *O. formigenes* (Allison et al. 1985; Baetz and Allison 1989, 1992). Yield of cells per mole of oxalate metabolized is higher than expected in *O. formigenes* given the small change in free energy (−26.7 kJ/mol oxalate) for oxalate metabolism. It has been proposed that the energy derived from decarboxylation of oxalate is conserved by mechanisms other than substrate level or electron transport phosphorylation. In this regard a membrane-bound oxalate^{2-}-formate^{-} antiport protein and a cytoplasm decarboxylase may work together to create a proton gradient for ATP synthesis by coupling the exchange of oxalate and formate across the cell membrane with a proton-consuming decarboxylation reaction (Allison et al. 1985, Anantharam et al. 1989, Baetz and Allison 1989, 1992, Ruan et al. 1992). Alternatively, decarboxylation reactions that generate Na^+ ion gradients via sodium pumping decarboxylases may also consume energy (see Daniel and Drake 1993). Recently it has been demonstrated that both oxalate and glyoxylate are metabolized to acetate by the bacterium *Clostridium thermoaceticum* originally isolated from horse manure (Daniel and Drake 1993).

4.6. Fluoroacetate

Fluoroacetate (FA; CH_2FCOOH) is the most common of a range of organofluorine compounds that occurs in plants (McEwen 1978). It was first identified in the southern African shrub *Dichapetalum cymosum* (Marais 1944). Fluoroacetate occurs in a variety of Australian plants of the family Fabaceae, including various species of *Acacia*, *Gastrolobium*, and *Oxylobium* (Cheeke and Shull 1985). Despite their toxicity, some of the plants represent an important food resource for foraging animals. Poisoning of domestic animals is influenced by availability of other food sources and the growth cycle of the plants (Kellerman et al. 1988). There are indications that organofluorine compounds confer an advantage on plants producing them since they may provide defense against predation by insects, birds, or animals. In addition, sodium monofluoroacetate (compound 1080) is used extensively in different parts of the world for controlling various agricultural vertebrate pests such as rats, mice, and coyotes (Cheeke and Shull 1985).

FA is toxic to animals by means of a ''lethal synthesis'' in which it is converted to highly toxic fluorocitrate in the liver. Fluorocitric acid competitively inhibits aconitate hydratase (aconitase), resulting in a block in the TCA cycle (Goldman 1969). Thus glucose metabolism is impaired, accompanied by an accumulation of unmetabolized citrate, resulting in death. Another mode of action has been advanced which proposed that FA may inhibit citrate transport through mitochondrial membranes rather than inhibition of aconitate hydratase (Kun et al. 1978). This is due to the formation of thiol-ester bond with the sulfhydryl groups of two enzymes in the mitochondrial membrane which function in citrate transport.

Detoxification has been demonstrated in vivo in both adapted and nonadapted mammals (King et al. 1978, Oliver et al. 1979). However, recent research indicates that genetic tolerance to FA cannot be attributed to differing abilities to defluorinate FA. Differences in F^- accumulation patterns appear to reflect differences in metabolic rate between eutherians and marsupials and between large and small animals, rather than any differences in hepatic defluorinating abilities (King et al. 1981, Mead et al. 1985). It therefore appears that defluorination is a ubiquitous low-level detoxification mechanism by the liver, which is not a major means of circumventing FA toxicity in vivo in resistant mammals (Mead et al. 1979). However, no research has focused on ruminal detoxification of fluoroacetate. It is interesting to note that browsers such as eland (*Taurotragus oryx*) and kudu (*Tragelaphus strepsiceros*) appear to be more resistant to fluoroacetate poisoning than grazers such as springbok (*Antidorcas marsupialis*), and resistance may have developed in mammals with foregut fermentation that have had long exposure to these plants. It is also possible that livestock and wild animals adapt by avoiding ingestion.

Despite the great stability of the carbon-fluorine bond, enzymatic cleavage had been documented. Studies with a partially purified enzyme from a soil pseudomonad indicated degradation of FA as follows:

$$F\,CH_2COO^- + H_2O \rightarrow HOCH_2COO^- + HF$$

The enzyme was not general for cleavage of C-F bonds but specific for the dehalogenation of monohalogenated acetic acids (Goldman 1965, 1969). On the basis of this specificity, the enzyme was named haloacetate halidohydrolase. Cleavage of the C-F bond was shown to be faster than that of the other carbon halogen bonds; in addition, the K_m for chloroacetate (2×10^{-2} M) was 10-fold higher than for the FA. Two further halidohydrolases have also been isolated (Goldman et al. 1968) which catalyze the following reaction:

$$\text{L-RCHXCOO}^- + H_2O \rightarrow \text{D-RCHOHCOO}^- + HX$$

where X = Cl or I and R = H, CH_3, or CH_3CH_2. Although the organism could be grown aerobically, none of the reactions was found to require molecular oxygen.

There is also considerable research interest in the anaerobic reductive dehalogenation of haloaromatic compounds from anoxic ecosystems such as sediments and sewage sludge. The conditions required for degradative activity are obligate anaerobiosis, a microbial consortium (often methanogenic), and long residence times. Research to isolate anaerobic bacteria from the rumen of resistant animals capable of detoxifying fluoracetate has been unsuccessful thus far. Recently a haloacetate halidohydrolase gene from *Moraxella* species was used to construct a fluoroacetate dehalogenase expression vector (pBHf) in *Butyrivibrio fibrisolvens* (Gregg et al. 1994). Although none of the bacteria (*E. coli* or *B. fibrisolvens*) hosting the cloned gene approached the same levels of expression measured in *Moraxella* species, fluoroacetate (10 mM) added to the culture medium was defluorinated with a specific activity of 10 nmol/min/mg soluble cell protein. Stable expression of dehalogenase activity in *B. fibrisolvens* demonstrates the possibility of using genetically modified rumen bacteria to detoxify feedstuffs in ruminant animals (Gregg and Sharpe 1991).

4.7. Pyrrolizidine Alkaloids

Pyrrolizidine alkaloids (PAs) are widely distributed in the plant families Compositae, Leguminosae, and Boraginaceae, and the genera that cause toxicity in livestock include *Senecio*, *Crotalaria*, *Heliotropium*, *Amsinckia*, and *Echium*. These alkaloids are reactive basic substances which contain N in a characteristic bicyclic heterocyclic ring system. Common structural features of hepatotoxic PA are a 1,2 double bond in the B ring and esterification of the CH_2OH group to a carbon side chain (Fig. 15.6). The compounds require metabolism to pyrroles (dihydropyrrolizine derivatives) by hepatic mixed function oxidases before toxicity occurs, and the 1,2 unsaturated bouble bond facilitates hepatic metabolism to a pyrrole derivative (Culvenor et al. 1976). Pyrroles are highly reactive alkylating agents that bind with vital cellular components including DNA.

Susceptibility to PA toxicity and the syndromes produced are quite variable

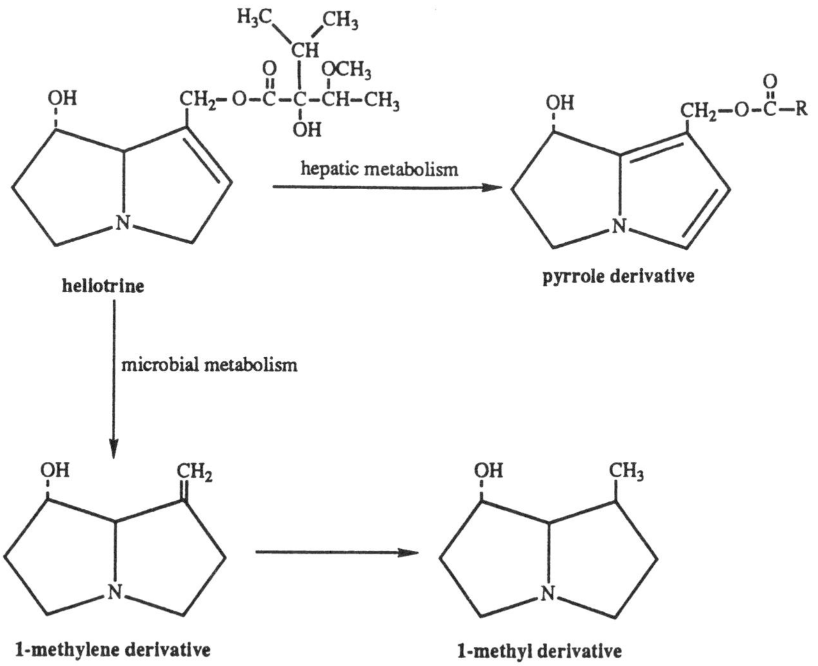

Figure 15.6. Ruminal metabolism of the pyrrolizidine alkaloid heliotrine.

among domestic livestock species (Hooper 1978). Of the foregut fermenters, cattle are highly susceptible while sheep and goats appear to be quite tolerant (Sharrow and Mosher 1982, Cheeke 1988). Resistance in sheep may be due to their ability to detoxify PA in the rumen coupled with hepatic detoxification of pyrroles. A ruminal bacterium *Peptococcus heliotrinereducens* has been isolated from sheep which metabolizes the PA, heliotrine, to nontoxic 7α-hydroxy-1α-methylene-8α-pyrrolizidine and 1-methylpyrrolizidine derivatives by reduction of the 1,2 double bond in the heterocyclic ring and cleavage of the ester-linked carbon side chain (Russell and Smith 1968, Lanigan 1976). Aliphatic and aryl esterase activity of microorganisms in the rumen is high and probably readily hydrolyzes the open esters of heliotrine, but other PAs such as retronecine, which have closed esters, would be more resistant to cleavage (Shull et al. 1976, Swick et al. 1983). Inhibition of methanogens in the rumen increases the rate of ruminal metabolism of heliotrine to nontoxic metabolites (Lanigan 1971, 1972). The response has been explained in terms of an increase supply of hydrogen for reduction of the PA; however, it is likely that in the absence of methanogens and increased partial pressure of H_2, rumen organisms are forced to reduce other substrates to dispose of electrons rather than producing hydrogen.

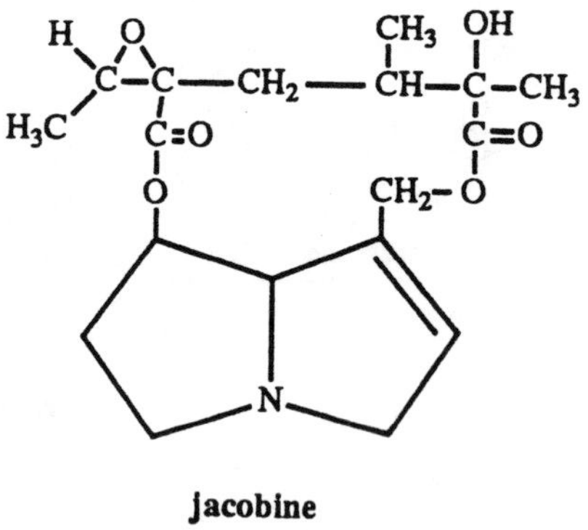

Figure 15.7. Structure of the pyrrolizioline alkaloid jacobine.

Biotransformation of tansy ragwort (*Senecio jacobaea*) PA, jacobine (Fig. 15.7), in the sheep rumen provides further evidence of the ability of sheep to detoxify PA by ruminal metabolism (Craig et al. 1992). Rate of transformation of these PA was approximately 10 times faster (19 to 26 μg/mL/h) in the rumen contents of sheep and goat compared with cattle (Wachenheim et al. 1992). The ruminal bacteria responsible for metabolism of jacobine have not been identified, nor have the transformation products been characterized, although ester hydrolysis of the dicarboxylic side chain and reduction of the double bond of jacobine are likely. Transformations of this type could reduce or eliminate the toxicity of the PA (Mattocks 1968, 1981). However, the importance of ruminal versus hepatic metabolism as the mechanism of resistance to PA toxicity is a contentious issue. Cheeke (1994) suggests that the evidence presented to support ruminal detoxification as the basis for resistance of sheep to PA toxicity is inconclusive and that differences in hepatic metabolism are the likely cause of protection.

4.8. Phenolics

Phenolics are widely distributed in the plant kingdom and are often present in the diet of herbivores. They occur as polymeric and simple compounds such as tannins, lignin precursors, flavonoids, and cell wall phenolic acids. Most of these compounds have no adverse affects on the animal, but toxicity and antinutritional affects have been attributed to tannins, and phenolic acids may be inhibitory to some ruminal organisms.

The two major groups of soluble polymeric phenolics are hydrolyzable and condensed tannins. Hydrolysable tannins (MW 500 to 3,000) are complex esters consisting of a core of glucose esterified with various combinations of mainly gallic (gallotannin), and ellagic acid (ellagitannin), which are trihydroxyphenolics

(Figs. 15.8, 15.9). Condensed tannins (MW 1,000 to 20,000) which are also referred to as procyanidins, are polymers of flavan-3-ols or flavan-3,4-diols such as catechin and epicatechin (Fig. 15.10).

HYDROLYSABLE TANNINS

Plants that are considered to be toxic to ruminants because of their hydrolyzable tannin content include *Terminalia*, *Acacia*, *Quercus*, and *Ventilago* species. Studies of the degradation and metabolism of these compounds in foregut fermenters have been confined mainly to the hydrolysable tannins from *Terminalia oblongata*. The main toxic principle in *Terminalia oblongata* appears to be punicalagin, an ellagitannin that is hydrolyzed to another toxic tannin, terminalin (Figs. 15.9, 15.11) (Doig et al. 1990). Terminalin could be produced from punicalagin by esterase activity of ruminal microorganisms or by acid hydrolysis in the abomasum.

The hydrolyzable tannin (HT), tannic acid which was prepared from *Rhus cotinus* or *Rhus coriaria* has been used as a model for HT toxicity in sheep (Murdiati et al. 1992). There are apparent similarities between tannic acid and HT from *T. oblongata*, although this can depend on the species of origin of the tannic acid (Zhu 1993). Murdiati et al. (1992) found that the major phenolic metabolites in the urine of sheep fed either tannic acid or *T. oblongata* were glucuronide conjugates of resorcinol and 2-carboxy-2′4′4,6,-tetrahydroxy diphenyl 2,2′-lactone with traces of unconjugated pyrogallol, resorcinol and phloroglucinol. Other urinary metabolites from tannic acid including gallic acid and the 4-O-methyl derivative have also been identified (Zhu 1993). Formation of the diphenyl lactone metabolite is in accord with known pathways for the anaerobic microbial degradation of gallic acid acting instead on hexahydroxydiphenic acid which had been liberated from the HT by ester hydrolysis in the rumen. The steps involved are partial decarboxylation, reductive dehydroxylation at the 4 position, and lactonization of the remaining carboxyl. Hexahydroxydiphenic acid is one of the principal acids found in HT along with gallic acid and *m*-digallic acid (3-galloylgallic acid). As a free acid hexahydroxydiphenic acid spontaneously cyclizes to the dilactone, ellagic acid which is stable and highly insoluble. On hydrolysis, punicalagin is converted to a phenolic product that consists of two gallic acid molecules carbon linked to an ellagic acid. Further metabolism of the compound in the rumen may lead to the dilactone product found in the urine, but this has not been confirmed.

Simple phenolic metabolites from HT including gallic acid, ellagic acid, pyrogallol, and resorcinol have also been identified in digesta of sheep fed tannic acid and *T. oblongata* (Murdiati et al. 1992, Zhu 1993). These metabolites are not particularly toxic unless fed in high concentration and some also undergo further metabolism in the rumen. Krumholz and Bryant (1986) have shown that gallic

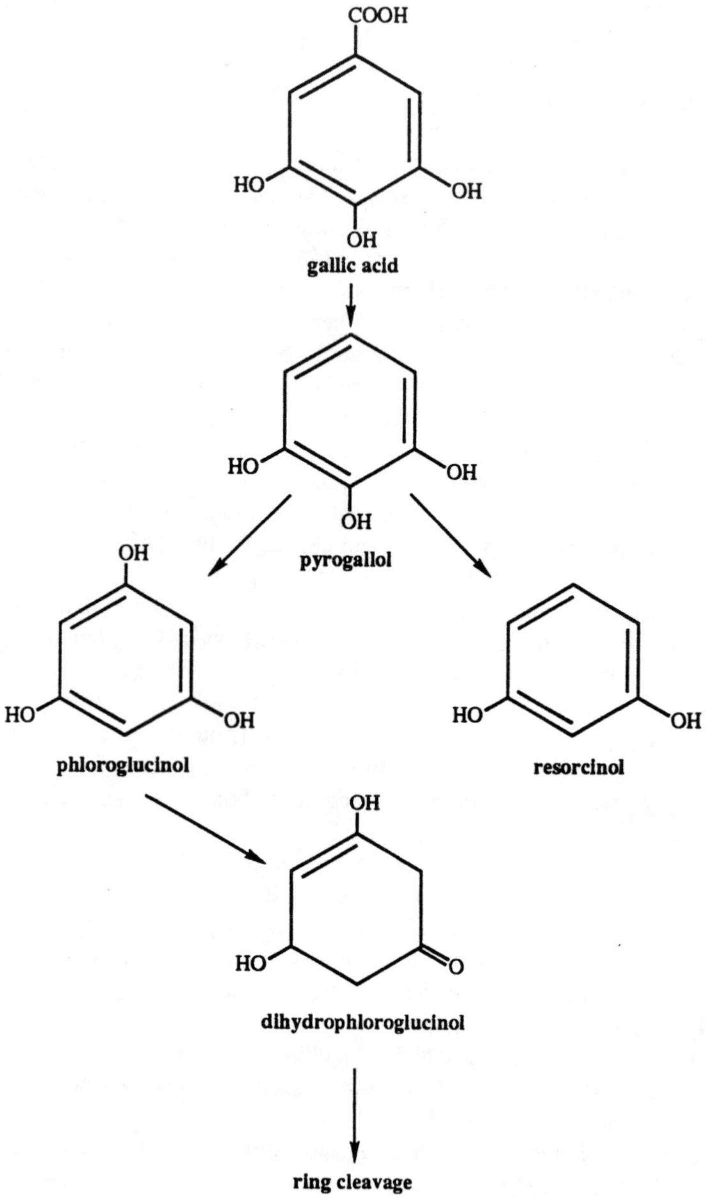

Figure 15.8. Ruminal metabolism of trihydroxybenzenoids.

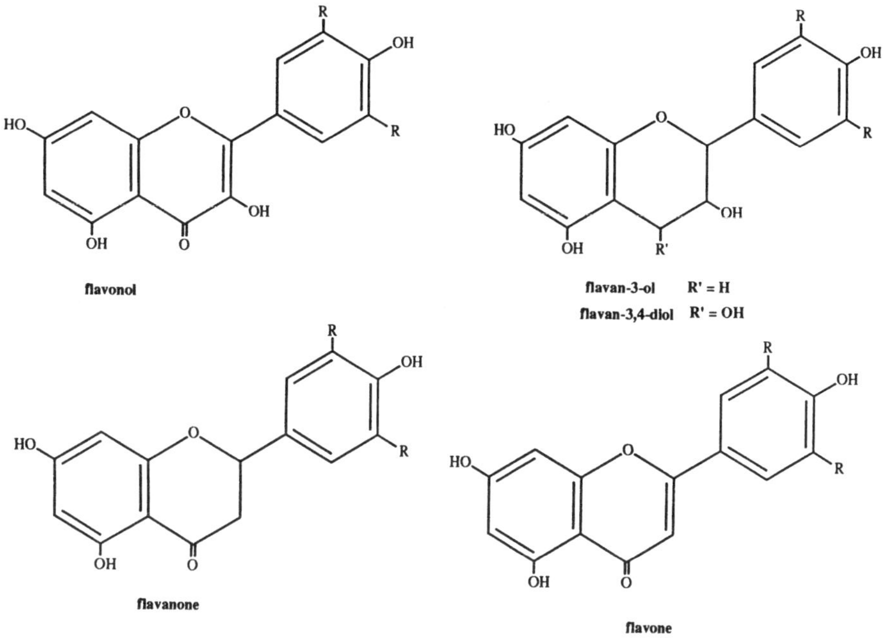

Figure 15.9. Structure of plant flavonoids and flavonols.

acid is decarboxylated to pyrogallol and converted to phloroglucinol before degradation of the phenolic ring to acetate and butyrate (Fig. 15.8) by the gram-positive ruminal bacterium *Eubacterium oxidoreducens*. This organism also degrades the flavonol quercetin to dihydroxyphenylacetate and butyrate. Hydrogen and formate were required as electron donors for these catabolic reactions. Ruminal bacteria, *Streptococcus bovis* and *Coprococcus* sp., are able to degrade phloroglucinol to acetate and carbon dioxide (Tsai and Jones 1975). Unidentified ruminal bacteria also produce resorcinol from gallic acid (Krumholz and Bryant 1986). It has been postulated that rumen, cecal, and large-bowel microbial ecosystems are unable to degrade monobenzenoid compounds having two or fewer hydroxyl groups such as resorcinol (Krumholz and Bryant 1986). Furthermore, the studies of Murdiati et al. (1992) inferred that the trihydroxybenzenoid gallic acid can be completely degraded in the rumen, but when fermented in the hindgut, the main end product is resorcinol. The initial rate of degradation of the phenolic components of HT in the rumen is slow but gradually increases as the ruminal microorganisms adapt during the first 1 to 2 weeks of exposure (Murdiati et al. 1992).

FLAVONOIDS AND CONDENSED TANNINS

Bacteria other than *Eubacterium oxidoreducens*, which degrade plant flavonoid glycosides such as rutin, quercitrin, and naringin, have also been isolated from

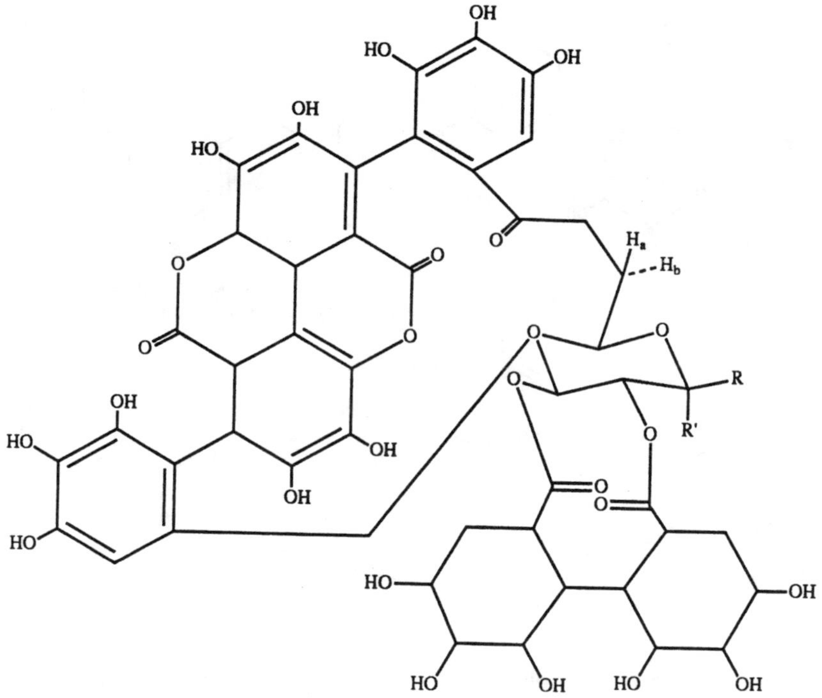

punicalagin

Figure 15.10. Structure of the hydrolyzable tannin punicalagin.

the rumen (Simpson et al. 1969, Cheng et al. 1969). Various strains of *Selenomonas* and *Butyrivibrio* were able to hydrolyze the glycoside and ferment the sugar but were unable to degrade the heterocyclic ring (Fig. 15.10), while another *Butyrivibrio* strain and a *Peptostreptococcus* sp. fermented both the sugar and heterocyclic ring. Cleavage of the heterocyclic ring occurred with the flavonol glycosides but not with the corresponding aglycone. However, mixed ruminal microorganisms were able to degrade two aglycones—quercetin, which is a flavonol and the flavanone hesperetin (Simpson et al. 1969). The products of flavonoid degradation in the rumen include acetate, butyrate, dihydroxyphenolic compounds, phloroglucinol, and possibly monohydroxyphenolics (Simpson et al. 1969, Krumholz and Bryant 1986). Phloroglucinol is probably released from the A ring of the flavonoid while the di- and monohydroxyphenolics would derive from the B ring. Cleavage of the heterocyclic ring of flavonoid glycosides seems

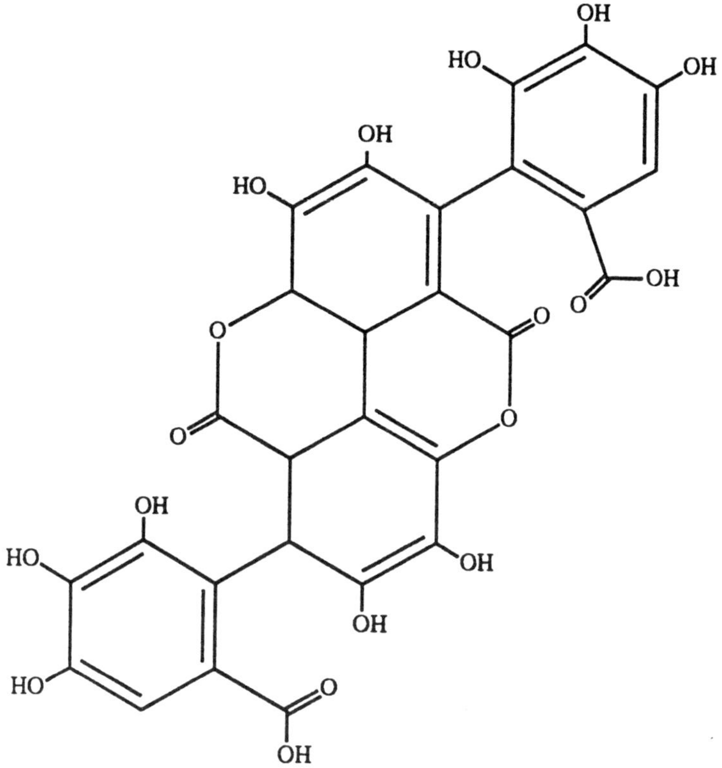

Figure 15.11. Hydrolysis product of punicalagin.

to occur when the sugar moiety is attached at carbon 3 or carbon 7, but the presence of a methoxy group in the 4 position on the B ring may prevent degradation. Although the flavonoid ring systems of the common plant flavonoids are readily degraded in the rumen, there is no evidence of cleavage of the heterocyclic ring system of flavan-3-ols and flavan-3,4-diols (Fig. 15.10) that are the subunits of condensed tannins. The ability to degrade condensed tannins would be beneficial in circumstances where these tannins have antinutritive properties such as complexing with plant protein, digestive enzymes and endogenous protein, and inhibiting microbial activity (Mangan 1988).

CELL WALL PHENOLICS

Phenolic monomers and dimers of hydroxybenzoic and hydroxycinnamic acids are a characteristic component of the cell walls of gramineaceous plants (Lowry

1990, 1993). Structurally, the polysaccharide and lignin in the cell wall are linked by hydroxycinnamic bridges that are attached through their carboxyl or phenolic groups to form ester and ether bonds, respectively (Iiyama et al. 1990; Kondo et al. 1990; Lam et al. 1990, 1992). Hydroxycinnamic acids are also esterified to hemicellulose or etherified to lignin without forming bridges (Helm and Ralph 1992, Susmel and Stefanon 1993).

Recent studies have shown that the rumen microbial population is well endowed with feruloyl and *p*-coumaryl esterase activities which release hydroxycinnamic acids from arabinoxylan (Akin and Benner 1988; Borneman et al. 1990, 1991, 1992). While the released phenolic acids are not generally regarded as toxic, they may be inhibitory when present at high concentration in the microenvironment of ruminal bacteria (Jung and Deetz 1993). The steps involved in the ruminal metabolism and excretion of cell wall phenolics are not well understood, but it has been proposed that phenolic cinnamic acids are hydrogenated in the side chain, demethylated, and dehydroxylated to 3-phenylpropionic acid in the rumen (Martin 1982a,b). We have observed that several ruminal microorganisms rapidly transform ferulic and *p*-coumaric acid by hydrogenation of the unsaturated carbon side chain, which explains why the unmodified acids were not observed in appreciable quantities in the rumen by Jung and co-workers (1983a,b). Phenylpropionic acid is absorbed from the gut and metabolized to hippuric acid by the liver (Cremin et al. 1995). Metabolic activity of the liver does not appear to be affected by the ruminal metabolites of cell wall phenolic acids at concentrations normally found in the diet of ruminants (Cremin et al. 1995). However, metabolism and excretion of these compounds by the liver as glycine conjugates may impose a nitrogen demand on the animal (Lowry et al. 1993).

4.9. Mycotoxins

Mycotoxins are secondary metabolites produced by fungi and normally have potent biological effects. Mycotoxins cause a broad spectrum of effects, both acute and chronic, and these are usually exerted on specific organs and tissues at the subcellular level resulting in a variety of biochemical changes which may or may not be accompanied by pathological changes (Shull and Cheeke 1983). It is important to note that many mycotoxins are produced by ubiquitous fungi which contaminate or invade common feeds such as maize, peanuts, rice, and other feeds utilized directly by livestock and humans. The importance of these mycotoxins cannot be overemphasized, especially in animal feedstuffs where a large proportion of an animal's diet in many parts of the world is comprised of products that are naturally infected with Fusaria and related fungal species.

Ruminants are generally less susceptible to mycotoxin poisoning than are monogastrics (Kurmanov 1977). Preparations of whole ruminal ingesta have been shown to degrade T-2 toxin (Kiessling et al. 1984), deoxynivalenol (King et al.

1984), and ochratoxin (Hult et al. 1976, Kiessling et al. 1984). Thus the rumen is probably responsible for conferring a degree of mycotoxin resistance in ruminants. Degradation of aflatoxins is negligible.

The trichothecenes are a chemically related group of biologically active metabolites of which the most important is T-2 toxin. The reactivity of trichothecenes is due to the 12,13-epoxide and 9,10-double bond in the trichothecane ring system and relative potency among the more than 40 trichothecenes is determined by the R-group substitution on the ring skeleton (Ueno 1984). A series of experiments was conducted to study the degradation of trichothecenes in the rumen. Preliminary experiments showed that T-2 toxin, HT-2 toxin, deoxynivalenol, and diacetoxyscirpenol were all degraded to varying extents by whole ovine ruminal contents. The rate of T-2 toxin degradation was 2.22 mg/L/h. From these results, it appeared that T-2 toxin was first deacetylated to HT-2 toxin which was then further deacetylated to T-2 triol. Toxin degrading activity was slightly higher in the protozoal than the bacterial fraction and absent from cell-free preparations. However, T-2 toxins cause an 83% decrease in protozoal protein over a 6-hour incubation period whereas bacterial protein increased by 14% over the same period (Westlake et al. 1989). This effect on eukaryotic cells is due to binding to ribosomes and inhibition of protein synthesis (Ueno 1984), whereas trichothecenes have little effect on bacteria.

Since the detoxification of T-2 toxin involved deacetylation, further experiments were carried out with pure cultures of rumen bacteria known to have esterase activity (Westlake et al. 1987a). T-2 toxin (10 μg/mL) had no inhibitory effect on lag time or growth rate of pure cultures of *Butyrivibrio fibrisolvens*, *Selenomonas ruminatium*, and *Anaerovibrio lipolytica*. Even at a concentration of 1 mg/mL there was no effect on growth or lag time of *B. fibrisolvens*. This organism was shown to hydrolyze T-2 toxin to HT-2 toxin, T-2 triol, and neosolaniol whereas the other two organisms produced only H-2 toxin and T-2 triol. This suggested enzymatic hydrolysis at C-3 and C-4 to yield HT-2 toxin and T-2 triol, while a single hydrolytic cleavage at C-8 would lead to production of neosolaniol (Westlake et al. 1987a). A membrane preparation isolated from *B. fibrisolvens* was able to degrade T-2 toxin to HT-2 toxin and T-2 triol. This protein fraction had an approximate molecular weight of 65 kDa and showed esterase activity (396 μmoles *p*-nitrophenol formed/mg protein/min with p-nitrophenylacetate as substrate). A certain degree of enzyme-substrate specificity is evident since this esterase preparation did not produce neosolaniol, and thus two different pathways are probably involved.

Further experiments with *B. fibrisolvens* showed that although aflatoxin B_1 was not inhibitory, there was no degradation. However, there was degradation of zearalenone, verrucarin A, diacetoxyscirpenol, deoxynivalenol, and acetyl T-2 toxin (Westlake et al. 1987b). Thus this predominant species of rumen bacteria plays an important role in increasing the resistance of ruminants to trichothecene toxicity. Proposed pathways for biotransformation of deoxynivalenol, diacetoxyscirpenol, and T-2 toxin by anaerobic bacteria from the rumen of sheep and cattle are presented in Figure 15.5. These reactions include both enzymic reduction and ester hydrolysis, resulting in less toxic compounds. For example, deepoxy T-2 was reported to be at least 400-fold less toxic than T-2 toxin in a rat dermal irritation bioassay (Swanson et al. 1988). Thus, microbial detoxification of deleterious compounds would be an important mechanism for assisting the survival of the host animal.

4.10. Selenium

Selenium accumulates in a diverse group of plants and causes toxicity in animals consuming them (James et al. 1992). The element is present in these plants in several forms, including proteinaceous and nonproteinaceous selenoaminoacids and low-molecular-weight organic and inorganic compounds. The solubility of selenium compounds in the digestive tract influences the availability and potential toxicity of ingested selenium. Animals can die acutely or develop a chronic condition called alkali disease (James et al. 1992).

All animals are susceptible to selenium intoxication, but foregut fermenters can probably tolerate higher levels of selenium intake as a result of microbial metabolism in the gut. In ruminants, water-soluble forms of dietary selenium are reduced and precipitated as elemental selenium by ruminal bacteria (Butler and Peterson 1961, Peterson and Spedding 1963). This has been demonstrated in ruminal bacteria from the genera *Selenomonas*, *Wolinella*, and *Prevotella*; sulfate-reducing species of the genus *Desulfovibrio*; and facultative anaerobes of the Enterobacteriaceae family (Hudman and Glenn 1984, 1985; Tomei et al. 1992; Rasmussen and James 1994). These microbial transformations reduce the net absorption of selenium from the gut and increase the tolerance to intoxication. This may explain why uptake of the element from the gut and susceptibility to selenium toxicity were higher in lambs than in adult sheep (Cousins and Cairney 1961). Although elemental selenium is precipitated in the rumen there is evidence that selenium is also incorporated into bacteria as an organic compounds including protein (Hidiroglou et al. 1968, 1974). However, a comprehensive investigation of the role of ruminal microorganisms in preventing selenium toxicity is yet to be undertaken.

5. Ruminal Digestive Disorders

Foregut fermenters have complex stomachs that harbor a stable population of microorganisms that derive energy mainly by fermenting the soluble and structural carbohydrates of plant tissues. The forestomach in domesticated ruminants has evolved to slow the passage rate of plant material through this fermentation chamber so that the microorganisms can obtain energy by digesting the fiber. Digestive disorder usually occurs in ruminants when their diet is suddenly changed from a fibrous forage to a concentrate or when they are fed excessive amounts of a readily fermentable diet. In these circumstances, an imbalance in microbial population develops or a discrete group of organisms proliferates which produce metabolic products that have an adverse effect on both digestion and the animal. This section will describe the changes in rumen microbial metabolism that result in characteristic digestive disorders following a change to a readily fermentable diet.

5.1. Concentrate Diet-Induced Indigestion

Lactic acidosis is the most common digestive disturbance which occurs in foregut fermenters when their diet is suddenly changed from forage to concentrate or when excessive amounts of a rapidly fermented diet are eaten. Other names for the syndrome include grain engorgement, grain overload, and acute indigestion. Lactic acid accumulates in the rumen owing to an imbalance in microbial populations, an increase in the rate of fermentation in the rumen, and a concomitant drop in pH. The acidic conditions in the rumen favors the absorption of undissociated lactic acid into the blood, which creates an acid-base imbalance. Symptoms of the disorder include anorexia, rumen stasis, rumenitis, diarrhea, dehydration, laminitis, and liver abscess (Huber 1976, Brent 1976).

The numbers and types of microorganisms and the pattern of fermentation end products produced by them in the rumen and intestines of sheep and cattle following feeding with concentrate diets have received considerable attention (Allison et al. 1964, 1975; Latham et al. 1971, 1972, Mackie et al. 1978, Mackie and Gilchrist 1979). Stepwise adaptation of the ruminant to a high concentrate diet results in an increase in the protozoal population and a shift in the bacterial population to amylolytic (*Streptococcus*, *Lactobacillus*, *Peptostreptococcus*, *Ruminobacter*, *Bacteroides*, and *Butyrivibrio*) and lactate-utilizing (*Megasphaera*, *Veillonella*, *Selenomonas*, *Anaerovibrio*, *Propionibacterium*) bacteria (Caldwell and Bryant 1966, Latham et al. 1971, 1972, Mackie et al. 1978, Mackie and Gilchrist 1979). In these experiments the percentage of amylolytic and lactate-utilizing bacteria increased steadily from 1.6% and 0.2% on a low concentrate diet to 21.2% and 22.3% on a high concentrate ration. However the transition in ruminal microbial population from a roughage to concentrate diet does not occur in an orderly fashion when the dietary change is abrupt. An abundant supply of carbohydrate that is rapidly fermented results in a decrease in the acetic/propionic acid ratio, the accumulation of lactic acid, and increased concentrations of glucose in the rumen with a decline in pH below 5.4 (Mackie et al. 1978). Together, these changes in chemical composition of the rumen represent a sudden change in fermentation pattern in the rumen and an imbalance in the ruminal population. Under these conditions, the ciliate protozoa population, which has a stabilizing effect on ruminal fermentation, declines because the ruminal pH is below the growth optimum for these organisms (Purser and Moir 1959). Protozoa have an important role in regulating the production of lactic and volatile fatty acids in the rumen (Kariya et al. 1989). In particular, entodiniomorphs ingest starch and soluble sugars and store it as amylopectin, as well as ingesting bacteria, thus reducing the rate of fermentation in the rumen (Williams and Coleman 1992). Protozoa not only regulate the rate of lactic acid production in the rumen but also are involved in the metabolism of lactic acid (Newbold et al. 1987). Thus, the presence of a stable and active population of ruminal protozoa is likely to reduce the incidence of lactic acidosis.

The rate at which a concentrate diet can be introduced into the rumen without upsetting the balance of the ruminal population thus partially depends on the rate at which the protozoa can increase their numbers in response to the diet, compared with the bacteria. The minimum doubling time for protozoa is 5.5 to 7.3 hours (Warner 1962, Hungate et al. 1971, Potter and Dehority 1973). In contrast, the lactate-producing bacteria *Streptococcus bovis* can increase its numbers in the rumen 2,000-fold in 6 hours—i.e., T_d of ca. 30 min (Hungate et al. 1952) while amylolytic bacteria such as *Ruminobacter* (*Bacteroides*) *amylophilus*, *Selenomonas ruminantium*, and *Anaerovibrio lipolytica* have minimal doubling times of 1.4 to 1.5 hours (Hobson 1965, Hobson and Summers 1967). When the availability of sugar is not limiting, *S. bovis* and *S. ruminantium* grow rapidly and produce lactic acid as an end product (Scheifinger et al. 1975, Russell and Baldwin 1979). Lactate dehydrogenase is activated to produce lactic acid by the accumulation of pyruvate in *Selenomonas ruminantium* (Wallace 1978) and by the combined effects of low intracellular pH and fructose 1,6-biphosphate concentration in *S. bovis* (Russell and Hino 1985). Furthermore, *S. bovis* is more tolerant of acid conditions than most other ruminal bacteria (Therion et al. 1982) and therefore is able to establish in high numbers until pH declines further and it is replaced by lactobacilli that are even more acid resistant. Ruminal microorganisms produce both the D and L isomers of lactic acid, and combined concentrations can approach 100 mM in the rumen during an acute disturbance (Allison et al. 1975).

The other critical factor in the development of the syndrome is the rate of increase in number of acid-tolerant, lactate-utilizing bacteria that grow in response to an increase in lactic acid concentration in the rumen at low pH. On a roughage diet the number of lactate-utilizing bacteria is low and may consist of mainly acid-sensitive strains such as *Veillonella* and *Selenomonas* rather than acid-tolerant species of *Megasphaera* and *Anaerovibrio* (Mackie et al. 1978). Consequently, animals that lack or have small populations of acid-tolerant, lactate-utilizing bacteria can be susceptible to lactic acidosis if they are abruptly changed to a high concentrate diet. In these animals, the ruminal pH may fall below the tolerance levels for growth of the acid-resistant, lactate-utilizing species and thus inhibit their growth before accumulated lactate can be utilized. The main route for lactate metabolism in the rumen appears to be via the acrylate pathway (Russell and Baldwin 1978, Counotte et al. 1981, 1983, Hino and Kuroda 1993). Lactate metabolism has been studied mainly in *Megasphaera elsdenii*, which ferments lactate to propionic acid via acrylyl-CoA (acrylate pathway) without ATP synthesis (Ladd and Walker 1965). Lactate racemase is the initial enzyme in the pathway and is induced by lactate (Hino and Kuroda 1993). This organism occupies an ecological niche as a lactate utilizer in the rumen because it is not subject to catabolite repression by glucose and maltose like other lactate utilizers.

5.2. Bloat

The excessive distension of the forestomach with gas is called bloat. The main gaseous products of fermentation in the foregut of herbivores are carbon dioxide and methane, which can amount to several hundred liters of gas production per day in large ruminants (Church 1979). Normally, bubbles of gas are produced in the digesta, rise, and accumulate as a pocket of free gas in the dorsal rumen before being expelled by the process of eructation. Bloat occurs when the rate of escape of gas from the rumen liquor is delayed owing to the formation of a frothy digesta which expands as a foam and subsequently impedes the eructation of any free gas. Severe inflation of the forestomach restricts breathing, interferes with cardiovascular function, and may result in death. Characteristically, bloat occurs in ruminants either eating lush legume pastures and legume hay or fed concentrate (grain) diets.

Susceptibility to bloat is highly variable and dependent upon dietary, animal, and microbial factors. There appears to be an unidentified genetic character that affects conditions in the rumen digesta and predisposes animals to legume bloat, but the role of animal factors in feedlot bloat is not well understood. Animals that are prone to one form of bloat are not necessarily susceptible to the other form. Soluble proteins in legumes, combined with the dispersion of chloroplast-membrane fragments and the rate of digestion of the legume are major predisposing factors in the development of foam in legume bloat (Howarth et al. 1986). The process of microbial digestion of bloat-causing legumes is more rapid than with bloat-safe forages. It can be divided into four distinct events: (1) greater bacterial colonization of the leaf surface; (2) faster bacterial penetration of the epidermal layer; (3) earlier maceration of leaf tissue; and (4) faster bacterial penetration of mesophyll cell walls (Cheng et al. 1980). *Lachnospira multiparus* is a pectinolytic ruminal bacteria that may contribute to maceration of legume leaves (Cheng et al. 1979). However, the role of ruminal microorganisms in the development of pasture (legume) bloat (Reid et al. 1975) is poorly characterized in comparison with grain (feedlot) bloat (Bartley et al. 1975).

Grain bloat usually occurs in animals that have been fed a high concentrate and low roughage diet for several weeks. Changes in the number or types of predominating ruminal microorganisms have not been definitively related to the occurrence of this disorder (Bryant et al. 1961, Mackie et al. 1978). However, it has been hypothesized that the production of a dextran slime by amylolytic bacteria is a major contributing factor to the formation of stable foam in grain bloat (Bartley et al. 1975, Cheng et al. 1976). *Streptococcus bovis* has been cited as the most likely slime producer (Bailey and Oxford 1958, Gutierrez et al. 1959), but its numbers in the rumen are not closely associated with the onset of bloat (Bartley et al. 1975). Several amylolytic ruminal bacteria including *S. bovis, Selenomonas ruminantium*, *Butyrivibrio fibrisolvens*, and *Megasphaera elsdenii* have been implicated as playing a role in grain bloat by degrading salivary mucin (Mishra et al. 1967, 1968). However the contribution of these organisms to slime production was not determined in these experiments. Large numbers of Lactobacilli are present in the rumen at low pH (< 5.5) following concentrate feeding, but grain bloat is not associated with a proliferation of these organisms (Gutierrez et al. 1959). Marked variations in the number and types of protozoa occur in ruminants fed high concentrate diets but their role in the development of bloat has not been established (Eadie and Mann 1970, Mackie et al. 1978). Bloat tends to reoccur after an initial incident unless the animal is returned to a roughage diet for several days. It has been suggested that this change in feeding regimen is required; otherwise, slime adhered to the rumen wall acts as a source of slime-producing organisms that will grow rapidly when a concentrate diet is reintroduced.

While many attempts have been made to define the microbial role in bloat and identify the microorganisms involved, it is still unclear whether susceptibility or occurrence of bloat is due directly to particular ruminal organisms. Sudden changes in diet and other undetermined factors also cause imbalances in or proliferation of microbial populations that result in excessive production of particular metabolic products and/or enzymes that do not normally predominate in the rumen. The following sections describe changes in rumen microbial metabolism that contribute to the pathogenesis of characteristic disease syndromes.

5.3. Ruminal Induction of Pulmonary Edema and Emphysema

Acute bovine pulmonary edema and emphysema (ABPE; fog fever) is a naturally occurring respiratory disease of cattle which can be induced under experimental conditions in small ruminants. The disease is characterized by the rapid onset of respiratory distress after a change in feed, particularly a sudden change from dry to lush green pasture (Blake and Thomas 1971, Selman et al. 1974, Breeze et al. 1976). It is generally regarded that the pathogenesis of ABPE involves the ruminal metabolism of L-tryptophan (TRP) to indoleacetic acid (IAA)

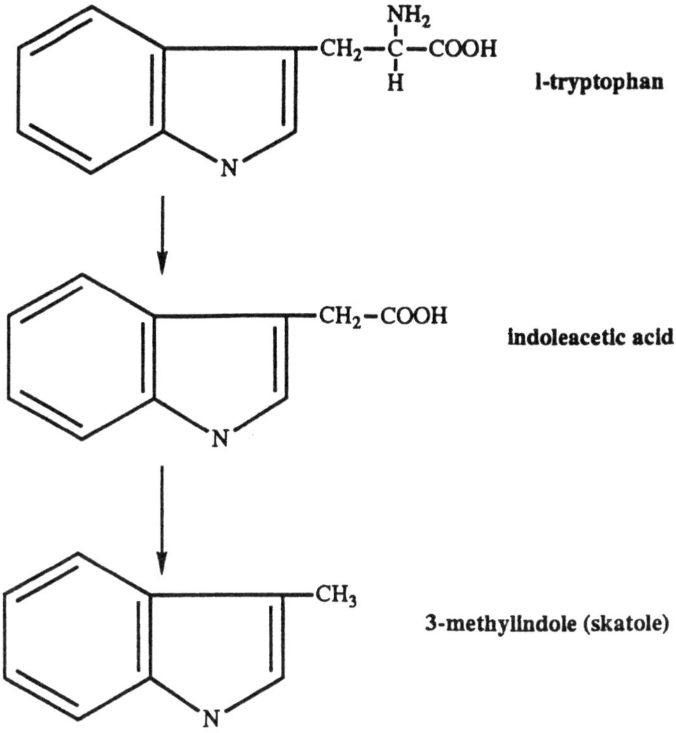

Figure 15.12. Ruminal metabolism of L-tryptophan.

and then decarboxylation to 3-methylindole (3MI, skatole) (Fig. 15.12), which causes pulmonary disease (Selman et al. 1976, Carlson and Dickinson 1978, Hammond et al. 1979). The major microbial metabolites of TRP in the gastrointestinal tract are indolepropionic acid, tryptamine, indoleacetic acid, 3MI, and indole. Metabolism of tryptophan in the gastrointestinal tract has been studied mainly in monogastric animals and humans (see Yang and Carlson 1972), while ruminal metabolism has received limited attention. Lewis and Emery (1962a,b,c) found that the rate of deamination of TRP in the rumen was slow, and the predominant end products were indole and 3MI. However, conversion of TRP to indolepropionic acid and indoleacetic acid by ruminal bacteria has been observed (Lacoste and Lemoigne 1961). Furthermore the ruminal bacterium *Ruminococcus albus* can synthesize tryptophan from IAA (Allison and Robinson 1967). Ruminal organisms responsible for the conversion of TRP to IAA have not been identified but a *Lactobacillus* species which metabolize IAA to 3MI has been isolated from the bovine rumen (Yokoyama and Carlson 1974, Yokoyama et al. 1977).

Production of 3MI by this *Lactobacillus* species appears to be inhibited by the presence of maltose, sucrose, or mannose in the growth medium. The 3MI-producing *Lactobacillus* also utilizes metabolites from other aromatic amino acids including tyrosine and phenylalanine. *p*-Hydroxyphenylacetic acid, a microbial metabolite of tyrosine, is decarboxylated to 4-methyl phenol (*p*-cresol).

Indole is usually a major metabolite of TRP irrespective of the diet of the animal, which probably reflects the widespread ability of bacterial species to produce indole from TRP (Demoss and Moser 1969, Roth et al. 1971). The main enzyme involved is tryptophanase which uses pyridoxal phosphate as a cofactor to cleave the C-C bond in TRP and yield indole, pyruvic acid, and ammonia (Newton and Snell 1964). Generally, tryptophanase is induced by TRP and repressed by glucose (Demoss and Moser 1969). Factors that influence the pathway of TRP metabolism in the rumen are relatively unidentified, and the benefits to the bacterium of 3MI production are not readily apparent. The physiological significance of catabolism of TRP to indole is obvious in that pyruvate becomes available for ATP synthesis. Skatole (3MI)-producing bacteria cannot utilize IAA as a sole energy source for growth, and the decarboxylation reaction provides little energy. However, 3MI has an inhibitory effect on some gram-negative bacteria, which may assist 3MI producing bacteria to compete in the diverse ruminal ecosystem.

The key to understanding the pathogenesis of ABPE is to determine more precisely the conditions under which ruminal fermentation is upset, thus favoring 3MI production rather than indole as the end product of TRP metabolism.

5.4. *Ruminal Thiaminase Production*

A distinct neurological syndrome known as either cerebrocortical necrosis (CCN) or polioencephalomalacia (PEM) is associated with abnormal rumen fermentation. This nervous disorder has been reported in cattle, sheep, goats, and deer. In the etiology of CCN, thiamin (vitamin B_1) deficiency plays an essential role. The condition appears to be due to thiamin-destroying enzymes; thiaminase I, which is a methyl transferase; and the hydrolyase thiaminase II (Edwin and Jackman 1973, 1982). These enzymes are produced in the rumen by different bacteria although factors influencing thiaminase production are unknown. Concise reviews of the disease and its etiology have been published (Brent 1976, Edwin and Jackman 1982, Brent and Bartley 1984, Rammell and Hill 1986).

It is generally accepted that the ruminal microbiota synthesize sufficient quantities of B-vitamins to satisfy the requirements of the host animal. However, calculations based on thiamin concentrations in gastrointestinal contents of sheep indicate that microbial thiamin synthesis may be barely adequate and sometimes inadequate to meet the needs of the host animal. Thiamin deficiency has been reported when diets consisted mainly of hay (Phillipson and Reid 1957, Buzaissy and Tribe 1960), straw (Buzaissy and Tribe 1960, Porter 1961), and maize silage (Candau and Massengo 1982). Microbial thiamin synthesis was shown to be inadequate in the rumen of sheep adapted to a protein-free diet (Naga et al. 1975, Breves et al. 1980). Results suggest that the thiamin status of feedlot cattle may be influenced by the level of protein in the diet (Grigat and Mathison 1983). Alternatively, high thiaminase activities in ruminal and intestinal fluid can lead to excessive destruction of dietary or microbially produced thiamin (Edwin and Jackman 1973, Edwin et al. 1968a,b). There is now little doubt that the disease results from a progressive thiamin deficiency induced by the action of bacterial thiaminases in the gut and rumen. However, the precise role of thiaminase has been difficult to establish, and the factors that lead to the development of the disease remain obscure.

The ''thiaminase 1 hypothesis'' has been advanced to explain the involvement of thiaminase enzymes in CCN (Brent and Bartley 1984). Thiaminase 1 catalyzes a base exchange reaction that substitutes a N-containing ring, or ''cosubstrate,'' for the thiazole ring. This not only results in a loss of enzyme activity but also creates a thiamin analog composed of the pyrimidine ring of the original thiamin and the cosubstrate. The analog can then be absorbed and inhibit thiamin requiring metabolic reactions. Since the brain derives its energy from glucose utilization via glycolysis and thiamin pyrophosphate (TPP) is an essential cofactor in decarboxylation, the CNS might be expected to show symptoms before other major organ systems (Brent and Bartley 1984).

Clostridium sporogenes and *Bacillus thiaminolyticus* have both been isolated from the rumen of animals with CCN (Boyd and Walton 1977, Edwin et al. 1979). However, these bacteria are not part of the autochthonous microbiota of ruminants, and attempts to establish them in the rumen have failed. This fact, together with differences in biochemical activity between these bacteria and thiaminases isolated from CCN cases, makes it likely that the source of thiaminase 1 in CCN has not yet been found. The ingestion of bracken or nardoo can also result in thiamin deficiency although this must be regarded differently since the thiaminase is exogenous instead of endogenous (Edwin et al. 1979). CCN is frequently associated with high-energy diets, but no dietary regimen appears to be completely free from periodic episodes of thiaminase excretion. Dietary changes followed by upsets in delicate balance of microbial types favoring the proliferation of undesirable thiaminase-producing bacteria might lead to the disease (Brent and Bartley 1984). Animals that were changed very rapidly to high concentrate diets developed high levels of ruminal thiaminase (Sapienza and Brent 1974). Thus lactic acidosis seems to establish ruminal conditions conducive to development of CCN. However, many questions remain unanswered. For example, is acidosis a cause or an effect of thiaminase activity? Why are thiaminase bacteria selected and encouraged preferentially, or are thiaminase production and activity induced or enhanced in the autochthonous microbiota? It is also worth noting that rumen motility and mixing of rumen contents was markedly decreased when thiamin deficiency symptoms developed in sheep (Naga et al. 1975). This may explain the frequent association between free-gas bloat (i.e., gas released through a stomach tube) and animals fed feedlot diets (Howarth 1975).

One factor that has gained prominence in describing the pathogenesis of PEM is dietary sulfur. Inorganic sulfate ingestion at a concentration of 2% to 3% (0.66% to 0.99% S) in order to limit feed intake resulted in a high incidence of PEM in cattle (Raisbeck 1982). A further study (Gooneratne 1989) showed that high dietary sulfur led to depletion of blood thiamin and Cu and the occurrence of PEM. James et al. (1992) have also speculated that the consumption of high-sulfate waters may be the cause of "blind staggers" in rangeland cattle. Although it is dangerous to postulate a causal relationship, it is possible that SO_4 or a ruminal metabolic product of SO_4 (SO_3^{2-}, S^{2-}) results in ruminal thiamin degradation by cleavage of the methylene bridge, or SO_4 inhibits biosynthesis of thiamin and/or alters the balance of ruminal microbiota resulting in reduced thiamin biosynthesis. Sagar and co-workers (1990) have shown that in calves with dietary-induced PEM, elevated rumen sulfide concentrations were not associated with depleted rumen or plasma thiamin concentrations. Thus it is possible that this type of PEM is a form of sulfide toxicity since H_2S is an inhibitor of cellular respiration (Beauchamp et al. 1984).

6. Conclusion

The gastrointestinal microbiota can be considered as the most metabolically adaptable and rapidly renewable organ of the body which plays a vital role in the normal nutritional and physiological functions of the host animal. We need to increase our understanding of the basic mechanisms involved so that the ruminal microbiota can be successfully manipulated and fully exploited to the ultimate benefit not only of the ruminant animal but also of mankind. This requires a sustained research effort using novel techniques and approaches to elucidate the mechanisms involved. The areas of research covered in this chapter represent some of the advances in ruminal detoxification of phytotoxins and mycotoxins and digestive disorders that have an enormous potential impact on nutrient productivity and performance.

References

Adler JH, Weitzkin-Neiman G (1970) Demethylation of methyl-O-coumestrols by rumen microorganisms in vitro. Refu Vet 27:51–55.

Akin DE, Benner R (1988) Degradation of polysaccharides and lignin by ruminal bacteria and fungi. Appl Environ Microbiol 54:1117–1125.

Allison MJ, Reddy CA (1984) Adaptations of gastrointestinal bacteria in response to changes in dietary oxalate and nitrate. In: Klug MJ, Reddy CA, eds. Current Perspectives on Microbial Ecology, pp. 248–256. Washington, DC: American Society for Microbiology.

Allison MJ, Robinson IM (1967) Tryptophan biosynthesis from indole-3-acetic acid by anaerobic bacteria from the rumen. Biochem J 102:36–37P.

Allison MJ, Bucklin JA, Dougherty RW (1964) Ruminal changes after overfeeding with wheat and the effect of intraruminal inoculation on adaptation to a ration containing wheat. J Anim Sci 23:1164–1171.

Allison MJ, Robinson IM, Dougherty RW, Bucklin JA (1975) Grain overload in cattle and sheep: changes in microbial populations in the cecum and rumen. Am J Vet Res 36:181–185.

Allison MJ, Cook HM, Dawson KA (1981) Selection of oxalate-degrading rumen bacteria in continuous cultures. J Anim Sci 53:810–816.

Allison MJ, Mayberry WR, McSweeney CS, Stahl DA (1992) *Synergistes jonesii*, gen. nov., sp. nov.: A rumen bacterium that degrades toxic pyridinediols. Syst Appl Microbiol 15:522–529.

Allison MJ, Dawson KA, Mayberry WR, Foss JG (1985) *Oxalobacter formigenes* gen. nov., sp. nov.: oxalate-degrading anaerobes that inhabit the gastrointestinal tract. Arch Microbiol 141:1–7.

Anantharam V, Allison MJ, Maloney PC (1989) Oxalate:formate exchange: the basis for energy coupling in *Oxalobacter*. J Biol Chem 264:7244–7250.

Anderson RC, Rasmussen MA, Allison MJ (1993) Metabolism of the plant toxins nitropropionic acid and nitropropanol by ruminal microorganisms. Appl Environ Microbiol 59: 3056–3061.

Angermaier L, Simon H (1983) On the reduction of aliphatic and aromatic nitro compounds by Clostridia, the role of ferredoxin and its stabilization. Hoppe-Seyler's Z. Physiol Chem 364:961–975.

Baetz AL, Allison MJ (1989) Purification and characterization of oxalyl-coenzyme A decarboxylase from *Oxalobacter formigenes*. J Bacteriol 171:2605–2608.

Baetz AL, Allison MJ (1990) Purification and characterization of formyl-coenzyme A transferase from *Oxalobacter formigenes*. J Bacteriol 172:3537–3540.

Baetz AL, Allison MJ (1992) Localization of oxalyl-coenzyme A decarboxylase, and formyl-coenzyme A transferase in *Oxalobacter formigenes* cells. Syst Appl Microbiol 15:167–171.

Bailey RW, Oxford AE (1958) A quantitative study of the production of dextran from sucrose by rumen strains of *Streptococcus bovis*. J Gen Microbiol 19:130–145.

Bartley EE, Meyer RM, Fina LR (1975) Feedlot or grain bloat. In: McDonald IW, Warner ACI, eds. Digestion and Metabolism in the Ruminant, pp. 551–562. Australia: University of New England Publishing Unit.

Batterham TJ, Hart, NK, Lamberton JA, Braden AWH (1965) Metabolism of oestrogenic isoflavones in sheep. Nature (Lond) 206:509.

Beacham IR (1987) Silent genes in prokaryotes. FEMS Microbiol Rev 46:409–417.

Beauchamp RO, Bus JS, Popp JA, Borreiko CJ, Andjelkovich DA (1984) A critical review of the literature on hydrogen sulfide toxicity. CRC Crit Rev Toxicol 13:25–97.

Bell EA (1973) Aminonitriles and amino acids not derived from proteins. In: Toxicants Occurring Naturally in Foods, pp. 153–169. Washington, DC: National Research Council, National Academy of Sciences.

Berenbaum M, Feeny P (1981) Toxicity of angular furanocoumarins to swallow tail butterflies: escalation in a coevolutionary arms race. Science 212:927–929.

Blake JT, Thomas DW (1971) Acute bovine pulmonary emphysema in Utah. J Am Vet Med Assoc 158:2047–2052.

Borneman WS, Hartley RD, Morrison WH, Akin DE, Ljungdahl, LG (1990) Feruloyl and p-coumaroyl esterase from anaerobic fungi in relation to plant cell wall degradation. Appl Microbiol Biotechnol 33:345–351.

Borneman WS, Ljungdahl LG, Hartley RD, Akin DE (1991) Isolation and characterization of p-coumaroyl esterase from the anaerobic fungus *Neocallimastix* strain MC-2. Appl Environ Microbiol 57:2337–2344.

Borneman WS, Ljungdahl LG, Hartley RD, Akin DE (1992) Purification and partial characterization of two feruloyl esterases from the anaerobic fungus *Neocallimastix* strain MC-2. Appl Environ Microbiol 58:3762–3766.

Boyd JW, Walton JR (1977) Cerebrocortical necrosis in ruminants: an attempt to identify the source of thiaminase in affected animals. J Comp Pathol 87:581–589.

Braden AWH, Hart NK, Lamberton JA (1967) The oestrogenic activity and metabolism of certain isoflavones in sheep. Aust J Agric Res 18:335–348.

Brent BE (1976) Relationship of acidosis to other feedlot ailments. J Anim Sci 43:930–935.

Brent BE, Bartley EE (1984) Thiamin and niacin in the rumen. J Anim Sci 59:813–822.

Breeze RG, Pirie HM, Selman IE, Wiseman A (1976) Fog fever (acute pulmonary emphysema) in cattle in Britain. Vet Bull 46:243–251.

Breves G, Hoeller H, Harmeyer J, Martens H (1980) Thiamine balance in the gastrointestinal tract of sheep. J Anim Sci 51:1177–1181.

Bryant MP, Robinson IM, Lindahl IL (1961) A note on the flora and fauna in the rumen of steers fed a feedlot bloat-provoking ration and the effect of penicillin. Appl Microbiol 9:511–515.

Butler GW, Petersen PJ (1961) Aspects of the faecal excretion of selenium by sheep. NZ J Agric Res 4:484–491.

Buzaissy C, Tribe DE (1960) The synthesis of vitamins in the rumen of sheep. 1. The effect of diet on the synthesis of thiamin, riboflavin and nicotinic acid. Aust J Agric Res 11:989–1001.

Caldwell DR, Bryant MP (1966) Medium without rumen fluid for nonselective enumeration and isolation of rumen bacteria. Appl Microbiol 14:794–801.

Candau M, Massengo J (1982) Evidence of a thiamine deficiency in sheep fed maize silage. Ann Rech Vet 13:329–340.

Carlson JR, Dickinson EO (1978) Tryptophan-induced pulmonary edema and emphysema in ruminants. In: Keeler RF, Van Kampen KR, eds. Effect of Poisonous Plants on Livestock, pp. 251–259. New York: Academic Press.

Carlson JR, Breeze RG (1984) Ruminal metabolism of plant toxins with emphasis on indolic compounds. J Anim Sci 58:1040.

Cheeke PR, Shull LR (1985) Other plant toxins and poisonous plants-Fluoroacetate (1080). In: Natural Toxicants in Feeds and Poisonous Plants, pp. 375–377. Westport, Conn: AVI.

Cheeke PR (1988) Toxicity and metabolism of pyrrolizidine alkaloids. J Anim Sci 66: 343–2350.

Cheeke PR (1994) A review of the functional and evolutionary roles of the liver in the detoxification of poisonous plants, with special reference to pyrrolizidine alkaloids. Vet Hum Toxicol 36:240–247.

Cheng KJ, Jones GA, Simpson FJ, Bryant MP (1969) Isolation and identification of rumen bacteria capable of anaerobic rutin degradation. Can J Bacteriol 15:1365–1371.

Cheng KJ, Hironaka R, Jones GA, Nicas T, Costerton JW (1976) Frothy feedlot bloat in cattle: production of extracellular polysaccharides and development of viscosity in cultures of *Streptococcus bovis*. Can J Microbiol 22:450–459.

Cheng KJ, Dinsdale D, Stewart CS (1979) Maceration of clover and grass leaves by *Lachnospira multiparus*. Appl Environ Microbiol 38:723–729.

Cheng KJ, Fay JP, Howarth RE, Costerton JW (1980) Sequence of events in the digestion of fresh legume leaves by rumen bacteria. Appl Environ Microbiol 40:613–625.

Church DC (1979) Rumen fermentation of natural feedstuffs. In: Digestive Physiology and Nutrition of Ruminants, Vol. 1, p. 295. Portland: Oxford Press.

Clark PH (1984) The evolution of degradative pathways. In: Gibson DT, ed. Microbiological Degradation of Organic Compounds, pp. 11–27. New York: Marcel Dekker.

Counotte GHM, Prins RA, Jansen RHAM, de Bie MJA (1981) Role of *Megasphaera elsdenii* in the fermentation of DL-[2-^{13}C] lactate in the rumen of dairy cattle. Appl Environ Microbiol 42:649–655.

Counotte GHM, Lankhorst A, Prins RA (1983) Role of DL-Lactic acid as an intermediate in rumen metabolism of dairy cows. J Anim Sci 56:1222–1235.

Cousins FB, Cairney IM (1961) Some aspects of selenium metabolism in sheep. Aust J Agric Res 12:927–942.

Cox RI, Braden AWH (1974) A new phyto-oestrogen metabolite in sheep. J Reprod Fertil 36:490–493.

Craig MA, Latham CJ, Blythe LL, Schmotzer WB, O'Connor OA (1992) Metabolism of toxic pyrrolizidine alkaloids from tansy ragwort (*Senecio jacobaea*) in ovine rumen fluid under anaerobic conditions. Appl Environ Microbiol 58:2730–2736.

Cremin JD Jr, McLeod KR, Harmon DL, Goetsch AL, Bourquin LD, Fahey GC Jr (1995) Portal and hepatic fluxes in sheep and concentrations in cattle ruminal fluid of 3-(4-

hydroxyphenyl) propionic, benzoic, 3-phenylpropionic, and trans-cinnamic acids. J Anim Sci. 73:1766–1775. Submitted.

Culvenor CCJ, Edgar JA, Jago MV, Outleridge A, Peterson JE, Smith LW (1976) Hepato- and pneumotoxicity of pyrrolizidine alkaloids and derivatives in relation to molecular structure. Chem-Biol Interact 12:299–324.

Daniel SL, Drake HL (1993) Oxalate and glyoxalate-dependent growth and acetogenesis by *Clostridium thermoaceticum*. Appl Environ Microbiol 59:3062–3069.

Davies HL (1987) Limitations to livestock production associated with phytoestrogens and bloat. In: Wheeler JL, Pearson CJ, Robards GE, eds. Temperate pastures: their production, use and management. pp. 446–456. Sydney, Australia: Australian Wool Corporation/CSIRO.

Dawson KA, Allison MJ (1988) Digestive disorders and nutritional toxicity. In: Hobson PN, ed. The Rumen Microbial Ecosystem, pp. 445–459. London: Elsevier.

Dawson KA, Allison MJ, Hartman PA (1980a) Isolation and some characteristics of anaerobic oxalate-degrading bacteria from the rumen. Appl Environ Microbiol 40:833–839.

Dawson KA, Allison MJ, Hartman PA (1980b) Characteristics of anaerobic oxalate-degrading enrichment cultures from the rumen. Appl Environ Microbiol 40:840–846.

Demoss RD, Moser K (1969) Tryptophanase in diverse bacterial species. J Bacteriol 98: 167–171.

De Vries W, van Wijck-Kapteyn WMC, Oosterhuis SKH (1974) The presence and function of cytochromes in *Selenomonas ruminantium*, *Anaerovibrio lipolytica* and *Veillonella alcalescens*. J Gen Microbiol 81:69–78.

Doig AJ, Williams DH, Oelrichs PB, Baczynskyj L (1990) Isolation and structure elucidation of punicalagin, a toxic hydrolysable tannin, from *Terminalia oblongata*. J Chem Soc Perkin Trans 1:2317–2321

Dominguez-Bello MG, Stewart CS (1990) Degradation of mimosine, 2,3-dihydroxy pyridine and 3-hydroxy-4(1H)-pyridone by bacteria from the rumen of sheep in Venezuela. FEMS Microbiol Ecol 73:283–289.

Eadie JM, Mann SO (1970) Development of the rumen microbial population: high starch diets and instability. In: Phillipson AT, ed. Physiology of Digestion and Metabolism in the Ruminant, pp. 335–347. Newcastle-Upon-Tyne: Oriel Press.

Edwin EE, Jackman R (1973) Ruminal thiaminase and tissue thiamine in cerebrocortical necrosis. Vet Rec 92:640–641.

Edwin EE, Jackman R (1982) Ruminant thiamine requirement in perspective. Vet Res Commun 5:237–250.

Edwin EE, Lewis G, Allcroft R (1968a) Cerebrocortical necrosis: a hypothesis for the possible role of thiaminases in its pathogenesis. Vet Rec 83:176–178.

Edwin EE, Spence JB, Woods AJ (1968b) Thiaminases and cerebrocortical necrosis. Vet Rec 83:417.

Edwin EE, Markson LM, Shreeve J, Jackman R, Carroll PJ (1979) Diagnostic aspects of cerebrocortical necrosis. Vet Rec 104:4–8.

Fowler ME (1983) Plant poisoning in free-living wild animals—a review. J Wildl Dis 19:34–43.

Goldman P (1965) The enzymatic cleavage of the carbon-fluorine bond in fluoroacetate. J Biol Chem 240:3434.

Goldman P (1969) The carbon-fluorine bond in compounds of biological interest. Science 164:1123.

Goldman P, Milne GWA, Kiester DB (1968) Carbon-halogen bond cleavage. III. Studies on bacterial halidohydrolases. J Biol Chem 243:428.

Gooneratne SV, Olkowski RG, Klemmer GA, Kessler GA, Christensen DA (1989) High sulfur related thiamine deficiency in cattle: a field study. Can Vet J 30:139–146.

Gregg K, Sharpe H (1991) Enhancement of rumen microbial detoxification by gene transfer. In: Tsuda T, Sasaki Y, Kawashima R, eds. Physiological Aspects of Digestion and Metabolism in Ruminants, pp. 719–735. New York: Academic Press.

Gregg K, Cooper CL, Schafer DJ, et al. (1994) Detoxification of the plant toxin fluoroacetate by a genetically modified rumen bacterium. Bio/Technol 12:1361–1365.

Grigat GA, Mathison GW (1983) A survey of the thiamine status of growing and fattening cattle in Alberta feedlots. Can J Anim Sci 63:715–719.

Gustine DL, Moyer BG, Wangsness PJ, Shenk JS (1977) Ruminal metabolism of 3-nitropropanoyl-D-glucopyranoses from crownvetch. J Anim Sci 44:1107–1111.

Gutierrez J, Davis RE, Lindahl IL, Warwick EJ (1959) Bacterial changes in the rumen during the onset of feed-lot bloat of cattle and characteristics of *Peptostreptococcus elsdenii* n. sp. Appl Microbiol 7:16–22.

Hammond AC, Bradley BJ, Yokoyama MT, Carlson JR, Dickinson EO (1979) 3-Methylindole and naturally occurring acute bovine pulmonary edema and emphysema. Am J Vet Res 40:1398–1401.

Hammond AC, Allison MJ, Williams MJ, Prine GM, Bates DB (1989) Prevention of *Leucaena* toxicosis of cattle in Florida by ruminal inoculation with 3-hydroxy-4-(1H)-pyridone-degrading bacteria. Am J Vet Res 50:2176–2180.

Hegarty MP, Schinckel PG, Court RD (1964) Reaction of sheep to the consumption of *Leucaena glauca* benth. and to its toxic principle mimosine. Aust J Agric Res 15: 153–167.

Helm RF, Ralph J (1992) Lignin-hydroxycinnamyl compounds related to forage cell wall structure. 1. Ether-linked structures. J Agric Food Chem 40:2167–2175.

Hidiroglou M, Heaney DP, Jenkins KJ (1968) Metabolism of inorganic selenium in rumen bacteria. Can J Physiol Pharmacol 46:229–232.

Hidiroglou M, Jenkins KJ, Knipfel JE (1974) Metabolism of selenomethionine in the rumen. Can J Anim Sci 54:325–330.

Hino T, Kuroda S (1993) Presence of lactate dehydrogenase and lactate racemase in *Megasphaera elsdenii* grown on glucose and lactate. Appl Environ Microbiol 59: 255–259.

Hobson PN (1965) Continuous culture of some anaerobic and facultatively anaerobic rumen bacteria. J Gen Microbiol 38:167–180.

Hobson PN, Summers R (1967) The continuous culture of anaerobic bacteria. J Gen Microbiol 47:53–56.

Hodgson RE, Knott JC (1936) The composition and apparent digestibility of the flat pea *Lathyrus silvestrus* (*wagneri*). J Dairy Sci 19:531–534.

Holtenius P (1957) Nitrite poisoning in sheep, with special reference to the detoxification of nitrite in the rumen. Acta Agric Scand 7:113–163.

Hooper PT (1978) Pyrrolizidine alkaloid poisoning-pathology with particular reference to difference in animal and plant species. In: Keeler RF, Van Kampen KR, James LF, eds. Effects of Poisonous Plants on Livestock. New York: Academic Press.

Howarth RE (1975) A review of bloat in cattle. Can Vet J 16:281–294.

Howarth RE, Cheng KJ, Majak W, Costerton JW (1986) Ruminant bloat. In: Milligan LP, Grovum WL, Dobson A. eds. Control of Digestion and Metabolism in Ruminants, pp. 516–527. Englewood Cliffs, NJ: Prentice-Hall.

Huber TL (1976) Physiological effects of acidosis on feedlot cattle. J Anim Sci 43: 902–909.

Hudman JF, Glenn AR (1984) Selenium uptake and incorporation by *Selenomonas ruminantium.* Arch Microbiol 140:252–256.

Hudman JF, Glenn AR (1985) Selenium uptake by *Butyrivibrio fibrisolvens* and *Bacteroides ruminicola.* FEMS Microbiol Lett 27:215–220.

Hult K, Tieling A, Gatenbeck S (1976) Degradation of ochratoxin A by a ruminant. Appl Environ Microbiol 32:443.

Hungate RE, Dougherty RW, Bryant MP, Cello RM (1952) Microbiological and physiological changes associated with acute indigestion in sheep. Cornell Vet 42:423–449.

Hungate RE, Reichl J, Prins R (1971) Parameters of rumen fermentations in a continuously fed sheep: evidence of a microbial rumination pool. Appl Microbiol 22:1104–1113.

Iiyama K, Lam TBT, Stone BA (1990) Phenolic acid bridges between polysaccharides and lignin in wheat internodes. Phytochemistry 29:733–737.

Jakoby WB, Ziegler DM (1990) The enzymes of detoxification. J Biol Chem 265: 20715–20718.

James LF, Allison MJ, Littledike ET (1975) Production and modification of toxic substances in the rumen. In: Digestion and Metabolism in the Rumen, pp. 576–590. Armidale, Australia: University of New England Publishing Unit.

James LF, Panter KE, Molyneux RJ (1992) Selenium poisoning in livestock. In: James LF, Keeler RF, Bailey EM, Cheeke PR, Hegarty MP, eds. Poisonous Plants. Proceedings of the Third International Symposium, pp. 153–158. Ames: Iowa State University Press.

Jones GA (1972) Dissimilatory metabolism of nitrate by the rumen microbiota. Can J Microbiol 18:1783–1787.

Jones RJ (1981) Does ruminal metabolism of mimosine explain the absence of *Leucaena* toxicity in Hawaii? Aust Vet J 57:55.

Jones RJ, Megarrity RG (1983) Comparable toxicity responses of goats fed *Leucaena leucocephala* in Australia and Hawaii. Aust J Agric Res 34:781–790.

Jones RJ, Ford CW, Megarrity RG (1985) Conversion of 3,4-DHP to 2,3-DHP by rumen bacteria. Leucaena Res Rep 6:3–4.

Jones RJ, Megarrity RG (1986) Successful transfer of DHP-degrading bacteria fron Hawaiian goats to Australian ruminants to overcome the toxicity of *Leucaena*. Aust Vet J 63:259–262.

Jones RJ, Lowry JB (1984) Australian goats detoxify the goitrogen 3-hydroxy-4(1H)pyridone (DHP) after rumen infusions from an Indonesian goat. Experientia 40:1435–1436.

Jung HG, Deetz DA (1993) Cell wall lignification and degradability. In: Jung HG, Buxton DR, Hatfield RD, Ralph J, eds. Forage Cell Wall Structure and Digestibility, pp. 315–346. Madison, Wisc: ASA-CSSA-SSSA.

Jung HG, Fahey GC, Garst JE (1983a) Simple phenolic monomers of forages and effects of in vitro fermentation on cell wall phenolics. J Anim Sci 57:1294–1305.

Jung HG, Fahey GC, Merchen NR (1983b) Effects of ruminant digestion and metabolism on phenolic monomers of forages. Br J Nutr 50:637–651.

Kariya R, Morita Z, Oura R, Sekine J (1989) The in vitro study on the rates of starch consumption and volatile fatty acid production by rumen fluid with or without ciliates. Jpn J Zootech Sci 60:609–613.

Kellerman TS, Coetzer JAW, Naudé TW (1988) Plant Poisonings and Toxicoses of Livestock in Southern Africa. Cape Town: Oxford University Press.

Kelly RW, Lindsay DR (1975) Change with length of feeding period in the oestrogenic response of ovariectomized ewes to ingested coumestants. Aust J Agric Res 26:305–311.

Kiessling KH, Pettersson H, Sandholm K, Olsen M (1984) Metabolism of aflatoxin, ochratoxin, zearalenone, and three trichothecenes by intact rumen fluid, rumen protozoa, and rumen bacteria. Appl Environ Microbiol 47:1070.

King DR, Oliver AJ, Mead RJ (1978) The adaption of some Western Australian mammals to food plants containing fluoroacetate. Aust J Zool 26:699.

King DR, Oliver AJ, Mead RJ (1981) *Bettongia* and fluoroacetate: a role for 1080 in fauna management. Aust Wildl Res 8:529.

King RR, McQueen RE, Levesaue D, Greenhaigh R (1984) Transformation of deoxynivalenol (vomitoxin) by rumen microorganisms. J Agric Food Chem 32:1181.

Kondo T, Mizuno K, Kato T (1990) Cell wall-bound p-coumaric and ferulic acids in Italian ryegrass. Can J Plant Sci 71:495–499.

Krumholz LR, Bryant MP (1986) *Eubacterium oxidoreducens* sp. nov. requiring H_2 or formate to degrade gallate, pyrogallol, phloroglucinol and quercetin. Arch Microbiol 144:8–14.

Kun E, Kirsten E, Sharma ML (1978) Catalytic mechanism of citrate transport through the inner mitochondrial membrane: enzymatic synthesis and hydrolysis of glutathione-citric acid thioester. In: Azzone GF, ed. The Proton and Calcium Pumps, pp. 285–295. Amsterdam: Elsevier.

Kurmanov IA (1977) Fusariotoxicosis in cattle and sheep. In: Mycotoxic Fungi, Mycotoxins and Mycotoxicosis, Vol. 3, pp. 85–110. New York: Marcel Dekker.

Lacoste AM, Lemoigne MM (1961) Degradation du tryptophane par les bacteries de la panse des ruminants. C R Acad Sci 252:1233–1235.

Ladd JN, Walker DL (1965) Fermentation of lactic acid by the rumen microorganism *Peptostreptococcus elsdenii*. Ann NY Acad Sci 119:1038–1045.

Lam TBT, Iiyama K, Stone BA (1990) Distribution of free and combined phenolic acids in wheat internodes. Phytochemistry 29:429–433.

Lam TBT, Iiyama K, Stone BA (1992) Cinnamic acid bridges between cell wall polymers in wheat and phalaris internodes. Phytochemistry 31:1179–1183.

Lanigan GW (1971) Metabolism of pyrrolizidine alkaloids in the ovine rumen. III. The competitive relationship between heliotrine metabolism and methanogenesis in rumen fluid in vitro. Aust J Agric Res 22:123–130.

Lanigan GW (1972) Metabolism of pyrroliziodine alkaloids in the ovine rumen. IV. Effects of chloral hydrate and halogenated methanes on rumen methanogenesis and alkaloid metabolism in fistulated sheep. Aust J Agric Res 23:1085–1091.

Lanigan GW (1976) *Peptococcus heliotrinereducens*, sp. nov., a cytochrome-producing anaerobe which metabolizes pyrrolizidine alkaloids. J Gen Microbiol 94:1–10.

Latham MJ, Sharpe ME, Sutton JD (1971) The microbial flora of the rumen of cows fed hay and high cereal rations and its relationship to the rumen fermentation. J Appl Bacteriol 34:425–434.

Latham MJ, Storry JE, Sharpe ME (1972) Effect of low-roughage diets on the microflora and lipid metabolism in the rumen. Appl Microbiol 24:871–877.

Lewis D (1951a) The metabolism of nitrate and nitrite in the sheep. 1. The reduction of nitrate in the rumen of the sheep. Biochem J 48:175–180.

Lewis D (1951b) The metabolism of nitrate and nitrite in the sheep. 2. Hydrogen donators in nitrate reduction by rumen microorganisms in vitro. Biochem J 49:149–153.

Lewis TR, Emery RS (1962a) Relative deamination rates of amino acids by rumen microorganisms. J Dairy Sci 45:765–768.

Lewis TR, Emery RS (1962b) Intermediate products in the catabolism of amino acids by rumen microorganisms. J Dairy Sci 45:1363–1368.

Lewis TR, Emery RS (1962c) Metabolism of amino acids in the bovine rumen. J Dairy Sci 45:1487–1492.

Long TA, Washko JB, Palmer WL (1977) Flat pea hay promising as ruminant forage. Sci Agric 24:711.

Lowry JB (1990) Metabolic and nutritional significance of the cell-wall phenolic fraction. In: Akin DE, Ljungdahl LG, Wilson JR, Harris PJ, eds. Microbial and Plant Opportunities to Improve Lignocellulose Utilization by Ruminants, pp. 119–126. New York: Elsevier.

Lowry JB, Tangendjaja M, Tangendjaja B (1983) Autolysis of mimosine to 3-hydroxy-

4–1(H)pyridone in green tissues of *Leucaena leucocephala*. J Sci Food Agric 34: 529–533.

Lowry JB, Sumpter EA, McSweeney CS, Schlink AC, Bowden B (1993) Phenolic acids in the fibre of some tropical grasses, effect on feed quality, and their metabolism by sheep. Aust J Agric Res 44:1123–1133.

Mackie RI, Gilchrist FMC (1979) Changes in lactate-producing and lactate-utilizing bacteria in relation to pH in the rumen of sheep during stepwise adaptation of sheep to high concentrate diets. Appl Environ Microbiol 38:422–430.

Mackie RI, Gilchrist FMC, Robberts AM, Hannah PE, Schwartz HM (1978) Microbiological and chemical changes in the rumen during the stepwise adaptation of sheep to high concentrate diets. J Agric Sci Camb 90:241–254.

Mackie RI, McSweeney CS (1991) Microbiology of foregut and hindgut fermentation. In: Ho YW, Wong HK, Abdullah N, Tajuddin ZA, eds. Recent Advances on the Nutrition of Herbivores, pp. 189–197. Kuala Lumpur: Malaysian Society of Animal Production.

Majak W, Cheng KJ (1984) Cyanogenesis in bovine rumen fluid and pure cultures of rumen bacteria. Can J Anim Sci 59:784.

Majak W, Clark LJ (1980) Metabolism of aliphatic nitro compounds in bovine rumen fluid. Can J Anim Sci 60:319.

Majak W, Pass MA (1989) Aliphatic nitro-compounds. In: Cheeke PR, ed. Toxicants of Plant Origin. Vol. II, Glycosides, pp. 143–159. Boca Raton, Fla: CRC Press.

Majak W, Cheng K-J (1981) Identification of rumen bacteria that anaerobically degrade aliphatic nitrotoxins. Can J Microbiol 27:646–650.

Majak W, Cheng K-J (1983) Recent studies on ruminal metabolism of 3-nitropropanol in cattle. Toxicon 3(suppl):265–268.

Majak W, Clark LJ (1980) Metabolism of aliphatic nitro compounds in bovine rumen fluid. Can J Anim Sci 60:319–325.

Majak W, Cheng K-J, Hall JW (1982) The effect of cattle diet on the metabolism of 3-nitro-propanol by ruminal microorganisms. Can J Microbiol 62:855–860.

Majak W, Cheng K-J, Hall JW (1986) Enhanced degradation of 3-nitro-propanol by ruminal microorganisms. J Anim Sci 62:1072–1080.

Mangan JL (1988) Nutritional effects of tannins in animal feeds. Nutr Res Rev 1:209–231.

Marais JSC (1944) Monofluoroacetic acid, the toxic principle of Gifblaar, *Dichapetalum cymosum* (Hook) Engl Onderstepoort. J Vet Sci Anim Ind 20:67.

Martin AK (1982a) The origin of urinary aromatic compounds excreted by ruminants. 2. The metabolism of phenolic cinnamic acids to benzoic acid. Br J Nutr 47:155–164.

Martin AK (1982b) The origin of urinary aromatic compounds excreted by ruminants. 3. The metabolism of phenolic compounds to simple phenols by ruminants. Br J Nutr 48: 497–507.

Mattocks AR (1968) Toxicity of pyrrolizidine alkaloids. Nature (London) 217:723–728.

Mattocks AR (1981) Relation of structural features to pyrrolic metabolites in livers of rats given pyrrolizidine alkaloids and derivatives. Chem Biol Interact 35:301–310.

McEwen T (1978) Organo-fluorine compounds in plant. In: Keeler RF, Van Campen KR, James LF, eds. Effects of Poisonous Plants on Livestock, New York: Academic Press.

McSweeney CS, Allison MJ, Mackie RI (1993a) Amino acid utilization by the ruminal bacterium *Synergistes jonesii* strain 78–1. Arch Microbiol 159:131–135.

McSweeney CS, Mackie RI, Odenyo AA, Stahl DA (1993b) Development of an oligonucleotide probe targeting 16S rRNA and its application for detection and quantitation of the ruminal bacterium *Synergistes jonesii* in a mixed-population chemostat. Appl Environ Microbiol 59:1607–1612.

Mead RJ, Oliver AJ, King DR (1979) Metabolism and defluorination and fluoroacetate in the brush-tailed possum (*Trichosurus vulpecula*). Aust J Biol Sci 32:15.

Mead RJ, Moulden DL, Twigg LE (1985) Significance of sulfhydryl compounds in the manifestation of fluoroacetate toxicity to the rate, brush tail possum, woylie and Western grey kangaroo. Aust J Biol Sci 38:139.

Mishra BD, Fina LR, Bartley EE, Claydon TJ (1967) Bloat in cattle. XI. The role of rumen aerobic (facultative) mucinolytic bacteria. J Anim Sci 26:606–612.

Mishra BD, Bartley EE, Fina LR, Bryant MP (1968) Bloat in cattle. XIV. Mucinolytic activity of several anaerobic rumen bacteria. J Anim Sci 27:1651–1656.

Mohn WW, Tiedje JM (1992) Microbial reductive dehalogenation. Microbiol Rev 56: 482–507.

Morris MP, Garcia-Rivera J (1955) The destruction of oxalates by rumen contents of cows. J Dairy Sci 38:1169.

Murdiati TB, McSweeney CS, Lowry JB (1992) Metabolism in sheep of gallic acid, tannic acid and hydrolysable tannin from *Terminalia oblongata.* Aust J Agric Res 43: 1307–1319.

Naga MA, Harmeyer JH, Hoeller H, Schaller K (1975) Suspected B-vitamin deficiency of sheep fed a protein-free urea containing purified diet. J Anim Sci 40:1192–1198.

Nakamura Y, Tada Y, Shibuya J, Yoshida J, Nakamura R (1979) The influence of concentrates on the nitrate metabolism of sheep. Jpn J Zootech Sci 50:782–789.

Newbold CJ, Williams AG, Chamberlain DG (1987) The in vitro metabolism of DL-lactic acid by rumen microorganisms. J Sci Food Agric 38:9–18.

Newton WA, Snell EE (1964) Catalytic properties of tryptophanase, a multifunctional pyrodoxal phosphate enzyme. Proc Natl Acad Sci USA 51:382–389.

Nilsson A (1961) Demethylation of the plant oestrogen biochanin A in the rat. Nature 192:358.

Nilsson A (1962) Demethylation of the plant estrogen formononetin to daidzein in rumen fluid. Ark Kemi 19:549–550.

Nilsson A, Hill JL, Davies HL (1967) An in vitro study of formononetin and biochanin A metabolism in rumen fluid from sheep. Biochim Biophys Acta 148:92–98.

Oliver AJ, King DR, Mead RJ (1979) Fluoroacetate tolerance, a genetic marker in some Australian mammals. Aust J Zool 27:363.

Peterson PJ, Spedding DJ (1963) The excretion by sheep of ^{75}selenium incorporated into red clover (*Trifolium pratense* L.): The chemical nature of the excreted selenium and its uptake by three plant species. NZ J Agric Res 6:13–23.

Phillipson AT, Reid RS (1957) Thiamin and niacin in the rumen. J Anim Sci 59:813–822.

Porter JWG (1961) Vitamin synthesis in the rumen. In: Lewis D, ed. Digestive Physiology and Nutrition of the Ruminant, pp. 226–234. London: Butterworths.

Potter EL, Dehority BA (1973) Effects of changes in feed level, starvation, and level of feed after starvation upon the concentration of rumen protozoa in the ovine. Appl Microbiol 26:692–698.

Provenza FD, Pfister JA (1991) Ingestion of plant toxins on food ingestion by herbivores. In: Ho YW, Wong HK, Abdullah N, Tajuddin ZA, eds. Recent Advances on the Nutrition of Herbivores, pp. 199–206. Kuala Lumpur: Malaysian Society of Animal Production.

Purser DB, Moir RJ (1959) Ruminal flora studies in the sheep. IX. The effect of pH on ciliate population of the rumen in vivo. Aust J Agric Res 10:555–564.

Raisbeck MF (1982) Is polioencephalomalacia associated with high-sulfate diets. J Am Vet Med Assoc 180:1303–1305.

Rammell CG, Hill JH (1986) A review of thiamine deficiency in ruminants. Vet Ann 25: 71–77.

Ramos JL, Timmis KN (1987) Experimental evolution of catabolic pathways of bacteria. Microbiol Sci 4:228–237.

Rasmussen MA, Foster JG, Allison MJ (1992) *Lathyrus sylvestris* (flatpea) toxicity in sheep and ruminal metabolism of flatpea neurolathyrogens. In: James LF, Keeler RF, Bailey EM, Cheeke PR, Hegarty MP, eds. Poisonous Plants. Proceedings of the Third International Symposium, pp. 377–381. Ames: Iowa State University Press.

Rasmussen MA, Allison MJ, Foster JG (1993) Flatpea intoxication in sheep and indications of ruminal adaptation. Vet Hum Toxicol 35:123–127.

Rasmussen MA, James LF(1994) Selenium metabolism in the rumen. In: Colegate SM, Dorling PR, eds. Plant-Associated Toxins: Agricultural, Phytochemical and Ecological Aspects. Wallingford, U.K.: CAB International.

Reid CSW, Clarke RTJ, Cockrem FRM, Jones WT, McIntosh JT, Wright DE (1975) Physiological and genetical aspects of pasture (legume) bloat. In: McDonald IW, Warner ACI, eds. Digestion and Metabolism in the Ruminant, pp. 524–536. Armidale, Australia: University of New England Publishing Unit.

Ressler C (1964) Neurotoxic amino acids of certain species of *Lathyrus* and vetch. Fed Proc 23:1350–1353.

Rosenthal GA, Janzen DH (1979) Herbivores: Their Interactions With Secondary Plant Metabolites. New York: Academic Press.

Roth CW, Hoch JA, Demoss RD (1971) Physiological studies of biosynthesis indole excretion in *Bacillus alvei*. J Bacteriol 106:97–106.

Ruan Z-S, Anantharam V, Crawford IT, et al. (1992) Identification, purification, and reconstitution of Oxit, the oxalate:formate antiport protein of *Oxalobacter formigenes*. J Biol Chem 267:10537–10543.

Russell GR, Smith RM (1968) Reduction of heliotrine by a rumen microorganism. Aust J Biol Sci 21:1277–1290.

Russell JB, Baldwin RL (1978) Substrate preferences in rumen bacteria: evidence of catabolite regulatory mechanisms. Appl Environ Microbiol 36:319–329.

Russell JB, Baldwin RL (1979) Comparison of maintenance energy expenditures and growth yields among several rumen bacteria grown on continuous culture. Appl Environ Microbiol 37:537–543.

Russell JB, Hino T (1985) Regulation of lactate production in *Streptococcus bovis*: a spiraling effect that leads to rumen acidosis. J Dairy Sci 68:1712–1721.

Sagar RL, Hamar DW, Gould DH (1990) Clinical and biochemical alterations in calves with nutritionally induced polioencephalomalacia. Am J Vet Res 51:1969–1974.

Sapienza CA, Brent BE (1974) Ruminal thiaminase vs. concentrate adaptation. J Anim Sci 39:252. Abstract.

Sapiro ML, Hoflund S, Clark R, Quin JI (1949) Studies on the alimentary tract of the Merino sheep in South Africa. XVI. The fate of nitrate in ruminal ingesta as studied in vitro. Onderstepoort J Vet Sci 22:357–372.

Sayler GS, Hooper SW, Layton AC, King JMH (1990) Catabolic plasmids of environmental and ecological significance. Microbiol Ecol 19:1–20.

Scheifinger CC, Latham MJ, Wolin MJ (1975) Relatonship of lactate dehydrogenase specificity and growth rate to lactate metabolism by *Selenomonas ruminantium*. Appl Microbiol 30:916–921.

Selman IE, Wiseman A, Pirie HM, Breeze RG (1974) Fog fever in cattle: clinical and epidemiological features. Vet Rec 95:139–146.

Selman IE, Wiseman A, Breeze RG, Pirie HM (1976) Fog fever in cattle: various theories on its aetiology. Vet Rec 99:181–184.

Sharrow SH, Mosher WD (1982) Sheep as a biological control agent for tansy ragwort. J Range Man 35:480–482.

Shull LR, Cheeke PR (1983) Effects of synthetic and natural toxicants on livestock. J Anim Sci 57(suppl 2):330.

Shull LR, Buckmaster GW, Cheeke PR (1976) Factors influencing pyrrolizidine (*Senecio*) alkaloid metabolism: Species, liver sulfhydryls and rumen fermentation. J Anim Sci 43:1247–1253.

Simpson FJ, Jones GA, Wolin EA (1969) Anaerobic degradation of some bioflavonoids by microflora of the rumen. Can J Microbiol 15:972–974.

Stanier RY (1951) Enzymatic adaptation in bacteria. Annu Rev Microbiol 5:35–56.

Susmel P, Stefanon B (1993) Aspects of lignin degradation by rumen microorganisms. J Biotechnol 30:141–148.

Swanson SP, Helaszek C, Buck WB, Rood HD, Haschek WM (1988) The role of intestinal microflora in the metabolism of trichothecene mycotoxins. Food Chem Toxicol 26:823.

Swick RA, Cheeke PR, Ramsdell HS, Buhler DR (1983) Effect of sheep fermentation and methane inhibition on toxicity of *Senecio jacobaea*. J Anim Sci 56:645–654.

Takahashi J, Masuda Y, Miyaga E (1978) Effect of pH and level of nitrate on the reduction of nitrate and nitrite in vitro. Jpn J Zootech Sci 49:1–5.

Takahashi J, Masuko T, Endo S, Dodo K, Fujita H (1980) Effects of dietary protein and energy levels on the reduction of nitrate and nitrite in the rumen and methemoglobin formation in sheep. Jpn J Zootech Sci 51:626–631.

Therion JJ, Kistner A, Kornelius JH (1982) Effect of pH on growth rates of rumen amylolytic and lactolytic bacteria. Appl Environ Microbiol 44:428–434.

Tomei FA, Barton LL, Lemanski CL, Zocco TG (1992) Reduction of selenate and selenite to elemental selenium by *Wolinella succinogenes*. Can J Microbiol 38:1328–1333.

Tsai CG, Jones GA (1975) Isolation and identification of rumen bacteria capable of anaerobic phloroglucinol degradation. Can J Microbiol 21:749–801.

Turnbull KE, Braden AWH George JM (1966) Fertilization and early embryonic loss in ewes that had grazed oestrogenic pastures for 6 years. Aust J Agric Res 17:907–917.

Ueno Y (1984) General toxicology. In: Ueno Y, ed. Trichothecenes: Chemical Biological and Toxicological Aspects. New York: Elsevier Scientific.

Van der Meer JR, de Vos WM, Harayama S, Zehnder AJB (1992) Molecular mechanisms of genetic adaptation to xenobiotic compounds. Appl Environ Microbiol 56:677–694.

Van Etten CH, Miller RW (1963) The neuroactive factor alpha-gamma diaminobutyric acid in angiospermous seeds. Econ Botany 17:107–109.

Wachenheim DE, Blythe LL, Craig AM (1992) Characterization of rumen bacterial pyrrolizidine alkaloid biotransformation in ruminants of various species. Vet Hum Toxicol 34:513–517.

Wallace RJ (1978) Control of lactate production by *Selenomonas ruminantium*: homotrophic activation of lactate dehydrogenase by pyruvate. J Gen Microbiol 107:45–52.

Warner ACI (1962) Some factors influencing the rumen microbial population. J Gen Microbiol 28:129–146.

Watts PS (1957) Decomposition of oxalic acid in vitro by rumen contents. Aust J Agric Res 8:266–270.

Westlake K, Mackie RI, Dutton M (1987a) T-2 toxin metabolism by ruminal bacteria and its effect on their growth. Appl Environ Microbiol 53:587.

Westlake K, Mackie RI, Dutton M (1987b) Effects of several mycotoxins on the specific growth rate of *Butyrivibrio fibrisolvens* and toxin degradation in vitro. Appl Environ Microbiol 53:613.

Westlake K, Mackie RI, Dutton M (1989) In vitro metabolism of mycotoxins by bacterial, protozoal and ovine ruminal fluid preparations. Anim Feed Sci Technol 25:169.

Williams MC, Barneby RC (1977) The occurrence of nitro-toxins in North American *Astragalus* (Fabaceae). Britonia 29:310–326.

Williams AG, Coleman GS (1992) In: The Rumen Protozoa. New York: Springer-Verlag.

Yang JNY, Carlson JR (1972) Effects of high tryptophan doses and two experimental rations on the excretion of urinary tryptophan metabolites in cattle. J Nutr 102: 1655–1666.

Yokoyama MT, Carlson JR (1974) Dissimilation of tryptophan and related indolic compounds by ruminal microorganisms in vitro. Appl Microbiol 27:540–548.

Yokoyama MT, Carlson JR, Holdeman LV (1977) Isolation and characteristics of a skatole producing *Lactobacillus* sp. from the bovine rumen. Appl Environ Microbiol 34: 837–842.

Yung HG, Deetz DA (1993) Cell wall lignification and degradability. In: Jung HG, Buxton DR, Hatfield RD, Ralph J, eds. Forage Cell Wall Structure and Digestibility, pp. 315–346. Madison, Wisc: ASA-CSSA-SSSA.

Yung HG, Fahey GC, Garst JE (1983a) Simple phenolic monomers of forages and effects of in vitro fermentation on cell wall phenolics. J Anim Sci 57:1294–1305.

Yung HG, Fahey GC, Merchen NR (1983b) Effects of ruminant digestion and metabolism on phenolic monomers of forages. Br J Nutr 50:637–651.

Zhu J (1993) Tannin toxicity studies in mice and sheep. Ph.D. thesis, University of Queensland, Brisbane, Australia.

Index

Note: *Italicized* page numbers locate figures; pages followed by *t* locate tables.